# AUTOCANNON

A HISTORY OF

AUTOMATIC CANNON AND THEIR AMMUNITION

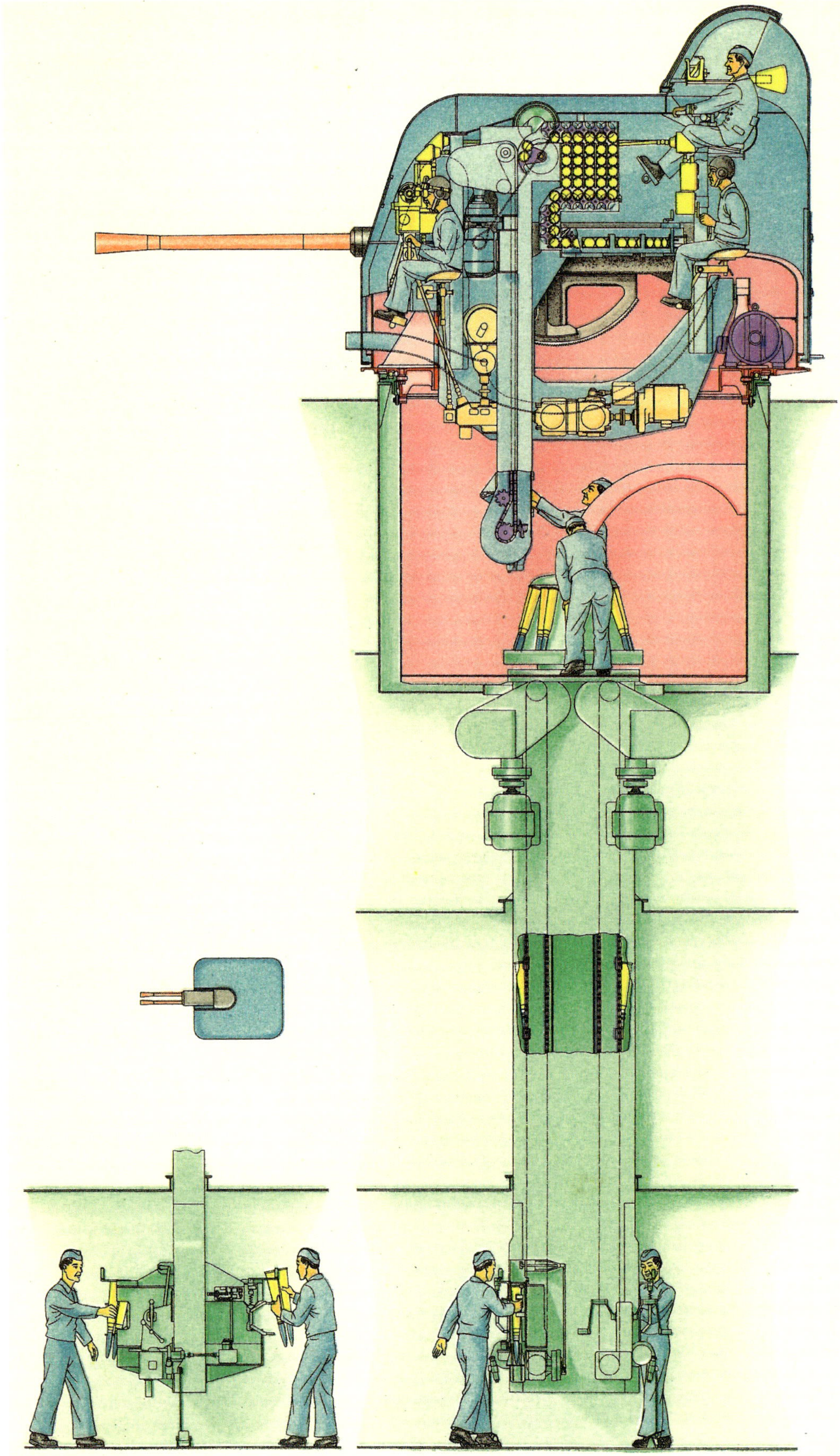

*57 mm Bofors M/50 twin mounting*

# AUTOCANNON

## A HISTORY OF AUTOMATIC CANNON AND THEIR AMMUNITION

Anthony G. Williams

THE CROWOOD PRESS

First published in 2022 by
The Crowood Press Ltd
Ramsbury, Marlborough
Wiltshire SN8 2HR

enquiries@crowood.com

**www.crowood.com**



**British Library Cataloguing-in-Publication Data**
A catalogue record for this book is available from the British Library.

ISBN 978 1 78500 920 4

Typeset by Simon and Sons

Cover design by Blue Sunflower Creative

Printed and bound in India by Parksons Graphics

# CONTENTS

# ACKNOWLEDGEMENTS

This book could not have been written without the help of the large number of people who have contributed information and illustrations in the twenty years since my first book on this subject – *Rapid Fire* – was published. I am very grateful to all of them.

Many of them have sent material to the *ECRA Bulletin* (European Cartridge Research Association) of which I was the editor for fourteen years. Others have been active on specialist internet forums, especially the IAA (International Ammunition Association), BOCN (British Ordnance Collectors Network), and my own forum, Military Guns and Ammunition. The web addresses are included in the Sources section at the end of the book, along with those of other sites which have provided useful information and a small selection of the books and other documents I have learned from. My education in this subject has also benefited from thirteen years spent as an editor of the Jane's ammunition reference, which has been through various name changes and is currently *Jane's Weapons: Ammunition*.

Crediting all of the people who have helped me would be an impossible task, but I would like to name several individuals whose help was particularly significant, or have reviewed and/or made major contributions to sections of this book, as follows:

**General**
Richard Jones, formerly of the Ministry of Defence Pattern Room and the National Firearms Centre
Steve McGregor, ECRA
And some who are no longer with us, including: Peter Labbett (ECRA), William Woodin (IAA) and Herbert Woodend (MoD Pattern Room)

**Ammunition feeding arrangements (magazines, belt links, clips etc)**
Koen Bakker

**The following national sections**

| | |
|---|---|
| France: | Claude Jacquinet |
| Germany: | Alex Diehl |
| Italy: | Viligelmo Ballabio |
| Japan: | Yasufumi Kunimoto |
| Russia: | Evgeny Aranov, plus Maxim Popenker and Denis Estavyev |
| Sweden: | Sigbjörn Hellström |
| Switzerland: | Robert Gerber |
| UK: | Tim Gurney |
| USA: | Robert Leiendecker |

# INTRODUCTION

The term 'cannon' in British use originally referred to large-calibre muzzle-loading artillery. This book is concerned with modern cannon; while there is no agreed single definition, these are commonly regarded as being fast-firing high-velocity automatic guns with calibres (projectile diameters) between 20 mm and about 57 mm.

The precursors of such guns were developed in the late 19th century, in the form of larger versions of hand-cranked multi-barrel machine guns (MGs) like the Gatling and the Nordenfelt, which were made in calibres of up to one inch (25.4 mm) and fired solid projectiles. These were followed by the Hotchkiss *cannon revolver* with a rotating cluster of five barrels, also hand-cranked but in much larger calibres of 37 mm, 40 mm, 47 mm and 53 mm, which fired hollow shells filled with explosives.

The first truly automatic cannon came near the end of that century, in the form of the Vickers-Maxim One Pounder, so-called from the nominal weight of the projectile (1 lb or 454 g). This was a scaled-up version of the company's machine gun and used the recoil energy from firing each shot to power the firing cycle. This famous gun was nicknamed the 'pom-pom' from the distinctive sound of its firing, a designation which was also applied to its successors, particularly the British 2-pounder naval anti-aircraft (AA) gun of World War II.

World War I saw the first extensive use of cannon in a variety of roles, most notably against the growing threat of aircraft attacking ground or naval targets. Some were also mounted on aircraft, for firing at each other or at ground targets. The guns were still in their infancy at this stage and developed only slowly in the interwar years. Only in the late 1930s was there an accelerating rush of interest in acquiring such weapons, and World War II saw such famous names as Oerlikon, Bofors and Hispano become synonymous with cannon, principally for use against, or by, aircraft.

After World War II the development of cannon continued for a variety of roles, but from the late 1950s they became increasingly challenged by guided missiles. However, experience revealed that the theoretical performance of the early missiles was rarely achieved in combat, so cannon remained in use as primary weapon systems against aircraft for longer than expected. Cannon also found other roles, particularly for use by and against light armoured vehicles; their principal use today. Their versatility has also led to their continued use: by combat aircraft (as a back-up weapon, or for use against targets for which a missile is not appropriate); in the ground-based anti-aircraft role (to supplement missiles); and also installed on warships, for dealing with small boats and/or as a last-ditch defence against incoming anti-ship missiles, in which role they are known as Close-in Weapon Systems (CIWS). The latest application of cannon is to adapt the naval CIWS to the C-RAM role (Counter-Rocket, Artillery and Mortar), for the protection of military land bases.

Alternative weapons keep emerging, in the form of smaller and cheaper missiles or entirely different technologies such as lasers, but it is likely that there will be roles for cannon for the foreseeable future.

*2-barrel 1-inch Nordenfelt (4 barrel versions also saw service)*

*37 mm Hotchkiss 'cannon revolver'*

## GUN CALIBRE

The choice of the optimum calibre for any particular purpose is inevitably the result of various compromises. Other things being equal (an important qualifier in any statements comparing guns and ammunition), large calibre weapons provide greater destructive power and a longer effective range, but are bigger and heavier, generate more recoil, have a lower rate of fire (RoF), and the bigger ammunition restricts magazine capacity.

The popularity of different calibres has varied depending on the tactical needs of the time. However, the first true automatic cannon emerged in the nineteenth century in 37 mm calibre for legal reasons. In the middle of that century explosive projectiles were introduced in rifles. These were felt to be inhumane when used against soldiers, so an international convention was held that led to the St. Petersburg Declaration of 1868, which stated:

> *The Great Powers agreed to renounce, in case of war among themselves, the use of any explosive projectile of less weight than 400 grams (14 ounces avoirdupois) or one charged with fulminating or inflammable substances.*

At that time, a 400 g shell was typically of 37 mm calibre so this became an international standard which was honoured for the next half century. By the start of World War I, except for a few obsolete weapons there was a huge calibre gap between rifle-calibre (6.5 to 8 mm) MGs and 37 mm cannon.

The pressure of war eventually led to a demand for smaller cannon calibres, partly for the anti-tank (AT) and anti-aircraft roles and partly to provide more firepower for aircraft without burdening them with large cannon. Aircraft weapons in 20 and 25 mm were developed which, although initially available only with solid shot, used high explosive (HE) projectiles by the end of the war.

After World War I was over, it was decided that the legal position of explosive shells should be clarified so in 1922 a Committee of International Jurists met at the Hague to consider a set of rules for aerial warfare. They decided that the use of tracer, incendiary or explosive projectiles by or against aircraft was not prohibited. This draft Convention was never formally ratified, but everyone thereafter behaved as if it had been, albeit with a degree of nervousness in some quarters.

In the interwar period, 20 mm became generally regarded as the smallest calibre whose projectiles were capable of carrying a useful quantity of HE. It therefore became a standard calibre with the great majority of nations and has remained so up to the present time. The 37 mm calibre remained popular and was joined by 40 mm weapons from Britain and, most notably, Bofors in Sweden. This remained the largest calibre to see significant service in rapid-firing weapons until the middle of the 20th century, with the exception of the early Hotchkiss manually-cranked 'canon revolver' (actually a rotary gun in modern terminology), produced in 47 mm and 53 mm calibres.

Before World War II there were few weapons in calibres intermediate between 20 and 37 mm, the most significant being 25 mm AA guns used by Sweden, France, Japan, Russia and, just in time for the war, the American 1.1-inch (28 mm) naval gun (whose odd calibre indicates that this was probably the last cannon to have been designed to adhere to the St Petersburg Declaration, since the shell weight was 416 g). More emerged during the war, with 23 mm weapons becoming established in the Soviet Union and 30 mm guns, initially for aircraft, in Germany, with Japan following suit right at the end. Attempts were made in various countries to produce fast-firing automatic weapons in 50–57 mm calibre, but cannon of this size were not notably successful until after the war.

After the lessons of the war had been absorbed, NATO broadly settled for automatic cannon in 20, 30 and 40 mm calibre (with 35 mm, 27 mm and 25 mm being added later) plus 57 mm in naval guns (land and aircraft versions not lasting long). NATO cartridges are allocated STANAG (STANdardisation AGreement) numbers, which specify characteristics required to ensure interoperability in different weapons. The Warsaw Pact (and by extension China and nations supplied by these countries), used weapons in 23, 25, 30, 37, 45 and 57 mm calibres. It should be noted that NATO countries currently produce or use weapons chambered for three different 20 mm cartridges, two 25 mm and four 30 mm (not including variations using different case materials or priming) as well as 27 mm, 35 mm, and soon to be two, or even three, 40 mm, and a 57 mm. In contrast, the complete and considerable range of Russian automatic cannon currently in production is focused on the 30 × 165 cartridge (albeit in both percussion and electrically-primed versions), although the 57 mm seems to be enjoying a minor revival of interest.

## DEFINITIONS

The current usage of the term 'cannon' has only become widely accepted since World War II. Before then, the term 'machine gun' was used to describe a relatively small-calibre weapon normally firing solid projectiles while larger-calibre versions were generally called 'automatic guns'. The situation changed in British practice with the selection for the Royal Air Force (RAF) of the French Hispano Suiza HS 404 *moteur-canon* in the late 1930s. Anglicized as 'cannon', the name became adopted for the Hispano and subsequently for all other automatic shell-firing guns of 20 mm or more calibre. The plural of cannon can be either cannon or cannons – the former will be used in this book.

The distinction between MGs and cannon is an arbitrary one that has varied in different countries and at different times. As far as projectiles are concerned, the accepted convention has been that MGs are designed to fire bullets: ball, incendiary and armour piercing (AP) projectiles, with or without tracers; that are usually 'jacketed'; the core being covered by a relatively soft, usually copper-based, alloy which can be formed by the pressure of firing to fit the rifling grooves. Cannon projectiles – generally known as shells – are normally of steel with separate, softer driving or rotating bands to take the rifling, and are typically filled with HE and/or incendiary materials plus a fuse to initiate them, or have a hard metal core for armour piercing (AP shot). The dividing line between bullets and shells is generally accepted to be 20 mm calibre, exceptions to this general rule being the 13 mm Rheinmetall-Borsig MG 131 and 15 mm Mauser MG 151 aircraft heavy machine gun (HMG) rounds, which used cannon-type projectiles; conversely, the AP shot for the 25 mm Hotchkiss cannon was made just like an MG bullet.

Different nations have had different naming practices. For example, in World War II German automatic weapons of up to and including 20 mm were known as machine guns (*Maschinengewehr*) with larger calibres being known as cannon (*Maschinenkanonen*), leading to the MG and MK prefixes. In addition, designations based on function were used for particular applications, whether or not the weapons were automatic. A gun intended for the anti-aircraft role was called *Fliegerabwehrkanone* (sometimes given as *Flugzeugabwehrkanone* or *Flugabwehrkanone*), for the anti-tank role *Panzerabwehrkanone*, and for mounting in armoured fighting vehicles *Kampfwagenkanone*; Flak, Pak and KwK for short. The BK (*Bordkanone*) designation was used for very large calibre (37+ mm) airborne cannon intended for ground attack, although currently that term is applied to the 27 mm Mauser BK 27 cannon.

Nomenclature also varied between services. In Sweden, the Air Force called their 12.7 mm and 13.2 mm m/39 guns *automatkanon*, presumably because they were larger than usual for aircraft guns at the time, while their Navy called the 25 mm M/32 a *kulspruta*, (which translates as machine gun), presumably because this was a small weapon by naval standards. In the US services, naval weapons are generally given Mk (mark) designations, army guns M (model). The numbers that form part of such designations have in most cases related to the date of introduction or acceptance into service (e.g. M/32 introduced in 1932) but in other cases model numbers are allocated on a different and not always obvious basis, as is typical of current US and UK practice.

Since World War II the definition of a cannon as any fast-firing, high-velocity automatic weapon of 20 mm or more has become commonly accepted; the upper calibre limit is less clearly defined, but is generally taken to be around 57 mm as, broadly speaking, this is the largest calibre of automatic weapon to include multi-purpose guns that were fitted to aircraft, naval and land mountings. The next step up in automatic weapon calibre is to 75 mm, which are usually specialist single-purpose guns.

Calibre designations have also varied. The UK and related countries used to designate guns of 37+ mm calibre (whether automatic or not) by 'pounder' measurements (abbreviated to Pr or Pdr, often written $^{PR}$ or $^{PDR}$), after the nominal weight of the standard projectile in pounds. The British also used inch measurements, but by the 20th century mostly for guns from 76.2 mm (3-inch) calibre upwards. The USA used inch measurements as standard for some weapons (e.g. the 1.1-inch naval gun) but from World War I onwards also used metric measurements, having adopted 37 mm, 75 mm, 105 mm and 155 mm artillery from France during that war. Only in the last half-century have metric calibre measurements almost entirely taken over from inch or pounder measurements in Britain and the USA. Even within the metric group, practices have varied; in wartime Germany and certain other countries, heavy weapon calibres of 20 mm or more were usually described in centimetres rather than millimetres.

The cartridges fired by the guns are also described using what has become accepted as the standard metric measurements of calibre and cartridge case length, plus any non-metric designations they have been known by. The metric measurements may not have been used at the time (e.g. the USN 1.1-inch cartridge was only ever known by that name; the metric 28 × 199 SR is a retrospective designation for comparison purposes). Metric measurements are therefore used for all calibres and cartridges in this book, alongside any other designations which may have been used. In some instances two or more cartridges may have the same measurements, so these are distinguished by additional letters which may refer to the rim design (R, SR and RB or RR for rimmed, semi-rimmed or semi-rimless, and rebated – with a small-diameter rim) or some other identifying letters, notably B to describe a belted case. The worst case of this lies with 20 mm cartridges with 110 mm case lengths: six in total (including experimentals), none of which are interchangable. So as well as 20 × 110 RB and 20 × 110 R there are three named for the gun: 20 × 110 HS (Hispano Suiza), 20 × 110 Scotti, 20 × 110 AP-20; and one named for the user organisation, the 20 × 110 USN. Again, these designations may not have been officially used, but are applied now to distinguish between them. A third metric to help with identification is the rim diameter; this is shown in the ammunition tables.

A final note on ammunition designations: in larger-calibre artillery in which the projectile and propellant are loaded separately, a 'round' of ammmunition refers to the projectile, the 'cartridge' is the separate propellant charge. However, in small arms and cannon, in which the projectile is firmly secured to the cartridge case during assembly, the complete unit of ammunition – cartridge case, propellant, projectile and primer (means of igniting the propellant) – may be known as either the 'cartridge' or the 'round' of ammunition.

## BOOK STRUCTURE

This work is structured into four parts: first, a brief history of cannon and their use; second, the design of cartridges and guns; third, the cartridges described in calibre order; and fourth, the cannon arranged by country. Finally there are data tables on the cartridges and guns for quick reference purposes, and a glossary of terms.

Part I provides a history and analysis of the use of automatic cannon by armies, navies and air forces of all nations from the end of the 19th century to the present day. It is focused mainly on those guns which made it into service, and describes how their use changed over time.

Part II considers the main factors affecting the design of cartridges and cannon, including the different types of projectile used, a brief section on ballistics and consideration of unconventional systems.

Part III includes all of the cannon cartridges to have seen service, plus many which did not.

Part IV gives the same treatment to the cannon.

It should be noted that the focus in this book is sharply on the guns and ammunition, with some information about mountings but nothing about sights or fire control. These omissions are not due to any disregard of their importance – in fact, they are vital to the effective performance of the weapon systems – but they are complex subjects of their own and deserve a much more detailed treatment than there was space for in this book.

# DEVELOPMENTS BEFORE 1932

The late 19th century was a period of rapid technological development in guns, dramatically increasing their volume of fire. The key to this was the introduction of the self-contained cartridge or round of ammunition, loaded with 'smokeless powder' which did not foul up mechanisms in the same way as black powder (gunpowder), and which was sturdy enough to survive the violent handling needed to achieve high rates of fire. The precision engineering needed to mass-produce the many small mechanical parts of a fast-firing gun also developed.

The first self-powered machine gun was the Maxim, which entered service in 1886. This used the recoil force from firing each cartridge to push the breechblock back, extracting and ejecting the fired cartridge case then removing the next round from a fabric belt before pushing it into the chamber ('chambering' it) to be fired.

The Maxim machine gun was adopted by many armies, usually firing the same ammunition as the army's rifle. However, interest grew in a bigger model to compete with the hand-cranked multi-barrel Hotchkiss revolving cannon. Maxim selected the smallest of the Hotchkiss 37 mm rounds, the 37 × 94 R, as the basis of their large machine gun.

Designed in the late 1880s, the first versions entered service in the 1890s and the gun was adopted by the British Army in 1900 as the 'Ordnance QF 1 pounder Mk. 1' (QF stood for Quick Firing, an artillery term which meant that the propellant was contained within a brass cartridge case). However, it was popularly known as the 'pom-pom'. The explosive shells were not only valued for their destructive effects, but also because at long range the puffs of smoke indicated where the shells were landing, allowing the aim to be corrected as required.

The pom-poms were widely used by both sides throughout World War I, although they became obsolete in ground and naval fighting. In ground fighting, they were outranged by modern fast-firing artillery, and in naval use they were no longer effective against larger torpedo boats and torpedo-boat destroyers. However, when installed on high-angle mountings they found a new role against low-flying aircraft. The first installation of a pom-pom on a high-angle mounting for anti-aircraft (AA) fire was actually approved for the British Army before World War I, in February 1914.

Maxim's basic recoil-operated cannon design was scaled up and down, and modified to accommodate the use of different cartridges from the Vickers 1-inch gun (25.4 mm) to the Vickers 2-pounder (40 mm) naval AA gun, as well as versions retaining the 37 mm calibre but with smaller or larger cartridge cases to meet specific needs. In World War I Britain briefly made use of the 37 mm Vickers 1½-pounder installed in warships for the AA role (some also equipping the Austro-Hungarian navy), quickly replaced by the 2-pounder; while the small 1-pounder Mk III aircraft gun also saw very limited use. The USA developed the One Pounder 'Heavy' 37 mm Mk. 6 naval gun, while Krupp produced

*German Maxim 37 mm captured in southern Africa (NJR ZA – Wikimedia Commons CC BY-SA 3.0)*

*Naval 2-pounder pom-pom (Wikimedia Commons, public domain)*

the *3.7 cm Luftschiff Flak* gun, better known as the *Sockelflak*.

Not all such guns were inspired by Maxim's design. In France, Hotchkiss produced *Le canon automatique de 37 mm Hotchkiss Mle 1900/1901*, an enlarged version of the company's gas-operated machine gun, and this achieved some sales. Gas operation was also adopted by Samuel McClean in the USA for his 37 mm cannon which was first tested in 1904. The gun proved too unreliable to pass US tests but a couple of hundred were sold to Russia during World War I and turned up in many places thereafter.

Long recoil was the method of operation chosen for the British 1½-pounder C.O.W. gun (named after the Coventry Ordnance Works where it was developed). This gun used powerful ammunition, but only just got into service at the end of World War I and, after various attempts to arm aircraft with it (some aircraft were even designed around it, but did not see service) the surviving guns saw their last use in the AA role in World War II.

The need to ignite the hydrogen which provided the lift for observation balloons and airships led to the introduction of explosive/incendiary bullets even in MGs, which were clearly in breach of the St Petersburg Declaration (described in the Introduction). However, after World War I lawyers argued that the purpose of the Declaration was to prevent explosive bullets being used against personnel, and using them solely against aerial craft of any type complied with the spirit if not the letter of the agreement. This argument was generally accepted thereafter.

High-Explosive/Incendiary (HEI) MG bullets were really too small to be effective, so interest grew in smaller cannon, firing shells large enough to hold a useful quantity of HEI. The 1-inch Vickers (made in the Italian Terni works) and the similar 25.4 mm Revelli-FIAT did fire HEI ammunition at least later, if not initially, but saw little use in World War I, mostly flexibly-mounted in Italian bomber aircraft. In the early 1920s the Vickers-Terni was sold to Sweden for arming submarines and remained in service to the late 1930s.

Much more significant was the German 20 mm Becker, which was first designed and tested just before World War I. The Becker had a unique operating system, now called Advanced Primer Ignition (API) Blowback. It was very simple, with few moving parts, and had the effect of softening the recoil blow, so was particularly suitable for use in aircraft. Germany made several hundred Beckers during World War I, about half arming aircraft, the others on ground mountings for both AA and anti-tank (AT) use.

After the War, Germany was prohibited from making such armaments, so further development was transferred to Switzerland and undertaken first by SEMAG and later by Oerlikon. By the mid-1930s these guns, in three versions (all 20 mm calibre but with three sizes of cartridge: 20 × 72 RB, 20 × 101 RB and 20 × 110 RB) were thoroughly developed and available on the open market. However, relatively little attention was paid to them until World War II created a huge demand for aircraft and AA cannon.

Oerlikon was not the only manufacturer developing cannon in the early interwar period. Rheinmetall of Germany nominally transferred its prohibited cannon development work to its Swiss subsidiary, Solothurn. This was based on the experimental Ehrhardt recoil-operated 20 mm gun of World War I, and while the 20 mm calibre was retained, the cartridge size was substantially increased, from 20 × 70 RB to 20 × 105 B and 20 × 138 B. However, very few sales initially resulted.

The other arms manufacturer offering 20 mm cannon at this time was the Danish company, Madsen. They had made their name with a light MG early in the century, which had many sales worldwide and remained in production until at least the 1950s, but was never adopted by a major army. Much the same could be said of their 20 mm cannon, which like many early guns of this type was marketed as a dual-purpose weapon: for AT purposes on a low mounting and AA on a high-angle one. With a 20 × 120 cartridge it was more powerful than most, and a cyclic rate of fire (RoF) of 400 rpm was par for the course at that time. It saw service with various armed forces, was still being marketed by the company in the 1950s, and ammunition was still being made for decades thereafter.

*2 cm Becker (left) with vertical box magazine, alongside a belt-fed 7.9 mm MG 14 Parabellum (Courtesy Philip Jarrett)*

# 1933–1940

In 1933 automatic cannon were still a rarity in service. Most armies didn't see the point of them, most aircraft struggled to carry them, and only navies had really identified a clear need for them. They were very aware of the threat to their ships posed by bomb and torpedo-carrying aircraft and wanted to be able to shoot them down before they could release their weapons. That meant a layered gun defence, with manually loaded guns of 76 to 127 mm calibre (3 to 5 inch) as the outer layer, something in the 25 to 40 mm calibre range (preferably in multi-gun mountings) for the middle layer, and MGs (initially) or 20 mm cannon (eventually) mainly on unpowered mountings as the inner layer.

## NAVAL AA

The principal issues in selecting naval guns for the middle layer of defence at this time were: the calibre (range versus volume of fire); the choice between a high-velocity manually-loaded gun or a lower-performing automatic; the ship impact of the mounting – size and weight (linked to the number of barrels required per mounting); what kind of ammunition feeds were more suitable for extended engagements; and whether barrel cooling, power operation or even stabilisation of the mounting, were worth the added complication.

It should be noted that the tactical situations of navies required different solutions from the armies: navies expected their warships to be subjected to successive waves of air attacks, putting a premium on the ability to sustain continuous fire for an extended period, something which required water cooling or very quick-change barrels. On the other hand, weight was generally less critical aboard ship, so large, multi-barrel mountings were attractive, particularly when coupled to more sophisticated sighting systems.

The method of aiming was an important factor; MGs and light cannon used open sights so their accuracy was greatly dependent on the gunner's skill. The larger automatic cannon mountings commonly featured director control, which added to the mounting and manning demands.

Powered mountings offered much faster training and elevation rates and (if coupled with stabilisation to compensate for ship rolling and pitching) greater accuracy, but were more expensive and vulnerable to loss of power through action damage.

Also of concern was the crew required to man the weapons (as warships were commonly overcrowded in wartime) and the maintainability of the equipment; World War II experience led to simplicity and reliability becoming highly valued.

The most prominent cannon, in terms of service with a major nation, was World War I-vintage British 2-pounder pom-pom naval AA weapon in a single mounting, nearly 800 being made for the RN, and many export sales being achieved. The pom-pom was unusually advanced for the time in having a belt feed for the ammunition (initially fabric), plus a water-cooled barrel to sustain a high rate of fire. This was supplemented by a modified gun in an eight-barrel (and subsequently four-barrel) mounting from 1930 but production was initially very slow. Furthermore, even as the octuple mounting began to enter service there was some concern about its adequacy. However, an attempt to replace the 2-pounder with a more modern system failed; instead of this, a high-velocity loading of the pom-pom was developed.

One other naval AA cannon was developed in Britain in the 1930s: the high-velocity Vickers 25.4 mm gun. The only user was Argentina, and the only ship it was fitted to was a training cruiser launched in 1938, so it was of no practical significance.

Italy took the needs of the navy more seriously than most, the 37 mm Breda 37/54 being an automatic weapon available from the start in a M32 twin mounting with water-cooled barrels (although later models were air-cooled) and, very unusually for a system in this class, a limited stabilisation system.

*Quad Pom Pom mounting, showing the steel-link ammunition belts used by the later Marks (RAN)*

*Two quad 1.1-inch mountings on the USS Enterprise (USN)*

*This view of a triple mounting for the Japanese 25 mm Type 96 AA gun (originally naval, this one in land service on Guadacanal, October 1942) shows the three vertically-mounted 15-round box magazines (USMC Photo)*

The USN was slow to introduce an equivalent system; the 1.1' Mk 1/1 gun was first tested in 1933 but was not available in quantity until 1940. The system proved to require expert maintenance, the rate of fire was low for the calibre, and the projectiles were also rather small for a gun in this class, so it was rapidly eclipsed in popularity by the 40 mm Bofors as soon as that became available.

The other major navy to select a calibre smaller than 37 mm for its mid-layer defence was Japan, which bought the 25 mm Hotchkiss to replace the Vickers 2-pounder. The gun was designated the Type 96 (indicating that it entered service in 1936) and was fitted into single, twin and triple mountings. Vast numbers were made, but there were criticisms: the air-cooling and magazine-feeding greatly limited the sustained rate of fire, the calibre was rather small, and there were problems with vibration, slow training and elevation rates, and inadequate sights.

*The Bofors factory in 1936, showing the production line for naval 40 mm twin mountings, with water-cooled barrels*

Ironically, the French navy did not order the 25 mm Hotchkiss until too late (although a few were fitted to French ships after the German invasion). Their existing systems, the 37 mm Mle 1925 (single barrel) and Mle 1933 (twin barrel), used manually-loaded air-cooled guns firing high-velocity ammunition, as did the German 37 mm SKC/30 and the Russian 45 mm 21-K; all were limited to firing at about 30 rpm. These nations all attempted to replace their systems with automatic cannon, with varying results. France held a competition which resulted in the 37 mm Mle 1935 being chosen, in a twin mounting, but it was still at the trial stage when Germany invaded in 1940. Germany developed the little-known but apparently Bofors-inspired 3,7 cm SK C/36, but few were built. Its unique ammunition as well as general design survived to be used in the Flak M42, which only entered service late in the war, followed by the Flak M43, which used the same gun and ammunition as the army Flak 43. Only the Russian navy did the obvious thing from the start and fitted the 37 mm M1939 developed for the army, which served alongside the 45 mm guns from 1941. However, these army-pattern guns relied on air-cooling, limiting their sustained firing rates.

This brings us to the most successful of the medium-calibre cannon developed in this period: the 40 mm Bofors. The first model was introduced in 1932 but sales didn't really get going until the middle of the decade with the M36 twin naval mounting. The navies who were the biggest users of the Bofors in World War II were the RN and the USN, but production was initially unable to keep up with demand so they did not get the guns into service until the 1941–50 period covered in the next section. Bofors also produced guns in 25 mm and 20 mm calibre in this period, but they were nowhere near as successful. The 25 mm M/32 did see service in the Swedish navy and also inspired the development of a Russian naval system in this calibre, introduced later in the war. The 20 mm m/40 was mainly used by the army.

Now to the inner-layer naval systems, or 'last ditch defence'! The MGs fitted prewar proved inadequate in this role, being steadily replaced with shell-firing cannon of 20 mm calibre, almost invariably in a simple, free-swinging, single-barrel unpowered mounting with open sights.

Barrels were air-cooled, and magazine feed was almost universal. Several models were available, from Madsen, Breda, Lahti, and Rheinmetall, but by far the most popular was the largest and most powerful of the Oerlikons, the Type SS, which was made and used in vast numbers by both the RN and USN.

The 20 mm Rheinmetall mounting was the L/30, using their C/30 gun (the same as the land-based Flak 30), which weighed

The gun crew of a 20 mm Oerlikon aboard a Greek warship in the Mediterranean (Wikimedia Commons / public domain)

*The naval experience of the British with the Hispano-Suiza proved the superiority of the Oerlikon. After the fall of France supplies of Oerlikons from Switzerland were cut off, and as a stop-gap Hispano guns were installed on several escort vessels. The results were uniformly bad. Ships reported frequent stoppages from the unsatisfactory design of the cradle, salt water corrosion around the breech block, unlocked tappets, and the collapse of the magazine springs… when water got into a hot gun barrel stoppage was likely at the reopening of fire. The guns were so unreliable that one vessel reported that she preferred not to open fire with her Hispanos in the hope that the Germans would think that she was not a warship. Opposed to this, British experience with the 100 Oerlikons at sea in November of 1940 was uniformly excellent.*

420 kg; this became the L/38 when fitted with the Mauser C/38 gun, which first appeared at the very end of this period. Italy used the same Breda 20/65 M35 as the army, in various mountings including a complicated twin in a diagonal mounting (the RM35, featuring power operation and limited stabilisation, and weighing 2,330 kg) in which one gun was above and to one side of the other, plus two single mountings; the little-used M39 and the free-swinging M40 of 312 kg.

The multi-purpose Madsen cannon saw some use on naval mountings, and a few of the Finnish Lahti L34 were fitted to small naval craft.

Some use was also made of the Hispano aircraft cannon in this role, but it was not a success. The USN BuOrd (Bureau of Ordnance) history, as quoted in Friedman's 'US Naval Weapons', recounts the following concerning the Oerlikon and Hispano as naval AA guns:

*'…the Oerlikon offered greater reliability because of [its] greater reserve of power. It was operated by maximum gas pressure, about 23 tons per square inch, and by springs which were correspondingly powerful. On the other hand the Hispano-Suiza was operated by reduced gas pressure of about 8 tons per square inch and by springs which were similarly weak… the forces at work in the Oerlikon were consequently greater than those in the Hispano, and hence such factors as friction, the effect of different elevations, cold weather, rain and the like, were too small in proportion to the operating forces to cause stoppage. Primarily as a result of this reserve power, the Oerlikon did not require skilled service maintenance. One of the greatest advantages of the Oerlikon… was [that] it had a barrel which could be replaced in about 30 seconds. Since an airplane gun fired very short bursts, and was mounted in the wings of a plane moving through the air at a high rate of speed, it did not present much of a cooling problem. The Hispano was therefore designed without any thought of permitting a change of barrels during action; in fact, it was hardly possible under such conditions. At sea, however, against multi-plane dive-bombing attacks, prolonged firing was necessary… The magazine employed with the Oerlikon was superior… the Oerlikon magazine could be kept fully loaded without any tension on the springs while in the Hispano the spring was compressed as long as the magazine was loaded, a condition which brought about occasional spring collapse. Secondly, the magazine had a greater reserve because of the easier task of feeding at 490 rpm as against the Hispano's 690. Thirdly, the Oerlikon magazine had a straight tangential lead through its mouthpiece, whereas the Hispano had a 90 degree bend…'*

## LAND AA

Most armies showed very little interest in AA cannon until the end of the 1933–1940 period. The typical approach was to field manually-loaded 75+ mm guns, backed-up by MGs for close defence. What changed this attitude was the Spanish Civil War (1936–39), an intense conflict in which several countries became involved, mainly for ideological reasons but they also took the opportunity to test how their latest weapon systems and tactics worked in practice. The Republicans (loyal to the existing democratic government) were supported by Russia, while the eventually successful Nationalists led by General Franco were supported by Germany and Italy. France, the UK and the USA stayed officially neutral but their militaries observed developments with professional interest.

One of the lessons of this conflict was the devastating potential of air attacks and the inadequacy of existing AA systems to deal with them: the 75+ mm guns were too slow-firing, and their elevation and training rates too low, to engage fast-moving planes, particularly at low level, while MGs were not destructive enough. The result was a sudden interest in cannon for AA purposes. The initial focus was primarily on medium-calibre guns (34 to 40 mm) which were in many cases similar to those being sought for middle-layer naval defence. However, the requirements of land forces were somewhat different from the naval ones. Mobility was far more important, so minimising weight was a priority and single-gun mountings with air-cooled barrels plus manual training and elevation were the norm. The lack of sustained fire capability which resulted was not regarded as much of a problem, as it was expected that air attacks on ground forces would be more sporadic and dispersed than attacks on ships, with engagement opportunities for the guns being brief and widely separated.

The British Army initially took the logical step of adopting a version of the naval 2-pounder pom-pom, but when the 40 mm Bofors gun was evaluated in 1937 its superiority in terms of mobility, ballistics and reliability was so obvious that the Army changed tack and ordered the Bofors instead, the few 2-pounder which had been delivered being used for static defence before ultimately being handed to the RN. Between 1937 and 1939, orders for more than 500 guns were placed with Bofors, with licensed domestic production beginning in mid-1939. The typical towed Bofors mounting for land use weighed around 2,000–2,500 kg.

The US Army was the other major, large-scale user of the Bofors (although many other armies bought it in smaller quantities), but they used it alongside their Browning-action 37 mm M1.

Russia used the 37 mm M1939, derived from a Bofors design. Germany had the benefit of Rheinmetall's considerable resources and experience, so they had the 37 mm Flak 18 cannon in service in 1935; this was rapidly replaced by the Flak 36 and then the 37, but the changes concerned the mounting and sights, not the gun. Italy had the Breda 37/54. All of these guns had a comparable performance in terms of ballistics and rates of fire, and all of them used magazine or ammunition strips for feeding, but the quality of their mountings and sights (crucial issues in determining effectiveness) is more difficult to evaluate.

One of the earliest army AA cannon to be developed was the French Schneider 37 mm Mle 1930, using its own unique and quite powerful ammunition. Interest at this time was negligible, however, and the gun did not achieve any significant sales until the French Army realised that war was looking increasingly imminent, and their preferred 40 mm Bofors would not be available for some time. An interim order for the Schneider was placed, but only a handful of guns made it into service. A 37 mm Hotchkiss saw even less use.

A very interesting system in this class was the Swiss 34 mm Flab K38. This was unusual in two respects (in addition to its calibre): it had belt feeding and a water-cooled barrel, thereby enabling a much higher sustained rate of fire than most other guns in this class. Even the cyclic rate was much superior to other pre-war systems in this class, as it fired at 250–270 rpm instead of the more usual 120–160 rpm. Only the Swiss acquired it, but it stayed in service until the late 1960s.

This takes us the 'inner layer' cannon, which were of 20 to 25 mm calibre. Before the fall of France, the 25 mm guns in service were mainly confined to a couple of hundred M1940 twin Hotchkiss mountings with the French army. The Russian army adopted the Bofors-based 25 mm M1940 which entered service in the following year.

*40 mm Bofors in Dutch Indonesia, post-World War II*

*Oerlikon 20 mm gun on an interesting anti-tank mounting*

The 20 mm guns were generally marketed as dual-purpose AA and AT guns (in the 1930s the AP ammunition of many of them could penetrate most tanks). This was of course dependent on the mounting, a high-angle mount being needed for the AA role, a low tripod for AT; some were convertible between the two. All cannon were air-cooled. The great majority of the guns were fed by magazines: either box magazines typically of 20 rounds (mounted horizontally or vertically) or drums (typically of 60 rounds, but both smaller and larger were used).

Some used strip-feed, in which clips of cartridges, held together at the base by metal strips, were placed on trays feeding into the side of the gun (permitting continuous feed without the need to pause for a magazine change – but the ammunition was exposed to the elements instead of being protected). Only a small minority were belt-fed.

First in the field were Oerlikon, with variants of their Type L and Type S; photos exist of an anti-tank model on a tracked trailer, for towing behind a small tracked vehicle. Madsen were marketing their 20 mm Model 1935 with some success, and in the German invasion of 1940 the Danish army knocked out several light AFVs with them. The Swedish army adopted the Bofors 20 mm m/40. The Swiss had their 20 mm Flab K38 which again differed from most other guns in this class in using a belt feed rather than magazines.

In Finland, Lahti introduced the M40 but the designation is somewhat misleading: the prototype was commissioned in 1939 but due to the lack of German *Flakvisier* sights, it did not get into service until much later. Italy had two 20 mm AA cannon from different designers: the Breda 20/65 M35 came first, followed in 1938 by the Scotti 20/77. These two guns, and the Lahti, all used Rheinmetall's 20 × 138B ammunition.

Rheinmetall put a great deal of effort into developing effective 20 mm cannon, initially marketed through their Swiss subsidiary, Solothurn, as a result of treaty limitations on German arms production following World War I. The first into service was the Flak 30, in 1935. From 1940 this was supplemented in service by a modified version by Mauser, the Flak 38, which had a higher rate of fire (420–480 rpm rather than 280 rpm). There was a lightweight mounting for mountain and airborne troops (the 2 cm *Gebirgs Flak*) and a successful quad mounting, the *Flakvierling*, much feared by Allied close-support fighter-bombers. Different versions of these cannon (the KwK 30 and 38) were mounted as light AFV armament. All of these were fed 20 × 138 B ammunition via horizontal magazines.

The Imperial Japanese Army (IJA) adopted 20 mm as their standard light AA cannon calibre. Their first attempt was a massive 20 × 157 cartridge in a modified Rheinmetall/Solothurn gun designated Type 94, but this was unsuccessful so quickly replaced by the Type 98, firing 20 × 143 ammunition. Despite an effort to copy the Bofors they never did manage to get a larger-calibre AA cannon into service, although later in the war they acquired some 25 mm Type 96 mountings from the navy.

*20 mm Madsen in the AT role; a 40-round drum magazine is fitted, belt-feed was available*

Poland and Russia, as well as Germany, mounted 20 mm cannon in some of their light AFVs for ground fighting.

## AIRCRAFT CANNON

The French built on their World War I experience of manually-loaded 37 mm calibre *moteur canon*, this time focusing on 20 mm cannon. One benefit of such a centrally-mounted position was that the projectile trajectory remained in line with the sights at all ranges, without the lateral harmonisation issue of wing-mounted guns (although the sights still needed to be adjusted for vertical harmonisation at some specified range). Other advantages of *moteur canon* were that the weight was concentrated along the centreline rather than spread along the wings, which helped with the aircraft's roll-rate; the gun could be rigidly attached to the crank case to help with recoil control (some strengthening of the crank case was required); there was likely to be room for a bigger magazine in the fuselage; and (in some cases) it permitted the use of a bigger gun than would be desirable in the wings. Cowling or wing-root mounted guns also enjoyed some of these benefits, but they needed to be synchronized with propeller rotation to avoid the projectiles hitting the blades, which added complication and reduced the rate of fire. These remarks are of course concerned with conventional front-single-engined fighters with tractor propellers; even small twin-engined fighters could group several cannon in the nose (e.g. the Westland Whirlwind), and there were a few unconventional single-engined fighters with pusher props but these were not successful.

In the early 1930s Hispano-Suiza developed a liquid-cooled V12 engine with a 20 mm cannon firing through the propeller hub. It was almost essential to design or at least considerably adapt the engine specifically for such an installation, because the space between the cylinder banks (through which the gun barrel would need to run) would normally be filled with items such as intake ducting, carburettors or fuel injection systems, superchargers and other essential impedimenta. The propeller needed to be designed with a hole in the centre of the hub and its location also needed to be adjusted so that the hole lined up with the gun barrel. And of course, there was no hole through the centre of a radial engine, so such installations were impossible with this engine type. Evidently, it would not have been feasible to take an existing fighter designed for wing armament and fit it with an engine gun, without some major redesign.

Hispano-Suiza's choice of *moteur canon* was the biggest of the Oerlikon cannon which it licence-manufactured under the HS 7 and HS 9 designations (the differences being minor). The engine and cannon combination was fitted into the Dewoitine D 501 fighter, which entered service in 1934, and subsequently into the D 510. The Oerlikon gun family continued to be developed even after this, with the guns becoming steadily lighter and faster-firing. By 1935 they were close to their ultimate in performance, in the FF, FFL and FFS, the last of these being a lightweight version of the Type SS AA gun. The smaller aircraft guns were primarily used by Germany as the MG-FF (the Oerlikon FF with minor modifications) and Japan, as the naval Type 99–1 (FF) and Type 99–2 (FFL). The earliest user, however, seems to have been Poland, which mounted two of the AF or FF cannon in the PZL P.24 from 1934 (service entry 1936).

Meanwhile, two inventors tried to improve on the Oerlikon API blowback design by introducing a locked breech which enabled the reciprocating bolt to be lighter and thereby faster-moving, to increase the rate of fire. The first of these was the Italian designer Alfredo Scotti, followed by the Swiss Marc Birkigt. Scotti developed more than one aircraft gun but his only success was the 20 mm AA cannon which made it into service. Birkigt, working at the French branch of Hispano Suiza, developed the 20 mm HS 404 aircraft gun whose test-bed performance was so impressive that it was not only adopted by France (it entered production in 1938, and in service with the M.S.406 and D 520) but also by the UK and the USA. However, it proved a troublesome weapon to develop, with various installation issues which took time to resolve, making it too late for the Battle of Britain. Like the Oerlikon guns, it was initially fed by a 60-round drum magazine, with belt feed entering service in RAF planes in 1941.

The Russian Polikarpov I-16 Tip 17 was another early user of contemporary cannon; this variant entered service in 1937 armed with a pair of 20 mm ShVAK in the wings; these enjoyed the benefit of belt feed, permitting a significant increase in ammunition load. Russia also acquired a licence to build the Hispano V12 engine, and became a prominent user of engine-mounted cannon.

Germany initially tried the French motor-cannon approach by fitting the very powerful 20 mm Rheinmetall-Borsig MG C/30L into a prototype He 112 fighter which was sent to Spain for combat testing. It was found to be effective against vehicles, but too slow-firing at 300–350 rpm for aerial combat, as well as too heavy. The ammunition feed was of interest: a 100-round drum mounted under the gun. The cannon were later put into use as light AA weapons, fitted with shorter barrels (the aircraft version had a very long barrel, to fill the space all the way to the propeller hub). While this aircraft installation was unsuccessful, the motor-cannon concept was later revived.

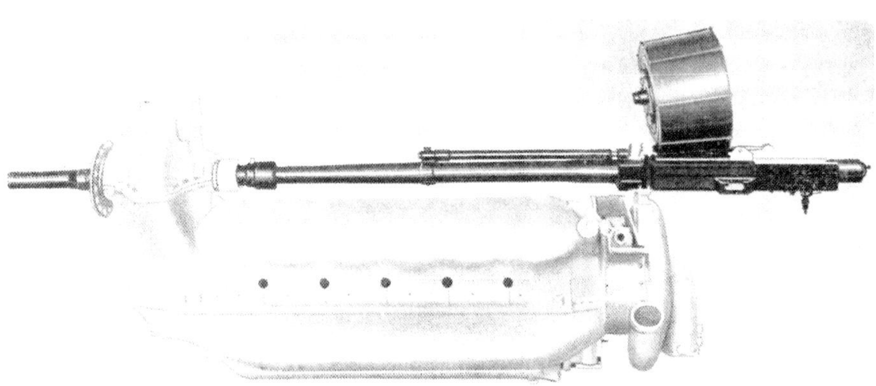

*HS 404 mounted on engine block as designed*

Rheinmetall-Borsig developed other aircraft guns at this time. A lot of effort went into evolving 20 mm cannon using less powerful ammunition than the MG C/30 L, resulting in the MG 204 of which only a small number was built and deployed by the *Küstenfliegergruppen* (coastal air service). More successful was the company's 30 mm MK 101, firing very powerful ammunition. This was primarily used in ground attack, first being fitted to a dozen Messerschmitt Bf 110C-6 in 1940. Both guns were magazine-fed.

Other aircraft cannon developed in this period included:

- Madsen 23 mm: identical to the 20 mm except for the increased calibre: belt feed standard. Tested by various countries (including the USA) but only a small number were sold in South America.
- 20 mm Furrer FMK 38: aircraft version of the AA gun described above, saw Swiss service both in engine-mounted installations and in twin-engined planes.
- 37 mm M4, developed from a Browning design, which went on to see service in the Bell P-39 Airacobra.

Various French aircraft cannon were developed in the 1930s in 33 mm, 25 mm and 23 mm calibres, but apart from these being installed in a handful of aircraft mainly for test purposes, they saw little or no use.

Last but far from least, the MG 151/20 entered production in 1940 and went on to be the *Luftwaffe's* principal aircraft cannon in World War II.

# 1941–1950

This section encompasses the period of the most urgent development of cannon at the height of their importance. Their principal use was against aircraft, for which purpose they were fitted to warships, ground vehicles, towed mountings, and other aircraft. Some use was also made of cannon for arming AFVs for ground fighting (although this developed to a far greater extent later), and for arming aircraft for ground or naval attack.

The primary concern was initially to obtain enough cannon which were suited to these purposes as well as meeting reliability requirements (particularly for installing in aircraft, when repairs and maintenance in flight were not feasible). Even with competent and experienced cannon engineers (a rare species) the process of designing, developing, debugging and manufacturing a new design took years. The cannon which had most impact in World War II were therefore those which were either already in production at the start of the war, or for which a great deal of development work had already taken place. Development during the war was mainly focused on incremental improvements to existing designs or, in some cases, a concentration on simplifying the manufacturing process to aid mass production.

In the years after World War II some radical innovations such as revolver and rotary cannon were developed, but with the loss of wartime urgency (and reduced resources) these were not introduced into service until the 1950s.

## GROUND AND NAVAL ANTI-AIRCRAFT SYSTEMS

With a few exceptions, the cannon used in the AA role were multi-purpose, being used in land or naval mountings as required. Most of them had already emerged before World War II and were mentioned in the previous section.

Wartime experience revealed the practical superiority of the 20 mm Oerlikon and 40 mm Bofors over their contemporaries with their principal development emphasis being on simplifying manufacture for mass production (both guns initially required a lot of hand-finishing). Both the UK and USA were involved in this process; production time required for the Bofors dropped by around 50% while that for the Oerlikon was reduced by no less than 80%. The USN formally decided to adopt the 20 mm Oerlikon to replace the .50 Browning on 7 August 1941, at the same time as they decided to adopt the 40 mm Bofors to replace the 1.1-inch system.

A variety of naval mountings for the Oerlikon gun were developed for both the RN and USN. Most were single, unpowered, free-swinging mountings, but there were also twin mountings, both unpowered and powered. Some of the British powered Mk V twins were converted to take one 2-pounder Pom Pom, others to take a 40 mm Bofors gun (a system designated 'Boffin') as a result of the advent of kamikaze attacks, as they stood a much better chance of destroying the aircraft. Both navies also developed powered quad mountings, the Mk 15 for the USN (which saw limited service) and the Mk XIV, using belt-fed Oerlikons, for the RN (which did not, as the RN decided that fitting a pair of Bofors guns made a better use of the mounting).

The Polsten, a drastically simplified version of the Oerlikon, was used by the British army rather than the navy (presumably to avoid any logistics muddle) and was fitted to a variety of wheeled and vehicle mountings, including double, triple and quadruple (Canada having developed and manufactured the quad). It was in service between March 1944 and the 1950s.

The naval Bofors mountings saw considerable development in UK and US service. The RN equipment ranged from army-pattern air-cooled guns on single naval mountings (which made up most of the installations) to twin mountings featuring water-cooled barrels and in some cases stabilisation and even an on-mount radar.

While the 20 mm Oerlikon remained in USN and RN service in vast numbers to the end of the war, the British Army showed less interest in the light cannon,

*Bofors STAAG mounting (NavWeaps)*

preferring improved models of 40 mm Bofors. This view was probably influenced by operational research into the relative effectiveness of the 20 mm Oerlikon and 40 mm Bofors in dealing with low-flying coastal raiders. It was noted that these were not usually spotted until 1,500 yards (1,370 m) away, so the Bofors could not use its range advantage. Engagement times were only about 13 seconds (from first sighting the target to the last shot) and the Bofors only had time to fire seven or eight rounds to the Oerlikon's fifty. Even so, in most circumstances the Bofors was more effective.

Calculations were made about the probability of success (immediate break-off, or eventual crash) against a Junkers 88. Using simple sights only, the Oerlikon's chances were estimated at 3% against a crossing target, 12% against one approaching head-on; the Bofors' 18% and 20% respectively. With more advanced sights, performance was much better. Using predictor control, the Bofors' success rate rose to 35% and 39%. The addition of a Triple Gyro sight had an even more dramatic effect on the Oerlikon's performance – to 17% and an astonishing 76% respectively.

In USN service, the Bofors accounted for 33% of all Japanese aircraft shot down by AA fire during World War II, compared with 28% for the 20 mm, 30% for 5 inch guns, and 3% for MGs. The Bofors was credited with 50% of the total between October 1944 and March 1945, showing how its importance increased as the war progressed. In terms of the number of rounds fired for every plane shot down, the Bofors averaged 1,713; the 1.1-inch 1,932; the Oerlikon 5,287; HMGs (.50-inch cal/12.7 mm) 11,143; and MGs (.30' cal/7.62 mm) 28,127. At the other end of the scale, the 5' guns firing time-fuzed ammunition managed 654 rounds per plane; when firing proximity (VT) fuzes this improved to 340 rounds.

While generally considered inferior to the Oerlikon in the AA role, at the end of the war the Hispano HS 404 was being proposed for such use by the USN, for example in 'Assembly 334' with quad guns as an anti-kamikaze weapon capable of 2,800 rpm, but this was too late to see action in World War II.

Although Japan had criticisms of the 25 mm Type 96, it was nonetheless rated by users as the most reliable Japanese AA weapon of World War II. Late in the war it saw some use on land, with a two-wheel carriage being designed for it. The maximum effective range was 2,000 m, with the system being most effective at under 1,000 m. It took an average of 1,500 rounds to shoot down an aircraft at a range of 2,000 m and an altitude of 1,000 m.

Japan captured numerous 40 mm Bofors guns from both the Netherlands and the UK early in the war, manufactured the ammunition and tried to reverse-engineer the guns, but too late to have any effect.

Germany saw significant development of AA cannon during the war, although new guns arrived too late to see much service with the exception of the 2 cm Flak 38 which was just beginning to enter service in 1941. In the second half of the war the formidable *Flakvierling* emerged as a major threat to Allied fighter-bombers.

Others which entered service were the 3,7 cm Flak M42 naval gun (which seems to have been an improved version of the unsuccessful SK C/36) and the 3,7 cm Flak 43 intended for both land and naval use which fired the standard 37 × 264 B, but neither was in time to make a difference. The same could be said of the 3 cm MK 103 aircraft gun when installed in ground mountings: the Flak 103/38 was an unsuccessful combination of the gun with the 2 cm Flak 38 mounting. Another high-performance 30 mm AA gun was being developed in Czechoslovakia under the designation MK 303; this eventually entered service in that country as the M53.

Germany was the only country which put a larger-calibre automatic AA gun into (limited) service, in the form of the 5 cm Flak 41. The gun part apparently worked well enough, but the mounting and sights were unsatisfactory. Late in the war a second attempt was made, in the form of the 5,5 cm *Gerät* 58, but this did not have time to get into service. The Russian 57 mm S-60 just squeezes into this section, entering service in 1950, as does the Bofors 57 mm M/50 naval gun.

Only a few years after World War II, both Oerlikon and Bofors began to replace their wartime guns with faster-firing models designed around more powerful, high-velocity ammunition; a process driven by the much briefer firing opportunities provided by high-speed jet aircraft. Oerlikon introduced two different high-performance general-purpose cannon, the Type 5 TG (later known as the KAB) and the Type 204 GK (KAA), both using a new 20 × 128 round. Bofors introduced the L/70 development of their 40 mm gun firing 40 × 365 R rather than 40 × 311 SR ammunition. Hispano-Suiza became involved in the AA field as a direct rival to Oerlikon, in new 20 mm (HS 820) and 30 mm (HS 830) calibres, while the Russian 25 mm benefited from a major redesign to achieve a significantly higher rate of fire. Mountings and sights also needed to become more sophisticated to provide any chance of a successful engagement.

The older Bofors L/60 guns were not forgotten, with improvements being made to the huge existing stocks of guns. Immediately after the war the British army introduced the Bristol Bofors, which had a powered version of the single towed mount, able to be laid on target by one man instead of two.

Self-propelled anti-aircraft guns (SPAAGs) became more common; initially consisting of little more than standard land or naval mountings fitted to the back of a truck, some became better protected by mounting them in turrets on AFV chassis, with powered elevation and traverse and gyro sights adding to their effectiveness. The high volume of fire made these systems popular for use against ground targets also; e.g. the US M19 with twin 40 mm Bofors, which was too late to see action in World War II but extensively used in Korean ground fighting in the early 1950s.

Purpose-designed cannon for ground fighting remained rare at this time, the German 20 mm KwK 30 and 38 already mentioned being the main exceptions, but Poland also developed a 20 mm gun which saw action in a light tracked chassis. The Russians fitted a modified version of the 20 mm ShVAK aircraft cannon to some light tanks.

*'Wirbelwind' self-propelled anti-aircraft gun on the chassis of the PzKpfw Ausf H tank with the Flakvierling (Narodowe Archiwum Cyfrowe / Wikimedia Commons / Public Domain)*

## AIRCRAFT CANNON

By far the most intensive cannon development in this period concerned the armament of fighter aircraft and, to a lesser extent, ground attack planes (see separate section below). The former was driven by the rapid development of aircraft in the pre-World War II period; in the decade from 1935, combat aircraft in service went from being primarily biplanes with fabric-covered wings to all-metal jet-powered aircraft capable of three times the speed. These developments not only made aircraft much harder to hit, their stronger construction also made them far harder to damage, requiring guns of ever-increasing destructive effect.

These developments did not apply equally to all theatres in World War II. The Japanese favoured light, manoeuvrable aircraft with a very long range, so saved as much weight as they could and did not fit much in the way of armour protection until very late in the war. Most of their aircraft therefore remained vulnerable to machine gun fire, and the batteries of heavy machine guns (HMGs) of 0.50-inch / 12.7 mm calibre carried by US fighters were more than adequate to deal with them.

In the European theatre the USAAF was primarily concerned with combating fighters which were easier to damage than bombers, so again their .50 cal HMGs proved adequate. However, the USN was more worried about bomb and torpedo-carrying planes attacking their warships, so wanted the more reliable destructive effect of 20 mm cannon. Unfortunately, they selected the Hispano which proved even more problematic for the USA than it was for the UK. As mentioned, the USA did develop the Bell Airacobra fighter with a 37 mm M4 or (later) M10 cannon, but Russia was the main user of these.

The RAF primarily faced medium bombers and fighters, and after the Hispano's teething problems were surmounted, their eventual standard fighter armament of four 20 mm cannon proved more than adequate to deal with these targets (the Spitfire usually had only two cannon, plus MGs or HMGs).

After early-war experience with the MG-FF (which remained in service for a long time), Germany adopted the Mauser MG 151, initially in 15 mm calibre but soon modified to 20 mm to be able to use the MG-FFM's high-capacity *Minengeschoss* shells. The resulting MG 151/20 was certainly one of the best – if not *the* best – all-round aircraft guns of World War II. The icing on the cake was that it was also available in an electric-primed version ideally suited to synchronized installations, so it could be fitted into wing-root and cowling mountings (the Hispano, like the Oerlikons, could not be synchronized – at least until the electric-primed M24 version was made by the US post-war).

Russia started the war with the 20 mm ShVAK cannon, one of the most fast-firing guns whose main problem was its relatively light and ineffective projectiles for the calibre. The 20 mm Berezin, based on the 12.7 mm HMG from the same designer, maintained the performance of the ShVAK with a considerable weight saving and much better reliability, but did not come into use until the latter part of the war.

The Japanese had the most varied and interesting collection of 20 mm cannon in service, although the initial focus was more on MGs. As ever, the Imperial Japanese Army and Navy (IJA and IJN) went their different ways. The IJN displayed a great enthusiasm for Oerlikon 20 mm cannon; the Type 99–1 (Oerlikon FF) was in service at the start of the Pacific war, the more powerful Type 99–2 (Oerlikon FFL) partly replacing it later. The IJA showed a similar enthusiasm for Browning MG designs, scaling them up to 20 mm (Ho-5), 30 mm (Ho-155) and 37 mm (Ho-204) calibre. They also adopted the powerful 20 mm Ho-1 and Ho-3, based on a self-loading anti-tank rifle.

In the latter part of the war, Germany and Japan shared a problem which the other nations in World War II escaped: they were both attacked by massed fleets of heavy bombers. Against such targets, the 20 mm cannon were found wanting. In one calculation, the *Luftwaffe* estimated that it took, on average, 20 hits to down a heavy bomber, and average shooting accuracy was such that only around 5% (at best) of shots fired hit the target. This meant that about 400 rounds needed to be fired for each bomber shot down, and that was more 20 mm ammunition than most planes carried.

The German response was to focus on developing 30 mm guns in two flavours: a low-velocity weapon which was light and compact enough to be carried by small fighters (the MK 108); and a high-velocity one for arming the largest and most powerful fighters (the MK 103 – although in practice this was mostly used against ground targets). The Japanese also developed 30 mm aircraft cannon – three of them (the IJA Ho-155 mentioned, plus the IJN Type 2 and Type 5) – although they were too little, too late to make a difference.

Neither country stopped at 30 mm calibre for aerial combat: Germany fitted the 50 mm BK 5 (based on an automated anti-tank gun) to some heavy fighters and, at the end of the war, were working on low-velocity (MK 112) and high-velocity

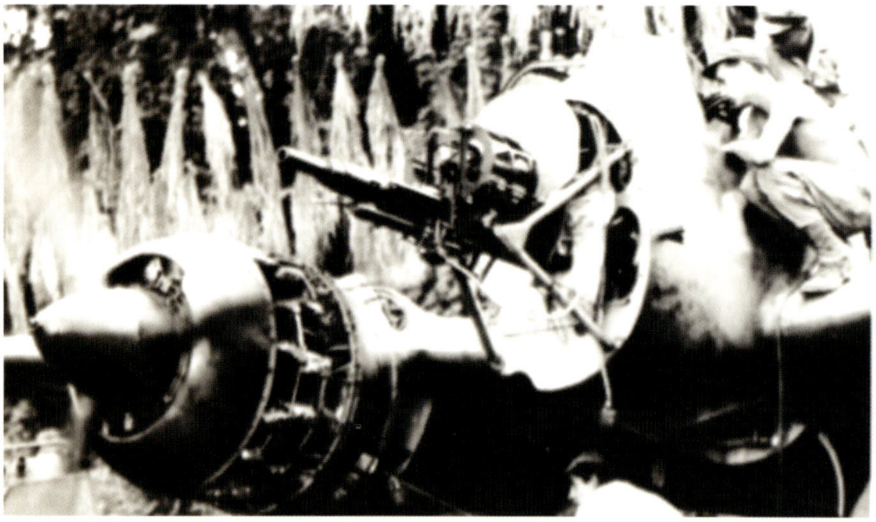

*Japanese 37 mm Ho-203 in the nose of a Ki-45KAI aircraft*

(MK 412) guns in 55 mm calibre, on the basis that one hit from such shells could be expected to bring down a heavy bomber. Japan fielded low-velocity 37 mm (Ho-203), 40 mm (Ho-301) and 57 mm (Ho-401) cannon as well as the more powerful 37 mm Ho-204 already mentioned. Such guns had a negligible impact on the fighting.

Russia may not have had heavy bomber fleets to deal with, but the aircraft of their German opponents were well-protected. A high-velocity cannon (the VYa-23) was developed for use in both aerial combat and ground-attack roles, and by the end of the war a lower-velocity gun in the same calibre (the NS-23), but smaller, lighter and faster-firing, was becoming available. Guns in 37 mm calibre were also fielded, the similarly dual-purpose high-velocity NS-37 saw some wartime service (as an engine cannon in the Yak-9K fighter) with the lower-velocity equivalent, the N-37, in use in the post-war period. Various cannon in 45 mm and 57 mm calibre were tested but not adopted.

## GROUND ATTACK

Any 20 mm cannon could be very useful in the ground-attack role against unarmoured targets, but for the specific purpose of punching holes in tanks, something more powerful was needed. It should be pointed out that airborne AT cannon had two advantages over their ground-based equivalents; the speed of the aircraft added to the velocity of the projectiles, increasing their effectiveness, and the pilot could choose to attack the side, rear or top of the vehicles, where the armour was much thinner than over the frontal arc. Airborne AT cannon therefore did not have to be as powerful as the army anti-tank guns.

Ground attack aircraft of World War II can be divided into three categories: more-or-less modified fighters; light bombers; and purpose-designed close-support planes armoured against ground fire. The modified fighters were most common in the USAAF and RAF, because these forces had no previous requirement for the ground attack role (it was seen by some as making the air force too subservient to the needs of the army). Furthermore, fighters were already available in large numbers, and had the added advantage of versatility, as they were better able to defend themselves against enemy fighters. It was also an attractive way of making use of planes which had been superceded in their designed role. Fighters modified for ground attack missions typically received better armour protection of vital areas but were usually limited in terms of their ability to carry cannon powerful enough to penetrate tank armour. As a result, rocket projectiles (RPs) became popular in both the RAF and USAAF; it was easier to modify planes to carry them, and they could be omitted if the planes were needed in the fighter role.

Unfortunately, RPs were far less accurate than cannon (and bombs even more so). Even when practising in ideal conditions, the RAF hit rate with RPs against tank-sized targets was around 5%; British Operational Research teams discovered that in the stress of combat, this dropped to 0.5% (better sights improved this somewhat). In contrast, the one US/UK fighter which saw action with heavy AT cannon – the Hawker Hurricane (in IID and IV versions, carrying two 40 mm Vickers S guns underwing), achieved around a 25% hit rate, with much less difference between practice and combat. However, these guns became less useful in Europe as tank armour thickened up.

The light bombers proved too vulnerable to both ground fire and fighters, and did not survive for long, except in the general case of carrier-borne naval attack planes which were obviously limited in size. There was another exception: the Ju-87G, which was adapted from its designed dive-bomber role to carry two powerful anti-tank cannon underwing; the 37 mm BK 3,7, a modified AA gun. This achieved considerable success, the development of tungsten-cored *Hartkernmunition* helping to keep the guns effective.

Germany did have available another powerful aircraft cannon which was developed pre-war and used primarily in the ground attack and anti-shipping roles; the 30 mm MK 101. It saw little use until it found its niche from 1942 onwards in the Henschel Hs 129B-1 described below. It was out of production by then, so the gun was gradually replaced by the 30 mm MK 103.

This takes us to the specially-designed ground attack planes. Only two of these saw extensive service, and it is no surprise that these were with Germany and Russia, which were both far more concerned with providing maximum support to their armies and therefore developed aircraft well-suited to the role.

Russia built vast numbers of the Il-2 from 1941 to the end of the war. This was heavily armoured and most of the planes carried a pair of 23 mm VYa-23, but in the later Il-2M of which about 1,000 were built, this was changed to the powerful 37 mm NS-37. The heavy recoil of these guns did cause aiming problems,

*Hans-Ulrich Rudel's Ju-87G (Bundesarchiv, Bild 101I-655–5976–04 / Grosse / CC-BY-SA 3.0)*

however: the Il-2 suffered from being single-engined, using an engine not designed to take a motor cannon, so the guns had to be carried underwing.

This problem did not affect the German equivalent, the compact and equally well-armoured Hs 129. This had twin engines in the wings, leaving the fuselage free to carry a wide variety of armament without any concerns about synchronisation. This layout meant that one really big cannon could be carried instead of two smaller ones underwing; the central gun location made it easier to adjust the sights and also to absorb recoil, and provided more space for automatic loading systems and large-capacity magazines. The most common versions of the Hs 129 carried one 30 mm MK 101 or 103, but some variants carried the BK 3,7 instead, or even the huge BK 7,5 which was essentially an automated version of the standard Pak 40 75 mm anti-tank artillery! The Hs 129 was a highly impressive design, almost perfect for its role although serviceability was not good and its low performance made it vulnerable to fighters.

The British attempted something similar with a version of the DH Mosquito high-speed light bomber, the FB Mk XVIII 'Tsetse', which was armed with a 57 mm Molins gun (an autoloading version of the army 6 pounder AT gun). This combination proved exceptionally accurate and effective in testing. However, while the plane carried additional armour there was clearly some concern about its vulnerability to the formidable German light Flak, and with the RAF deciding to use RPs for the AT role instead, the Tsetses were handed over to Coastal Command without ever being used against tanks. The small number in service achieved modest success in the anti-submarine and anti-ship roles.

## THE AFTERMATH

The end of World War II saw the production of most of the cannon of the Axis powers cease (although some equipment remained in service with smaller armed forces for decades). The two notable exceptions were from Mauser: the MG 151/20 and the MG 213C. The MG 151/20 saw continued service for some years with France and is still made in South Africa as the GA-1. The fact that it is significantly lighter than most other 20 mm cannon, and with a lot less recoil, has enabled it to find a niche arming light vehicles and helicopters. The MG 213C was the original of the multi-chamber but single-barrel revolver cannon; the UK and France continued the development of the 30 mm version which entered service as the ADEN and the DEFA series respectively. Switzerland, the USA and Russia produced more heavily modified versions of the principle.

The development of conventional cannon was not forgotten, however. The Hispano-Suiza HS 404 was taken over by the Swiss branch of the firm and renamed HS 804. Bofors developed their own recoil-operated 20 mm aircraft guns around the Hispano's ammunition; these only saw service with Sweden. They also produced the rather odd medium-velocity 57 mm m/47 aircraft gun, intended for the anti-shipping role, which saw service in a twin-engined light bomber. In the late 1940s Russia introduced the faster-firing NR-23 to replace the NS-23, firing the same ammunition, and also the N-37 in 37 × 155 calibre, a faster-firing, lower-velocity but much smaller and lighter gun than the NS-37.

# 1951–1969

Development during this period was a combination of technological improvements on the one hand, and on the other, doubt about the future of cannon in the light of the promise of guided missiles. For example, the French were initially unwilling to spend too much on the 30 mm DEFA aircraft revolver cannon as they expected that it would only be a temporary solution. For the AA role, the Royal Navy never adopted the 40 mm L/70 Bofors gun, putting their money into the Seacat manually-guided short-range AA missile instead (they ended up keeping the basically World War II-era 40 mm L/60 Bofors for longer than the Seacat). The nuclear-powered missile cruiser USS *Long Beach* was designed in the 1950s without any guns at all; it was well-equipped to counter major warships, warplanes and even submarines, but had nothing to engage other kinds of target, for example small craft or coastal defences, or to fire warning shots. Some guns were added later.

The faith which some users placed in the guided missile was initially misplaced. Success rates for missiles achieved in testing under ideal conditions turned out to have no bearing on the actual effectiveness in combat, particularly for air-to-air missiles which suffered from being carried externally and battered by the elements; hit rates in the 1960s typically dropped to around 10%. Aircraft designers had been happy to omit guns, which cause problems with recoil, vibration, and propellant chemicals affecting the airframe and being sucked into the engines (potentially causing flameout). However, once the lessons of combat had been learned, air forces scrambled to fit guns to their fighters once again.

For much of this period the main driver affecting AA and aircraft cannon development was again the need to shoot down faster and stronger planes, calling for an increased rate of fire and/or more destructive ammunition. There were (and still are) three principal sources of cannon design and production, which took different approaches: the USA, Russia, and Western Europe. The last of these was originally composed of a diverse group of companies, mainly in Switzerland (Oerlikon and Hispano-Suiza), Germany (Rheinmetall and Mauser), France (DEFA, later GIAT) and to a lesser extent the UK (Royal Small Arms Factory at Enfield Lock); Bofors in Sweden concentrated on improvements to the 40 mm and 57 mm systems introduced after World War II. Occasional mergers and closures have reduced the number of firms since then.

## ANTI-AIRCRAFT SYSTEMS

As mentioned in the previous chapter, new 20 mm AA guns were introduced in Western Europe, faster-firing and using higher-velocity ammunition. These were all based around two new cartridges of virtually identical performance: the 20 × 128 Oerlikon and 20 × 139 Hispano-Suiza. The 20 × 139 has been more widely used, not just in the original HS 820 but also in the Rheinmetall Rh 202 and French 20 F-1 and F-2 (later designated 20M693) also made in South Africa as the G1–2. The Oerlikon round was used in various products of this company, namely the Type 204GK (now KAA), Type 5TG (now KAB), and Type FZK (now KAC).

Despite developing the 37 mm T249 Vigilante SPAAG (designed around the T250 six-barrel rotary cannon) in the 1950s, the USA did not adopt any new cannon specifically for AA. Instead, they adapted their 20 mm Vulcan aircraft cannon (see below) for AA use as the M168 gun, in two mountings: towed (M167) and self-propelled (M163).

Doubts began to be raised about the adequacy of the 20 mm AA guns in dealing with jet aircraft. The Russian army already had a very powerful HMG – the 14.5 mm KPV – and in the cannon class preferred AA guns of larger calibre than 20 mm. They selected 23 mm and designed new guns around the powerful 23 × 152B cartridge, an uprated version of the one used in World War II VYa-23 aircraft guns. The guns were the 2A14 (used in the ZU-23 towed twin mounting) and the 2A7, a water-cooled version in a quad mounting for the ZSU-23–4 *Shilka* tracked SPAAG.

The Russian navy retained the 25 mm calibre (albeit with a modified 25 × 218 cartridge), introducing first the 110-PM in 1950, then the faster-firing M-110 in the late 1950s; the twin gun mountings were designated 2M-3 and 2M-3M respectively. From 1960 onwards these

*ZSU-23–4 Shilka (Bukvoed – Wikimedia Commons – CC-BY-SA-3.0)*

were replaced by 30 mm guns in the form of the NN-30 revolver cannon, installed in a remotely-controlled, radar directed, twin mounting designated AK-230.

At the end of World War II Germany was trying to get 30 mm AA guns into service, with negligible success. One project, the MK 303 gun developed in German-occupied Czechoslovakia, was continued by that country after the war and adopted in the 1950s as the M53. Initially this was in a towed twin mounting but it is better known mounted in a 6x6 SPAAG; the M53/59, and modified M53/70. It was distributed to several other countries friendly with the Warsaw Pact, but is now probably only held in reserve.

Another high-velocity 30 mm intended primarily for AA use was a Hispano-Suiza project dating from around 1940. Originating in the French branch as the HS 411, this was evaluated (unsuccessfully) in the UK in 1942 and was taken over by the Swiss branch as the HS 830 (later modified and redesignated HS 831). During the 1950s it was acquired by France for use in towed mountings, with the AMX-13 DCA tracked SPAAG following in the late 1960s. It saw wider use in naval mountings, both single and twin, but with few sales until the 1970s.

Oerlikon skipped the 30 mm calibre for AA guns, focusing instead on larger and more powerful cartridges. After an unsuccessful attempt to introduce the 42 mm 421 RK revolver system (paralleled by a British effort in the same calibre) Oerlikon concentrated on the 35 mm calibre.

*Oerlikon Twin Field AA Gun GDF-005 (Oerlikon Contraves)*

The KD series guns were developed to use the resulting 35 × 228 round, the KDA being for use in SPAAGs and naval mountings, its earliest and best-known application being the German Gepard tracked SPAAG. Other models (KDB and KDC) were used in towed mountings.

During World War II the principal Russian naval AA gun in larger cannon calibres was a manually-loaded 45 mm, and after the war an automatic gun was developed to fire considerably more powerful ammunition in the same calibre. This was the 45 mm SM-21, which only saw limited service before the Russians decided that, with 30 mm and 57 mm AA guns in use, there was no need for something in between. The ammunition for the Russian naval 57 mm guns was the same as the army's S-60, but a different design of gun was selected, the SM-15, and used in various single, twin and quad mountings in the ZIF series. The gun evolved with the mountings, a series of improvements being made over its production life.

As mentioned in the previous chapter, Bofors introduced a 57 mm naval gun in 1950; the gun, feed arrangements and mounting were all gradually improved, a process which goes on today. A towed version was also developed, but had had much less success and a shorter life.

One other 57 mm AA gun is worth mentioning, the Czech R10 (also known as the PLK). This was developed as a rival to the Russian S-60 but was much less successful, only a couple of hundred being bought by the Czech government, for mainly political reasons. It used its own ammunition.

## AIRCRAFT GUNS

In the 1950s a new generation of aircraft guns began to enter service, most notably including revolver and rotary cannon, all delivering far higher rates of fire than those in service in World War II. The Mauser MG 213C revolver cannon, first designed in 20 mm calibre but soon modified to 30 mm to increase its destructive effect against heavy bombers, was developed to production in the UK and France. The resulting ADEN and DEFA guns comfortably exceeded the original Luftwaffe target rate of fire of 1,000 rpm, achieving 1,200–1,400 rpm. The ammunition for these guns only developed a medium muzzle velocity, no higher than the previous generation. Oerlikon produced a series of aircraft revolver cannon in 20 mm, 25 mm and 30 mm calibres, mostly using their own high-velocity ammunition, but only one of these saw service, the big 30 mm KCA, which was installed in some versions of the SAAB Viggen.

After achieving little in World War II US cannon designers were working overtime, with a wide range of development projects involving a variety of operating systems. Only one of the successful ones was closely related to anything done in World War II: the USN's MK 12, based on the Hispano only firing more powerful ammunition, the 20 × 110 MK 100 series with a muzzle velocity of over 1,000 m/s, at a rate of fire of around 1,000 rpm. This remained the navy's standard cannon for this period. Interestingly, it met the German late-war target for a 'million point gun' without the complication of a revolver mechanism.

The US did introduce into service two very different revolver cannon: one was the relatively conventional M39 firing the new 20 × 102 ammunition. The M39 was the USAF's standard fighter gun in the late 1950s/early 1960s and would have disappeared from view shortly afterwards were it not for its use in the widely popular Northrop F-5 light fighter. The other was the USN's wildly unconventional MK 11, a twin-barrel revolver used only in a gunpod, and using the Navy's 20 × 110 MK 100 ammunition.

In parallel with revolver cannon US engineers were also working on multi-barrel rotary guns. These were highly modified versions of the original Gatling hand-cranked rotary MG of the latter half of the 19th century, and promised both a very high rate of fire and a relatively smooth system of operation, to the benefit of reliability and longevity. What became both the USAF's and subsequently the USN's definitive fighter gun, the six-barrel 20 × 102 M61, appeared in the 1950s and is still in service in updated form as the M61A2. The rate of fire has remained at around 6,000 rpm since the advent of the linkless feed M61A1 in the late 1950s. The main downsides are that the system is bulky and heavy; the ammunition feed as well as the gun. It was decided to power the gun and linkless feed with an electric motor, which has the advantage that it continues working even if a round fails to fire, and fired cases can be returned to the magazine instead of being ejected from the aircraft.

The basic rotary design was modified to meet other needs: the GAU-4, intended for gunpods, was gas-operated to save weight; and the M197 is a helicopter version with three barrels and a rate of fire reduced to 1,500 rpm.

It is notable that the US has retained the 20 mm calibre for fighter guns, when Europe was using 30 mm (and more recently 27 mm) and Russia started with 23 mm guns, rapidly followed by 37 mm and 30 mm cannon. The US felt that the high rate of fire of their rotary cannon in particular compensated for their less destructive ammunition. However, by the end of this period they were working on a replacement for the M61 in a larger calibre: the GAU-7.

Russia initially followed a different development path, of steady evolution based on existing design principles. The NR-23 of 1948 had already replaced the NS-23 in the early 1950s was in turn replaced by the AM-23 using the same 23 × 115 ammunition, which fired even faster (1,250 rpm instead of 800–950 rpm). However, it was decided to develop a gun intermediate in calibre between the 23 mm and 37 mm, resulting in the NR-30 of 1955. This used powerful 30 × 155B ammunition, of only medium velocity similar to the ADEN/DEFA guns but firing much heavier projectiles. For a basically simple recoil-operated design weighing only two-thirds as much as a revolver cannon it managed an impressive 850–1,000 rpm.

This does not mean that the Russian engineers were closed to new ideas: they did adopt the revolver mechanism for one aircraft gun (as well as a naval AA gun as described); the remarkable R-23 of 1964. This was specifically designed for mounting in a remote-controlled defensive turret and adopted a front-loading design to keep the ammunition feed as close as possible to the centre of gun mass, near the trunnions. This also meant conventional ammunition could not be used, and the gun saw service in only one type of bomber. Another novel operating system was adopted for the far more successful GSh-23, adopted a year later and used in a wide variety of installations. This is the twin-barrel Gast type firing the standard 23 × 115 ammunition, with the operating mechanism firing the barrels alternately. Very compact and light, this fires at 3,000–3,400 rpm.

# 1970 ONWARDS

This period is mainly notable for the following:

- A new class of weapon, the naval Close-In Weapon System (CIWS) designed to shoot down incoming anti-ship missiles, which have been subsequently doubling-up in the land-based Counter-Rocket, Artillery, Mortar (C-RAM) role for base protection.
- The design and evolution of AFV cannon for ground fighting, now a major application; similar guns are also being used to protect naval vessels against small-boat attacks.
- The appearance of new gun mechanisms, most notably externally-powered, especially the US Chain Gun series.
- The development of an entire family of Russian 30 mm cannon based on the 30 × 165 cartridge case, using various operating systems, to replace previous models for all purposes.

It was also a period of consolidation in the European cannon market, with Oerlikon acquiring Hispano-Suiza in 1971, Rheinmetall taking over Mauser in 1995 and Oerlikon Contraves in 1999, leaving the French company DEFA/GIAT/NEXTER as the only other major group designing and developing new cannon. US firms have also been through many changes in this period; a process which is continuing, as described in the USA section of the national summaries.

## ANTI-AIRCRAFT GUNS

The importance of cannon in the AA role has been declining for some decades due to the steady improvement in effectiveness of surface-to-air guided missiles (SAMs), from shoulder-fired up to vehicle-mounted short-range systems. Those gun systems which survive are now mainly from 23 mm calibre upwards and are frequently found sharing a mounting with SAMs; missiles usually have a minimum range under which they don't have time to acquire the target and steer towards it, so guns help to fill that close-in gap to give comprehensive coverage.

The oldest of the current systems seeing significant use is the towed ZU-23 mentioned in the previous chapter. The survivors have seen various upgrades, and mountings are still being developed, especially in Poland. First, in 1988, came the ZUR-23–2S *Jod* with improved electro-optical sights and a twin launcher for *StrelA-2M* AA missiles. In 2002, the Polish Army ordered 44 examples of the more heavily modified ZUR-23–2 KG which has two 23 mm guns and two *GROM* MANPADS, plus a more advanced sight. Rosboronexport of Russia also offers a ZU-23–3 upgrade, with powered elevation and traverse, new sights, plus an optional twin-SAM pod (9M333 – SA13). NORINCO of China is also offering upgrades.

Naval versions of the twin-23 mm have also been developed in Poland (the ZU-23–2M *Wróbel*).

Upgrades of the *Shilka* SPAAG system are popular, for example: the Ukraine has offered '*Donets*' on a T-80U chassis with improved radar and FCS plus four SA-13 SAMs; a Russian upgrade adds SA-16/SA-18 SAM pods to the turret; and the Netherlands offers an upgrade to the radar and FCS.

As well as updating the FCS, improvements in effectiveness have been achieved by changing the ammunition. The best-known example is the 35 mm Oerlikon AHEAD which fires time-fuzed shells set to burst open and release a 'shotgun blast' of tungsten pellets in the path of the target. In smaller calibres Frangible APDS (FAPDS) ammunition consists of a sub-calibre tungsten projectile fired at high velocity, which fragments on striking the target.

The equivalent Western cannon to the Russian 23 mm is the Oerlikon KBA in 25 × 137 calibre, but this has only been used in one dedicated army AA system, the Italian SIDAM 25, a tracked SPAAG mounting a quad turret. Greater success is being enjoyed by the Chinese PG-87 (25 × 182B) which is in service in the Type 87A twin mounting (1,500 kg) and two tracked SPAAGs with quad mountings: the PGZ95 and the PGZ04. Oerlikon tried again with the more powerful KBB (25 × 174/181) but this was only sold as part of the Sea Zenith CIWS (see below).

Several 30 mm AA systems were developed from 1970. One of the first was an oddity: the A436 Model 1980 from Romania, a towed system utilising a pair of Russian NN-30 naval cannon, with various modifications. This was followed in the early 1980s by the Russian 2S6 *Tunguska* tracked SPAAG, mounting two twin-barrel (Gast type) 2A38M

*ZU-23–2M_Wróbel 2 (Łukasz Golowanow / Wikimedia Commons / Public Domain)*

SIDAM 25 (Oerlikon Contraves)

*25 mm Type 95 SPAAG turret (Max Smith – Wikimedia Commons – Public Domain)*

cannon (30 × 165), alongside SAMs. This has been replaced in production by the similarly-armed *Pantsir*. By the mid-1980s Greece had developed the Artemis towed system with two Mauser MK 30 Model F (30 × 173), related to the KUKA Arrow (KUKA having since been acquired by Rheinmetall).

In larger calibres, the long-established Bofors 40 mm L/70 remains in service. One famously abortive application was the US M247 Sgt York SPAAG (using two of the guns) in the 1980s, but in the early 1990s the gun was used in the *Luftvärnskanonvagn* (lvkv) a SPAAG based on the CV90 MICV. In the same decade, the Oerlikon Contraves Skyshield 35 towed system was revealed, based on the fast-firing 35 mm KDG revolver cannon. The combat weight is 3,600 kg for the gun mounting only; sensor units and a command post are required in addition. Finally, Thales revealed their RAPIDFire SPAAG in 2012, a wheeled vehicle mounting a turret with a CTAS 40 mm cannon and SAMs.

There has recently been some upsurge of interest in the potential effectiveness of SPAAGs against the proliferating miscellany of UAS – drones used by armed forces for tasks ranging from reconnaissance, through target laser illumination, to suicide types with an HE payload. Many of these are too small and cheap to be worth a guided missile (even assuming that the missile guidance system is capable of dealing with such targets) but an airburst or AHEAD type shell might be able to deal with them quite economically.

## CIWS

On 21 October 1967 the Israeli destroyer *INS Eilat* was sunk by three Russian P-15 *Termit* missiles (known to NATO as the SS-N-2 Styx) which had been launched from Egyptian Navy missile boats. Although ships had been sunk by German guided missiles in World War II, these had been launched by aircraft. The *Eilat* was the first victim of a ship-launched anti-ship missile, and this caused great concern among NATO navies, as existing AA systems had extreme difficulty in defending against such attacks. The Styx was relatively large next to most modern anti-ship missiles, but compared with manned aircraft it made a very elusive target: small, fast, and giving the target ship only a short time to react.

Work began on countermeasures in the USA, resulting in a prototype system in 1973 which was approved in 1977 and entered service in 1980. This was given the designation Mk 15 Phalanx, and at its core is a 20 mm M61 six-barrel rotary cannon. This is an automated system, the mounting having its own radar and a fire control system which is capable of tracking the trajectory of the gun's projectiles as well as the incoming missile and correcting the aim if necessary – referred to as a 'closed loop' system. The system is fully automatic; once switched on, it will detect any incoming aircraft or missiles, aim, and fire as many bursts as necessary to destroy the target. It is also quite compact and easy to install, as the 6 tonne mounting does not penetrate the deck but sits on it. It has therefore been fitted to a large number of warships in many navies and has been steadily updated during its life.

*Phalanx CIWS and Bofors 40 mm L70 Gun aboard ROCN Di Hua (Xuan Shisheng – Wikimedia Commons – CC-BY-SA-3.0 licence)*

Anti-ship missiles have been fired at Phalanx-equipped ships on only a handful of occasions. On 17 May 1987, the destroyer *USS Stark* was hit by two Exocet missiles while patrolling in the Persian Gulf; the Phalanx system was not switched on. On 14 July 2006 the Israeli corvette *INS Hanit* was hit by a missile. Again, the Phalanx system was not switched on. In February 1991 during the Persian Gulf War an Iraqi Silkworm missile was fired at the battleship *USS Missouri*, which was being escorted by *HMS London* and *HMS Gloucester* (both Type 42 destroyers carrying Sea Dart). The missile was destroyed by a Sea Dart fired from *Gloucester*, the first time a SAM had successfully engaged an enemy missile during combat at sea. The *Missouri* did have Phalanx systems but these did not engage as the missile was destroyed before getting close. A few weeks later, *Missouri* was involved in another incident in which a (mistaken) warning of a missile attack was given. The Phalanx system of the escorting frigate *USS Jarrett* was switched on, and fired at defensive chaff launched by *Missouri*.

The versatility of Phalanx has been increased in recent years by fitting a manual control coupled to an electro-optical system, enabling the controller to engage other targets such as small boats. On the other hand, the development of much faster anti-ship missiles, greatly reducing Phalanx's engagement window, has raised concerns about its effectiveness against such threats.

Phalanx dominates the Western fleets, but other systems have seen much more limited service. The most-used is the Dutch Goalkeeper (first ordered in 1983), which is similar in principle and capabilities to Phalanx, but is designed around the much more powerful GAU-8/A cannon in 30 mm calibre. As a result, it is bigger and heavier, using a deck-penetrating mount, and is more difficult to fit to existing warships. The gun mounting weighs 6.8 tonnes (loaded) and the remainder of the below-deck equipment 3.75 tonnes. Two other western systems are in service: the Oerlikon Contraves Sea Zenith, mounting four KBB cannon in 25 mm calibre, which has been acquired only by the Turkish navy, entering service in the late 1980s; and the Spanish Meroka, a unique 12-barrel volley gun in 20 mm calibre, which is slowly disappearing as the ships it is fitted to are retired from service.

*Meroka gun of the Spanish Navy aircraft carrier Príncipe de Asturias (Outisnn – Wikimedia Commons – CC-BY-SA-3.0 licence)*

*Goalkeeper CIWS (LA(Phot) Amanda Reynolds / MOD 45151583 / OGL v1.0 / via Wikipedia Commons)*

A more recent challenger from Rheinmetall in a slightly different category is the Millennium in 35 × 228 calibre, with a KDG revolver cannon firing AHEAD ammunition. This has separate gun and radar modules and is intended for both land and naval use. The AHEAD projectile is packed with tungsten pellets and has a small explosive charge in the base, initiated by a time fuze set as the projectile leaves the muzzle. This splits the projectile apart before it reaches the target and sends a 'shotgun blast' of pellets in the path of the target. The KDG fires at 1,000 rpm, so a short burst will shower the target with high-velocity pellets. This is effective against both aircraft and missiles, but also against other targets such as small boats or personnel on shore. Against larger vessels, the time fuze can be disabled so that the projectile penetrates the target before splitting up with considerable fragmentation effect. The Millennium system was developed in the mid-1990s and in service from 2003.

Russia having developed the world's first shipborne anti-ship missiles, it is not surprising that they were also first in the field with CIWS. The AK 630, mounting an AO-18 (later, the improved GSh-6–30) six-barrel rotary, replaced the AK 230 in production and has been in service since 1976, the first of the new family of guns in 30 × 165 calibre. This was followed in 1989 by the redoubtable 3M87 (better known as the *Kortik*, or *Kashtan* for the export version) which mounts two GSh-6–30P cannon plus eight guided missiles (with a power reloading system below deck). These systems, plus

*35 mm Millennium CIWS (MKFI – Wikimedia Commons – Public Domain)*

Kashtan 'Chestnut' (from manufacturer's brochure, courtesy Richard Jones)

the simpler and lighter *Palash* or *Palma* (export version), have been improved in service but do not feature closed-loop gun aiming.

The most recent Russian development is the *Pantsir-ME*, a naval version of the *Pantsir* system, announced in 2019. This is claimed to offer a considerable improvement in effectiveness over *Kashtan-M*.

China has so far stuck with gun-only CIWS mountings: the seven-barrel rotary H/PJ-12 (known as Type 730 in the West) which appears to be based on the GAU-8/A except it fires electrically-primed 30 × 165 Russian ammunition rather than percussion-primed 30 × 173. This was recently joined by an eleven-barrel version, the H/PJ-11 (or Type 1130) which reportedly fires at 10,000 rpm. Little is known about the performance of these systems.

As a result of experience in Iraq and Afghanistan, where military bases were attacked using indirect fire weapons, some of these naval CIWS have found a new role in the protection of such bases, known as C-RAM (Counter-Rocket, Artillery and Mortar). The first of these (and it appears the only one to have seen active service so far) was Phalanx in 2005, fitted to a wheeled trailer or truck-bed complete with its own generator and renamed Centurion. The US Army made extensive use of these systems which are credited with defeating more than 100 attacks. The British also deployed some Centurion systems, based on Phalanx 1B mountings borrowed from the RN (and since returned).

Other systems developed since then are the German MANTIS based on the Skyshield 35, and two Chinese systems; truck-borne versions of the naval Type 730 and 1130 known as the LD2000 and 3000 respectively.

## GENERAL-PURPOSE NAVAL CANNON

On 2 April 1982 Argentinian forces invaded the Falkland Islands, a British Overseas Territory in the South Atlantic. The invasion was repelled after intense fighting which concluded on 14 June. Casualties were heavy as were material losses, including several warships. The RN lost the Type 42 destroyer *HMS Sheffield* to an air-launched Exocet anti-ship missile, and another Type 42, the *Coventry*, plus two Type 21 frigates *Ardent* and *Antelope*, to bomb-carrying aircraft in confined waters. The *RFA Sir Galahad* was also lost to bomb attack, while the container ship *SS Atlantic Conveyor* was sunk by another Exocet attack.

Many lessons were learned from this conflict, but in terms of naval armament it was clear that close-in defences against air and missile attack needed to be increased. The RN acquired both

*See-through illustration of Oerlikon 30 mm Twin Naval Gun Type GCM-A03–3 (BMARC)*

*MK 44 cannon on British DS30M Mk 2 from MSI*

the Phalanx and Goalkeeper CIWS to address the anti-ship missile problem, but also opted for new automatic cannon to replace their basically World War II-era 20 mm Oerlikon SS and 40 mm Bofors L/60. These were the 20 mm Oerlikon KAA in a single, manually-operated GAM-BO1 mounting, and the 30 mm Oerlikon KCB in manually-aimed single (LS-30B) and twin (GCM-AO3) power-operated and stabilized mountings.

In the 1980s the Italian company OTO Breda also developed single and twin Sentinel remotely controlled mountings for the 30 mm Mauser MK 30F.

As time passed, so short-range missiles became more effective and were fitted as a matter of course to frigates and larger warships. The installation of AA guns therefore declined. In the case of the RN, the number of ships equipped with the 30 mm twin mount reduced steadily, although the 30 mm single mount became more popular as a replacement for both of World War II-era 20 mm and 40 mm guns. At this point, yet another incident caused a different view to be taken.

On 12 October 2000 the destroyer *USS Cole* was refuelling in Aden harbour when it was attacked by two suicide bombers in a small boat packed with explosives. Casualties were heavy and the ship was severely damaged, out of commission for over a year. This incident had a similar effect to the sinking of the *INS Eilat* by anti-ship missiles in 1967: navies worldwide reviewed their security procedures and defensive systems. One outcome was an increase in the number of light guns, from MGs to medium automatic cannon.

Compared with AA guns, there was an important difference in the nature of the cannon preferred for providing close-in defence against small boats; they did not require a high rate of fire. So guns which were designed for ground fighting, such as the Bushmaster family, were frequently chosen; these are claimed to offer lower running costs and greater accuracy, reliability and longevity.

A good example of this change is the RN's 30 mm gun: in the latest mounting, designated DS30M Mk 2, the 30 mm Oerlikon has been replaced by the 30 mm MK 44 (Bushmaster II) which has a cyclic rate of around 200 rpm rather than 650 rpm for the Oerlikon. Another difference between the mountings which is typical of current production is that the Mk 2 is designed for remote operation, by a gunner within the ship using electro-optical day/night sights and a joystick to adjust traverse and elevation.

Such remotely-controlled mountings (generally known as RWS for remote-controlled weapon system) are becoming the norm for all naval cannon. The most popular general-purpose cannon is probably the Bushmaster (25 × 137; known to the US Army as the M242) due to its adoption by the US Navy in the late 1980s in the MK 38 MGS (Machine Gun System). The original MK 38 (now designated the Mod 1) utilized a MK 88 single mounting, which was neither stabilized nor powered. In the mid 2000s, a powered, stabilized and remotely-controlled mounting was introduced based on the Israeli Raphael Typhoon Weapon System and designated MK 38 Mod 2. Various other options exist, including 40 mm MK 19 Grenade Launcher and 7.62 mm MG as coaxial armament alongside the Bushmaster. The Typhoon RWS, which can take a range of armament, is mostly associated with the 25 mm Bushmaster cannon and is in service with several navies. Many other similar mountings are now available, in calibres from 20 mm to 30 mm.

Not all such installations use slow-firing guns, however: since gun calibre is not particularly important, the opportunity has been taken to recycle some older cannon to defend against small craft. A good example is the French 20 mm Modèle F2 (20M693 gun; 15B mounting), and the smaller 20M621 (15A mounting);

*25 mm Mk 38 Mod 1 (USN / Wikimedia Commons / Public Domain)*

*25 mm MK 38 Mod 2 (USN / Wikimedia Commons / Public Domain)*

both mountings are manually-operated, and both guns are capable of around 750–800 rpm. More recently, the French Navy has adopted the Narwhal RWS which uses the 20M621 gun.

Interestingly, Rheinmetall introduced in 2015 a modified version of the Oerlikon KAA cannon (20 × 128) adapted to remote-control mountings and designated KAE. This is designed for use in the Searanger 20 RWS, along with new SAPPIE-T ammunition (Semi-Armour Piercing Pyrotechnically Induced Effect – Tracer) optimized for the close defence role.

The Republic of China (Taiwan) has single and twin remotely-controlled mountings (XTR-101 and XTR-102 respectively) fitted with the US M39 revolver cannon (20 × 102). The German Navy (and some others) have adopted Rheinmetall's MLG 27 using the Mauser MK 27 revolver cannon (27 × 145B); both the M39 and the BK 27 fire at about 1,700 rpm. Their use, however, may be more connected with the availability of surplus aircraft guns rather than any perceived advantages of such cannon in this role.

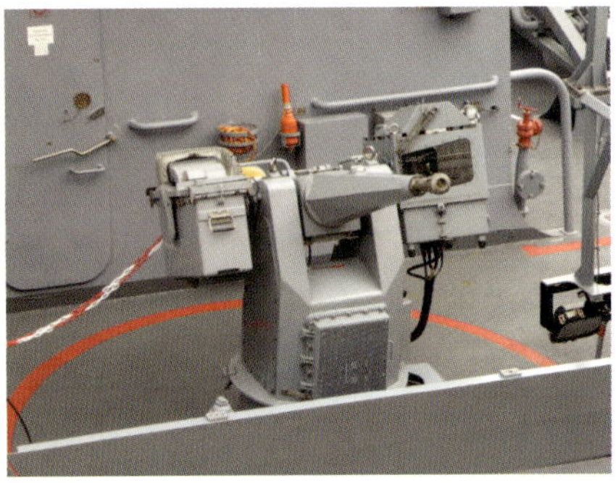

*MLG 27 on K130 Braunschweig class corvette F264 at DSEI 2015*

China offers three remotely-controlled single-barrel 30 mm mountings, the H/PJ-14, -15 and -17, but very little is known about them.

At the heavier end of the general-purpose naval cannon market come mountings for guns in 35+ mm calibre. These can be divided into those intended only for surface targets or helicopters, and those also suited for AA defence.

Oerlikon developed the 35 mm GDM-A naval twin mounting, introduced in the 1970s, which has both KDC guns mounted close together. Oerlikon also cooperated with OTO Melara in developing the GDM-C twin mounting, in which the KDC guns are mounted wide apart. More recently Denel has developed the GA-35 gun, which matches the performance of the KDC and is available in a twin mounting designated the Naval Air Defence System 35 mm Light Gun Mount (NADS 35 LGM) and has been in service on South African frigates since 2005.

China is offering two different naval guns in its own 37 × 240 calibre: the original Type 76 twin mounting available in two versions (H/PJ76A), which has an automatic fire control system, and a simpler manually-directed mounting, the H/PJ76F. In addition, there is a single manual or semi-automatic mounting, the AN3, which is available in either the 37 × 240 or Oerlikon 35 × 228 calibres.

The Bofors 40 mm L/70 has a long history as related in previous sections. It reached its peak of performance in the OTO Breda Dardo system, which comes in two versions, both with twin guns mounted close together; the Type A deck-penetrating mount, with 736 rounds in a seven-layer magazine, and the Type B with 444 rounds in four layers. This was developed in Compact and Fast Forty versions capable of 600 rpm or 900 rpm respectively. The Fast Forty is intended for the anti-missile as well as AA role and the guns have dual feed, with HE mainly used but APFSDS available for when missiles close to 1,000 m.

The latest development for the naval 40 mm Bofors are in a different direction, with lightweight, non-deck penetrating, remotely-controlled general-purpose mountings being developed by both BAE and OTO Melara. The BAE Bofors 40 Mk 4 weighs less than 2,500 kg complete with 100 rounds of ammunition (30 rounds in the primary magazine plus another 70 in an intermediate magazine which are loaded automatically into the primary magazine). The turret uses electric drive rather than hydraulics and is fully digitized and modular in the approved modern fashion. Local control is possible using a gun-mounted video sight. Rate of fire is variable between 30 and 300 rpm. Its 3P ammunition is basically pre-fragmented HE and includes a fuze which can be set electronically in various ways, including impact, time and proximity, with sub-options such as delayed-action impact and gated proximity (i.e. the fuze only becomes active when close to the target). Development was completed in 2015 and the launch customer was Brazil, with five mountings ordered for patrol vessels.

Oto Melara's Forty Light was developed a few years later than the Mk 4. Its performance appears to be virtually

*Denel 35 mm (NADS 35 LGM) (Brent Best – Wikimedia Commons – Public Domain)*

*BAE Systems Bofors 40 Mk 4 naval gun system (BAE)*

*OTO Melara Forty Light (Leonardo S.p.A.)*

identical, and it weighs 1,900–2,100 kg depending on the mounting: Type A (full remote-control slaved to a combat management system); Type B (includes a co-axial daylight TV camera to provide a reversionary autonomous mode); or Type C (fully autonomous, incorporating its own fire-control system based on an on-mount EOD equipped with a day-light TV, a cooled thermal camera, and a laser rangefinder). The mount carries 72 ready-to-fire rounds with a dual-feed system, and a newly-developed smart-fuzed programmable ammo is comparable with the 3P.

One new option not yet in service at the time of writing is the Thales RAPIDSeaGuardian, announced in 2016. This is based on the CTAI 40 mm CTAS (Cased Telescoped Armament System) developed jointly by France and the UK.

Finally, two larger guns which have, in various mountings, been around for about seven decades have been making a come-back. The first is the 57 mm Bofors, now adopted by the USN and USCG as the MK 110. The other is the Russian 57 mm A-220M, based on the army S-60.

The Bofors gun, now in the SAK 57 L/70 Mk 3 version, fires at 220–230 rpm, contains 120 rounds in the turret and a similar number below deck, and weighs around 7,000 kg. It is in Swedish service in the Visby class corvettes, and US service in the Littoral Combat Ships and Coast Guard cutters.

The A-220M has had its performance upgraded to fire at a cyclic rate of up to 300 rpm with the aid of a barrel cooling system. The 6,000 kg mounting holds 400 ready rounds below the deck. So far, no users have been identified.

## GUNS FOR GROUND FIGHTING

The use of automatic cannon for ground fighting in recent decades has almost entirely been restricted to mountings (usually in turrets) in light AFVs, with the main purposes of engaging other light AFVs and of providing general fire support for infantry attacks. The cannon have two origins; adapted versions of the automatic cannon also used in air-defence weapon systems, and specially developed weapons optimized for the ground fighting role.

The German Army was among the first to reintroduce the concept after World War II, with the adoption of the *Schützenpanzer* HS 30 (*SPz lang*) APC in the late 1950s. It was low, with well-sloped armour, and some carried a 20 mm Hispano-Suiza HS 820 cannon in a turret. In modern parlance, these were really Armoured Infantry Fighting Vehicles (AIFV) or Mechanised Infantry Combat Vehicles (MICV), as they were intended for fighting as well as carrying troops.

The HS 30 was replaced from the late 1960s by the much heavier Marder MICV with the *Spähpanzer Luchs* 8 × 8 Armoured Car (AC) of 1974 taking on the reconnaissance role. Both of these feature a Rheinmetall Rh 202 cannon, firing the same 20 × 139 ammunition as the HS 820, in a turret mounting.

During the 1960s Hispano-Suiza offered a range of cannon in 20 mm, 23 mm, and 30 mm calibres. These all used a recoil-operated mechanism and were intended for AFV use. However, they might have been ahead of their time as they failed to achieve any sales.

Many other nations have since adopted high-velocity small-calibre cannon for arming their light AFVs, a typical example being the South African Vektor (now Denel) 20 mm G1–2, a version of the French 20 F-2 which is fitted to armoured cars (ACs). While the Warsaw Pact's BMP-1 tracked MICV of the late 1960s featured a 73 mm low-velocity

*Bundeswehr Marder A5 with 20 mm Rh 202 cannon (KrisfromGermany / Wikipedia Commons / public domain)*

*BRDM with 23 mm KPVB in 2000 (Maxim Popenker)*

smoothbore cannon, the BMP-2 of the early 1980s switched to the 2A42 30 mm cannon. The gun was specifically designed for the AFV role and features a variable rate of fire (either 200–300 or 600–800 rpm). In the BMP-3, the Russians have covered all of the options, providing the turret with 7.62 mm, 30 mm 2A72, and a 100 mm cannon.

The requirements of the fighting in Chechnya stimulated the development of a variety of more or less informal cannon installations to provide heavy firepower for close-in fighting, ranging from a 30 mm 2A42 on an overhead mounting installed on a BTR-80, to installations of 23 mm guns – not just the 2A14 in 23 × 152B but even the GSh-23 twin-barrelled aircraft gun in 23 × 115, delivering 50 rounds per second. More formally, there is the BMPT Tank Support Combat Vehicle on a T-90 chassis, intended to protect tank formations from infantry wielding anti-armour weapons. This has a remotely operated overhead mounting with two coaxial 30 mm 2A42 plus a 7.62 mm PKTM plus 16 AT-16 supersonic ATGW, and two remote controlled 30 mm AGS-17D grenade launchers on the vehicle sides. One unusual Russian light AFV gun is the KPVB, which is the 14.5 mm KPV HMG rechambered to use 23 × 115 ammunition. This involved lengthening the action as the 23 mm round is longer than the 14.5 mm. The current status of the project is uncertain.

The Yugoslav Army went their own way with their locally-designed 30 mm Zastava M86 (single feed) or M89 (dual feed) cannon (also used in the AA role) in the BVP M80AK MICV, unusually chambered for the 30 × 210B ammunition from the Russian naval NN-30 revolver cannon. Armour penetration is claimed to be 60 mm at 1,000 m (projectile type and striking angle not specified). This gun continues to be developed, the latest version being the M97/12, offered by Yugoimport in either 30 × 210B or the 30 × 173 NATO chamberings.

During the 1980s the Bulgarians introduced the BMP-23 AIFV and BRM-23 reconnaissance variant armed with the 23 mm cannon from the ZU AA gun, the 2A14. The BMP-30 version uses the Russian BMP-2's turret and 30 mm 2A42 gun. The 2A14 cannon was also used by the Ukraine in a twin overhead mounting in their BTR-94, a modified Russian BTR-80 8 × 8 vehicle, 50 of which were made for export to Jordan and subsequently passed to Iraq. The Ukraine now offers two 30 mm cannon: the ZTM-1 (closely based on the Russian 2A72) in an external, overhead mounting in the upgraded MT-LB tracked AIFV; and the ZTM-2, based on the older Russian 2A42. This is installed in the Grom Universal Fighting Module.

The British Army also moved to automatic cannon in the series of LAFVs introduced in the early 1970s, with the Fox 4 × 4 AC and the Scimitar tracked reconnaissance vehicle both being equipped with the Royal Ordnance Factory (ROF) L21 Rarden cannon. This was a ground-breaking weapon in two ways; it introduced the 30 mm calibre to AFVs, in a gun which was specifically designed for AFV use. The gun is optimized for maximum single-shot accuracy rather than a high rate of fire.

The British originally backed both horses with the 1970s LAFVs, as the original version of the Scimitar, the Scorpion, was introduced with a modified version of the Saladin AC's 76 mm manually-loaded gun. This battle was evidently won by the Rarden, since the 30 mm gun was not only fitted to the MCV-80 Warrior but was also retrofitted to the Scorpion, using turrets from scrapped Fox ACs to create a new model called the Sabre (the recoil of the 76 mm having apparently caused structural cracking in the lightweight aluminium alloy Scorpion). Incidentally, this change also indicated that the tracked reconnaissance vehicle had triumphed over the wheeled armoured car in British Army thinking, followed through by the selection of the (much larger) GD ASCOD 2 as the Ajax, due to enter service in 2019/20. The Ajax is another ground-breaker, being the

*Scimitar with Rarden gun (MoD Pattern Room)*

*AJAX with 40CTWS (Andrew Linnett / MOD_45159445 / OGL v1.0)*

first application of the Anglo-French 40CTWS (40 mm Cased Telescoped Weapon System), along with the French Jaguar 6 × 6 AC.

The US Army was surprisingly slow to adopt the MICV concept. Some use was made of the 20 mm M139 (the US Army's term for the HS 820) installed on the M114 Command and Reconnaissance Carrier, but it was not found to be satisfactory as it needed to be kept clean and well-lubricated to function properly. The 1980s saw the introduction into US service of the M2 Bradley MICV, equipped with a turret-mounted 25 mm McDonnell Douglas M242 Bushmaster, popularly known as the Chain Gun. This weapon has become a standard fitting in a wide range of light AFV turrets (including the USMC 8 × 8 LAV), with some competition from the Oerlikon KBA, a conventional fast-firing cannon using the same ammunition, and from the externally-powered GIAT/NEXTER 25M811.

China has adopted a 25 mm gun for its ZSD-89-II (Type 89) MICV, presumably using the PG-87 high-velocity AA gun. Both manned turret and overhead gun mounts have been seen. A curious new vehicle has also been developed; the NORINCO FAV (Fast Attack Vehicle). It has an open, jeep-type body and is primarily of interest because of a 23 mm dual-feed cannon in 23 × 115 calibre mounted on the roll-over bars. This is said to be capable of 200 to 400 rpm and to be a Chain Gun type although ATK, which owned the rights to the name, professed no knowledge of this gun.

There is clear evidence of a classic gun-armour race taking place in MICVs. As their armour becomes tougher under the threat of attack from enemy MICVs, so the gun calibre goes up in order to defeat the thicker armour. The 25 × 137 ammunition, which was at one time the standard calibre for NATO light AFVs, is being replaced in new developments by the 30 × 173 round, either in the Mauser MK 30 or the Bushmaster II/MK 44 (redesignated XM813 in the latest AFV version, with a magazine attached to the gun). Apart from the German Puma MICV, the MK 30 has been adopted by Korea (in the KIFV), Austria (Ulan), and Spain. The Bushmaster II/MK 44 equips the Norwegian CV-90/30, the Swiss SPZ2000, and in its XM813 form, some of the US Army Stryker 8 × 8 fleet in an overhead mounting.

South Africa did consider the EMAK 35, an externally powered gun in 35 mm Oerlikon calibre, for arming future IFVs but in the end settled on a smaller version of the same thing in 30 mm; the GI-30 Cam Gun. This is intended to arm the Denel Badger 8 × 8, and has also been chosen by Malaysia, who selected the Denel LCT 30 two person turret armed with a stabilized GI-30 to arm their new 8 × 8 IFVs.

Japan chose the 35 mm Oerlikon KDE for its Type 89 IFV (which still appears to be in low-rate production) and the 35 mm Bushmaster III is in service in the CV90/35 used by Denmark, the Netherlands and Estonia. Sweden has a version of the classic Bofors 40 mm L70 in its CV90/40; to date, the most powerful automatic cannon to see service in a MICV. South Korea has followed suit with the K40 gun firing Bofors ammunition in the K21 MICV. The almost equally powerful 40CTWS is shortly to enter service, as may be the Super 40, a less powerful 40 mm gun with ammunition created by necking-out the 30 × 173 case (the XM813 requires little more than a barrel change to fire the new ammunition).

Even larger calibres are being considered. The US Army has expressed interest in a version of the Bushmaster III in 50 mm calibre, modified with a new barrel and stronger recoil management and renamed the XM913. The 'Super 50' ammunition was created by necking-up the 35 mm Oerlikon case. Russian companies are offering a modified version of the old 57 mm S-60 cannon in a AU-220M turret for ground fighting, and are developing new ammunition for it.

One reason for this increase in calibre can be understood by looking at comparative armour penetration figures. Depending on the range, striking angle and projectile type, a 20 × 139 shot will penetrate about 25 mm, a 25 × 137 about 30–40 mm, a 30 × 170 APDS 40–60 mm and 30 × 173 APFSDS around 80 mm, a 35 × 228 APDS 85 mm and APFSDS 110 mm, and a 40 × 365R or 40CTWS APFSDS c.140 mm. These figures also indicate why APFSDS has taken over from APDS as the preferred anti-armour munition.

Another reason for the calibre increase is the emergence of Air-Burst Munitions, also known as High Explosive Air Burst or Programmable Air Burst Munition (ABM/HEAB/PABM). These are HE/fragmentation shells (designed to generate a spherical pattern of small, high-velocity fragments) which are detonated by a time fuze set as the projectile leaves the gun, and designed to explode in the air directly over the target. Clearly, this requires a sophisticated and expensive sighting/fire control/fuzing system, so larger and more effective shells make better use of the costly fuze.

A different approach to the same problem is the Oerlikon AHEAD or KETF (Kinetic Energy, Time Fuzed), in which the shell's payload consists of tungsten pellets with a small burster charge in the base, with a time fuze set at the muzzle to explode the shell before it reaches the target. This generates an expanding cone of pellets mostly in a forwards direction, in the path of the target. HEAB is better at catching personnel hiding behind walls or in trenches, KETF delivers a much denser pattern of heavy pellets onto the target and could be effective against even MBTs by damaging their optical sensors.

Among the light automatic cannon, which use less-powerful cartridges in order to minimize weight and recoil, are the South African 20 mm GA-1 Cobra, which uses 20 × 82 MG 151 ammunition and weighs just 39 kg, the Israeli G360, firing the 20 × 110 Hispano cartridge, which weighs 48 kg, and the 56 kg M230LF, a longer-barrelled, belt-fed version of the M230 Chain Gun in service in the AH-64 Apache combat helicopters.

The conventional turret mounting of such weapons is also under threat in favour of external mounting in remote (controlled) weapon stations (RWS). This in effect reduces the 'turret' to the gun mounting, leaving the crew sitting down inside the vehicle. The advantage is that only the gun need be exposed when firing from cover, and more space is freed up in the compartment. This has only become feasible because of modern sighting arrangements, using low-light and/or thermal imaging electro-optical sights on the mounting, providing a day/night, all weather image which can be viewed on screens inside the vehicle.

Installation weight of modern RWS varies with both the calibre and the degree of armour protection provided. However, a typical 20 mm RWS (including a co-axial MG) weighs in at 300–400 kg, with those for the 30 mm M230LF weighing only a little more, at 350–450 kg. Most RWS for the 25 mm NATO guns are in the region of 800–1,000 kg, while those for the 30 mm cannon are around 1,500–2,000 kg. For 35 mm guns, the KMW RWS weighs in at 3,500 kg (with a two-man turret being closer to 5,000 kg), and a manned Nexter 40 mm CTWS is 4,200 kg; but the Lockheed Martin turret adopted by the UK weighs no less than 6,000 kg because of its extra armour protection.

Typically, IFVs are armed with cannon whose size and power is primarily determined by the need to punch holes in their enemy equivalents. This gives priority to armour penetration with the use of HE ammunition against other targets being a secondary, albeit very important, requirement. High velocity 30 mm cannon must now be considered the minimum to deliver adequate penetration against medium AFVs, with 35 mm or one of the more powerful 40 mm rounds providing a margin for future-proofing.

Armies with AFVs which are not intended to fight their opposite numbers might find RWS with lighter weapons, better suited to general fire support, more appropriate. In which case, the various lightweight mountings for the 30 mm M230LF / XM914 might well be of interest, especially to those armies using the AH-64 Apache helicopter. The standard ammunition for the AH-64 is a HEDP which has some armour penetration capability, but other HE rounds are being developed for the XM914.

Broadly speaking, in IFVs in which delivering cannon fire is a primary role, a manned turret still tends to be preferred as the gunner and commander are always facing in the direction the gun is pointing and the commander can put his head out of the turret to improve situational awareness, although given the risks of so doing and the development of 360 degree all-weather, day and night optronic coverage by cameras spread around the vehicle, the advantages of relying on the unaided eyeball may not be what they were.

It is not difficult to foresee the adoption of technical developments from the computer gaming industry which will see crew helmets fitted with visors incorporating video screens, with the view slaved to the head and eye position so that the crew can effectively look through the vehicle in all directions as if it weren't there. Something like this is in service with the F-35 Joint Strike Fighter.

No-one really knows how the pros and cons of the different gun and mounting options will work out until they are thoroughly tested in combat against sophisticated as well as irregular enemies.

*Armadillo Lemur RWS mounting a medium-velocity 30 mm M230LF / XM914*

## AIRCRAFT GUNS

There was a period in the 1950s and 1960s when most nations felt that guns would no longer be required, and several fighters were designed to use only missiles. Most of these, such as the USAF's F-102A and F-106A, were intended for intercepting strategic bombers. Naval fighters such as the RN's Sea Vixen and the USN's F-4 Phantom II were also introduced without guns.

The US realised its mistake in the Vietnam War and accordingly fitted the USAF's F-4E with an M61A1 cannon. This version achieved half of its kills with the gun rather than its missiles. Some F-106As were also modified to carry an M61A1 in the missile bay. Where guns could not be retrofitted, gun pods were used as an expedient, but they were not as rigid (and therefore as accurate) as internal cannon, besides increasing aerodynamic drag.

The experience of the Americans, Europeans and Russians in developing their current range of aircraft cannon has been markedly different. After an attempt at updating the M-39 (the Ford Tigerclaws, intended for the abortive F-20 Tigershark light fighter) the Americans have remained true to the rotary cannon.

In the late 1960s the Americans attempted to produce an advanced 25 mm rotary cannon (the Philco-Ford GAU-7/A being the chosen version) intended for the F-15 fighter. The projectile was buried within a cylindrical combustible case which resembled an oversized shotgun cartridge. This was defeated by technical problems and the 20 mm M61 has remained in US service ever since, with the latest M61A2 version (with lighter barrels to reduce both weight and spin-up time) being installed in the F-22 and F/A-18E/F. A three-barrel version of the M61A2 has also been developed for export; it is fitted internally to the TA-50 and FA-50 light combat variants of the South Korean KAI Golden Eagle, and the gun appears to be known as the A-50.

The USMC was the first of the US combat plane operators to mount a 25 mm gun; the five-barrel GAU-12/U rotary cannon fitted externally to the AV-8B (American Harrier). The new F-35 is the first tri-service (USAF, USN, USMC) combat plane to carry a 25 mm gun; in this case, the GAU-22/A which is a lightened, four-barrel version of the GAU-12/U. Only the USAF F-35A will carry the gun internally; for the USMC's F-35B and the USN's F-35C the cannon is optional, and when carried is installed in a conformal pod.

Two 30 mm cannon have seen USAF service, but only in the ground attack role. These are the massive seven-barrel General Electric GAU-8/A rotary cannon fitted to the Fairchild A-10 'tankbuster' (which was designed around the gun and its ammunition tank) and a four-barrel version of the same gun, the GAU-13/A, which has only been used in the GPU-5A gun pod, for fitting to aircraft with only an occasional need for such firepower. Like gunpods in general, this suffers from accuracy problems due to the relative lack of rigidity in the mounting.

In contrast, Western Europe has remained wedded to the revolver. In 1977 the JA 37 variant of the SAAB Viggen entered service, equipped with a powerful Oerlikon KCA 30 mm revolver cannon in a conformal pod under the fuselage. In addition to the ADEN, DEFA and Oerlikon developments, Mauser re-entered the market with the BK 27, which fires high-velocity 27 mm ammunition. Introduced in the Panavia Tornado and the Alpha Jet light strike aircraft in the 1970s, it is also fitted to single-seat versions of the SAAB Gripen and the Eurofighter Typhoon. Although it was originally expected that the Eurofighter would carry the Oerlikon KCA, this was changed to a version of the BK 27.

The French have been primarily concerned with the low muzzle velocity of their 30 mm cannon, which they have remedied by developing a longer case (which also differs slightly in other dimensional details) to create the 30 × 150B. The associated GIAT 30M791 seven-chamber revolver cannon is fitted to the Dassault Rafale; it is heavier than earlier models at 120 kg but can achieve a remarkable 2,500 rpm.

The British decided to develop a new weapon, the ADEN 25, to equip the final version of the RAF's Harrier, instead of staying with the very similar BK 27 which was already in RAF service. The ADEN 25 was essentially similar to the earlier models and could fit into the same mountings but was designed to use the standard 25 mm NATO ammunition. It experienced considerable development problems which led to its cancellation in 1999.

The 1970s saw the start of a rash of new cannon for all three services in Russia, with no fewer than four new aircraft guns entering service between the mid-1970s and mid-1980s. First up was the six-barrel rotary GSh-6–23, adopted in 1974, which achieved a remarkable 9,000 rpm. It was fitted to the Su-24 for ground attack, and also to the MiG-31 interceptor (for use against cruise missiles) but was withdrawn in the early 1980s due to various problems.

The other three guns were all designed around the new tri-service 30 × 165 Russian cartridge, originally developed for the naval AO-18 six-barrel rotary cannon. The GSh-6–30 (9A621) is an air-cooled version of the navy gun, developed in parallel with the GSh-6–23 and also accepted into service in 1974. From 1975 it equipped the MiG-27 ground attack fighter but was not popular, developing too much vibration and recoil in the relatively light aircraft, leading to some equipment literally falling to pieces; it was withdrawn from service in the late 1990s.

There is also a twin-barrelled gun, the GSh-30, which was fitted to the Su-25 ground attack aircraft from 1978. It was also originally intended for the new generation of fighters (the Su-27 and MiG-29 families) but preference switched to the lighter and more compact GSh-301.

The GSh-301 is a single-barrelled 30 mm gun weighing only 45 kg and firing at 1,500–1,800 rpm, continuing the Soviet tradition of combining remarkably low weight with high rates of fire. Amazingly, it has an 'old fashioned' recoil-operated linear action rather than the revolver mechanism originally attributed to it by Western sources, but handsomely beats all other aircraft guns in terms of hitting power for its modest weight and size. To make an historical comparison with World War II weapons, it is approximately the same size and weight as a 20 mm MG 151 but has the hitting power of a battery of four 30 mm MK 103s – which weighed twelve times as much! Such power did bring with it some initial problems with equipment damage in the MiG-29 due to vibration.

Incidentally, it is curious that in the 'collectivist' USSR, successful gun designers were generally honoured in the designations of their guns, as with Gryazev and Shipunov in the current 'GSh-' range, whereas in the 'individualist' West this tradition largely died with John Browning. It is tempting to speculate that this might have a bearing on the remarkable performance of Russian guns.

One typically Russian characteristic is the tendency to favour heavy projectiles at the expense of muzzle velocity. The 23 × 115 round fires a 170 or 200 g projectile, in comparison with the 102 g of the 20 × 102 M61 and the 180 g of the much bigger NATO 25 × 137. Similarly, the 390 g 30 × 165 projectile compares with around 270 g for the 30 × 113B (ADEN) and 30 × 150B (GIAT). If the

Russian cartridge were loaded with 270 g projectiles, muzzle velocity in the GSh-301 could be expected to increase, for the same muzzle energy, from 860 to around 1,030 m/s. The preference for the relatively low MV is surprising, as this is generally considered to be of paramount importance in air fighting because it minimizes the time of flight and therefore reduces the problems of deflection shooting. The Russians are apparently content to accept a lower velocity in return for greater destructive effect. A heavy projectile is also a benefit in low-altitude firing (especially ground attack) as it loses velocity at a slower rate than a lighter projectile of similar shape.

One interesting cannon developed by the Czech Republic is the ZPL-20, a version of the GSh-23 modified to fire the US 20 × 102 ammunition. This required significant changes to convert the gun from percussion to electric ignition. It is so far only available in the *Plamen* (Flame) PL-20 gun pod system, compatible with standard NATO pylons, which weighs 259 kg with 210 rounds of ammunition. It can fire at two rates: 780 rpm or 2,600 rpm.

Despite the retention of the fighter gun, it is worth noting that while its technology and performance has continued to develop (although in rather more subtle ways than in the past, concerned with better materials and careful design analysis) its relative fall in importance is reflected in reduced magazine capacities. A typical revolver cannon installation includes space for only about 150 rounds; enough for just five or six seconds' firing. The MiG-29 carries only 100 rounds. Even the apparently generous 570 rounds in the F/A-18 lasts for only six seconds, while the GSh-6-23 installation in the MiG-31 could expend its 260 rounds in less than two seconds. On the other hand, some modern fire control systems can prevent a gun from firing until a hit is certain, so wasted ammunition is far less of a problem than in the past.

In the specialized field of ground attack, the GAU-8/A of the American A-10 'tankbuster' has a clear lead in hitting power over any other, and also has a huge advantage in combat persistence. Its massive ammunition capacity lasts for about twenty seconds firing; the GSh-30 in the Su 25 just five seconds.

## HELICOPTERS AND GUNSHIPS

Arming helicopters with cannon started in France in 1957, with the testing of a 20 mm MG 151/20 attached to a Sikorski H-34 (S-58). The MG 151/20 was preferred to the French Hispano-Suiza HS 404 in this and other mountings because it is a significantly smaller and lighter gun and generates much less recoil. These installations were prompted by the wars of independence with French colonies in North Africa.

The next generation of helicopters, which entered service in the 1970s, saw more powerful cannon being used. Some of the French AS Gazelle were fitted with a fixed, forward-firing GIAT 20M621 cannon, while the equivalent MBB Bo 105M could be fitted with a fixed under-fuselage 20 mm Rheinmetall Rh 202 (or similar). Gun pods may also be carried, e.g. the Nexter NC 621 pod for the 20M621.

The French also produced manually-aimed doorway mountings, for example the 19A for the 20M621, and chin turrets: the 20 mm THL 20 (20M621) and the THL 30 (fitted with the externally-powered Nexter 30M781, in 30 × 113B calibre).

The French HAP (*Helicoptère d'Appui et de Protection*) version of the Eurocopter Tiger uses the THL 30. At one point the German version of the Tiger (now known as the UHT, for *Unterstützungshubschrauber* Tiger) was intended to be equipped with a new recoilless Mauser weapon, the RMK 30, but this was cancelled.

The US Army also experimented with cannon installations in the late 1950s, but it took the Vietnam War to stimulate the development of armed helicopters, with the first official cannon installation consisting of a forward-firing 20 mm M195 six-barrel rotary fixed to the port side of the AH-1G. Over 315 units were built, starting in 1968. Other fixed installations were also tried using the M24 (Hispano) cannon. The ultimate US solution turned out to be a steerable cannon in an under-nose turret fitted to a purpose-designed two-seat attack helicopter. Two different systems have been extensively fielded and remain in US service. The USMC chose the M97 turret, mounting a three-barrel 20 mm M197 rotary for installation in its SeaCobra (AH-1J, entering service in 1971). For its AH-64 Apache, the US Army selected the M230 Chain Gun which entered service in 1983.

Russia initially fitted HMGs to its big Mi-24 combat helicopters, but experience in Afghanistan indicated the need for longer range; the Mi-24P was given the powerful, fuselage-mounted twin-barrel 30 mm GSh-30K, with longer barrels than the GSh-30. However, the heavy recoil generated by the powerful ammunition and the high rate of fire apparently caused problems with keeping the guns on target, so the Mi-24VP was accordingly fitted with the

*19A door mounting for the 20M621 (Nexter)*

*30 mm M230 Chain Gun mounted in chin turret for AH-64 combat helicopter*

23 mm twin-barrel GSh-23 cannon in the NPPU-24 turret.

Among more recent Russian developments, the Mi-28 Havoc is fitted with a NPPU-28 turret mounting a 30 mm 2A42 cannon, In contrast, the KA-50 and –52 attack helicopters carry their 2A42 on the side of the fuselage; however, this is not rigidly fixed as it can be depressed by up to 37° and traversed by a few degrees to aid gun stabilisation. These weapons use the Russian 30 × 165 cartridge, significantly more powerful than the 30 × 113B used in Western helicopter guns, although it should be noted that the Army guns use percussion priming, the GSh-30K electric, so the ammunition is not interchangeable.

Other nations which have produced armed helicopters include South Africa, Italy, Romania and India. The South African Denel AH-2 Rooivalk attack helicopter is fitted with a 20 mm Vektor F2 (GIAT 20M693) dual feed gun in a chin turret. Some of the Italian A.129 Mangusta have a 20 mm M197 in a Lockheed Martin/OTOBreda chin turret. Romania has also produced a 'gunship' version of the licence-built IAR 330L Puma; 24 of these were fitted with a THL 20 chin turret under their SOCAT upgrade programme. Finally, the Indian HAL Light Combat Helicopter, which entered production in 2017, is fitted with a Nexter THL 20 chin turret with a 20M621 cannon.

Gunships are transport aircraft fitted with side-firing guns so that they can maintain fire continuously on a ground target by flying in a tight circle. A US development, they have proved very effective in combat against lightly armed irregulars (starting in Vietnam) but would be too vulnerable against an enemy equipped with any kind of air defence systems.

At first various aircraft were employed, fitted with multiple MGs, but ultimately the AC-130 Hercules became the aircraft of choice and cannon of various types were fitted in order to achieve a greater stand-off distance. The 20 mm M61 was used initially, 25 mm GAU-12/U later, with 40 mm Bofors L/60 guns and then a manually-loaded 105 mm howitzer. Initial attempts to use the 30 mm Bushmaster II were unsuccessful due to accuracy issues, but these have been overcome and the ultimate armament for the AC-130 is likely to be a combination of the Bushmaster II (under the designation GAU-23/A) and the 105 mm. For export purposes, a smaller and more lightly armed AC-235 is offered, armed with the M230LF in 30 × 113B calibre.

## AMMUNITION NATURES

The selection of the appropriate ammunition type to load into magazines or belts clearly depends on the expected targets. This might not be a simple choice in circumstances when the nature of the target can change rapidly; for example, from aircraft to armoured vehicles.

There are three possible solutions to this problem. The first is to provide the gun with a quick-change dual feed system, so it is possible to switch rapidly from HE to AP belts or magazines. This is desirable in vehicle or ship based guns but too bulky in aircraft. The second is to use multi-purpose ammunition as previously described, accepting some loss in absolute HE or AP performance. The third is to mix different types of ammunition in one belt, acknowledging that some of the rounds fired will always be less than ideally effective against any particular target.

The mixed belt solution was common in aircraft guns in World War II, with AP and HE or incendiary rounds mixed in varying proportions as required, usually with a tracer round added in every fifth or sixth position. It was less necessary in larger calibre aircraft cannon ammunition as the HE/I shells were usually powerful enough to penetrate aircraft armour and large enough to carry a tracer if required. However, it was common to cover the muzzles of cannon with cloth or rubber to prevent the blast of air from freezing the mechanism. In such circumstances, it was prudent to ensure that the first round fired was an AP shot in order to remove the cover without risk of explosion.

Traditional mixed belts are still used in Russian aircraft cannon, but most other nations have selected a multi-purpose round for their combat aircraft, in one of three flavours: perhaps most commonly some variation of the Raufoss Multipurpose 'fuzeless' projectile (such as the 20 × 102 PGU-28 series, standard in US fighters); a chemical-free FAP or PELE type; or (in the case of France) a traditional base-fuzed SAPHEI with a relatively low HEI content but a hardened steel body.

# CARTRIDGE DESIGN: CONVENTIONAL SYSTEMS

## CARTRIDGE ELEMENTS

A cartridge is a unit or round of ammunition consisting of a projectile, propellant which burns to generate gas which thrusts the projectile from the gun, a primer to ignite the propellant and a cartridge case to hold it all together. This section is concerned with conventional ammunition as shown below, which is designed with the projectile at the front, with its base gripped by the mouth of the cartridge case which contains the propellant, and is designed to be loaded from the rear into a chamber formed within, or (in the case of revolver designs) in line with, the barrel. Various alternative systems have been designed, with very little success until recently: these are covered in the section on Unconventional Systems.

For as long as it has been used, the cartridge has been the heart of all automatic weapons. It is almost invariably designed first and the gun configured around it. This particularly applies to gas or recoil operated weapons because their functioning depends upon the characteristics of the ammunition; not just the size and shape but also the gas pressure and recoil generated. These factors are less significant in externally powered weapons such as 'Chain Guns' and most multi-barrel rotary cannon.

While it is not uncommon for different guns to be designed around the same cartridge, it is therefore rare for a service cannon to be adapted to fire a new cartridge. The few exceptions are generally concerned with an alteration to the neck diameter of the cartridge case in order to take a projectile with a different calibre. The best-known example of this is probably the German Mauser MG 151 aircraft gun, which commenced life in the 1930s in 15 mm calibre but was increased to 20 mm during World War II (WW2). Current examples (not in service so far) are the 40 mm Super 40 version of the 30 mm Mk 44 round and the 50 mm Supershot and Super 50 versions of the 35 mm Bushmaster III.

Cartridges are designed to achieve particular effects. The official requirement might specify the ability to penetrate a given thickness of armour at a particular range, a maximum time of flight to a given range, or the effectiveness of incendiary or high explosive projectiles. There will also be a range of other criteria including acceptable levels of accuracy, consistency, chamber pressure, reliability and safety, which will need to be sustained over a wide range of climatic conditions and a long period of time. These only partly apply to blank cartridges and not at all to drill or dummy rounds.

For use in automatic weapons (except in the case of externally powered guns), it almost goes without saying that the ammunition must be able to operate the gun mechanism reliably. This might not be a simple matter given that the projectile weights and propellant powders can vary considerably in different loadings of any particular cartridge, and this can alter the recoil and pressure characteristics accordingly.

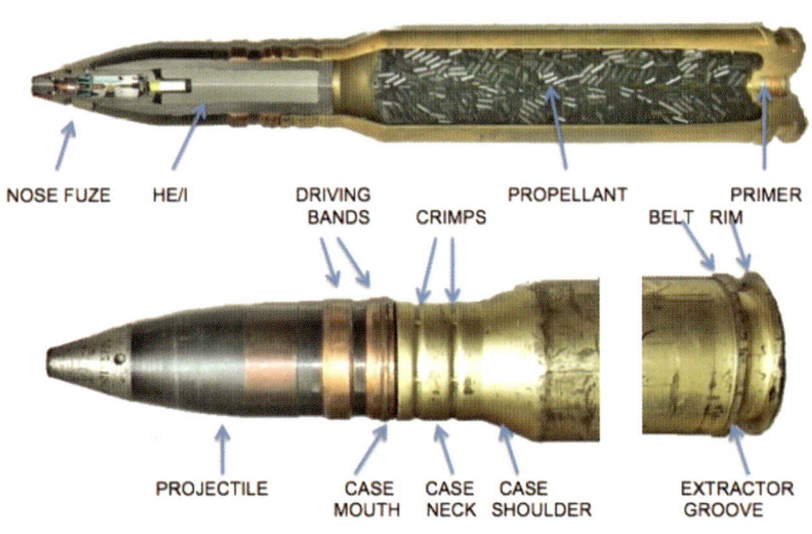


*Main features of a round of HEI cannon ammunition (photo from Derk Jaap Harms)*

## THE PROJECTILE

The projectile is what the weapon is all about. Everything else – the rest of the cartridge and the gun itself – is concerned with ensuring that the projectile hits the target at the desired velocity. The design of the projectile is therefore the most crucial aspect of ammunition design.

Projectiles are available in a wide variety of types, even ignoring those intended for practice or drill purposes. Service cannon ammunition can be classified into four broad categories: target practice (TP), also sometimes called 'ball' rounds; incendiary (I), high explosive (HE) and armour-piercing (AP) – any of which may also be equipped with tracers (T) so that gunners can observe the accuracy of their shooting. Other common designations are SAP (semi-armour-piercing), SD (self destruct), APDS (armour-piercing discarding-sabot) and APFSDS (armour-piercing fin-stabilized discarding-sabot). Recent additions include frangible projectiles designed to fragment after penetrating the target (giving rise to FAP and FAPDS designations), and high-explosive shells with time fuzes to air-burst over the target (HEAB or PABM – programmable air-burst munition).

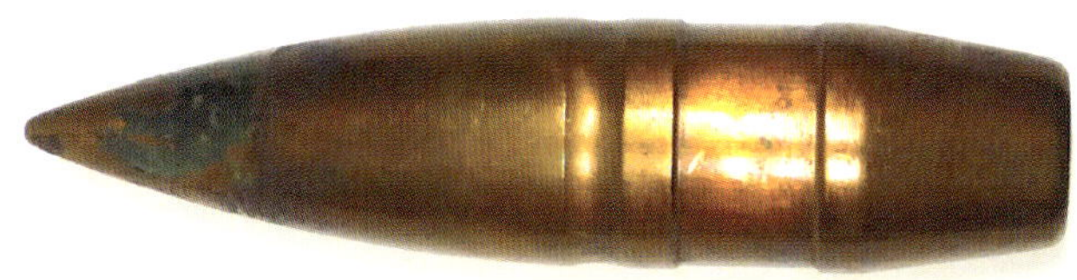

*A small-arms type jacketed AP bullet for the 25 mm Hotchkiss*

Some manufacturers use their own designations for proprietary types of projectiles, such as Nammo's MP (multi-purpose, with semi-armour piercing, explosive and incendiary effects), Diehl's PELE (penetrator with enhanced lateral effect) and Oerlikon Contraves' (now Rheinmetall) AHEAD (advanced hit efficiency and destruction) also known as KETF (kinetic energy, time fuzed); this is a modern version of the old shrapnel shell.

HEAT shells (high explosive, anti-tank, with a shaped-charge warhead) are in common use with tank guns, but cannon shells are generally too small for them to be worthwhile. There is one well-known example of HEDP (high explosive dual-purpose, HEAT with an added fragmentation effect for anti-personnel use); this is the M789 from ATK, the standard type for the 30 mm M230 cannon used in the AH-64 Apache helicopter.

Older designations for cannon projectiles which might be enountered include APHC, APCR, HVAP, and *Hartkernmunition* or *Pzgr. 40* for AP ammunition (to be explained later) and *Minengeschoss* or mine shell; a thin-walled, high-capacity HE shell used in World War II and for a while thereafter.

Cannon projectiles are usually painted according to their type for rapid identification purposes. Colour schemes have varied between nations (and sometimes services) and military blocs, and within such groupings over time, so this is not the place to attempt a complete listing. However, Western TP projectiles are nowadays commonly painted blue and AP ones black. Reds and yellows frequently indicate chemical contents: HE, incendiary or tracers. Combined projectiles often have bands of colour to indicate their particular mix of characteristics, so they can be quite colourful.

The method of construction of projectiles varies considerably. Cannon ammunition generally uses steel projectile bodies fitted with separate driving (UK) or rotating (USA) bands of greater than projectile diameter, which are gripped by the rifling in order to spin the projectile. Spinning the projectile ensures that it remains stable in flight, pointing towards the target, and greatly improves long-range accuracy. Originally driving bands were made of copper (still favoured by the Russians) but there are difficulties in making this strong enough to withstand the shearing effect of the rifling at very high velocities, so soft iron or steel driving bands are used on most Western high-velocity cannon.

After much experimentation US designers have perfected plastic driving bands, first successfully introduced in the 1970s for the 30 × 173 ammunition of the GAU-8/A aircraft gun. The advantage of plastic-banded projectiles is that barrel wear is only one-third that caused by metal driving bands. Plastic and iron bands also avoid the problem of 'coppering'; the coating of the bore with copper which has to be cleared from time to time with a 'decoppering' round.

Light-alloy projectile bodies are usually used in APCR shot, in which only the hard central core penetrates the target, the light-alloy body breaking up on impact. Small sub-calibre projectiles such as APDS and ADFSDS are nowadays commonly encased in plastic sabots (often with a light-alloy base) with larger calibres using light-alloy entirely.

There have been a few examples of cannon projectiles copying the characteristics of small arms (rifle and machine gun) bullets, with all-enclosing soft metal (usually copper-based) jackets designed to be easily engraved by the barrel rifling, but these have not been used since World War II.

*A USN 40 mm Bofors shell colour scheme interpreted: green body = HE; black band = Self-Destruct; white band = Tracer; red tip = Incendiary. So this is an HEI-T-SD*

## ARMOUR PIERCING PROJECTILES

The simplest form of cannon AP projectile consists of a pointed steel shot – the military term 'shot' (from the cannon ball era) meaning a solid or very thick-walled projectile without any explosive content. The steel will be hardened, at least at the point, and the shot will have a driving band. The best AP performance is obtained with a rather blunt point which generates considerable aero-dynamic drag, so the shot is sometimes given a 'ballistic cap' (UK) or windshield (USA); a hollow, pointed aerodynamic nose cone usually made out of light alloy or thin steel in order to minimize the added weight. Some projectiles were given a blunt cap of soft steel (under the ballistic cap where fitted) as this improves penetration against face-hardened armour, leading to designations such as APC (armour piercing, capped) and APCBC (armour piercing, capped, ballistic capped). This simple type of AP shot was usual in automatic cannon until after World War II, when it began to be replaced by more sophisticated composite projectiles. Other AP types were given a small HE filling with a base fuze (therefore becoming APHE), to enhance the behind-armour effects; always difficult to get right – the fuze tended to explode the shell on impact, or failed altogether.

*AP-T (20 × 138B)*

*APHE-T (37 × 264B)*

*APCR for ADEN gun (30 × 113B)*

*SAPHE (20 × 128)*

The first major improvement in cannon AP performance emerged during World War II – the composite rigid shot. It was probably the French who fielded it first, in the M1935 loading for the little 37 × 94R round still being used in some light tank guns, but this had a steel core in a magnesium sleeve. Germany improved the design by using a tungsten alloy core. Tungsten (usually in the form of tungsten carbide) is harder and much more dense than steel and therefore better at penetrating armour. Incidentally, reference to 'tungsten steel' may be encountered (especially in US documents); this is in fact steel, typically with only about 4% tungsten. When the shot hits the target, the sleeve breaks up and the narrow core penetrates. An example is shown above, for the 30 mm ADEN cannon; the ballistic cap has been removed to expose the pointed tip of the sub-calibre tungsten penetrator. This type of projectile is still in common use in cannon, particularly in aircraft guns such as the 30 mm GAU-8/A. In this, and in a few other cases, the core is of DU (depleted uranium), which is even more dense than tungsten and adds pyrophoric (spark producing) behind-armour effects after penetration. Dale Davis, in his *Historical Development Summary of Automatic Cannon Caliber Ammunition: 20–30 Millimeter*, argues that in late World War II Germany probably used some of their stocks of uranium (not depleted, so far more radioactive) as projectile cores, but no firm evidence has emerged.

When used with tank and anti-tank guns, the composite shot, known during World War II as APCR (armour-piercing, composite, rigid – an early document referred to this as 'composite, rigid, armour piercing' but there were presumably second thoughts about the acronym), in British service, HVAP – high velocity armour piercing – to the Americans and *Panzergranate 40 (Pzgr. 40)* or *Hartkernmunition* (hard-core projectile) to the Germans, was usually made much lighter than standard. This permitted a higher muzzle velocity and therefore better armour penetration. Such lightweight projectiles are less common in automatic weapons because of the need to match the trajectory and recoil characteristics of other ammunition types which may be mixed with AP rounds in the same belt. A current generic designation for this type of ammunition is APHC, for armour-piercing hard core.

The next stage of development of armour piercing ammunition – the squeeze or taper bore – has not seen service in automatic cannon (although various experimental guns were made) and therefore will not be discussed in detail. Essentially, it was the same as APCR except that the light sleeve was designed to be squeezed flat against the AP core by a tapered barrel (German Gerlich system) or barrel attachment (British Littlejohn adaptor) in order to reduce air resistance and thereby retain the APCR's performance out to a longer range. The main tactical problem is that firing a full-calibre HE shell is possible only by removing the Littlejohn adaptor, and is impossible with a Gerlich barrel. These systems were therefore very specialized, only seeing service in a few World War II anti-tank and armoured fighting vehicle (AFV) weapons, before being replaced by APDS.

APDS (Armour Piercing, Discarding Sabot) was initially designed in France before World War II, but was then

*Experimental 20 × 110 HS round with 15 mm squeezebore projectile*

Sectioned 35 mm APDS sabot, with the flight projectile next to it

*Sectioned 30 × 173 APFSDS*

developed in Canada and finally the UK, being issued for British 6[PR] and 17[PR] guns from mid-1944 onwards. It is now available in even the smallest cannon calibres.

APDS shot is similar to APCR except that the sleeve (sabot) is designed to break up and fall away as soon as the projectile leaves the barrel, leaving the dense, small-calibre AP core to continue alone. This provides a great improvement in the performance of APCR at longer ranges, in terms of both reducing the time of flight to increase hit probability and increasing the impact velocity (and therefore armour penetration), without the practical disadvantages of taper bores. It has accordingly seen widespread use post-war in automatic weapons; in smaller calibres the sleeve is nowadays mainly of plastic, typically with a light-alloy base for rigidity.

The most sophisticated type of AP shot is the APFSDS. This was developed as a result of the observation that a long, thin projectile has advantages for AP use, partly because it has less air resistance and therefore does not slow down as much on its way to the target, but also because it can punch through armour more easily since there is more mass and energy behind each square millimetre of frontal area. The problem is that there is a limit to the length of projectile which can be stabilized by being spun by barrel rifling – it can be no longer than about six times the calibre (an L/D or length/diameter ratio of 6:1). Long, narrow shot therefore have to be stabilized by fins and have sometimes been called arrow projectiles, notably by the Germans who developed this *Pfeilgeschoss* concept in World War II, although in practice they only applied it to HE projectiles in order to achieve very long range in large-calibre artillery weapons.

Saboted ammunition is not generally used in aircraft weapons because of the potential damage from bits of discarded sabot hitting the aircraft or being sucked into the engines. 'Sabot diverters' have been tried to circumvent this problem, but have not been adopted due to weight, bulk and reliability issues. The only current example of saboted ammunition being cleared for firing from aircraft is the 20 × 102 MK149 APDS for the gun in the AH-1 Cobra helicopter, presumably because the aircraft is slow enough to give the sabot fragments time to move out of the way, the gun is mounted well below the engine intakes, and downwash from the rotor will also help keep the plastic fragments away from the intakes.

*Different styles of APFSDS sabots: a Rheinmetall 30 mm at the top with two 25 mm; the photo below (from David Birkin) shows Nammo-pattern 30 mm. The plastic nose-cones to aid chambering are missing from the 25 mm examples.*

APFSDS penetrators with an L/D ratio of between 10:1 and 20:1 are in common use in smooth-bored tank guns, but less so in automatic weapons because their stability is upset by being spun by the rifling. Muzzle velocities of over 1,000 m/s can result in rotational rates of 100,000 rpm; spinning at these speeds, the slightest deviation from precise symmetry in the construction and location of the penetrator within the sabot will result in it bending out of shape, and the longer and narrower the penetrator, the more difficult that is to control. In these smaller calibres, some degree of rifling is needed (even if confined to a small section near the muzzle) to act as a 'sabot stripper', ensuring that the sabot separates cleanly from the penetrator with the petals being thrown clear, but the ideal rifling angle for this is much less than that needed to stabilize conventional projectiles.

The technical difficulties of firing APFSDS from rifled barrels have been overcome, mainly by fitting the sabot with slip rings which reduce the amount of spin provided to the projectile. APFSDS shot is therefore now available in automatic cannon ammunition, but the penetrators are often shorter and wider than those designed for smoothbore guns. Even so, 30 mm APFSDS can suffer from severe rifling-induced yaw in the early part of its trajectory, not entirely settling down until it has passed 400 m. To keep the overall length of the cartridge down to that of standard ammunition, the long penetrators are usually buried deep within the case and are fitted with 'puller sabots' which grip near the front of the penetrator and pull it up the barrel.

Both tungsten alloy and DU are used to make penetrators. DU is easier to work with and delivers more consistent results over a wider range of conditions than tungsten, as well as producing behind-armour pyrophoric effects. However, it has been criticized for creating toxic dust in and around the target, with obvious long-term health hazards.

One interesting recent cannon development is that of super-cavitating projectiles for shooting into water, with a tip designed to create a bubble of gas which surrounds the projectile, greatly reducing hydrodynamic drag. This has reached production in the Nammo/Crane 30 × 173 MK 258 Mod 2 APFSDS for the USN, which has been modified to have a flattened tip 2.3 mm in diameter. It is variously known as Hydroballistic or Swimmer round, and is used in close-defence naval guns where it has the advantage of being able to shoot through waves which may be hiding small boats.

## EXPLOSIVE/ INCENDIARY SHELLS

The term 'shell' is again of ancient origin and obvious meaning; a hard outer casing protecting something more vulnerable inside, in this case usually a high explosive and/or incendiary compound. It should be noted that military terminology is not always logical. All projectiles for the 20 mm Hispano, including AP shot, were officially referred to as 'shells' by the RAF.

In larger artillery calibres, shells can contain a wide variety of materials including smoke to obscure visibility, star-shell for illumination, chemicals, various antipersonnel rounds, and anti-tank sub-munitions (some capable of homing onto their targets) but these need not concern us here.

Some early shells were made from cast iron, but this was replaced by a process of drilling out a cavity of the appropriate shape in a steel shot to take the desired explosive and/or incendiary compound. This led to a relatively small capacity in smaller calibres however, so early in World War II, German technicians perfected the mine shell (*Minengeschoss* or *M-Geschoss*) which was made by stamping and drawing the shell from a thick disc of metal in the same way in which cartridge cases are made. This resulted in a shell with thin walls, approximately doubling the HE capacity. It also led to a lighter shell overall (HE being much lighter than steel), permitting a higher muzzle velocity and a shorter flight time to the target. This came at the expense of range, as lighter shells slow down more quickly. These characteristics made the mine shell particularly suitable for aircraft cannon, since in air-to-air combat ranges are short and a high velocity is important to maximize the hit probability.

Disadvantages of the original type of mine shell (designated *Ausf.A*) were its poor aerodynamic shape, lack of ability to penetrate structures, and lack of anywhere to add a tracer (except by sticking one onto the base). The aerodynamics were harmed by the blunt-nosed cylindrical body shape and the hemispherical base which did not provide a clearly defined air flow separation point, leading to instability and inaccuracy.

By the end of World War II the Germans had developed a modified form of mine shell, the *Ausf.C*, to achieve a better all-round compromise. This retained the thin shell body but used a much more streamlined nose shape and combined it with a thicker, squared-off

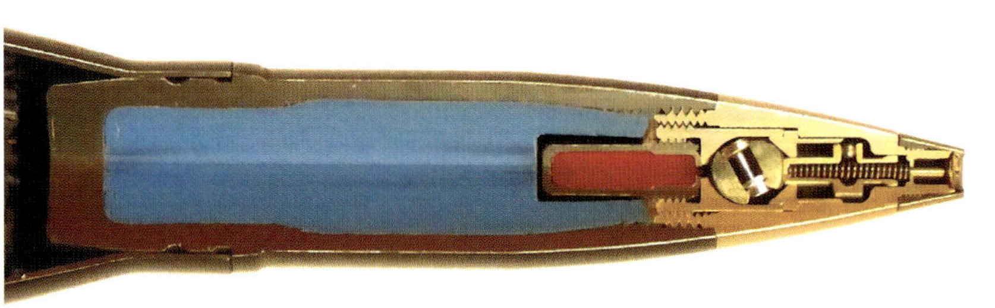

*Two nose-fuzed 20 mm HE shells: for World War II 20 × 138B (top) and the more modern 20 × 128. Note the much better aerodynamic shape of the modern one*

*Two examples of 30 mm Minengeschoss rounds for the Rheinmetall MK 108 cannon: the top one is an Ausf.A, the other is an Ausf.C tracer; note the more streamlined shape and smaller HE capacity (courtesy Derk Jaap Harms)*

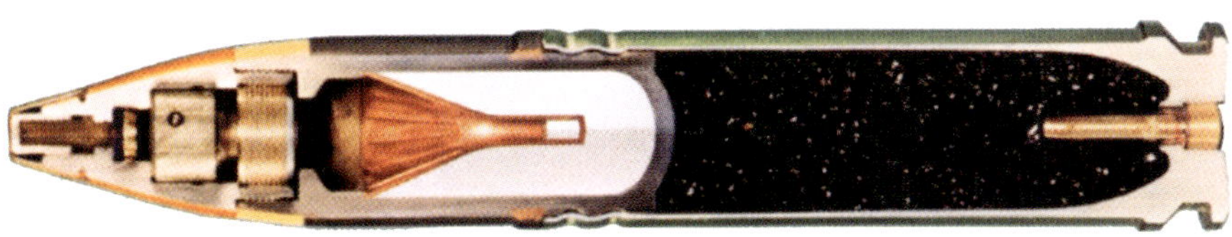

*M789 30 mm HEDP round, sectioned to show the cone-shaped liner (ATK)*

base which was more stable and provided space for a tracer.

Important but conflicting characteristics of HE fillings are that they must explode as violently as possible to make the most of the restricted quantity of compound that the shell can carry, yet be completely safe over a wide range of temperatures, despite the rough handling the ammunition may receive from the gun mechanism. During World War II, it was discovered that adding aluminium or magnesium dust to HE compounds not only improved the strength of the explosion but also had an incendiary effect. In consequence, pure incendiary cannon shells became less common. Two types of HEI were developed: one with a single compound containing both explosive and incendiary characteristics, the other with separate explosive and incendiary elements.

One specialized form of HE projectile is the HEAT (high explosive anti-tank) or shaped charge shell. In this, there is an empty cone-shaped hollow, lined with metal, at the front of the shell, with its open end forwards. An instant-acting fuze (often in the base) detonates the HE behind the cone on impact with the target. The explosion compresses the metal lining to form an elongated high-velocity jet whose tip can be travelling at 8,000–9,000 metres per second, with considerable armour-piercing capability. This type of projectile is standard in modern infantry anti-tank weapons (recoilless guns, rockets or guided missiles) and is used in some larger cannon. It was experimentally used in World War II in ammunition for German aircraft guns, but has not been generally used in smaller calibres because the effectiveness is largely determined by projectile diameter and is also lessened by being spun by barrel rifling. A rotation of only 150 revolutions per second is sufficient to reduce the penetration by 50%, yet a cannon shell will typically rotate considerably more. In larger calibres, such as the *Obus 'G'* (named after the inventor, Gessner) for the French 105 mm tank gun, this problem has been reduced by mounting the shaped charge on ball bearings within the case, greatly reducing the spin imparted to the explosive section, but this is not feasible with small-calibre cannon ammunition.

However, the problem has been tackled by careful design of the shape of the HE cone and metal lining, the latter being fluted to counteract much of the effect of spin. Despite this, pure HEAT projectiles are not currently used in cannon ammunition, but a variation is: the 30 mm M789 HEDP (HE Dual

*ATK PABM rounds sectioned, with common time fuze*

| Type | TNT | Picric Acid (Lyddite) | RDX (Cyclonite, Hexogen) | PETN | Pentolite 50/50 PETN/ TNT | CE (Tetryl) |
|---|---|---|---|---|---|---|
| Density (gram/cc) | 1.57 | 1.63 | 1.7 | 1.7 | 1.6 | 1.5 |
| Heat Produced (calories/gram) | 935 | 895 | 1,335 | 1,420 | 1,070 | 1,020 |
| Gases Generated (cc/gram) | 790 | 780 | 855 | 790 | 845 | 845 |
| Velocity of Detonation (m/s) | 6,900 | 7,250 | 8,400 | 8,100 | 7,455 | 7,300 |
| Calculated Power[1] | 106 | 100 | 166 | 161 | 130 | 123 |
| Brisance (shattering effect)[2] | 97 | 100 | 201 | 187 | 131 | 114 |
| Figure of Insensitivity[3] | 115 | 100 | 50–60 | 40 | 72 | 70 |

*1 = (calories/gram × cc/gram × 100)/(cal/gram of Picric Acid × cc/gram of Picric Acid)*
*2 = (power × velocity of detonation × density)/(velocity of detonation × density of Picric Acid)*
*3 = the higher the figure, the lower the sensitivity*

Purpose – anti-personnel as well as anti-armour), the standard type of ammunition used in the 30 mm M230 Chain Gun fitted to the AH-64 Apache attack helicopter, in which the angle of the rifling has also been reduced to the minimum level needed to achieve stability. The M789 is claimed to be capable of penetrating 50 mm of steel, as well as having a 4 m lethal radius against unprotected personnel.

Another recent variant of the HE shell is the timed air-burst type, known by different manufacturers as HEAB or PABM (HE Air-Burst, or Programmable Air-Burst Munition). These anti-personnel shells are intended to explode in the air directly over the target, and are designed to produce a spherical fragmentation pattern to ensure that even people hiding behind walls or in trenches may be hit. The timing of the detonation is by a fuze which is programmed as the gun is fired.

## TYPES OF EXPLOSIVE

The earliest explosive used in cannon shells was black powder (gunpowder) but at the end of the 19th century this began to be replaced by high explosives (HE), starting with picric acid, (known as Lyddite in the UK). This was in turn replaced during and after World War I (WW1) by TNT (TriNitroToluene), which is still widely used today; it is effective, inexpensive compared with more modern and powerful explosives, relatively insensitive and easy to pour. It also mixes well with other chemicals to provide different characteristics.

Aluminium (or sometimes magnesium) powder is often added to HE, as this both increases the brisance (shattering effect) and enhances the incendiary effect, as well as making the explosive pulse last longer. Wax of various types is sometimes added as a phlegmatizer to make the HE less sensitive and therefore safer to fire from a gun.

In both World Wars, TNT was mixed with up to 40% ammonium nitrate as an economy measure, to create Amatol; almost as effective, except for increased susceptibility to damp. In World War II this was supplemented by Minol (typically 40% TNT, 40% ammonium nitrate and 20% powdered aluminium).

By World War II, more powerful substances were being introduced into general service. One of these was PETN, also known as Penta or Penthrite. The problem was that this was generally felt to be too sensitive to use by itself, as it was inclined to be detonated by the shock of firing (although this did not deter the Japanese). It was accordingly usually desensitized by adding about 15% of Montan wax to produce Penthrite Wax, or Nitropenta. An alternative use was to mix PETN with TNT or Amatol to form one of the Pentolites (Pentolite being typically 50% PETN, 50% TNT).

Another more powerful explosive was Tetryl or CE (Composition Exploding), which had seen earlier use but became more widely used in World War II. Octogen, also known as HMX, is chemically related to RDX (see below) and is often mixed with 25–30% TNT to make it pour more easily, the result being Octol.

An important explosive, as powerful as PETN but less sensitive, is RDX (Research Department eXplosive), which is also known as Cyclonite or Hexogen. This has often been mixed with TNT, with or without aluminium powder or wax, for example: Composition B (60% RDX and 40% TNT); Cyclotol (70–80% RDX); Torpex (42% RDX, 40% TNT and 18% aluminium); Hexotonal (Torpex with 15% aluminium and 3% wax); and Composition H6 (44% RDX, 30% TNT, 21% powdered aluminium, 5% paraffin wax).

During World War II the Japanese used several types, with TNT, Pentolite and Cyclonite all being recorded, by themselves or in various combinations. Germany used HA.41, a mixture of 80% Cyclonite and 20% aluminium. The Soviets used a mixture of RDX and aluminium.

In recent decades attention has been paid to Insensitive Munitions (IM) which are much less likely to explode accidentally when subjected to high temperatures or violent impacts (e.g. from bullets or shell fragments). One common explosive of this type is plastic- or polymer-bonded explosive (PBX) consisting of explosive powder combined with a synthetic polymer. Various materials can be used, the first experiments in the early 1950s being with RDX and polystyrene plus a platicizer. HMX compositions use teflon-based binders, with other IMs being TATB (triaminotrinitrobenzene), FOX-7, TEX and IMX-101.

The characteristics of some of these explosives are shown above. The most powerful have not been much exceeded in destructive power since World War II, and World War II explosives, or variations on them, are still in common use today.

## COMBINED PROJECTILES

It is usual for projectiles to combine characteristics. The most common example of this is to add a tracer element to any type of projectile, leading to a '-T' at the end of the designation. These usually consist of a chemical in the base of the projectile that is ignited by the heat of firing and burns brightly, enabling the trajectory to be followed and the aim corrected as required. These are obviously more effective at night so daytime tracers (particularly in aircraft guns) have sometimes generated a smoke trail.

It is usual for the tracers used in ground fighting to burn without emitting light for the first part of their travel ('dark trace'), partly to avoid spoiling the gunner's night vision and partly to conceal the precise location of the gun. German night fighters in World War II used 'dim trace' ammunition visible to themselves, but less obvious to the bombers they were attacking (often from below). More recently, IR tracers have been developed which emit hardly any visible light and can only be seen through night vision devices.

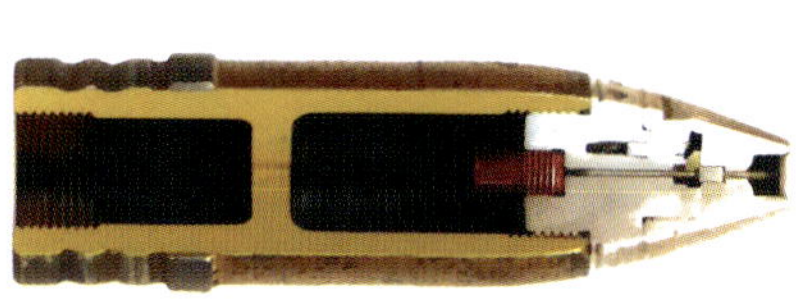

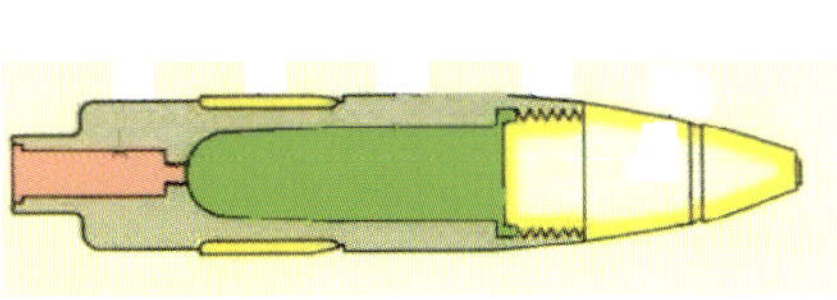

*Two HE-T shells from World War II: top is for the 37 mm German Flak, below is for the 1.1-inch USN (28 × 199SR). Note how the end of the tracer burn (dull pink colour) the HE (green) will be detonated, so this is an HE-T-SD shell.*

One undesirable characteristic of tracer rounds is the difficulty in matching their trajectory to non-tracers; the tracer is burned up in flight, thereby reducing the weight to some degree, but the gas generated at the base of the projectile can help to reduce the aerodynamic drag.

One advantage of tracers is that they can be used to achieve a simple self-destruct (SD) effect with HE shells, in that the final stage of tracer burn can be linked by a channel to the HE compartment. Otherwise, SD is achieved by including a timing element in the fuze. SD is particularly useful in AA ammunition, as it is undesirable to have live rounds which have missed their target hit friendly ships, vehicles or people. For the same reason, ammunition used in aircraft on home defence duties has been given SD fuzes, notably in Germany during World War II.

An alternative tracer proposal is OWL (One-Way Luminescent) tracers, which use materials in the projectile base which glow after being subjected to the heat and pressure of firing. Little space is used in the projectile and nothing is burned, so the ballistics are unaffected. Yet another proposal was for a reflecting material in the base of the projectile, illuminated by a laser from the firing aircraft or vehicle.

An originally unintended side-effect of tracers is psychological; the sight of a stream of tracers heading towards an attacking aircraft can distract the pilot. Clearly, developments aimed at minimising the chance of the enemy seeing the tracer may in some circumstances be undesirable as the distraction or suppressive effect will be lost. A further advantage of SD tracers in AA projectiles is that they create a bursting zone which might distract pilots even before they begin to fly through the tracers.

Other forms of combined projectile have been developed in order to maximize the effect on the target or to achieve reasonable effectiveness against a range of different targets, as we have already seen in the case of the HEI and HEDP shells respectively. The need for combined shells has been reduced to some degree by the use of dual-feed systems, which allow a gunner to instantly switch between two separate magazines containing two different types of ammunition (e.g. AP and HE), but combined shells still have uses, especially in aircraft guns which are all single-feed except for some Russian helicopter installations.

The term 'AP shell' (which should properly be APHE) is used to describe an essentially armour-piercing projectile which also contains some high explosive (usually only a very small percentage of the projectile weight), intended to detonate once the armour has been penetrated. It is generally only worth doing this in larger calibres, although national practices have varied. In World War II, for example, British AP shot for tank/anti-tank guns contained no HE, while the German equivalents did. This type of shell is now obsolete.

More of a compromise is the SAP or SAPHE (semi-armour piercing HE) shell, which is similar to an APHE shell but has thinner walls to provide more space for explosives, thereby losing some AP performance. A similar pattern of shell became popular in the late 19th century for naval guns, when it was known as CP (Common Pointed), although this originally used iron filled with gunpowder. Because of the necessity for a hard point to both APHE and SAPHE shells, the fuzes are fitted into the projectile base. SAPHEI ammunition is still made, most notably for the Nexter 30 mm aircraft cannon since the French consider this combination of characteristics to be optimal for a general-purpose aircraft gun round.

Conventional ammunition designations can therefore be as complex as SAPHEI-T. The problem with such multi-purpose rounds is obviously that they may not be very effective in any of their tasks. The Norwegian firm of Raufoss has been particularly successful in developing Multipurpose (MP) ammunition which combines a chemical impact fuze with incendiary, HE and sometimes armour piercing elements. This type has now been adopted by other countries for a wide range of ammunition, such as the American Olin PGU-28 round for the 20 mm M61 series cannon. While

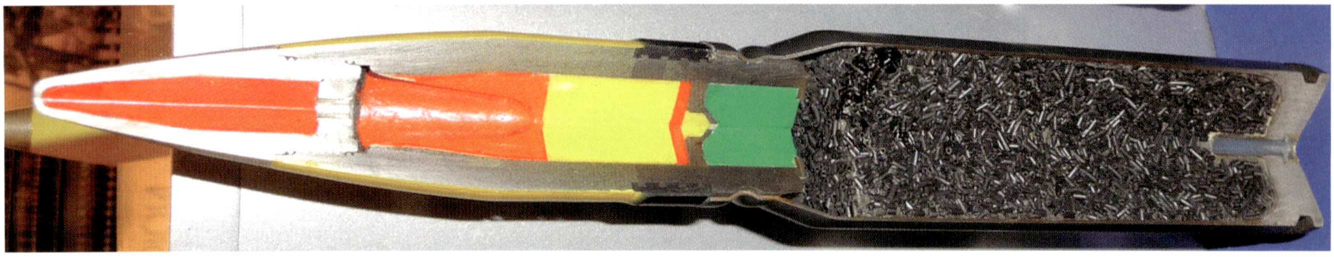

*30 × 173 Multipurpose: the incendiary in the tip ignites on impact, leading to the explosion of the HE element (yellow) after a short delay*

primarily for aircraft use, such projectiles have also been developed for rounds as large as the 40 mm Bofors.

Nammo has also developed a different form of multipurpose ammunition known as APEX (Armour Piercing EXplosive). This has a tungsten carbide penetrator for enhanced penetration, combined with HEI elements and with a delayed-action fuze. It is intended for aircraft use, being offered in 20 mm calibre and in 25 mm calibre for the F-35 fighter.

## OTHER PROJECTILES

Some projectiles are designed to break up into a hail of sub-projectiles, either on or immediately before impact with the target. The former are called 'frangible', leading to the 'F' designation. Some of these have been developed for training as mentioned in the section on Target Practice ammunition. Other uses are more warlike in the form of FAP, with the tungsten alloy projectiles being manufactured in such a way that they are relatively brittle. They are still able to penetrate armour almost as well as conventional tungsten penetrators but break up after penetration. There are incendiary effects as each fragment strikes a surface, and the cloud of fragments travelling at supersonic velocity generates a powerful shock wave, adding to the general mayhem. A bonus is that the lack of explosive chemicals means that it is subject to less restrictive transport and storage regulations than HE rounds.

A variation on this is FAPI, with a tip made from a pyrophoric material such as titanium or zirconium, which indicates a hit with a bright flash as well as generating incendiary effects after penetration.

FAP was developed by Rheinmetall for aircraft use, for which discarding sabots are unsuitable, but the same principle is used in APDS ammunition to create FAPDS which has a useful capability against both aircraft and light armour, with several advantages over other types. In AA (anti-aircraft) use the dense sub-calibre flight projectile has significantly better ballistics than conventional HE shells, providing a longer effective range and a higher hit probability at all ranges. A bonus is that it retains almost the same effectiveness as APDS against light armoured vehicles, so it's a dual-purpose round. The icing on the cake is that as the projectile is inert, it is safer to handle. What it means is that AA guns in these calibres can have their performance and usefulness substantially upgraded, just by adopting new ammunition (with amendments to the sights to allow for different ballistics). One variant of FAPDS is Rheinmetall's FMPDS (frangible missile-piercing discarding sabot), originally developed by NWM De Kruithoorn, which is loaded into the 30 mm cartridges used in the Goalkeeper naval anti-missile system.

Diehl of Germany has produced a rival to the FAP concept with the PELE (Penetrator with Enhanced Lateral Effect). In its original form this is the opposite of FAP in that it consists of a hard steel shell body with a lightweight (alloy or plastic) core and a light nose-cap, so designed that compression of the inner core on impact causes the steel shell to fragment and burst violently outwards after penetration. Projectile weights and muzzle velocities are the same as standard shells. Armour penetration is not as great as FAP but PELE will fragment after penetrating only 2 mm of Dural. Both types are available with the addition of pellets to enhance fragmentation, and the distinction between them is blurring because some PELE types now contain a tungsten penetrator. FAP and PELE are making inroads into sales of both the traditionally fuzed and Multipurpose HEI types, while PELE is also used in army guns as an alternative to HEI. Rheinmetall Waffe Munition (RWM) has acquired the rights to the PELE technology and is sharing development and production with Diehl.

Frangible rounds are also being produced for training, e.g. the Nammo TP-RRR (reduced ricochet risk – as opposed to RR for reduced range as already mentioned) is intended for aircraft use when firing at ground targets, to eliminate the risk of projectiles ricocheting off the ground and travelling far into the distance or even striking the firing aircraft as it flies past (a surprisingly common occurence).

A light cannon variant which has long been applied to Russian aircraft ammunition for ground attack purposes is the CC, or cargo carrying, projectile (also known as the multi-element projectile). This contains sub-projectiles, heavy metal pellets which are discharged from the body of the projectile at a set distance after firing in order to saturate the target area. The version for the 23 mm cartridge

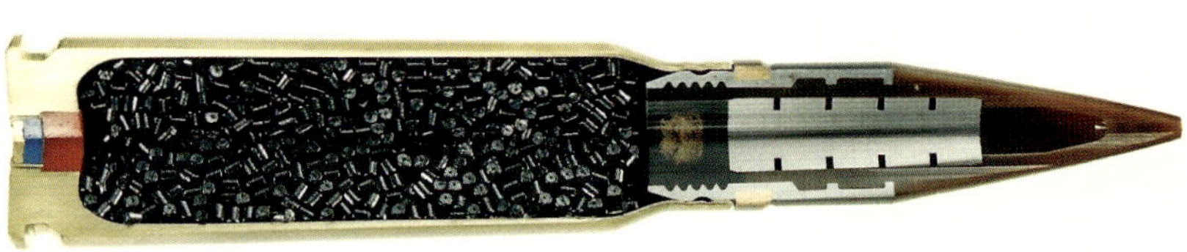

*20 × 102 Frangible Armour Piercing (FAP) (Oerlikon Contraves)*

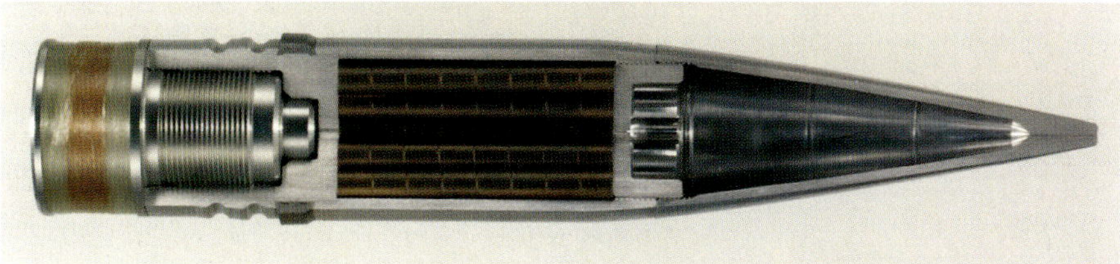

*35 mm Oerlikon AHEAD projectile sectioned to show the stacks of pellets in the body of the projectile, the base fuze and ejection charge, and the hollow nose cone (Oerlikon Contraves)*

carries 24 × 2 g sub-projectiles, that for the 30 mm guns 28 × 3.5 g. The current 30 mm round discharges at between 1,100 and 1,800 m after firing, the sub-projectiles forming a cone with an angle of 8°. Total projectile weights and muzzle velocities are similar to that of standard ammunition. Although promoted as being useful for attacking targets such as parked aircraft, they should clearly be very effective in the anti-personnel ground-strafing role for which aircraft guns have most commonly been used in recent conflicts.

A more sophisticated version of the Russian CC is represented by the Oerlikon AHEAD (Advanced Hit Efficiency And Destruction) anti-aircraft/anti-missile ammunition developed in 35 mm calibre. The fire control system measures the distance to the target, calculates the time needed for the AHEAD shell to reach it, and sets a time fuze electronically as the shell exits the muzzle. A small bursting charge is timed to detonate 25 m before the shell reaches the target, creating an expanding pattern of 152 tungsten sub-projectiles each weighing 3.3 g and travelling at high velocity due to a combination of the carrier projectile velocity and the force with which the sub-projectiles are ejected.

The original version of AHEAD has since been joined by a variant using a larger number of smaller sub-projectiles, for attacking personnel and other ground targets. For the last of these, intended for use in cannon mounted in armoured vehicles, both 35 mm and 30 mm versions are available and the ammunition is also sometimes designated KETF (Kinetic Energy Time Fuzed). In principle, KETF is similar to the nineteenth century Shrapnel shell, except that the timing method was then considerably less sophisticated. The disadvantage of KETF in the anti-personnel role is that the pellets are principally sent in a forwards direction, which means that they may not strike people hiding behind walls or in trenches, unlike the rival HEAB/PABM shells which are designed to do just that, despite delivering fewer and smaller fragments to the target area. While KETF has secured some sales, PABM seems to be winning in this contest, at least for the time being.

Another rival to AHEAD is the Bofors PFHE (pre-fragmented high explosive) shell available in 40 mm and 57 mm calibres as a component of the 3P system (prefragmented, programmable, proximity fuzed). This differs in being an HE shell lined (in 40 mm calibre) with 650 tungsten pellets. The 3P fuze can be set electronically immediately before firing to one of seven different modes, involving variations on and combinations of proximity, time and impact fuzing. This allows them to be set, for example, in the gated proximity mode (in which the proximity fuze is only armed when close to the target, to prevent it from being trigged by anything it passes on the way) or with a choice between instant or delayed-action impact fuzing. The effective radius of the exploding 40 mm shell is up to 7 m against aircraft and 3 m against sea-skimming anti-ship missiles. Compared with AHEAD/KETF, the main disadvantage is that the pellets are much smaller and, given the less directional fragmentation pattern, only a small percentage of them will be likely to strike the target. On the other hand, the HE blast means the fragments will be travelling a lot faster.

Rocket-assisted projectiles (RAPs) are used in artillery ammunition to extend the range but as the space required for the rocket reduces the HE capacity by about 50% they are not used in smaller-calibre weapons. However, an AP RAP (with the propellant wrapped around the penetrator) was experimentally developed for the 30 mm GAU-8/A aircraft gun in the 1970s and demonstrated the same penetration at 1,800 m as the standard APCR could achieve at 1,200 m.

Another abortive effort was a 1968/69 feasibility study for a 'Special Purpose' Multi-Flechette Round (MFR) awarded to Honeywell by Frankford Arsenal, intended for anti-personnel use in the Vehicle Rapid-Fire Weapons System – Successor (VRFWS-S), the planned replacement for the US Army's 20 mm M139 gun for light AFVs. The VRFWS-S eventually resulted in the 25 mm M242 'Bushmaster' Chain Gun, but the MFR was experimentally developed in 20 and 30 mm. Required characteristics were: projectile total weight 194 to 227 g, weight of each flechette 0.65 to 0.97 g, MV 914 to 1,220 m/s, maximum lethality range 600 m. Priorities were to minimize dispersion and tumbling, achieved by using a small charge to eject flechettes

*Bofors 57 mm PFHE (Bofors)*

from the base of projectile. The shell was loaded with 0.687 g SPIW flechettes; 88 of them fitted in a 30 mm projectile (it was estimated that a 20 mm shell would carry 28, a 25 mm 60).

Not all projectiles are designed for offence; some shells have contents designed to defend aircraft against missile attack. Some post-World War II Soviet aircraft cannon have been available with chaff-dispensing anti-radar ammunition (PRL-23 in 23 mm) and even infra-red decoy projectiles (IK-23). The 23 mm version of the IR decoy round ignites about one second after firing and burns for some four seconds. The 30 mm chaff dispenser projectile produces a cloud of radar-obscuring particles which grows from an initial 7–9 m$^2$ to a maximum of 14–18 m$^2$ in area. These were intended both for use in bomber defensive weapons and by fighters, which would fire a burst of such decoys then dive out of the way of any incoming missiles.

Finally, guided projectiles will probably emerge at some point in the future and these could transform the long-range AA performance of the larger calibre cannon. Various proposals for such guided systems have been made (notably for the Bofors and Russian 57 mm rounds) but none appears to be close to production, the smallest system of this type in service being for the OTO Melara 76 mm gun, which is outside the scope of this book.

One less ambitious version of the above is the course-correcting shell. For such a device to work, there need to be radar and communications systems plus a fire control system linked to the gun, so that the trajectory of the shell and its target can be constantly measured and correction signals sent to the shell. That hi-tech element is actually quite easy; what holds up its implementation is the hardware problem of steering the shell. There is very little room inside a cannon shell, so the method of steering needs to be as compact as possible. The most likely version of this to see service in the foreseeable future could be a 'one shot' system allowing just one change of course for each shell.

## TARGET PRACTICE AND OTHER NON-COMBAT ROUNDS

Target Practice ammunition is by far the most commonly used. It may be encountered both with (TP-T) and without a tracer and is designed to provide economical training and practice. It therefore needs to match the ballistics of the combat rounds while being inexpensive to make. Sometimes purpose-designed inert projectiles are used, in other cases the standard HE shell is given an inert filling and a dummy fuze.

A few TP projectiles contain chemicals designed to give a flash on impact as a visual indication of the striking point. Some more specialized types have been designed to fragment on impact, as described in the Other Projectiles section.

The very long range of which cannon ammunition is capable can be inconvenient as this limits the number of places where training can safely take place. Accordingly, some TP projectiles are shaped aerodynamically to follow the trajectory of service rounds for a certain distance, after which they become unstable and rapidly lose velocity, falling to the ground. These typically have the RR designation (reduced range) added to them, as in the Nexter 25 × 137 TPRR-T shown right, which has aerodynamic flutes formed in the shell body.

The most radical example of this is the 30 mm Rarden's DSRR (Discarding Sabot, Reduced Range) which fires a saboted tubular projectile that follows the trajectory of the APDS out to 1,500 m. At that range the velocity has fallen from 1,300 m/s to around 700 m/s and an aerodynamic shockwave effect chokes the flow of air through the tube, at which point it becomes aerodynamically unstable due to the modified pressure distribution. The range safety trace for this ammunition measures 4,500 m by 2,260 m, compared with 11,900 m and 5,800 m for the APDS.

In the past, TP ammunition has sometimes been used in combat, for example by the RAF for a time in World War II, when the 20 mm HE projectiles were suffering from fuzing problems. TP was

*Nexter 25 mm TPRR-T, with destabilising flutes*

*30 mm Rarden L15A1 DSRR tubular projectile with sabot*

then generally referred to as 'ball', the usual term for ordinary military bullets, named after the round lead or iron balls that were the standard projectiles until the nineteenth century.

Other ammunition used in training includes blank and drill or dummy rounds.

Blank rounds are cartridges without projectiles. In some applications they are used merely to make a noise to simulate real ammunition, as in small-arms training and the ceremonial firing of field guns. In automatic cannon they are generally used to practice firing the weapon without needing a large safety area to allow for a projectile flying down range. The construction of blank cannon rounds varies; some are based on a standard cartridge case with a 'rose crimp' (the neck folded inwards to seal the propellant); others are entirely made of plastic except for a metal base. By themselves, blank rounds do not develop enough gas pressure or recoil to function automatic mechanisms. One approach to resolving this is the use of 'break-up shot', thin-walled plastic shells filled with small metal balls or powder to approximate the weight of service projectiles, which disintegrate immediately after leaving the muzzle.

Drill or dummy rounds are used for manually cycling through the weapon and its ammunition feed (or powered cycling in the case of externally-driven guns). Drill rounds are used for safe training in gun and ammunition handling, dummy rounds are used to test the functioning of the mechanical ammunition handling process. Despite their different uses they basically have the same characteristics so the same rounds may be used for both purposes.

There is a wide variety of approaches to making drill or dummy rounds: some use conventional cases and projectiles but are completely inert (with no chemical components of any kind), however these may not be robust enough to withstand heavy use.

An alternative approach is to use a genuine cartridge case but with a special 'projectile' which is long enough to reach the base of the case, making the whole round much sturdier. A third variation is to make a once-piece round, usually machined from solid aluminium alloy to match the weight of conventional rounds. Finally, there are dummy rounds which are specially made in more than one piece, for production convenience or to use different materials – which may include synthetic materials, as in the French 30 mm round shown third from the left in the photo below left.

Finally, 'proof rounds' are made for all ammunition: these generate much higher chamber pressure than standard loadings and are intended for testing the strength of a gun before approving it for service. In some instances, special projectiles have been developed: in British use, these were sometimes cylindrical with a flat nose, so they could not be mistaken for anything else.

*Various types of drill and dummy rounds*

## FUZES

The triggering of incendiary compounds might not be too critical as they burn for a noticeable, albeit short, period of time. HE shells, however, have to be detonated at precisely the correct instant; and for this, a fuze is needed.

Every fuze has two distinct requirements. First, it must reliably detonate the explosive filling at the precise instant required: this might be the instant it hits the target; a fraction of a second later; at a preset point of its flight; or when it detects the target is close enough. Many fuzes are capable of being set to achieve more than one of these, or even all of them like the Bofors 3P fuze. Secondly, it must be completely safe when being handled (or mishandled) and fired. It is not easy to achieve these aims and fuzes are often intricate devices built with the precision of a Swiss watch, and therefore tend to be expensive (various fuzes can be seen in the photos of sectioned HE shells earlier in this chapter).

Fuzes are of several distinct types. The original version, applied to mortar shells in previous centuries, consisted of a length of fuze cord (rope soaked in a flammable compound) that could either be ignited just before firing or be lit by the propellant charge. It was a time fuze, which detonated after the cord had burned through to the filling, and could be adjusted by varying the length of the cord.

The first 'modern' time fuzes, which were used in both World Wars, were a more sophisticated version of the same principle. They contained a powder train which could be adjusted in length by turning the head of the fuze to achieve the required burning time. During World War II, Germany made much use of mechanical fuzes which were, in effect, clock timers (the British adopted these half-way through the war). These were much more precise than powder fuzes, being accurate to 0.5 per cent instead of 2 per cent. This was particularly important in anti-aircraft fire, in which time-fuzes were set to detonate shells at a specified altitude.

In more recent times many time fuzes, like clocks, have become electronic. These fuzes are mainly reserved for larger artillery, but are an essential feature of the Oerlikon AHEAD system in which the time is determined by the fire control computer and set electronically as the projectile leaves the muzzle. They are also used by HEAB/PABM ammunition, with ATK using a fuze which counts the number of projectile rotations to determine when to detonate the airburst shell.

Contact or impact fuzes may appear simpler to design – the general principle is the same as that used to fire a toy cap gun – but they have their own problems. The main one is to ensure that the shell will detonate at the instant of contact while remaining safe at all other times. Mechanical contact fuzes have elaborate safety measures, which usually involve the detonating mechanism being locked in a safe position by devices which are moved out of the way by the centrifugal force applied to the shell as it is spun by the barrel rifling. The performance of such fuzes is usually specified in terms of the minimum safe distance from the muzzle at which they become armed, and the maximum distance by which they will all be armed (typical safe/arm distances being 30/60 m).

The mechanism is further complicated by the usual requirement that there be a slight delay between contact and detonation. This is in order to achieve a more useful effect against vehicles as it is usually better for the shell to explode inside the target aircraft or vehicle rather than on the surface. This requirement can cause problems, for example Israeli pilots, in various actions during the 1960s, discovered that their 30 mm DEFA cannon fire was ineffective. It transpired that the shells were fitted with long-delay fuzes intended for use against bombers; against the much smaller fighters, they had passed right through before exploding. Similarly, the strafing of enemy personnel on the ground by combat aircraft may be relatively ineffective if the shells are burying themselves in the ground before exploding or fragmenting.

Another problem is concerned with ensuring that the fuzes do detonate at the right instant even when the target is struck a glancing blow. This is a particular problem in firing against aircraft, and despite the best efforts of fuze designers it is difficult to ensure reliable detonation. For example, the RAF discovered in the late 1950s that 20% of 30 mm ADEN strikes would ricochet off the target while a further 55–60% would explode on the aircraft skin instead of inside the structure, leaving just 20–25% of hits effective. Such problems are magnified where the target offers little resistance to penetration. One of the earliest contact fuzes specifically designed for aircraft use was the British No. 131 fuze of World War I, made extremely sensitive in order to achieve detonation when penetrating airship fabric. Somewhat surprisingly the Russians developed a modern equivalent for their 30 mm ammunition in the Cold War, intended to destroy unmanned reconnaissance balloons which sometimes drifted over the USSR.

Not all contact fuzes are mechanical. One exception is the 'air compression fuze', which relies on the violent compression on impact of a column of air to ignite the fuze; there is a hole in the tip. This operating principle saw some use in World War II by the Japanese, by whom it was known as the KUKI-SHINKAN. It was also use by the RAF; in the 20 mm Hispano (e.g. Fuze No. 254) and also tried in the early 30 mm ADEN LV. It appears to have only been used for aircraft guns, and the Japanese Army in particular valued it because their Ho-103 and Ho-5 guns fired HE shells from closed bolts, so it was important to use a fuze which was not ignited by heat in the chamber. Russia used them as well, in the 20 mm ShVAK.

Even simpler, but rather more dangerous, were those Japanese shells in World War II which were detonated by a nose fuze consisting of PETN explosive covered only by a thin copper cap, with no safety mechanisms at all.

Theoretically similar – but far safer – modern fuzes use chemicals which detonate on impact, notably the popular Raufoss (now Nammo) Multipurpose ammunition which is available in calibres of up to 40 mm. The characteristics of the incendiary fuze compound and its metal exterior are extremely carefully judged to ensure the necessary combination of reliable detonation with safety when mishandled. Put simply, the shell must never detonate when dropped from a height point-down onto a concrete floor, but must always detonate when it hits a target at high velocity.

One unusual fuze developed by Germany in World War II was the hydrostatic fuze *(Flüssigkeitszünder)* for aircraft cannon shells. This was designed to detonate as a result of the pressure rise on penetrating bomber fuel tanks. The fuze was designated the AZ 9501.

Fuzes which are detonated by the close proximity of the target are unsurprisingly known as proximity fuzes, although for security reasons during World War II (when they were first introduced) some disinformation was circulated to the effect that their randomly-allocated USN code letters – VT – stood for variable time. Some initial theoretical work was carried out in the UK in 1939–40, but this was transferred to the USA whose scientists overcame considerable difficulties in turning the concept into reality. These fuzes immediately achieved an improvement in AA successes of between three and five times in comparison with timed or contact fuzes. There are various methods of achieving detection but a small radar set is by far the most common in cannon shells, which limits its application to the larger calibres (originally 75 mm but now down to about 40 mm), partly because of the size of the fuze and partly because smaller shells don't contain enough HE or produce enough fragments to damage the target at a distance, so they need to penetrate the target.

Because of the difficulty of detecting other targets against background clutter, proximity fuzes for automatic cannon are used almost exclusively against aircraft or missiles. They can normally be set to detonate at different distances according to the nature of the target or of the environment, with a smaller distance being set when firing at targets close to the ground or sea to ensure that the fuze is not triggered accidentally. In recent years more sophisticated fuzes have been developed to deal with this problem.

Modern electronic fuzes can be extremely versatile, such as the Bofors programmable fuze that forms an essential part of the 3P ammunition already mentioned. As well as the normal proximity function, the fuze can operate in five other modes: gated proximity (fuze only activated close to the target to avoid premature detonation); gated proximity with impact priority (a slight delay in activation to provide the opportunity for a direct hit); time function (for airburst fire to provide a shrapnel effect against surface targets); and two impact functions with variable delay.

Some fuzes are associated with particular types of shell. For example mine shells produce few fragments, relying almost entirely on blast and incendiary effects, so need to explode inside the target. They are therefore used with delayed-action contact fuzes. Shells exploded outside the target by proximity fuzes cause most damage by means of fragments and therefore either contain separate pellets (as in the Bofors PFHE) or have a thick shell wall which is designed to break up into fragments of a specified range of sizes. The vast majority of fuzes are fitted to the points of projectiles. However, some types of shell make use of fuzes in the projectile base, as in the case of the SAPHE already mentioned, or mounted in the centre of the shell – e.g. with HEAB/PABM, to provide a spherical fragment distribution pattern.

There is the obvious risk that shells fired at an attacking aircraft might carry on to detonate on landing some distance away, possibly among friends. Contact and proximity-fuzed shells are therefore often fitted with a self-destruct device that detonates the shell after a certain period of time, when it can be presumed to have missed the target. Sometimes this is a function of the fuze, typically being triggered when the spin rate has fallen enough, but a simpler system used in smaller calibres is for the propellant to ignite a separate chemical fuze in the base of the shell; a reversion to the original form of time fuze. This may also be achieved at the end of a tracer burn.

Some explosives are difficult to initiate so the fuze might be designed to incorporate an explosive booster charge, or gaine, to magnify its effect.

## PROPELLANTS

The earliest machine gun cartridges were designed to use black powder (gunpowder) as a propellant. This is a simple mixture of saltpetre, sulphur and charcoal, and has been used for centuries. It is not ideal as a propellant, however, because of its inefficient burning characteristics and the large quantities of smoke and other residues generated.

Improvements in chemical science in the late nineteenth century led to the development of 'smokeless' powders, the first of these entering service in France in the late 1880s. Other nations soon followed suit, with the British version being known as 'cordite' because it was extruded in fine strings resembling cords, which were bundled together for insertion into the cartridge case.

There are three different types of smokeless propellant, known as single-base, double-base and triple-base. All are based on NC (nitro-cellulose) with NG (nitro-glycerine) or equivalent, and nitro-guanidine as other major components in the more complex versions, triple-base including all three. Cordite is a double-base type.

The increasing complexity of the propellants is due to the constant search for the ideal combination of characteristics. These include maximum power for a high muzzle velocity, moderate pressure and temperature to minimize the stresses and erosive effects on the gun, as little fouling and corrosion as possible, and a minimum of smoke and flash at the muzzle. Many of these desiderata are mutually exclusive so each propellant is a compromise. As with other chemical ammunition components, propellants must be insensitive to rough treatment, provide consistent performance over a wide range of climatic environments and be tolerant of storage for long periods in poor conditions.

Cannon propellants today are generally prepared in the form of grains or small pellets. The precise chemical composition, and the size and shape of the grains, will vary from one cartridge to another in order to provide the power and pressure characteristics suited to the gun. The nature of the projectile affects propellant choice: heavier projectiles need slower-burning powders to keep the pressure peak down, whereas light projectiles need faster-burning powders to accelerate them quickly enough to reach a high velocity.

More information about chamber pressures can be found in the Ballistics section.

Propellant technology continues to develop, with particular success being enjoyed by the products of Nitrochemie at their Wimmis plant in Switzerland. They developed the EI (Extruded Impregnated) 'semi-double base' propellant in 1985 (known generally as 'Wimmis propellant') that is especially temperature-stable, delivering consistent performance regardless of ambient conditions over a long period of storage. It has been widely adopted for Western cannon ammunition as it both increased muzzle velocity and considerably improved barrel life compared with earlier propellants.

Since then Nitrochemie (now majority owned by Rheinmetall) has continued development work and has introduced ECL (Extruded Composite Low-sensitivity) propellant, which compared with EI generates more energy at a lower burning temperature and with no increase in pressure. It also has enhanced IM (Insensitive Munition) characteristics, being less vulnerable to bullet impact and cook-off (self-igniting when a round is left in a hot gun chamber). ECL contains nitramine and other additions, including stabilizers. It omits the relatively unstable NC, providing three times the storage life.

Other manufacturers have also been working on advanced propellants, with Diehl BGT Defence and ICT (*Fraunhofer*

*Institut Chemische Technologie*) collaborating on DNDA (dinitro-diaza) temperature-independent gun propellants for IM ammunition. These combine RDX or a similar energetic with an NC binder and a DNDA plasticizer. GD-OTS St Marks Powder facility has been developing a Hybrid propellant which offers similar high-temperature stability and barrel wear advantages. They are also experimenting with higher loading densities by combining extruded and ball powder, increasing density by 19% and muzzle velocity by 11.5%.

Modern propellants like these typically generate a flame temperature of 2,800–3,000 degrees K and generate a force constant of 1,000–1,100 J/g (joules per gram). Self-ignition (cook-off) temperature is around 175–200 degrees. Some idea of the effect of ambient conditions even on such advanced propellants can be gained by the 27 mm DM83/93 PELE DNDA ammunition. Muzzle velocity varies from 983 m/s at –40°C to 1,033 m/s at +50°C and 991 at +70°C, while maximum pressures at the same temperatures are 372, 400 and 450 MPa respectively.

In cannon ammunition, the focus of IM development has mainly concerned explosive projectiles so far, but Low Vulnerability Ammunition (LOVA) is gradually being introduced, initially with tank gun and artillery propellants which are more vulnerable, being generally contained within combustible cases.

## PRIMERS

Primers are older than metallic cartridges. The first percussion primer was developed early in the nineteenth century and this invention soon replaced the flintlock in military service. It consisted of a small copper cap containing fulminate of mercury, which produced an intense flash when struck a sharp blow from a hammer or firing pin and thereby ignited the propellant charge.

Many of the early metallic cartridges departed from this arrangement in that the priming compound was contained within the rim of the cartridge case, and fired by the impact of the hammer on the rim. This proved inconvenient because it required the rim to be thin and soft, and also meant that the used case could not be reloaded. The rimfire principle soon fell from favour except in the small inexpensive cartridges still used for target shooting.

Rimfires were superseded by the return of the separate percussion cap, now housed centrally in a cavity, known as the primer pocket, in the base of the case (the cartridges therefore being called 'centrefire'). It is so arranged that when struck, the primer fires into the propellant via a central hole or holes in the cavity. Once fired, the primer can be knocked out of the case, which can then be reloaded. In larger calibres, the primer may ignite an intermediate igniter charge to boost its effect.

Little has changed in primer design since then. Fulminate of mercury has been replaced by less corrosive (and more recently, lead-free) compounds and the precise details of the design of the percussion cap have been refined. Smaller cannon rounds usually follow the practice of small-arms ammunition in using a simple push-fit of the primer in the case, but in larger calibres primers are often screwed into the case to provide a more secure fit (and in principle to make it easier to replace the propellant, the first element of ammunition to deteriorate after long storage). The only major departure has been the replacement in some cases of percussion ignition with electric.

Electric ignition is achieved by passing an electric current from the gun through a special primer which is heated to ignition point virtually instantaneously. Initially developed for artillery, it was first used in automatic weapons in Germany to facilitate the synchronisation of aircraft guns designed to fire through the propeller disc of a single-engined fighter. Previously, this had involved a complex mechanical or electromechanical linkage between the propeller and the gun firing mechanism to ensure that the gun fired only between the propeller blades. It was obviously simpler to design a direct electrical link between the propeller and the cartridge. Ironically, it came into general service in World War II just before the need for it disappeared because of the advent of jet engines.

Electric priming does have other advantages. It simplifies gun design because the firing pin and its associated spring and mechanism for releasing, withdrawing and recocking the pin are replaced by a simple electrical contact. A disadvantage is the risk that such primers may be accidentally ignited by powerful electrical emissions as found in some environments, especially warships. Nexter has produced alternative versions of some 30 mm rounds with different primer characteristics to overcome this problem.

It is usual for cartridges to be designed either for percussion or electric ignition but there are some cases where a particular cartridge has been available in both versions. They are difficult to tell apart visually, since only the internal design of the primer is different. However, in electric ignition the primer has to be insulated from the rest of the case and this insulation is sometimes visible by a concentric ring in the primer. Needless to say, percussion and electric versions of the same cartridge have to be used in the appropriate guns as they are not interchangeable. The best known examples are the 15 mm and 20 mm rounds for the MG 151, the 20 × 128 Oerlikon and the 30 × 165 Russian (which uses electrical priming for the naval and aircraft weapons, percussion for the AFV guns). There are others; the percussion primed 20 × 110 widely used in the HS 404 and related guns was produced in an electrically-primed version for the post-war American M24. It is possible to combine electric and percussion elements in the same primer, although it is not clear whether anyone has made use of this in service cannon rounds.

## THE CARTRIDGE CASE – DESIGN

The cartridge case has two major functions. First, it holds together all of the active components of the cartridge – the projectile, propellant and primer – in a waterproof container which is rugged enough to withstand rough handling, especially in automatic weapons. Second, when the gun is fired, the cartridge case is expanded by the gas pressure against the walls of the firing chamber, forming a gas-tight seal which prevents any propellant gas from seeping back into the gun mechanism – and possibly into the firer's face. Achieving this gas seal is known as 'obturation'. A less significant feature is that the metal case absorbs some of the heat generated by firing and removes this from the chamber when it is ejected.

Cartridge case design is a science in itself. There are certain basic requirements – the case must locate the projectile in precisely the correct position in the firing chamber (with the exception of advanced primer ignition – API – blowback guns which fire as the cartridge is moving forwards), it must be easy to extract once fired and it must function well in automatic weapons – but there are various methods of achieving these aims.

Before proceeding further it is well to be clear about the nomenclature of cartridge cases. The basic elements are shown in the illustration at the front of this Design section.

The base of the cartridge case usually has identifying symbols stamped around the primer pocket, which typically provide information about the cartridge, the manufacturer and the year of production. Somewhat confusingly, these symbols are popularly referred to as the 'headstamp' instead of the more logical 'base stamp'. Some cannon ammunition has similar code stamps on the projectiles, while stencilled data on the case side may also be found.

Rimless cases need an extractor groove and 'bottlenecked' cases also have a shoulder, where the case is reduced in diameter down to the neck in cartridges in which the calibre is significantly smaller than the case diameter. Some cases have a belt – an annular projection – just above the extractor groove.

Given the violent accelerations the ammunition for fast-firing cannon are subjected to, there needs to be an effective means of securing the projectile in the case otherwise it might fall out during chambering. This is normally provided by crimping; the case neck is pressed into recesses in the projectile, either all around the circumference (annular crimping groove) or at several points on it. Multiple rows of crimps are used in some cases, for extra security.

*Most headstamps just contain identification codes, but the AAC was famous for self-promotion! Note that the primer has its own stamp, showing it was made by the Western Cartridge Company in 1942*

Accurate location in the firing chamber is essential to the ignition of the primer; the cartridge needs to enter the chamber, but not too far or the firing pin (or electrical contact) will not touch the primer. It is also important in the internal ballistics of the gun, as the projectile must be in exactly the same position, relative to the conical section leading to the rifling, to achieve consistent pressure characteristics and thereby minimize the variation in muzzle velocity and trajectory. The distance between the face of the bolt and the part of the chamber that locates the cartridge is known as the 'headspace'.

The earliest method of achieving accurate location was by means of a rim around the head of the case. This remains outside the chamber or in a recess so that it is flush with the breech face. The rim also provides something for the extractor to hook on to in order to pull the fired case from the chamber. Rimmed cases are still satisfactory except that the rim can complicate ammunition feed arrangements in automatic weapons. In spring-loaded magazines the rims can foul each other if improperly loaded, while in belt-fed guns it is usually necessary for the cartridge to be withdrawn backwards from the belt before being pushed forwards into the chamber. This design has therefore largely been replaced for military purposes, although it is still encountered, for example in some Russian small arms and in larger cannon from 37 mm upwards.

By far the most common military case type is the rimless, in which the rim is reduced to the same diameter as the case so that the cartridges in magazines can be stacked on top of each other or be pushed forward from ammunition belts without risk of jamming. To give the extractor something to hook on to, an extractor groove is cut into the case. There have been a few examples of semi-rimmed (or semi-rimless) cases, in which the rim is only fractionally wider than the case and is combined with a small extractor groove, but these have not been a particularly successful compromise. Some Bofors cannon cases (25 mm and 40 mm L/60) combine a rim with what appears to be an extractor groove, but

*Case rim types (from left): rimmed; rimless; semi-rimmed; rebated; belted*

this is a clip groove to secure the rounds in the right place in the loading clip.

Rimless cases need a method of accurate location in the chamber and this is normally achieved by using a bottleneck case, in which the headspace is determined by the contact between the shoulder of the case and the matching part of the chamber. With straight or tapered cases this becomes more difficult to achieve so such cases often have a projecting belt just in front of the extractor groove to provide positive location.

A few cases have seen service with rebated rims, i.e. with a rim of smaller diameter than the case. This is essential in API blowback cannon that fire while the cartridge is still moving, because it allows the bolt and its extractor claw to follow the case into an extended or hooded chamber. This ensures that the cartridge is fully supported at the instant of firing.

All Oerlikon 25 × 137 and some 35 × 228 cases have an annular groove just below the shoulder. The purpose of this is to locate the type of belt links used with this ammunition.

Military cartridge designations are usually given via metric case measurements of the calibre and case length. The round for the 20 mm Hispano-Suiza HS 404 cannon is therefore described as the 20 × 110, as the calibre is 20 mm and the case length is 110 mm. Note that this is not the same as the overall cartridge length, which in this example is about 180 mm, since this can vary according to the length of the projectile. Note also that the calibre measurements are nominal, so it must not be assumed (for example) that a projectile from one 40 mm cartridge can necessarily be used in another.

Letters may be added after the numbers to denote some variation from the military standard rimless head or to aid identification. The 20 mm Becker is known as the 20 × 70 RB because it has a rebated rim (sometimes expressed as RR); the 30 mm Aden is called the 30 × 113B because the case is belted; the rimmed 20 mm ShVAK is known as the 20 × 99R; while semi-rimmed rounds have the SR suffix. Where two rounds have the same calibre, case length and case type, other letters may be added; for example the 20 × 110 USN which is not interchangeable with the 20 × 110 HS.

It should be noted that even in countries where metric designations are used, the case length may not form a part of the official designation. In this book metric measurements are applied to all rounds to aid identification, including older ones which used non-metric designations, so for example the semi-rimmed 1.1' US Naval round is listed in the Ammunition Table as the 28 × 199 SR.

The discussion in this chapter is about conventional cartridges. Various unconventional designs have been tried, almost entirely without success except for the new Anglo-French 40 mm Cased Telescoped system. This and less successful efforts are described in a later section.

## THE CARTRIDGE CASE – MATERIALS

Brass is still used for most small-arms and some cannon cartridge cases despite its weight and cost. Standard cartridge brass (70% copper, 30% zinc) has exactly the right characteristics to fulfill the functions described in the last section because it expands instantly to form a seal without being split open. Nowadays steel is normally used in larger cannon, particularly linear-action guns with a high rate of fire, because its extra strength is better at coping with the violent treatment the cartridge is given by the loading mechanism. It is also lighter and cheaper. However, it requires more protection against corrosion and, being less resilient, forms a less perfect seal on firing. This is a potential disadvantage in turret-mounted applications as it means that more propellant gas might seep back into the turret.

Aluminium alloy has even more attractions, as it saves between 25–38% of the total weight of a cartridge, with the greatest saving being with high-velocity rounds in which the large cartridge case makes up a high proportion of the weight. However, for a long time it proved very difficult to achieve satisfactory results with this material. It tends to split under the pressure of firing and, being flammable, can catch fire in a 'burn through' doing serious damage to the gun. So the cases need to be carefully made and chemically treated to minimize these risks. This is expensive so is only used when weight saving is critical.

The first aluminium-alloy case to see service is the 30 × 173 ammunition designed for the General Electric GAU-8/A rotary cannon fitted to the A-10 ground-attack aircraft. The difference in empty case weight is 143 g compared with 350 g for the Oerlikon KCA's dimensionally similar steel case, resulting in the weight of a complete HEI cartridge being about 670 g compared with 890 g for steel-cased rounds (around 25% saving – compared with a brass-cased version, the saving would be significantly greater). The A-10 has such a large ammunition capacity that the use of aluminium cases results in a weight saving of over 270 kg. Where the 30 × 173 cartridge has been adopted for ground guns such as the Mk 44 and the Rheinmetall MK-30/2, steel cases are now used.

The other cartridge type to use aluminium alloy cases is the medium-velocity 30 × 113B used in the M230 Chain Gun fitted to the AH-64 Apache helicopter. The brass-cased ADEN version weighs 500 g with a 270 g projectile; the light-alloy cartridge weighs 339 g with a 237 g projectile (a saving of around 130 g or 26%).

The earliest cartridge cases were made in more than one piece, typically an iron head to which was fixed a tube made from coiled brass. These were very much an interim measure, soon replaced by the one piece case which has been universal since the latter part of the 19th century. This is formed by discs, punched from a thick sheet of metal, being repeatedly 'drawn', that is mechanically pressed into a hollow shape through several stages until the desired case shape has been achieved. The final operation is usually the cutting of the extractor groove close to the base. Despite this, wrapped cases are still occasionally experimented with.

Composite cases may be returning in the future, in the form of polymer case bodies reinforced with metal bases. The metal is mainly required to strengthen the base of the case which is subject to great pressure on firing, but also usually forms the extractor rim since the extraction forces involved mean that an extractor claw could rip through polymer rims.

Such hybrid cases initially ran into problems with achieving sufficient strength and heat resistance. Only recently has materials technology advanced to the point where success appears to be in sight, initially with small-arms ammunition in which weight savings for the complete round, compared with brass-cased ammunition, range from 15–20% if a brass base is used to 22–30% for steel and/or aluminium alloy bases. If this proves successful the technology can be expected to be applied to cannon ammunition in due course, since it can offer other advantages apart from weight saving. For example, polymer is not a good conductor of heat so the gun chamber remains much cooler after heavy firing than with metal cases. The lack of heat transfer works the other way too; if left in a hot chamber, a live round is far less likely to 'cook off' (fire by itself) if the case is polymer rather than metal.

One problem with cases of conventional form in larger calibres is that a large projectile has considerable inertia, placing the thin and inherently weak neck of the case under great stress as the round is subject to violent accelerations during the firing cycle. An attempt to avoid this is the 'neckless' case, in which the projectile is heavily supported by much thicker polymer. The particular US example shown was developed by AAI and dates to 1973.

*An unusual throwback: an experimental coiled steel case for the Rheinmetall 50 mm Supershot: the base is made separately.*

*Hybrid polymer/metal cases have been tried from time to time since at least the 1960s. The example above, believed to be by AAI, is a polymer/aluminium 20 × 102 case which weighs 36 g compared with 120 g for the brass equivalent.*

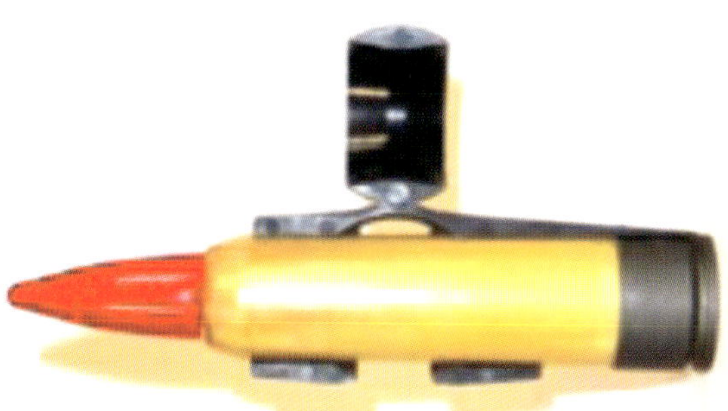

*AAI 25 × 145 neckless case with polymer belt link.*

# BALLISTICS

Ballistics is a complex and highly technical subject. In this book, it will be covered only in sufficient detail to enable the reader to grasp the basic essentials.

The subject is generally divided into internal and external (or interior and exterior) ballistics. The first deals with the passage of the projectile up the gun barrel, the second with what happens to the projectile between the gun muzzle and the target. There is a third category – terminal ballistics – concerning what happens when the projectile hits the target.

A key phrase must be borne in mind concerning all statements in this book which involve comparing different variables: 'other things being equal' (OTBE). There are many variables to take into account when making comparisons between cartridges, so it is assumed that only one variable at a time will be considered.

## CHAMBER PRESSURE

Some of the issues involved in internal ballistics have already been touched on. The relationship between the calibre and weight of the projectile and the weight, form and chemical composition of the propellant, together with the barrel length, will determine the pressure exerted on the chamber and gun barrel by the expanding propellant gas, the rate of acceleration of the projectile up the barrel, and the velocity achieved at the muzzle.

As soon as the primer ignites the propellant, gas is generated which rapidly builds up a considerable pressure as a result of initial resistance, caused first by the force required to separate the firmly crimped-in projectile from the case, immediately followed by the driving band hitting and being engraved by the rifling. Peak chamber pressure therefore occurs within the first few centimetres of projectile travel (which is why the barrel steel is thickest at this point) then steadily reduces as the projectile travels up the barrel, releasing more volume for the gas to expand into.

The pressure is still significant when the projectile leaves the muzzle, resulting in a rapid expansion of gas into the open air at supersonic velocity, causing the characteristic 'report' of a gun firing. This final expansion, coupled with the end of the friction between the projectile and the barrel, results in a final boost to the projectile so its maximum velocity is attained just beyond the muzzle (although 'muzzle velocity' – MV – is usually measured at several metres past the muzzle anyway).

The mean or maximum chamber pressures that a particular gun and ammunition combination is designed for can be measured in different ways; the older 'copper crusher' system measures the mechanical effect of the gas pressure allowed to escape through a small hole in the cartridge case and chamber wall (the location of the hole matters). Today, a more precise piezo-electric system is commonly used, and that can result in figures significantly higher than before.

The performance of a given cartridge depends upon the maximum pressure which the gun is designed to accept. This is usually determined by the strength of the mechanism which locks the bolt or breechblock to the breech at the instant of firing. The design of the cartridge case is also a factor, some being stronger than others. Cannon ammunition is generally loaded up to the highest practical pressure level, taking into account barrel wear, the risk of a case being stuck to the chamber and other potential problems.

Chamber pressure can be expressed in different units of measurement. The UK traditionally used 'tons per square inch', the USA 'pounds per square inch' (psi), the metric world 'kilograms per square centimetre' (kg/cm$^2$), with 'bar' (multiples of standard atmospheric pressure) also featuring. Today the international standard is megapascals (MPa), as used throughout this book, although all of the others (except tons per square inch) are still in use and may be quoted by ammunition manufacturers. The relationship between them is as follows:

1 MPa = 10 bar = 10.2 kg/cm$^2$ = 145 psi = 0.0647 tons per square inch.

Cannon have generally had chamber pressures in the 275–415 MPa range, which is to say (somewhat rounded): 2,750–4,150 bar; 2,800–4,200 kg/cm$^2$; 40,000–60,000 psi; or 18–27 tons per square inch. Chamber pressures for modern high-performance cannon are typically in the range 380–450 MPa (55,000–65,000 psi).

## MUZZLE VELOCITY AND ENERGY (MV AND ME)

Muzzle velocity is the speed at which the projectile leaves the gun barrel. For any particular gun and cartridge the MV can vary significantly depending on the projectile being fired; sub-calibre projectiles like APDS and – even more so – APFSDS are lighter than full-calibre projectiles so are launched at a higher MV. A more useful basis for comparing the performance potential of cartridges is therefore the muzzle energy (ME).

The ME developed by a cannon is measured in joules (J) and is calculated as follows: the projectile weight in grams is multiplied by the square of the muzzle velocity in metres per second (m/s), then the result is divided by 2,000. So a 400 g projectile fired at 800 m/s will generate (400 × 800 × 800) / 2,000 = 128,000J (or 128 kJ). Note that MV is much more important than projectile weight in determining the ME: doubling the MV of a projectile quadruples its ME, whereas doubling the projectile weight only doubles its ME.

The ME which is generated depends on the calibre of the gun as well as the chamber pressure; the calibre determines the area on which the gas pressure can work. In a gun of 20 mm calibre, the projectile has a base area of 3.15 cm$^2$. If the pressure is 4,000 kg/cm², the gas will apply a force equal to 12,600 kg to the base of the projectile. However, a 40 mm gun has a projectile base area of about 12.5 cm², so the same chamber pressure will apply 50,000 kg. This explains why, in cartridges with the same propellant capacity and chamber pressure but of different calibres, the larger calibre will develop more ME.

A high MV is almost always regarded as desirable in direct fire weapons (i.e. used against targets which can be seen from the gun position, as opposed to indirect fire such as mortars and artillery). In ground fighting, a high MV provides a flatter trajectory plus a shorter flight time, making it easier to hit distant targets (especially if they are moving). The armour penetration of AP projectiles is also, other things being equal, directly related to the striking velocity. In aircraft or anti-aircraft applications, the reduced flight time is even more valuable than in ground firing, simplifying aiming at targets moving rapidly in three dimensions.

The problem is that a high MV carries penalties. OTBE, achieving a significant increase in velocity requires considerably more propellant, which results in a larger and heavier cartridge case and a bigger and heavier gun. A longer barrel is required to give the projectile time to accelerate to its higher speed. Recoil, muzzle blast and barrel wear will all be worse, and in an automatic weapon the rate of fire is likely to be less than in a lower-velocity gun of the same type, as the mechanism will have to go through a larger range of movements in handling the larger ammunition.

There are also limiting factors. The law of diminishing returns applies, and using bigger cartridge cases holding more propellant will achieve ever-smaller increases in MV from the extra propellant. A cartridge which is so big as to be unable to use all its propellant efficiently is described as 'over bore'. Such cartridges generate high levels of muzzle flash and blast, and usually wear out barrels quickly.

In most cases, an MV of around 1,100–1,200 metres per second (m/s) is the practical maximum for conventional full-calibre projectiles. Sub-calibre projectiles such as APDS and APFSDS can achieve more – typically around 1,400 m/s for APDS and up to 1,600 m/s for APFSDS.

It is worth noting that the degree of barrel wear affects the internal ballistics. Most muzzle velocities are quoted from new barrels, a but few sources also include typical muzzle velocities from part-worn barrels. For each gun there is a maximum number of rounds before the barrel has worn to an unacceptable degree and must be changed.

The barrel length in comparison with the calibre is obviously an important factor in determining MV. This is expressed as the 'calibre length', which is simply the length of the barrel divided by the calibre. For example, the current Bofors 40 mm gun has a barrel 2.8 m long, and therefore has a calibre length of 70, normally expressed as L/70. World War II Bofors had a less powerful cartridge and needed a calibre length of only L/56. The shortest calibre-length cannon barrel in current use is probably for the medium velocity 30 mm M230 Chain Gun, at just under L/36 (107 cm).

Any loading of any cartridge has a ballistic optimum barrel length, at which the highest velocity is generated. Shorter barrels will lose velocity because the propellant gas has had insufficient time to accelerate the projectile, and some of the energy will therefore be lost in excess muzzle blast and flash. Longer barrels will lose velocity because the gas pressure begins to drop too low to overcome the friction of the projectile travelling up the barrel (a projectile will eventually grind to a halt given a barrel of sufficient length). In practice, most guns are built with a barrel shorter than the ballistic maximum, because the fall-off of velocity with reductions in length is initially quite small, and a shorter barrel provides a lighter and handier gun. A good example of this is provided by the current Russian 30 × 165 ammunition: three different MVs are quoted for these rounds, because the guns which fire them come in (at least) three different barrel lengths. The MVs for full-calibre loadings are 860 m/s, 890 m/s and 960 m/s in barrels of 150 cm, 162 cm and 240 cm respectively (L/50, L/54 and L/80).

## RECOIL

Two factors determine recoil: the cartridge ballistics and the characteristics of the gun.

The recoil impulse generated by firing a cartridge has two components; the momentum of the projectile, and the 'rocket effect' of the gas escaping from the muzzle. The simple formula for calculating this is as follows:

$$(\text{projectile mass} \times \text{muzzle velocity}) + (\text{propellant mass} \times \text{gas escape velocity})$$

Therefore a cartridge firing a 100 g projectile at 1,000 m/s should have the same projectile momentum as one firing a 200 g projectile at 500 m/s. Note that this is a different calculation from ME, as projectile weight and MV are of equal value.

The recoil caused by the escaping gas is much more difficult to calculate because it depends on the relationship between the burning characteristics of the propellant and the length of the barrel. If you assume two guns firing the same cartridge, one with a barrel of optimum length and the other with a much shorter barrel, the optimum length one will produce the higher muzzle velocity and therefore the greater recoil through projectile momentum. However, in the short-barrel gun the gas will be at a higher pressure when the projectile leaves the muzzle, and will therefore expand more violently, causing more muzzle blast and flash and generating a stronger 'rocket effect'. In this case, a higher proportion of the recoil will be generated by the expanding gas than with the optimum barrel.

For this reason, there is no simple ratio which will tell you exactly what proportion of the recoil is generated by the escaping gas as opposed to the projectile. However, a good approximation can be made, based on the weight multiplied by the velocity of the propellant compared with the weight multiplied by the velocity of the projectile. For cannon, the velocity of the gas escaping from the muzzle is typically in the region of 1,200–1,500 m/s. Rather than pick one figure, an approach which may be a little more sensitive to variations between cartridges is to assume that the gas escape velocity is 50% higher than the projectile's muzzle velocity.

It is therefore fairly simple to work out what proportion of the recoil impulse is generated by the escaping gas. The calculation goes like this, for a gun firing a

This muzzle brake is for the 30 mm MK 103. By the time the projectile leaves the brake, most of the propellant gas will have escaped sideways, contributing little to recoil.

100 g projectile at 1,000 m/s, using 40 g propellant (for this calculation, the units of measurement do not matter as long as they are consistent):

Projectile momentum: 100 × 1,000 = 100,000. Propellant gas momentum: 40 × 1,500 = 60,000. So the total recoil momentum is 100,000 + 60,000 = 160,000, of which the gas produces 60,000 / 160,000 × 100 = 37.5%.

The only way of reducing the recoil force generated by a cartridge while maintaining the ME, is to reduce the effect of the escaping gas by deflecting some of it, either to the side or (preferably) as far to the rear as is feasible. A device to achieve this is known as a muzzle brake. The extent to which a muzzle brake can reduce recoil obviously depends upon the proportion of the recoil impulse generated by the propellant gas – it gives the greatest benefit in very powerful, high-velocity weapons. One text on military cannon states that an efficient muzzle brake can reduce the recoil impulse by up to 30%. Higher figures are possible, but only by using brakes which are so large that they would be impractical. A degree of caution needs to be applied to efficiency figures claimed for muzzle brakes: sometimes a high figure of 75% or more may be claimed, but this will probably be referring only to the recoil generated by the escaping gas, omitting that generated by the projectile momentum.

A disadvantage of a muzzle brake is that the deflected gas may kick up dust, revealing the weapon's position and affecting the gunner's view. The deflected blast may also be extremely uncomfortable for anyone near the gun.

A recoilless gun is a specialized design which discharges most of the gas out of the back of the weapon, so in this case the 'rocket effect' acts in the opposite direction, more or less balancing the projectile momentum. However, this requires special ammunition containing several times as much propellant as a conventional round of the same muzzle energy, so the ammunition is bulky and expensive and the firing signature is substantial. The need to have a clear path for the gas from the chamber out to the rear also places considerable restrictions on gun design, making it far more difficult to design an automatic weapon. That has not stopped some designers from trying: e.g. the Rheinmetall-Borsig 55 mm MK 115 of World War II, and more recently the Mauser RMK 30 and RMK 35 series (see the Unconventional Systems chapter).

The brake to the right is fitted to a twin-barrelled GSh-30 installed on a Su-25 (Photo courtesy Ken Duffy). The vanes within the brake deflect the gas downwards and slightly to the rear, providing a small counter-recoil effect

So far only the recoil impulse generated by the cartridge has been discussed. However, the gun's recoil also depends on the design of the weapon and of the mounting. The weight of the weapon has a considerable effect. Momentum works both ways, equally (Newton's law of equal and opposite reactions must be obeyed), so the rearwards momentum of the gun matches the forwards momentum of the projectile plus the escaping gas. It therefore follows that the heavier the weapon, the more slowly it will move backwards under recoil, giving a smooth push rather than the sharper kick of a lighter weapon firing the same ammunition. The recoil momentum is the same, but it is delivered in a different way; the lighter weapon, recoiling more quickly, has the same momentum but higher recoil energy.

The mechanism chosen also has an effect. As described below, long-recoil and API blowback designs both inherently have a 'recoil-smoothing' effect, averaging out the recoil pulses so that the peak recoil blow is substantially lower. This makes it possible to use lighter gun mountings.

The rate of fire is a major factor. Obviously, doubling the rate of fire will double the recoil force, but it gets worse: a doubled RoF provides only half the time for the gun to recoil and return to battery, so that makes each recoil kick much sharper, multiplying the problem. The keenness of modern designers to limit the RoF where possible is understandable.

The nature of the mounting can make quite a difference. In all but the lightest cannon, the mounting allows the gun to recoil backwards between shots (there is commonly some kind of buffer). Detail design can make a big difference; the US Edgewater mounting used on both US and UK 20 mm aircraft cannon mountings in World War II substantially reduced the peak recoil blow.

An even more effective way of doing this is with a mounting, in which the barrel and action are held back in the full recoil position before firing. Pressing the firing button releases the barrel and action to run forwards; a round is chambered, then the weapon is fired while the barrel and action are travelling at maximum forward velocity, just before they get 'into battery'. The recoil force therefore has to overcome the considerable forward momentum of the barrel and action before it can start pushing them back again. This is a highly effective way of smoothing out the recoil pulses and is commonly used in modern automatic AA cannon. It is important to have reliable ammunition, as a 'hang fire' in which a cartridge ignites slightly late, after the barrel has come to a stop, would have unpleasant effects. This system is variously known as constant recoil, differential recoil, and floating firing.

Some automatic gun mechanisms have a greater recoil-smoothing effect than others. One of them is the long-recoil type, in which the barrel recoils a considerable distance between shots. This makes for a slow-firing gun and is mainly used in large-calibre weapons. It was commonly used in large aircraft cannon in World War II, such as the US 37 mm and the British 40 mm and 57 mm guns – the big 57 mm Molins had a peak recoil blow similar to that of the 20 mm Hispano (although the total recoil thrust was obviously many times greater). The other is the advanced primer ignition blowback (APIB) type, as pioneered by Becker in World War I and much used by Oerlikon and similar cannon in World War II. In this, the gun fires from an open bolt, which means that when the firing button is pressed, the bolt moves forwards and chambers a cartridge before firing. The key point of the API blowback is that the cartridge is fired while the bolt is still moving forwards at maximum velocity, so there is a kind of internal differential recoil effect. This was only possible because the Becker/Oerlikon guns used special ammunition with cartridge cases featuring a rebated rim (of smaller diameter than the body), enabling the extractor to hook over the rim while remaining within the overall diameter of the cartridge. This allowed the entire bolt to slide inside an extended firing chamber.

## EXTERNAL BALLISTICS

Internal ballistics provide only the starting point in achieving the desired effects on the target: external ballistics are of equal importance. Just two key factors determine the external ballistics of a projectile; the muzzle velocity and the ballistic coefficient (BC). The BC is significant because it determines the rate at which the projectile slows down, and in conjunction with the MV this decides the maximum range (at any given elevation) and the time of flight to any particular distance. The time of flight in turn decides the amount by which the projectile drops downwards as this happens at a constant rate due to gravity. The curved path of the projectile which results from the MV, the BC and gravity drop is called the trajectory.

In most types of long-range shooting a short time of flight is considered desirable because it maximizes the hit probability by flattering the trajectory and minimizing the effect of target movement. It also results in the projectile striking the target at a high velocity and therefore with greater effect. The main exception is when artillery fires in the 'upper register' (above 45 degrees elevation) to achieve plunging fire.

Some key external ballistics data for four common NATO cannon cartridges are shown in the table below left.

| HE loadings for: | 20 × 102 | 25 × 137 | 30 × 113B | 30 × 173 |
|---|---|---|---|---|
| MV m/s | 1,030 | 1,085 | 719 | 1,021 |
| ToF to 1.5 km | 2.93 | 2.07 | 4.10 | 1.92 |
| ToF to 2.5 km | 7.15 | 5.05 | 10.00 | 4.04 |
| BD to 1.5 km | 19.07 | 10.67 | 39.16 | 9.97 |
| BD to 2.5 km | 67.46 | 32.8 | 129.48 | 23.14 |

*MV = muzzle velocity. ToF = Time of Flight in seconds. BD = Ballistic Drop in mils.*

The advantages of a high muzzle velocity in reducing the time of flight are self-evident. So are the disadvantages as described above: more propellant is required, the barrel will need to be longer, the gun will be heavier and (in the case of a mounted weapon) so will be the mounting to cope with the greater recoil. In an automatic weapon, the rate of fire is also usually lower. As we have seen, there is also a practical limit to how high the velocity of any given projectile can be pushed. To make the most of the muzzle velocity, we need to achieve a high ballistic coefficient.

There are two elements which decide the ballistic coefficient (BC); the sectional density (SD) and the form factor (FF). The SD is a simple calculation as it is the ratio between calibre and projectile weight. The formula is:

For metric measurements: multiply the projectile weight in grams by 1.422, then divide the result by the square of the calibre in millimetres. So for a 20 mm projectile weighing 100 g: (100 × 1.422) / (20 × 20) = an SD of 0.355. The higher the SD figure, the better the velocity retention (assuming equal FFs).

What the SD measures is the weight (or momentum, when moving) behind every square millimetre of the projectile calibre (i.e. the frontal area of the projectile). If projectiles were solid cylinders then for a given SD figure they would all be the same length regardless of their calibre. In practice, of course, the length varies with the calibre; a 40 mm projectile will be about twice the length of a 20 mm, and will therefore have about double the SD figure. This explains why artillery shells travel much further than rifle bullets, no matter how fast or streamlined. OTBE, the bigger the calibre, the longer the range and the shorter the flight time to any given range.

Other things are of course far from equal, which is where the form factor comes in. The FF measures the aerodynamic efficiency of the projectile's shape, and is much more complicated to calculate; without access to manufacturers' data or live-fire testing on an instrumented firing range, only approximate estimates can be made. It is obvious that a projectile with a pointed nose will have much less air resistance than a simple cylinder, and it will therefore have a better FF, but problems arise when you try to become more specific.

The first problem is that the FF is different at subsonic and supersonic velocities, because shapes which work best at subsonic speeds are not the best at supersonic velocities. At subsonic speeds, the drag caused by the low-pressure area created at the back or base of the projectile is significant, and major reductions in drag can be made by tapering this to some extent (streamlining or boat-tailing). At supersonic speeds, it is the nose shape that is critical; finely pointed noses are needed, but the back end doesn't matter so much. Some taper towards the base may be useful, but the optimum taper angle is less than that at subsonic velocities.

The development of the 25 mm Bofors cartridge described in Part 3 provides an interesting example of the value placed on the FF in achieving a worthwhile improvement in performance.

A further factor affecting military projectiles is the addition of tracer elements. These generate gas which helps to fill the low-pressure area at the base, reducing drag. This gives them a different trajectory by comparison with non-tracer rounds, not helped by the fact that as the tracer burns up the weight of the projectile reduces, thereby worsening its SD. Tracers can therefore never achieve a perfect match with other projectiles and can only ever be an approximate guide to their trajectory.

An important aspect of external ballistics is the stability of the projectile. If an ordinary cannon shell were fired from a smoothbored gun, it would soon start to tumble, ruining its aerodynamics and accuracy. The answer was found to be to cause the projectile to spin rapidly by cutting spiral grooves into the barrel, called rifling. The effect of the spinning is to even out any asymmetries and keep the projectiles travelling point-first. Cannon shells are given a driving (or rotating) band near the base of the projectile, which is larger in diameter than the shell and is gripped by the rifling. The driving bands are traditionally copper but this is too soft for modern high-velocity cannon which normally use soft iron driving bands instead. The 30 mm projectiles for the US GAU-8/A cannon are unusual in using plastic driving bands. It appears surprising that more use of these is not made, as they reduce the rate of barrel wear by about two-thirds.

With large-calibre, high-velocity cannon there is some risk of the shock of impact with the rifling 'stripping' the driving band. To combat this, some weapons have progressive or gain-twist rifling, in which the rifling grooves start out parallel then gradually increase in twist down the barrel. This is particularly important in revolver cannon, as the projectile has longer to travel before hitting the rifling so is already travelling at high velocity.

Rifling permits a high degree of accuracy over the maximum range of a weapon. There is a relationship between the rifling twist (the angle of the rifling to the barrel) and the length of the projectile. For a given calibre, the longer (i.e. heavier) the projectile, the steeper the twist has to be in order to stabilize it. Clearly, with a particular rifling twist some light projectiles will be very stable, some medium-weight ones marginally stable and some heavy ones not stabilized at all. This can have consequences for the terminal as well as the external ballistics, as we shall see. As projectiles become steadily longer so rifling becomes ineffective, and long, thin projectiles such as APFSDS (armour-piercing fin-stabilized discarding-sabot) require fins, like arrows, at the back to keep them pointing, nose-first, in the right direction. Such projectiles are disturbed by rifling and work better from an (almost) smoothbore barrel (some degree of spin is considered useful in effecting clean sabot separation). Although APFSDS was originally developed for tank guns which are now mostly smoothbored, automatic cannon with rifled barrels also use APFSDS. The spinning can cause the projectiles to yaw (fail to point straight ahead) for some distance before their flight stabilizes.

Despite this problem, the advantages of 'sub-calibre' projectiles are such that they are extensively used in ground fighting and anti-aircraft roles. For a given calibre, a heavy projectile has more momentum and is therefore slowed down less by wind resistance. Heavy AP shot also has better armour penetration, OTBE, and as it slows down less, this advantage will increase with range. The disadvantage of heavy projectiles is that they require much more work to accelerate up to a high muzzle velocity. These conflicting demands explain the great success of sub-calibre ammunition such as APDS (and even more so APFSDS), which is extremely light for its calibre in the barrel (i.e. the calibre of the sabot) but extremely heavy for its very narrow calibre once the sabot has dropped away. There is a penalty in that the energy used to accelerate the sabot is wasted, so this is made as light as possible.

The advantages of these sub-calibre projectiles for external ballistics are considerable. For example, one modern high-velocity 25 mm HE shell has a maximum range of 9,000 m and will travel about 2,300 m in three seconds (a reasonable time limit for effective AA fire). The APDS version will reach over 3,400 m in the same time, and will carry to a maximum theoretical range of over 18,000 m. These figures explain the attraction of FAPDS (frangible) rounds for AA fire, in which time of flight to the target is of crucial importance to the hit probability.

The fin-stabilized APFSDS shot tend to be less accurate in theoretical terms than conventional spin-stabilized projectiles, but in practice this is more than compensated for by their flatter trajectory and shorter time of flight, which combine to increase greatly the hit probability, especially against moving targets.

Range tables for World War II-era 20 mm Madsen make interesting reading in indicating the ballistic performance of this type of ammunition at that time. At a range of 3,000 m (gun elevation 7° 48') the velocity of the HE shell fell from 840 m/s to 170 m/s and the 50% zone of dispersion was 4 m wide and 46 m long; the time of flight was 11.06 seconds. At the maximum range of 4,700 m (elevation

39° 40') the shell's terminal velocity had increased to 250 m/s due to the steep angle of descent (80° 19') while the 50% zone of dispersion increased to 14 m wide and 122 m long; the time of flight was 29.39 seconds.

## TERMINAL BALLISTICS – PENETRATION

This brings us onto the complex subject of armour penetration. It is obviously useful to be able to compare the thickness of armour which can be penetrated by different armour-piercing (AP) ammunition, but there are several variables to take into account, the two major ones being range and impact angle:

Range: penetration is directly linked to impact velocity (OTBE), which is in turn affected by the range to the target and by the rate at which the projectile loses velocity (expressed as its ballistic coefficient, or BC).

Impact angle: typically, a shot striking at 60° has only 80% of the penetration of one striking at 90°, at 45° this reduces to 66% and at 30° to 45%, but these figures can vary significantly depending on the projectile design. An unwelcome complication is that there are two different conventions for describing striking angle; from the horizontal or from the vertical. The US and British preference is to measure the angle from the vertical; the optimum for penetration, perpendicular to the armour surface, being 90° (as used in this book). Conversely, the continental preference is to measure from the horizontal, the optimum impact angle being given as 0°.

The above variables are normally dealt with in the standard way of describing penetration, which is to list the armour thickness penetrated at a given impact angle and range, as in 75 mm/60°/500 m. However, there are other variables which will affect the results:

Armour quality: the characteristics of armour plate used in penetration tests can vary considerably, not just in the composition of the steel or other material used, but in the way it is treated, e.g. steel armour can be 'face hardened', which requires different projectile characteristics to best defeat it.

Definitions of penetration: there have been different definitions of 'penetration' used in different places and times: whether the projectile passes completely through the armour or merely makes a hole in it; and whether all of the projectiles must penetrate, or only a percentage of them (and if the latter, what percentage).

Climate conditions during the test: ammunition performance is affected by ambient temperatures and pressure.

An added complication can occur with AP ammunition fired from aircraft guns; do the penetration figures take account of aircraft speed, and if so, are the speeds always the same?

The measurement variables described above mean that penetration statistics can only be taken as indicative and really don't mean very much in isolation. Reliable comparisons can be made only when it is known that the results being compared were obtained in exactly the same circumstances – and preferably in the same place and at the same time.

However, certain broad principles still hold generally true:

The higher the striking velocity the better, at least until the velocities are so high that the projectile is more likely to shatter than penetrate. For steel penetrators, this happens at impact velocities much over 1,000 m/s. Tungsten alloy penetrators are about 50% better than steel in their penetration capabilities.

A high SD is desirable both in retaining velocity and in penetrating armour. This is an advantage for APFSDS, which loses velocity relatively slowly even at long range.

A blunt nose is best for penetration, although this is often concealed by a pointed ballistic cap or windshield to reduce aerodynamic drag.

APCR is now uncommon in cannon calibres, as its penetration is much inferior to discarding sabot ammunition, so it is only used in circumstances when APDS can't be – which mainly limits it to aircraft guns. One well-known example of APCR is the 30 mm API used in the GAU-8/A cannon fitted to the A-10 aircraft; this is also unusual in having a depleted uranium core.

The relative AP performance of the different types of discarding sabot ammunition can be judged from data quoted for the 25 mm NATO cartridge (Oerlikon-Contraves production); FAPDS 26 mm (at 60°/1,000 m), APDS 31 mm and APFSDS 37 mm.

Comparing products from different sources is much less reliable, but broadly speaking the performance of modern AP rounds will be proportional to their muzzle energies, giving the following very approximate ranking of current western cartridges:

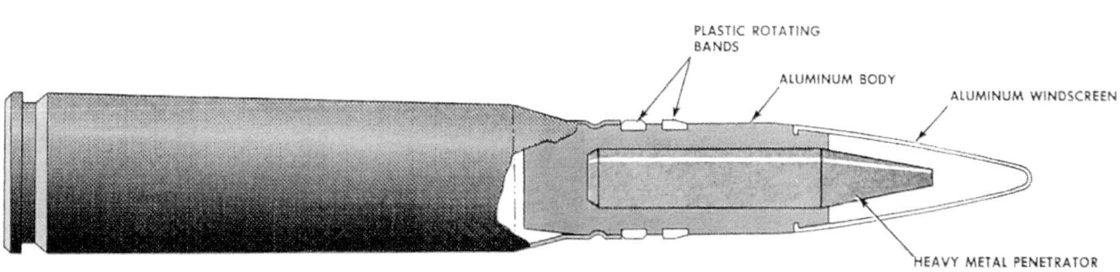


*30 × 173 PGU-14/B (APCR)*

| | 20 × 139 | 25 × 137 | 30 × 173 | 35 × 228 | 40 × 180 | 40 × 255 | 40 × 365 |
|---|---|---|---|---|---|---|---|
| ME kJ | 75 | 105 | 200 | 400 | 340 | 500 | 500 |
| AP mm | 25 | 37 | 80 | 110 | 95 | 140 | 140 |

# GUN ELEMENTS

## GUN ELEMENTS

A gun is in principle a simple device. Its essential components are a chamber shaped to fit the cartridge and strong enough to resist the pressure generated by the propellant gas; a barrel to guide the projectile and contain the propellant gas for long enough to permit the projectile to be accelerated to its maximum velocity; a method of firing the cartridge at the required instant; and some means for holding the cartridge in the chamber on firing, removing the fired cartridge case and inserting a new cartridge ready to fire the next shot.

In breech-loading weapons (which includes all automatic cannon) the last requirement results in the need for a break, or breech, in the rear of the chamber to enable cartridges to be inserted or withdrawn. The back of the chamber is known as the breech face. The part of the gun which is pressed against the open rear of the chamber during firing, and is moved out of the way to permit reloading, is known as the bolt or breech block; generally the term 'bolt' is used if it moves to and fro along the same axis as the barrel, and 'breech-block' if it is moved vertically or sideways, or is tilted away from the breech. This usually contains the mechanism to fire the cartridge (either mechanical or electrical). In all except revolver mechanisms and the unique 40 CTWS (40 mm Cased Telescoped Weapon System), the chamber is formed in the rear section of the barrel. The barrel may be extended rearwards in order to provide the locking surfaces for the mechanism; this part is unsurprisingly known as the barrel extension. The body or chassis of the gun, into which the barrel, bolt and other mechanisms (generally known as the 'action') are fitted, is known as the receiver.

Other components of a cannon system are: some means of holding an ammunition supply ready for loading; an appropriate method of mounting the gun (which may be fixed or movable, and will usually require some mechanism to absorb recoil); and a sighting device to give some expectation that the projectiles will hit their target.

## GUN CHARACTERISTICS

There are several key characteristics which provide the basis for comparison between guns and determine what they will be useful for. A quick reminder of one vital point: in making any such comparisons, the qualifier 'other things being equal' (OTBE) applies. there are many different variables affecting performance, so it is important when discussing one point of comparison that all other variables must remain the same. OTBE must therefore be assumed to apply, unless otherwise stated.

The key characteristics of cannon are: calibre; muzzle energy (ME); rate of fire (RoF); recoil; size, and weight.

Calibre: the larger the calibre, the heavier the projectile, the greater the distance it will travel, the more HE it will carry, and the more armour it will penetrate. However, the bigger and heavier the gun and its ammunition, the greater the recoil, and the lower the RoF is likely to be.

Muzzle energy: This is dealt with in more detail in the Ballistics section. The ME represents the power of a single shot, a higher ME usually meaning increased gun and ammunition size and weight, more recoil and heating effects, and often a lower RoF.

Rate of Fire (RoF): this is usually expressed as the number of rounds fired per minute (rpm). This does not mean that a gun can fire that many rounds in one minute (it will probably run out of ammunition and/or overheat the barrel before getting there); it is the cyclic rate at which it fires. Sometimes (mainly with AA cannon) a 'practical RoF' is given as well, a lower figure allowing for changing ammunition magazines, and possibly also allowing for cooling-off time between bursts. With some World War II fighter aircraft, cannon which were synchronized to fire through the propeller disc also experienced some reduction in RoF as each shot was timed to avoid hitting a propeller blade. Most cannon instantly hit their RoF, but externally-powered multi-barrel rotaries can take up to 0.5 seconds to 'spin up' to their maximum rate. A high RoF is particularly valuable when attacking targets which provide only brief firing opportunities, such as aircraft and (even more so) missiles; the downsides include much faster gun heating and barrel wear (for any given number of barrels) plus a more rapid expenditure of ammunition.

Recoil: there are two aspects to recoil: the recoil energy generated by one shot (which is obviously related to the ME of the cartridge, although not calculated in the same way) and the total recoil energy of a burst of fire, calculated by multiplying the recoil energy by the RoF. Again, there is more about this in the Ballistics section.

Size & weight: the size and weight of both the gun and its ammunition are clearly of considerable significance to aircraft armament, and have always acted as a restraint on the calibre of the guns fitted. While less important in other installations, there are still costs involved; for example, for the weight of a single-barrel AA gun mounting in 40 mm calibre, it might be possible to have a quad 20 mm mounting with up to ten times the RoF; which is preferred will depend on the purpose of the system.

## THE GUN BARREL

All cannon barrels are rifled (the only military guns which are not being most mortars and most modern tank guns, which are smoothbored and fire fin-stabilized projectiles). The rifling consists of spiral grooves inside the barrel which grip the projectiles by their driving or rotating band and spin them to ensure they are gyroscopically stable in flight. Without this, the projectiles would be likely to tumble in the air instead of flying point first. The degree of rifling twist needed to achieve stabilisation depends on various factors, including the length of the projectile the gun is designed for; the longer and heavier the projectile, the steeper the twist. There is a risk that the impact of hitting the rifling might strip the driving band, so some cannon have been designed with progressive rifling; that is, the grooves start out being parallel with the barrel to offer the minimum resistance, then gradually increase their twist until they achieve the desired rate.

The length of the barrel is commonly described in multiples of the calibre, as the 'calibre length' (usually expressed as L/xx or Lxx). So a 100 cm long barrel with a calibre of 2 cm is described as an 'L/50'. This figure is useful to know as it provides a means of comparing the characteristics and performance potential of guns of different calibres. In practice, cannon barrels have varied between L/15 to more than L/100: most fall within the range L/40 to L/60. Generally speaking, cannon of around L/30 or less will be considered 'low velocity' guns, those of about L/70 or more 'high velocity'; these are relative terms.

One point to be aware of is that there are two different ways of measuring barrel length: one measures the whole barrel from the muzzle to the breech face (not including any barrel extension); the other just measures the rifled portion of the barrel, i.e. from the muzzle to the front of the chamber (this necessarily applies to revolver cannon and the 40 CTWS). Also, muzzle attachments such as muzzle brakes and flash hiders may sometimes be included in the barrel length, which is misleading as they do not add to the velocity.

Another factor affecting the design of the chamber and barrel is the chamber pressure: the peak pressure developed by the propellant gas, which occurs as the projectile begins its journey up the barrel and engages the rifling (a couple of terms worth noting: the 'throat' is the part of the chamber immediately in front of the neck of the case, and the 'leade' is the transition zone between the chamber and the barrel, where the rifling begins). The higher the chamber pressure, the greater the thrust applied to the base of the projectile and the more rapidly it will accelerate. But the higher the pressure, the stronger (and therefore heavier) the chamber, barrel and action need to be.

As noted in the section on Ballistics, cannon have generally had chamber pressures in the 275–415 MPa range. This subject is covered in more detail in that section.

## BARREL COOLING & GUN SYNCHRONISATION

Still further descriptors concern the method of ensuring that barrels do not burn out or melt through overheating: they may use water or some other liquid, evaporative or air cooling.

The burning of the propellant, and (less significantly) the friction of projectiles accelerating down the barrel and being gripped and rotated by the rifling, combine to generate considerable heat in gun chambers and barrels. The higher the RoF per barrel, the more heat is generated and the less time there is available for it to be dissipated. The problem is also exacerbated by high pressure, high velocity cartridges which use a lot of propellant.

Heat generation causes two problems. The first is that after firing, the heat of the chamber might be enough to ignite, sometime later, the propellant in a cartridge left in the chamber, with potentially disastrous results. This is known as 'cooking off' and is overcome in most heavy weapons by ensuring that the chamber is left empty after each burst of fire. In linear weapons, this is achieved by designing the gun to fire from an open bolt, i.e. the bolt is held back after each burst (an inherent feature of API blowback weapons anyway). In the case of revolver or rotary cannon, the mechanism is generally arranged to ensure that the multiple chambers are cleared at the end of each burst. In the larger American rotary cannon this is achieved by putting the drive motor into reverse in order to load unfired cartridges back into the ammunition feed.

There was a problem with open-bolt firing in the past, in the case of aircraft guns that were synchronized to fire through a propeller disc. The accurate timing necessary to avoid shooting off a propeller blade meant that the instant of firing had to be under precise control, which could be achieved by firing from a closed bolt or, where open-bolt firing was retained, by separating the initiation of the firing cycle (by pressing the firing button to chamber a cartridge) from the ignition of the cartridge (controlled by a mechanical or electrical link from the propeller). Synchronized guns were effectively semi-automatic, in that the timing of each shot was deliberately controlled. Synchronisation mainly applied to MGs, but during World War II cannon of 20 mm (and even 30 mm, with less success) were synchronized in some German, Japanese and Soviet installations. The RoF was reduced to a degree which depended on the specifics of the gun mechanism and type of synchronisation system. In the best examples (German cannon which used electrical priming) the RoF was reduced by only about 10%; less efficient systems could reduce the RoF by up to 40%.

One other rare application of synchronisation was the four and eight-barrelled mountings of the British 2-pounder 'pom-pom' used in World War II. It was noted that if the guns were left to fire freely, they would fire at different rates and from time to time would all fire together, generating a heavy recoil blow which could damage the mounting. To avoid this, a hand-crank device was added to operate the firing gear. It should be emphasized that the crank did not operate the extraction/ejection/chambering cycle as it did in 19th century guns like the Gatlings and Nordenfelts – that was achieved by recoil operation as usual – but just the timing of the firing pin release. Later this feature, known by the RN as 'controlled firing', was operated by an electric motor in the four-barrel mountings and in 1939 abandoned altogether as unnecessary in the eight-barrel mounts. The effect on the rate of fire was relatively small: from 115 rpm free-firing to 95–100 rpm controlled.

The second problem arising from heat generation is more difficult to overcome. In extended firing, the barrel and chamber can reach such high temperatures that they glow visibly. Initially, this will result in a much increased rate of wear of the rifling. In extreme cases, the mechanism loses its integrity and becomes jammed.

This is not always a problem. Aircraft guns are normally used in short bursts and have a high-speed airstream to keep them cool. Aircraft also return to base between missions, where the gun barrels can be changed if necessary. Indeed, in the case of the latest revolver cannon, combining a single barrel, high MV and an extremely high RoF, the use of long bursts of fire can reportedly require the barrel to be changed almost as often as the magazine is reloaded. Even so, restrictions on weapon firing have been applied. For example, World War II US fighter pilots were told to fire 20 mm bursts of no more than 20 rounds and preferably 10, with 37 mm figures being 15 and 5 respectively.

Ground and ship-based guns have to find other means of overcoming the heat problem. It has always been common for shipboard automatic guns of all calibres to have water-cooled barrels, as the added weight and complication are less of a problem than with aircraft or mobile land systems. Russian armament firms, exercising their characteristic inventiveness, have even managed to provide water-cooling to rotary cannon. In the past, water-cooled guns have been easily recognisable because of the large-diameter water jackets. Modern versions use pumped circulating systems which have slimmer jackets. At least one of the Russian rotary cannon plus their 30 mm 2A38M twin-barrel AA gun, as well as the single-barrel GSh-301 aircraft gun, use transpiration cooling, which relies on the evaporation of a cooling fluid.

Water-cooling has sometimes been used in ground-based guns in the past – the Maxim 37 mm pom-pom and the Swiss 34 mm FMK 38, for example – but has since fallen out of favour as it is more suited to a static battle where arrangements can be made for a reliable supply of water. Instead, heavier barrels with greater capacity to absorb heat may be fitted, sometimes finned or fluted to increase the surface area and thereby assist cooling, and the barrels are commonly lined with Stellite or some similar heat-resistant material. The development of new alloys for gun barrels which are more resistant to overheating problems is still an active research field.

## GUN CONFIGURATIONS

The great majority of cannon to have seen service have had (and still have) a single barrel with a single chamber formed within the breech end of the barrel. This is the simplest and most compact type of layout, and meets most requirements. Since World War II, more complicated gun configurations have been introduced in order to achieve higher rates of fire, mainly for aircraft or anti-aircraft use.

A substantial increase in the RoF can be achieved by designing a twin-barrel mechanism in which the two bolts are linked by a yoke so that when one bolt is in battery (forwards and locked) the other is at the back of its stroke. The recoil of one bolt therefore pushes the other forwards, acting as an accelerator. Introduced in the German Gast machine gun of World War I, it has only otherwise seen service in three post-World War II Russian cannon: the highly successful 23 mm GSh-23 aircraft gun (which fires at up to 3,000–3,400 rpm), and the related 30 mm GSh-30 aircraft and 2A38M anti-aircraft guns which fire at 3,000 rpm and around 2,000 rpm respectively. More recently, the ZPL-20 – a version of the GSh-23 adapted to fire NATO 20 × 102 ammunition – was developed in the Czech Republic. Given the combination of compactness and high rate of fire, it is surprising that more use of this configuration has not been made; it may simply be that the major makers or users were already focused on alternative configurations.

An alternative approach to speeding the rate of fire is to separate the barrel from the chamber, to enable several chambers to be contained in a cylinder behind the breech, designed to rotate so that each chamber is lined up with the barrel in turn. Modern cannon of this type were first developed in Germany at the end of World War II, and the similarity in layout to the revolver handgun has led to such guns being called 'revolver cannon'. The advantage is that ammunition can be handled in stages, with one chamber empty, the next with a new round loaded, the third (lined up with the barrel) being fired, and the fourth with the fired case being ejected. Four chambers is the smallest number to see service in a revolver canon, with five being the most common and both seven (with a single barrel) and eight (with two barrels) also being used. As ammunition handling is the main limitation on gun performance, dividing this into stages permits revolver cannon to fire at very high rates: in the 1,000–2,500 rpm range depending on the calibre and the number of chambers, with the calibres of service guns being from 20 mm to 35 mm. There are disadvantages: the need to start and stop the rotation of the cylinder for each shot; the need to form and then remove a seal between the chamber and barrel for each shot (usually involving a sleeve being pushed forward to bridge the gap); increased barrel wear partly due to the high rates of fire generating high temperatures, and partly due to 'balloting' – the tendency of the projectiles to wear the barrel unevenly.

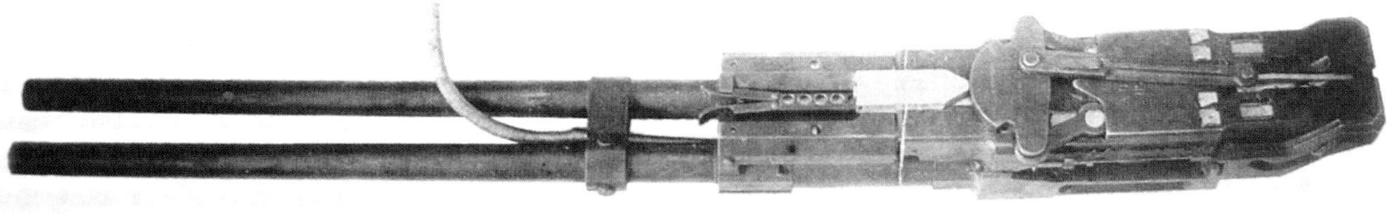

*1957 prototype of the Russian AO-9 (GSh-23) (Courtesy Maxim Popenker)*

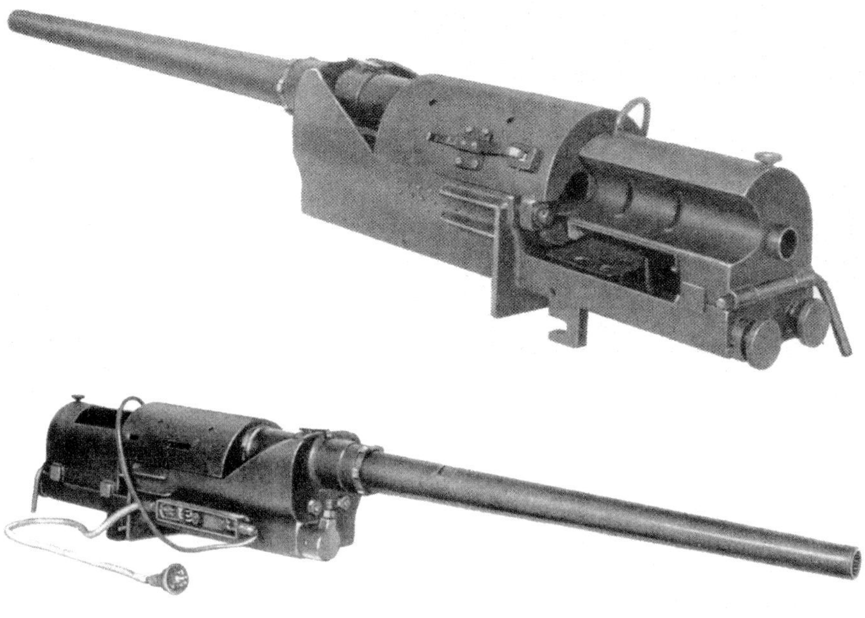

German prototype of the Mauser MG 213C revolver cannon (Chinn)

There is another cannon which has the chamber separate from the barrel for different reasons: the 40 mm Cased Telescoped Armament System (CTAS) from CTA International, a Franco-British joint project approaching service status at the time of writing. In this case, the breech with the cylindrical, open-ended chamber pivots sideways enabling a new round of ammunition to be inserted at right-angles to the barrel, pushing out the fired case as it does so. The main advantage of this system is that the ammunition and its feed system are very compact, enabling them to be installed in smaller turrets. The disadvantages are that the complex ammunition is more costly to make, and the separation of the chamber from the barrel can increase barrel wear. This is described in more detail in the chapter on Unconventional Systems.

The final type of gun configuration to see general use is the multibarrel rotary cannon, developed in the USA after World War II, starting with the idea of a motorized Gatling (the first attempt to fit an electric motor to a Gatling gun actually occurred around the end of the 19th century, but that went no further). The ammunition handling is in stages, similar to that for the revolver cannon described above but instead of one barrel, there is in effect one attached to each of the chambers in the cylinder, so the entire barrel group rotates. This has three significant advantages: the chambers are integral with the barrels; each barrel deals with only a small fraction of the RoF (e.g. a six-barrel gun firing at 6,000 rpm means that each barrel is firing at only 1,000 rpm) so overheating and barrel wear are less of a problem, and the barrel group spins continuously without having to stop to fire each shot. All of these characteristics combined can result in a great increase in the practical RoF and reliability.

Another advantage shared with other externally-powered mechanisms is that they operate independently of ammunition quality, pressure or recoil characteristics; if a cartridge fails to fire, the gun simply ejects it and loads the next one. Such a failure in a recoil or gas operated gun will normally stop the mechanism until the dud cartridge has been manually ejected, although some modern self-powered weapons have pyrotechnic systems to eject dud cartridges automatically. The Oerlikon KCA revolver cannon has four such charges; the Russian GSh-301 uses a small explosive charge, directed through a small hole in the chamber wall, to penetrate the cartridge case and ignite the propellant directly – a brutal but apparently effective war-only system.

The number of barrels can vary considerably; most of those in service have six but others have 3, 4, 5 or 7 barrels, and a recent addition from China has eleven. The maximum firing rates vary accordingly, but are typically in the range 600 to 1,200 rpm per barrel. Rotary cannon in 20 mm, 23 mm, 25 mm and 30 mm calibres are in service. The disadvantages are obviously the much greater bulk and weight of the gun, and the fact that it takes up to half a second for the heavy barrel group to spin up to the maximum rate, especially when externally driven although the latest version of the M61, the A2 used in the F/A-18E/F and F-22 fighter aircraft, has been redesigned to achieve quicker acceleration – mainly by reducing the weight of the barrels – but still takes 0.25 seconds to reach maximum speed. In aerial combat the opportunities for bursts of fire may be very brief, so that a slow initial rate of fire can be a disadvantage – although in some designs at least, the barrels can be spun-up before opening fire if the pilot has enough warning.

There is also a problem in achieving high levels of accuracy with a rotary gun, not just because of the need to align several barrels to the same point of impact (the alignment of the barrels

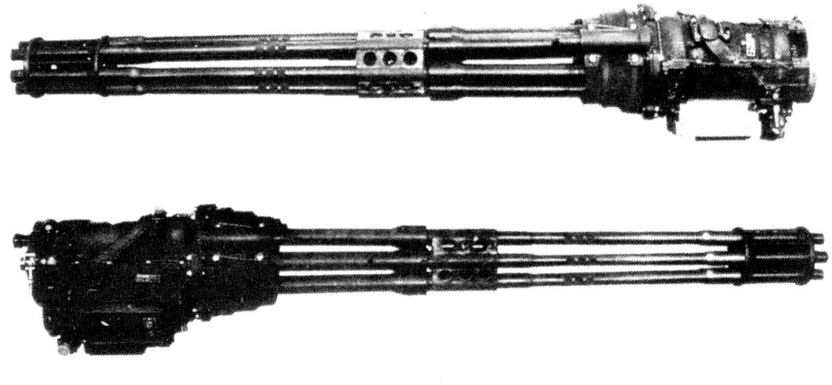

Prototype Vulcan, Model D (Chinn)

can be adjusted), but also because the projectiles are thrown sideways by the rotary motion of the gun, to a degree which varies as the gun accelerates to its maximum rate of fire. This delays the firing of the first effective shot. Having said that, the M61 has a reputation for high accuracy once it is spinning at peak rate. Rotary cannon also make great demands upon the ammunition feed system, both in terms of the quantity required and rate at which the cartridges have to be supplied to the gun. Finally, along with other externally-powered automatic cannon they are vulnerable to 'hang fires', or delayed ignition; it could be disastrous if a cartridge fired after its barrel had moved from the normal firing position.

Maximum RoFs in modern single-barrel, single-chamber cannon are generally in the region of: 1,000 rpm for 20 mm guns; 700 rpm for 30 mm; 550 rpm for 35 mm; 350–450 rpm for 40 mm; and 200–300 rpm for 57 mm. There are exceptions; the Russian 30 mm GSh-301 aircraft gun manages an astonishing 1,500–1,800 rpm. Some guns are deliberately designed to fire at much lower rates, especially those fitted to modern light armoured fighting vehicles (LAFVs) because high rates are not needed, and lower-rate guns are generally more reliable, more accurate, do not overheat or wear out so quickly, and conserve ammunition. Rates of fire of older guns tended to be much lower anyway: in World War II the typical rates for 20 mm guns were up to 450 rpm (for AA guns) and up to 800 rpm (aircraft guns), with 37–40 mm cannon firing at about 140–220 rpm.

## AMMUNITION HANDLING AND GUN LOCKING SYSTEMS

In automatic weapons the most complex and troublesome areas are those concerned with ammunition handling; the mechanisms for loading, locking, firing, unlocking, extracting, and ejecting. These all need to function reliably at very high speed and potentially in unfavourable circumstances: environmental (sand/dust or climatic) or gravitational, especially in the case of violently manoeuvring aircraft.

The mechanisms by which automatic guns function vary considerably, and form the main technical differences between different types of cannon. There are three principal variables, one of which has already been considered above: the configuration of the gun (single-barrel with integral or separate chamber, single barrel with multiple chambers, twin-barrel or multiple rotating barrel). This section is concerned with the method of locking (and unlocking) the bolt or breechblock to the breech, in those mechanisms which use locked breeches. The gun operating systems will be considered later.

## LOCKING SYSTEMS

A key design element of the locked-breech gun action is the mechanism for locking the bolt or breech block to the barrel for the instant of firing, to prevent the bolt from being hurled backwards by the chamber pressure and thereby losing most of the high-pressure gas out of the breech. Some mechanical means of locking and unlocking the bolt and breech at the appropriate moments are required, and variations in how this is achieved are an important distinguishing feature of the different designs to have seen service.

There are four main classes of locking mechanism used in recoil-operated guns: the toggle lock; rotary locking systems; tilting or pivoting locks; and sliding locks.

## TOGGLE LOCK MECHANISM

The toggle lock mechanism was the first into service, introduced in the Maxim machine gun. Hiram Maxim, an American living in England, designed the machine gun named after him and gave its first firing demonstration in 1884. The same toggle lock system was scaled-up for use in the first Maxim cannon, the 1-pounder or One Pounder (the nominal weight of the projectile), which used the existing 37 × 94R Hotchkiss ammunition. Variations on this design firing other sizes of 37 mm ammunition were developed in the UK (37 × 69R, 37 × 124R), the USA (37 × 136R) and Germany (37 × 101SR) , and the ultimate derivative was the Vickers 2-pounder naval AA gun in 40 × 158R calibre, which entered service in the First World War and saw extensive service in Second World War.

Several Swiss cannon of the late 1930s in 20 × 139, 24 × 138 and 34 × 239 calibre also used a variation on the Maxim mechanism. These were designed by Colonel Adolf Furrer and manufactured by *Eidgenössische Waffenfabrik* of Berne. They represented the final development of this type of action, with the last of them not leaving service until well into the 1960s. It is of historical interest that the famous Luger pistol also uses a toggle lock which is similar in general principle.

The toggle lock is similar in principle to an elbow joint. The bolt is hinged in the middle so that it can bend, but when the gun is ready to fire the bolt is straight, like an outstretched arm. On firing, the bolt and barrel recoil together until an unlocking cam frees the bolt from the barrel and forces the joint to bend; upwards, downwards, or to one side, depending on the design. At the same time, the rearward movement of the barrel is stopped. As the joint bends, the bolt head is pulled away from the breech face, extracting the fired case from the chamber. The recoil or return spring then pushes the bolt forwards, straightening it and chambering a fresh cartridge before the bolt is locked again as it returns to battery.

## ROTARY LOCKING SYSTEMS

Most modern small arms (whether recoil or gas operated) use rotary-locking bolts. The head of the bolt has two or more lugs which lock the bolt into recesses in the barrel extension when the bolt is rotated, just as in a bolt-action rifle. This system, which provides a very strong action, is also used in cannon calibres. During the rearward movement of the barrel/bolt assembly, the bolt is forced to rotate by cams until it is unlocked. A variation on this is the rotating bolt head design which is exactly the same except that, as the name suggests, only the head of the bolt rotates.

This system was used in various cannon, particularly from Germany and the USSR. Mauser used it for their 20 mm MG 151/20 cannon and the 20 mm Flak and KwK 38 guns. Rheinmetall used a rotating lock in most of their large

37 mm Flak family. Soviet cannon with rotating locks include the 23 mm NS-23 and NR-23 aircraft guns, plus the 37 mm NS-37 and 45 mm NS-45. Hispano Suiza of Switzerland developed (but failed to sell) various rotary-locked AFV cannon in the 1960s in 20, 23 and 30 mm calibres.

Another rotating mechanism is the Solothurn lock invented by Louis Stange early in the century. The firm of *Waffenfabrik Solothurn* was established in Switzerland in 1929 as a subsidiary of the German firm of *Rheinische Metallwaren-und-Maschinenfabrik*, better known by its trade name of Rheinmetall, as a way of evading the limitations on weapon development imposed by the Treaty of Versailles. These nominally Swiss weapons therefore featured significantly in German rearmament plans. Two years after the acquisition of August Borsig GmbH in 1933 the firm became Rheinmetall-Borsig, a designation which lasted until 1956, after which it reformed as Rheinmetall.

The Solothurn lock consists of a rotating collar fitting around the barrel extension and locking this to the bolt by means of an interrupted thread. As the barrel/bolt assembly moves rearwards, the collar is rotated by cams in the receiver, thereby disengaging the lock and enabling the bolt to be separated. This mechanism was used in various infantry weapons developed between the wars, and in the 20 mm MG 204 and 30 mm MK 101 aircraft cannon.

## TILTING OR PIVOTING LOCKS

An alternative system uses a pivoting lever or plate, fixed to the barrel extension and incorporating lugs which fit into recesses in the bolt to hold them together. The initial movement of the barrel pivots the lever, lifting the lugs and releasing the bolt. An early example of this was invented by Louis Schmeisser in the early 1900s and used in the Dreyse machine gun and, in enlarged form, the 20 mm Ehrhardt cannon under development in 1918. Ehrhardt was one of the key figures in Rheinmetall, so this mechanism was incorporated in various Rheinmetall weapons thereafter.

Strenuous efforts at the end of World War I to keep the Ehrhardt secret from the Allies were largely successful, with the surviving examples offered around 1930 by HAIHA *(Hollandsche Artillerie Industrie en Handelsmaatschappij)*, a Dutch company jointly established by Rheinmetall and Steyr of Austria. Rheinmetall then passed the development of this mechanism onto its Swiss subsidiary, Solothurn, who modified it to take more powerful 20 × 138B ammunition and marketed it as the S5–100 series. This later evolved into the Rheinmetall-Borsig Flak 30.

## SLIDING LOCKS

Another variation is the sliding lock, with that invented by the prolific American designer John M. Browning being the most successful. In this case the barrel and bolt are connected by a locking piece which is vertically cammed out of engagement as the assembly recoils. The Browning short-recoil machine gun emerged during World War I in rifle calibre as the M1917, later refined into the M1919, in which form it remained in service until well after World War II. The same basic design was scaled-up to a new 0.50-inch (12.7 mm) calibre after World War I, eventually emerging as the M1921. Subsequently refined to the M2, this was extensively used by all three services in various versions during World War II and has remained in service ever since.

The Americans never fielded any Browning short-recoil guns in cannon calibres (the 37 mm Brownings used long-recoil). However, during World War II, the Japanese were great enthusiasts of this action and fielded a variety of weapons using the design, including the 20 mm Ho-5, 30 mm Ho-155, and 37 mm Ho-204.

A different type of sliding lock is one in which the breechblock slides horizontally or vertically to expose the breech face. A different part of the mechanism then performs the extraction, ejection and loading cycle. These were commonly used in large-calibre manually-loaded guns but less often in automatic mechanisms.

## GUN OPERATING SYSTEMS AND POWER SOURCES

There are four different power sources used to drive the mechanism in automatic cannon: recoil; gas tapped from the barrel; blowback (the cartridge case, driven by gas pressure, pushes the breech open); and external drive, which may be electric, hydraulic, or ram-air (the last of these only feasible for guns installed on high-speed aircraft). There are also hybrid mechanisms which use more than one of these power sources. Each of these major categories has subgroups, for example recoil-operated weapons may have short or long recoil mechanisms, and blowbacks have four: simple, retarded, delayed and API blowbacks. These are described below.

## RECOIL-OPERATED MECHANISMS

The first successful self-powered automatic gun mechanism to appear, in the Maxim machine gun of the 1880s, was powered by recoil. Originally made in rifle calibres, the Maxim mechanism was soon scaled up to take cannon ammunition, the first such gun, in 37 mm calibre, emerging before the end of the 19th century.

The next significant development in recoil-operated cannon was the 20 mm Ehrhardt which was entering production at the end of World War I. This did not see service but fathered a range of Rheinmetall cannon which saw much use in the next war; most German cannon used various types of recoil operation at this time. During the interwar period, the Danish firm of Madsen had some success with their 20 mm cannon; Bofors developed 25 mm and 40 mm guns; and some US 37 mm guns based on Browning designs saw service in World War II. Russia also fielded various recoil-operated guns (some based on Bofors actions), and Japan made use of Browning designs for some aircraft cannon.

After World War II, gas operation became more popular, particularly in small-calibre weapons. The best-known recoil-operated small-arms survivor is now the .50 Browning heavy machine gun, the original version of which was designed a century ago. However, recoil operation is still proving to be suitable in some types of large-calibre cannon and accordingly remains in common use today.

In recoil-operated systems, the energy to drive the reloading cycle comes from the recoil force generated by firing the gun, in accordance with Newton's Third Law: for every action there is an equal and opposite reaction. So as the propellant gas drives the projectile forwards, the gun is driven backwards. At the moment of firing, the barrel and bolt or breechblock are locked to each other (using one of the systems described above) and initially recoil together within the receiver before being unlocked and separated to enable the fired case to be extracted and ejected. The recoil movement compresses a spring which is used to return the bolt/breechblock into battery (i.e. in its forwards position), thrusting a fresh cartridge from the ammunition feed into the breech before finally locking the action again. The key point is that the bolt or breechblock remains mechanically locked to the breech until the projectile has left the barrel, thereby ensuring that the breech cannot open until the gas pressure in the barrel has dropped to a safe level.

Recoil-operated mechanisms may be of short-recoil or long-recoil types. The difference between them is that in the long-recoil type, the barrel travels all the way back with the bolt before being unlocked. The bolt is then held to the rear while the barrel is pushed forward by a spring, extracting and ejecting the spent case. The bolt is released and pushed forward by a second spring once the barrel has returned to battery, collecting and chambering a new cartridge as it does so, and is locked to the breech again before the next shot is fired. In the short recoil mechanism, a cannon barrel travels only a few centimetres before the bolt is unlocked; the barrel is then stopped and the bolt continues its rearward movement alone, compressing the return spring and extracting and ejecting the fired case before reaching the rearward limit of its travel. The return movement of the bolt chambers a fresh cartridge and relocks the bolt as in the long recoil system. In both long and short recoil types the barrel has to be free to slide to and fro while its support is engineered closely enough to maintain acceptable accuracy.

As a general rule, long-recoil guns have an advantage over short-recoil of reducing the peak recoil force by spreading it out over a longer period. However, the need for the entire barrel/bolt unit to recoil a considerable distance, followed by a delay in commencing reloading while the barrel returns to the firing position, gives long-recoil mechanisms a relatively low rate of fire. Despite this disadvantage, the long and relatively soft recoil push puts less strain on the mechanism and makes accuracy easier to achieve while avoiding the need for the whole gun to recoil in its mounting. It also means that by the time the breech opens, the residual gas pressure is very low so little escapes from the breech – a positive benefit when the gun is in a manned turret. Some actions are intermediate between long and short recoil, with the barrel being unlocked and stopped partway through the recoil stroke.

Where a high rate of fire is required, guns with short-recoil mechanisms may incorporate an accelerator; typically, a pivoting lever which is struck by the barrel at the end of its rearward travel, transferring that energy into kicking the bolt backwards at a higher speed. Sometimes, a second accelerator is fitted to kick the bolt forwards again.

New short-recoil designs are now very uncommon, having been largely replaced by gas operated or externally powered designs. One surprising exception is the current Russian aircraft cannon which arms the MiG-29 and Su-27 families; the GSh-301.

## GAS-OPERATED GUNS: LINEAR MECHANISMS

A few years after Maxim introduced the short-recoil gun, John Browning invented a different principle of operation for his first machine gun. After observing that the propellant gases escaping from the muzzle still had considerable force, he devised a simple mechanism to trap some of the gas and use it to drive the loading cycle. In his first practical machine gun, the gas was tapped from the barrel before the projectile left the muzzle. A prototype was sold to Colt by 1890 and the gun was adopted by the US Navy in 1895. The inventor cannot have been entirely satisfied as most later Browning machine guns and cannon used recoil mechanisms. In Europe, Hotchkiss was an early exponent of gas operation.

In principle, gas-operated linear-action machine guns have worked in the same way ever since. There are variations in how far up the barrel the gas port is sited, and in whether the gas drives a piston which is connected to the action or drives the action directly, but these are detailed refinements. The only significant variation is the nature of the mechanism the gas is driving; linear, revolver or rotary. Gas operated weapons were exclusively linear until the invention of the revolver cannon which did not see service until after World War II. Gas operated rotary cannon are even more recent.

Gas operated linear actions have proved particularly popular in small arms, and the vast majority of the current generation are of this type. Gas operation has two advantages over recoil; as the barrel remains fixed to the weapon, it is more likely to be accurate; and it is easy to regulate the amount of gas operating the mechanism in order to ensure that the weapon functions properly regardless of climatic or ammunition variations. A disadvantage of gas operation is that it is usual with such systems to rely in part on blowback to help with extracting the fired case – which means that the pressure will still be high when the breech opens, making such guns unsuitable for use in manned turrets. The means of locking the bolt to the barrel has as many variations as in the case of recoil operated guns, with rotary bolts, lifting locking pieces and tilting or sliding breechblocks all being employed.

In automatic cannon, gas operation has constantly rivalled the recoil type. The Maxim 1-pounder was rapidly followed by the Hotchkiss Mle 1900/1901 in the same calibre, and shortly after that by the US McClean, both gas-operated. During the interwar period, the French Hotchkiss Mle 38/39 in 25 mm and the Italian Breda M35 in 20 mm were developed. The Russian 20 mm ShVAK and the later 23 mm VYa also used gas operation.

World War II Germany focused on recoil-operated guns but still developed a few gas-operated cannon: the Rheinmetall-Borsig 30 mm MK 103 hybrid and the 37 mm Flak 4, as well as the unsuccessful 50 mm Flak 41 and the experimental 55 mm Flak 58 by the same firm.

Post-war, gas operation has featured in several linear guns in which a high rate of fire combined with light weight is required. Russian aircraft cannon such as the NR-23, NR-30 and N-37, and several light AA guns; the Soviet 23 mm ZU, French 20 mm GIAT 20M621 and 20M693, and various Oerlikon and Mauser models. However, for reasons explained later it is arguable whether many of the modern guns should be described as gas-operated or gas/blowback hybrids.

The ZU-23 gun uses a vertically-sliding breechblock combined with a separate cartridge rammer and extractor, all powered by gun gas. The French firm of GIAT has produced the 20 mm 20M693 which, like many modern AA cannon, has a floating firing mounting to reduce recoil.

The Oerlikon KA, KB, KC and most of the big KD series are all gas-operated or gas-unlocked, with blowback assistance to aid extraction. The KD series (with the exception of the long-recoil KDE and the KDG revolver), combine the usual features of the modern series including floating firing and selective ammunition supply, which in this case can be via belts or a linkless feed system. It should be noted that the full range of Oerlikon cannon is more complex than is indicated here, with many sub-types developed for specific applications.

Mauser produced a range of guns based on a mechanism which uses gun gas not just to power the firing cycle but also to drive the ammunition feed, which may be linkless or via a disintegrating belt. Three calibres were offered with calibre designations; the MK 20, MK 25 Mod E and MK 30 Mod F, of which only the largest has been successful.

## GAS-OPERATED ROTARY CANNON

Rotary cannon are usually externally driven. However, Russian armament firms have developed this type of gun with gas operation, despite the problems of achieving this when the barrels are spinning round at about 1,000 revs per minute. The advantage, apart from not needing an external power source, is quicker acceleration to the maximum rate of fire, an important consideration in aircraft weapons. The six-barrel GSh-6–23 and GSh-6–30 have achieved extremely high rates of fire for their calibres.

A self-powered version of the American M61A1 described below, the GAU-4/A, was developed in the 1960s for installations with limited available power. This version still needs an electric motor to start it firing (although a powerful spring was developed as an alternative), after which gun gas takes over. It was fitted to the SUU-23/A gun pod, which also carried 1,200 rounds of ammunition.

## BLOWBACK MECHANISMS

The simplest form of automatic operation is the blowback. In this, the cartridge is held in the chamber by a heavy bolt, which is in turn kept pressed forward by a strong spring. When the cartridge fires, the expanding gases push the projectile up the barrel and the cartridge case back against the bolt. The inertia of the bolt holds it in place for the fraction of a second necessary to allow the projectile to reach the end of the barrel, by which time the bolt is being pushed back, ejecting the fired case and compressing the return spring as it goes. This leaves the rear of the chamber open so that the bolt, now being pushed forward again by the return spring, can pick up another cartridge from the ammunition feed and chamber it in the gun. During this cycle the bolt also recocks the firing mechanism.

The problem with the simple blowback is that it only works well with low-powered ammunition such as small pistol cartridges. Cannon cartridges are so powerful that the bolt would need to be prohibitively heavy to hold the cartridge case in place until the projectile had cleared the barrel. The penalty for premature opening of the breech is likely to be a burst case, with high-pressure gas escaping from the breech. Various methods of modifying the simple blowback have therefore been developed and many of these are in service today.

Some of the modifications involve engineering a delay to the movement of the bolt. This may be arranged by locking the bolt to the barrel at the instant of firing, then unlocking it as soon as pressures have dropped to safe levels and allowing the blowback cycle to proceed (delayed blowback); or arranging for the bolt to push some mechanism out of the way, engineered to provide maximum resistance for the first few millimetres of travel (retarded blowback). Delayed blowback has been frequently used in heavy automatic weapons (see 'The Hybrids') but retarded blowback has been restricted to small arms. Even less common is the gas-retarded blowback, in which gas tapped from the barrel is used to hold the bolt against the breech face until the gas pressure in the barrel has dropped enough for the breech to open safely. The simplest variation in heavy weapons, however, uses advanced primer ignition.

## API BLOWBACK

The advanced primer ignition (API) blowback is mechanically almost identical to the pure blowback, with two principal differences; first, it fires as the bolt is moving forwards and just before the cartridge is fully chambered; and secondly, an extended or hooded chamber is used within which the cartridge can slide while remaining supported. The effect of these differences is that the recoiling cartridge case has to arrest the forward movement of the bolt before it can begin to push it back again. The result is that a much lighter bolt is required; it is commonly stated that for a powerful API cannon, the bolt is only one-tenth the weight it would need to be if the gun used a pure blowback system. The reduced inertia of the relatively light bolt permits a much higher RoF than would be possible with a pure blowback.

All simple and API blowbacks have in common a bolt which is unlocked (i.e. not mechanically locked to the chamber) at the instant of firing, and all of the blowback family suffer from a common problem; a weakness in the positive extraction of the fired case. This is blown out of the chamber by gas pressure in the barrel (hence the 'blowback' designation) rather than positively pulled by the extractor as with most other mechanisms, and there is a risk that the pressure will instead stick the case to the walls of the chamber. At best this stops the gun from firing until the case has been cleared by the charger mechanism; at worst, the case may be pulled apart by the tension, leaving a section of case in the chamber and putting the gun out of action for some time.

The most effective solution to this problem is to wax or oil the cases before they are fired, but this carries its own problems such as collecting grit and carrying it into the mechanism. A later approach which works in some circumstances (actually developed in the 1930s) is to engrave shallow grooves or flutes into the chamber, leading from the case neck back to the body of the cartridge case. This provides a route for a thin film of propellant gas to seep back around the case, thereby preventing it from sticking. The fired cases exhibit longitudinal powder streaks from the mouth to close to the base.

Another characteristic of blowback weapons is that some cartridge shapes work better than others. Straight cases work well because they remain fully supported by the chamber walls as they slide backwards through the chamber after firing. Tapered or bottlenecked cases work much less well, as the case shoulder is never supported by the chamber. The gas pressure will instantly expand that part of the case to fill the gap, and may split the case or pull it apart. The only successful API blowback cannon with a (slightly) bottlenecked case is the 20 × 110RB Oerlikon S. Even so, the neck and shoulder of the fired case are considerably expanded from the unfired shape.

API case designs also need to take account of the fact that the cartridge is deep within the extended chamber when it is fired. All of the cartridge cases in cannon calibres therefore have a rebated (reduced diameter) rim to enable the bolt, with extractor claw hooked over the rim, to follow the cartridge into the chamber. The extractor is not required to extract fired cases because they are blown out, but it is necessary to pull a complete round out of the chamber if it fails to fire. The cases also often feature a considerable thickness of metal near the head, to reduce the risk of the unsupported head of the case splitting open as it is extracted.

Despite the apparent simplicity of the API blowback mechanism, the relationship between cartridge power, bolt weight and rate of fire is both subtle and critical. It is essential for the case to remain within the extended chamber until the projectile has left the barrel, thereby allowing the chamber pressure to fall to a safe level. This requires a certain minimum bolt weight (the inertia of the bolt being far more important at this point than the strength of the recoil spring). A powerful, high-velocity cartridge requires a long barrel to achieve the velocity. This inevitably entails a longer delay before the projectile clears the muzzle, which in turn requires a heavier and slower-moving bolt to keep the fired case within the chamber for that period. The consequence of this is a lower rate of fire than with short-barrelled low-velocity guns. To some extent, this is true of all automatic weapons, but in locked-breech types the effect is less severe as the mechanism only needs to allow for a fractionally longer delay before opening the breech; the weight of the bolt does not have to be increased.

A consequence of this is an unusual sensitivity to different types of ammunition. It is particularly difficult to balance the pressure characteristics of different cartridge loadings with the mechanism. In extreme cases the need to fire variant loadings featuring significantly different recoil characteristics required the gun mechanism to be rebalanced with different bolt weights and spring strengths. The best-known example is the change from the German 20 mm MG-FF (heavy projectiles at low velocity) to the MG-FFM (lighter projectiles generating less recoil), as described in more detail later.

A requirement of the API system is that the gun can only fire from an open bolt, i.e. with the bolt held fully back. In some respects this is desirable because it means that the chamber is left open between bursts of fire and therefore has a chance to cool down, and there is no risk of a cartridge 'cooking-off' in a hot chamber. However, it also means that there is a significant delay between pressing the trigger or firing button and the first round being fired. This system was therefore never suitable for synchronized mountings in piston-engined fighter aircraft, which required each shot to be precisely timed to avoid shooting off the propeller.

A result of the API principle is that the peak recoil forces generated are relatively low, because much of the recoil energy is absorbed in stopping the forward movement of the bolt before it can be thrust to the rear. As a result there is little surplus energy to drive a belt feed without slowing down the rate of fire, so most examples have been drum-fed. The main exceptions have been those such as the MK 108 whose design permitted the use of a very light bolt, the energy required to drive the belt feed being balanced by a reduction in bolt weight. The Japanese also managed to develop belt feeds for later models of their 20 mm Type 99–1 (Oerlikon FF) and Type 99–2 (Oerlikon FFL) aircraft guns, and the UK also developed, but did not field, a belt feed for the 20 mm Oerlikon Type SS. The light recoil push typical of API blowbacks also means that the complication of recoiling mountings can be avoided in all but the largest calibres.

The API blowback principle is used in many sub-machine guns, but as the cartridge cases used are the conventional rimless type, the cartridges are fired only a few mm before the bolt hits the breech face, so the effect is limited. Its use in heavy automatic weapons has mainly been restricted to a chain of development originating in a German gun, named after Reinhold Becker who patented the design in 1913. This simple weapon of modest performance saw limited service in both aircraft and ground fighting roles towards the end of World War I, but its significance lies in its post-war descendants, particularly the products of the Oerlikon factory which were extensively used in World War II (as described in the Switzerland section), and further developed in Germany and Japan. Larger cannon using this system were also developed, most notably the Rheinmetall-Borsig MK 108 firing 30 mm ammunition, and even a 55 mm version which saw further development in the USA post-war.

Despite its popularity in World War II due to its simplicity, ruggedness and reliability, the shortcomings of the Oerlikon system were becoming apparent by the end of the war. Its characteristics limited both the MV and the RoF, with little room for further improvement, so the next generation of Oerlikon cannon used locked-breech designs.

## THE HYBRIDS

Many automatic cannon combine elements of more than one operating system in their design. Probably the most common of these are the gas/blowback hybrids, typified by the famous Hispano-Suiza HS 404 aircraft cannon.

The 20 mm HS 404 was developed by Marc Birkigt at the French branch (HSF) during the mid-1930s. It is a linear-action weapon usually described as gas operated, but that is not strictly correct. In a purely gas operated mechanism, the gas tapped from the barrel is used to drive the entire reloading cycle. In the Hispano, this gas only unlocks the bolt from the breech; the remainder of the cycle is driven by the recoiling cartridge case being blown out of the chamber by residual gas pressure, just as in a blowback weapon. The entire gun recoils about 20–25 mm on firing and this energy is used to drive (via an intermediary spring) the ammunition feed in belt-fed versions, so the gun system incorporates elements of three different operating principles; gas, blowback and recoil.

One consequence of blowback case extraction is that fired cases are not positively pulled from the chamber, and so might sometimes stick. The first HS 404s used oiled cartridges to prevent this, but the British did not like this system and used instead a hard wax coating which melted on firing; it was forbidden to oil the ammunition, as this might remove the wax. Most guns of this type now have fluted chambers which allow some gas to seep around the outside of the case, balancing the pressure inside the case, so it is surprising that the USN's Mk 16 (a modified HS 404 on a naval mounting, still in service with the US Coastguard in the 1990s) was equipped with a built-in cartridge oiler.

Birkigt's gas-unlocked blowback mechanism has the virtues of light weight and relative simplicity so it has survived in the post-war Hispano range, examples of which were subsequently adopted by Oerlikon after the long-lasting rivalry between the firms was ended when Oerlikon absorbed Hispano-Suiza in 1971. During World War II the HS 404 itself was further developed in the UK and, in several forms, in the USA. Post-World War II, the Swiss branch of Hispano-Suiza took over production as the HS 804, which is still available from UK firm AEI Systems. The design was further developed into the USN MK 12 aircraft cannon, chambered for the more powerful 20 × 110 USN or Mk 100 series round.

The gas/blowback principle was also featured in the Italian Scotti system (which actually preceded the HS 404), except that a rotary lock was employed. An interesting aspect of the Scotti system (at least as it applies to small arms) is that while it is in theory a delayed blowback, in practice it seems to be more biased towards gas operation. On firing, the gas piston knocks the bolt carrier to the rear, an action which not only unlocks but also pulls back the bolt-head and therefore provides some positive extraction of the fired case. Scotti designs were made during World War II in 20 mm calibre.

The Japanese Army 20 mm Type 97 anti-tank gun, together with its associated Ho-1 and Ho-3 aircraft cannon and the closely related 20 mm Type 98 AA gun, used a different form of lock, in which a gas piston cams a locking piece down, leaving the breechblock to recoil.

Most modern gas-operated linear cannon could be described as gas/blowback hybrids. This is because their very high firing rates necessitate the breech opening very early, while gas pressure in the chamber is still relatively high, so this residual pressure is utilized to help accelerate the bolt and blow the fired case out of the chamber. It is therefore arguable as to which section to describe them under. As with other weapons using blowback extraction, fluted chambers are commonly used to ensure that the fired cases do not stick. One disadvantage to this solution is that the cannon are unsuited for use in manned turrets, because of the quantity of gas escaping out of the breech and into the turret.

The first of the post-war guns in the field in 20 mm calibre was the Hispano-Suiza HS 820, based on a powerful new 20 × 139 cartridge developed in the late 1940s. The gun, which uses percussion priming, is available in magazine, drum or belt-fed versions. Since Oerlikon took over the Hispano company, the HS 820 has been redesignated the Oerlikon KAD. The design was further developed in France, first as the F2 and then as the current 20M693, in which form it is made in South Africa as the Vektor G1–2.

In 1940 the French branch of Hispano-Suiza was working on a more powerful version of the gas-unlocked blowback design, the 30 mm HS 411. The Swiss branch continued to develop it, and introduced the gun immediately after the war as the HS 830. This was subsequently modified as the HS 831, known since the 1970s as the Oerlikon KCB.

The Hispano/Oerlikon family have not been the only ones to use a gas-unlocked blowback mechanism. The Rheinmetall Rh 202 is a particularly successful gun which uses the same 20 × 139 cartridge as the HS 820. Like many modern cannon, it features a floating firing or differential recoil mechanism to reduce recoil and a dual belt feed, enabling a rapid switch between HE and AP ammunition. It has been widely adopted in AA and AFV roles.

The French GIAT 20M621 is a modern lightweight gas/blowback gun designed around the US electrically-primed 20 × 102 round.

Not all hybrids have used gas pressure to unlock the bolt; it is equally possible for a delayed blowback weapon to use recoil unlocking. These are much more difficult to distinguish from short-recoil mechanisms, however, as the initial rearward movement of the barrel gives some impulse to the bolt before unlocking occurs and blowback takes over. The distinction is that in a short-recoil gun, the principal driving force is the momentum of the bolt assembly, which therefore needs to be reasonably heavy. In a recoil/blowback hybrid, the principal driving force is blowback, so the bolt can - and indeed, needs to be – as light as possible. In practice, the difference between short-recoil and recoil-unlocked blowback is one of degree rather than kind and is related to the rate of fire. As the bolt is lightened in order to increase the firing rate and open the breech more quickly, blowback becomes steadily more important in powering the firing cycle.

These hybrids can achieve higher rates of fire than other linear mechanisms, due to the fact that the maximum potential rate of fire is directly linked to bolt speed, which in turn depends on bolt weight, and delayed blowbacks can utilize the smallest and lightest bolts. This is because they neither need much momentum nor carry the additional weight of a gas piston assembly. Of the types of hybrid, the short-recoil/blowback has the potential for the fastest rate of all, partly because the initial barrel movement gives the bolt a 'flying start', partly because the design lends itself to the use of a bolt accelerator to give a further impulse to the bolt. There is another incidental advantage of the short-recoil type; it is easier to engineer the design so that the bolt is pulled slightly away from the breech face during the initial barrel movement, providing some positive initial extraction and reducing the risk of sticking cases.

These advantages may well explain why the Russians adopted recoil operation for the remarkable GSh-301 aircraft cannon, which has such a high rate of fire for a single-chamber gun (1,500–1,800 rpm) that blowback forms an important element of its operating cycle. Gas pressure remaining in the barrel when the breech is opened is so high that the fired cartridge case is ejected at a 'breech velocity' of 100 m/s. A 'return accelerator' is also used to return the breechblock as quickly as possible. The barrel is somewhat shorter than that of most of the Russian 30 mm guns, which reduces the time the projectiles take to exit the muzzle and thereby permits a fast breech opening time.

Not all hybrids have relied on blowback. One of the notable exceptions was the Rheinmetall-Borsig 30 mm MK 103 aircraft gun, in which barrel recoil operated the belt feed and initiated bolt motion, while a conventional gas system unlocked and accelerated the bolt. A primary concern of the designers appeared to be to ensure that the bolt did not unlock too early, so unlocking was delayed until the chamber pressure had dropped to the point at which blowback was not a dominant feature. It seems that the main reason for the barrel recoil was to absorb the recoil pulses without the necessity of allowing the whole gun to recoil in its mounting. The designers then presumably reasoned that the recoil movement might as well be put to good use. This is an interesting example of the difficulties that can exist in trying to place gun mechanisms into neat categories.

The experimental French 20 mm Delauney-Belleville anti-tank cannon of 1918 used a mixture of gas and long-recoil operation.

Recoil was the force driving a famous twin-barrel gun. During World War I, the German firm of Gast devised a different form of mechanism for an aircraft RCMG, in which the recoil of each barrel operated the loading and firing mechanism of the other, leading to a high rate of fire of around 1,600 rpm.

This principle was adopted in revised form much later, in the Russian GSh-23 and GSh-30 aircraft guns and the 2A38 AA gun, all of which display the usual national characteristics of compactness, light weight and a high rate of fire, and the experimental American 25 mm GE 225 (also offered in an externally-powered version). These all use gun gas to operate the mechanism and are therefore properly hybrid guns.

More recently, the Russians developed the 2A42 AFV/helicopter cannon which first emerged in the early 1980s in 30 × 165 calibre. It is perhaps surprising that it is not externally powered, given the modest rate of fire requirements, but it is quoted as using a mixture of gas operation and barrel recoil in the design.

Hybrid power sources can relate to the ammunition feed as well as the gun action. Externally-powered ammunition feeds are normally associated with externally-powered guns, but there have been rare exceptions, e.g. the GZ 1-FF electrically-driven belt-feed for the MG-FF API blowback cannon.

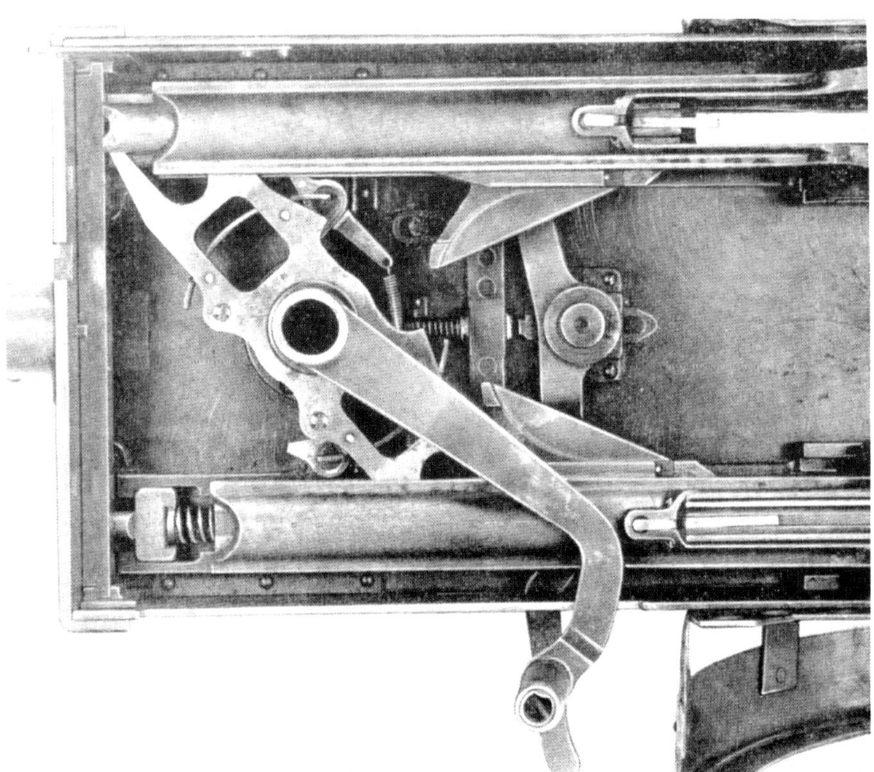

*Mechanism of the Gast twin-barrel gun*

## REVOLVER CANNON

Gas operation is usually the motive force for revolver cannon. The first practical example of these was invented in Germany towards the end of World War II by a team led by Anton Politzer of Mauser, who were seeking to improve the rate of fire of aircraft cannon. The main limiting factor was the length of time it took for the fired case to be removed and a fresh cartridge inserted into the chamber.

Politzer conceived the idea of achieving this movement in several stages rather than all at once, thereby effectively slowing down the process without slowing the rate of fire. He did this by designing a rotating cylinder, with its axis parallel to the barrel, which contains several chambers, just like a giant revolver handgun. When a chamber is in line with the barrel, the preceding chamber is in the case ejection position, while two are in the loading position and the fifth remains empty. With each shot, the cylinder is rotated to move the next chamber in line with the barrel. As it rotates, so the fired case is extracted and a live round loaded, in gradual stages; how gradual depending on the number of chambers, originally five but varying between four and seven in current weapons. The original design provided for blowback extraction, but this was soon replaced by mechanical means.

An obvious problem is the loss of propellant gas through the gap between the cylinder and the barrel. This is usually overcome by fitting each chamber with a sliding sleeve which bridges the gap at the instant of firing, forming a gas-tight seal, or by pressing the cylinder against the rear of the barrel. A further alternative is to make the cylinder shorter than the cartridge, so that the cartridge case extends into the barrel, sealing the gap. Another difficulty is that unless the mechanism is externally driven (normally gun gas is used) there has to be some means of starting the process. Special pyrotechnic cartridges are often used to generate the gas to initiate the cycle.

The Mauser guns, the MG 213C in 20 mm (later designated MG 213/20) and the MG 213/30 in 30 mm, were at an advanced stage of development by the end of the war and inspired the subsequent development of the principle. Politzer and other members of the Mauser team went to various western firms to continue their work; Anton Politzer went to France, Otto von

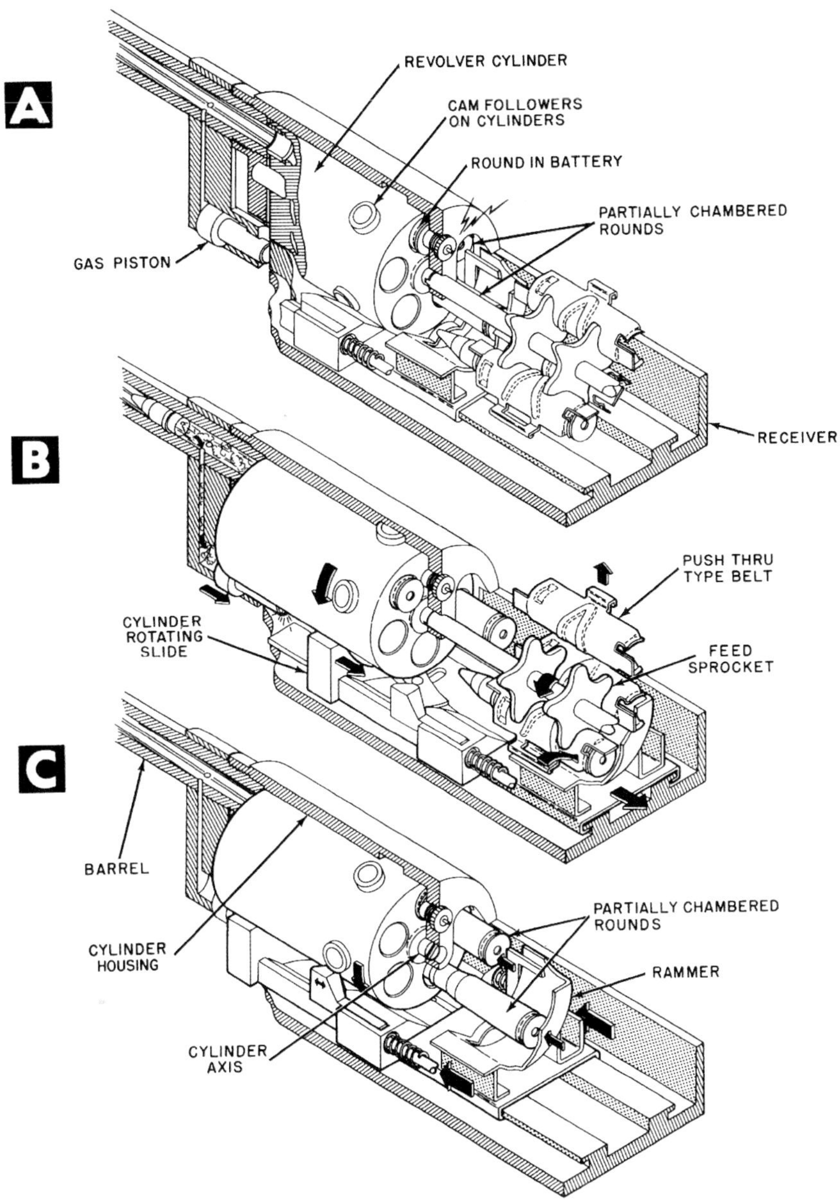


*The action of the MG 213C revolver cannon (Chinn)*

Lossnitzer to the USA, Fredrick Linder to Switzerland and Werner Jungemann to the UK. Without such aid, it took the Soviet Union years to catch up.

The best known of the revolver cannon were developed directly from the MG 213/30: the five-chamber British ADEN and the French DEFA, later GIAT, now Nexter. The most obvious changes have been to cartridge case length, which has increased from 85 mm to 111–113 mm in service weapons in order to increase the muzzle velocity, although the weapon development programmes have been rather more complex than this suggests. The most recent of the Nexter series is the 30M791 which fires a much more powerful 30 × 150B cartridge. The gun achieves a very high (2,500 rpm) RoF by using seven chambers.

The Americans adopted the revolver principle soon after the war (in fact the idea was patented in 1905 by Charles Clarke of Philadelphia, but then forgotten), but unlike the British and French chose a 20 mm cartridge, later also used in their M61 rotary cannon. This led to the Pontiac M39, introduced in the early 1950s, which over thirty years later was improved to create the lightweight, faster-firing Ford Tigerclaws intended for the abortive Northrop F-20 Tigershark.

Other calibres of revolver ranging up to 37 mm were built but not adopted.

The Swiss Oerlikon firm, another recipient of a Mauser engineer, developed a range of revolvers commencing with the 20 mm 206 RK, which used an electrically-primed version of their new 20 × 128 cartridge, and the 30 mm 302 RK. Other variants followed, perhaps the most unusual being the abortive 42 mm 421 RK AA gun developed for the UK (and designated 'Red King' by them), which featured two barrels loaded from a single seven-chamber cylinder to achieve 450 rpm. This led to the British 'Red Queen' weapon, with a single water-cooled barrel and a different 42 mm cartridge, which was tested secretly in the early 1950s before being abandoned.

Later versions of the Oerlikon aircraft guns differed from the original Mauser design in using four rather than five chambers. However, none was adopted for service until the big KCA (initially developed as the 304 RK), firing the powerful 30 × 173 cartridge, was selected by Sweden in the 1970s. The 304 RK was also chosen by the USA as a back-up in case of the failure of the GAU-8/A rotary cannon for the A-10 aircraft, slightly modified and dubbed GAU-9. In the mid-1990s Oerlikon-Contraves announced a scaled-up version of this gun; the 35 mm 35/1,000 (now designated KDG).

Mauser have appropriately reclaimed their invention in the five-chamber BK 27 aircraft gun, which fires their unique 27 × 145B cartridge.

An unusual revolver cannon was the American Mk 11 designed by Frank Marquardt, chambered for the 20 × 110 USN (Mk 100) cartridge. Like the Oerlikon 421 RK this featured two barrels, in this case fed by one eight-chamber revolver cylinder from two separate belts, with firing from both barrels simultaneously. The gun used both gas and recoil to power its operation. Prototype versions were even more peculiar as they used a unique ammunition feed system which enclosed the cartridges within linked 'projectors' (like large cartridge cases) from which the cartridges were fired into the chambers (see the section on Unconventional Systems for more detail). Production versions used a more conventional belt feed.

At the time of its introduction, the revolver cannon offered almost double the rate of fire of comparable linear-action weapons, i.e. around 1,200–1,400 rpm. Since then it has been doubled again, up to 2,500 rpm. This extremely high rate of fire has mainly limited the applications of the revolver cannon to aircraft, which only need to fire very short bursts. At this rate of fire, the single barrel would clearly overheat and be worn out very quickly. In recent years, however, the BK 27 has been adopted for use in naval mountings, and Oerlikon now offer the 35 mm KDG for land and naval AA and anti-missile mountings.

The Soviet Union has made little use of revolver cannon, with only two very different types seeing service. Best known is a powerful naval AA weapon, the four-chamber NN-30, introduced in 1959 and firing a 30 × 210B cartridge. This has been adopted in China as the NORINCO Type 69 and was adapted by the Romanian army as a towed AA gun, the A436. In addition there was the Rikhter R-23, a 23 mm gas-operated aircraft gun developed in the 1950s which is unlike all other service revolvers in being front-loaded; that is, the ammunition feed arrangements are underneath the barrel and the cartridges are fed backwards into the chambers. These characteristics were designed to meet the needs of a turret-mounted bomber defensive gun. There are more details about this cannon in the Russia section of the book.

The experimental American T168 developed at the same time and for the same purpose used a similar loading system; the 30 × 136RB or 30 × 120RB cartridge also featured a case tapering down to an extremely rebated rim (just 17.3 mm in diameter) but the projectile was exposed in the usual way.

The front-loading concept was revived and combined with recoilless operation and combustible cases in the abortive Mauser RMK 30 and 35 series (see Unconventional Systems).

## EXTERNALLY POWERED GUNS

The earliest machine guns of all were externally powered. The external power source was manual labour, in that the guns were made to operate by the gunner cranking a handle. The rate of fire was therefore largely dependent on the energy of the gunner, and tended to slow down as the gunner tired. The most successful large-calibre weapons of this type were the Gatling (up to 25.4 mm calibre) and Hotchkiss (from 37 to 53 mm) multi-barrel rotary guns and the Nordenfelt (again, up to 25.4 mm) which had fixed barrels (usually two or four). They were all developed in the latter part of the 19th Century, and by the end of it had been reduced to obsolescence by the new self-powered weapons, commencing with the recoil-operated Maxim.

The first experiments with coupling an electric motor to a Gatling gun actually began in 1890, and eventually resulted in a RoF of 3,000 rpm – a remarkable achievement. Nobody could think of a use for this until close to the end of World War II, when the demands of air fighting placed a premium on a high RoF and prompted the American Project Vulcan.

Project Vulcan resulted in the famous M61 Vulcan rotary cannon, essentially similar to the Gatling. In 20 × 102 calibre, with six barrels and a rate of fire of between 4,000 and 7,200 rpm (normally set at 6,000 rpm), this has been the USAF's standard air combat gun since the early 1960s. Cannon of this type are also in service in 25 and 30 mm calibres, with between three and seven barrels. The external power source can be electric, hydraulic, pneumatic or even (in aircraft pods) from a turbine driven by the slipstream.

Current service models are the six-barrel M61 in 20 × 102 (now in the A2 version), the three-barrel M197 version and the South Korean A-50 in the same calibre, the five-barrel GAU-12/U in 25 × 137 and four-barrel GAU-22/A in the same calibre, the seven-barrel GAU-8/A in 30 × 173 and the four-barrel GAU-13/A gunpod version also in 30 × 173. All of these normally use external power, and the demands are considerable. The M61A1 has an engine of 26kW, the GAU-8/A requires 52kW – enough to drive a small car at 160 km/h. In the SUU-16/A gun pod, a ram-air turbine is used to power the M61A1, a neat solution with the disadvantages of a slower spin-up time (around one second) and a rate of fire which reduces below 6,000 rpm if the airspeed drops below 650 km/h. Guns can also be powered by gas bled from the engine, a solution adopted for the GAU-12/U fitted to the AV-8B.

At least one externally-powered version of the Russian rotary cannon is also available (the AO18L lightweight naval gun) but most of them are gas-operated and are described in the previous section.

Oerlikon briefly joined the group of rotary cannon developers with the KBD, a seven-barrel gun in 25 × 173/181 calibre, which was intended for the naval CIWS (close-in weapons system) role but attracted no sales.

Finally, China has adopted seven-barrel and eleven-barrel power-driven rotaries in 30 × 165 Russian calibre, initially for CIWS mountings.

More recently still, external powering has been introduced in a range of single-barrel machine guns and cannon. The best-known of them are the Chain Gun family. These are simple linear weapons, named after the chain which drives the reciprocating bolt mechanism. Unlike the Vulcan, the purpose of this design is not rate of fire but reliability, combined with the ability to vary the rate of fire depending on the tactical needs of the moment. Chain Guns also have a very compact mechanism, a major advantage in AFVs in which turret space is always at a premium. Also useful in manned turrets is that breech opening can be delayed until the gas pressure has dropped, minimising fumes entering the turret.

The Chain Gun concept was originated by Hughes Helicopters in the USA (it is now the property of Northrop Grumman Innovation Systems) and has been applied to weapons in a variety of calibres. The most important cannon-calibre versions are the M242 'Bushmaster' in 25 × 137 calibre; the M230 in 30 × 113B; and the Bushmaster II (plus the Mk 44 and XM813 versions), chambered for the 30 × 173 round. The Bushmaster III in 35 × 228 is also in service. Power demands are much lower than with rotaries; the 30 mm M230 requires a motor of only 5 kW.

An interesting side-effect of external powering is its insensitivity to ammunition type. The Bushmaster II can quickly be adapted, by exchanging a few parts, to fire 30 × 170 Oerlikon (KCB) or Rarden ammunition. A related disadvantage is that they are prone to be vulnerable to 'hang fires'; cartridges which fire a fraction of a second late. These don't affect self-powered guns because the mechanism stops if a cartridge does not immediately fire. However, an externally-powered gun mechanism will open the breech and extract the cartridge regardless whether it has fired or not – and the results of the cartridge firing after the breech is opened could be catastrophic. With Chain Guns, this can be countered by building in a recoil sensor which detects any failure to

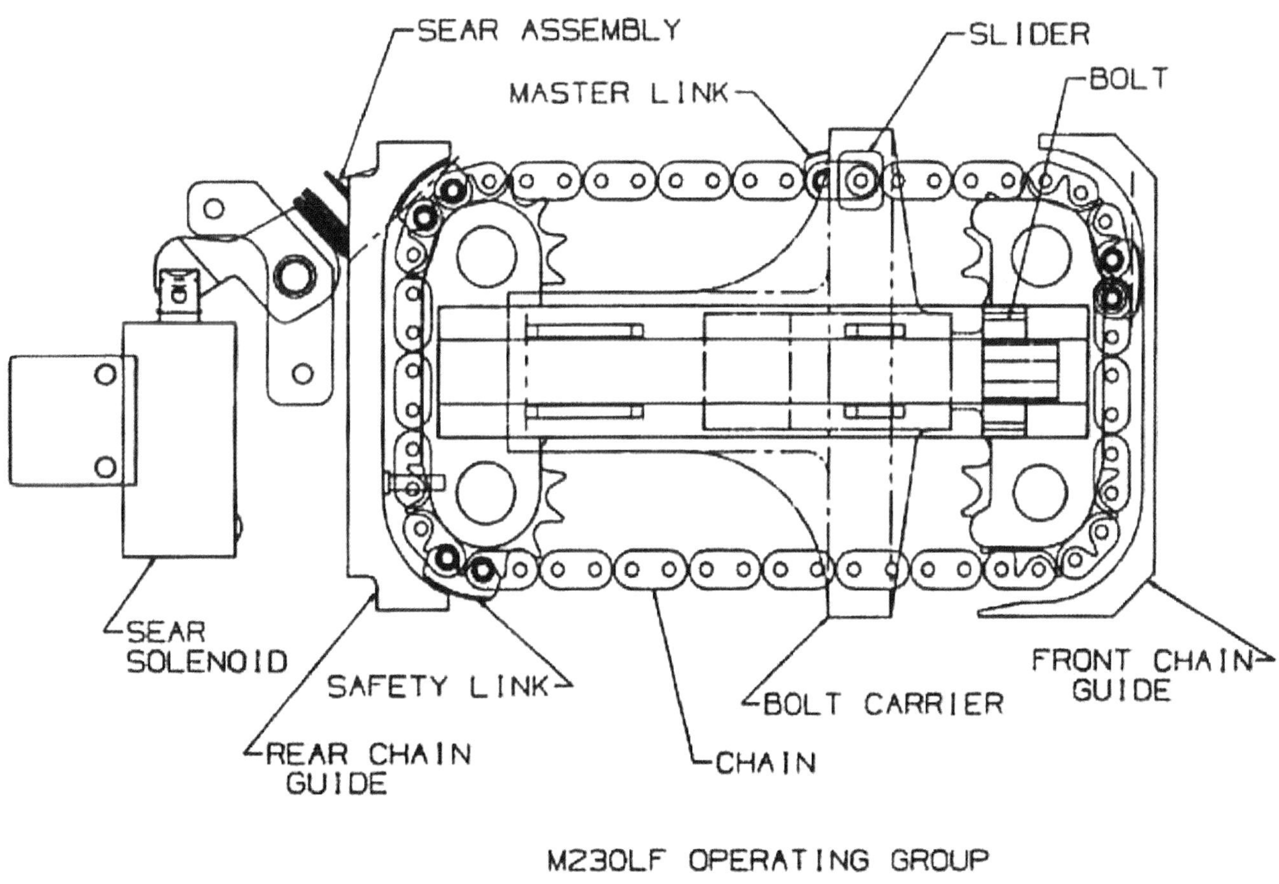


*M230 Chain Gun mechanism*

fire and stops the gun action. This only works up to about 400 rpm. However, that is enough for most purposes.

Externally powered weapons have also been developed in France. GIAT offers two such weapons; the 25M811 gun in 25 × 137 and the 30M781 in 30 × 113B. In Germany, Rheinmetall developed the Rh 503 in both 35 × 228 and 50 × 330 'Supershot' calibres in the 1980s, only to see sales opportunities disappear with the end of the Cold War. Yet another weapon in 35 × 228 is the South African EMAK 35 (since replaced by a 30 mm version, the GI-30 CamGun). China has also fielded a lightweight 23 mm externally-powered gun, intended for helicopter or ground vehicle use. As will be seen, externally-powered guns are beginning to dominate the AFV and helicopter weapon fields, the main exception being Russian equipment. By no means all externally-powered mechanisms are chain driven. Others use a rotating element, such as a drum with cam tracks engraved in the surface, to convert their rotary motion into a reciprocating one to operate the bolt.

A unique application of external powering is used by the Spanish Meroka CIWS. This mounting contains twelve externally-powered guns, using electrically-primed 20 × 128 ammunition, which are closely packed together in two rows of six, each row with its own belt feed. Reloading utilizes a combination of compressed air and the recoil of the breechblock. Each belt is moved on six places to reload. The guns fire in salvos of twelve shots, the slightly staggered firing giving a burst rate of fire of 9,000 rpm, with a 0.2 second pause before the next volley. It is a modern revival of the 'volley gun' concept, made obsolete in the 1870s by the first manually-driven machine guns, now using modern technology to provide an impressive performance. Its only real drawback is that its weight limits the system to naval applications.

*A selection of belt links, clips, strips and magazines (Koen Bakker)*

## FEED SYSTEMS

Another aspect of cannon design is the method of holding the ammunition ready to be loaded into the gun, which is a complex subject of its own.

Ammunition may be contained in a flexible belt originally made mainly of fabric, but since the interwar years typically made of metal links (pulled into position for reloading by another mechanism driven by the gun, or more rarely separately powered). The links may remain connected after the cartridges are removed for loading, but more commonly they are 'disintegrating link' belts, designed to fall apart. Alternatively, ammunition may be held in a box magazine (with a spring used to ensure that the cartridges are held in position) or in a drum or pan magazine (which may be gun powered, spring-driven or use some other form of external power). Strip or clip feeds have been used in some weapons, in which several rounds held together by a rigid metal strip or clip (usually along their bases) may be fed into loading trays. Gravity has been used to aid feeding, either by itself (using guides or hoppers) or in conjunction with other systems, particularly clip feeds; this is now generally restricted to large AA weapons. The most advanced methods use various forms of linkless feed, with the ammunition transferred to the gun by some kind of power-operated conveyor system.

Identical cannon can be used in different weapon systems which may have a different method of feeding. Some weapon systems have double or even triple feeds giving the user the option of immediately or consecutively firing different types of ammunition in the system. In general all ammunition has to be removed from the belt, clip, link, magazine (box or drum) or strip before firing and removal is therefore part of the ammunition feeding process.

The above list of possibilities is not comprehensive but does give the major categories, examples of which are described and illustrated below.

## GRAVITY FEED

The earliest hand-cranked cannon generally relied on gravity to feed ammunition to the weapons. Typically, there would be some form of vertical guide above the gun into which the cartridges would slot, properly aligned with the chamber. The lowest cartridge would be suspended just above the bolt, or pressed against it, so that after recoiling to the rear (extracting and ejecting the fired case as it does so), the bolt would push the next cartridge into the chamber on its return. This form of gravity feed remained popular in larger cannon calibres for many decades, as it is simple and reliable. The reloading process was speeded up by connecting several rounds by clips, so they could be handled as one unit. Ammunition for the original 40 mm Bofors gun, for example, was held in groups of four by a metal clip fitting around the base of the cases.

40 mm Bofors L/60 with extended ammunition guides (Celsius)

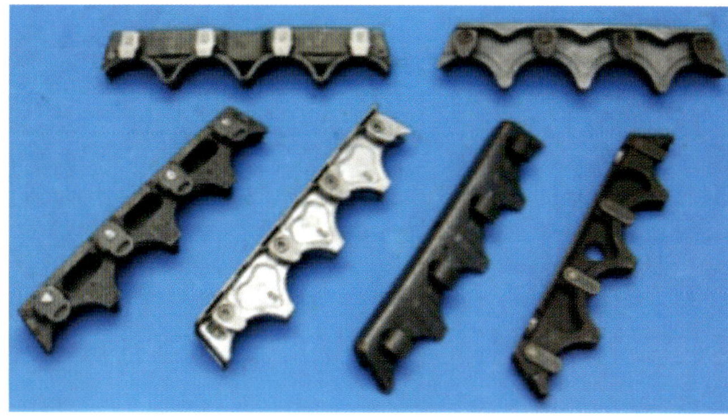

40 mm Bofors clips holding 4 rounds for cannon L60 and L70 (Koen Bakker)

A filled 37 × 250R clip holding 5 rounds for cannon M1939 61-K (Koen Bakker)

A quad 1.1-inch mounting on USS Ranger during an alert in December 1942. All eight of the open frame-type ammunition clips are on the guns, with more being held ready – see bottom right of picture. (US National Archives)

Each group was manually dropped into the guides above the gun (which could take two or more groups) as required.

In modern versions of the Bofors, large-capacity power-operated magazines are available together with automatic clip handling, but gravity is still utilized to aid feeding.

One variation of a clip feed was adopted for the USN 1.1-inch gun of World War II which used a rectangular frame holding eight rounds, the clip being replaced when empty; a kind of half-way house between a clip and magazine. Two clips were fitted to each gun side by side, permitting continuous firing.

The main problem with the basic version of this method is that only a small quantity of ammunition is ready to fire at any given moment. Continuous fire requires several loaders as shown in the above photo. A second problem is that the location of the feed rails keeps

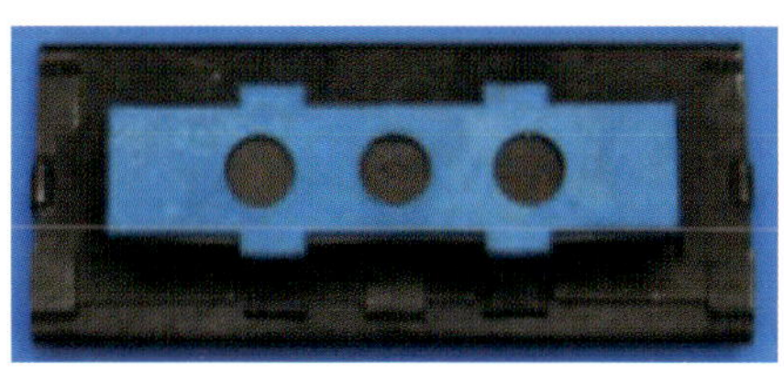

30 mm × 170 training ammunition for Rarden cannon

30 mm × 170 HE ammunition for Rarden cannon

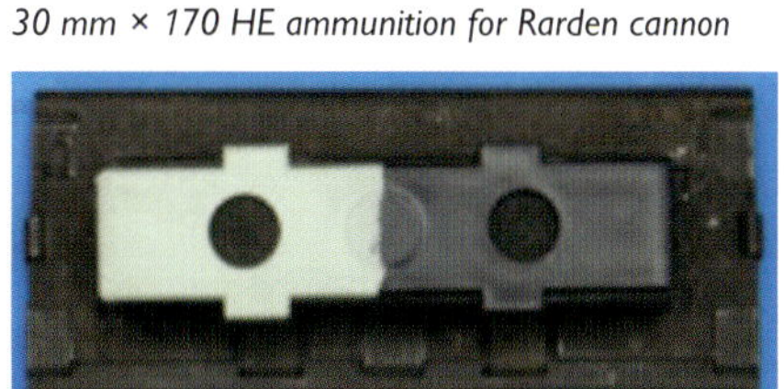

30 mm × 170 mixed ammunition for Rarden cannon

30 × 170 clip holding 3 rounds for the Rarden cannon. The different colour of the plastic see p. 82 indicates the type of clipped ammunition; in poor light conditions the number of holes in the plastic can also be used to determine the type of clipped ammunition (Koen Bakker)

changing with gun elevation, which can hinder rapid reloading. It is therefore mainly appropriate for circumstances in which the gun is stationary and there is plenty of room around it for the gun crew to operate without much fear of them being gunned down; in other words, in land or ship based anti-aircraft weapons.

One exception, and a curious survival of the small-capacity clip feed, is used by the British 30 mm L21A1 Rarden AFV cannon; just three rounds are held in each clip, with room for two clips in the feed above the breech.

The many and various means of providing an extended period of uninterrupted fire account for the different types of ammunition feed currently in use. In larger calibres, this has been achieved by retaining gravity feed but arranging a device which holds several stacks of cartridges side by side, from which rounds are moved to the breech in turn. This method was used in the 57 mm Class M Molins Gun carried by Coastal Command's de Havilland Mosquito FB Mk XVIII anti-submarine aircraft (also known as 'Tsetse'), in which gravity was assisted by a spring-powered arm on top of each bank of five rounds, with

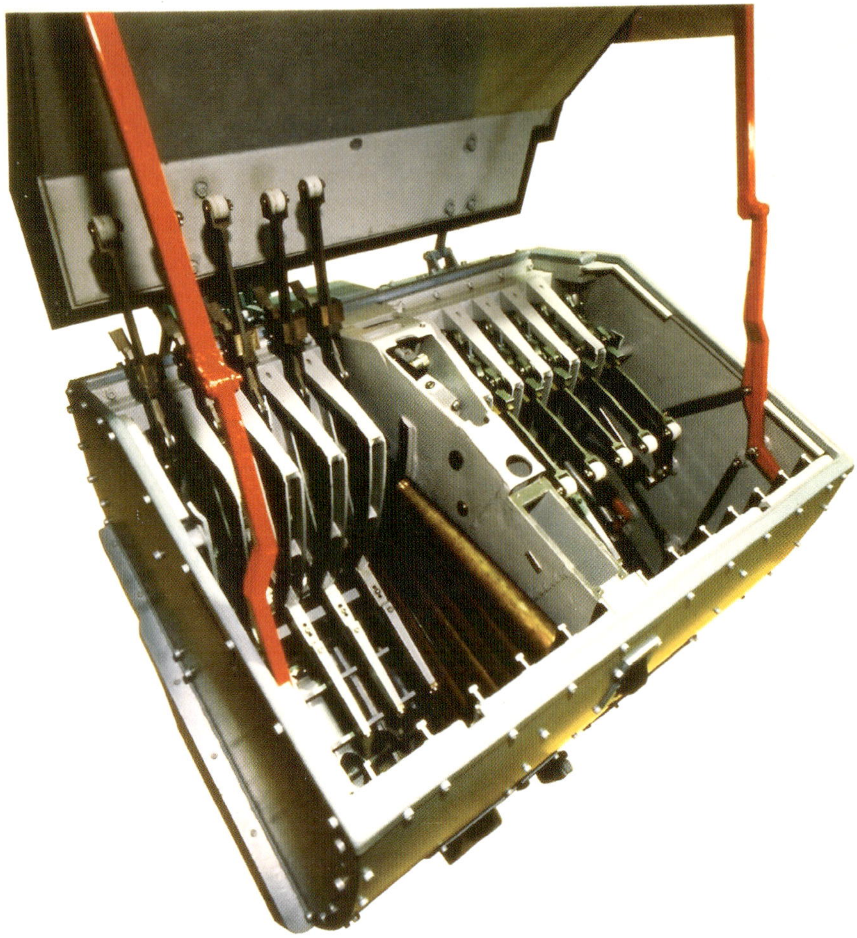

*40 mm Bofors 101-round magazine (AB Bofors)*

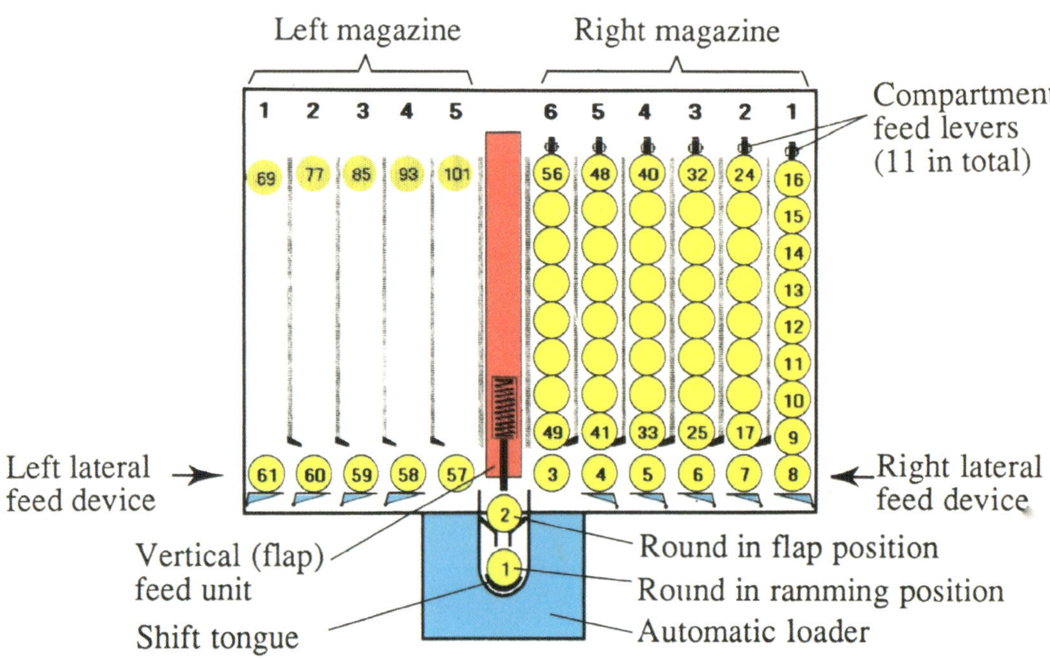


*101-round magazine for the Bofors SAK 40/70 E naval gun. The numbers on the rounds indicate the feeding order. Either the left or right magazine can be used, allowing two different ammunition types to be loaded. The compartment feed devices are cocked manually during loading. The lateral and vertical feed devices and the rammer are all recoil operated (AB Bofors)*

electrical power being used to shift the four banks into position. The total capacity was 23 rounds, including rounds in the feedway.

A modern example is the Bofors SAK 40/70E naval gun, which has a 101-round magazine consisting of ten columns.

## STRIP FEED

Some cannon use strip feeds, in which a number of rounds (limited by the need for easy handling, so the number decreases with size and weight) are connected by a strip of metal shaped to hold the rounds in place. This is usually rigid although some examples are articulated and a few disintegrate. The strips are typically placed in a horizontal tray, the strip being pulled through the feed mechanism as the gun fires, rather like a belt feed. Rapid reloading can be facilitated if the design places the ammunition feed at the centre of mass of the gun (i.e. the trunnions, around which the gun elevates); some designs have combined this with oversized trunnions with the feed passing through them, so that the feed trays remain in the same location (albeit varying in their angle) regardless of gun elevation changes.

Examples of this feed type were used in Germany in World War II; the 3,7 cm Flak 18, 36, 37 normally used 6-round or 8-round strips while the 5 cm Flak 41 used 5-round strips. They are less common now, although some

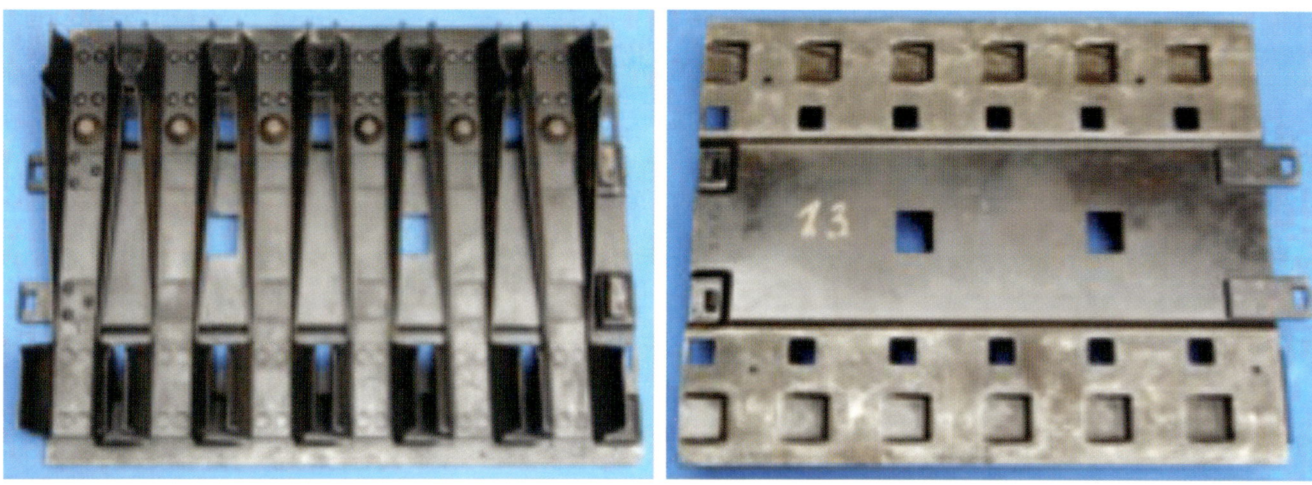


*37 × 264B strip holding 6 rounds for cannon Flak 18, 36 and 37; the strips can be connected with each other (Koen Bakker)*

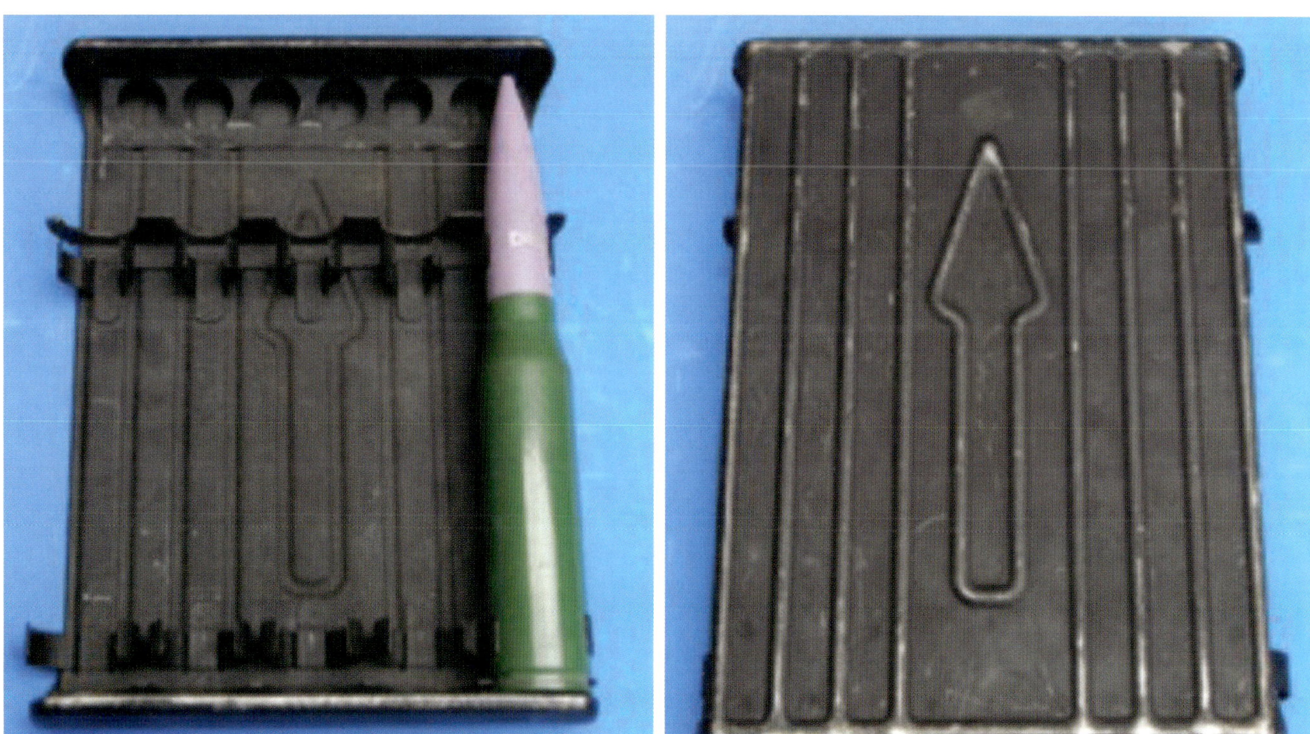

*30 × 170 strip holding 5 rounds for Hispano-Suiza 831 series (Oerlikon KCB) cannon (Koen Bakker)*

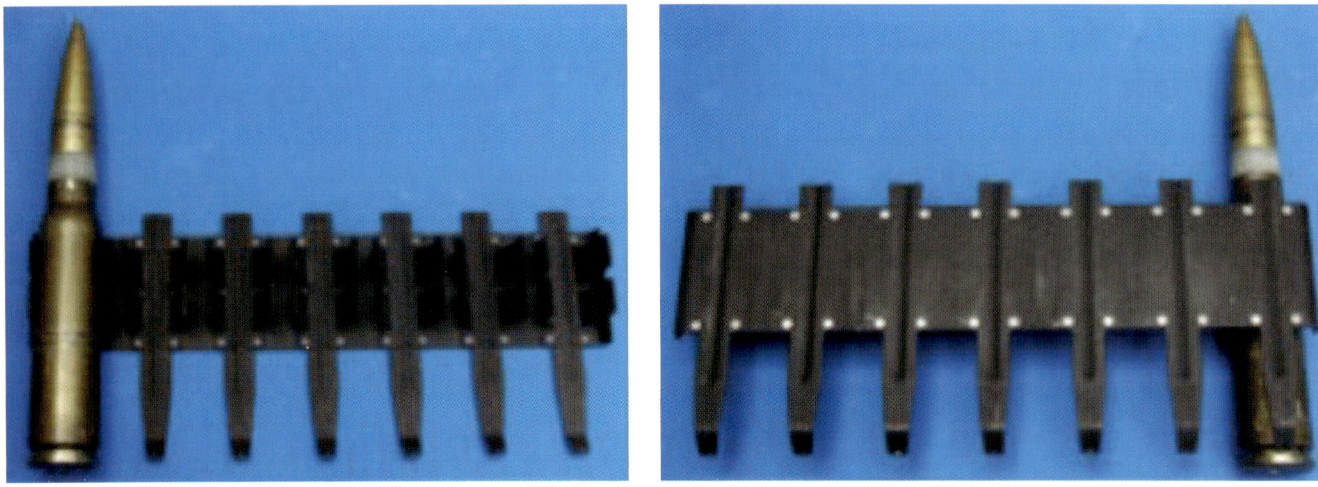

*30 × 173 strip for GAU-23/A (Bushmaster II) cannon installed in AC-130U gunship. The rounds are positioned in the strip by the extractor groove; comparable with a metal variant of the link (Koen Bakker)*

post-World War II cannon have used them, as shown.

Strips have the benefit of being significantly lighter than magazines (with the disadvantages of exposure to dirt and damage), and more limited round capacity. Some examples of empty weights, all German of World War II vintage:

- For 3 cm Flak M44 (disintegrating): 10 rounds capacity, 1.8 kg
- For 3,7 cm Flak 18: 6 rounds capacity, 3 kg
- For 3,7 cm Flak 43: 8 rounds capacity, 3 kg
- For 5 cm Flak 41: 5 rounds capacity, 7.5 kg

## MAGAZINE FEED

In smaller cannon calibres, detachable box, drum or pan magazines have often been used. In a box magazine the cartridges are stacked on top of each other, in one or two interlocking rows (the latter almost halving the length of the magazine); they may be mounted vertically, above or (in a few cases) below the gun, or horizontally to feed in from the side. One very rare example of a four-row box magazine ('quad-stack') was that for the 20 mm Polsten.

A pan magazine is circular but flat, with the cartridges arranged radially, all pointing inwards; there may be one or more rows (large multi-row pans may also be referred to as drums).

A drum magazine is a cylindrical container which holds the rounds parallel to each other, aligned with the cylinder axis.

All of these types usually have an internal spring which keeps the cartridges pressed against the open end of the magazine; they are stopped from falling out by lips which only permit them to slide out lengthways.

Advantages of magazines are that the ammunition is protected to some extent from damage and collecting dirt, and the feed will work at any angle or even in a violently manoeuvering aircraft as it maintains a fixed relationship between the magazine and the gun. The main disadvantage is that their capacity is necessarily small, both because of their bulk and because there is a limit to the amount of work a spring can reliably be expected to do (especially if working against gravity, as with a bottom-mounted box magazine). Unlike clip and strip feeds, there is also a pause in the firing whenever the magazine needs to be changed.

The simplicity and low cost of the box magazine has led to its becoming the standard ammunition feed used in virtually all small arms today. However, they rarely hold many rounds in larger calibres simply because they would otherwise be too bulky and heavy to manhandle, and maintaining a good rate of fire calls for an energetic loading crew.

One of the most common examples is the German 2 cm Flak 30 and Flak

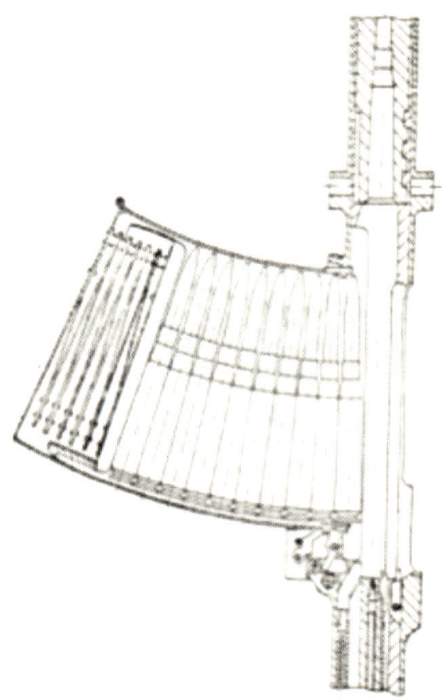

*2 cm Flak 30 magazine, horizontally mounted*

38 – the standard wartime light AA weapons – which normally used side-mounted 20-round boxes, although larger ones were also developed.

Some idea of the empty weight of box magazines can be gained from World War II German examples:

- 2 cm KwK 30/38 (20 × 138B): 10 rounds capacity, 2.1 kg
- 2 cm Flak 30/38 (20 × 138B): 20 rounds, 3.4 kg
- 2 cm Flak 38 (20 × 138B): 30 rounds, 5.3 kg

25 × 163 double stack magazine holding 15 rounds for the Hotchkiss Mle 1938 and variants, including Japanese naval 25 mm Type 96 (Koen Bakker). The magazine was top-mounted, feeding downwards, so is shown upside-down here. The holes in the magazine sides permit the crew to check the number of rounds remaining; the flat spring which presses the rounds against the feed lips at the open end of the magazine can be seen (Koen Bakker)

- 2 cm Flak 38 (20 × 138B): 40 rounds, 7.0 kg
- 3 cm MK 101 (30 × 184B): 6 rounds, 6.2 kg

A development of this type was the multiple box magazine as used by the French 25 mm Hotchkiss experimental aircraft cannon just before World War II, which consisted of four box magazines stacked vertically, with a total capacity of 35 rounds. The Russians used a similar idea in the limited-service 23 mm MP-6, for which the magazine contained nine clips of ammunition, each holding nine rounds for a total of 81, resulting in a bulky box-shaped magazine.

The radial pan magazines were at one time popular in LMGs, but not in larger calibres except in special circumstances, the necessarily large diameter normally being a deterrent. A rare example was developed for the Finnish 20 mm L34.

Not all 'pan' magazines were what they seemed to be: one example of a 'false pan' was installed under the wings of some World War II German fighter aircraft to feed their Mauser MG 151 cannon: the ammunition container was pan-shaped, but actually enclosed a coiled belt with about 140 rounds in three layers.

A less common variation with drum or pan magazines is for them to be rotated by the gun's mechanism or by some other external power source. Some of the USAF ammunition feeds for aircraft cannon have used a large-capacity power-operated helical drum with the rounds pointing inwards, although the trend now is to fit more compact magazines with a power-operated linkless feed.

The largest-calibre examples were fixed, power-driven types that were side-mounted in some experimental AA guns of the 1950s such as the US 37 mm

20 × 139 single stack magazine for Rheinmetall Mk 20-01 (Koen Bakker)

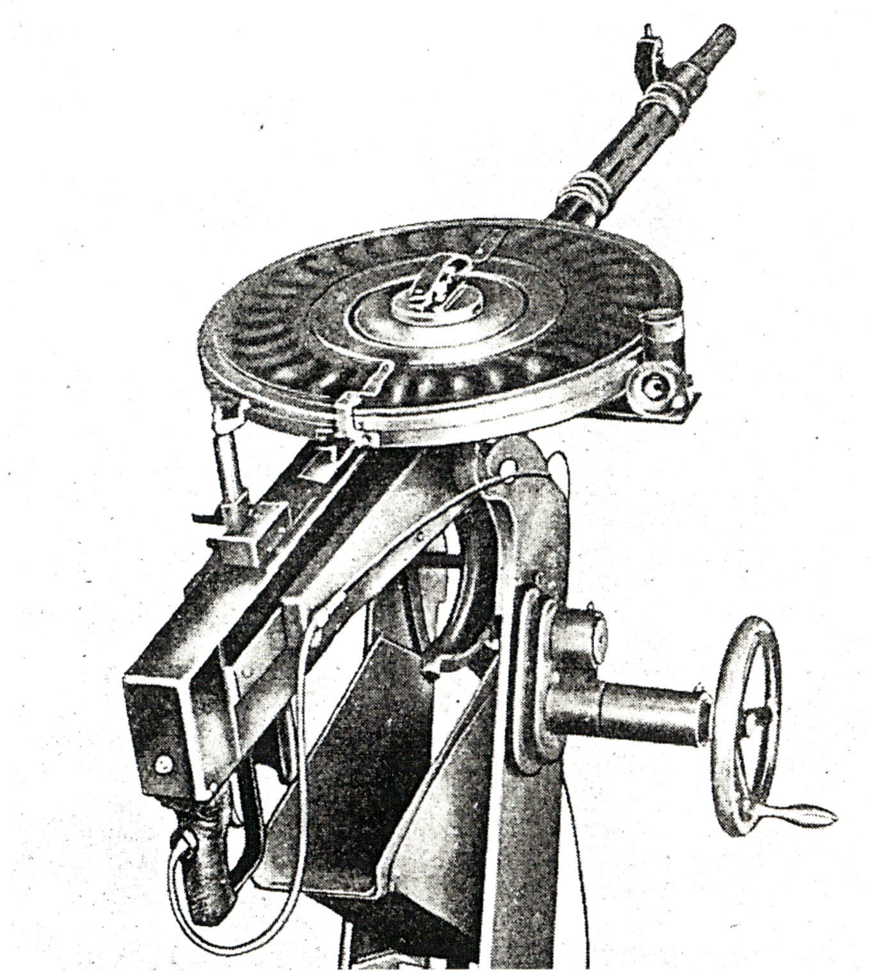

A 32-round pan magazine on the Finnish Lahti L34 20 mm cannon

Vigilante (192 rounds) and the Anglo-Swiss 42 mm Red King and Red Queen (75 rounds).

Parallel drum magazines have been far more common; they represent the most practical way of packing a large ammunition supply within compact dimensions. The best known is probably for the original Oerlikon Type S 20 mm cannon, used by the Royal Navy and the USN from World War II, which uses a 60-round drum. The earliest marks of 20 mm Hispano aircraft cannon of the 1930s also used a drum of this capacity, as did most of the Luftwaffe's Oerlikon-based 2 cm MG-FF. There were some interesting differences between the Oerlikon and Hispano type drums: The Hispano drum is symmetrical, with the axis of rotation directly over the gun breech, so that the drum overhangs the gun to an equal extent on both sides. The Oerlikon drum is asymmetrical, with the magazine to one side of the gun (see photos below).

Two versions of the Oerlikon drum were made, with the overhang being on the left or right side. This complication provided two advantages: in twin mountings, the guns could be mounted closer together with the right-hand gun having a magazine with a right-hand overhang, and the left hand gun with a left-hand feed. A more subtle advantage of the Oerlikon pattern was that the spring force was pushing the rounds directly downwards into the feed, whereas in the Hispano the spring provided rotational force only; the rounds were forced to change from a horizontal movement to a vertical one just by the shape of the feed path, which proved less reliable. Finally, it was possible to load the Oerlikon magazine without tensioning the magazine spring, but in the Hispano a loaded magazine always had the spring under tension, causing it to lose strength if left too long.

Smaller-capacity drums were used to some extent, particularly in confined spaces such as AFV turrets or in flexibly-mounted aircraft installations where the large drums were awkward to handle and blocked the gunner's view.

However, while the need to change drums at frequent intervals was acceptable in a warship, the capacity soon proved inadequate in fighter aircraft use where reloading in combat was not possible. Larger drums of 90–100 rounds were developed (particularly in Germany and Japan), but their bulk counted against them and attention turned to belt feeds.

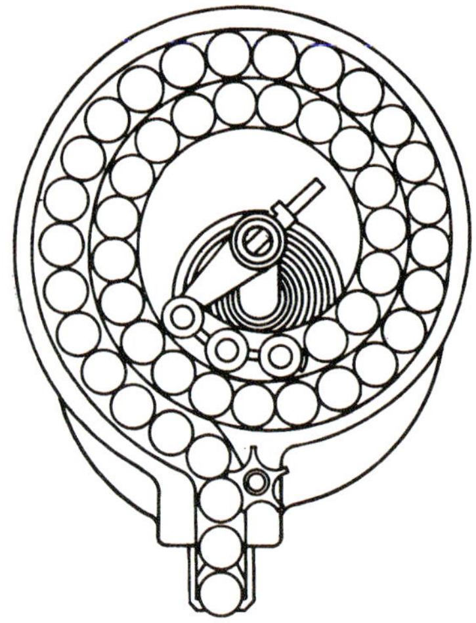

*Drum magazine for Oerlikon 5TG (Chinn)*

One rare variation in drum magazines was the saddle drum: this had two small drums, one each side of the gun, whose feed paths came together in such a way that the gun fired one round from the right drum followed by one from the left and so on. This was mainly intended for flexibly-mounted aircraft guns, to keep the magazine out of the gunner's line of sight.

Examples were the German 20 mm MG 204 which had a 20-round capacity, and the Japanese Ho-1 (Flexible) and Ho-3 (Fixed) which had 15 and 50-round capacities respectively.

Examples of the weight of empty drum magazines from World War II are as follows:

- 2 cm Oerlikon L (20 × 99RB): 15 rounds, 2.0 kg
- 2 cm MG-FF (20 × 80RB): 90 rounds, 11.5 kg
- 2 cm Lb 201/204 (20 × 105B): 20 rounds, 2.5 kg
- 2 cm Ls 201/204 (20 × 105B): 100 rounds, 32 kg
- 20 mm Hispano (20 × 110): 60 rounds, 10.8 kg
- 20 mm Oerlikon S (20 × 110RB): 60 rounds, 14.5 kg (29 kg filled)
- 2 cm Flak 38 (20 × 138B): 50 rounds, 8.5 kg
- 2 cm MG-C 30/L (20 × 138B): 100 rounds, 37.5 kg
- 3 cm MK 101 (30 × 184B): 30 rounds, 20.5 kg

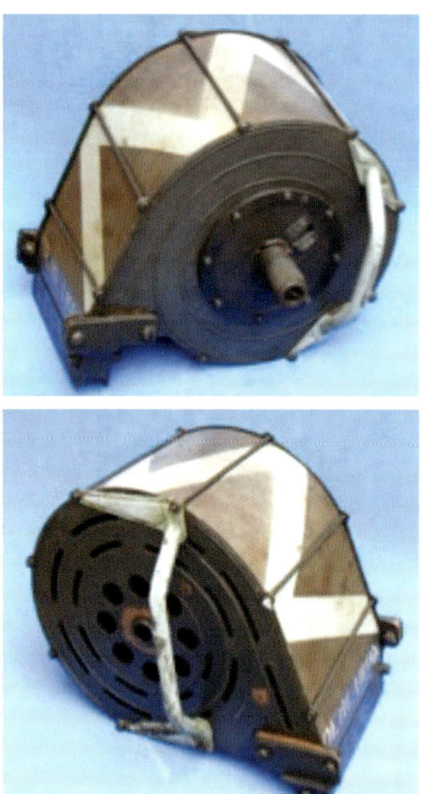

*Drum magazine holding sixty 20 mm × 110RB rounds for the Oerlikon 20 mm Type S (Left Hand Feed) (Koen Bakker)*

*Drum magazine holding 15 rounds, for the 20 mm Madsen M/38 (Koen Bakker)*

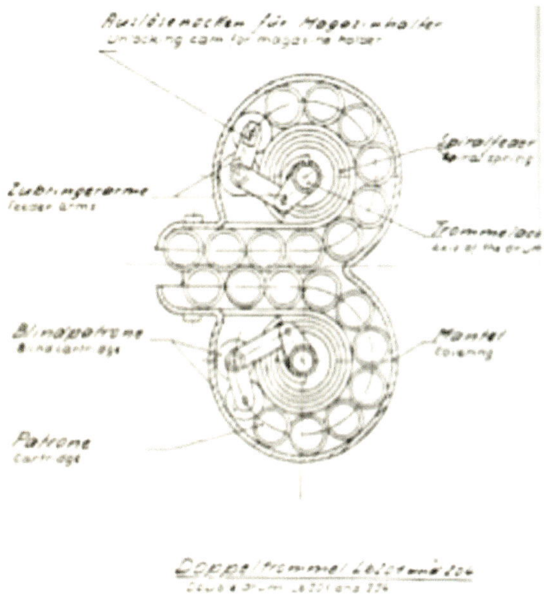


*Double drum for 2 cm MG 204*

One advantage of magazine systems compared with belt feeds is that the ammunition feed maintains a fixed relationship with the gun breech, which saves on the space needed to allow the belt feeds to flex and also enhances reliability. Conversely, the disadvantage (particularly in cramped AFV turrets) is that the bulky ammunition feed depresses as the gun elevates, demanding a lot of space around the breech and limiting the round count available for immediate fire. Well-known examples include the 40 mm Bofors gun in the Swedish CV90 (Combat Vehicle 90), and the Oerlikon 35 mm KDE used in the Japanese M89 Infantry Fighting Vehicle. In these AFV systems, the entire magazine moves with the gun as it elevates. A similar system has been adopted for the ATK XM813, an AFV version of the belt-fed 30 mm MK44 (Bushmaster II) Chain Gun, and the 30 mm South African GI-30.

The CV90 has 24 rounds in the feed mechanism (divided into three sections, so that the gunner can rapidly switch between three different ammunition types) while the Oerlikon KDE has 34 in two separate magazines, one each side of the breech. Obviously, far more ammunition than this is stored within the vehicle, but the feeds have to be reloaded by hand. The XM813 benefits from its smaller ammunition and has two 90-round magazines, one on each side of the gun, but this requires a lot of space to be made available within the turret.

## BELT FEED

Belt feeds were developed very early in the life of machine guns, being utilized by the original Maxim. In this weapon, as with the Vickers derived from it and the scaled-up one pounder (1 pdr) pom-pom version firing 37 mm ammunition, the belt was of canvas with metal reinforcements, and fitted tightly around each cartridge case. As the early military cartridges were rimmed, this meant that they had to be withdrawn to the rear by the loading mechanism before being pushed forward into the chamber. Canvas also had the problem of being affected by moisture, and a wet belt could freeze rigid at low temperatures. Finally, the existence of the emptied belt flapping about clearly proved a nuisance in aircraft applications, prompting the introduction of the disintegrating link belt.

As its name suggests, the disintegrating link belt is formed of individual steel links that clip the cartridges together. The cartridges form an integral part of the belt, and the links are designed to fall free of each cartridge as it is chambered. Initially devised in Germany for aircraft MGs, this was quickly introduced into British service in 1916 (the Air Service Mk I for .303-inch ammunition). The first application of a disintegrating steel link belt to cannon appears to have been the Mark II*C version of the British 2-pounder pom-pom naval AA gun, probably some time in the 1920s, which came in belt lengths of 14 rounds (earlier versions of the gun used a 25-round fabric belt). Not all steel belts have been of the disintegrating type, and some (e.g. for the German MG 151) were designed to be convertible between disintegrating and non-disintegrating depending on the installation requirements. Disintegrating-link belts can be extended by clipping on additional sections, non-disintegrating generally cannot. Belts may be fitted with special start and end tabs.

Belt design has to take into account a number of factors. The clips must be strong enough to maintain their grip on the cartridges while not being so tight that cartridge extraction is hindered (not always achieved, particularly with aircraft gun belts subjected to high-G forces in violent manoeuvering). The belt must obviously be flexible in the vertical plane but usually also needs to permit some degree of curvature in the horizontal plane to allow for the needs of different installations and for some gun movement relative to the ammunition container (the 2-pounder links shown below clearly did not permit such flexibility). This was of particular importance in aircraft turret mountings, in which ammunition was often contained within the fuselage, so the belt feed had to allow for both gun elevation and turret traverse. It is common for belt feed systems to incorporate belt guides or chutes to ensure that the belts are fed smoothly into the gun.

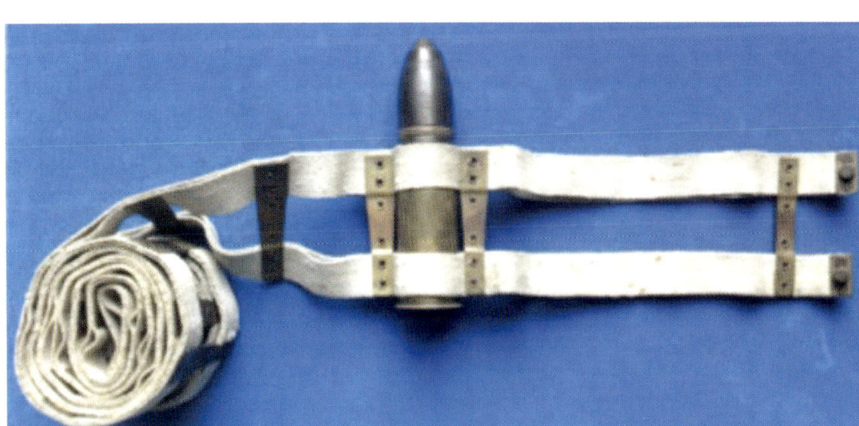

*1-pounder (37 × 94R) fabric belt with brass spacers for 20 rounds (Koen Bakker)*

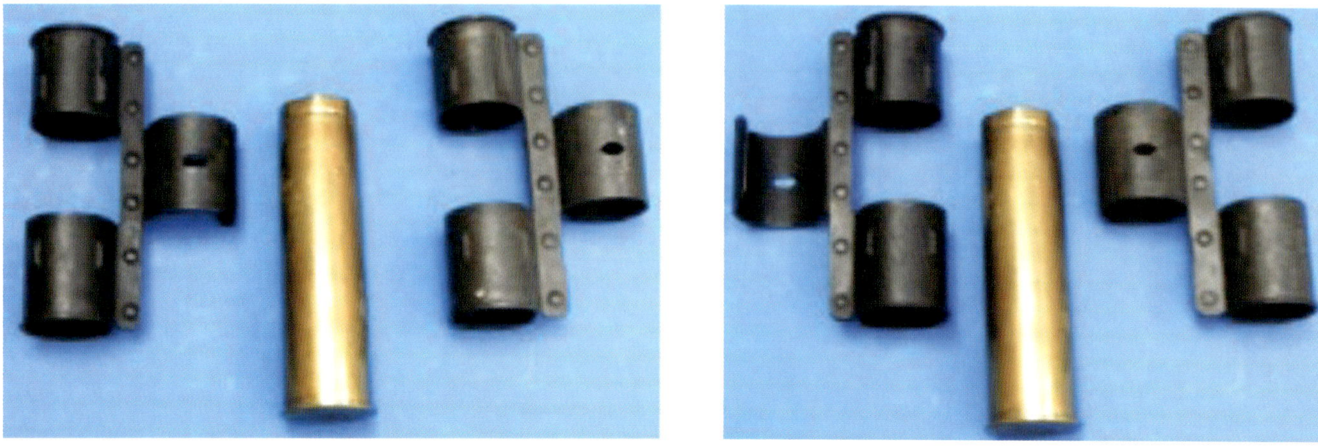

*2-pounder pom-pom link and endlink: 40 × 158R (Koen Bakker)*

Some data on the flexibility of belts can be gauged from the German data on the 2 cm MG 151/20:

With the belt straight, the rounds are spaced 30 mm apart
With the belt lying flat, maximum horizonal curvature with the cartridges base inwards results in the bases forming a circle 330 mm in diameter
With the belt lying flat, maximum horizontal curvature with the projectiles nose inwards results in the projectile noses forming a circle of 210 mm diameter
With the belt twisted (warped), a belt of 10 rounds will warp at 90 degrees

Joining individual links and ammunition into a linked ammunition belt is achieved in different ways, with the following options:

- link and link with round;
- link in link with round;
- link to link connection (e.g. pins, rings etc.);
- hook in eye with round.

Examples of these are shown below.

Another aspect of using links is the positioning of the round in the individual link:

- using a groove in the cartridge case;
- using the extractor groove of the cartridge case;
- using a groove on the projectile;
- using the shape of the round;
- using linking equipment.

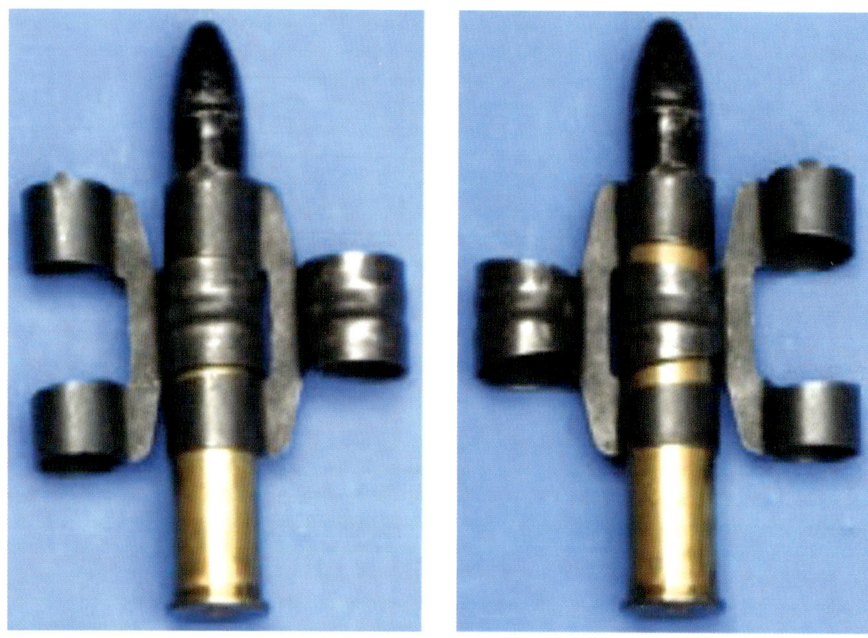

*ShVAK (20 × 99R): link and link with round; rounds are positioned in the link using the shape of the round (Koen Bakker)*

Delinking before firing can be done by pushing the round forwards, backwards or downwards. The method of delinking influences the design of the individual link.

For practice purposes lightweight rounds made from plastic are sometimes connected using plastic links.

Knowing the weight of belt links provides a basis for comparison with the magazine weights provided earlier in this section:

2 cm MG 151/20 (20 × 82): 0.0175 kg
3 cm MK 103 (30 × 184B): 0.12 kg
3 cm MK 108 (30 × 90RB): 0.11 kg
5.5 cm MK 212 (55 × 175RB): 0.5 kg

A couple of other data points, from Russia: 20 mm ShVAK link = 0.038 kg, 23 mm VYA = 0.094 kg.

The power to move the belt through the gun can be provided in several ways. In smaller cannon it is often provided by the movement of the bolt, which drives a mechanism to pull the cartridges into place. Alternatively, in cannon fitted to recoiling mountings the recoil movement is often used. Another variation, demonstrated by the Rheinmetall-Borsig MK 103, involved the drive being taken from the recoiling barrel. Gas can be tapped from the barrel to drive the belt (as distinct from the gun mechanism), or

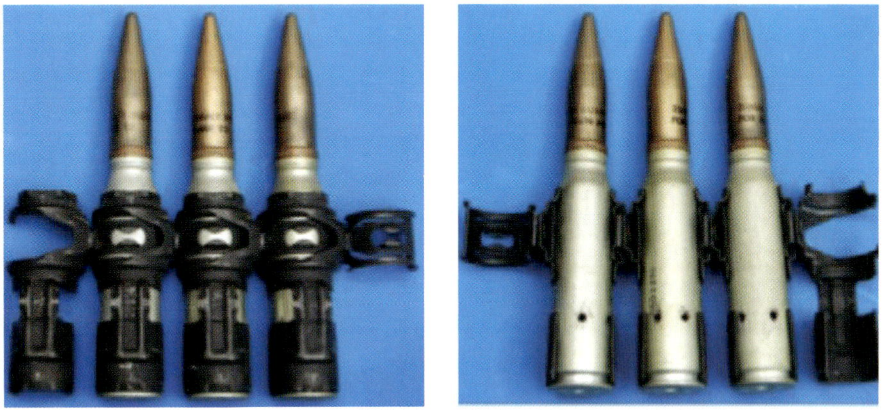

*USN Mk 12 (20 × 110) Mk 2 series link; link in link with round; rounds are positioned in the link using the extractor groove (Koen Bakker)*

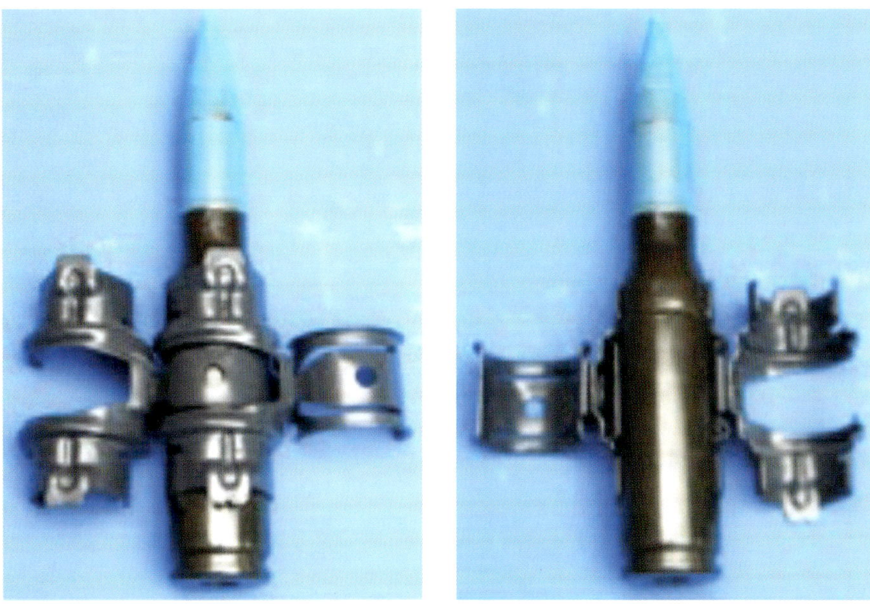

*Oerlikon KBA (25 × 137): link in link with round; rounds are positioned in the link using a groove in the cartridge case, just below the shoulder (Koen Bakker)*

*Links for M10 and M9 (37 × 145R and 37 × 223SR respectively): link in link with round; the links fit around the neck of the case and are positioned using a groove in the driving band. This allows the same link to be used for two very different cartridges (Koen Bakker)*

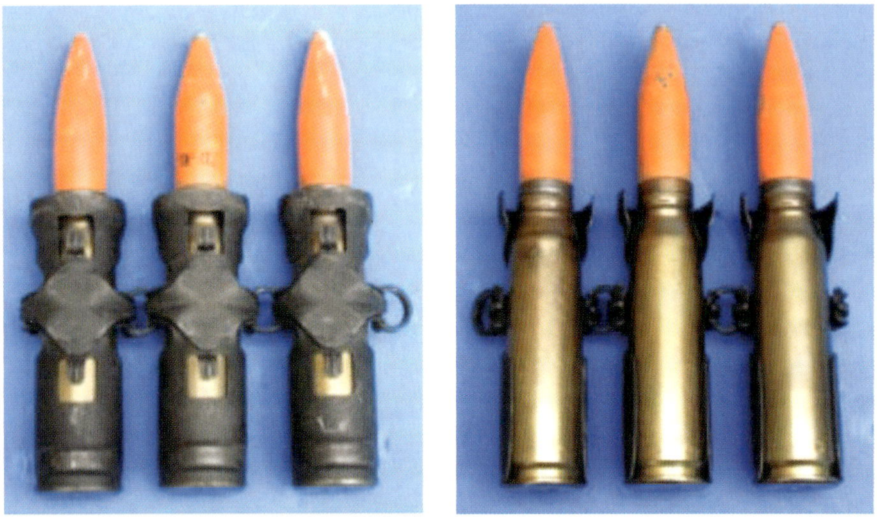

*GIAT AME 621 F1 (20 × 102) link to link connection; rounds are positioned in the link using the extractor groove. This link is compatible with the M12 for the M39 revolver cannon and variants (Koen Bakker)*

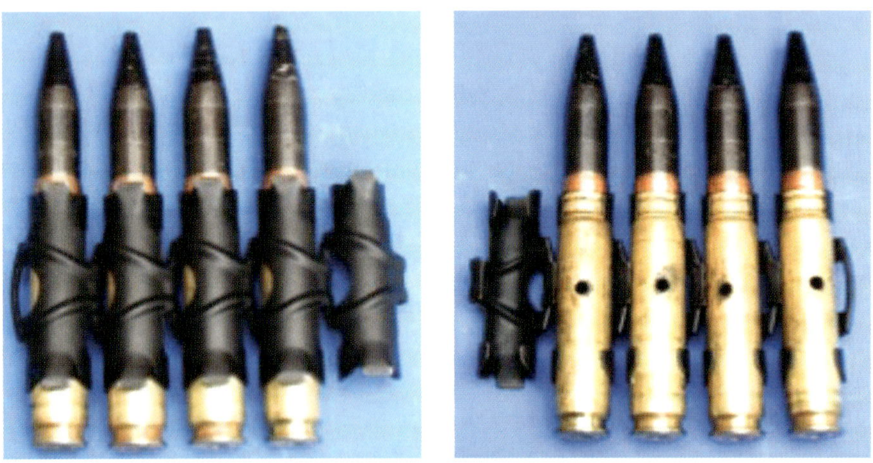

*NR-23 (23 × 115): hook in eye with round; round are positioned in the link using the shape of the cartridge (Koen Bakker)*

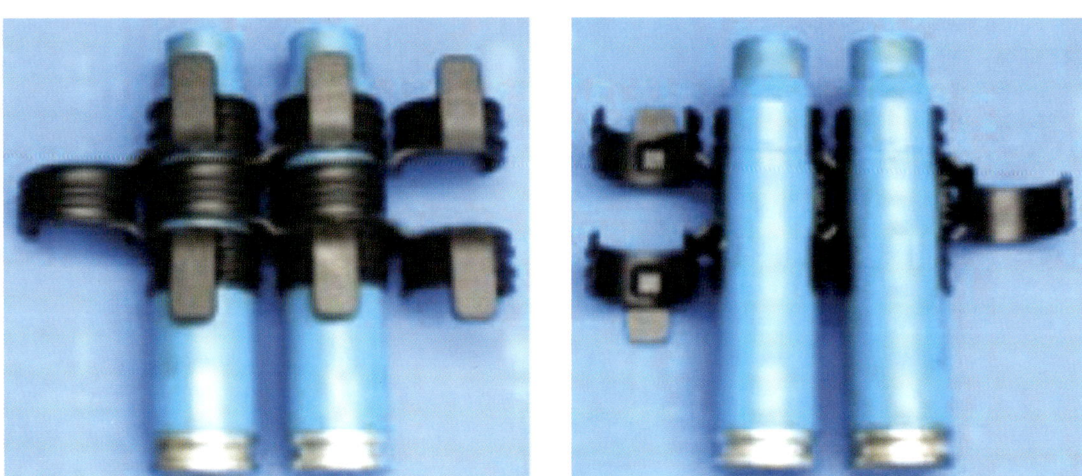

*Bushmaster II / Mk 44 (30 × 173) plastic link variant: link and link with round; rounds are positioned in the link using a groove in the cartridge case. Ammunition and link were developed by Nammo Bakelittfabrikken (Koen Bakker)*

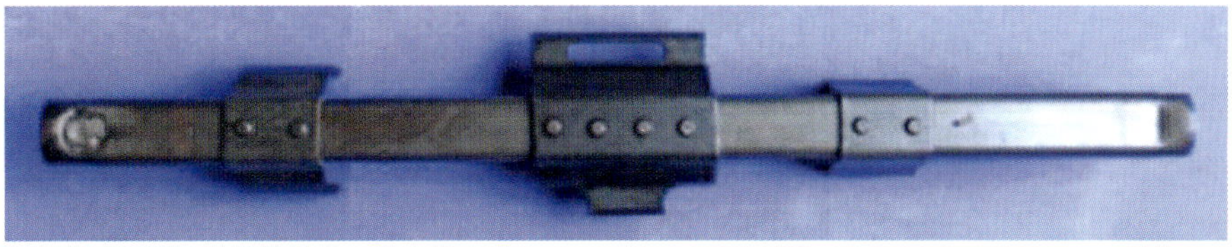

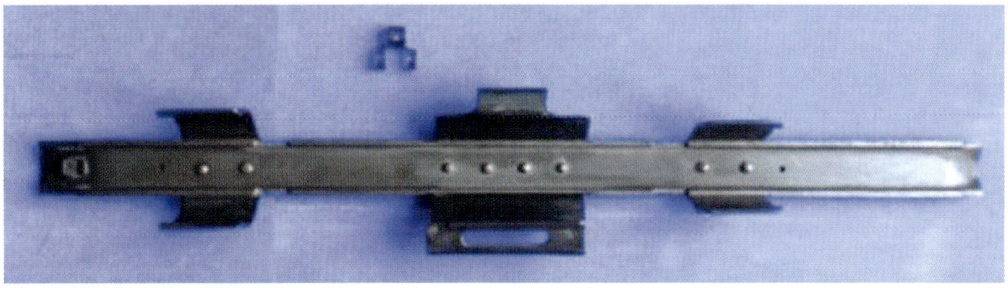

*Mauser MK 214A aircraft cannon (50 × 420R): the huge link needed for the massive ammunition is emphasized by the 7.62 × 51 M13 link in the lower picture (Koen Bakker)*

a separate electric or hydraulic motor can do the work. An interesting hybrid was the Chatellerault-based drive for the 20 mm Hispano; this was driven by a pre-wound spring, the tension of which was maintained by the gun recoil movement. The advantage of this complexity was a smooth belt-pull, unaffected by the jerkiness of the gun motion. One unusual method, tried (apparently unsuccessfully) in the Ikaria MG-FF, was to utilize the force of the gas escaping from the muzzle.

The advantage of belt feeding is that large quantities of ammunition can be held ready to fire, limited only by the space available to store it and by the power available to overcome the belt's inertia. This has made belt feeding by far the most common method used by the kind of heavy weapons described in this book; but there are also disadvantages which have led to the use of alternatives in certain circumstances.

*30-round M6 endless belt for 37 mm M4*

The main problem in systems which use the gun's action to drive the mechanism is that this places certain limitations on the arrangements for holding the belt. The power which can be tapped from the gun mechanism cannot be expected to pull a heavy belt very far, and even in the best of circumstances the rate of fire of a belt-fed gun is likely to be lower than that of an equivalent magazine-fed weapon. There are also the problems in aircraft installations of disposing of the belt links and fired cartridge cases (the bulges underneath the nose of the 1950s Hawker Hunter jet fighter were containers for discarded links), and in some cases of the belts jamming or even pulling apart under G forces (which the 20 mm Mk 12 installation in the USN F-8 Crusader was noted for). Additional problems occur in installations where the belt feed has to cope with a wide range of gun elevations, which means that the belt has to flex over a considerable arc, which can result in significant feeding problems.

In the largest calibres of automatic weapon a belt would simply be too unwieldy, although a variation in the form of a continuous loop or captive belt (i.e. forming an endless chain) was used in installations such as the USAAF's 37 mm M4, the IJA's 37 mm Ho-203 and 57 mm Ho-401, and the autoloading versions of the 5 cm PaK 38 anti-tank gun developed for the Luftwaffe.

## LINKLESS FEED

These difficulties have led to the development of linkless feeds for use with heavy automatic cannon or where very high rates of fire are required. These are typically power driven, the best examples being those for the rotary cannon used by many American fighter aircraft. These have mainly taken the form of a large helical pan or drum with the cartridges pointing inwards, and have had capacities of up to 2,048 rounds (an

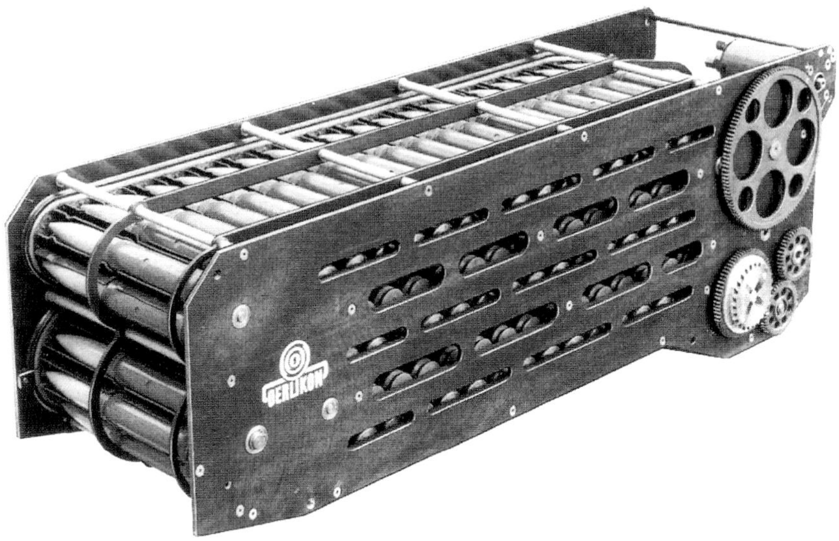


*Linkless feed for BK 27 cannon installed in Eurofighter Typhoon*

optional 20 mm M61A1 installation for the General Dynamics F-111 aircraft) or 1,174 rounds for the integral 30 mm GAU-8/A gun which has the Fairchild A-10A built around it. The ammunition feed is driven by the same motor that powers the gun, ensuring that the cartridges are fed into the weapon at the appropriate rate. Because of the difficulty of disposing of empty cartridge cases from a high speed aircraft, the ammunition feed slots the fired cases tidily back into the magazine. More compact linkless feed magazines have since been introduced, not just in fighters but also for the 1,075-round capacity 30 mm M230 Chain Gun in the AH-64 Apache helicopter.

Such large magazine capacities pose obvious problems in reloading. These have been tackled in USAF fighter guns by the Universal Ammunition Loading System which is a mobile unit specifically designed to reload M61 magazines (and remove the fired cases). It contains its own pneumatic drive and can reload a 515-round magazine in four minutes.

An alternative approach to reloading has been adopted for the M230 Chain Gun in the AH-64D Apache Longbow helicopter. The 'Sideloader' unit is a mechanical system built into the aircraft that enables one man to load 1,075 rounds, in 11-round strips, in less than 15 minutes. This compares with the AH-64A which is reloaded by three men, using ground support equipment, taking more than half an hour.

Ammunition feeding is complicated by the fact that the majority of heavy automatic weapons are designed to recoil in their mountings in order to spread out the recoil force and minimize vibration. A clip, or a box or small drum magazine, can recoil with the gun, but it is not feasible to arrange this for belt-drives or other large ammunition capacity systems. It is therefore necessary for designers to find some way of transferring ammunition from a stationary supply to a moving gun.

One unique feed system is that developed for the Anglo-French 40 mm CTWS, designed initially for use in AFVs. The cylindrical cartridges are stored in the turret sideways, at right-angles to the direction of fire. They are fed through the gun trunnions and into the chamber, which then rotates through 90 degrees to line up with the barrel for firing. Since the trunnions are the one part of the gun which does not move with changes in elevation, the power-driven feed is fixed in place which should make it very reliable.

It should be noted that while some guns have been developed around one specific type of feed system, others have used more than one type of system, depending on the particular gun application.

# UNCONVENTIONAL SYSTEMS

Almost all ammunition to have seen service is conventional in layout, with the projectile at the front, followed by the propellant, with a metal cartridge case containing the propellant and gripping the projectile close to its base. Various attempts have been made to develop alternatives to this arrangement. The motivations for this include achieving lighter and/or more compact ammunition, and/or repackaging the components to achieve a different shape more suited to a novel design of gun mechanism plus, in the case of caseless/combustible case rounds, eliminating the need to extract and eject a fired cartridge case.

As such ammunition is invariably associated with an unconventional gun design, the guns will be considered alongside the ammunition in this chapter.

Many different approaches have been tried, but until recently have met with very little success. The most prominent of these can broadly be categorised as follows:

1. Telescoped or cased telescoped (CT) systems
2. Hughes Lockless system
3. Folded ammunition
4. Ammunition for open-chamber guns: the Tround
5. Split-breech guns: the Nutcracker
6. Reverse-loading systems
7. Multi-barrel salvo-firing guns
8. Two-piece ammunition: SLAmmo
9. Electrothermal-chemical propulsion
10. Caseless or combustible cased ammunition
11. Liquid propellant
12. Recoilless ammunition
13. USN Mk 11 original type.

The above variants are not all mutually exclusive, and they can occur in combinations: e.g. telescoped rounds can have combustible cases, and/or incorporate recoilless characteristics.

## TELESCOPED AMMUNITION

From the mid 1950s to the 1990s the US in particular carried out much research into telescoped cannon ammunition. This has the projectile entirely contained within the cartridge case and with most of the propellant wrapped around the projectile instead of behind it. The resulting cartridge shape is usually a plain cylinder, notably shorter but somewhat fatter than a conventional round. This ammunition, also known as 'cased telescoped' or CT, achieves more compact dimensions to save magazine space. In some examples, the case is combustible or even dispensed with altogether.

All such ammunition faces a technical problem, in that the projectile has to travel a greater distance to enter the leade (the part of the chamber immediately before the barrel rifling starts) so there is the risk of gas escaping past the projectile and up the barrel before the projectile can get there and form a gas-tight seal. It is therefore usual to have two-stage propellant ignition: a small quantity of powder occupies a space between the primer and the projectile, which on firing drives the projectile out of the case and into the leade. As the projectile moves forward, the burning gases ignite the main propellant.

Some of the advantages of this kind of configuration were spelled out in a report

*Telescoped rounds: 20 mm AGT; 30 mm AMCAWS; 40 mm CTAI; 20 mm Tround; 25 mm GAU-7/A with projectile; 30 mm Hughes Lockless; 30 mm Folded*

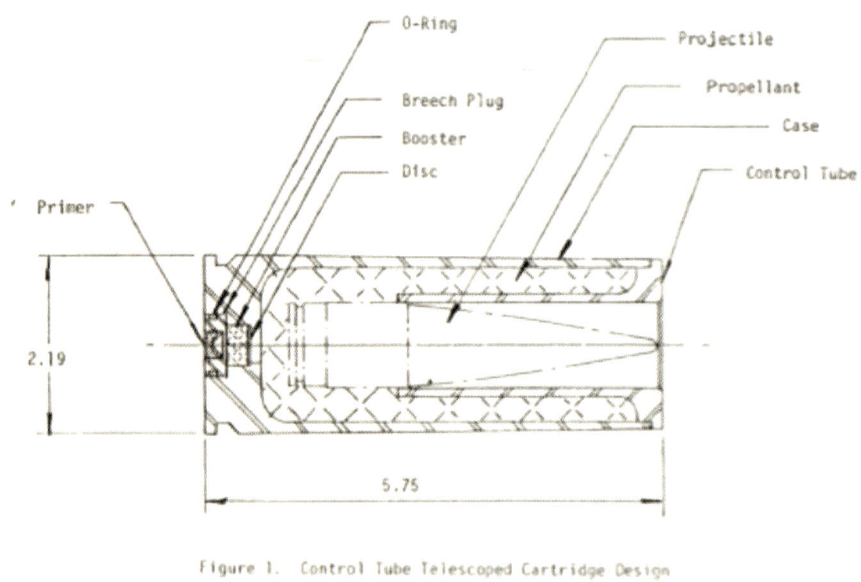


*TK*

on the Control Tube Telescoped Cartridge (CTTC) development programme conducted by GARD Inc, a subsidiary of the GATX Corporation of Illinois, in 1976–77. Work was carried out in conjunction with the US Air Force Armament Laboratory at the Eglin AF base.

Five key design features of CTTC listed were: a metal case, to provide strength and chamber sealing; a control tube (or inner case) which positively guides the projectile into the barrel and prevents any possibility of gas blow-by before the barrel is obturated by the projectile; a standard percussion or electric primer; and usually loose rather than moulded propellant as the latter is subject to fractures and variable burning rates.

Some advantages given for the CTTC: the projectile does not require a crimp groove, so its wall thickness can be reduced; the driving band may be located further to the rear, reducing the risk of balloting; less of the projectile is exposed to the chamber pressure, so the projectile body can be thinner. This all adds up to a reduced projectile body weight (with more room for HE); the reduced overall projectile weight permitting an increase in MV, a shorter flight time, and a greater hit probability.

In addition, a fully telescoped round is shorter and more compact, reducing the volume needed for the gun mechanism, the feed system and ammunition storage by up to 40 per cent. The shorter round also leads to shorter ramming and extraction strokes, so a higher RoF becomes possible, plus several gun components can be shortened and lightened, reducing overall gun weight.

Following the problems experience by the GAU-7/A programme (see Combustible Case systems below) the CTTC contract objective was to develop a fully-telescoped round in which ballistic performance was reproducible at temperatures between –54° and +71°. Specific goals when fired through a 25 mm calibre barrel with a length of 213 cm were a mean MV of 1,189 m/s (3,900 fps) at a peak chamber pressure less than 448 MPa (65,000 psi).

The specification of the CTTC round included: the 25 mm GAU-7/A TP projectile (194 g); 122 g double-base propellant similar to that used in GAU-8/A round; an overall length of 146 mm and a diameter of 55.6 mm. The case had an extractor groove and was slightly tapered, but this could be altered if required. After various development changes, this round successfully met the goals.

An alternative approach to solving the GAU-7/A's problems was undertaken by Brunswick, also in the early 1970s. A 1975 report on an experimental telescoped plastic case showed moderate success in a single-shot test barrel, the case notably withstanding chamber pressures of up to 690 MPa (100,000 psi). The critical element – the gas seal between the chamber and the barrel – failed at pressures greater than 550 MPa (80,000 psi).

Not all CT rounds have been perfectly cylindrical, the AMCAWS-30 being an exception. This 30 × 166 round was developed for the Army Materiel Command Automatic Weapon System 30 mm gun (also known as the Advanced Medium Caliber, or AMC-30) project of the mid-1970s. It was unusual among telescoped ammunition for two reasons: it used a tapered case, smaller at the base (51 mm) than the mouth (54.7 mm) and designed to be pushed backwards into the separate chamber; and it was designed to operate in a 'stop mode'.

The stop mode addressed a basic problem with telescoped ammunition, which is to get the projectile into the barrel before igniting the main propellant charge, both to create space for combustion and to prevent the escape of propellant gases past the projectile. The AMCAWS-30 only

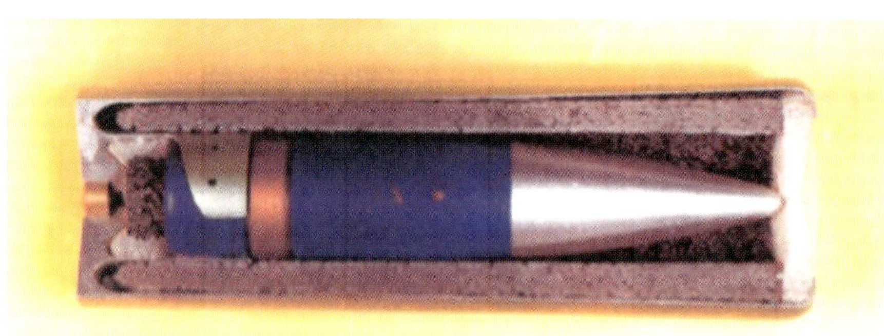

*Sectioned 30 mm AMCAWS/AMC-30 CT consolidated charge round: on the left can be seen a small initial charge in front of the primer, to get the projectile moving up into the barrel. Around the base of the projectile up to the copper driving band is a short silver-coloured tube, partly cut away. This locates the projectile in the case and only allows combustion to reach the compressed main charge (which forms a lining within the case) once the projectile has begun to move forwards (photo courtesy of Robert Leiendecker)*

gave the projectile enough initial impulse from the primer/booster to propel the projectile into the barrel where it would stop. The main charge then ignited and set the projectile in motion again.

Such behaviour was generally avoided in other CT rounds (in which projectile movement was continuous) because it resulted in long and uncertain action times, which were unacceptable for high-RoF externally powered guns. The Army was only interested in low-RoF self-powered guns so this idea appeared promising. The prototype AMCAWS-30 weapon fired ten-round bursts at a nominal rate of 120 rpm, with the second prototype having a nominal rate in excess of 400 rpm. Unfortunately the problem of getting the projectile to enter the barrel and stop consistently turned out to be too difficult so the idea was abandoned. The aluminium case held 194 g of consolidated propellant, generating an MV of 1,120 m/s at a chamber pressure of 360 MPa; ballistic characteristics which were slightly better than the GAU-8 rounds.

In 1996 the Inspector General of the US Department of Defense issued a report on DoD Cased Telescoped Ammunition and Gun Technology Programmes. This covered the following projects:

1. 20 mm AGT (1982–1992): USAF Advanced Gun Technology: 20 × 140, 42 mm diameter; total weight 453 g; propellant 112 g, projectile 90.7 g; MV 1,500 m/s, ME 102,000J.
2. 25 mm CTCA (1991–93): USAF Cased Telescoped Combat Ammunition: 25 × 152, 46.2 mm diameter, 179 g propellant pushing 149 g projectile at 1,500 m/s, ME 168,000J.
3. 30 mm COMVAT: 30 × 207 mm, 54 mm diam, 1,043 g cartridge, 304 g propellant, 245 g projectile at 1,460 m/s, ME 260,000J.
4. 45 mm COMVAT (1986–91): 45 × 305 mm, 70 mm diameter, 3,175 g cartridge, 615 g propellant firing 755 g projectile at 1,350 m/s, ME 688,000J.

The Inspector General's report concluded that cased telescoped ammunition is ballistically inefficient with inherent performance, weight, volume, and cost disadvantages. The report argued that CTC cases are wider so contain far more propellant generating a very high MV – this is ballistically inefficient, more complex and costly, adds weight and bulk, and brings heating and rapid barrel wear problems. Barrel obturation is a major problem, leading to blowby of the propellant gases, balloting of the projectile, and muzzle debris. He stated that CTC is not necessary; conventional cannon are satisfactory and much cheaper to develop.

While some of the Inspector General's conclusions have some validity (the technical issues over obturation and balloting, and the higher cost) most of them do not. There is nothing inherent about the considerable quantities of propellant used in the listed cartridges (the cause of the other complaints); that was the choice of the designers, and could be avoided simply by dialling back the high-velocity requirement, thereby reducing the quantity of propellant needed and resulting in slimmer cases, avoiding the bulk, inefficiency, over-heating and barrel wear (problems which would also occur with conventional systems of the same performance).

It should be noted that some apparently telescoped ammunition actually isn't. The best example of that to see (limited) service was the Russian Rikhter R-23, designed for a reverse-loading revolver in which ammunition was fed backwards into the chamber. (The USA had a more conventional – but still strange – approach to the same reverse-feeding requirement, in the 30 mm T168 revolver.) The R-23 ammunition looks peculiar, partly because it is tapered the 'wrong' way, and partly because the case is extended forwards to enclose the projectile. However, the propellant is all contained behind the projectile and the overall length of the round is the same as a conventional alternative, so it doesn't meet the definition of 'telescoped'. The same can be said of the British experimental 'Nutcracker' ammunition for split-breech guns (see below).

### 45 × 305 COMVAT

The COMVAT project (Army Combat Vehicle Armament Technology Programme) formed part of US CT research from 1986–1991 and was intended to result in a more powerful replacement for the 25 mm M242 Bushmaster cannon fitted to LAFVs. Initially a 30 mm round was developed, but this was superseded by the 45 mm version. COMVAT followed on from an earlier programme called Combat Vehicle Armament System Technology (CVAST). ATK was the prime contractor, working with ARES who developed the 45 mm XM295 cannon.

*40 mm and (sectioned) 45 mm CTAS, courtesy MoD Shrivenham*

Dimensions of the cylindrical case are 70 mm diameter and 305 mm long; the case weighs 1.991 kg and the round weight is 3.175 kg (with long projectile). Two different APFSDS loadings are known, with short and long projectiles:

Short projectile: Flight projectile length 228 mm; diameter 11.6 mm; sabot length 133.5 mm. Total launch mass 562 g; plastic bands 4 g; flight projectile 320 g; 3-petal sabot weight 238 g.
Long Projectile: Flight projectile length 264 mm; diameter 13.3 mm; sabot length 151 mm. Total launch mass 750 g; plastic bands 4 g; flight projectile 478 g; 3-petal sabot weight 268 g.

The long projectile was fired at 1,350 m/s, giving a muzzle energy (including sabot) of 688,000J.

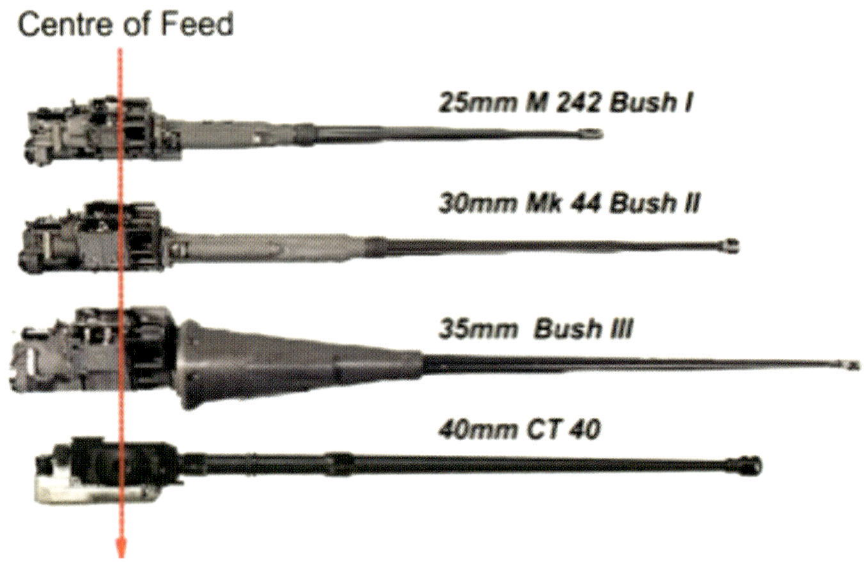


*A comparison between the CT 40 and rivals, showing the very compact mechanism (CTAI)*

## 40 × 255 CTAI

The Franco-British 40 mm CTAS (Cased Telescoped Armament System) is set to make history as the first cased-telescoped system to see service, having been selected by the British Army (for the Ajax reconnaissance vehicle) and the French Army (for their new *Engin Blindé de Reconnaissance et de Combat*). The development history of the CTAS is covered in the International section of Part 4, with details of predecessors in the US (COMVAT) and France (45M911) sections.

The ammunition uses a cylindrical steel cartridge case measuring 65 mm in diameter and 255 mm long, with the 40 mm projectiles completely contained within the case and surrounded by Nitrochemie ECL propellant. This configuration makes the CTA rounds smaller in volume by a factor of about 35% compared with conventional rounds. The case contains a telescopic insert at each end. As the round is fired, these telescopic inserts expand outwards slightly to form a complete seal with the barrel. The projectile is supported at the base of the case, and also at the front by an obturating sabot in combination with an annular aligning ring, all made of moulded plastic.

*Two options for ammunition feed arrangements (CTAI)*

One of the advantages of cylindrical CT ammunition is that it permits different loading mechanisms, as exploited by the 40 CTAS: the chamber is separate from the barrel and pivots horizontally through 90° after firing; the next round is then pushed into the chamber, simultaneously pushing out the fired case; and the chamber then pivots back into line with the barrel for firing. This simple system avoids the usual need for the fired case to be hauled out of the chamber by an extractor, exerting enormous tension on the case rim which therefore has to be very strong and heavy. The CT case can be more lightly constructed. On the other hand, such separate-chamber mechanisms run into another problem (shared with conventional revolver cannon); sealing the gap between the chamber and the barrel, which requires careful attention being paid to the design of the breech and the front of the case to create a seal for the brief instant of firing.

Up to seven different loadings have been proposed so far for the CTA family:

- APFSDS-T, with an MV of 1,500 m/s and an RHA penetration of greater than 140 mm at 1.5 km.
- GPR-PD-T (General-Purpose Round Point-Detonating Tracer), fitted with a base impact fuze, with a weight of 1,000 g and a muzzle velocity of over 1,000 m/s, generating a muzzle energy of 500,000J. The tracer element burns for more than six seconds.
- GPR-AB-T (General-Purpose Round Air-Burst Tracer) is similar to the above but with a programmable

Sectioned 40 mm CTAS ammunition

Sectioned 40 mm CTAS AAAB round

base-mounted electronic fuze, designed to achieve a fragmentation pattern biased to the sides, in order to engage troops behind cover (in defilade). The nose of the GPR projectiles is domed to maximize the HE content and increase fragmentation in airbursts.

- AAAB, or A3B (Anti-Aerial Air-Burst) which contains more than 200 cylindrical tungsten pellets with a detonating booster and appears from the description to be comparable with the Oerlikon AHEAD.
- TP-T (Training Practice Tracer), ballistically matched to the GPR-PD-T.
- TPRR-T (Target Practice Reduced Range Tracer), designed to have a range of less than 4,500 m.
- KETR-T (Kinetic Energy Training Tracer), ballistically matched to the APFSDS-T at 1,500 m and safety-limited to 4,500 m.

## CT SYSTEMS FROM OTHER COUNTRIES – RUSSIA

In the mid-1980s an experimental 45 mm CT gun was developed in Russia as a potential replacement for the 30 mm 2A42 and 2A72. The case dimensions are not known, but an HE round fired a 1.3 kg projectile (containing 170 g HE) at an MV of 850 m/s; an APDS round fired a 670 g projectile (with a 420 g penetrating core) at 1,640 m/s, penetrating up to 150 mm armour plate. The RoF was 150–200 rpm.

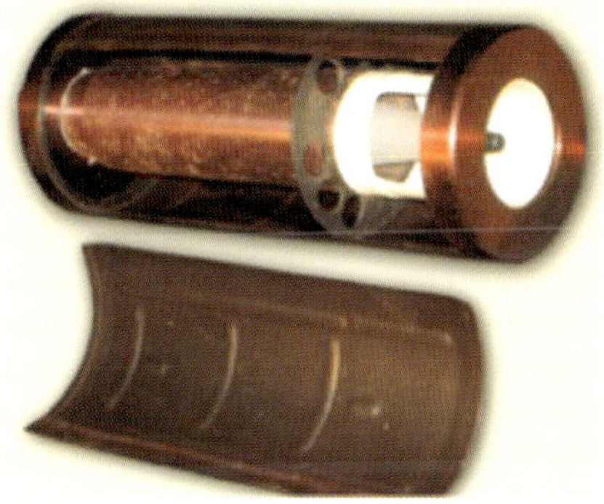

Two photos of Russian CT ammunition from the ECRA Bulletin

## CT SYSTEMS FROM OTHER COUNTRIES – JAPAN

During the period 2006–2011 a 40 mm cannon firing what was claimed to be a new CT round was developed in Japan. Little is known of the gun and ammunition except for a YouTube video. The cartridge is approximately 330 mm long and 55–60 mm wide, with the 40 mm projectile contained in a sabot. It barely qualifies as CT, as nearly all of the propellant is behind the projectile. The Japanese case appears to have very similar dimensions to the experimental 50 × 330 Supershot round developed in Germany in the 1980s, which used a necked-out and lengthened 35 × 228 Oerlikon case and was designed to be fired from suitably rebarrelled and modified 35 mm guns. Other web sources mentioned that a 50 mm version of the Japanese gun was considered, which further supports the idea that this new round may be derived from the Supershot. If this is correct, it therefore seems likely that the Japanese gun is a modified Rheinmetall (Oerlikon)

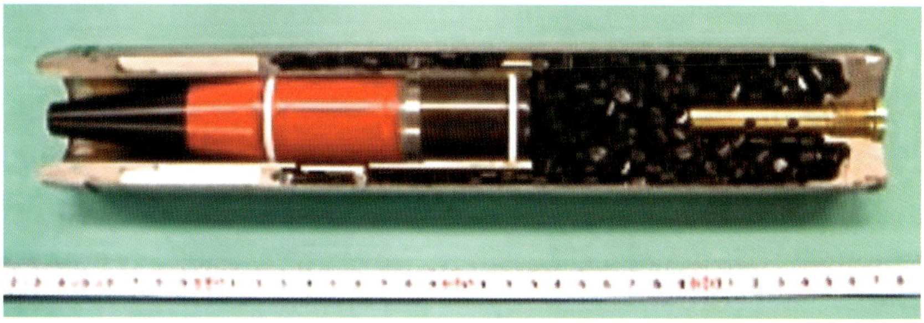

Sectioned Japanese 40 mm round, from internet video

Photo of Chinese CT rounds from the Internet

40 mm Lockless compared with BK 27 (courtesy of David Birkin)

35 mm gun: the rate of fire measured from the video is 560 rpm, close to the 550 rpm of the standard Oerlikon KDA models (which Japan has in service in the twin-gun Type 87 SPAAG).

### CT SYSTEMS FROM OTHER COUNTRIES – CHINA

In November 2016 NORINCO of China showed a cased telescoped weapon system of 40 mm calibre. All that is known about the ammunition at the time of writing is that the HE loading (designated CS/BTE3) has an MV of 1,000 m/s while the APFSDS (CS/BAA3) achieves 1,450 m/s and is claimed to penetrate 130 mm RHA at 1,000 m.

Compared with the Franco-British CTAI 40 mm ammunition the proportions are different; these rounds appear to be longer and/or more slender. Judging by the gun photos, though, a similar sideways-loading configuration appears to be used.

It is not known whether this is intended for the Chinese Army, for export, or just as a technology demonstrator.

## HUGHES LOCKLESS SYSTEM

Another unusual ammunition configuration which has failed to achieve service status was the Hughes Lockless (dating from the early 1970s), based on a cased-telescoped round which took the form of a flat polymer box in order to suit a gun design that involved the rounds being slotted in sideways. The Lockless system was so-called due to its unique mechanism, which did not have the usual bolt or breechblock locking to the barrel. When a round was pushed sideways into the fixed breech, in line with the barrel, an external sleeve slid forwards to enclose the breech and support the round as it was fired. The sleeve was then retracted, permitting the next round to push the fired case sideways out of the gun.

The interior of the box-shaped cartridge case has three longitudinal compartments: the inner one contains the projectile, while most of the propellant is in the compartments to either side; as in cylindrical telescoped rounds, there is a small initial charge behind the projectile. When the gun is fired, the primer ignites the propellant behind the projectile, sending it out of the case. As the projectile leaves the case, the main charges are ignited, driving it up the barrel.

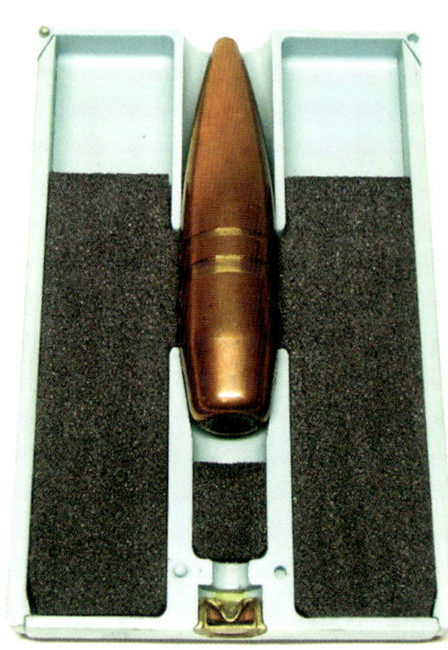

Sectioned 12.7 mm Lockless

Various calibres from 5.56 mm to 40 mm were made (the small ones were dubbet 'chicklets' after the chewing gum sweets) with most effort being put into the 30 mm system, but none was adopted. The initial 30 mm Mod 1 cartridge used loose propellant and fired a 324 g projectile at 1,067 m/s. The Mod 2 used consolidated propellant to drive a 356 g projectile at the same velocity (generating 203,000J ME), despite having a slightly shorter case; 157 mm long, 78 mm wide and 34 mm deep.

The 40 mm was sponsored by the Air Force Armament Laboratory, Eglin AFB. This configuration allowed a 136% increase in energy per pound of ammunition over the standard 40 mm Bofors with no increase in cartridge volume. Firing was done with a single shot test figure using a standard Bofors barrel. The 40 mm round measures 228 mm long by 133 mm wide and 44 mm deep.

The 30 mm gun weighed 113 kg, was 280 cm long, and used a long-recoil method of operation with a pneumatic feed. Its nominal rate of fire was 350 rpm, although in tests it managed 200–230 rpm.

## FOLDED AMMUNITION

This was a project to minimize the length of the gun action by making the ammunition as short as possible, achieved by locating the propellant in a container next to the projectile, with a channel for propellant gas from the base of the propellant chamber across to the base of the projectile. The inventor of the concept was Andrew Grandy, working at the Frankford Arsenal in the early 1970s. He was primarily interested in 5.56 mm applications but Folded Ammunition was made in a wide range of calibres up to 40 mm.

As the propellant burns, the gas passes through the channel before reaching the base of the projectile, in theory permitting more complete combustion. In the 30 mm example examined, the primer pocket is midway between the projectile and the propellant container, on the channel between them. Combustion therefore begins with the propellant which is closest to the projectile, rather than at the other end of it as with a conventional round.

An advantage Folded Ammunition had compared with Cased Telescoped is that the projectile was positioned in the correct place in the chamber before firing, like a conventional round, so that it did not have to 'jump' from the cartridge case to the barrel. The gun used a push-rod to extract the cartridge case instead of a conventional extractor, thereby eliminating potential extractor breakages.

In the 30 mm version, the goal was to launch a 324 g projectile at 1,100 m/s with a peak chamber pressure of just 380 MPa. Honeywell developed the ammunition and ARES the gun system. A comparison with the 30 × 173 GAU-8/A round showed the following reductions: length 41%, volume 24% and weight 12%. The shape of the case evolved during development, with a folded version of the 30 mm WECOM being further reduced in volume by giving the propellant container a crescent-shaped rather than circular cross-section and partly wrapping it around the projectile.

The configuration of the cartridge facilitated the development of a recoilless variant, as recoil-balancing gas could be allowed to escape rearwards directly from the propellant container, and the percentages of the gas used for propelling the projectile and balancing the recoil could be adjusted by varying the size and shape of the channel.

## OPEN CHAMBER DESIGNS: THE TROUND

During the 1950s there was much interest in the 'open chamber' principle, with various designers and manufacturers becoming involved in investigations. The best-known inventor was David Dardick whose novel gun and ammunition system was dubbed the Tround (short for triangular round, after the cross-sectional shape). The gun had several chambers in a revolving cylinder, similar in this respect to a revolver cannon, but the chambers were open on the outside of the cylinder. As the cylinder rotated, the open chambers enabled rounds to be inserted sideways with spent cases removed in the same way. Support for the plastic cartridge during firing was maintained by a sleeve-like 'breech' around that part of the cylinder which was in line with the barrel.

The advantages of this system are its mechanical simplicity and the very short ammunition handling movements, giving the potential for a high rate of fire. The difficulty is in achieving the required support to the cartridge (case splits were a significant issue during development). Other difficulties were that it shares the problem of all telescoped ammunition

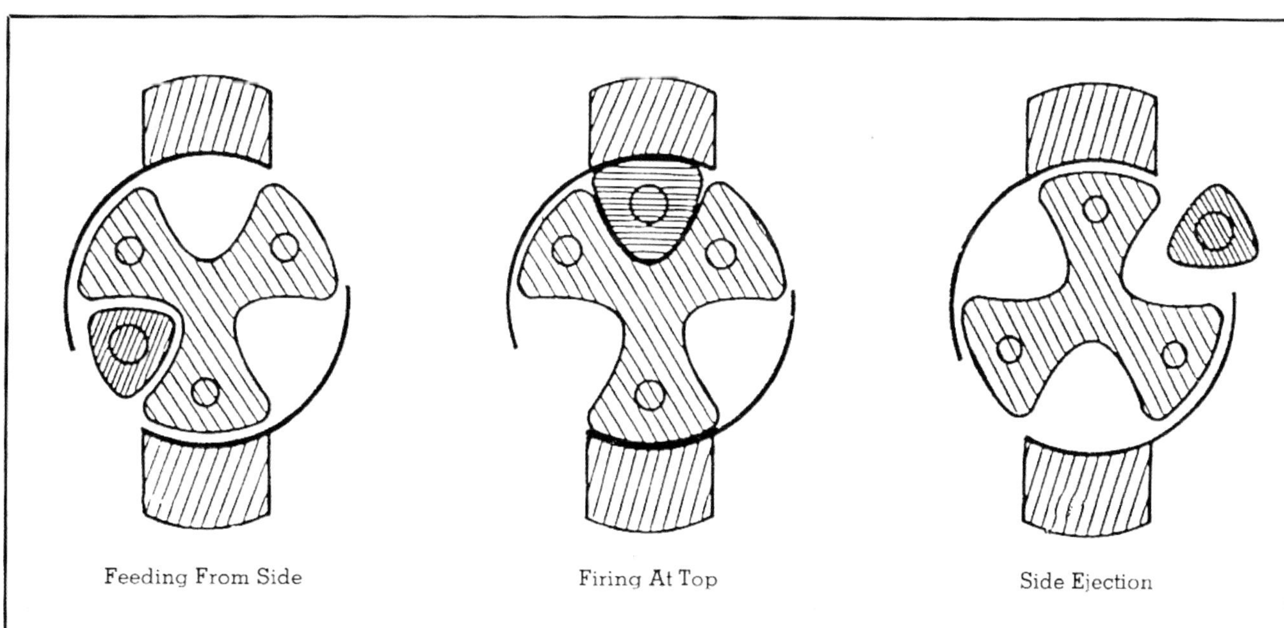


*The open chamber principle (Chinn)*

that the projectile has to make the jump to the barrel before the gun gas can flow past it, and it shares with revolver cannon the need for the cylinder to cease rotating for a fraction of a second to fire each shot, to ensure that the projectile cleanly enters the barrel. That is with a single-barrel gun; in principle there seems no reason why the open-chamber system should not be applied to a multi-barrel rotary version.

The cartridge might be described as semi-telescoped; the projectile is enclosed within the case, but most of the propellant is behind it, with only a small proportion extending to the front of the case. It is not actually triangular in cross-section, since the outer side of the case must exactly follow the inner curvature of the breech sleeve to maintain full support during firing; the three sides of the case are therefore rounded, a shape described as 'convex triangular'. Early work had asymmetrical case cross-sections, but it was soon realised that making them symmetrical simplified ammunition handling. The 20 mm version of the Tround is 170 mm long, with each side of the case measuring 42 mm (around the curve). The empty plastic case weighs 130 g.

Various designs of gun and ammunition were tried as part of the open chamber investigations, but the principal interest came from the USAF, whose 1954–56 T154 programme led by the Hughes Tool Company sought a very compact gun suitable for use in bomber defensive mountings (see the reverse-loading 30 mm T168 below for a different approach to this).

## SPLIT-BREECH GUNS

Another system which bears a superficial resemblance to the Tround is the split-breech system, which was the subject of experiments in the UK in the 1950s as part of research into future aircraft guns. This had a breech consisting of two cylinders, each with several deep semi-cylindrical grooves along their outsides, which were held side by side and rotated in opposite directions. The rotation was so arranged that grooves from each cylinder came together to form a perfect cylindrical shape; the gun's chamber. The cartridges were dropped between the rotating cylinders and collected by the grooves which, for an instant, formed an all-enveloping chamber, at which point the cartridge was fired. This action led to the nickname of 'Nutcracker'.

The advantages of the Nutcracker design were smoothness and simplicity of operation, with the potential for an extremely high rate of fire. The disadvantages should also be apparent. Each chamber was only formed for a brief instant as the cylinders rotated; any mistiming in a cartridge firing would leave it partly unsupported, leading to splitting along its length. There would also have been the usual revolver problems of sealing the gap to the barrel, and excessive barrel wear. One complicating factor was that the cylinders had to move closer together to form the breech, then pull apart slightly to enable rotation to continue.

It seems likely that (as with revolver cannon) the rotation needed to be stopped at the instant of firing, partly to ensure that the cartridge was supported and partly to keep it lined up with the barrel until the projectile had completed the transfer from the cartridge case to the barrel. This would have slowed down the rate of fire and spoiled the smoothness of the action. The temptation must have been to minimize any such interruptions in the interests of a high rate of fire, but as the experiments failed it must be presumed that this dilemma was never satisfactorily resolved: reportedly case splits were not uncommon.

There is no known official source of information about this system, but a few cartridges survive. The cartridge was necessarily tube-shaped (although the tube could vary in diameter along its length), but was not telescoped as all of the propellant was behind the projectile. Since the diameter of the case was fixed by that of the cartridge neck (i.e. only fractionally larger than the calibre) the cartridge was very long. Various calibres and lengths were envisaged, but a 30 mm cartridge with a 330 mm case length exists.

There was one weapon using this system which entered service, although in a very different form: the US Honeywell 40 mm Mk 18 grenade launcher, of which about 1,200 were made between 1965 and 1968. It fired standard low-velocity grenade ammunition and the mechanism was hand-cranked; the cartridges remained in their fibreglass-tape belt as they were fired. The low rate of fire and low pressure cartridges combined to make this weapon work, but the split-breech system has not been repeated.

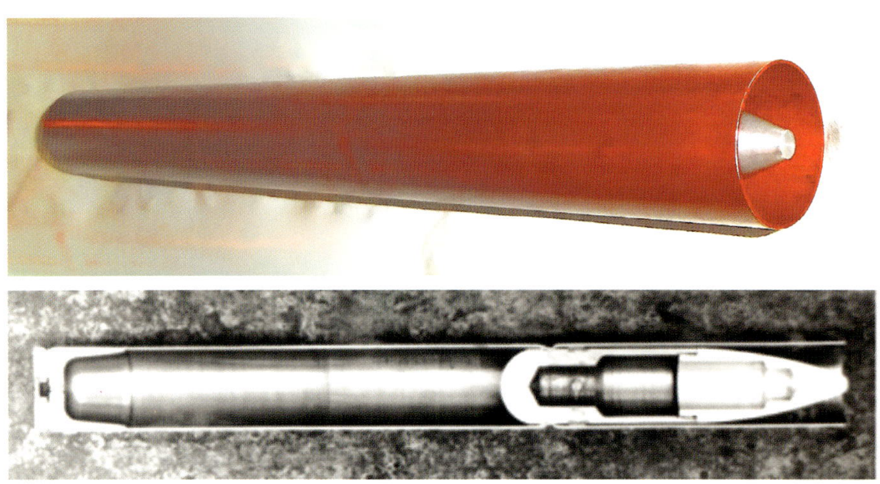

*30 mm 'Nutcrackers': lower photo from Geoff Tillotson*

## REVERSE-LOADING SYSTEMS

***30 mm T168:*** **30 × 120RB AND 30 × 136RB AMMUNITION.**
For some time after World War II air forces with strategic bombers were still thinking in terms of fitting defensive guns in turrets, including remotely-controlled ones. Conventional cannon were not well-configured for that purpose: they

had the gun action and ammunition feed at the back of the gun and a long barrel which created a lot of aerodynamic drag, especially at jet aircraft speeds. It also meant that the ammunition feed had to allow for considerable elevation movements, introducing complexity and potential unreliability. As a result, in 1951 the Springfield Armory came up with the concept of a reverse-loading four-chamber revolver, and United Shoe Machinery Corp. were given the contract for the initial feasibility study. The concept of reverse-loading had already been patented by Mauser in World War II as a part of the MG 213 programme. Instead of the cartridges being pushed nose-first into their chambers from the back, they are inserted base-first into the front. This means that the ammunition feed is close to the centre of the gun therefore does not move much with elevation. It also shortens the gun overall, providing a more compact installation.

To make such a gun possible, the electric-primed ammunition (designated T268 when loaded with a T328 TP projectile) needs to be repackaged so that it tapers in a reverse direction, with the base being the narrowest part of the case: the rim is tiny at just 17 mm diameter. Cases were made in brass, steel and aluminium. The first case developed was the 120 mm FAT32 (T222 to T225 series cartridges): subsequently this was replaced by the 136 mm FAT32E1 (T226 to T269 series). Overall length with the 120 mm case is 195 mm and weight around 450 g when fitted with the 250 g T328 target practice projectile. MV was 850 m/s, but Davis says the loading was similar to the conventional T204–208 series (Aden 30 × 113B) round (207 g T328 projectile at 823 m/s).

Test firing started in 1952, and a programme of testing and modification (with various cartridge case lengths and loadings) continued until 1956 when the project was closed down. The ultimate version was the T168E2, a lightweight 'receiverless' gun.

Russian designers faced the same problem and addressed it in the same way with the Rikhter R-23, which did see service.

***23 mm R-23:*** **23 × 260 AMMUNITION.** The Russian air force had discovered problems with using the turreted NR-23 as a bomber defence gun because when firing on the beam the pressure of the slipstream against the barrel affected the functioning of the gun. As this was experienced even with the propeller-driven TU-4 it would clearly be a much bigger problem with supersonic jets, so a new gun was needed which would not experience these problems. This was developed in the early 1960s.

OKB-16's designer Rikhter decided on a radical configuration to meet the requirement, selecting a four-chamber revolver mechanism (with fluted chambers) but with the cartridges loaded backwards into the chambers. This meant that the ammunition belt feed (into the right side of the gun) is sited under the barrel, close to the centre of gravity of the gun, and there is very little of the gun behind the revolving cylinder. As much as possible of the gun barrel is therefore contained within the gun mounting, with only a short length protruding into the slipstream. Altering the elevation of the gun also has a minimal effect on the ammunition feed as it is very close to the trunnions. The gun is gas-operated, with three different gas systems: one to drive the revolver cylinder and feed mechanism, one to blow cartridges into the chambers and a third to blow fired cases out again.

Loading the cartridges backwards means that conventional ammunition cannot be used, as it tapers from the base to the tip; a configuration which is tapered towards the base is needed, and this results in a strange, pipe-shaped 23 × 260 case within which the projectile is completely concealed. The case has a tapering base without a rim or extractor groove. Note that this is not a telescoped cartridge as all of the components are in a line as usual so it is no shorter than a conventional round.

The overall round length is the same as the case length, at 260 mm and the weight is 509 g. The empty case weight is 267 g and contains 67–70 g of propellant. There were two combat loadings: a nose-fuzed HEI (designated OFZ) containing 17.4 g A-XI-2 and a base-fuzed HEI (also OFZ) containing 14.5 g of the same mix, with a flat nose covered by a short ballistic cap. Both weigh 174–175 g and were fired at 850 m/s.

The gun did see limited operational service in bomber defensive armament. It was initially designated the 261-P but on adoption in 1964 it was named the R-23. It originally fired at 2,400–2,600 rpm, but in service this was reduced to 1,800–2,000 rpm to improve reliability. The gun weighs 58.5 kg, is 147 cm long with an L/50 barrel and has a life of 3,000 rounds. It was only ever fitted in a tail mounting to certain variants of the TU-22 supersonic bomber: the A, B, K and R subtypes.

This cartridge should be compared with the experimental US 30 × 120RB cartridge for the T168 cannon described

*Rikhter R-23 with belt link (top) with 30 × 120RB T168 (collection Jean-François Legendre)*

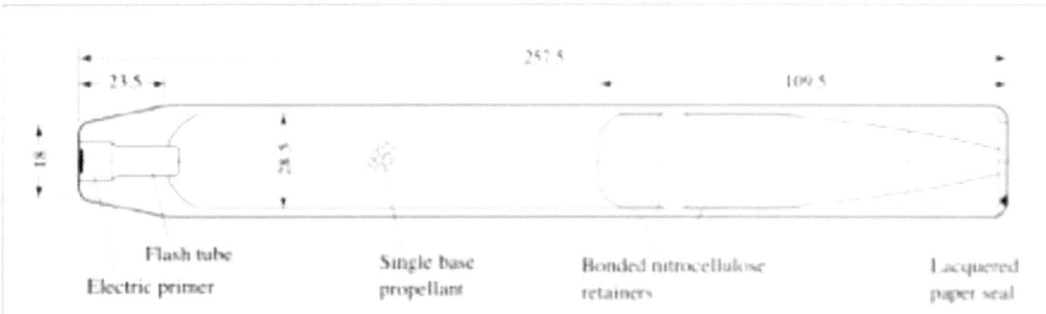


*Sectioned R-23 round (Courtesy Dr J R Crittenden Schmitt)*

above, which used a similar loading system for the same reasons. Other reverse-loading systems are described elsewhere in this part: the 30 mm AMCAWS (cased telescoped section) and the Mauser RMK 30/35 series (recoilless section).

## MULTI-BARREL SALVO-FIRING GUNS

Before the invention of the hand-cranked machine guns such as the Gatling, Nordenfelt and Hotchkiss rotary guns (which all depended on unitary cartridges to work), there was a brief vogue for multi-barrel guns in which each barrel was loaded with one shot. Cranking the firing lever fired the barrels, either all together or in rapid sequence, giving a very high rate of fire until the barrels had all been fired – at which point there was a very long pause while the barrels were individually reloaded by hand. Some of these volley-gun systems survived into the cartridge era, which enabled reloading to be accelerated to some degree, but the system was considered inferior to the hand-cranked machine guns, which were themselves displaced by self-powered automatic guns.

Despite this, such 'salvo-firing' guns have never quite vanished from the imagination of gun designers, whose efforts finally bore fruit in the shape of the Spanish Meroka CIWS. This features twelve 20 mm barrels in one cluster, with a single breechblock for all of the barrels which are simultaneously reloaded by external power. The Meroka has been in service for some decades with the Spanish Navy and is described in detail in the Spain section of Part 4. Long before this emerged, however, there was a little-known attempt to revive the concept in Germany in the 1930s for use in short-range AA systems.

The German systems were known as *Salvenmaschinenkanonen* (salvo-firing machine guns) and *Mehrlaufwaffe* (multi-barrel weapons). Four systems of this type were listed in a 1945 Allied inventory of all German weapons: two of them, by Mauser and Gustloff, were of 16 mm calibre, but two others (both by Rheinmetall-Borsig) were 20 mm systems, designated *Mehrlaufwaffe SMK 18 V1* and *V2*. Several guns of this type were found (in varying degrees of completeness) in an outdoor museum in Moscow in 1996, although matching them to the descriptions in the inventory proved difficult. All of them are pre-World War II, being referred to in German documents dated 1936 and 37.

The most obvious difference between the V1 and V2 (known as Type 1 and Type 2) is that the former's eight barrels are laid out horizontally, in one row, whereas the V2's are in two groups of four, each group stacked vertically. From the available description, the gun fired from a locked breech and had a short-recoil mechanism. The RoF per barrel is given as 80 rpm, giving an overall RoF of 640 rpm. The main technical problem was said to be that all eight barrels had to fire at the same instant, or the breechblock could jam.

No details appear to be available of the ammunition, except the 20 mm calibre and an MV of 900 m/s. These figures point to the usual 20 × 138B adopted for other Flak guns at the time; the 20-round magazines also look similar.

## TWO-PIECE AMMUNITION (SLAMMO)

The 30 mm Separate-Loading Ammunition High Performance Automatic Cannon (SLAMMO-HYPAC) was developed in the late 1970s by Pulsepower Systems of San Carlos, California. They submitted an unsolicited proposal to make a 25 mm revolver gun in which the cartridge and the projectile were loaded separately. After initial research the Automatic Cannon Advanced Concepts Team decided in 1979 to investigate the possibility of a 30 mm version of SLAmmo and selected a reverse-tapered case shape. The gun features dual feeding paths: the projectile is rammed into the bore, while the cartridge is simultaneously rammed rearwards into the separate chamber, which is then rotated in line with the bore and pressed forwards against the integral lip seal to fully seal the chamber.

The characteristics of the ammunition were: projectile (from the GAU-8/A) weight 356 g and length 140 mm; case length 148 mm and diameter 31.5 mm (base) and 38.6 mm (mouth); case weight 60 g (aluminium) or 175 g (steel); propellant weight 143 g; maximum chamber pressure 414 MPa; muzzle velocity 1,110 m/s; total cartridge weight (minus projectile) 209 g (aluminium) or 324 g (steel); total ammunition weight per round: 566 g (aluminium) or 680 g (steel).

A comparison with the GAU-8/A and the AMCAWS-30 was made. While all three are similar in their muzzle energy, the SLAmmo provides reductions in ammunition weight of 7% and 16% compared with the AMCAWS-30 and GAU-8/A respectively, and reductions in volume of 22% and 38%.

## ELECTROTHERMAL-CHEMICAL PROPULSION

A different approach to ammunition technology is the electrothermal-chemical system. This uses a (relatively) conventional cartridge case and propellant, but the primer is replaced by a plasma generator extending into the propellant. When a strong pulse of electricity is applied to the generator, temperatures of up to 10,000°K can be developed. More importantly, the temperature and therefore the burning rate of the propellant are variable, enabling a much better control of the pressure being generated. In a conventional cartridge, it is difficult to achieve a theoretically ideal pressure curve; the plasma generator allows a careful rise in initial pressure to start the projectile safely on its way, building up to achieve the maximum temperature, pressure and velocity which the gun can accept. Test firings were demonstrated by Bofors around the 1990s using a modified 40 mm gun, with the cartridge case being made of synthetic material to achieve electrical isolation. Replacement of the conventional propellant with a material whose gases have a lower molecular weight would permit even higher velocities. The main problem is the electrical power needed by the plasma generator, not as great as with a rail gun but still too much for current military use.

## CASELESS AMMUNITION SYSTEMS

The most drastic change involves doing without the cartridge case altogether. It will be noted that most large-calibre artillery (e.g. the NATO 155 mm guns) use separate-loaded propellant charges. Traditionally, the propellant increments were kept in cloth bags, but since the 1980s combustible cartridge cases have increasingly been used. Obturation is achieved in such artillery by the design of the breech-block. The effects of environmental conditions on propellant performance are well understood and can be compensated by making suitable adjustments to the gun elevation or propellant increments. This all works perfectly well, but artillery breech designs and loading systems which are adequate to cope with rates of fire of a few rounds per minute are not suited to coping with over a thousand rpm.

*Caseless ammunition: from the left, 20 mm USAF; 30 mm US (repA); 40 mm Ho-301 (repA); above, base of Ho-301 showing vent holes for propellant gas*

Viable caseless cannon ammunition can theoretically be achieved in various ways, although only one (rather unsuccessful) cannon example was ever adopted for service: the Japanese 40 mm Ho-301 aircraft gun of World War II described below. In this, the propellant was inserted into the rear of the projectile, which was driven forward by gas pressure escaping from holes in the base of the projectile. This may sound rather like a rocket, but there are important differences: the Ho-301 projectile was driven forward by the pressure of the gases trapped in the chamber and barrel, (a tube-fired rocket generates little chamber pressure); and the propellant was all burnt before the projectile left the muzzle (a rocket carries on accelerating). While this solution removes the need to extract and eject the fired cartridge case, the obvious problem is that there is very little space available within a projectile for propellant, so the muzzle velocity will be very low. This technology is best suited to low-pressure, low-velocity ammunition such as projected grenades, and is used in certain Russian and Chinese grenade launchers.

A second method is to form the propellant into a hard case or solid block, moulded around the base of the projectile. There are many problems which result from this. The loss of the mechanical strength provided by the case leaves the propellant blocks relatively fragile and unsuited to violent mechanical handling. Also, the lack of a case means that some alternative means must be found for obturation – the sealing of the breech against gas escape. Furthermore, the propellants are fully exposed to variations in environmental conditions that can cause considerable differences in pressures and velocities. In the case of aircraft guns in particular, the environmental variations can be severe since the ammunition can be at +50°C on take-off and –50°C a few minutes later at high altitude, with equally drastic variations in humidity. As a result, caseless cannon ammunition has never achieved service status and there seems to be no prospect that it will.

### 20 × 87

The 20 mm round shown on the left in the group photo above was developed in Spain in the mid-1950s by the Belgian firm of SIDEM International under contract to the European Office of the US Air Force Research and Development Command. It was referred to as a 'combustible' or 'self-consuming' round. The 20 × 87 case was formed from ball propellant and a mixed cellulose binder, and was filled with loose propellant. The electric primer was also combustible. Overall length of the complete round is 158 mm and the weight is 175 g including an M99 TP projectile weighing 130 g. This round was intended to match the ballistic performance of the M99 cartridge (20 × 110 Hispano), but proved a failure: at room temperatures the MV was about 36 m/s less than the M99; at extreme temperatures, 58% of the rounds misfired; much of the case was not consumed; and the case was much too fragile to withstand handling by an automatic gun mechanism.

### 30 mm CASELESS TELESCOPED

A round like the 20 mm one above with most of the projectile exposed, may be particularly prone to suffer from the projectile becoming detached during the loading cycle. This can be avoided by enclosing most or all of the projectile within the propellant, to form telescoped ammunition as shown with the 30 mm round above, but that brings other difficulties as already described in the CT section. This particular dummy round, reportedly developed by Hughes, is typical of many similar experiments; it is 150.5 mm long overall and 47.5 mm in diameter.

### 40 mm Ho-301 CL

Technically the most interesting of the IJA aircraft cannon, this API blowback was limited to firing very low-velocity caseless ammunition as described above. Another problem was the box magazine capacity of just ten rounds. The gun was light at 40 kg and compact, with an L/20 barrel, and was easily able to be mounted in fighter wings. It had a high rate of fire; British tests averaged 433 rpm (391–478 min/max) although that meant that the magazine was emptied after one burst of 1.3 seconds.

In appearance the Ho-301's cartridge resembled a small finless rocket, but it differed in principle in that the propellant (expelled through twelve 3.8 mm diameter holes in the base of the projectile, surrounding the primer) burned out within the barrel, generating a recoil thrust to drive the blowback mechanism.

Overall length (fuzed) was 130 mm, of which the rear 30 mm was devoted to propellant. The rounds weighed 559–585 g (sources differ) of which 52–57 g was the HE load of TNT or TNT + picric acid (another source gives 63.7 g).

The propellant charge weighed 10 g and in British tests, involving three guns and 1,000 rounds of ammunition, generated an average of 120 MPa (17,200 psi) chamber pressure, but with minimum and maximum pressures of 100 and 135 MPa. This variation led to MVs of between 213–256 m/s (average 238 m/s) resulting in the size of 10 shot groups at 180 m ranging from 0.5 × 0.7 m to 1.2 × 2.0 m.

## COMBUSTIBLE CASE

One way of obtaining the advantages of caseless ammunition while avoiding the worst problems is to use a combustible case, made of some material which is strong enough to survive moderate handling and to provide some protection to the propellant, but which is burned up in the chamber. This not only saves weight compared with a metal cartridge case but avoids the need to extract and eject the fired cases, a common source of problems.

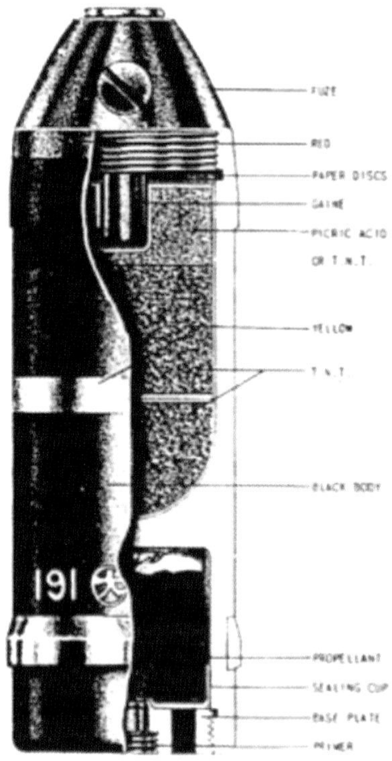

*TK*

Bofors explored the possibility of combustible cases while developing their famous 40 mm L/60 in around 1930. The biggest problem they were finding was how to make the loading mechanism function properly. At one point they tried to speed up the cyclic rate of fire by using electrically primed zinc cases with very thin walls, the idea being that they would vaporize when fired and thus eliminate the time needed to extract the spent case. Unfortunately the zinc fumes would condense inside the barrel, leaving a zinc residue that was not healthy for the bore.

Part-combustible cases are successfully used in current 120 mm smoothbore tank guns, albeit with metal bases to provide obturation and facilitate the removal of unfired rounds from the chamber. Despite being tried late in World War II in the Rheinmetall-Borsig 55 mm MK 115 (which combined an autoloader firing at 300 rpm with a part-combustible case plus recoilless operation!) neither combustible nor part-combustible cases have so far been adopted in automatic cannon; presumably they are not strong enough to withstand violent handling.

The best known combustible-case project, which was intended to arm the USAF's then-new F-15 fighter, is the 25 mm GAU-7.

### 25 × 152 GAU-7/A

The GAU-7/A project, which commenced in the late 1960s, was a daring attempt to achieve a radical leap forward in aircraft weapon technology and was the intended armament of the F-15 fighter which was then on the drawing board. The gun was a multi-barrel rotary but the powerful ammunition was telescoped (with the projectile completely buried inside it) and also used caseless or a combustible case technology. This meant that there was no fired case to extract or eject, simplifying the firing cycle; the ammunition was also significantly lighter and more compact.

The two gun companies which produced prototypes for testing were Philco-Ford (working primarily with Brunswick on ammunition development) and General Electric (working mainly with Hercules). Philco-Ford won the competition and were awarded a further development contract.

In appearance the Philco-Ford/Brunswick cartridge is a simple cylinder, 152 mm long and 40 mm in diameter. The case weighs only 33 g (of which the outer shell, made of felted nitrocellulose, acrylic fibre and HYCAR resin, weighs 25 g) and the long-nosed, finely tapered projectile weighs 192 g; it has a broad plastic driving band. The propellant was divided into a forward charge of 91.5 g and a rear charge of 38.4 g, giving a total round weight of c.360 g. Percussion priming was used. The maximum chamber pressure was 448 MPa, the muzzle velocity of 1,200 m/s producing a muzzle energy of 138,000J.

The GE/Hercules cartridge is also cylindrical, 151 mm long and 38.5 mm in diameter, and the complete round weighs c.350 g. Unlike the combustible-cased Brunswick round, the Hercules one is caseless. The main propellant charge weighs 124 g and was moulded into a solid shape to surround the 192 g projectile; there is a 28 g black-powder booster charge ignited by the percussion primer, which then initiated the main charge. Protection of the propellant was provided by a polyester tape covering. Designed performance characteristics matched those of the Brunswick round.

Unfortunately, packing so much unprotected propellant into the magazine was considered a major fire risk in combat, so it was decided that each round should have a fireproof sheath which had to be stripped off before chambering, significantly adding to the complexity of the system. The fact that the propellant was not sealed off from the atmospheric conditions was also a problem (especially for the Hercules round), because the performance could vary depending on the ambient temperature and humidity. As a result, the GAU-7/A failed to generate repeatable chamber pressures, muzzle velocities and action times over the desired temperature range. In the end, the problems proved insoluble and the project was cancelled.

The Mauser RMK30/35 series of combustible case systems appears in the recoilless gun section below.

## LIQUID PROPELLANTS

An interesting variation on the caseless theme is the use of liquid propellant. With such systems only the projectiles are mechanically handled, the propellant is contained in a separate tank from which it is squirted into the chamber in the same way as a fuel injection system for an internal combustion piston engine.

There are many theoretical advantages with such systems, including much faster mechanical handling, more efficient storage of propellant, and precise control of ballistics by varying the quantity injected. However, these have been more than outweighed by practical problems.

The first problem is that a tank full of very volatile and potentially explosive propellant is not what is wanted in a combat system. Accordingly, nonhypergolic bipropellants (in which the fuel and oxidizer are stored in separate tanks and only combined in the gun chamber) were preferred instead of monopropellants, at the cost of added complexity. Incidentally, hypergolic bipropellants are those in which mixing the fuel and oxidizer instantly causes spontaneous combustion, which while satisfactory in a rocket is not what is wanted in a gun, thereby explaining the 'nonhypergolic' bit.

The second problem concerns the considerable energy required to pump the propellant(s) at very high pressure into the chamber. A separate source of power for this purpose was found to be impractical, leading to the development of regenerative injection, in which combustion chamber pressure is amplified by a piston system and used to force the fuel into the chamber, at the cost of yet more complexity.

A great deal of theoretical and practical work was done on such systems from the end of World War II onwards, resulting by the early 1970s in a fast-firing 25 mm aircraft cannon using nonhypergolic bipropellants being made and tested in the USA. While impressive ballistics were obtained during the long period of experimentation, the variation in muzzle velocities which occurred was too great to meet military requirements and there were also some catastrophic failures which destroyed the guns.

In the early 1980s several French companies became involved in working on a 25 mm liquid-propellant system as part of the ABF Program aimed at developing an aircraft gun to replace the 30 mm DEFA. Work was begun in three places: EFAB (*Etudes et Fabrication d'Armements*) at Bourges; SEP (*Société d'Etudes pour la Propulsion*, later absorbed by SNECMA, then GIAT and now Nexter, part of the Safran Group); and ONERA (*Office National pour la Recherche en Aéronautique*). It appears likely that this project later became part of the Franco-German GEC Program for caseless rounds, Rheinmetall developing the gun and EFAB the ammunition. Research ended in 1984 but the results were transferred to the MCF Program (Medium Calibre, Future) that later evolved into the Anglo-French 40 mm CTA.

The most recent attempt to use liquid propellants was for the US Army's 155 mm Advanced Field Artillery System (AFAS) project in the 1990s, which ultimately resulted in the Crusader self-propelled gun. The liquid propellant system was however abandoned in favour of solid propellants in 1996 because the technical risks were expected to result in a protracted development timescale, too long to meet the Army's requirements. The solid propellant version of the Crusader was also cancelled, in 2002.

Proposals have also been made for gas-powered guns, injecting inflammable gas into the chamber instead of liquid. These seem unlikely to fare any better.

## RECOILLESS AMMUNITION

The first recoilless (RCL) gun to see use was developed by the American Cleland Davis in 1910, using the principle of balancing the recoil resulting from the projectile being fired forwards by simultaneously firing a load of lead shot

*25 mm GAU7/A cartridge*

of the same weight rearwards. Guns were developed in 2-pounder (40 mm), 6-pounder (57 mm) and 12-pounder (76 mm) calibres and saw some use in World War I, with trials continuing afterwards. The system worked well enough but the guns were extremely long (two barrels, back to back) and slow to reload. The lead shot being fired to the rear was also a hazard to friendly troops (or to the structure of an aircraft firing one).

In the 1930s it was discovered that the same effect could be achieved by loading considerably more propellant into the case and diverting most of the resulting gas rearwards via venturi to provide a 'rocket effect'. Clearly, some careful juggling with the design is necessary to ensure that exactly the right proportion of the gas is diverted to balance the recoil generated by the projectile and the gas propelling it. This approach almost halved the length and weight of the guns, and made reloading easier. The back-blast was still a potential hazard, but guns using this system remain in service today – notably in the form of the Swedish 84 mm Carl Gustav. However, such weapons are manually loaded and have a relatively low MV.

Two basic approaches to channelling the gas to the rear have been followed. One has a 'blow-out' cartridge base (usually made from plastic or some similar weak material), allowing the gas to escape directly to the rear; the other has a case with many perforations in the body sides which are sealed by an internal liner until firing, when the gas bursts through into a chamber which is slightly wider than the case, from where it is ducted to the rear.

Whichever approach to channelling the gas is taken, the need to do this complicates the design of automatic weapons. That has not prevented many attempts, with top marks for effort going to Russian designers, particularly Kondakov who was actively involved in the 1930s. The projects in our calibre range (some were 76 mm guns) were as follows:

**ARKON** (*Avtomaticheskaya Reaktivnaya Kondakova*): initially in 45 mm and then 37 mm calibre. This was a gas-operated repeater with a 'blow-forward' type of action, the barrel moving forwards by 450 mm on firing to eject the fired case, loading another as it returned to battery. The electrically-primed cartridges were brass-cased with plastic blow-out bases. The 45 mm was fed by a six-round clip, the definitive 37 mm by a five-round magazine. The development of the 37 mm was complete by 1936. It weighed 40 kg and a charge of 175 g fired the 674 g projectile at 545 m/s. Firing tests were carried out on an R-6 aircraft, but the project was then cancelled.

**RPTR**: Kondakov's final effort was the RPTR, a 37 mm recoilless gun with similar characteristics and ballistic performance to the ARKON, except that it had a cyclic rate of 30 rpm. The gun weighed 63 kg, was 165 cm long with a 155 cm barrel, and had a chamber pressure of 240 MPa. Development was completed in 1937, but all work on the recoilless weapon programme was then cancelled.

**GK-37 and GK-45**: curiously, Kondakov reverted to the Davis principle with these two guns which were developed in parallel with the RPTR; two barrels were used, back to back, with a shared chamber. Both barrels fired projectiles (one inert), and gas was tapped to separate the barrels and reload. Recoil was therefore perfectly balanced, at the cost of a very long and cumbersome weapon. The guns were used in firing trials in 1937, but were caught up in the general cancellation of work on RCLs. Their characteristics were:

GK-45: gun weight 340 kg; barrel weight 2 × 94.2 kg; a five-round clip weighed 31.3 kg; projectile weight was 1,425 g and the MV 758 m/s; RoF was 60 rpm (on trial).

GK-37: gun weight 140 kg; barrel weight 2 × 37.2 kg; a three-round clip weighed 8.9 kg (12 round clip also available); projectile weight was 600 g and the MV 940 m/s; RoF was 140 rpm (theoretical).

The other country to have persevered with autoloading RCLs is Germany, with two very different projects taking place, one in the mid-1940s and the other in the 1990s.

### MK 115 (55 × 318B)

This round, which was under development by Rheinmetall-Borsig at the end of World War II, has two claims to fame: it used a part-combustible case and was intended for a recoilless gun. The shape of the case looked conventional, but there was a steel base 96.5 mm long, the rest being made from nitrided cardboard. The overall length of the round is 500 mm, the weight 3,000 g (of which the cartridge case weighs 1,000 g, the propellant 500 g). The rim and body diameters were both 67 mm.

This fired the same 1,480 g *M-Geschoss* shell at the same 600 m/s MV as the conventional MK 112 (55 × 175RB), but used far more propellant, the extra gas generated being diverted into a tube and expelled to the rear of the gun in order to balance the recoil (apart from a proportion tapped off to drive the belt feed). There are no known survivors of this round, not surprising given the case materials, the one shown below being a replica.

The MK 115 weighed 180 kg, was gas-operated and belt-fed. The gun was 330 cm long, with an L/24 barrel. Despite the large case, the rate of fire was a planned 300 rpm.

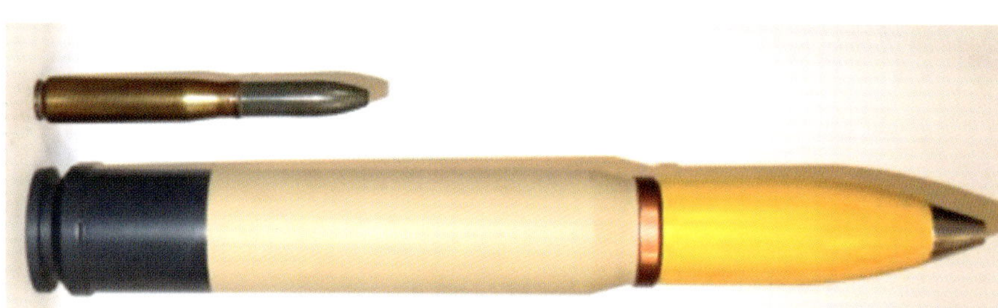

*A replica of the 55 × 318B round for the MK 115 from Rheinmetall-Borsig, under development at the end of World War II (20 × 110RB for scale)*

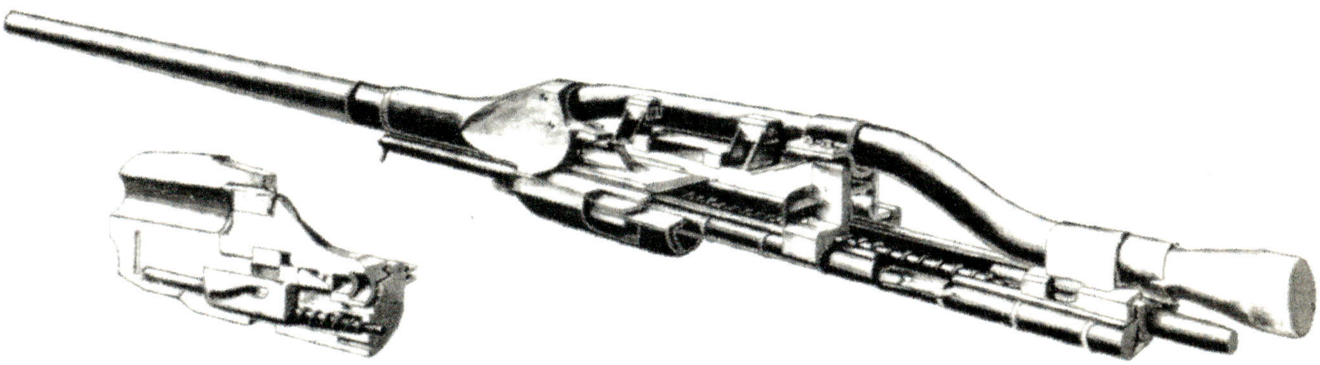

*The MK 115 (Chinn)*

***MAUSER RMK 30 / 35 SERIES:***
30 × 230 / 30 × 280 / 35 × 300 / 35 × 350 AMMUNITION.

Perhaps the ultimate attempt at combining desirable characteristics was made by Mauser in the 1990s, with the RMK (*Rückstoßfreie Maschinenkanone* = Recoilless Machine Cannon) family of 30 mm and 35 mm weapons and ammunition. The cartridges for all of this family were telescoped within a completely combustible cylindrical case with an inductive primer, the latter also being combustible, and the weapons were both automatic and recoilless.

The RMK rounds were developed for a technically very interesting series of cannon, developed by Mauser from the 1990s but shelved in the 2000s. Two 30 mm versions were made with different round lengths: the RMK 30/1 (30 × 280) was first demonstrated to the German armed forces firing from a Wiesel light airborne tank in November 1996. The RMK 30/2 (30 × 230), with linkless feed, was originally intended to be the main gun armament for the German version of the European Tiger attack helicopter. Subsequently two 35 mm versions were proposed. The dimensions and performance of the ammunition were as follows:

- 30 × 230 RMK 30/2: 280 g at 1,050 m/s = 154,000J: diameter 44 mm, mass 500 g
- 30 × 280 RMK 30/1: 280 g at 1,350 m/s = 255,000J: diameter 50 mm, mass 655 g
- 35 × 300 RMK 35/1: 400 g at 1,250 m/s = 312,000J: diameter 50 mm, mass 900 g
- 35 × 350 RMK 35/2: 430 g at 1,480 m/s = 470,000J: diameter 63 mm, mass 1,170 g

The RMK incorporated several innovations in the field of automatic cannon. The most important, as the name suggests, was that they were recoilless, the recoil being balanced by diverting a large part of the propellant gases through a nozzle at the rear of the gun. This is not unusual in single-shot weapons, but simultaneously allowing for rapid automatic fire is a different matter, especially given that unlike service RCLs, the ammunition was to deliver velocities and energies comparable with conventional cannon. Mauser's solution to this problem involved further unusual approaches: separate chambers which could be loaded from the front (already patented by Mauser in World War II as a part of the MG 213 programme), plus cylindrical, telescoped and caseless (or combustible cased) ammunition using an inductive primer.

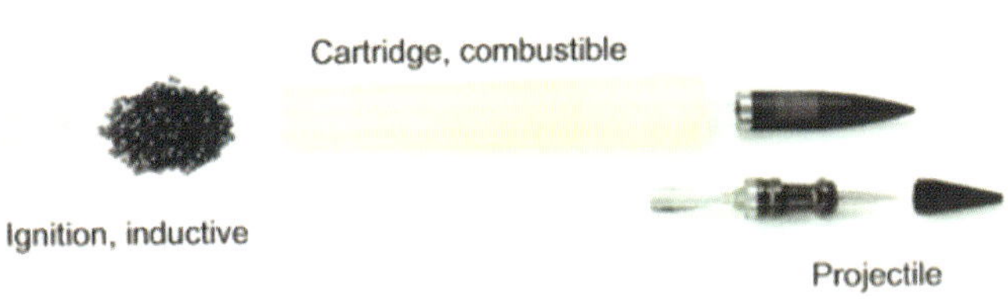


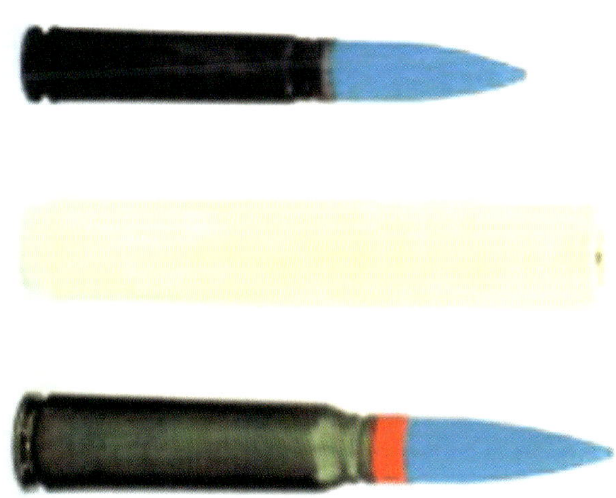

*RMK30 flanked by a 27 × 145B (top) and 30 × 173 (images from Mauser)*

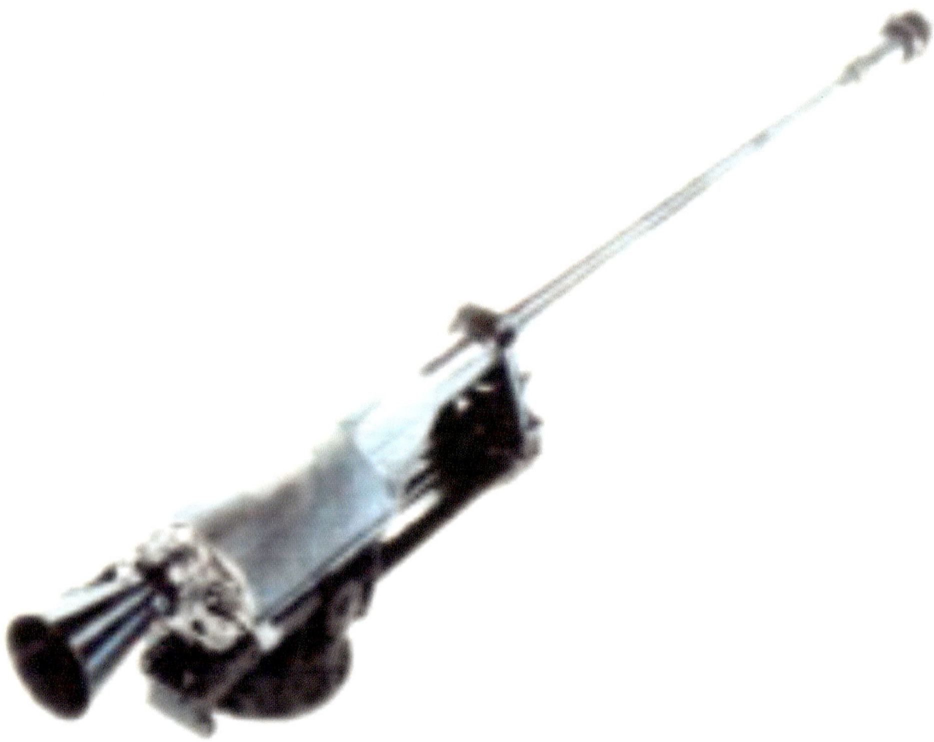

*Mauser RMK 30/1*

The advantage of this system was a considerable reduction in the vibration which usually accompanies cannon firing, as well as zero recoil whether in single shot or automatic mode. These characteristics led to a low shot dispersion (reportedly only half that of a conventional cannon) and permitted the installation of powerful cannon on vehicles (aircraft, naval or ground) which would normally be considered much too light to mount them.

Four different guns using four different rounds of ammunition were proposed, as listed below. All were multi-chamber revolvers (three chambers in the 30/1 and 35/1, four in the 30/2) , except for the 35/2 which had a single rotating chamber; and all were power-operated, with a variable rate of fire up to 300 rpm (200 rpm for the 35/2) and powered linkless ammunition feed:

RMK 30/1 30 × 280 ammunition: weight 125 kg; barrel/total length 240/330 cm

RMK 30/2 30 × 230 ammunition: weight 95 kg; barrel/total length 170/220 cm

RMK 35/1 35 × 300 ammunition: weight 152 kg; barrel/total length 280/320 cm

RMK 35/2 35 × 350 ammunition: weight 280 kg; barrel/total length 280/350 cm

The first, and possibly only, gun to be built and tested was the RMK 30/1. It is unclear what development stage the other models reached.

### *THE 20 mm MK 11 (PROTOTYPE)*

The original prototype of the USN's MK 11 cannon must be one of the strangest cannon to have been built. This was the culmination of a long design process by naval technician Frank Marquardt, starting in 1945. All manner of revolver-type arrangements were considered to achieve a very high firing rate, including twin drums, and loading the projectile (from the front) separately from the propellant (from the back). The first firing prototypes in 1949 had an ammunition feed consisting of 'projectors' connected by wire links, from which special cartridges were 'fired' into the chambers, to avoid the need for heavy mechanical rammers. The spent cases (which have no extractor groove) were blown out at high velocity by compressed air, back into the projectors which remain wired together. Empty chambers were flushed with carbon dioxide to cool them.

The production MK 11 was modified to use the conventionally belted 20 × 110 USN ammunition developed for the MK 12 cannon.

*The original form of ammunition for the USN 20 mm MK 11*

# PART III: THE CARTRIDGES

This part consists of descriptions and illustrations of all of the cannon cartridges which have seen service, plus a large number which were not adopted. They are grouped into five sections: 20 mm, 22–28 mm, 30 mm, 33–37 mm, and 40–57 mm. Within each section, they are arranged by calibre and case length. The headings for each cartridge are given in the standard metric measurements, although for cartridges whose formal designation was given in inches or 'pounder' units, these are given first with the metric conversion in brackets. Cartridges which have seen service have their headings in larger font and underlined.

Each calibre group contains group photos to enable the appearance of the rounds to be compared; for service rounds and (separately) for experimentals. A 20 × 110RB Oerlikon cartridge is included in each group to aid comparisons.

Examples of all of the experimental rounds are not available, so replicas have been made of many of these.

Photo captions may include the following terms: (repP), (repC), (repA), indicating 'replica projectile', 'replica case' or 'replica all' respectively. The 'replica all' category includes some examples of cases and projectiles which were made separately at different times, but in other examples complete replica rounds, mostly based on factory drawings, have been machined out of solid aluminium. The 'replica projectile' category includes a few which are genuine but have been repainted to represent a different nature, e.g. HE instead of TP. In an ideal world, all paint schemes would be original but with older rounds this is frequently not the case. Also, all projectiles would ideally be crimped into the cases by the factory, but in practice much of the cannon ammunition in collections consists of fired cases which have had projectiles added later.

The basic data for the rounds, including further dimensional information, are summarized in the Ammunition Table in the Appendix.

Unconventional ammunition, such as cased-telescoped, caseless, and recoilless, are dealt with in Part 2 and will not be listed here.

# 20 mm CARTRIDGES

Cannon in 20 mm calibre first saw service during World War I but achieved their peak of success in World War II, mainly as aircraft and short-range anti-aircraft weapons but also in some light armoured vehicles and small naval craft. Since the 1950s the popularity of this calibre has declined, as it has gradually been replaced in most of its roles either by larger-calibre cannon or by guided missiles. However, 20 mm cannon continue to equip some fighter planes (mainly of US origin), remain popular for the short-range defence of warships and have found new roles in arming helicopters.

A wide range of different cartridges was developed for 20 mm cannon, varying greatly in size and power. Only one saw service in World War I, but by World War II some seventeen different rounds were being manufactured; no fewer than five of them by Japan and four by Germany. Many of these disappeared soon after World War II, and today only four cartridges from that period remain in use – 20 × 82, 20 × 110RB, 20 × 110 Hispano and 20 × 99R ShVAK – the last being used only for secondary roles rather than in automatic cannon).

However, another four were introduced in the 1950s and these also remain in service today – 20 × 102, 20 × 128, 20 × 139 and 20 × 110 USN – although the last of these is obsolescent and no longer in production.

## 20 mm SERVICE CARTRIDGES

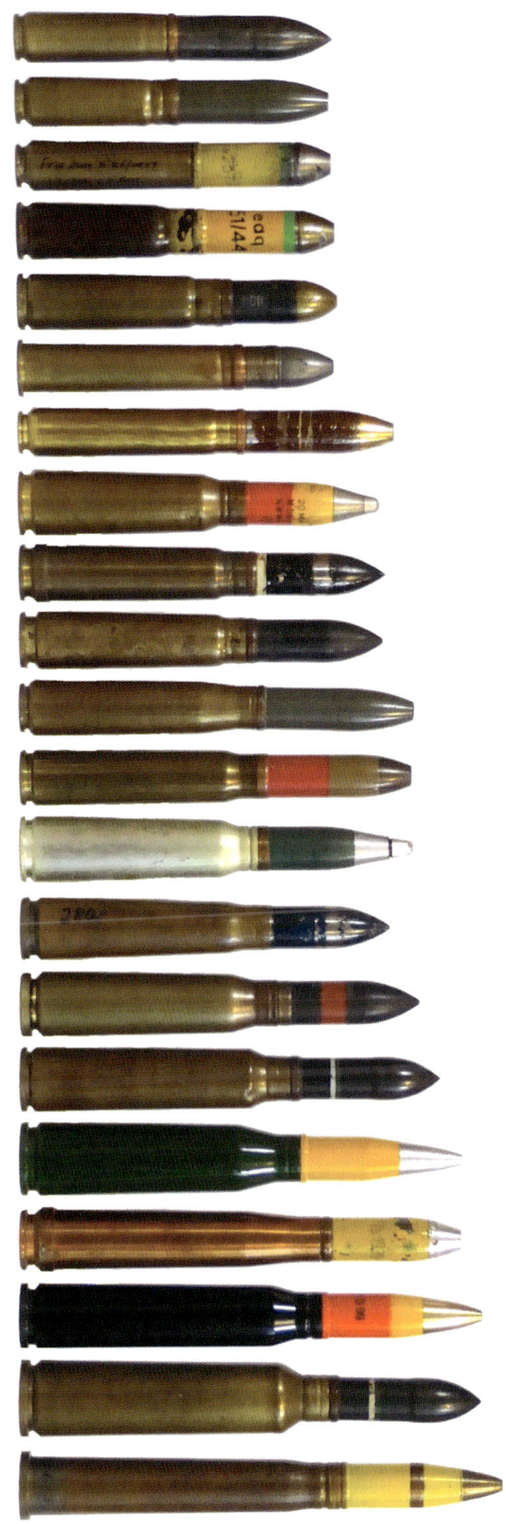

*20 × 70RB Becker*
*20 × 72RB Oerlikon FF*
*20 × 80RB MG-FF*
*20 × 82 MG 151/20*
*20 × 94 Type 2 / Ho-5*
*20 × 99R ShVAK*
*20 × 101RB Oerlikon FFL*
*20 × 102 US*
*20 × 105B Rheinmetall*
*20 × 105 MG 204*
*20 × 110RB Oerlikon Type S/SS*
*20 × 110 Hispano HS 404/804*
*20 × 110 MK 100 USN*
*20 × 113 Lahti*
*20 × 120 Madsen*
*20 × 124 Ho-1 / Ho-3*
*20 × 128 Oerlikon KAA*
*20 × 138B Rheinmetall*
*20 × 139 HS 820 / Oerlikon KAD*
*20 × 143 Type 98*
*20 × 144R Bofors*

## 20 mm EXPERIMENTAL CARTRIDGES

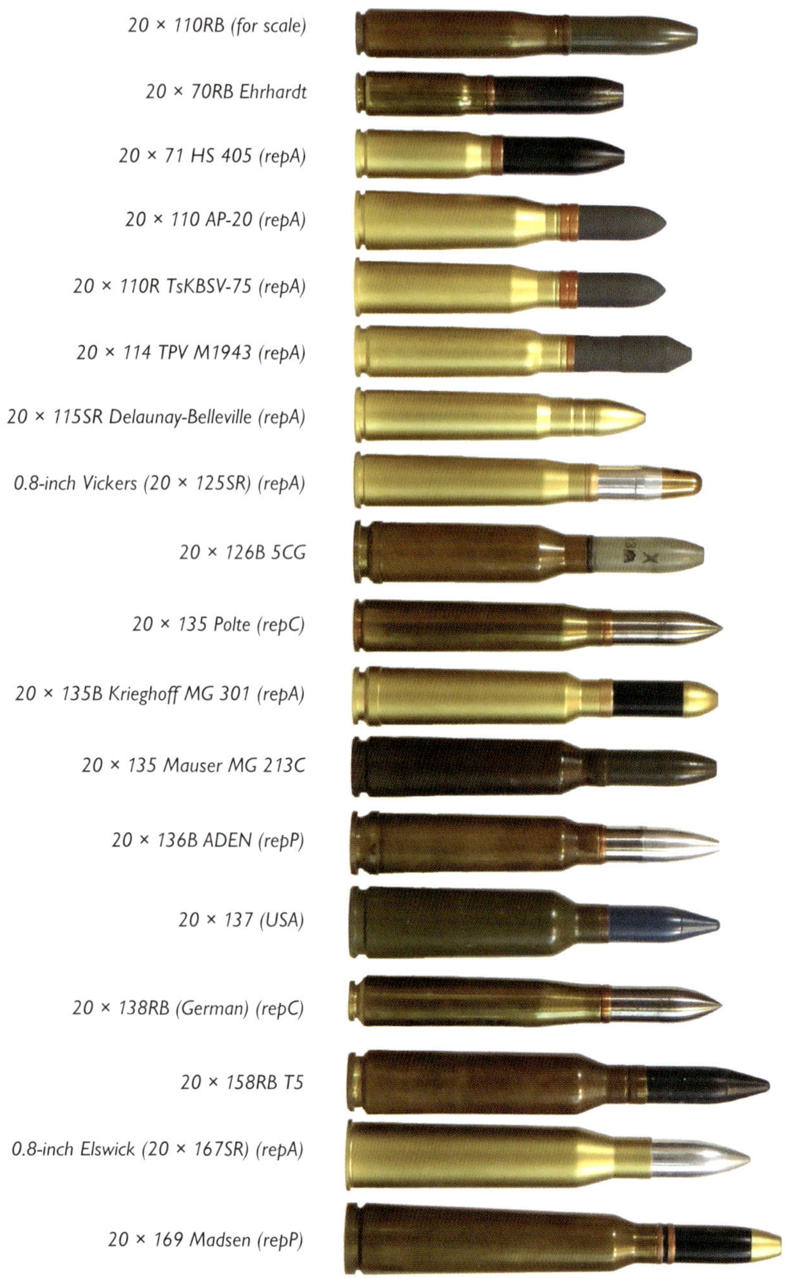


## 20 × 70RB BECKER

This was the ancestor of all 20 mm cannon cartridges, being introduced (along with the API blowback operating principle) in the German Becker cannon developed during World War I. The brass cartridge case had a rebated rim, as required for cartridges designed for API actions. The original (Type 1) round has a straight case; the Type 2 used a very slightly bottlenecked case that is sometimes mistaken for the round for the experimental Ehrhardt gun (see below), possibly accounting for some confusion over the 'Type 3' Becker; as far as is known, there were only two types of Becker case. Overall length of the complete round is 147 mm and the weight is c.180–200 g. The propellant load was 12.5 g of Gew.BLP plus 1.3 g of ignition compound for all loadings; MV is quoted as 490 m/s, without specifying with which projectile.

Three different types of shells were made; HE, incendiary and tracer. The unpainted HE shell has a very prominent plunger-type nose fuze, contained 0.5 g picric acid plus 5.5 g melted tolite, and weighs 137 g with a brass fuze, 128 g with a zinc fuze. The incendiary shell, painted dark red and weighing 137 g, has a longer body containing 8 g of white phosphorus and drilled with twelve side holes, and a simple nose impact fuze. The tracer shell (apparently the only one used in aerial combat) came in two versions: a pointed nose, painted black, weighing 145 g and containing 7 g of tracer compound; and a flat nose, painted green, weighing 144 g and containing 11 g of tracer compound.

## 20 × 70RB OERLIKON TYPE F / 20 × 72RB FF / TYPE 99–1

After World War I the development of the Becker was continued, first by SEMAG of Switzerland and, at the end of the 1920s, by Oerlikon who renamed the gun the Type or Model F. In 1934 the FF aircraft cannon was introduced; lighter and faster-firing than the Type F. At the same time the case was modified; the rim diameter was reduced (to 19 mm rather than 20 mm) and the extractor groove lengthened by 2 mm, increasing the case length to 72 mm. The FF saw service in various air forces during the 1930s, but was most famous for its adoption by the Japanese Navy air force

*Becker Type 2 (left) and Ehrhardt cases*

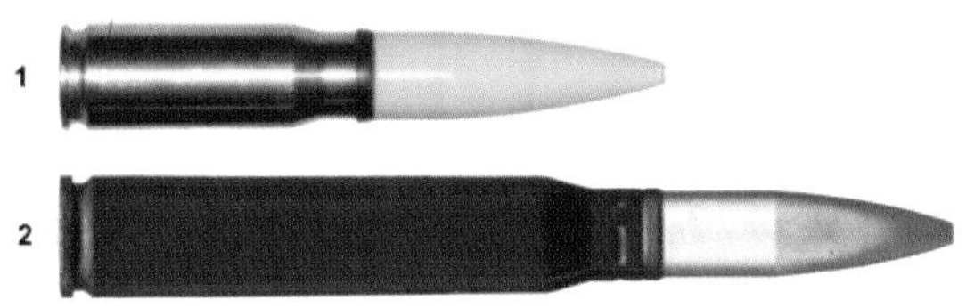

*HS 821 rounds: 20 × 70 and 20 × 139 (courtesy Robert Berger)*

under the designation 20 mm Type 99–1; their standard aircraft cannon in the early years of World War II.

The brass cartridge case has a rebated rim with a slight bottleneck and contains 13.5 g of propellant, giving a muzzle velocity of 605 m/s. Projectiles, which have copper driving bands, were secured in the case by three or six stab crimps in the case neck. Overall length of the complete round is 140–147 mm and the weight is 180–200 g. Principal combat loadings (Japan) were HE (128 g, inc 10.2 g PETN+TNT), HE-T (128 g, inc 5.6 g HE), HEI (120 g, inc 5.7 g white phosphorus and 5.0 g TNT), Tracer (128 g), and API (131 g).

### 20 × 70RB EHRHARDT

This is for the German Ehrhardt *FlzK (Flugzeugkanone,* or aircraft cannon) which entered production by the end of World War I but did not make it into service.

The brass cartridge case has a slight bottleneck and a rebated rim, and is often confused with the Becker Type 2, but the base and rim are different, about 1 mm more in diameter and with a less-rounded base.

Overall length of the complete round is 143 mm and the weight is 193 g. Details of the loadings are not known, but projectiles were the same as used in the Becker rounds and a muzzle velocity of 670 m/s has been quoted. This cartridge was also apparently selected for the experimental Scotti M35 aircraft gun of the mid-1930s. The Ehrhardt was the ancestor of the Rheinmetall/Solothurn 20 mm cannon developed through the 1930s.

### 20 × 70 HS 821

A 1960 project for a Heavy Infantry MG saw a new 20 mm cartridge proposed for it. After various experiments, a bottlenecked 20 × 70 cartridge was selected with an overall length of 144 mm and a weight of 161 g. The projectile weighed 90 g, contained 18 g HE and was fired at 650 m/s. This was dropped at an early stage in favour of a longer and more powerful gun firing standard 20 × 139 ammunition.

### 20 × 71 HS 405

Hispano Suiza France developed this shortened version of their 20 × 110 round for the HS 405, a shorter and lighter version of the HS 404 intended for use in aircraft turrets and flexible mountings, but it was still at the development stage in 1940. It fired the same range of projectiles of around 123 g but at only 600 m/s. Overall length is 145 mm.

## 20 × 80RB MG-FF/M

The German *Luftwaffe* decided in the late 1930s to adopt a light cannon to equip its fighters and selected the Oerlikon FF, which the firm of Ikaria (Oerlikon's German subsidiary) made under licence in Berlin as the MG-FF. However, they made some changes to the gun and ammunition, including lengthening the cartridge case (although the overall round length remained the same). During 1940 they introduced the *'Minengeschoss'* or *M-Geschoss* (mine shell), which was made by drawing the projectile body in the same way as a cartridge case rather than drilling a cavity for explosives. This enabled very thin projectile walls to be made, doubling their HE/I capacity while also reducing their weight. The recoil characteristics were different, causing the gun to be modified and subsequently given the designation 'MG-FFM'. Initially, the design did not lend itself to fitting a tracer.

The straight cartridge case is brass or steel and has a rebated rim. It contains 13.5 g of propellant for the heavy projectiles and 15.5 g for the Minengeschoss, developing about 280 MPa chamber pressure. Projectiles, which have steel driving bands, are secured in the case by a crimping groove in the case neck. Overall length of the complete round is c.145 mm and the weight is c.180 g with the heavy projectiles, c.160 g with the *M-Geschoss*. Combat loadings include:

- *Sprenggranatpatrone L'spur mit Zerleger*: the principal loading used with the original MG-FF: 134 g HE-T SD including 3.7 g HE; MV was 600 m/s. For use with the MG-FFM the recoil had to be reduced; this was achieved by using an aluminium alloy rather than brass fuze (reducing the weight to 115 g) and reducing the muzzle velocity to 585 m/s.
- *'M' Geschosspatrone*: thin-walled high-capacity shell without tracer, for use with the MG-FFM; 92 g, including 20 g HEI; MV 700 m/s. Fuzes available with or without a self-destruct function.
- *Panzerbrandgranatpatrone (Ph)*: pointed shell of thick, hardened steel, with a phosphorus incendiary capsule inserted from the base; for use with the MG-FFM. Weight 115 g, MV 585 m/s.

## 20 × 82 MG 151 (20 × 83.5)

The Mauser MG 151 gun first entered service in late 1940 as a replacement for the interim MG-FF. Initially it was made in 15 × 96 calibre, but the cartridge case was soon shortened and neck-out to take the high-capacity 20 mm *Minengeschoss* (mine shells) developed for the MG-FFM, creating the 20 × 82 round. The gun designation was changed to MG 151/20, and this remained the Luftwaffe's standard aircraft cannon from 1941 to the end of World War II. Two versions were made: the standard gun used percussion priming but for use in synchronized installations an electrically-primed version was developed, along with the associated ammunition. The cartridges looked identical except for the primer so care had to be taken to ensure that no errors occurred, as the electric-primed rounds would not work in percussion guns and vice versa.

The MG 151/20 was also used by France (latterly as a helicopter gun) until at least the 1970s, and they made brass-cased ammunition for it. A slightly modified version of the gun (using steel

*20 × 82, from left to right: M-Geschoss (HE-T SD); sectioned M-Geschoss with dummy fuze; French TP; South African TP*

cases and percussion priming) is still made in South Africa as the Denel Land Systems (originally Vektor) GA-1. This version of the round is often referred to as the 20 × 83.5 but while such case lengths exist that is not consistent: some South African-made ammunition measured by the author uses cases of around 82 mm. The light weight, small size and low recoil of this gun suit it for fitting to light vehicles and small helicopters. The cartridge is also chambered in the first version of the Mechem NTW-20 anti-materiel rifle.

In the years following World War II the French briefly developed the MG 250, a small revolver cannon chambered for electric-primed MG 151 ammunition in both calibres, but only a few prototypes were made.

The original cartridge case is of brass or steel, rimless with a slight bottleneck. It contains 19 g of propellant, developing a chamber pressure of 285–310 MPa (current production: 275 MPa average, 325 MPa maximum). Projectiles are secured in the case by stab-crimps in the case neck. Overall length of the complete round is 147 mm and the weight is c.180 g for M-Geschoss loadings, c.200 g for other World War II projectiles (current production 205 g).

Combat loadings from World War II included:

- *Sprenggranatpatrone L'spur mit Zerleger*: 115 g HE-T SD including 3.7 g HE; MV was 710 m/s.
- *'M' Geschosspatrone*: thin-walled high-capacity shell without tracer; 92 g, including 20 g HEI; MV 800 m/s. Fuzes available with or without a self-destruct function.
- *Panzerbrandgranatpatrone (Ph)*: pointed shell of thick, hardened steel, with a phosphorus incendiary capsule inserted from the base. Weight 115 g, MV 710 m/s.
- *'MX' Geschosspatrone*: extra-long version of the M-Geschoss to increase the HEI load to the maximum; 105 g, including 25 g HEI; MV 640 m/s; SD fuze. This required a different barrel as the rifling twist had to be changed and the ballistics were different; which also meant that the characteristics of other ammunition types had to be adjusted to match.
- *Brandgranatpatrone M.44*: thick-walled incendiary M-Geschoss fitted with hydrostatic fuze for detonating inside bomber fuel tanks; 104 g; MV 745 m/s.

Current Denel combat loadings are the base-fuzed M1A2 SAPHEI and nose-fuzed M1A3 HEI (an HEI-T is also available). Projectiles have soft iron driving bands, weigh 110 g and are fired at 720 m/s. The SAPHEI is claimed to penetrate 15 mm at 100 m.

## 20 × 94 TYPE 2 / Ho-5

This round was developed during World War II for the Japanese Army air force's Ho-5 cannon, also known as the 20 mm Type 2. It offered a useful combination of effectiveness and light weight, and was the Army's standard aircraft cannon at

*20 × 94: AP-T (medium); Ma 202*

the end of the war. The brass cartridge case is rimless with a slight bottleneck and contains c.20 g of propellant. Overall length of the complete round is 148 mm and the weight is 212–255 g depending on the loading. Principal projectiles were HEI (79 g, inc 3.4 g HE); Ma 202 Special Incendiary (fuzeless, 79 g, inc 3.1 g PETN and 8.8 g incendiary); and Ma 205 Special Incendiary (79 g, with air-column fuze).

AP-T was also used in three versions made of soft, medium and hard steel: these weigh 112, 116 or 120 g respectively and are coloured plain black (soft); black with a green band (medium); or black with green and white bands (hard). Muzzle velocity was stated to be 750 m/s, although the US measured the performance during the war and found it to be 730 m/s with the HEI shell, 700 m/s with AP-T. This performance was very modest for the size of cartridge case and at one time it was thought that the original loading was more powerful, but had to be downloaded because of wartime production difficulties with the gun. However, this was contradicted by later research; the performance was low from the start.

### 20 × 96 HS 421 /422 / MAC Mle 1949 / 2CG

This round was one result of a French project immediately following World War II to develop new aircraft guns to replace the HS 404. Two calibres based on the same cartridge case were made by SFM – 15 × 113 and 20 × 96 – and both HSF and MAC designed guns to fire them (the HS gun is usually designated 421 but there is a 20 × 96 cartridge case drawing headed HS 422 – possibly the 421 is the 15 mm version?) . The cartridge case was based on that of the Soviet 14.5 × 114 anti-tank rifle round. Overall round length is 156 mm for both calibres. Muzzle velocity of the 20 mm is quoted as 863 m/s (compared with 1,125 m/s for the 15 mm) with the projectile weights not known. These were cancelled in favour of the 30 mm DEFA programme (designated 3CG).

## 20 × 99R ShVAK

This was developed in the late 1930s for the Soviet ShVAK aircraft cannon. In fact, the gun was first produced in

12.7 × 108R calibre, but that saw very little use, the cartridge soon being necked out to 20 mm calibre. The performance of the cartridge was not impressive, as the short and light shells held only a small quantity of HE. As well as being extensively used in Soviet aircraft during World War II, a version of the gun was fitted to some light tanks. At the end of the war, the Berezin B-20 appeared, a lighter gun chambered for the same cartridge. The cartridge has survived to this day for use in sub-calibre training devices, and rather curiously was also selected for the Finnish Helenius anti-material rifle of the early 2000s.

The brass cartridge case is slightly tapered and rimmed (unusual for the calibre), and contains 18 g of propellant, developing 300 MPa. Projectiles, which have steel driving bands, are secured in the case by a crimping groove in the case neck. Overall length of the complete round is 145 mm and the weight is 175–180 g. Principal cannon loadings were HE / HEI (91–96 g, of which 3.7–5.6 g was HEI filler) and API (96.6 g, with a hardened steel core and 2.8 g incendiary composition). The use of tungsten carbide-cored API ammunition has been mentioned, with which the maximum armour penetration was stated to be 35 mm. Muzzle velocities were in the range 750–815 m/s, mainly depending on barrel length – this varied significantly for different installations.

## 20 × 99RB SEMAG TYPE L / TYPE 94 FLEX: 20 × 101RB OERLIKON FFL / TYPE 99–2

When SEMAG took over the development of the Becker, they also introduced a more powerful cannon using a longer-cased 20 × 99RB cartridge. Oerlikon continued its development as the Model or Type L. The Japanese Army acquired a few of these guns under the designation 20 mm Type 94 Flex in 1932–33, for fitting to heavy bombers (previous reports that these used a unique 20 × 89RB round are now known to be incorrect). In around 1935 Oerlikon introduced the lighter and faster-firing FFL aircraft gun, with a modified cartridge case: as with the 20 × 72RB, the rim diameter was reduced (to 19 mm rather than 20 mm)

*20 × 101RB Type 99–2 below the 20 × 72RB FF for comparison*

and the extractor groove lengthened by 2 mm, increasing the case length accordingly to 101 mm. At the same time, the propellant load was increased to 21.4 g, raising the muzzle velocity from 670–700 m/s to 750 m/s.

These guns would have seen very limited use were it not for the adoption of the FFL by the Japanese Navy air force as the Type 99–2. This supplemented the smaller Type 99–1 and was the IJN's standard aircraft cannon at the end of World War II.

The brass cartridge case is very slightly bottlenecked with a rebated rim. Projectiles, which have steel driving bands, are secured in the case by a crimping groove in the case neck. Overall length of the complete round is 173 mm and the weight is c.215–220 g. Japanese projectiles are similar to those of the Type 99–1: HE (128 g, inc 10.2 g PETN+TNT), HEI (120 g, inc 5.7 g white phosphorus and 5.0 g TNT), Tracer (128 g), and API (131 g).

## 20 × 102 US

This American cartridge is probably the most widely used 20 mm round in service today, due to the fact that it is chambered in the M61 (Vulcan) six-barrel 'Gatling' rotary cannon fitted to US fighter aircraft and to the naval Phalanx CIWS (close-in weapon system) which is also used on land as the Centurion C-RAM (Counter-Rocket, Artillery, Missile) base protection system. Three-barrel versions of the gun are also made: the M197 helicopter gun and the A-50 made for the fighter version of the South Korean KAI T-50 jet trainer. The round is also chambered in: the US M39 revolver cannon (as fitted to many 1950s USAF fighters, still used in the F-5, and made in Taiwan for land and naval AA mounts); the current French Nexter 20M621; and the Czech ZPL-20, a version of the Russian 23 mm GSh-23 twin-barrel aircraft gun rechambered for the NATO round.

The history of this cartridge goes back to World War II when a very powerful .60 inch anti-tank rifle was developed. This never saw service but the 15.2 × 114 ammunition was used in various experimental aircraft guns developed during World War II, notably the T17 which was a modified Mauser MG 151 but also the T18 which was a modified Hispano. After the war the .60 cartridge remained the USAAF's preferred basis for new aircraft guns, and the first versions of what became the M39 and the M61 were chambered for it. However, the advantages of explosive shells were finally accepted by the USAF so the case was necked out to 20 mm and slightly shortened to keep the overall length the same.

Many different loadings have been developed for this round. The original USAF ammunition was the M50 series which has relatively light projectiles of indifferent aerodynamic shape intended for high-altitude combat (projectile weights are c.100 g and muzzle velocities are 1,030–1,050 m/s); such projectiles lose velocity quickly in the denser air of lower altitudes. The Air Force Armament Laboratory investigated new loadings in the late 1960s; one was a ground-attack series using heavier as well as better-shaped projectiles to maximize velocity retention and armour penetration (136 g at 878 m/s, or 162 g at 808 m/s with a heavy-metal core). The other was optimized for aerial combat and featured better-shaped but lighter projectiles to maximize hit probability (PGU-17/B HEI and PGU-18/B TP); the target ballistics were a weight of 78–84 g at an MV of 1,130–1,160 m/s. The requirements for both of these variants were dropped.

The M50 series was largely replaced by the more streamlined, 'Multipurpose' PGU-28/B SAPHEI from the late 1980s. At this point it should be mentioned that 'PGU' = Projectile, Gun, Unit; the number which follows is a unique designation for a particular projectile; the '/' means the first version; and 'B' means 'aircraft, expendable'. A designation finishing 'A/B' means it's the second version; the differences between the PGU-28/B and the PGU-28A/B are minor, the most obvious being a change from a copper alloy driving band to an iron one.

Following a series of in-bore explosions, PGU-28/B was restricted to emergency wartime use in 2000, pending the introduction of a lower-pressure version of the loading. The M940 is a self-destruct version of the Multipurpose design for use in the Centurion C-RAM system. The latest developments for aircraft use are: the tungsten-cored FAP (Frangible Armour Piercing) PEA 280 from RWM; Nammo's APEX (Armour Piercing EXplosive) which combines a Multipurpose design with a tungsten penetrator; and Diehl's PELE, designated M-DN 203 (or 293 with a tracer), which can penetrate 10 mm armour plate at 500 m.

A family of APDS rounds has been developed for the Phalanx, starting with depleted uranium projectiles which were replaced in the 1980s with tungsten alloy. The plastic sabots within which the projectiles are housed are colour-coded: black (Mk 149 Mod 0 and 1) and subsequently orange (Mk 149 Mod 2) for DU loadings, white for the tungsten Mk 149 Mod 4 and red for the latest Mk 244 Mod 0. There was also a green version which remained experimental.

The Mk 149 Mod 4 used in the original Phalanx fires a 12 mm diameter tungsten alloy projectile weighing 70 g. The Mk 244 used in the latest Mod 1B of Phalanx is loaded with a significantly bigger projectile, 14 mm in diameter and weighing 104.5 g. The longer barrels of the M61 fitted to the Mod 1B, together with the use of more energetic propellant, allow the Mk 244 to maintain a high muzzle velocity despite the heavier projectile.

The brass cartridge case used in service loadings is rimless and bottlenecked (steel, aluminium and plastic cases have been tried). It contains 38 g propellant, developing a maximum chamber pressure of 420 MPa. All service versions of the 20 × 102 use electric priming (there was an experimental percussion-primed variant, the M150 series, made in the 1950s for the T220 AA gun). Projectiles have steel driving bands and are secured in the case by a crimping groove in the case neck. Overall length of the complete round is 168 mm and the weight is c.260 g.

Principal combat loadings (full calibre) are: M56 HEI; M53 API; PGU-28 SAPHEI; M940 MP-T-SD; FAP; APEX; and PELE.

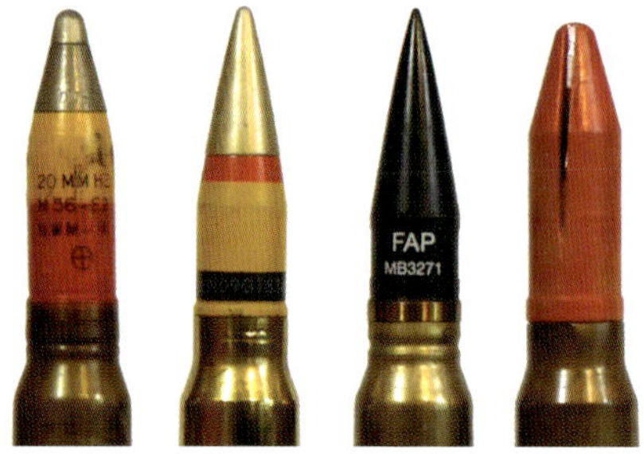


*20 × 102 projectiles, from left to right: M56 HEI; PGU-28/B; FAP; Mk 244 APDS*

Sub-calibre: Mk 149 Mod 4 APDS (94 g fired at 1,120 m/s); Mk 244 Mod 0 APDS (126.5 g fired at 1,100 m/s – from long-barrelled M61 in Phalanx Mod 1B only).

## 20 × 105B

This cartridge was developed in the interwar years for Rheinmetall. As it is the smaller of the two 20 mm rounds identified with Rheinmetall's Swiss Solothurn operation it is commonly known as the 'Short Solothurn'. It was developed in conjunction with the Lb 204 aircraft cannon which never seems to have been adopted, but did see service with various armies in the Solothurn S18–100 series semi-automatic anti-tank rifles. It was therefore generally loaded with AP projectiles, of which there are various types.

The belted cartridge case is of brass or steel and bottlenecked. It contains 30 g of propellant for a muzzle velocity of 725 m/s when firing the heavy 146 g APHE-T shells (760 m/s with 134 g HE). Projectiles, which have iron driving bands, are secured in the case by a crimping groove in the case neck. Overall length of the complete round is 170 mm and the weight is 270–280 g. Commercial loadings were given female name codes, e.g. Anna for the practice/ball round, Liesel for the HE-T and Pauline for the APHE-T.

## 20 × 105 MG 204

This Rheinmetall-Borsig cartridge was developed from the 20 × 105B for the Lb 204 (see above), the change apparently being made to improve the reliability of ammunition feeding. The 20 × 105B was in its turn developed from the 20 × 108 for the Lb 201a and Lb 202 (see below). The 20 × 105 was used in the MG 204 aircraft gun of the late 1930s, which saw only very limited service in some Luftwaffe seaplanes.

The brass or steel cartridge case (weighing 111 or 102 g respectively) is rimless and bottlenecked and contains 27.5 g of propellant giving a muzzle velocity of 760 m/s with the HE shell, 725 m/s with the APHE. Projectiles are secured in the case by a segmented crimping groove in the case neck.

Overall length of the complete round is 170 mm and the weight is 270 g. The principal loadings used a nose-fuzed HE-T SD shell weighing 134 g and containing 3.7 g of Penthrite wax plus a large tracer compartment (which initiated the HE content on burn-out), and a base-fuzed APHE-T weighing 146 g including 1.4 g Penthrite wax.

### 20 × 108 Lb 201a / Lb 202

This was the first iteration of the ammunition for Rheinmetall's compact 20 mm aircraft gun, used in the models Lb 201a and Lb 202. It introduced the range of projectiles which were subsequently loaded into the 20 × 105B Lb 204 and 20 × 105 MG 204, but had a lower muzzle velocity of around 700 m/s with the 134 g HE shells. The dimensions were similar to those of the 20 × 110 Hispano, the most obvious recognition point being the much longer extractor groove, also used on the 20 × 105 MG 204 (see photo below).

## 20 × 110RB (OERLIKON)

After Oerlikon took over the development of the SEMAG guns in the 1920s they developed an even larger version of the gun around this cartridge. It is known as the Oerlikon Type S or SS (or FFS in its later, aircraft version) and is probably the mostly widely used 20 mm cannon ever made. As an aircraft gun it saw only limited use with the French air force in the 1930s, but it was adopted by the RN and USN and became their standard light AA gun in World War II. It lasted in RN service certainly until the late 1980s if not later, and remains in use around the world to this day, although now primarily for use against light surface craft.

*20 × 105B (left) compared with 20 × 105*

The cartridge case is normally brass (although some steel cases were made, particularly in the USA in late World War II) and has a rebated rim with a tapering shoulder. It contains 28.5 g of propellant, developing a chamber pressure of 310 MPa (maximum 360 MPa). Projectiles, which have copper driving bands, are secured in the case by a crimping groove in the case neck. Overall length of the complete round is 181 mm and the weight is 240 g.

Many variations in projectile types have been used over the years, initially weighing around 128 g, but the standard HE types now weigh 120–125 g and are fired at around 830 m/s. Two current exceptions are: an API from Soloboda in Serbia which weighs 142 g, has an MV of 845 m/s and penetrates 25 mm armour at 100 m; and the M74 HEI (and M72 HEI-T) from CBC of Brazil which weigh only 100 g and are fired at 875 m/s. These appear to use M50 series projectiles from the 20 × 102.

### 20 × 110 SCOTTI

Very little is known about this cartridge, of which few examples survive. It appears to be an interim stage between the 20 × 110RB Oerlikon and the 20 × 110 Hispano, since it combines the sloping shoulder of the former with the rimless base of the latter. This accords with the only known headstamp being dated 1933 (and helpfully stamped '20 mm Scotti' as well). It was intended for an experimental gun (probably the M32) developed by Scotti in Italy. Oerlikon was involved in assisting with development, so the cartridge used standard Oerlikon projectiles fired at around 850 m/s.

## 20 × 110 HS (HISPANO)

This cartridge was developed in France in the mid-1930s for use in the Hispano-Suiza HS 404 aircraft gun. It was apparently developed from the 20 × 110 Scotti described above, using a less sloping shoulder. The gun was adopted by the RAF as the 20 mm Hispano, remaining in service until replaced by the 30 mm ADEN in the 1950s, and was also used by US aircraft during and after World War II as the AN-M2 and M3. The last version of the USAF's gun, the M24, had electric rather than the percussion priming of all of the others. The HS 404, and the post-war HS 804 version by the Swiss branch of the company, have been widely used as aircraft and AA guns, and the ammunition remains in production; the guns are also still available. The cartridge was also used in the post-war Bofors m/45 and m/49 aircraft guns.

The cartridge case is rimless and bottlenecked, and usually made from brass (although steel cases may be encountered in post-war US production). It contains 32.5 g of propellant, developing an average chamber pressure of 335 MPa. Projectiles, which have steel driving bands, are secured in the case by a crimping groove in the case neck. Overall length of the complete round is 184 mm and the weight is 257 g.

HE projectile designs and weights were originally very similar to those for the 20 × 110RB described above, but fired at a higher muzzle velocity of 850–880 m/s (depending on barrel length). Initially, the nose fuze was too sensitive, exploding the shell on the surface of the target, so the inert ball round was used in a combat role until a delayed-action fuze was developed in 1941. The HE was replaced by an HEI weighing 130 g and containing a total of 10.75–11.4 g of HEI compounds; this went through several Marks, with minor variations between them.

British AP projectiles were generally heavier, e.g. the wartime British Mk 2.z

*20 × 110 Scotti case (ECRA Bulletin)*

weighed 140 g and was fired at 811 m/s. This was a two-piece shot; a pointed, hardened steel body with a bakelite ballistic cap with a blunt tip. Performance of the Mk 2.z against High Hardness plate was as follows:

Range 200 yards (183 m), penetration: 27 mm/0°; 24 mm/20°; 19 mm/40°
Range 400 yards (366 m), penetration: 24 mm 0°; 19 mm/20°; 15 mm/40°

The AP Mk 3.z was an APCR type with a tungsten core inside a duralumin body. This projectile weighed only 96.5 g (of which 57.5 g was the core) and was fired at 985 m/s. At 200 yards (183 m) it penetrated 67 mm/0° or 48 mm/30°; at 400 yards the penetration was 51 or 38 mm respectively (these figures assumed an additional 107 m/s velocity from the aircraft speed). Despite this impressive performance, accuracy was poor. The A.P. Mk 4.z was a much simpler design, a steel one-piece 141 g projectile withg a cavity filled with inert material.

The AP (and especially the Mk 3) were primarily intended for attacking armoured vehicles. For aerial targets, the RAF used the SAPI (semi-armour piercing incendiary), which was an HE shell with a hardened steel nose instead of a fuze; the incendiary contents were ignited by the shock of impact. This weighed 134 g (including c.12 g of incendiary material) and was required to penetrate 20 mm armour plate at 0°/200 m, then ignite petrol in cans placed behind the armour.

By the late war, the typical RAF belt loading was two HEI followed by two SAPI.

The US adopted the British HEI design as the Mk 1; this fired a 130 g shell (containing 11.3 g filling – 7 g of tetryl plus 4.3 g incendiary) at 850 m/s. The US M75 AP-T was their own design, a monolithic solid steel shot weighing no less than 165 g and was propelled at 777 m/s; this could penetrate 17.8 mm of homogenous plate or 16.5 mm of face-hardened plate at 500 yards (457 m) and 20°.

Post-war US designations included the M95 AP-T, M96 Incendiary and M97 HEI (containing 7.7 g HEI), all weighing c.130 g and were fired at 832 m/s. In the final generation, the M97 was replaced by the M58 which was lighter (c.120 g) but contained 13.6 g HEI. Both percussion and electric primed US ammunition used the same designations with the addition of 'Electric' where appropriate. Modern loadings of the same HE shell types tend to match those of the 20 × 110RB, but are typically fired at 845 m/s.

British 20 × 110 HS: HEI; SAPI; AP; Ball

### 20 × 110 (AP-20 / ATSLEGA)

This Russian cartridge was developed in 1937 following criticism of the relatively weak 20 × 99R ShVAK round. As well as the Atslega AP-20, it was reportedly used in the DShAK from OKB-2 and the initial prototype of the Berezin B-20 of 1941. The rimless cartridge case was larger than the ShVAK's in all dimensions resulting in an increase in muzzle velocity to 940 m/s despite the heavier projectile, which was based on the Oerlikon's and weighs 129 g. Overall length is 168 mm.

### 20 × 110R (TsKBSV-75)

This was a rimmed equivalent of the Atslega AP-20 round described above, intended for an AA version of the ShVAK cannon which was designed to work with rimmed cases. The rim diameter is the same as the AP-20 case, but the body diameter was reduced from 31 to 27 mm. This reduction in case capacity would presumably have reduced the muzzle energy. The projectile weight is believed to be 129 g, but the muzzle velocity is not known.

## 20 × 110 USN Mk 100 SERIES

The USN did not initially share the USAAF/USAF's interest in new revolver and rotary guns. Instead, they improved the performance of the Hispano by speeding up its rate of fire and rechambering it for a powerful new round, the 20 × 110 USN, known as the Mk 100 series ammunition (the initial designation was T130 series). This was developed by taking the base dimensions of the 20 × 102, but stretching the case length to the maximum that the Hispano could handle. It hardly seems worth the effort as the performance is only marginally better than the 20 × 102. The modified Hispano was designated the Mk 12 and was the USN's standard aircraft cannon from the mid 1950s until the 1970s when it was finally replaced in new aircraft by the M61. As well as the Mk 12, the new round was chambered in a remarkable twin-barrel revolver cannon, the Mk 11, which was used only in the Mk 4 gunpod. This ammunition is no longer made, has been steadily going out of use as the planes that carry the guns are retired (the A4 Skyhawk, still in service in various countries, being probably the last surviving platform), and is therefore obsolescent.

The steel cartridge case is rimless and bottlenecked. It contains 41 g of propellant, developing a maximum chamber pressure of 400 MPa. Like its shorter brother the 20 × 102, it uses electric priming. Projectiles, which have steel driving bands, are secured in the case by a crimping groove in the case neck. Overall length of the complete round is 185 mm and the weight is 270 g. Combat loadings are the Mk 106 HEI (nose fuzed), Mk 107 API (hardened steel penetrator in the base, with incendiary material in the nose cap) and Mk 108 AP-T: they all weigh 110 g and are fired at 1,030 m/s.

## 20 × 113, 20 × 113B LAHTI

These are among the least-used service cannon rounds. The L-34 'boat gun' was developed in Finland in the mid-1930s by Aimo Lahti. It was a light automatic cannon made by VKT and fitted to ten patrol boats of the Finnish Coastguard. The cartridge case for this gun is brass, rimless and tapered as well as slightly bottlenecked. Projectiles, which have brass driving bands, are secured in the case by a segmented crimping groove in the case neck. Overall length of the complete round is 177 mm and the weight is around 300 g with a 148 g AP projectile, fired at 700 m/s; a lighter HE loading was also made. When firing AP shot penetration was 30 mm/300 m/90° or around 20 mm/300 m/60°.

Subsequently Tampella took over production and developed a belted version of the cartridge. A brochure for this gun gives an overall round weight of 275 g, with both HE and AP-T projectiles weighing 136 g and fired at 800 m/s; a significantly more powerful loading than the original.

Lahti also developed a number of other 20 mm rounds. The complete list of all known Finnish 20 mm cases is below. The purpose of the experimentals (in italics) is unclear.

*20 × 75B (rim 26.0 mm, body 25.1 mm)*
*20 × 81 (rim 23 mm, body 23.5 mm)*
*20 × 97B (rim 25.4 mm, body 26.2 mm)*
*20 × 98B (rim 27.5, body 26.0 mm)*
*20 × 99B (rim 25.0 mm, body 25.6 mm) very close to 20 × 97B – may be the same*
20 × 113 (rim 26.1 mm, body 26.1 mm) early version of beltless ammo for L-34 gun?
20 × 113 (rim 27.9 mm, body 28.0 mm) later (more common) version of L-34 beltless?
20 × 113B (rim 27.5 mm, body 26.0 mm) Tampella version of L-34, also aircraft cannon prototypes L-35 and L-36?

### 20 × 114 TPV MODEL 1943

This was the last attempt by Russia to field a more powerful 20 mm cannon during World War II. The cartridge case was based on the 14.5 × 114 anti-tank rifle round, necked-out to the larger calibre. Rim and body diameters are therefore 27 mm, and the overall length with an HE projectile is 182 mm. Performance data is not known, although the ME would probably have been in the region of 40–44,000J, resulting in an MV with a notional 129 g projectile of 800–820 m/s.

*20 mm Lahti: 20 × 81; 20 × 113 (26 mm rim); 20 × 113 (28 mm rim); 20 × 75B; 20 × 98B; 20 × 97B; 20 × 99B; 20 × 113B (Woodin Laboratory)*

### 20 × 115SR DELAUNAY-BELLEVILLE

This was a late-World War I project for an anti-tank machine gun designed by the luxury car-making firm of Delaunay-Belleville to meet a French Army requirement. The company also designed the ammunition for it; essentially a shortened and semi-rimmed version of an experimental rimmed 'Tube-Canon' round from 1900, which had been designed for sub-calibre practice with naval guns. Armour-piercing projectiles of various types were tried, some with separate driving bands and others brass-jacketed, varying in weight from 80 to 126 g, but following comparative trials in September to October 1918 a 102 g brass-jacketed type was selected. With a 30 g propellant charge of Ballistite this was fired at 895 m/s but signs of overpressure caused the charge to be reduced to 28 g. Testing went on until May 1919, after which the project is no longer mentioned.

## 20 × 120 MADSEN

This round was developed in Denmark for the Madsen automatic cannon, offered from the 1930s in various mountings for aircraft, AA and anti-tank use, although its primary use was in the AA role. It was never the primary armament of any major army, but was widely used in World War II and remained in service for decades afterwards. The ammunition for the gun was made in several countries, most notably in the UK by Kynoch, as was usually the case for Madsen ammunition as they had no ammunition production facilities of their own. Apart from Madsen's own cannon it appears likely that the experimental Norwegian Brøndby M/1940 self-loading anti-tank gun was also designed around this cartridge, as Brøndby obtained his ammunition from Madsen.

The brass cartridge case is rimless and bottlenecked and has a notably thick rim. This rim was the result of the tilted-block gun action, which requires a large headspace in order that the face of the block will clear the breech as it pivots; the cases accordingly need a thick rim and head to withstand the pressures on the unsupported part of the case. Overall length of the complete round is c.185 mm and the weight is around 300 g (both varied depending on the loading). The ammunition evolved over the production period. Principal loadings were a 154 g AP shot fired at 780 m/s and a 127 g HE at 840 m/s, but other loads mentioned used incendiary and HE shells of 113 g (fired at up to 880 m/s), AP-T of 146 g, and both tracer and APHE (with a bursting charge of 3.5 g) of 136 g. The APHE was also known as the APES (Armour Piercing Explosive Shell) which was fuzeless, relying on the shock of impact to ignite the contents. In tests held in 1940 at the US Aberdeen Proving Ground the MV of the 146 g AP-T was measured at 820 m/s and of the 134 g APES at 828 m/s.

Madsen experimented with more powerful 20 mm ammunition. In the 1930s a version of the standard case was produced, lengthened to 134.4 mm. In the 1950s came two further designs, both using a wider than standard case (35.5 mm rather than 29.5 mm rim); one case design was 144.3 mm long, the other was 169 mm. Only the last of these is known to have survived physically, in the form of an unfired case (shown in the experimental group photo with a replica projectile). No ballistic data are known.

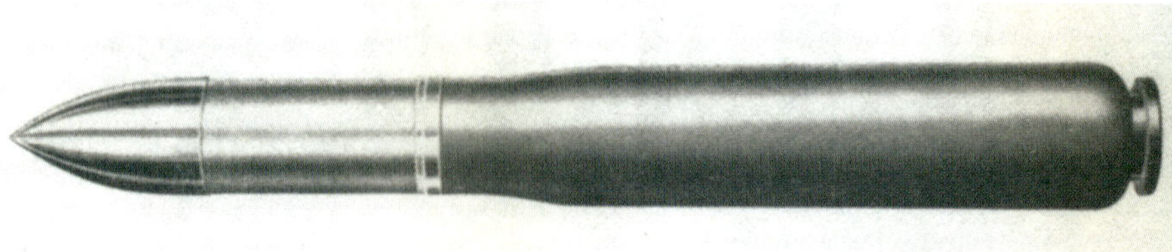

*20 x 120RB Simoneta*

*20 × 126B (bottom) compared with the 20 × 122 case*

### 20 × 120RB SIMONETA

Virtually nothing is known of this round which was briefly advertised by FN in the interwar years, and even less about the gun which it was designed for, but judging by the shape of the case it might have been an Oerlikon-type API blowback. Examples of the brass cartridge case survive; it is bottlenecked with a sloping shoulder and a markedly rebated rim.

### 20 × 122 5CG

Developed as part of an unsuccessful 1950s French programme for a light AA gun. German projectiles were used for test purposes; see the 20 × 126 B below. The brass cartridge case is rimless and bottlenecked. Projectiles, which have steel driving bands, are secured in the case by a crimping groove in the case neck. Overall length of the complete round is 188 mm and the weight is c.300 g.

## 20 × 124 Ho-1, Ho-3

This Japanese Army cartridge was first used in the Type 97 anti-tank rifle introduced in the late 1930s, but during World War II the mechanism of this semi-automatic gun was adapted to create an automatic aircraft cannon. This saw limited use in two versions; the Ho-1 for flexible mountings and the Ho-3 for fixed use. The brass cartridge case (sometimes described as 125 mm – it is actually 124.5 mm) is rimless and bottlenecked and contains c.37 g of propellant. Overall length of the complete round is c.195 mm and the weight is c.320 g. Principal loadings were the Type 98 HEI-T (136 g with 4.0 g Cyclonite), Type 100 HEI-T SD (132 g with 4.5 g Cyclonite), and Type 100 AP-T (162 g). There was also a 109 g Ma 201 Special Incendiary, stated by some to be fuzeless (with 4.3 g PETN burster in the nose and 8.8 g incendiary mix in the body, giving a total load of 13.1 g), but other sources say that it was designed not to be triggered by the aircraft metal skin but only after penetrating to the fuel tanks; possibly both are correct. Muzzle velocity was 792 m/s with the AP-T ammunition, around 820 m/s with HE.

### VICKERS .8-inch (20 × 125SR)

This was developed for the .8-inch Vickers Automatic Gun of 1931; the precise measurements are 20.3 mm calibre and 124.7 mm case length. It seems likely that this was the same as the VA '0.8-inch anti-tank gun' mentioned in the Small Arms Committee Minutes of December 1930, in which the projectile weight was stated to be 128.5 g and the MV 884 m/s. Both API and nose-fuzed HE loadings were available; the surviving ammunition is fitted with a fuzed projectile.

### 20 × 126(B) 5CG

Developed as part of an unsuccessful 1950s French programme for a light AA gun. German projectiles were used for test purposes; see the 20 × 122 above. The bottlenecked brass cartridge case was made both with and without a belt. Projectiles, which have steel driving bands, are secured in the case by a crimping groove in the case neck. Overall length of the complete round is 192 mm and the weight is 300 g.

## 20 × 128 OERLIKON

This is one of the two European 20 mm rounds to have been introduced since World War II. It was developed by Oerlikon in the 1950s for their KAA cannon, intended as a successor to the famous 20 mm Type S. This cartridge is far more powerful, and the gun has a much higher rate of fire. It has not matched the level of success of the older Oerlikon, although it has achieved numerous sales, primarily as a naval gun. The cartridge was also adopted, in electric rather than percussion-primed form, for the unique Spanish Meroka volley gun, used as a naval CIWS.

The steel cartridge case is rimless and bottlenecked. It contains 53–58 g propellant developing a chamber pressure of 330 MPa (initially) to 420 MPa (current loadings). Projectiles are secured in the case by a segmented crimping groove in the case neck. Overall length of the complete round is 204 mm and the weight is 340–360 g. Projectile weights for full-calibre HEI/SAPHEI loadings are 125–128 g and muzzle velocities are 1,050 m/s (KAA) or 1,100 m/s (longer-barrelled KAB). Projectile types in combat loadings are:

- HEI SD (nose-fuzed): contains 10 g hexal (Tracer version contains 5.6 g).
- HEI (base fuzed): contains 8.9 g hexal/incendiary, optimized for AA use.
- SAPHEI (base fuzed, contains 4.7 g hexal (Tracer version also).
- API-T: contains tungsten carbide penetrator.

Initially an AP-T loading was also offered, weighing 115 g and with an MV of 1150 m/s (KAB).

The APDS loading for the Meroka CIWS has a tungsten penetrator 12 mm in diameter in a light alloy and plastic sabot, a total weight of 310 g, a projectile weight of 75 g (102 g including sabot), 60 g of propellant for an MV of 1,215–1,290 m/s (sources vary), and penetration of 30 mm at 1,500 m.

In 2015 Oerlikon introduced a new SAPPIE-T (Semi-Armour Piercing Pyrotechnically Induced Effect – Tracer) loading for use with the new KAE cannon designed for remotely-controlled weapon stations. This round combines armour-piercing characteristics with a 'cascading pyrotechnic impact' inside the target, yet contains neither explosives nor fuse.

### 20 × 135B KRIEGHOFF MG 301

The development path of this obscure round is not entirely clear. Krieghoff produced a series of experimental gas-operated 20 mm cannon, starting in the late 1930s, all of which appear to have used this cartridge. Unterlüss reports indicate that it was for an early attempt at the 'million point gun' (see below) with an MV of 1,000 m/s. However, in 1942 the Luftwaffe commenced a project to develop '*HT Geschosse*' for use against armoured targets (i.e. protected bombers) up to about 200 m range. These were very long projectiles of various sizes and types (up to 162 mm long) which were set deeply into the case, limiting the propellant capacity – and therefore the muzzle velocity to around 500 m/s. The projectiles were developed by Hasag for this cartridge. Clearly, they could not have relied on spin-stabilisation as they were too long; it is not clear how stability was achieved, as little information about this project seems to have survived.

The cartridge case was the precursor of the 20 × 135 MG 213 described below, and has a more tapered case as well as a belt.

*20 × 128: sectioned HE and SAPHE; Meroka*

### 20 × 135 MAUSER MG 213/20

This cartridge was developed for the Luftwaffe's 'million point gun' project, for an aircraft gun (designated MG 213) which would achieve a muzzle velocity of 1,000 m/s and a rate of fire of 1,000 rpm. Various types of gun mechanism were tried, the successful one being the Mauser MG 213C. This was the original modern revolver cannon, first developed around this high-velocity round before attention switched to a 30 mm version. The two versions were then designated MG 213/20 and MG 213/30. A few test prototypes in both calibres were made by the end of World War II.

The large, steel 20 × 135 cartridge case is rimless and bottlenecked, electric primed, and weighs 154 g. The first version of the cartridge for the MG 213A has a tapered case, but the MG 213C is more cylindrical. Projectiles, which have iron driving bands, are secured in the case by a crimping groove in the case neck. Overall length of the complete round is 199.8 mm and the weight is c.340 g. The standard HE-T projectile weighs 117 g (the tracer burned to 1,250 m) and 63 g of Nz.R.P. propellant gave a muzzle velocity of 1,000 m/s. The Unterlüss report mentions three other loads:

- *MX-Geschoss*, using the lengthened version of the M-Geschoss (also used in the 20 × 82): weight 104 g; HE weight 24.5 g; MV 1,000 m/s.
- *Incendiary-tracer* (by Hasag): weight 115 g; MV 960 m/s.
- *Incendiary* (*Brandgranate 44*): weight 107 g; MV 990 m/s.

After the war, an example of the MG 213/20 arrived in the USA where it was carefully analysed under the designation T74, and various modifications made. The 20 × 135 ammunition was retained for these analyses, but made with brass rather than steel cases (designated T7). It was probably also used in the experimental ARF T33E3 gun. The loading for this ammunition reportedly used a conventional 104 g projectile (T61E1 – a shortened version of the 20 × 110 Hispano M99 TP projectile) at an MV of 1,150 m/s. Chamber pressure was 338 MPa.

### 20 × 136B ADEN

The UK took over the development of the MG 213C post-war (as did various other countries) but was primarily interested in the 30 mm version. The 20 mm round for the '30/20 mm ADEN' was experimented with between 1948 and 1952 but probably only in ballistic test barrels; there is no record of a gun being made in this calibre. Unlike the German 20 × 135, the brass cartridge case was belted (a feature retained from the 30 mm version). Overall length of the complete round is 199 mm, as with all ammunition intended to be used in guns closely based on the MG 213C. A unique range of projectiles was designed for this round, weighing 83–95 g (with light alloy fuzes: brass fuzes weighed 13 g more). The projectiles have a low-drag nose design but a hemispherical base that is not aerodynamically satisfactory. Muzzle velocities were in the range 1,100–1,200 m/s at a chamber pressure of 350 MPa.

### 20 × 137

Simply the 25 × 137 necked down to 20 mm. The purpose and performance could not be determined, but it was most probably a US experiment aimed at exploring possibilities for future fighter guns following the failure of the GAU-7/A.

*20 × 138B from left: Flak HE; U-boat SAP; Pzgr 40 (repP); Italian HE; Finnish AP*

*20 × 139 FM-K.38 / Flab-K.38 (courtesy Robert Gerber)*

## 20 × 138B RHEINMETALL

This was the larger of the 20 mm cannon cartridges developed by Rheinmetall in the 1930s, with the earliest date-stamped round being 1933. It is popularly if inaccurately known as the 'Long Solothurn' round as the Swiss company was used as a legal 'front' for those of Rheinmetall's activities which were banned under the Treaty of Versailles. It proved far more successful than the 20 × 105B described above, being primarily used in the Flak 30 and 38 AA guns that were extensively used by German and other Axis forces in World War II and remained in service in some nations for decades after the war. It was also used in the KwK 30 and 38 light AFV guns, the Solothurn S18–1000 series anti-tank rifles, and the MG C/30L aircraft (and subsequently AA) gun, which following experience with the Legion Condor in Spain was given modified ammunition with a booster charge above the primer. As well as these German and Swiss weapons, it was used in the Finnish Lahti L39 anti-tank rifle and L40 AA gun, and in the Italian Breda M35 and Scotti AA cannon. The ammunition was examined in the USA and cartridge cases made in 1941 under the designation 20 mm T1.

The belted cartridge case is of brass or steel, bottlenecked with a tapered shoulder, and contains a variable quantity of propellant, depending on the projectile weight: 147 g AP-T used 36.5 g to give an MV of 800 m/s, 115 g HE-T used 40.5 g to give 900 m/s, and the 101 g APHC *(Pzgr 40)* used 44 g to achieve 975 m/s.

Overall length of the complete round is 203 mm and the weight varies with the loading and case material between around 280 g for the HE-T to 340 g for a Finnish-made AP. A very wide range of ammunition was developed for this gun, including *M-Geschoss* shells and *Pzgr 40* tungsten-cored APHC *(Hartkernmunition)*. Projectiles used by Germany in World War II included:

- *Panzergranatpatrone L'spur*: a fuzeless AP-T weighing 143 g with an empty central cavity or 148 g with a phosphorus capsule (thereby an API-T). Armour penetration was 31 mm/60°/100 m or 25 mm/60°/300 m.
- *Panzersprenggranatpatrone L'spur mit Zerleger*: a base-fuzed APHE-T SD weighing 121 g (including 4 g HE) in which self-destruction was achieved via tracer burnout (naval use).
- *Sprenggranatpatrone L'spur mit Zerleger*: a nose-fuzed HE-T SD weighing 115 g and containing 7 g HE.
- *Brandsprenggranatpatrone ohne L'spur mit Zerleger*: a nose-fuzed HEI SD weighing 117 g and containing 20 g HEI (the omission of a tracer left far more space for HEI).
- *'MX' Geschosspatrone L'spur mit Zerleger*: a nose-fuzed Minengeschoss SD shell weighing 109 g and containing 24 g HEI. The tracer was a separate element attached to the base of the shell, instead of being integral. MV was 950 m/s. Experimental only, no evidence of use.
- *Panzergranatpatrone 40 L'spur*: a tungsten-cored *Hartkernmunition* principally used in the KwK 30 and 38 cannon fitted to armoured cars and light tanks. Various projectile designs were made. Armour penetration was 49 mm/60°/100 m or 37 mm/60°/300 m.

## 20 × 139 FM-K.38 / Flab-K.38

This was developed in Switzerland in the late 1930s for the FM-K.38 aircraft gun *(Flugzeugmotor Kanone Modell 1938)*, which saw limited use, and the Flab-K.38 AA gun *(Fliegerabwehr-Kanone Modell 1938)*, which remained in use for some time after the war.

The cartridge case is rimless and bottlenecked and contains 53.3 g of propellant (HE loading) or 50.8 g (AP-T). Overall length of the complete round is 205 mm and the weight is 336 g with the HE shell. Principal loadings were a 131 g HE shell containing 4.3 g HE and fired at 1,010 m/s, and a 147.5 g APT fired at 930 m/s, which could penetrate 300 mm/60°/200 m. This cartridge is very similar to the post-war 20 × 139 cartridge for the Hispano-Suiza Hs 820 (see below), but the rim is thinner and the cases are made of brass rather than steel.

## 20 × 139 HISPANO

This one of the two post-war European 20 mm rounds, and a close rival for Oerlikon's 20 × 128. It was based on the existing (now obsolete) 20 × 139 for the Swiss Flab Kan 38 and FMK 38 (see above), but features a steel rather than

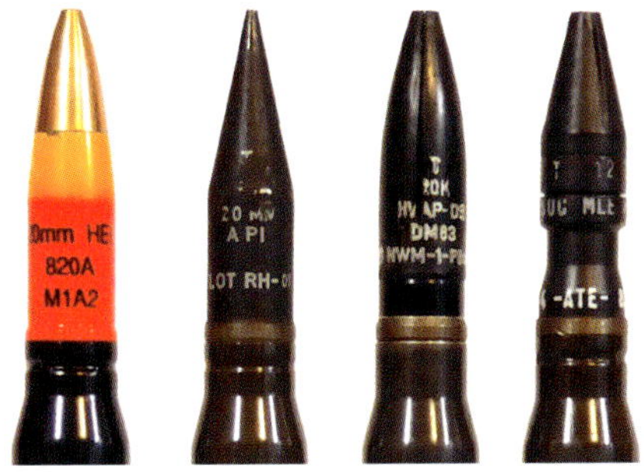


*20 × 139 HS: HEI; API; APDS; French APDS (20 × 141)*

brass case, a thicker rim, and different loadings. It was originally introduced by Hispano-Suiza in the 1950s for their HS 820 cannon, which they saw as the natural successor to the HS 804. Hispano-Suiza was taken over by Oerlikon in 1971, after which the HS 820 was renamed the Oerlikon KAD. Oerlikon is now a part of Rheinmetall Waffe und Munition. The cartridge has been more successful than the 20 × 128, mainly because of its use in the popular Rheinmetall Rh 202 cannon. It is also chambered in the GIAT F2 / 20M693, which is made in South Africa as the Vektor (now Denel) G1–2. These guns have seen various uses, as naval and land-based AA, light AFV armament and in helicopters. They are less used now as the focus shifts to 25 mm or 30 mm calibres for most purposes.

The steel cartridge case is rimless and bottlenecked. It contains 52 g of propellant, developing an average chamber pressure of up to 420 MPa (450 maximum); this is for Denel production, other sources quote 360–405 MPa. Projectiles, which have steel driving bands, are usually secured in the case by a crimping groove in the case neck. Overall length of the complete round is 213 mm and the weight is 315 g.

A wide range of different loadings has been developed by several manufacturers. Full-calibre projectiles typically weigh 120–125 g and are fired at a muzzle velocity of 1,045 m/s, such as the HEI (DM 101: 6.5 g HEI) and the HEI-T SD (DM 81) originally made by RWM. Two variations in loadings from the same source were the API-T (DM 43: 111 g at 1,100 m/s: penetrates 32 mm armour at 1,000 m) and the APDS-T (DM 63: tungsten carbide penetrator, 108 g; 1,150 m/s).

France developed an APDS loading with a slightly extended case (141 mm). Nammo makes an NM 75 Multipurpose round with standard ballistics (MP-T-SD in their terminology, more conventionally a SAPHEI-T SD); this is capable of penetrating 12.7 mm/60°/200 m and will be initiated by 1.5 mm dural at up to 1,000 m range. A more recent development is Diehl's PELE fragmenting ammunition, designated M-DN 213. It also has standard ballistics and can penetrate 12 mm armour plate at 500 m.

## 20 × 140B WZÓR 1938

Poland tested various 20 mm guns from Solothurn (Rheinmetall), Madsen and Oerlikon but, having decided to develop their own cannon, they settled on Rheinmetall's 20 × 138B round as the basis for ammunition development. The resulting cartridge bears a very close resemblance to the 20 × 138B but the case is slightly larger in key dimensions: 139.6 mm long, with rim, belt and body diameters of 28.0, 29.6 and 28.0 mm, compared with 27.0, 28.5 and 27.0 mm for the German case. In consequence, the Polish ammunition would not fit into the German guns, whereas the German ammunition could be (and reportedly was) fired in the Polish guns, albeit possibly with some loss in performance. It is unclear whether this advantage was deliberate or just a happy accident.

The overall length of the round remains the same at 204.5 mm, and the overall weight (with a 134 g projectile) is 320 g. Muzzle velocity with this projectile was 856 m/s. The capacity of the case was increased from 52 mm$^3$ to 55 mm$^3$, providing marginal increases in propellant charge, muzzle velocity and armour penetration. Solothurn projectile designs were used, to save time, and the following rounds were developed:

- PWS: APHE-T with a base fuze;
- PZS: API-T
- PC and PCS (TP and TP-T for AP).
- LWS: HE-T
- LC and LC-S (TP and TP-T for HE)

About 90,000 AP rounds were ordered before the war.

Polish AP ammunition showed the following performance:

| Armour penetration: | type of armour, | angle of impact, | thickness / distance |
|---|---|---|---|
| Carbonated armour plates at 0°: | 20 mm / 300 m; | 15 mm / 500 m | |
| Carbonated armour plates at 30°: | 20 mm / 100 m; | 15 mm / 400 m | |
| Homogeneous armour plates at 0°: | 40 mm / 200 m; | 25 mm / 800 m; | 15 mm / 1500 m |
| Homogeneous armour plates at 30°: | 25 mm / 300 m; | 20 mm / 500 m; | 15 mm / 600 m |

## 20 × 143 TYPE 98

This was the ammunition used for the Japanese Army's Type 98, their standard light AA gun from the late 1930s to the end of World War II. This powerful round was also credited with some anti-tank capability and so came with AP as well as HE projectiles. The large brass cartridge case was based on that of the 20 × 157 Type 94 (see below) but modified to use the same, more conventional, projectiles as the 20 × 124. The case is rimless and bottlenecked and contains c.58 g of propellant. Overall length of the complete round is 214 mm with the AP projectile (208 mm with HE) and the weight is 430 g (404 with the HE-T shell). Principal combat loadings were the Type 98 HEI-T (136 g with 4.0 g Cyclonite), Type 100 HEI-T SD (132 g with 4.5 g Cyclonite) and Type 100 AP-T (162 g). Muzzle velocity was 950 m/s with the HE-T SD loading; the AP-T would have been somewhat less.

## 20 × 145R BOFORS

This was a Bofors development, originally for an unsuccessful aircraft gun which

*20 x 154B AME 20M571*

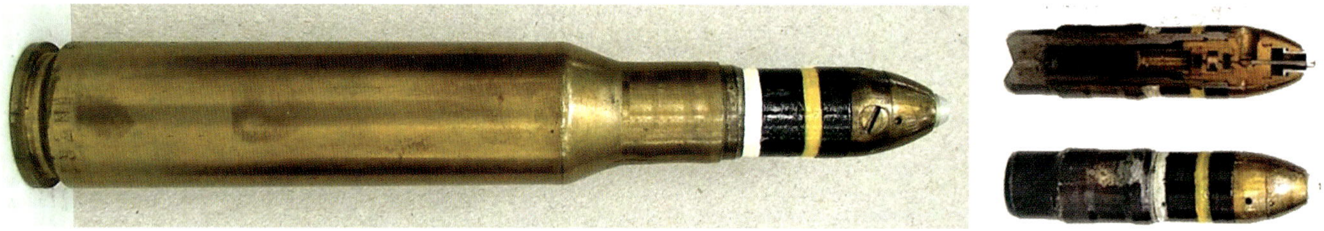

*20 x 157 Type 94 (photos courtesy of Ken Elks from his book on Japanese Ammunition 1880–1945)*

was later developed into the Army's m/40 AA and anti-tank gun. It was in effect a scaled-down version of their famous 40 mm, but did not achieve anywhere near the same success. The brass cartridge case is tapered with a slight bottleneck and has a large rim. It contains 35 g of propellant, generating a chamber pressure of 330 MPa. Overall length of the complete round is c.220 mm and the weight is 290–300 g, depending on the loading. Principal loadings were AP-T and HE-T, weighing 139–145 g, with muzzle velocities of 800–810 m/s. The tracers burned for 3–3.5 s and the HE content weighed around 14 g. The ammunition was subsequently used in sub-calibre training systems.

### 20 × 154B AME 20M571

An experimental French revolver cannon project from AME in 1957. It was clearly an attempt to squeeze the maximum MV from a 20 mm case which was based on the standard 30 × 113B. The greatly extended case length of 154 mm resulted in a significantly increased propellant capacity compared with the 20 × 135B ADEN, but only left space for a very short and stubby projectile. Somewhat astonishingly, there was also a 20 × 163B version, with only 30 mm of projectile protruding from the case mouth. No performance data for this ammunition (or the gun) have been located.

### 20 × 157 TYPE 94

The Type 94 AA gun and its unique ammunition saw only trial use with the Japanese Army before being replaced by the Type 98 described above. The Type 94 gun was based on a Solothurn cannon, modified to take a big new purpose-designed cartridge, but it proved unsatisfactory.

The huge brass cartridge case is rimless and bottlenecked, and is unusual in having a very long neck, with almost half of the long HE projectile inside the case when loaded (the driving band being halfway up the projectile), so the overall length of the complete round is more or less the same as the Type 98 at about 215 mm. No source has been found for any data on the ballistics for the Type 94 round, but performance was probably similar to the Type 98.

### 20 × 158RB T5

This massive cartridge was developed in the USA in the late 1940s for the unsuccessful T33 aircraft gun. At that time the USAF was concerned to achieve the highest possible muzzle velocity to maximize the chance of hitting enemy aircraft, but this gun and ammunition were dropped by the early 1950s in favour of the more efficient weapons designed around the 20 × 102. The brass cartridge case (designated 20 mm T5) is bottlenecked and has a rebated rim. It contains 58 g of propellant. Projectiles, which have brass driving bands, are secured in the case by a crimping groove in the case neck. Overall length of the complete round is 230 mm and the weight is 375 g with a 130 g M99 TP projectile. Calculated muzzle velocity was 1,150 m/s at a chamber pressure of 338 MPa (maximum 380 MPa). In the Colt T24, 56 g of 4,879 I.M.R. powder generated 1,070 m/s, at a pressure of 355 MPa.

### .8-inch ELSWICK (20 × 167SR)

The .8-inch Elswick (actually 20.3 mm calibre) was developed by the British Design Department in 1927 as the basis for an infantry anti-tank weapon (DD/L/3468). Elswick developed an automatic gun to fire it, and trials took place between 1928 and the early 1930s. It fired a 146 g bullet at 869 m/s; a great deal of experimentation was carried out to determine the optimum case size and chamber pressure, although the case seems very large for the performance.

# 22–28 mm CARTRIDGES

This class of ammunition has never been as popular as the 20 mm or 30 mm rounds, with few cannon in these calibres being widely used. Up to the end of World War II, the Russian 23 × 152B VYa and 25 × 218R M1940, French 25 × 163 Hotchkiss (mainly used by the Japanese Navy) and US Navy 1.1-inch (28 × 199SR) were the only ones to see significant use. Since then, the most successful have been the Russian 23 × 115 aircraft gun round and 23 × 152B ZU, and the 25 × 137 NATO used mainly in ground and naval systems. However, the 23 mm guns are now obsolescent in first-line armed forces (with the exception of the recent Chinese 'chain gun' in 23 × 115, used in helicopter and light vehicle mountings) and the previous dominance of the 25 × 137 in NATO and associated countries is being challenged by the 30 × 173. The most recent introductions in this calibre range whose use is still growing are the German 27 × 145B for the BK 27 cannon and the Chinese 25 × 182B Type 87.

## 22–28 mm SERVICE CARTRIDGES

*20 × 110RB (for scale)*
*23 × 106 Madsen*
*23 × 114 NS-23*
*23 × 152B ZU*
*23 × 260 R-23 (repA)*
*24 × 139 Pzw.Kan 38*
*1-inch Vickers (25 × 87R)*
*25 × 87 Revelli-FIAT*
*25 × 137 NATO*
*25 × 159.5 DCAN Mle 1935 (repA)*
*25 × 163 Hotchkiss Mle 1938*
*25 × 173/181 Oerlikon KBB*
*25 × 182B PG87 (repA)*
*25 × 187SR Bofors M/38*
*25 × 205SR Bofors M/32 (case only)*
*25 × 218 2-M3*
*25.4 × 189 Vickers*
*27 × 145B Mauser BK 27*
*1.1-inch USN (28 s 199SR)*

## 22–28 mm EXPERIMENTAL CARTRIDGES

### 22/23 × 138B

In 1940 Rheinmetall examined a proposal to improve the destructive effect of the 20 × 138B round in the MG C/30 by necking the case up to take a 22 mm projectile. It is not known whether this was actually made, but factory drawings of three different nose-fuzed HEI-T loadings survive; one with a 170 g shell containing 21 g HE and a large tracer (MV unspecified); one with a 146 g shell containing 24 g HE and a small tracer, fired at 905 m/s; and a slight variation on this weighing 150 g, containing 26 g HE and fired at 880 m/s. Overall cartridge length is 228 mm, significantly more than the 20 × 138B suggesting that the action and magazines would have required lengthening. The cartridge shown here is a replica.

An almost identical 23 mm version of the above also exists, as does a different 23 mm version (possibly by Polte?) in which the visible length of the projectile is considerably less to keep the overall round length to 202 mm, the same as the 20 × 138B, thereby making the conversion of existing 20 mm guns much easier. The cartridge shown here is a modern mock-up of this round using a replica projectile.

Sectioned 23 mm Madsen HE-T

## 23 × 106 MADSEN

This used a cartridge case formed in the 1930s by necking-out and shortening the case of the popular 20 × 120 Madsen round; the guns were identical, except for the barrel. The idea was to trade velocity for high-explosive content to increase the effectiveness of the gun in its aircraft applications; the HE shell contained nearly twice the load of tetryl as the 20 mm. The 23 mm Madsen featured in various experimental installations in several countries (including the USA), but only seems to have made it into service with Chile (fitted to Breda Ba.65 bombers) and – just possibly – in a couple of Hawk 75 planes bought by China, so it just about squeaks in as a 'service' rather than 'experimental' round.

The brass cartridge case is bottlenecked and had a typically thick Madsen rim. Overall round length was 186 mm and weight was 340 g, including 34.5 g nitro-cellulose propellant, generating a chamber pressure of 260 MPa. There were several loadings, all with projectiles weighing around 174 g and fired at a muzzle velocity of 720 m/s. These were: HE; Tracer; HE-T SD; Incendiary; and an inert 'Service Bullet'. As was usual with Madsen ammunition, these were made in the UK (by ICI).

### 23 × 110B HS 804

This post-World War II cartridge was developed to increase the power of the Hispano-Suiza HS 804. The calibre was increased from 20 to 23 mm; this meant that the cartridge case was straight, without any shoulder to use for headspacing, so HS added a belt to the case. In principle it would only have taken a barrel change to existing HS 804 guns, but it found no takers; the round shown in the Experimental Group photo is a replica. Performance data was not available, but it is reasonable to estimate an increase in the ME from 50,000 to c.57,000J; enough to drive a notional 135 g projectile at 920 m/s.

## 23 × 115 RUSSIAN

This was introduced in the NS-23 aircraft gun that emerged at the end of World War II, and was made simply by necking-up the 14.5 × 114 anti-tank rifle case (later used in the KPV heavy machine gun) to take projectiles from the 23 mm VYA round. The NS-23 was soon replaced by the faster-firing NR-23, then the even faster AM-23 (using modified ammunition), before the famous twin-barrel GSh-23 made its appearance in the late 1950s. In the 1970s, the cartridge was retained for the GSh-6–23 six-barrel rotary gun, which saw little use. All of these were designed as aircraft weapons. More recently, China has introduced a single-barrel 'chain gun' in this calibre; it is fitted into attack helicopter chin turrets and in light vehicle mountings. Ammunition is still widely made in several different countries.

Cases may be of brass or steel, filled with around 33 g (originally) or 37 g (later) of propellant developing 300 MPa chamber pressure, and percussion priming is used. The first group of loadings differed from later ones intended for the AM-23, GSh-23 and GSh-6–23 in using heavier projectiles loaded to a slightly lower muzzle velocity, but the final generation reverted to an intermediate weight.

Overall round length is 200 mm and weight is about 350 g with a brass case and original (200 g) projectile. As well as the standard loadings listed below, FZ anti-balloon rounds (using the 175 g HEI shell with a flat-nosed, sensitive fuze) and PRL chaff-dispensing anti-radar loadings have been made.

- HEI (OZ) for NS-23, NR-23: 201 g; nose fuze; contains 15.6 g A-IX-2 (RDX and aluminium); MV 690 m/s.
- HEI-T (OZT) for NS-23, NR-23: 196 g; nose fuze; contains 11 g A-IX-2; MV 690 m/s.
- API (BZ) for NS-23, NR-23: 199 g; hardened steel core plus 6 g incendiary mix within ballistic cap; MV 690 m/s.
- HEI (OZ) for AM-23, GSh-23 (early): 175 g; nose fuze; contains 17.4 g A-IX-2; MV 710 m/s.
- API (BZ-A) for AM-23, GSh-23 (early): 175 g; solid steel body with a tracer and a ballistic cap containing 5.2 g incendiary mix; MV 710 m/s.
- HEI (OFZ) for AM-23, GSh-23 (late): 184 g; nose fuze; contains 19 g A-IX-2; MV 700 m/s.

*23 × 115: OFZ; BR; PRL*

- HEI-T (OFZT) for AM-23, GSh-23 (late): 184 g; nose fuze; contains 19 g A-IX-2; MV 700 m/s.
- APHEI (BR) for AM-23, GSh-23 (late): 182 g; base fuze; contains 4.7 g A-IX-2; MV 700 m/s.
- API (BZA, BZAT) for AM-23, GSh-23 (late): 176 g; solid steel body (with or without a tracer) with a ballistic cap containing 4.4 g incendiary mix; MV 710 m/s.
- Multi-Element (ME) for AM-23, GSh-23 (late): 184 g; shell contains 24 tungsten-carbide pellets each 2 g; internal fixed-time fuze; MV 700 m/s.

### 23 × 122 HS 406 / 407

These guns were being developed by Hispano-Suiza's French branch at the start of World War II. They were essentially scaled-up versions of the 20 mm HS 404 which entered RAF and US service during the war and the 23 mm round was powerful, matching the performance of the bigger (and later) VYa cartridge. The gun was at the testing stage when France was defeated. The rimless brass case is bottlenecked, with a crimping groove around the neck. Overall round length is 203 mm and weight is c.450 g. The 200 g projectiles were fired at 900 m/s.

### 23 × 133 HS 827

This was for one of the Hispano Suiza range of AFV guns marketed in the 1960s (other calibres being 20 × 139, 30 × 138 and 30 × 170). They failed to sell. This one fired a 160 g projectile at 1,050 m/s; the HEI SD shell had a filling of 19 g, the HEI-T 16 g. The overall length of the round is

213 mm and the weight 420 g. The cartridge case weighs 195 g, propellant was 65 g, chamber pressure 353 MPa.

There was an APDS loading which weighed 365 g overall; the penetrator weighed 60 g, the sabot 40 g and the propellant 70 g, developing a muzzle velocity of 1,300 m/s. The projectile took just 0.82 secs to reach 1,000 m, at which range it penetrated 27 mm of armour at 45°.

The cartridge case appears to be the same as that of the US 23 × 133 AMRON experimental of the late 1960s.

*23 × 152 MP-6 (bottom) and 23 × 152B VYa (Courtesy Evgeny Aranov)*

**23 × 139 HS 823**

This was for a 1950s experimental Hispano Suiza AA gun. Although slightly longer than the later 23 × 133, the case is more slender as it is just the 20 × 139 necked-up to the larger calibre. The overall length is 213 mm and the weight 280 g (without propellant or HE); the projectile weighs 135 g (inc 19 g HE) and had an MV of 1,085 m/s. A probably identical cartridge was also offered as an option for the Mauser 20 mm Model B.

**.90 cal T2 (23 × 139RB), .90 cal T3 / T4 (23 × 139SR)**

Several rounds were developed for the US .90 cal aircraft gun projects of the late 1930s. All except the T1 had basically the same case shape and performance, with only the rim type changing to suit the different guns being made for them. The T2 features a rebated rim. There is also a T2E1 version which is rimless. Two other variations were the T3 and T4, which are both semi-rimmed; they are apparently identical apart from the headstamps, but were for different guns. In all of the T2, T3 and T4 models, the brass cartridge cases have a very gentle shoulder angle, a kind of cross between a bottlenecked and tapered case. Overall round length is 215 mm and weight is c.440 g. Projectile weight is 204 g and the MV was 823 m/s.

The T4 was later tried as a tripod-mounted anti-tank rifle, and was claimed to penetrate 32 mm armour at 91 m and 20° striking angle.

**.90 cal T1 (23 × 146SR)**

For the US .90 cal aircraft gun projects of the late 1930s. Unlike the later T2-T4 series, the T1 was designed to achieve a higher muzzle velocity so the brass, semi-rimmed case is bigger. It was based on the USN's 1.1-inch case, shortened and necked-down with a particularly long and tapered shoulder. Overall round length is 222 mm and weight is c.530 g. Projectile weight is 204 g and MV was 869 m/s.

## 23 × 152B VYa, 23 × 152 MP-6

The powerful 23 mm VYa (Volkov and Yartsev, the designers) aircraft gun entered service in World War II, being used mostly in the ground attack role. This went out of use not long after the war, but different loadings of the ammunition continued in use for sub-calibre training, remaining in production for many years.

The production round has a belted, bottlenecked brass case containing 60 g propellant and featuring double crimping grooves in the neck. Overall round length is 236 mm and weight is around 475 g. Combat loads were:

- HEI (OZ): 201 g; contained 15.6 g A-IX-2; nose fuze; MV 890 m/s.
- HEI-T (OZT): 196 g; contained 11 g A-IX-2; nose fuze; tracer; MV 890 m/s.
- API (BZ): 199 g; contained steel penetrator; incendiary mix within ballistic cap; MV 890 m/s; penetration was given as 25 mm at 400 m.

Another gun to use similar ammunition was the MP-6 or PTB-23 which preceded the VYa-23 and was replaced in production by it; the rounds were rimless, with the belt being added to the case later. Initially, the projectiles weighed 184–190 g and were fired at 870–890 m/s.

It is interesting that development of these powerful 23 mm cartridges started early in 1938, one year after Russia had purchased production rights to the Hispano-Suiza HS 406/7 which used equally powerful 23 × 122 ammunition. Possibly the performance of the French ammunition inspired the Russian designers to match it. However, the various Russian guns developed to use the ammunition bore no relationship to the HS guns.

## 23 × 152B ZU

The 23 mm ZU AA gun uses basically the same 23 × 152B case as the VYa above, but made of steel, with a different primer and different loadings so it is not interchangeable. The chamber pressure must also be higher as the muzzle velocity is increased significantly. The gun was the Warsaw Pact's standard light AA gun, mostly in land mountings but also on some naval vessels, and is still in common use with many nations, although Russia has replaced it with the 30 × 165 and China is doing so with the 25 × 182B. Russia has stopped developing the ammunition but modern loadings are available from other makers.

In addition to manufacture in China and Warsaw Pact nations, ammunition has also been made by Pretoria Metal Pressings of South Africa (now PMP Denel) for use in captured weapons. This features an additive that reduces the flash signature while remaining completely compatible in handling and ballistics with other 23 mm rounds. PMP produce 23 mm HEI and API-T rounds, which form part of the New Generation family of ammunition. The API-T is claimed to penetrate 50 mm of armour at 200 m.

As well as the Soviet loadings listed below, APDS or FAPDS rounds have been developed by Arsenal of Bulgaria, Mesko of Poland, and RWM (the PMA

*23 × 152B ZU: BZT; OFZT; S. African HEI*

276, which is no longer advertised). Mesko advertised two different APDS-T loads: one with a 103 g projectile fired at 1,220 m/s, the other 130 g at 1,170 m/s.

Overall round length is 236 mm and weight is c.450 g with full-calibre projectiles. The cartridge case contains c.76 g propellant that develops a maximum chamber pressure of 314 MPa. Standard Soviet loadings are:

- HEI SD (OFZ): 184 g projectile containing 18.5 g A-IX-2; nose fuze with self-destruct feature; MV 980 m/s.
- HEI-T SD (OFZT): 188 g; contains 10.7 g A-IX-2; nose fuze with self-destruct feature; tracer burns for 5+ s; MV 980 m/s.
- API-T (BZT): solid steel body with a tracer, and with a ballistic cap containing 4.4 g incendiary mix; MV 970 m/s.

## 23 × 260 R-23

The 23 × 260 was developed for the Soviet Rikhter R-23 reverse-loading aircraft revolver cannon of the early 1960s. The cartridge is extraordinary, looking like a length of plumbing pipe, with the projectile concealed within it and a tapering base without a rim or extractor groove. It did see limited operational service in bomber defensive armament. Rikhter cannon also saw service on board some Soviet spacecraft as a weapon for shooting down enemy satellites, but the exact calibre of that version is unclear. There are more details in the Unconventional Systems section. This cartridge should be compared with the experimental US 30 × 120RB cartridge for the T168 cannon, which used a similar loading system.

## 24 × 139 Pzw Kan 38 AND Tb 41

The Swiss 24 × 139 round (also described as 24 × 138) was introduced in 1938 in the Pzw Kan 38 self-loading gun for arming light AFVs. In 1941 the gun was adapted for use in the Tb 41 light anti-tank system. It is a compact and powerful round but the AT gun did not stay in service with the infantry for very long, since tank armour became too thick for it to cope with.

It lasted longer in fortifications and for arming small patrol boats. The round was retained in service for decades longer in a sub-calibre training system, for which special marking projectiles were used (tracer and flash/smoke impact).

The round uses a brass, bottlenecked, rimless case. Overall round length is 210 mm and the weight is 460 g. Both AP (Pz-G.V.) and HE (St-G) projectiles weigh 225 g and were fired at a muzzle velocity of 860 m/s (initially) and 900 m/s in the final, most produced, Type III. The AP could penetrate 13 mm/1,000 m/60°, and was still travelling at 315 m/s at 2,000 m. More realistically, it could penetrate 30 mm armour plate at 200 m/30° or 36 mm at 500 m/0°.

### 25 × 116 OERLIKON 204 GK

This cartridge was developed as an alternative to the 20 × 128 round for which Oerlikon's 204 GK was designed. The GK (*Gurtkanone* = belt fed) was not a revolver cannon, but a conventional gas-operated gun intended for arming light AFVs. The 25 × 116 case was created simply by shortening and necking-out the 20 × 128 case. Interestingly, this resulted in a case body diameter identical to the contemporary 25 × 116B described below, so this GK round could be regarded as effectively being a beltless version of the RK round.

The overall weight of the cartridge with an HEI shell was 380 g; the shell weighed 165 g and contained 20 g HEI; 58 g propellant generated 330 MPa and an MV of 1,050 m/s. With an AP-T shot weighing 143 g, the propellant charge was 61 g and the MV 1,120 m/s.

*24 × 138: AP; sub-calibre*

### 25 × 116B OERLIKON 251/25 RK

This interesting cartridge was offered in the early 1960s in one version of the Oerlikon 251 RK aircraft revolver cannon (the others being in 20 × 128 and 30 × 113B). It basically used the 30 × 113B case necked-down to 25 mm, with the overall round length kept the same at 199 mm. Weight of the dummy round shown is 300 g (the service round would presumably have weighed c.380 g, the same as the 25 × 116 above). The 165 g HE projectile was designed to contain 32 g HE (considerably more than the beltless version above, evidently an *M-Geschoss* type for aerial combat) and was fired at 1,040 m/s.

This was potentially the best overall compromise for an aircraft gun of the period, combining a muzzle velocity comparable to 20 mm rounds with a significantly greater destructive effect, but the timing was wrong – NATO nations were by then committed to the 30 mm ADEN/DEFA or the 20 mm M39 or M61.

### 25 × 135.5 CANON de 25 mm MODÈLE 1937B

This was a shortened version of the 25 × 159.5 round described below, with both the case and projectile being reduced in length. Presumably the gun was a suitably modified version of the *Canon de 25 mm Modèle 1937A*, but no

*25 × 140 GAU-8/A dummy with GAU-7 projectile (courtesy Tade (Chip) Orr, Woodin Laboratory)*

details have been identified concerning either weapon. The overall round length is 207 mm. The projectile weight is 200 g, the muzzle velocity 875 m/s with a maximum chamber pressure of 240 MPa.

### 25 × 136 GAU-8/A

This used a shortened and necked-down version of the 30 × 173 case with the overall round length reduced to 241 mm. It was intended for use in a modified version of the 30 mm GAU-8/A aircraft gun, for installation in the F-15 fighter as a conventional alternative to the failed 25 mm GAU-7 gun (see 25 × 152 below). It was designed to fire a 194 g HEI projectile (from the GAU-7 series) at an MV of 1,220 m/s, from an L/90 barrel (it was calculated that an L/98.5 barrel would be required to match the expansion ratio of the 30 × 173). Chamber volume was 127 cm$^3$ with a chamber pressure of 400 MPa. Very few of these rounds seem to have been made, but several case lengths were tried between 127 mm and 144 mm. The length of 136 mm is taken from a drawing in the official report on the project.

## 25 × 137 OERLIKON

The most successful of all of the 25 mm cartridges, the 25 × 137 was originally developed from 1964 by Oerlikon Contraves (which is now Rheinmetall Waffe und Munition – RWM – Schweiz AG, a branch of Rheinmetall Defence) for the American TRW cannon developed to meet the US Army's requirement for a more powerful light vehicle cannon to replace the M139 (Hispano HS 820). When the TRW project was abandoned Oerlikon developed their own KBA general purpose AA/AFV/naval cannon to fire it. It has since become a standard NATO round used in various other guns, most notably the ATK M242 Bushmaster Chain Gun (which won the US Army contract) but also in the Nexter 25M811, the GAU-12/U five-barrel rotary and the new GAU-22/A four-barrel version for the F-35 fighter (for which new ammunition natures are being developed). Two interesting guns designed around this ammunition failed to achieve service status: the twin-barrel (Gast type) GE 225 of the 1980s and the ADEN 25 revolver cannon.

*25 × 137: HEI-T; APC; APDS; APFSDS*

Ammunition is available from a large number of makers with a wide range of different loadings featuring every type of modern projectile from multi-purpose SAPHEI to APFSDS. The bottlenecked, rimless case is of steel and percussion priming is used. The case has an annular groove below the shoulder to provide a secure location for belt links (a characteristic shared with some of the 35 × 228 cases). Overall round length is 223 mm and weight is 490 g. The case holds 90 g propellant generating 360 MPa (full-calibre loadings), or 410 MPa (sub-calibre).

The principal US combat loadings are:

- HEI-T (M792): nose-fuzed, 184 g, containing 30–32 g HE with 6 s tracer burn; MV 1,100 m/s; time of flight to 2,000 m is 3.6 s.
- APDS-T (M791): 102 g tungsten-alloy core in plastic/light alloy sabot, total projectile weight 132 g; MV 1,345 m/s; time of flight to 2,000 m is 1.7 s; penetrates 25 mm RHA/60°/1,300 m.
- APFSDS-T (M919): 98 g DU long-rod penetrator in plastic/light alloy sabot, total projectile weight 132 g; MV 1,390 m/s; time of flight to 2,000 m is 1.56 s (remaining velocity 1,149 m/s).
- API (PGU-20/U; for AV-8B): 150 g DU core in 215 g projectile; MV 1,000 m/s.
- SAPHEI-T (PGU-32/U; for AV-8B): Multipurpose type fuzeless projectile with nose incendiary initiating the HE in a thick steel body; standard HE ballistics.

Other different and significant combat loadings (from RWM unless otherwise stated) are:

- SAPHEI-T SD: hardened steel body with ballistic cap and base fuze initiating 11 g Hexal; self-destructs after 5 to 11 s.
- FAPDS-T (PMD 098): 150 g projectile, MV 1,310 m/s; time of flight to 1,000/2,000 m is 0.83/1.8 s; penetrates at 60° 25 mm RHA/1,000 m, 19 mm/2,000 m. Purchased by Canada for anti-helicopter use in M242.
- APFSDS-T (PMD 090): 130 g projectile with tungsten alloy penetrator; MV 1,405 m/s; time of flight to 1,000/2,000 m is 0.75/1.57 s; penetrates at 60° 39 mm/1,000 m, 32 mm/2,000 m. Purchased by Canada and the Netherlands for use in M242.
- FAP (PMB 352): 250 g projectile with frangible tungsten core in aluminium body with plastic ballistic cap; MV 950 m/s. Intended for GAU-22/A; being developed in conjunction with the Netherlands; USMC also interested in FAP to replace PGU-20/U.
- APEX (by Nammo): Multipurpose round with tungsten carbide penetrator within the body, plus HEI initiated by delayed-action fuze. Intended for GAU-22/A; being developed in conjunction with Thales Australia.

In addition to the above, a PELE round was under development at the time of writing (M-DN 283), capable of penetrating 13

*Recently discovered in the Hungarian Military History Institute and Museum, this unknown round seems likely to be the 25 × 90.5RB round designed for the 25 mm SEMAG cannon. It is 146 mm long overall, features a Becker type rebated rim only 20 mm in diameter, and is head stamped (by hand) 1933.*

mm/60°/1,000 m, and HEAB was being developed by Nexter.

In the late 1970s a non-compliant loading was developed for the experimental Oerlikon KBC aircraft revolver cannon. This utilized a longer, heavier and more aerodynamic projectile (similar to the KBB's) which increased the overall cartridge length to 244 mm. Projectile weight was 230 g and muzzle velocity 1,000 m/s from the 90 g propellant load. TP, HEI and SAPHEI loadings were proposed, but the gun did not go into production.

### 25 × 145 AAI

This hybrid round, developed in the early 1970s by AAI as part of the US Bushmaster programme, is unusual in that it was specifically designed for a polymer case with a light-alloy base.

More details are provided in the Unconventional Ammunition section.

## 25 × 159.5 D.C.A.N.

Along with the 23 × 122 HS 406/7, this was the most significant of a series of 23 mm and 25 mm French aircraft gun projects of the 1930s. This one was a naval project, and the cartridge case is much slimmer than that of the 25 × 163 of the Hotchkiss Mle 38/39 AA guns (see below) at 33.8 mm instead of 42.6 mm. The case length was initially 161.5 mm and the gun designated the Mle 1933 but was modified to 159.5 mm before adoption as the Mle 1935. Projectile weights tested for the Mle 1933 varied from 265 g to 313 g, with MVs of around 830 m/s. For the M1935, the standard HE-T-SD shell weighed 289 g including 37 g of T4 explosive; muzzle velocity was 825 m/s at a maximum chamber pressure of 320 MPa. Some cartridges and a drawing (showing a different projectile) are known to survive: the cartridge measures 270 mm long overall, the drawing gives a length of 263.4 mm.

A small number of *Canon de 25 mm Modèle 1935* were made and three were fitted to Aeronavale flying boats which saw service in World War II. The *Canon de 25 mm Modèle 1937A* was also made using this cartridge case and projectile, but the ammunition was downloaded to 775 m/s at 240 MPa chamber pressure and the gun lightened to make it more suitable for flexible mounting.

## 25 × 163 HOTCHKISS / TYPE 96

Two different 25 mm Hotchkiss guns using two different cartridges saw French service in the 1930s: the 25 × 163 for the Mle 1938 automatic AA gun (adopted by the Japanese Navy as the 25 mm Type 96) and the 25 × 194R for the Mle 1934 manually-loaded anti-tank/light tank gun.

The brass case is bottlenecked and rimless, and contained 160 g of propellant. Overall round length is 233 mm and weight is around 735 g. The nose-fuzed 250 g HE shell contained 23 g HE; there was also an HE-T of the same weight that traced to 2,000–3,000 m and contained 15–20 g HE. Both were fired at 900 m/s. The AP loading uses the same projectile as the 25 × 194R, which is unusual in that it consists of a hardened steel projectile enclosed within a gilding metal jacket, just like a big AP rifle bullet. This weighs 300 g and was fired at about 875 m/s.

An experimental aircraft version of the automatic cannon is stated by one source to have fired a 250 g shell at 850 m/s. It is not absolutely certain that this was standard 25 × 163 ammunition, as a different source quotes the propellant charge as 118 g.

The 25 × 163 case seems to have been the parent of the 30 × 170 HS 411 Hispano that started development in around 1940, since the base dimensions are the same.

## 25 × 173/181 OERLIKON KBB

The 25 × 137's big brother, developed from 1966, uses a lengthened version of the same case in two versions with different neck lengths: 181 mm with Anti-Missile Discarding Sabot (AMDS) and other discarding sabot rounds, and 173 mm for full-calibre TP-T and HE loadings (see photo below). To complicate matters further, this round was for a long time commonly referred to as the 25 × 184; at least one example with this case length is known, so it may have been tried in development.

Although the KBD rotary cannon was developed to fire it, the only service use is in the Oerlikon KBB single-chamber gun, which itself has only been used in a naval CIWS which was purchased only by Turkey. Although it remains in service, it is being gradually replaced and it seems unlikely to enjoy a long

*TK*

future. The ammunition is no longer advertised.

Overall round length is 288 mm and weight is 620 g. Combat loadings are:

- HEI SD: 230 g hardened steel shell with ballistic cap, contains 20 g Hexal plus incendiary mix; base fuze with self-destruct function; MV 1,160 m/s.
- APDS-T: 190 g projectile with 14.5 mm diameter, 156 g tungsten alloy penetrator in a plastic/light alloy sabot; MV 1,285 m/s; penetrates 34 mm/60°/1,000 m (also AMDS without tracer, and FAPDS; same ballistics).

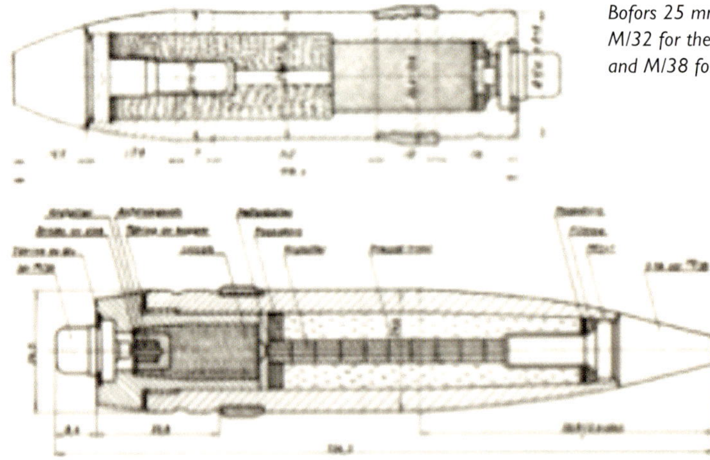

*Bofors 25 mm HE projectiles: M/32 for the 25 × 205SR (top) and M/38 for the 25 × 187SR*

## 25 × 182B PG 87

This Chinese 25 mm was introduced in 1987 as a more powerful replacement for the 23 mm ZU anti-aircraft guns which the Chinese army had in service. In fact, the cartridge bears a striking resemblance to the 23 × 152B, except for being larger in every dimension. The case carries 114 g propellant generating a chamber pressure of 314 MPa (except for the APDS-T: 121 g and 338 MPa). The round has so far been seen only in one type of gun, fitted to towed and self-propelled AA mountings.

Overall round length is 285 mm and weight is 690 g. Combat loadings are:

- HE SD: 250 g nose-fuzed shell containing 30 g HE; MV 1,050 m/s. (Also HE-T version with 19 g HE).
- API-T: 250 g shell containing penetrator; MV 1,050 m/s; penetrates 20 mm/60°/1,000 m.
- APDS-T: 135 g projectile; MV 1,350 m/s; penetrates 25 mm/30°/1,000 m.

### 25 × 193 HOTCHKISS

A few examples of this ammunition (known headstamp dates are 1933 and 1934) plus a factory drawing from 1930 survive. It bears a close resemblance to the rimmed 25 × 194R for the manually-loaded anti-tank gun but it has the same rimless base as the 25 × 163 described above. A US source indicates a projectile weight of 317.5 g and an MV of 823 m/s. Total round weight was 816 g.

The associated gun was marketed as an 'anti-tank machine gun' and a photo exists of one on a low-angle mounting. The development sequence is unclear but the earliest known dates on Hotchkiss factory drawings of the three case types are:

25 × 193: *mitrailleuse Hotchkiss*: March 28 1930

25 × 194R: *canon anti char Mle 1934* and *canon SA Mle 1938*: September 18 1931

25 × 163 : *mitrailleuse Hotchkiss* 25A -J case aviation type: June 13 1934

## 25 × 187–205SR BOFORS

The 25 mm Bofors M/32 was introduced into service in 1934. Initially this fired a 25 × 205SR M/32 round, but in the late 1930s the guns were modified to take a shorter-cased 25 × 187SR round designated M/38 (the guns retained the M/32 designation). The case was shortened by 18 mm to permit the use of longer and more streamlined projectiles to improve the effective range, while keeping the overall round length the same at 265 mm. The M/32 round fired a 250 g projectile at 900 m/s, the M/38 used a 280 g shell at 850 m/s. The use of these weapons seems to have been confined to the Swedish Navy, although the gun and ammunition inspired various Russian weapons.

## 25 × 218SR M1940

The 25 mm Russian rounds have a complex history. The inspiration seems to have been the Bofors long-recoil 25 mm M/32 AA gun and its 25 × 205SR ammunition, but the M40 adopted for Russian naval service used a modified gun design with the cartridge also altered to a 25 × 218SR with a longer neck but a slimmer case. This was replaced after the war by a new gun firing a modified, rimless version of that cartridge (described below).

The cartridge follows Bofors practice in having a wide, thick rim with a clip groove immediately above it. Overall round length is 291 mm and weight including 100 g of propellant is c.660 g. There were two combat loadings: a nose-fuzed HEI-T SD (OZR-132) weighing 288 g and containing 24 g TNT + incendiary, fired at 910 m/s; and an AP-I (BR-132) weighing 295 g and fired at 900 m/s, which had a blunt-nosed steel projectile covered by a short ballistic cap and with a large tracer compartment which burned for a minimum of 5 s. Both projectiles have three driving bands, the lowest of them being concealed within the case.

## 25 × 218 RUSSIAN

After World War II a rimless version of the 25 mm cartridge was adopted to replace the 25 × 218R described above, with a different gun designed for it. The gun and ammunition are now obsolescent, having been replaced in Russian service decades ago by weapons chambered in 30 × 165, with other nations gradually following suit. These weapons have only ever been mounted on warships.

The 25 × 218 closely resembles the 25 × 218R in appearance and dimensions except for the lack of the case rim and a sharper shoulder. It has a segmented crimp groove in the neck; presumably the higher rate of fire of its gun, and the more violent ammunition handling associated with this, required more secure

*25 × 218R M1940 and 25 × 218 (top) (collection Jean-François Legendre)*

fixing of the projectile in the case. Overall round length is 293 mm and weight is 640 g, including 95 g of propellant. There are three variations of nose-fuzed HEI-T SD projectiles, the first bearing a close resemblance to (and weighing the same 288 g as) the one used in the earlier 25 × 218R, the other two being of a different design with two rather than three copper driving bands (the lower still concealed within the case) and weighing 280 g. They contain 19.1 g A-IX-2, are fired at 900 m/s, and are identical except for the fuze. There is also an AP-T loading with a steel projectile which retains the three driving bands, has a slightly more aerodynamic shape than the one in the rimmed case and is slightly lighter at 281 g. MV is the same at 900 m/s.

## VICKERS 1-inch (25.4 × 87R)

This was designed by Vickers for a 1-inch aircraft gun in 1914 (the only known survivor still exists in the National Firearms Centre) but this saw no British use, being only manufactured in Italy by Vickers-Terni. This saw limited service during World War I in some Italian bombers and was purchased by the Swedish navy in 1922 for arming submarines; the only known official adoption. Ammunition still seems to have been made for Sweden in the late 1930s, with the case length extended slightly to 95 mm: such a round exists, with the headstamp **1935 25.4 mm VA** (Vickers Armstrong). There are photos showing the guns in AA use in Chinese hands in the Sino-Japanese War of the early 1930s, but the origin of these is not known.

The slightly tapered and rimmed brass case appears to have been based on that of the old 1-inch Nordenfelt (an adapted version of this round then still being in service in the 'One Inch Aiming Rifle' subcalibre training system). Overall round length is 162 mm and weight is 307 g. The 200 g pointed steel projectiles were fired at 470 m/s, the 12.5 g of propellant generating a chamber pressure of 223 MPa.

## 25.4 × 87 REVELLI-FIAT

The Revelli-FIAT was inspired by the Vickers-Terni, but perhaps too closely as production was curtailed when Vickers threatened legal action over patent violations. The cartridge case was modified to a more modern, sharp-shouldered, rimless type. The gun was in action on board aircraft in World War I, but was not popular due to the tendency of the HE shells to explode prematurely, so it soon disappeared. As with the Vickers 1-inch, overall round length is 162 mm and weight is around 300 g. The round used the same 200 g projectiles as the Vickers, with the muzzle velocity slightly less at 440 m/s.

### 25.4 × 148B S19–100

The Solothurn S19–100 was a 1933 project for a selective-fire anti-tank and AA gun. The belted cartridge resembles a scaled-up version of their 20 × 105B round, and is 251 mm long overall, weighing 620 g with a 280 g HE shell or 640 g with a 300 g APHE projectile. The example shown in the group photo is a replica (no actual rounds are known to exist).

## 25.4 × 189 VICKERS

The 25.4 mm Vickers AA gun was developed in the 1930s around a slender rimless cartridge. The brass, bottlenecked case contained 87 g propellant. Overall round length is 257 mm and weight is 570 g. The nose-fuzed HE projectile weighs 260 g and the thick steel body contained only around 10 g of Trotyl - presumably the aim was to maximize the fragment size rather than the blast effect. The APC projectile weighs 250 g and was fired at 914 m/s, the resulting muzzle energy being comparable with the contemporary 25 × 163 Hotchkiss and the current 25 × 137 NATO. The gun saw service in twin mountings in the Vickers-built Argentinian training cruiser *La Argentina*, commissioned in 1938; later, some were transferred to another Argentinian training ship.

### 26 × 130 COLT CR-26

This was a late 1960s project for a light AFV gun (the CR standing for Constant Reaction, a reference to the low-recoil operating mechanism), a role eventually filled by the Bushmaster M242 in 25 × 137. The greenish lacquered steel cases were made at Altdorf, the Swiss government ammunition factory, which Colt loaded with various projectiles for testing; aluminium alloy cases were also made, by General Impact Extrusion Ltd. of Canada.

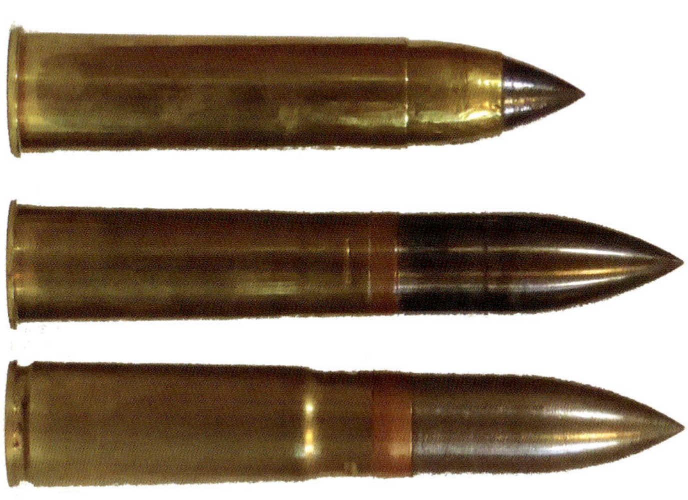

*From top to bottom: 1-inch Nordenfelt; 1-inch Vickers; 25.4 mm Revelli-FIAT*

Two service loadings were developed: an HE weighing 195 g (including 28 g HE filler) and fired at 945 m/s; and an APFSDS-T weighing 112 g at 1,265 m/s (penetrator weight 65 g). Both had a maximum chamber pressure (piezo) of 345 MPa.

The AP projectile is often quoted as APDS but this is misleading; in the US at this time it was not unusual to refer to APFSDS as APDS. The stated penetrator length of 130 mm and diameter of 6.35 mm – an L/D ratio of 20:1 – clearly means this was APFSDS. Another APFSDS version examined had a length of 152 mm and a diameter of 5.5 mm, giving an L/D ratio of 27.5:1. Fin stabilisation was used, but the projectile was also spun by the rifling at 1,360 rps (1,040 rps for the HE-T). Penetration was stated to be 26.7 mm/2,000 m/60°. Overall round length is 185 mm and the weight of the APFSDS-T round is 258 g with an aluminium alloy case and 380 g with a drawn steel case: weights of the HE-T are 328 g and 449 g respectively.

### 27 × 70B T150 VULCAN

The 27 mm Vulcan rounds (sometimes also referred to as 27 mm Winchester, as they were developed by Olin Mathieson which owned Winchester) were an attempt to pack the maximum destructive effect into a cartridge with the same overall dimensions as the 20 × 102, so that the M61 Vulcan cannon could be rechambered to take the ammunition. Priming is electric. The first attempt has a straight, belted case (shown in the group photo) and resembles a scaled-down 30 × 86B low-velocity ADEN round. It fired a 250 g shell at 610 m/s. Overall round length is 169 mm and weight is 340 g. Several types were made: HEI (designated T144); Target Practice (T142); Dummy (T143); and High Pressure Test, but in relatively small quantities.

This round was felt to have too low a muzzle velocity so the 27 × 96B was developed (see below).

### 27 × 96B T151 VULCAN

This was the 20 × 70B (described above) only with the cartridge case lengthened to take more propellant and the projectile accordingly shortened to keep the same 169 mm overall length. The projectile weight was reduced to 200 g and the MV increased to c.840 m/s. This still had double the projectile weight of the 20 × 102 and probably would have been quite effective, but the USAF presumably considered the 20 mm's high muzzle velocity to be more important. Very few were made. Overall round weight would have been slightly lighter than the 20 × 70B.

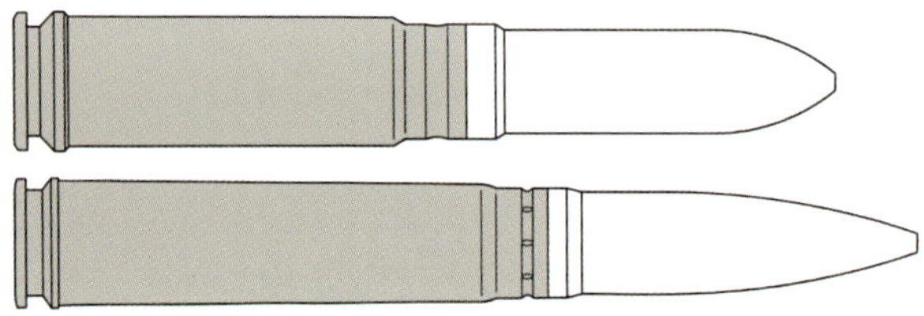

*27 × 123.5B STRADEN (top) compared with 27 × 145B BK 27 (Collection Jean-François Legendre)*

### 27 × 123.5B STRADEN

This was developed as an alternative to the 30 × 135.5B (see below) as part of the programme of improvements to the ADEN cannon in the 1970s. As with the 30 mm version, the round length was increased from 199 to 222 mm while rim and body diameters were also increased over those of the standard ADEN case. It seems odd that the 27 mm projectile was longer than the 30 mm one; it also had a very old-fashioned high-drag shape compared with that of the Mauser BK 27 projectiles. Possibly the projectile was only developed for function testing. Performace data have not been found.

## 27 × 145B BK 27

The 27 × 145B was developed by Mauser for their BK 27 revolver cannon. This is now the standard European aircraft gun, is also fitted to naval mountings, and is offered in a land mounting. The ammunition is rather more powerful than the rival 25 × 137, with the well-shaped, heavier projectiles providing a longer effective range. Aircraft ammunition is full-calibre with a range of loadings being available, but FAPDS is standard for naval use.

The steel, belted case has a slight bottleneck, is electrically primed and contains around 85 g of propellant that generates a chamber pressure of 460 MPa. Wimmis EI propellant is used, except in the PELE round which uses advanced DNDA propellant from Diehl BGT Defence and ICT (*Fraunhofer Institut Chemische Technologie*). Overall round length is 243 mm and weight is in the 500–520 g range depending on the projectile.

All of the full-calibre loadings are ballistically matched, with projectiles weighing 260 g and fired at 1,025 m/s. The original loadings include nose-fuzed HE shells designated DM 21, DM 31 and DM 11 (the last with a self-destruct fuze). There were also AP (DM 43), APHE (DM 13), APHE SD (DM23), SAPHE (DM 53) and SAPHE SD (DM 43).

More advanced rounds which have since been developed and are replacing the more traditional types are:

- FAP (DM 103) with a frangible tungsten alloy core supplemented by tungsten balls;
- Multipurpose (SAPHEI), a fuseless loading developed by Nammo under contract to the UK and qualified as the M90. The incendiary nose element is initiated on striking dural skin as thin as 0.5 mm, detonating the HE after penetrating c.300 mm;
- PELE (DM83), which is in service with Germany and Hungary aboard their Typhoon and Gripen. It will penetrate 30 mm/60°/100 m. There is also a tracer version, the DM93.

There is one sub-calibre round to see service: FAPDS (DM63) has been developed as the standard round for the MLG 27, the naval version of the Mauser BK 27 now in German Navy service and also sold for export. This weighs 180 g and has a muzzle velocity of 1,100 m/s. Flight

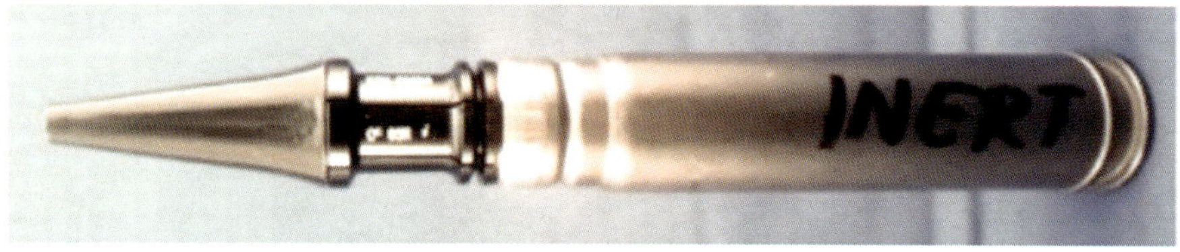


*27 × 145B in DM63 FAPDS loading*

times to 1,000/2,000 m are 0.98/2.14 s.

## 27 × 173 MAUSER SCORE

This one really doesn't belong here, but is included only because it looks as if it does, so it may puzzle people who find one! It is made by necking-down the 30 × 173 case, but was never intended to be used in automatic cannon, and never has been. It is used in a manually-loaded sub-calibre barrel insert which fits inside a 105 mm or 120 mm tank gun barrel for economical training (known as the Score system). It is loaded with a TPFRAN-T projectile (Target-Practice Frangible with Tracer) that is ballistically matched to tank HEAT rounds out to 1,800 m. Overall cartridge length is 272 mm and weight 735 g. The 255 g projectile is fired at 1,180 m/s.

### 27.5 × 160 AND 27.5 × 140 BUSHMASTER

These were manufactured by Amron Aerojet in 1968/69 for the GE-127 Light AFV gun under the US Army's Bushmaster programme. The rimless, bottlenecked steel case was initially made in two versions with the same case volume of 133 cm$^3$: one was 140 mm long with a 44.8 mm diameter rim (weight 276 g), the other was 160 mm long with a 39.8 mm diameter rim (293 g). About 200 of the short, fat cases were made in 1968, but around 15,000 of the long, slim version were made in 1969. Development continued into the early 1970s.

The 27.5 × 160 AP has an APDS-T shot in a plastic sabot that leaves the nose of the projectile exposed. Total cartridge length is 242 mm and weight 533 g, of which 206 g is the penetrator. MV was 1,100 m/s from a 208 cm barrel and chamber pressure 390 MPa. The tracer functioned out to 2,000 m. There is also an HE loading which is 247 mm long and weighs in at 555 g (without HE). Ballistic performance with the 140 mm-long case was presumably the same as the 160 mm one.

An alternative source of data on the 27.5 × 160 APDS-T gives 630 g as the total weight, 277 g for the zinc chromate plated steel case, and 254 g for the projectile of which 168 g was the 100 mm-long penetrator. The propellant weight was 120 g and the MV was 1,128 m/s, the chamber pressure being 393 MPa (57,000 psi). The two sources presumably illustrate different stages of the development process.

## 1.1-inch USN (28 × 199SR)

The US Navy adopted the 1.1-inch AA gun in the 1930s and it saw service in a four-barrel mounting during World War II. The big 28 × 199SR cartridge was powerful, but the gun was gradually replaced by the much superior 40 mm Bofors during the war with few remaining by 1945. The round is interesting in being probably the last which was deliberately designed to comply with the historic St Petersburg lower limit of 400 g for explosive/incendiary projectiles.

The tapered, bottle-necked, semi-rimmed brass case contained 120 g NC propellant, generating a chamber pressure of 250 MPa and a muzzle velocity of 820 m/s in a new gun, or about 790 m/s on average. The standard AA projectile was a 416 g HE-T containing 16.8 g Explosive D, with a super-sensitive nose fuze. Later development added a self-destruct feature by creating a channel between the end of the tracer compartment and the shell interior, so that the tracer initiated the HE as it burned out; HE content was slightly reduced to 15 g. The tracer burned for about 2,750 m, thereby defining the maximum effective range of the SD shells. Overall round length was 309 mm and weight was 850 g.

The case was the basis for the Cal . 90 inch T1 (23 × 146SR – see above), for an experimental automatic cannon. The USN also experimented with two other cartridges based on the semi-rimmed 1.1-inch case necked out to 30 mm and shortened to 183 mm (designated 30 mm EX 4) and 155 mm (30 mm EX 5). No further information is available; they are known only through a drawing of the cartridge cases. Less directly, the 1.1-inch was also the parent case for the first type of 30 mm WECOM round, the bottle-necked 30 × 110.

### 28/20 × 210B MK 8202

The German Army did have in service in World War II the 2,8 cm PzB 41 manually-loaded light anti-tank gun using 28/20 × 187R ammunition. This was a Gerlich type squeezebore with a calibre of 28 mm at the chamber but with a conical barrel forcing this down to 20 mm at the muzzle. The tungsten-cored 131 g projectile provided a very high MV of 1,400 m/s and excellent penetration.

Mauser designed the MK 8202, an automatic cannon to fire ammunition of similar performance but this used a different case: longer and belted rather than rimmed.

### 28 × 216B GERÄT 411

This was developed by Krupp in the early 1940s for an aircraft motor cannon, which in this case means one driven by the engine. The cartridge resembled a scaled-up 20 × 138B and fired a 300 g projectile at 900 m/s. Very few examples are known, but both brass and steel-cased versions survive in the Woodin Laboratory (the latter is shown below). The brass cartridge case has the headstamp 'aux 41 067' and the lacquered steel case has the headstamp 'avu 1a 41'. The steel case has a Luftwaffe acceptance marking. The round is 295 mm long overall.

There is some confusion between this round and a 'MK 411', which was of 30 mm calibre. No further details of this are known.

*27 x 173 Mauser Score*

*28 x 216B Gerät 411 (Woodlin Laboratoy)*

# 30 mm CALIBRE

Apart from the discovery of the mysterious 30 × 107RB described below, cartridges of 30 mm calibre are a relatively recent development. They only entered service during World War II with two German (30 × 90RB and 30 × 184B) and three Japanese (30 × 92RB, 30 × 114, 30 × 122) cartridges, all of which were abandoned after the war. In the 1950s NATO attention focused on low/medium velocity 30 mm rounds, initially for the ADEN and DEFA revolver cannon and subsequently for helicopter guns. The Soviet Union chose larger and more powerful ammunition; the 30 × 155B for aircraft guns and the massive 30 × 210B for naval AA guns. The Czechs also introduced the even bigger 30 × 211 for the M53 AA gun.

In the West, the 1950s saw the introduction of the powerful 30 × 170 in a Hispano AA gun, followed by the 30 × 173 Oerlikon in different forms for two very different aircraft guns. Since then, the USSR/Russia has adopted an 'all purpose' 30 × 165 round while the 30 × 173 is being increasingly used in various different guns for AA, AFV and naval purposes. France has gone its own way, adopting a new 30 × 150B round for its 30M791 aircraft cannon.

As a result of these developments, 30 mm is now effectively the international 'standard calibre' for automatic cannon for most purposes.

## 30 mm SERVICE CARTRIDGES

*20 × 110RB (for scale)*
*30 × 86B ADEN LV*
*30 × 91RB MK 108*
*30 × 97B DEFA 540*
*30 × 111–113B ADEN*
*30 × 115 Ho-155 (repA)*
*30 × 122 Type 5 (repA)*
*30 × 150B 30M791 (repA)*
*30 × 155B NR30*
*30 × 165 Russian*
*30 × 170 KCB*
*30 × 170 Rarden*
*30 × 173 Oerlikon KCA*
*30 × 173 GAU-8*
*30 × 184B MK 101/103*
*30 × 210B NN-30*
*30 × 211 M53*

## 30 mm EXPERIMENTAL CARTRIDGES

*20 × 110RB (for scale)*
*30 × 85B MG 213/30*
*30 × 91 MK 212 (repC)*
*30 × 92RB Type 2 (repA)*
*30 × 99B MK 255*
*30 × 100B WECOM*
*30 × 110 WECOM*
*30 × 110B MK 253 (repA)*
*30 × 120RB T268*
*30 × 126B T239*
*30 × 136 HS 825*
*30 × 136B HS 825*
*30 × 138 HS 836 (repA)*
*30 × 158 HS 603 / 604 (repA)*
*30 × 164 GAU/8*
*30 × 170 Oerlikon KCC*

### 30 × 85B

This cartridge was developed at the end of World War II for the Mauser MG 213C/30 cannon. The gun was originally designed for a high-velocity 20 × 135 round, but the need to shoot down heavy bombers led to the addition of a much more destructive 30 mm low-velocity version with a straight, belted case. The 330 g *Minengeschoss* shell was from the MK 108, but muzzle velocity was slightly increased to 530–540 m/s. Total round weight is 500 g, of which the case weighs 141 g and the propellant 29 g. The belt links weigh 104 g, so the total weight of 100 rounds of belted ammunition is 60.5 kg. Only a few prototype guns were completed before the end of the war, but the British (among others) continued the development of the gun and ammunition design that, with various minor modifications, became the first version of the ADEN cannon in 30 × 86B calibre.

## 30 × 86B ADEN 3M

This was developed for the first generation of British ADEN aircraft revolver cannon, known as the 'low velocity' (LV) or 3M type. It bears a close resemblance to the 30 × 85B cartridge being developed at the end of World War II for the Mauser MG 213C/30 cannon, which is hardly surprising since the gun and ammunition were closely based on it. The 30 × 86B briefly saw RAF service in the mid-1950s before being replaced by the 30 × 113B. The HE Mk 1.z projectile was of the *Minengeschoss* type with a hemispherical base, weighing 273 g and filled with 75 g HE. It had a muzzle velocity of 604 m/s.

The USA developed some guns to use this ammunition; the guns were the T121 revolver cannon and the T4 (rechambered Hispano type) and T5 (rechambered 20 mm T33 type). The ammunition designations were T158 (TP), T159 (dummy), and T160 (HE), firing a 270 g T241 shell at 640 m/s. The final version of the T158 series used a case lengthened to 102 mm, designated FAT 33, and fired its 250 g projectile at 830 m/s.

### 30 × 91 MK 212

This cartridge was for Mauser's MK 212 which was the rival to the MK 108 (see below), but lost the competition. The gun was a conventional locked-breech type so the cartridge case was a rimless, bottlenecked type rather than the rebated-rim straight-cased round of the MK 108. It fired a 300 g Minengeschoss shell at 575–580 m/s; this differed from the usual MK 108 shell most obviously by having the driving band set higher up the shell body, resulting in a reduced overall length of 185 mm. Cartridge weight is 583 g, of which the case weighs 250 g.

## 30 × 91RB MK 108

This first emerged in the latter part of World War II and was only ever used in the Rheinmetall-Borsig MK 108 cannon, a compact, light and fast-firing aircraft gun, most famously installed in the Me 262 jet fighter. The low-velocity ammunition uses a straight, rebated-rim steel cartridge case 90–91 mm long weighing 124.5 g; overall weight is 480 g. Propellant is 30 g of Nz.Bl.P and electric priming was used. Normal chamber pressure was 280 MPa with a maximum of 330 MPa.

It was almost exclusively loaded with Minengeschoss HEI shells weighing 330 g, of two different designs: a round-based, blunt-nosed Ausf.A type filled with 84 g HTA 41 explosive; and a flat-based streamlined Ausf.C filled with 72 g HEI. Muzzle velocity was 505 m/s. A longer, high-capacity (MX-Geschoss) 470 g shell fired at 440 m/s was adopted at the end of the war but it is not clear whether it was ever used. The low muzzle velocities meant that the weapon was principally of value in the anti-bomber role. It became obsolete at the end of the war.

*Three loadings of the MK 108 round, from top to bottom: M-Geschoss Ausf.A; M-Geschoss Ausf.C; and Ubungsmunition (practice)*

### 30 × 92RB TYPE 2

This was developed in mid-World War II for the 30 mm Type 2 aircraft gun of the Imperial Japanese Navy. Like the MK 108, this was intended as a bigger version of the 20 mm Oerlikon-pattern guns, but while a few fighter aircraft armed with this gun did see combat it was never formally adopted. The cartridge case was very similar to the 30 × 91RB but that appears to be a coincidence rather than a copy; the Japanese gun was different in layout to the MK 108, and the ammunition used much shorter and lighter projectiles fired at a higher muzzle velocity. The case shape was also different, having a slight bottleneck and a more rounded transition to the extractor groove. Actual case length was 91.5 mm, with overall round length being 167 mm. The main combat rounds were a 265 g HEI containing 20 g WP plus 3.5 g Pentolite, and a 273 g HE-T with 21 g PETN. Other loadings mentioned in documents are an HE and an APHE. MV was 710 m/s.

## 30 × 97B DEFA 540

The French also copied the Mauser MG 213C but after experimenting with 99 mm and 116 mm case lengths for their first attempt (designated MK 255) they chose a 97 mm case for the DEFA 540 series cannon, adjusting the exposed projectile length to maintain the same overall length. Subsequently the French adopted the British 30 × 113B cartridge dimensions for the DEFA 550 series, in the interests of standardisation.

The straight, belted cartridge case was originally made of brass but this changed to steel as a result of technical problems with the cases. Complete rounds are 199 mm long and weigh 435 g (steel-cased). Chamber pressure was 245 MPa. Only one combat load was adopted, the flat-based mine shell, containing 70 g of hexal, which weighed 297 g and was fired at 600 m/s.

### 30 × 100B M552 SERIES (WECOM)

During the 1960s, the US Army began a programme for a new cannon of 30 mm calibre, the XM140, intended for the abortive Cheyenne attack helicopter. Development work lasted for about 13 years, with well over 1,000,000 rounds being made, in at least ten variations. The programme was known as the 30 mm WECOM (from the Army's WEapons COMmand).

The first version of the round (rarely seen) uses a fat, bottle-necked rimless 30 × 110 case with 45 mm rim and body diameters and 177 mm overall length, based on the semi-rimmed 1.1-inch US Naval case (but with the body diameter slightly increased to the same diameter as the rim) and proposed by the Picatinny Arsenal. This was soon replaced by the 30 × 100B round described here, which resembles a shortened 30 × 113B ADEN/DEFA round but differs slightly in every dimension (e.g. rim and body diameters 34.5 and 32.5 mm compared with 33.3 and 32.3 for the ADEN/DEFA round). It is 178 mm long and weighs c.410 g. Both brass and aluminium cases were made. The 195 g HE shell was fired at 670 m/s; the chamber pressure was only c.200 MPa. Unlike the ADEN/DEFA rounds, the WECOM used percussion ignition.

The WECOM ammunition was known as the M552 series. The cartridge case was the XM193, with XM639 the TP round and XM640 the TP projectile it fired. The first version of the HEDP round was designated XM552, with the projectile being the XM553. This HEDP projectile was also used in the XM757 round, which used an aluminium case, the XM193B1E1. Other rounds using the XM193 case were the XM554 Practice, XM555 Spotting, and the XM250 and XM251 HPT. There were also high-velocity loadings using a modified case, the XM211, which was redesigned internally to resist higher pressures: these rounds were the XM719 TP, firing the XM719 projectile (220 g, at 760 m/s), and the XM717 HEDP firing the XM553 projectile (as mentioned above).

*30 x 107RB (Photo by Tade (Chip) Orr, at the Woodin Laboratory)*

*Oerlikon 30 × 110B for the 301 FK (courtesy Robert Gerber)*

The WECOM ammunition programme was terminated in 1976 in favour of using ammunition compatible with the NATO standard 30 × 113B.

Three rival guns were developed around the WECOM ammunition: the unique XM140, the three-barrel rotary XM188 and the first version of the M230 Chain Gun. The M230 won the competition, was adapted to use the 30 × 113B, and is currently in service in the AH-64 Apache combat helicopter.

### 30 × 107RB (UNKNOWN)

The only evidence concerning this cartridge is that one round exists, now in the Woodin collection. It was dug up in northern France and resembles a scaled-up 20 mm Becker so has been dated to World War I. The headstamp consists of the *Artillerie Werkstatt Spandau* monogram plus the letter I. The overall length of the round is 216 mm. The gun it was intended for is unknown.

### 30 × 110B 301 FK

The Oerlikon 301 FK aircraft gun project dates from 1948. The gun was not made, but some ammunition was manufactured for testing in a pressure barrel. The cartridge case is slightly wider than the usual MG123/30/ADEN/DEFA type, so is bottlenecked (but not quite as much as the MK 253 described below). The cartridge weighs 550 g of which the projectile, carrying a 70 g HE load, weighs 300 g (a belt link weighs another 100 g). Muzzle velocity was 650 m/s. Length is 205 mm.

### 30 × 110B MK 253

This is a shortened version of the 30 × 184B as used in the Rheinmetall-Borsig MK 103 aircraft gun, intended for use in the MK 253, a shortened version of the gun. It was developed in France after the end of World War II. Unlike the 30 × 97B and 30 × 113B rounds for DEFA revolver cannon, the 30 × 110B case is bottlenecked, providing a significantly greater propellant capacity. Overall length is 225 mm. No further details are known. The round shown in the Experimental Group photo is a replica.

## 30 × 111–113B ADEN, NEXTER AND LW30

The origin of the 30 × 113B family of cartridges goes back to the latter part of World War II, when a team of Mauser engineers led by Anton Politzer designed a new aircraft gun to meet an official specification for a 20 mm calibre weapon which could fire at 1,000 rpm and achieve a muzzle velocity of 1,000 m/s (the so-called Million Point Gun: 1,000 × 1,000). This project was given the designation MG 213, and Politzer's design, one of four competing projects, was designated MG 213C (later renamed MG 213/20). This was the first of the modern revolver cannon, loading each 20 × 135 cartridge in stages into chambers within a revolving cylinder, and it easily met the specification.

The Luftwaffe soon decided that a 30 mm calibre version should take priority because of the urgent need to destroy heavy bombers, so a new cartridge designed to use the same projectiles as the existing MK 108 was designed. This was electrically primed, had a case length of 85 mm (necessary to keep the overall round length the same 199 mm as the 20 × 135) and featured a straight case with a belt in front of the extractor groove to locate the round in the chamber. Compared with the 20 mm version, the hitting power was greatly increased but the muzzle velocity was halved. This new version was designated MG 213/30.

At the end of the war, Mauser's revolver cannon was examined with great interest by the Allies, although very few prototypes had been completed. Guns based on this principle were produced in the US (20 mm M39) and in Switzerland (various Oerlikon designs, culminating in the 30 mm KCA), but the French and the British copied the Mauser design more closely. In the case of the British ADEN gun (Armament Development ENfield), the cartridge was initially almost identical, a 30 × 86B with only minor variations in dimensions (see entry above). The French chose a longer case (30 × 97B) for the DEFA 540 series (also see above). All retained electric ignition.

*Ammunition developed from the MG 213/30 (not a complete list): 30 × 85B; 30 × 86B; 30 × 99B; 30 × 97B; 30 × 111B ADEN; 30 × 113B DEFA; 30 × 113B LW30 (repP); 30 × 126B*

During the 1950s the 30 × 86B and the 30 × 97B were replaced by the 30 × 113B, the subject of a NATO Standardisation Agreement (STANAG 3231) to standardize chamber dimensions (although ADEN cases are actually 111 mm long). The 30 × 113B retains the standard 199 mm overall round length. This change was made to increase the muzzle velocity to a level considered more suitable for air combat, by increasing the propellant capacity at the cost of smaller shells. Unfortunately this agreement did not extend to other aspects of performance, such as the pressure curve, electrical impulse for primer actuation, or belt link strength and flexibility, so ammunition for the ADEN (Mk 4, or HV – high velocity) and DEFA (550 series) is not generally interchangeable, although Oerlikon has made some dual-function ammunition compatible with both. A more obvious difference is that the DEFA (later GIAT, now Nexter) ammunition is steel-cased, while ADEN rounds were invariably brass-cased in British production, although some foreign makers have used steel cases. While the Nexter guns are still widely used the ADENs are becoming obsolescent, mainly surviving in the Jaguar aircraft in service abroad.

The USA also experimented with a very similar round (with a 114 mm case) in the T182 revolver. Ammunition designations were T204 (TP), T205 (dummy), T206 (HEI), T207 (High Pressure Test), and T208 (incendiary).

A different version of the French 30 × 113B is used in the externally-powered Nexter 30M781 gun fitted to the European Tiger helicopter. This version of the ammunition, also used in modified 550-series guns fitted to naval *Super Étendard Modernisé* strike planes, has a modified priming system so that the rounds will not accidentally be fired by the strong electrical currents which they are likely to encounter in their operating environments.

The most recent version of this cartridge was developed in the USA for the M230 Chain Gun (as used on the AH-64 Apache helicopter). This is designated the LW (lightweight) series. ATK production uses an aluminium-alloy case, although GD-OTS uses a lightweight steel case. Another important difference is that the standard loading for the LW30, designated M789, uses a HEDP projectile (High-Explosive Dual-Purpose) which is the only current automatic cannon service round to incorporate a HEAT shaped-charge warhead, for use against light armour as well as against personnel. Other designations are the M788 (TP), M848 (dummy) and M883 (High-Pressure Test).

Earlier production for the 30 mm ADEN gun used in the AV-8A/C aircraft used a different projectile, the M799 HEI. An HEI shell is now being reintroduced for use with the M230LF gun.

The most recent developments for the Nexter and LW30 ammunition include the adoption of improved propellants and explosives designed to meet IM (insensitive munition) standards. A percussion-primed version of the M230LF and its ammunition has been developed, for any users (presumably naval) who might prefer that, but seems to have attracted no sales so far.

Many different loadings have been developed for the ADEN and GIAT rounds: the ones listed below are the most significant.

- ADEN HE SD Mk 6.z: the standard RAF round for nearly 30 years; overall weight 500 g; 270 g nose-fuzed shell with hemispherical base, contains 48 g Torpex; MV 790 m/s; fuze self-destructs in 6–12 s.
- ADEN AP Mk 1.z: APCR round with tungsten carbide core; 270 g flat-based projectile; MV 790 m/s; penetrates up to 25 mm armour.
- ADEN MPT-LD RO 379: Multi-Purpose Tracer – Low Drag; Nammo Raufoss type fuzeless projectile with incendiary filling initiated on impact, plus Hexal explosive; final service round, with improved shape for better ballistics.
- GIAT OMEI Type F 7572 HEI SD: a thin-walled, flat-based mine shell weighing 245 g and containing 48 g Hexal; complete round weight 452 g; MV 810 m/s.
- GIAT OAPEI type F5272 Air to Ground thick wall HEI weighting 275 g and containing 48 g Hexal; complete round weight 442 g; MV 775 m/s
- GIAT OPIT Type F 5970 API-T: intended for ground attack of hard targets; contains 20 g incendiary material and has a hard steel penetrator covered by a ballistic cap; projectile weight 275 g; round weight 490 g. MV 775 m/s; penetrates up to 20 mm armour plate.
- Nexter OSPEI Type F 7670/7671 SAPHEI SD: current standard French type; overall round weight 480 g; 50 g propellant generating 294 MPa chamber pressure; thick-walled high-fragmentation 275 g projectile with ballistic cap, contains 18 g Hexal and a delayed-action base fuze; MV 775 m/s.
- LW30 M789 HEDP: current standard type used in AH-64 helicopter; round weight 339 g; 50 g propellant generating 280–310 MPa; projectile weight 237 g, contains shaped-charge warhead with 27 g PBXN-5 HE; MV 805 m/s; penetrates 25 mm/50°/500 m RHA.

A version of the LW30 ammunition fitted with a proximity fuze has been developed by Northrop Grumman under the designation XM1211 HEP (High Explosive Proximity). It is intended for use by AFVs to improve effectiveness against both ground (personnel) and aerial (UAS) targets. In a separate development, the XM1198 HEDP features a self-destruct fuze.

## 30 × 115 Ho-155

A late-war cartridge for the Imperial Japanese Army's Browning-type Ho-155 aircraft gun, which only had time to see very limited service (although it was intended to take a major role had the war continued). The brass case is rimless and bottlenecked with a crimping groove; the cartridge is 190 mm long overall and weighs 520 g. The conventional nose-fuzed HEI shell weighs 235 g and was fired at a muzzle velocity of 700 m/s; there is also a fuzeless Ma 301 incendiary loading. This was a relatively low performance for the size of cartridge, possibly the result of a weak gun action due to the poor steel quality that had delayed service entry.

### 30 × 120RB AND 30 × 136RB T168

These strange-looking cartridges were designed for the US T168 front-loading revolver cannon of the early 1950s, in which the belt feed was in front of the cylinder and the cartridges were pushed backwards into the chambers. As a result the cartridge has reverse taper, with the base of the case tapering down to a tiny rim just 17 mm in diameter. These cartridges should be compared with the Russian 23 × 260 round for the Rikhter 23, which used a similar loading system. There are more details in the Unconventional Systems section.

## 30 × 122 TYPE 5

The third of World War II Japanese trio of 30 mm aircraft gun rounds, this time for the Navy's 30 mm Type 5 cannon. Like the Army's Ho-155, this emerged so late in the war that it saw hardly any service. The brass cartridge case is rimless and bottlenecked and contains 70 g propellant. Overall cartridge length is 211 mm and projectiles weigh 350 g, with a muzzle velocity of 750m/s. Combat loads were HE (containing 39 g pentryl); HEI (Type 1 filled with 27.5 g white phosphorus plus a bursting charge, Type 2 with 43 g barium nitrate and magnalium); HET (with 29 g pentryl) and APHE.

### 30 × 120.5B ADEN C

As with the US T239 described below, this round used an extended version of the usual ADEN 30 × 111B cartridge case in order to provide more room for propellant. This came at the cost of a shorter and lighter projectile as the overall length remained the same at 199 mm. Details of the performance have not been found but it would probably have been similar to the T239 described below.

### 30 × 126B T239

This round used an extended version of the usual ADEN/DEFA 30 × 113B cartridge case (retaining the same 199 mm overall length) for a version of the experimental 1950s US T182 revolver cannon. The gun was intended as a tail-mounted bomber defense weapon. The round is usually found loaded with a black-painted 250 g T328 practice projectile, in which form the cartridge is designated T239 (about 45,000 were made); the HEI shell was designated T306 and the complete HEI round the T206. All had square rather than hemispherical bases. Muzzle velocity was intended to reach 914 m/s with the pressure not exceeding 275 MPa, but in practice only around 850 m/s was achieved. The overall round weight was c.500 g. Other ammunition designations were the T252 dummy and the T253 HPT. Initial production had brass cases but these proved too soft and deformable so were replaced by steel.

### 30 × 135.5B STRADEN

This was developed (along with a 27 mm version – see that calibre group) as part of a programme to develop an advanced ADEN gun, dubbed STRADEN for 'stretched ADEN'. Various improvements were made to the gun, but to boost the ballistic performance more powerful cartridges were developed in 27 mm and 30 mm calibres. Their overall length was increased from 199 to 222 mm, and the rim and case body widths were also increased, from 33.3 and 32.3 to about 36.2 and 34.6 mm respectively. These dimensions were scaled from unofficial cartridge drawings. Unfortunately, no other details of the STRADEN guns or their ammunition have been located.

### 30 × 136(B) HS 825

The Hispano Suiza Type 825 aircraft cannon almost made it into service; it was selected for various Swiss aircraft projects (plus one Swedish one) that were eventually cancelled, and was rejected by the British Fleet Air Arm after testing. The case is a bottlenecked type. Early production rounds, as made in the UK, have a rimless brass case, but later Swiss-made ones have a belted steel case. Overall length is 209 mm and the round weight is c.600 g. The loadings were optimized for high-altitude performance against aircraft, featuring a relatively light 225 g HE shell fired at the high velocity of 1,050 m/s. In the dense air of low altitudes, the shell lost velocity relatively quickly.

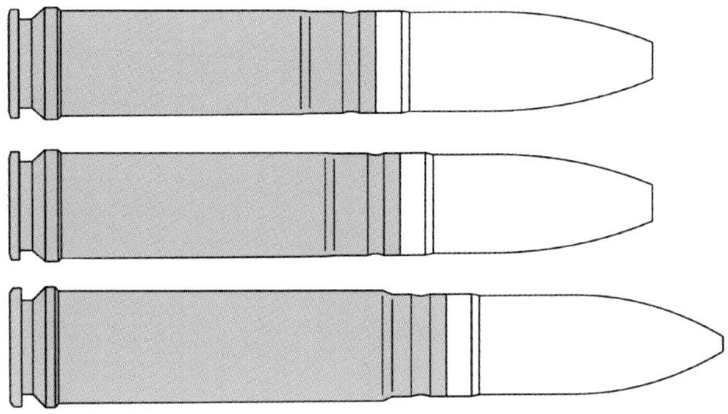

*From top to bottom: 30 × 111B ADEN; 30 × 120.5B ADEN C; 30 × 135.5 STRADEN (Collection Jean-François Legendre)*

### 30 × 138 HS 836

This cartridge was developed for a 1960s Hispano Suiza AFV gun (dubbed 'Spitfire') but failed to achieve any sales. The cartridge case (which weighs 300 g and contains 120 g propellant) is rimless and bottlenecked and the complete round is 230 mm (HE) or 210 mm (APDS) long. It fired a 230 g HE shell (containing 46 g HE) at 1,150 m/s or a 110 g APDS (170 g including sabot) at 1,350 m/s. Chamber pressure was 350 MPa. Cartridge weight is 650 g (HE), 590 g (APDS). Time of flight of the APDS: 0.8 sec to 1,000 m; penetration at 1,000 m/60° was 28 mm, dropping to 21 mm at 1,500 m.

## 30 × 150B 30M791

This is the most recent introduction in 30 mm calibre, developed from the 1980s for the Giat (now Nexter) 30M791 aircraft revolver cannon. This gun is (so far) only installed on the Dassault Rafale fighter, which entered service in 2001. The new ammunition is based on experience gained with the 30 × 113B DEFA range, but uses a longer cartridge case to accommodate the increased propellant load of 90 g to obtain the high muzzle velocity of 1,025 m/s. It therefore resembles a lengthened 30 × 113B round, although the base dimensions are slightly different, and it is easily mistaken for the 27 × 145B. The case is made of dark-grey lacquered steel and the priming is electric. Overall length is 250 mm and weight 520 g.

The only combat loading is an all-purpose base-fuzed 275 g SAPHEI SD. The projectile has a single iron drive band and is secured to the lacquered steel cartridge case by a single crimping ring. The fuze has a delay element, and will self-destruct the projectile 5–17 seconds after firing. The projectile body is coloured deep yellow, with a red band just below the black fuze. Black projectiles are also used. A safety priming device is used that renders them insensitive to electromagnetic interference.

## 30 × 155B NR-30

This cartridge has only been used in one gun, the Nudelman-Rikhter NR-30, also made in China as the NORINCO 30 mm Type 1. This was a standard Soviet aircraft cannon from the mid-1950s onwards particularly in the Su-17, MiG-19 and the F-6 (the much-modified Chinese version of the MiG). It is now obsolescent as the last aircraft that carry the gun are leaving service; North Korea is believed to be the last operator of the MiG-19/F-6, although the Su-17 family remain in service with various forces. Ammunition was made in Egypt and Pakistan as well as the USSR (and other Warsaw Pact countries) and China.

The ammunition is based on a brass, bottlenecked case with a narrow belt above the extractor groove. It is loaded with 95–100 g of propellant. Projectiles are firmly held in the case by means of two segmented crimping grooves. Overall length of the rounds is 266 mm, total weight c.840 g. Two main generations of loadings were developed for the

NR-30. The original HE types have projectiles with a long, mainly parallel-sided, body and a conical nose fuze, whereas the more modern types have a shorter parallel section and a more gradually tapered nose in the interests of improving the aerodynamics; these are very similar to the HE shells developed for the new 30 × 165 ammunition. The shape of the AP-T projectiles also varies. However, projectile weights of 400–410 g and the muzzle velocity of 780 m/s remained the same throughout. Specialist ammunition types such as anti-radar (chaff), multi-element (shrapnel) and anti-balloon loadings were also developed. Like most Soviet and Russian cannon ammunition, the projectiles are usually left as uncoloured grey steel, although some are painted black; fuzes are usually in natural aluminium colour.

The principal combat loadings are as follows:

First generation:

- HEI SD (designation OFZ): contains 40 g A-IX-2 HEI; delayed action / self-destruct nose fuze; (there was also a High Capacity variant of more streamlined shape which contained 60 g HEI)
- APHEI SD (BR): thick-walled shell with a ballistic cap and a base fuze; contains 15 g A-IX-2 HEI (there was also modified type with a more streamlined shape, which contained 25 g HEI)

Second generation:

- HEI (OFZ): contains 48.5 g A-IX-2G filler and a DA nose fuze
- APHEI (BR): description as for old generation but contains 25 g A-IX-2 HEI

### 30 × 158 HS 603/4/6

Using a slightly shortened version of the 30 × 170 case, this cartridge featured in two Hispano-Suiza guns of the 1950s, designated HS 603 (a short-barrelled version for the air force), and HS 604 (a lengthened barrel for AA use). It fired a 350 g HE shell at an MV quoted as 960 or 1,000 m/s (for the 603 and 604 respectively), or a 400 g shell at 900 m/s (HS 604). Overall length is 264 mm.

See also 37 × 158B for HS 606.

### 30 × 164 PHILCO-FORD

This was developed for an unsuccessful rival to General Electric's GAU-8/A. Like GE's 30 × 173, this used an aluminium alloy case and a plastic driving band, but had a purpose-designed, shorter and fatter case. Ammunition development was subcontracted to Honeywell, with case production by Harvey Aluminum, Amron, and IVI of Canada. Overall length is 278 mm and weight 630 g. Performance was similar to the 30 × 173 GAU-8/A rounds, with an MV of 975–1,065 m/s.

## 30 × 165 RUSSIAN

Developed in the 1970s and originally known as the AO-18 cartridge after the gun it was developed for, this was meant to be the USSR's attempt at a tri-service cannon cartridge to replace all other types. In a sense it has, since guns in this calibre now seem to be the only automatic cannon made in Russia, with the possible exception of the new versions of the 57 mm S-60 (should they be ordered). The weapons chambered for this round have included the twin-barrel GSh-30 (Su-25 aircraft) and GSh-30K (Mi-24 helicopter), single-barrel GSh-301 (current fighter gun in MiG-29 and Su-27 families), six-barrel rotary GSh-6–30 (variants used in naval CIWS mountings and formerly in the MiG-27 aircraft), twin-barrel 2A38 (army SPAAGs), and single-barrel 2A42 and 2A72 (army AFV and helicopter guns).

However, the three services developed their own versions of the ammunition. Army ammunition uses percussion priming (except for the GSh-30K helicopter gun), and while the navy and air force rounds share electric priming their projectiles vary in overall length due to different fuzes so may not be interchangeable. Army projectiles are uncoloured as usual for Russian cannon ammunition, but naval ones are painted green, and air force ones from non-Russian sources are sometimes blue. One distinctive projectile is the HE-T for army and naval guns, which differs from other full-calibre rounds in having a 'waisted' shape, possibly for rapid identification when making up ammunition belts.

Russian armoured vehicles, aircraft and naval gun mountings equipped with guns in 30 × 165 calibre have achieved numerous export sales, leading to ammunition being made (not necessarily for all three services) in China, India and Iran as well as various former Warsaw Pact countries. This ubiquity has also attracted the attention of foreign ammunition companies; Mecar developed the M929 APFSDS, while RWM advertised the PMC 304 FAPDS and were working on an APFSDS round, but both have since been cancelled.

The steel cartridge case is rimless and bottlenecked, and contains 120–133 g of propellant (depending on the loading) developing a chamber pressure of 353 MPa. There are two crimping grooves in the neck. With the exception of saboted sub-calibre rounds, muzzle velocities are mainly determined by the barrel lengths: army guns and the GSh-30K helicopter gun have 240 cm barrels, aircraft guns 150 cm and naval guns 162 cm. Complete full-calibre rounds are 291 mm (army and

*30 × 165: aircraft HEI; naval dummy; aircraft TP; drill round; TP-T; AP dummy*

navy) or 284 mm (aircraft) in length and weigh 830–850 g.

Most of the ammunition made in 30 × 165 has traditional HEI or AP loadings, with or without tracers, and with copper driving bands. The HEI compound is designated A-IX-2. APDS ammunition has also been developed for the 2A42 and 2A72 guns. The first generation was known as *'Trezubka'*; the entire projectile (including sabot) weighed 390 g and was fired at 870 m/s to achieve a penetration of 25 mm/60°/1,000 m, the latest is known as *'Kerner'*, described below. Bulgaria has produced an APCR-T round known as 'STING', which contains a frangible tungsten penetrator in a steel body with a long ballistic cap; weight is 400 g; penetration 25 mm/60°/1,500 m; tracer burn 3.5 s. China has developed an APFSDS round with an MV of 1,250 m/s for use in the LD3000 11-barrel gun (and presumably in the naval Type 1130 version). APFSDS for AFV guns has also been developed in Europe, with Nammo involved in one project.

The principal loadings are as follows:

Army Guns (MV 960 m/s except APDS-T):

- HEI SD (Russian designation OFZ): nose-fuzed; 390 g; filled with 48.5 g HEI
- HEI-T SD (OT): waisted shape with large tracer compartment (9–10 s burn); nose-fuzed; 388 g; filled with 11.5 g HEI
- AP-T (BT): solid steel body with tracer compartment (3.5 s burn) and long ballistic cap; 400 g; penetration of 18 mm/60°/1,000 m (at which distance velocity is 750 m/s)
- APDS-T (BP): tungsten penetrator in aluminium-alloy sabot with copper driving band and plastic ballistic cap; 304 g (complete); MV 1,120 m/s; penetrates 25 mm/60°/1,500 m

Naval Guns (MV 890 m/s):

- HEI SD (UOF-84): as for Army OFZ except electric primed and with MG-32 delayed-action fuze instead of A-670M
- HEI-T SD (OR-84): as for Army OT except electric primed and with MG-32 delayed-action fuze instead of A-670M

Aircraft Guns (MV 860 m/s):

- HEI (OFZ): as for Army OFZ except electric primed and with AG-30 or AG-30D delayed-action fuze instead of A-670M
- HEI-T (OFZ-T): uses similar body to OFZ rather than waisted army/naval types; 387 g; contains 43.3 g HEI compound; tracer burns for 1.5–4 s
- APHEI (BR): blunt-nosed thick steel body with ballistic cap, contains 14.6 g HEI and base fuze; penetrates 15 mm/60°/1,300 m
- AP-T (BT): as for Army BT except electric primed
- CC (ME): Cargo-Carrying (Multi-Element) ground-attack round containing 28 tungsten-carbide sub-projectiles (each 3.5 g) and a bursting charge which is detonated by a fixed-time fuze after the projectile has travelled 800–1,300 m, forming an 8 degree dispersion cone; projectile has yellow band around body.

In the mid-2010s, Nammo was working with Nitrochemie Wimmis and Arcus to develop modern 30 × 165 ammo for the 2A42 and 2A72 guns in Finnish service. Arcus made the cases, Wimmis provided the propellant and Nammo the projectiles. Natures being developed were MP-T/SD, APDS-T, APFSDS-T and TP-T. It was discovered that while the APFSDS worked in the 2A42, the projectile was too light to develop enough recoil to operate the 2A72, which had to revert to the APDS.

### 30 × 167 M.A.T.

A French design from the 1930s by *Manufacture d'armes de Tulle*, a state factory, for a powerful aircraft gun. The 455 g shells were fired at more than 950 m/s, generating 205,000J ME. No other information is available.

## 30 × 170 HISPANO / RARDEN / KCC

This round has a complicated history. The gun was originally a French design, the HS 411, work on which started in 1939/40. It is probably no coincidence that the base dimensions of the cartridge case are the same as the Hotchkiss 25 × 163 that was in French service at that time. Data on the gun and ammunition was transferred to Hispano Suiza's British subsidiary BMARCo (British Manufacturing and Research Company) at around the same time, since the company then built an example of the gun and proposed its adoption by the UK. Several Lots of ammunition were made in Britain in 1942. The case headstamps for all of the known examples have the following elements in common: **30 mm RLB 1942 LOT (#) (WD arrow)**. The only variation is in the Lot numbers, which range from 1 to 5. RLB indicates manufacture by the Royal Ordnance Factory, Birtley.

The British rejected the gun but the Swiss branch of Hispano Suiza continued to develop it; after World War II it was renamed the HS 830 and later HS 831, and it eventually entered service in the 1950s. With the HS 831L of the late 1950s the brass case was dropped in favour of steel, and the copper driving band for sintered iron. The original HE loading was for a 420 g shell fired at 920 m/s but the HS 831L saw the introduction of a new loading, with a more aerodynamic 360 g shell fired at 1,080 m/s. This has remained the ballistic standard ever since.

The gun was renamed the Oerlikon KCB after the companies merged in 1971. It has exclusively been used in the anti-aircraft role, in both land-based and naval mountings.

In the late 1960s HS developed a range of cannon for mounting in AFVs, one of which (the HS 837 'Thunderball') was designed around the 30 × 170 cartridge. An APDS loading was developed for this purpose, the projectile weighing 220 g (150 g penetrator, 70 g sabot) and being fired at 1,330 m/s; it was claimed to penetrate 32 mm/1,000 m/60° but was not put into production.

At about the same time, the HS 831 case design was adopted for the British Army's Rarden AFV cannon, but the case material reverted to brass, possibly because this provides a better gas seal and thus reduces gas leakage into the turret. The cartridge cases for the 1942 BMARCo round and the current Rarden are therefore almost identical, the main difference being in the primers.

Rarden and KCB ammunition are not qualified as interchangeable, although in

*30 × 170:1942 RLB; two variations on HS 831A; Oerlikon KCB HEI*

*30 × 170 Rarden: APSE-T; TP; HE; APDS; DSRR*

practice they seem to be (at least, the Rarden can fire both types). Both HE and APDS loadings (as well as training ammunition) are in use with the Rarden; the APDS has different characteristics from the experimental HS round developed for the 'Thunderball'.

A long and difficult search for a more effective APFSDS round culminated in a competition between Oerlikon (now RWM) and Nammo, using modified versions of projectiles developed for their 30 × 173 rounds in the 30 × 170 case of the Rarden gun. RWM won the contest in 2007, and its 30 × 170 APFSDS was officially qualified as the L21A1 (somewhat confusingly, the same designation as the Rarden gun). The British Army did not order it, but RWM offered it in kit form for Rarden users to assemble.

A final version of the 30 × 170 cartridge was that developed for the experimental KCC gun of the 1980s. This was based on the KCA aircraft revolver cannon with the ammunition type changed from 30 × 173 to 30 × 170, and from electric to percussion priming. The intended purpose was to fit two such guns in a naval AA mounting; the calibre was chosen to be attractive to navies which already used the KCB. As it happened, the cartridge case needed modifying to work satisfactorily in the much faster-firing KCC, and was given a stronger, two-row crimp and a longer extractor groove. These changes were however backwards-compatible with the KCB, and it was intended that the KCC ammunition would replace the KCB type in production, but the KCC was cancelled.

The cartridge case is rimless and bottle-necked, and contains 160 g of propellant generating a chamber pressure of 420 MPa (Oerlikon). It may be of steel (KCB) or brass (Rarden). Complete rounds in HE loadings are 285 mm long and weigh 870 g (Oerlikon) or 904 g (Rarden). Percussion priming is used. The principal loadings are:

Oerlikon: all current loadings have matching ballistics, with 360 g shells fired at 1,080 m/s. Flight times: 1,000 m/1.08 s; 2,000 m/2.61 s; 3,000 m/4.93 s.

- HEI SD: contains 40 g Hexal P30; delayed-action nose impact fuze
- HEI-T SD: contains 25 g HEI; tracer burns for 4 s
- SAPHEI SD: thick-walled blunt-nosed steel shell with aluminium ballistic cap and delayed-action base fuze; contains 20 g HEI

Rarden (current warshots):

- HE-T (L13A2): 360 g shell contains 25.6 g Torpex; MV 1,070 m/s; tracer burns for 2.6 s
- APDS-T (L14A2): 217 g tungsten carbide penetrator with pyrophoric cap (produces flash on impact) in a plastic sabot with a light alloy base; complete projectile 300 g; MV 1,170 m/s; penetrates 40 mm/40°/1,500 m; tracer burns to 2,000+ m.
- APFSDS-T (L21A1): 160 g tungsten alloy penetrator in a plastic sabot; complete projectile 235 g; MV 1,325 m/s

Also of interest is the DSRR (L15A1): a reduced-range training round with a tubular projectile which matches the initial trajectory of APDS but then loses stability; safety range for this is 4,500 m compared with 11,900 m for APDS. Formerly two other warshots were issued: APSE-T (L5A2) had a hardened steel shell with an aluminium ballistic cap and no fuze (the HEI mix was ignited through friction on impact); and APHC (L6A1) with a tungsten carbide core.

## 30 × 173 OERLIKON KCA / GAU-9/A

This was developed by Oerlikon in the 1950/60s for their 302/303/304 RK experimental aircraft revolver cannon which resulted in the KCA. The first

*30 × 170 KCC: note the double row of crimps and the longer extractor groove*

*30 × 173: Oerlikon RK 302; KCA; KCA drill; GAU-9/A; PGU-14; PGU-13; PGU-15; MPDS; RWM APDS; Nammo APFSDS; drill round; experimental loading with DEFA projectile*

version of the cartridge case measured 178 mm and this as well as the second version had brass cases and electric priming. For the KCA a steel case was adopted. The KCA only saw service in some versions of the SAAB Viggen fighter, which is no longer in use, and it was also offered in the Hughes Model 34 gunpod. It is now effectively obsolete, although curiously the KCA is still (2020) advertised by Rheinmetall as suitable for use in remote-controlled weapon stations. The cartridge dimensions were copied for the US GAU-8/A cannon, although with significant technical changes.

The bottle-necked rimless cartridge case contains 160 g of propellant. Complete rounds are 290 mm long and weigh 890 g. Projectile weight is 360 g and muzzle velocity 1,030 m/s. The principal loadings are: nose-fuzed HEI, base-fuzed SAPHEI and AP.

The KCA was tested in the USA under the GAU-9/A designation, for a twin installation in the A-10 aircraft, as insurance against the failure of the GAU-8/A. The ammunition adopted an aluminium-alloy case but retained the electric ignition and metal driving band. The case was coated internally with a urethane elastomer to prevent burn-through. About 50,000 rounds were made, mostly with a TP projectile weighing around 300 g and fired at about 1,036 m/s.

## 30 × 173 GAU-8/A AND AFV

In the late 1960s the Oerlikon KCA's cartridge design (see above) was adopted by General Electric for their version of the American GAU-8/A rotary cannon, as fitted to the A-10 'tankbuster' aircraft and subsequently to the Goalkeeper naval CIWS. GE subcontracted ammunition development to Aerojet, with most (if not all) of the cases being made by Harvey Aluminum.

The GAU-8/A ammunition differed from that of the KCA in three respects: the case material was changed from steel to aluminium alloy; the priming from electric to percussion; and the projectile driving bands from metal to plastic (after initial work with a gilding metal band). As a result, the rounds are not interchangeable. Later still the GAU-8/A cartridge was adopted for the Mauser MK 30F (now Rheinmetall MK 30–1) cannon which is in widespread use as an AA gun, and then for the American Bushmaster II Chain Gun.

Some manufacturers reverted to making the cheaper and stronger steel-cased ammunition (but still percussion-primed) for the MK 30F used in land or naval mountings, a trend which has accelerated since the growth in use of weapons in this calibre for arming AFVs. The Rheinmetall MK 30–2 version is specifically designed for this purpose, as is the latest version of the Bushmaster II/Mk 44 designated XM813. It is also used in the South African GI-30 CamGun.

The 30 × 173 is now the most common Western cannon cartridge in this calibre and is available in a wide variety of loadings from several different manufacturers, principally ATK and GD-OTS in the USA, Nammo and RWM in Europe. Ammunition is also made in other countries such as South Africa, South Korea, Greece and Poland.

In order to attack tanks, the A-10's ammunition was required to penetrate 57 mm RHA (Rolled Homogenous Armour) at a 30° angle at a range of 1,200 m, assuming that the aircraft was travelling at 250 knots (460 kph) – the aircraft speed increased the effective velocity of the projectile. Discarding sabot ammunition was ruled out because of the risk of the sabot petals being sucked into the engines, while HEAT was also rejected as the spin rate of the projectiles significantly degraded their effectiveness (technological developments have partly solved this problem since). So the projectile type chosen was an API with a Depleted Uranium (DU) penetrating core within a mostly aluminium body (the Incendiary effect arising from the pyrophoric nature of DU). This loading was given the designation PGU-14, which is distinguished by a black and red projectile. Both rocket-assisted AP and APFSDS loadings (the latter for use with a sabot shredder at the gun muzzle to avoid the engine-damage problem) were developed later but not adopted. The standard ammunition load for the A-10 in its intended anti-tank role consists of one PGU-13 HEI for every five of the PGU-14 API. More recently, counter-insurgency anti-personnel use has seen a switch to all-HE loads.

A special Missile Piercing Discarding Sabot (followed by a frangible version – the FMPDS) was developed by NWM (now part of RWM) for use with the Goalkeeper CIWS. This has a red plastic sabot the same shape as the full-calibre projectiles. A further experimental development of HEI-T ammunition by Aerojet took place in the early 1970s for use in the abortive Army Radar Gun Air Defense System (ARGADS) programme based on the GAU-8/A. While maintaining interchangeability with existing ammunition, this focused on minimising the time of flight through an improved ogive shape plus a tracer/fumer in the base, increased propellant load, and longer (254 cm) gun barrels. The projectile, which contained approximately 50 g HE, took 2.33 s to reach 2,000 m.

The newer steel-cased ammunition for AFVs has seen the introduction of

FAPIDS, HEAB/PABM, MPLD (multi-purpose low-drag SAPHEI), PELE and APFSDS loadings, plus an anti-personnel KETF shrapnel round, which was selected by the German Army for the Mauser MK 30–2 gun in its Puma MICV. Various target practice rounds are also available.

The initial 30 mm APFSDS was found to suffer from severe rifling-induced yaw in the early part of its trajectory, not entirely settling down until it had passed 400 m. Nammo accordingly elected to use a slip-ring driving band to minimize the spin imparted to the projectile, although RWM has selected a solid sabot for their loading. Both were offered for the Mk 44 gun in the USMC's now cancelled amphibious EFV, and no doubt will also be competing for use in the XM813, which has already been fitted to some Stryker 8x8 vehicles.

One unusual application of 30 × 173 APFSDS shot was the RAMICS round (Rapid Airborne Mine Clearance Programme), which was intended for firing vertically downwards from a helicopter to detonate or sink sea mines. It has a blunt tip to achieve a supercavitating effect, and could sink mines detected by a blue-green LIDAR at depths of up to 60 m. This interesting project was stopped in May 2011, although the ammunition remains available as the MK258 Mod 1 and has been dubbed the 'Swimmer', being marketed as enabling warship self-defence guns to shoot through waves to hit small boats.

The cartridge case is rimless and bottlenecked, and contains c.150 g propellant developing a maximum chamber pressure of 420 MPa. Complete rounds with HE projectiles are 290 mm long and weigh c. 830 g with steel cases or c. 670–690 g with light-alloy.

The principal loadings are:

GAU-8/A:

- API (PGU-14): full-calibre projectile with DU penetrator, weight 390 g; MV 1,013 m/s.
- HEI (PGU-13): 378 g; contains 56 g HEI and impact nose fuze. MV 1,020 m/s.
- MPDS/FMPDS (PMC 342 / 345): for use only in Goalkeeper CIWS; complete projectile with tungsten-alloy flight projectile/penetrator and sabot is 312 g; MV 1,150 m/s; flight time to 1,000 / 2,000 m is 0.94 / 2.07 s.

Mk 44 and MK 30:

- APFSDS-T (PMC 287 / MK268; Nammo similar): 235 g projectile with sabot; 160 g tungsten alloy flight projectile/penetrator; MV 1,405 m/s; flight time to 1,000 / 2,000 m is 0.75 / 1.56 s; penetration of armour plate at 60° is 53 mm/1,000 m and 46 mm/2,000 m.
- FAPIDS-T (PMC 283): 235 g projectile with sabot; MV 1,405 m/s; flight time to 1,000 / 2,000 m is 0.76 / 1.66 s; penetration of armour plate at 60° is 31 mm/1,000 m and 25 mm/2,000 m.
- HEI (MK266): generally similar to PGU-13 (see above) with tracer added.
- MPLD-T: 363 g Nammo Raufoss Multipurpose SAPHEI shell; MV 1,100 m/s.
- PELE (PMC 381 / 384): 363 g steel shell with plastic core (without / with tungsten penetrator); MV 1,100 m/s; penetrates 25 / 45 mm armour at 100 m.
- PABM (MK310): 423 g HE/frag with time fuze; MV 980 m/s.
- KETF (PMC 308): 362 g projectile containing 162 tungsten pellets each weighing 1.24 g; MV 1,100 m/s; flight time to 1,000 / 2,000 m is 1.07 / 2.64 s.

The proximity fuze already announced for the 30 x 113B LW30 has also been adapted for use with 30 x 173 ammunition. The HEP (High Explosive Proximity) is intended for use by AFVs to improve effectiveness against both ground (personnel) and aerial (UAS) targets.

## 30 × 184B MK 101 AND MK 103

This powerful round was developed by Rheinmetall-Borsig of Gemany in the mid-1930s, for use in their heavy magazine-fed MK 101 cannon. A private venture, this saw little use until it was fitted to the Hs 129 'tankbuster' plane during World War II. This demonstrated the usefulness of a powerful cannon and it was succeeded by the faster-firing and belt-fed MK 103 gun firing (almost) the same ammunition. By the end of World War II, the MK 103 was being developed for the AA role in addition to various aircraft installations.

One important difference between the two guns was that the MK 101 used percussion-primed ammunition, the MK 103 electric, so they were not interchangeable. A variety of different HE and AP loadings was developed but the best known are the *Minengeschoss* HEI for aerial combat, and the tungsten-cored *Hartkernmunition* for anti-tank use. The belted and bottlenecked steel cartridge case weighs 320 g and contains c.97–113 g of propellant (depending on the loading). Chamber pressures were typically 330–340 MPa for MK 101 and all AP loadings, but were reduced to 280 MPa for the MK 103 HEI loading to preserve gun life. Complete rounds are 298 mm long and weigh 778–936 g depending on the loading.

*30 × 184B: M-Geschoss Ausf.A; Ausf.C; Hartkern TP*

The principal loadings were:

- API (MK 101): 500 g thick steel pointed shell containing *Elektron-Thermit* incendiary capsule; MV 700 m/s; penetration 25 mm/90°/300 m
- APHEI-T: 450 g shorter thick steel pointed shell with base fuze and ballistic cap, containing 6 g *Elektron-Thermit* and 5.2 g Nitropenta; MV 760 m/s; penetration at 90° 32 mm/300 m, 27 mm/600 m; penetration at 60° 27 mm/300 m, 21 mm/600 m.
- APCR-T (*Hartkernmunition*): 356 g projectile with 231 g tungsten-carbide penetrating core, 85 mm long and 16 mm in diameter; MV 960 m/s; penetration at 90° 75–95 mm/300 m; penetration at 60° 42–52 mm/300 m.
- HEI-T (MK 101): 440 g thick-walled nose-fuzed shell containing c.30 g HEI mix; MV 800 m/s
- HEI (*Minengeschoss*): 330 g thin-walled nose-fuzed mine shell containing 85 g (Ausf.A) or 72 g (more

*30 × 210B: UBR-83; drill round*

streamlined Ausf.C); MV 900 m/s (MK 101), 860 m/s (MK 103).

## 30 × 210B NN-30

This was developed in the late 1950s for the Russian navy's NN-30 anti-aircraft revolver cannon as fitted in the widely-used AK-230 mounting, also adopted by China. A modified version of the gun was more recently developed by the Romanians for their land-based AA twin mounting designated the A436 or Model 80. The cartridge was also chosen by Zastava of Yugoslavia for their M86/M89 AFV guns which, it appears, are still featuring as an option in Serbian developments (the latest version being the Zastava M97/11).

It is one of the largest 30 mm rounds to see service, exceeded only slightly by the Czech 30 × 211 described below. Production of the system ended in the USSR in the early 1980s but ammunition is or has also been made in China, India, Romania, Serbia and Poland.

The brass cartridge case is bottle-necked and belted, and there are two crimping rings to secure the projectiles, which have two copper driving bands. The case holds 190 g propellant that generates 314 MPa chamber pressure. Complete rounds are 306 mm long and weigh 1.06 kg. Common to almost all of the loadings are a projectile weight of a 358 g and a muzzle velocity of 1,050 m/s. Soviet loads were as follows:

- HEI SD (UOF-83): nose-fuzed shell filled with 39 g (30 g for OF-83D) of A-IX-2 HEI compound
- APHEI SD (UF-83): base-fuzed shell containing 30 g A-IX-2 HEI compound; blunt steel nose covered by a ballistic cap
- AP-T (UBR-83): 328 g mild steel projectile with a large tracer that burns for 8 s. Loaded in ammo belts at a ratio of one to every 14 OF-83 rounds.

Romanian loads for the A436 differ, and carry an additional 'R' behind the ammunition code:

- The OR-83R is a HE-T projectile weighing 360 g, it is filled with A-IX-2 high explosive and carries the MG-30 nose fuse. The projectile is 135 mm long and a tracer burning time of at least 9 seconds. Muzzle velocity is 1050 m/s, the round weighs 1,076 g in total.
- The BZR-83R is an API-T projectile weighing 360 g, it is 119 mm long and the tracer burning time is a minimum of 4 seconds. This projectile consists of a short-pointed solid shot with a rear tracer cavity and has a long ballistic cap. Muzzle velocity is 1,050 m/s, the round weighs 1073 g in total. The projectile is able to penetrate 20 mm of armour at 1,000 m.

## 30 × 211 M53

This round was first designed at the Czechoslovak Brno factory, then developed under World War II German occupation for the MK 303 anti-aircraft gun. After the war, the Czechs decided to continue with its development and it emerged along with the M53 cannon. The ammunition was also produced in Serbia. It is unique to the M53 gun.

The steel-cased rimless cartridge is the biggest and most powerful of all of the service 30 mm rounds, and has often been designated the 30 × 220 (it may also be designated 30 × 210). The steel cartridge case is rimless and bottle-necked, initially weighing 750 g (brass) or 530 g (steel); late production weighs 490 g and contains 195 g of propellant generating 314 MPa chamber pressure. There are either one or two crimping grooves in the neck. The overall round length is 332 mm and the total weight around 1.14 kg.

The standard loading is a nose-fuzed 435 g HE-T SD fired at 1,000 m/s; there was also a conventional AP-T with matching ballistics, said to penetrate 55 mm of conventional steel armour at 500 m/0°. Yugoimport marketed an HE-T designated M69 (made by Sloboda Chachak), fitted with a super-quick action impact fuze with mechanical self-destruct; the fuze burned for four seconds. Yugoimport also marketed the M78 action-testing break-up round loaded with a 250 g plastic projectile filled with iron powder; this broke up after leaving the muzzle.

Some interesting experimental loadings were developed for this cartridge. One, dating to the late 1960s, was a fin-stabilized HEAT. Another was an APFSDS round that was still in the development stage when it was curtailed by the end of the Cold War.

The HEAT round, an unusual choice in such a small calibre, was developed for the late-1960s Vydra project, a cannon-armed APC. The VK-30 cannon was intended to engage both aerial and light armoured ground targets and had a pneumatic ammunition feed holding 120 rounds. The existing 30 × 211 cartridge was chosen as the basis, and a 30 mm HEAT projectile was developed to penetrate about 90 mm armour. To avoid the usual degrading effect of spinning a HEAT projectile, it was designed not to be spun by the barrel rifling; stabilisation was achieved by fins. The project was abandoned in the early 1970s in the interests of Warsaw Pact standardisation.

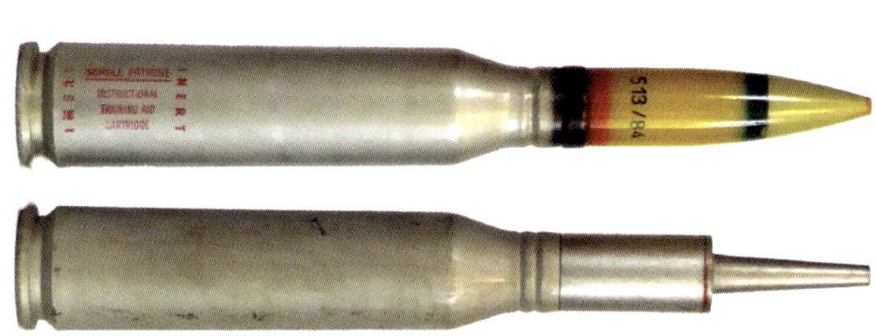

*30 × 211 rounds: HEI-T above, Vydra below*

# 33–37 mm CARTRIDGES

With the exception of the Swiss 34 × 239 round and its more successful replacement, the 35 × 228, service automatic weapons in this category have been monopolised by the 37 mm calibre. The reason for this goes back to the Declaration of St Petersburg of 1868, the result of an international conference called to discuss the use in warfare of the new explosive rifle bullets. It was agreed to ban all explosive projectiles weighing less than 400 grammes, and at that time the appropriate minimum calibre of shells designed to exceed that weight was 37 mm. Most countries subsequently retained this calibre out of sheer convenience – it meant that when designing a new cartridge it was easier to use existing projectile and barrel-making machinery. While the success of the Bofors guns has made the 40 mm calibre much more popular since World War II, the Chinese adopted a new 37 × 240 cartridge as late as 1976.

The cartridge which became the international standard 37 mm round at the start of the automatic cannon era was the 37 × 94R Hotchkiss 'one-pounder' (the nominal weight of the projectile, usually abbreviated to $1^{\text{PDR}}$ or 1-pounder). Although developed for a Hotchkiss manually operated gun, this was also chosen for the first self-powered automatic shell-firing gun, the Maxim 1-pounder 'pom-pom' which emerged in the 1890s. These guns and their ammunition were hugely successful in the international market, with a wide variety of loadings being made by many different firms in several countries.

Attempts to improve on the performance of guns in 37 mm calibre in the first third of the 20th Century saw cartridge cases gradually increasing in length and width in order to take more propellant, and the length and weight of the projectiles also grew significantly. 37 mm cannon of various types saw extensive service with several nations in World War II, mostly for AA use. Despite all of this enormous development activity, probably greater in total than that for any other calibre of cannon ammunition, only two 37 mm cannon rounds remain in service today, and one of those is obsolescent.

## 33–37 mm SERVICE CARTRIDGES 1

## 33–37 mm SERVICE CARTRIDGES 2

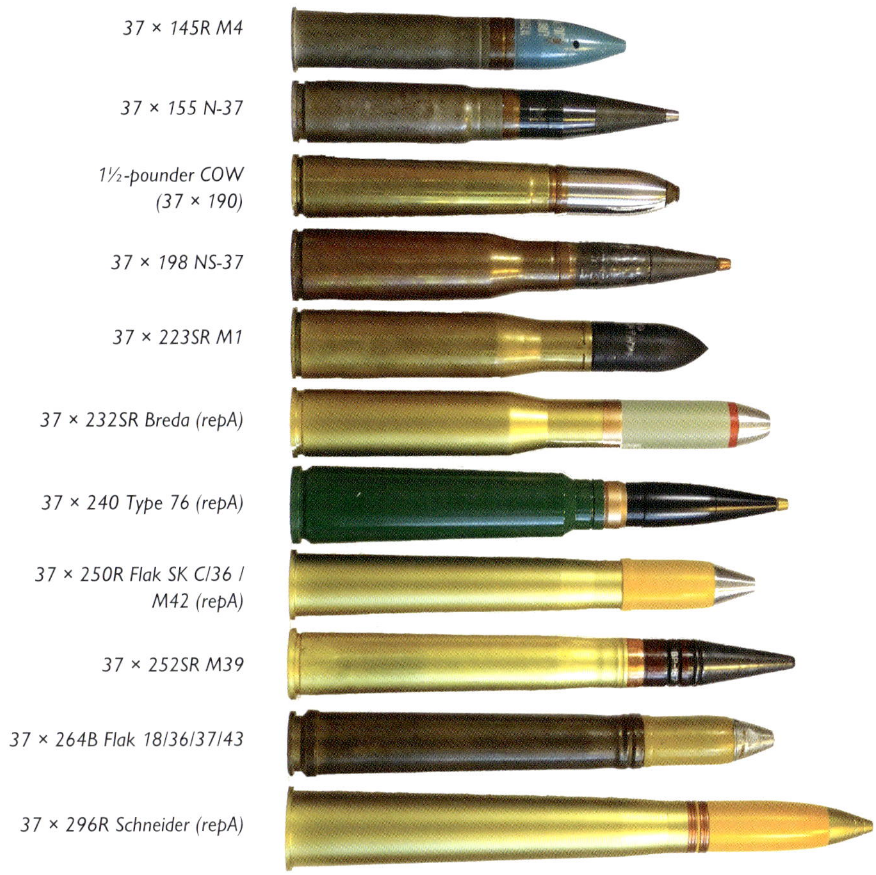


## 33–37 mm EXPERIMENTAL CARTRIDGES

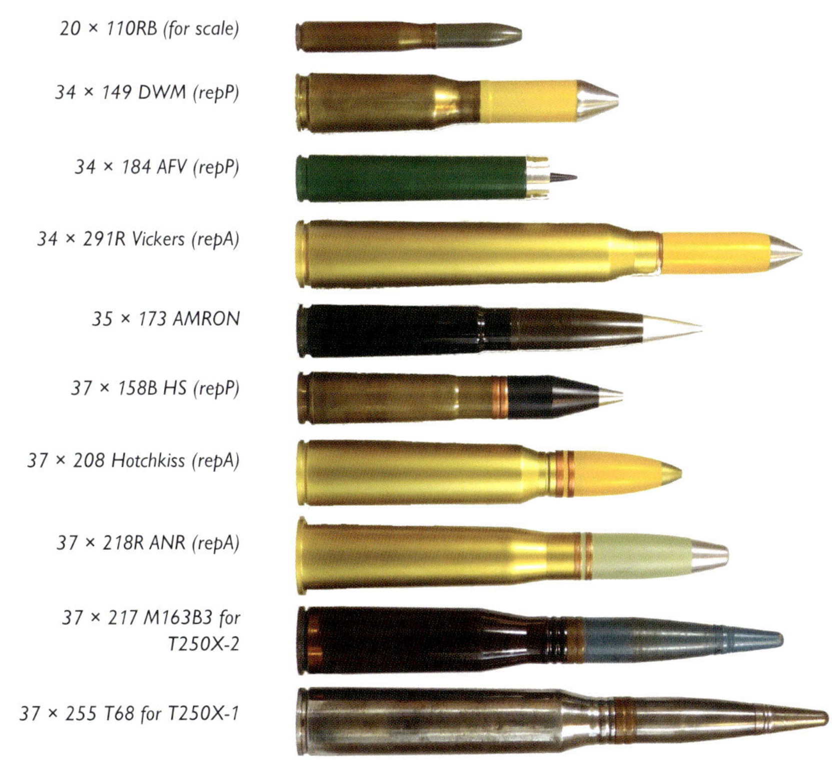


### 33 × 190R A.P.X.

Ordered to provide a *moteur canon* for French aircraft, A.P.X. developed this at the Puteaux Arsenal from 1927 to 1933, reaching the aircraft test stage in 1932. Known characteristics were: 33 mm calibre, with a rimmed case initially tested at 217.5 mm, then 188–192 mm (and generally described as 190 mm length) firing a 630 g shell containing 72 g HE at 650 m/s (at 225 MPa) or 687 m/s (from 162 g of US3 propellant generating 245 MPa).

### 34 × 149 DWM

Several cases and projectiles of this ammunition survive (the case in the Experimental group photo has a replica projectile which, like the original, is of the *M-Geschoss* type). This round was intended for an aircraft gun about which nothing is known, but it is said to have been developed by the German company DWM in 1941. The overall length of the round is 263 mm.

### 34 × 184 RHEINMETALL

This Rheinmetall round (which may also be described as 35 × 185) is known from some surviving cases and a couple of diagrams in a 1985 magazine. It was based on the same case as that used for the 25 × 173/181 Oerlikon KBB round, but without being necked down. The idea was to keep the overall length of the 34 mm ammunition the same as the 25 × 137, so that 25 mm guns could be rebarrelled to take the larger calibre round. While no performance data have been unearthed, the diagrams reveal that the 34 mm had a much bigger APFSDS penetrator than the 25 mm, while the HE shell was very long, leaving little space for propellant, and could therefore only have been fired at a low velocity.

The project was probably inspired by the 50 mm Supershot designed during the same period, which used a 35 × 228 Oerlikon case, straightened and extended to 330 mm and with the projectiles almost entirely buried in the case. The 34 mm round was presumably intended to be a 'mini-Supershot', to increase both the armour penetration and HE/fragmentation effect of suitably modified 25 mm AFV guns.

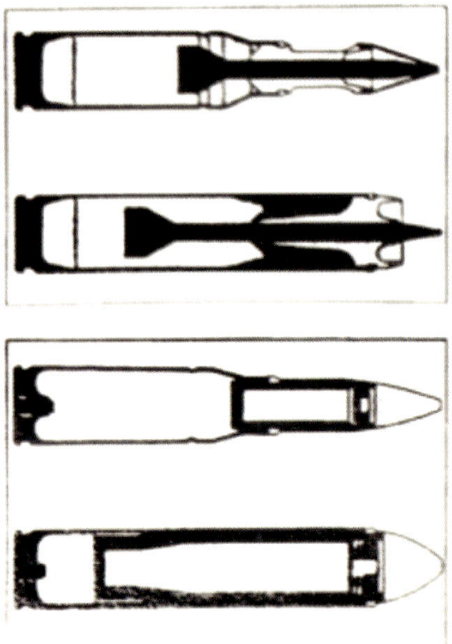

*34 x 184 rounds compared with the equivalent 25 x 137 rounds*

## 34 × 239 FK 38

This cartridge was developed for the Swiss 34 mm Flab Kan 38 AA gun adopted in 1938. The cartridge case is brass, rimless and bottlenecked. Projectiles have a single broad driving band of copper and are fixed to the case by two segmented crimping rings. The warshots are a nose-fuzed HE-T and an AP-T, the latter being a full-calibre steel projectile with a ballistic cap. Projectiles weigh 720 g with a muzzle velocity of 900 m/s for the HE-T, 930 m/s for the AP-T. The round is 362 mm long overall and weighs around 1.6 kg. The maximum feasible range was 4,500 m (which the projectiles reached in 8.8 sec) but the more practical effective range was the tracer burn-out distance of 3,000 m (reached in 4.8 sec).

In the early 1950s a new gun designed by the Swiss branch of Hispano Suiza, the HSS 834, was intended to use this ammunition, but it remained experimental. This used a modified loading, firing a 630 g shell (containing 70 g HE) at 1,000 m/s; Round length was 365 mm and weight 1.46 kg. The FK 38 remained in Swiss service until replaced by the 35 mm Oerlikon in 1968, but never seems to have been used by anyone else.

### 34 × 291 VICKERS CLASS T

Only drawings survive of this ammunition (which was probably never made – the Experimental group photo shows a replica) that was designed in the late 1940s by C.E.A.D. (Chief Engineer Armament Development) for a short-range AA gun system. The large case (chamber capacity 30 cubic inches, or 490 cc) was to contain 369 g propellant, and both semi-rimless and rimless versions were considered. Overall round length was 410 mm, and rim/body diameter was 55.9 mm. No ballistics are known, but this cartridge should clearly have been considerably more powerful than the smaller Swiss 34 mm Flab. Kan.38 (234 mm case, 47.8 mm diameter), which generated in the region of 300,000J muzzle energy. For the sake of comparison, the 35 mm Oerlikon has a case length of 228 mm, a rim diameter of 54.5 mm and generates 380,000–400,000J (depending on the loading). So the Class T cartridge should presumably have generated in the region of 400,000J.

Vickers-Armstrong were asked to prepare a design study for a gun to fire this ammunition, which they designated Class T, but this never proceeded beyond a paper study.

### 1½-pounder Mk V (35 mm)

The 1½-pounder Mk V was an experimental British naval AA system being developed in the mid-1930s as a replacement for the 2-pounder pom-pom. It was cancelled in 1937, mainly because the eight-barrel mounting ended up weighing more than the 2-pounder, but also no doubt because of the planned introduction of the HV version of the pom-pom, which reduced the main criticism of that system; the low muzzle velocity. All that is known about the Mk V ammunition is that it was of 34.54 mm calibre and fired a 680 g shell at 790 m/s. No other data or illustrations of the round have been located.

### 35 × 173 AMRON

This interesting US round was created by necking-out the 30 × 173 GAU-8/A case to take the standard 550 g HEI projectile from the Oerlikon KD series 35 mm ammunition. It is unclear what gun it was intended for – the increased overall length would have prevented its use in guns designed for the 30 × 173 round – but it could have been a formidable performer in a purpose-designed aircraft revolver gun, probably capable of achieving a muzzle velocity in the region of 900–1,000 m/s.

## 35 × 228 OERLIKON

This round was developed by Oerlikon Contraves, which is now Rheinmetall Waffe und Munition (RWM) Schweiz AG, a branch of Rheinmetall Defence. It was designed following extensive research into medium calibre AA needs, which saw comparisons being made between calibres ranging from 30 mm to 50 mm.

*35 × 228: APDS; HEI-T; AHEAD; drill round (other colour schemes also used for AHEAD)*

The optimum choice was whittled down to 35 mm or 50 mm, with 35 mm ultimately succeeding. In Swiss service, the 35 mm systems replaced the 34 × 239 used in the Flab Kan 38, but unlike the earlier system, the 35 mm has enjoyed considerable international success.

Along with the 35 mm round came a new range of high-velocity, fast-firing AA guns in what became the Oerlikon KDA, B, C and F, plus the slower-firing KDE AFV gun. Current emphasis is on the KDG revolver cannon in the Millennium system for AA, CIWS and C-RAM roles, and the Bushmaster III Chain Gun for use in AFVs. Oerlikon 35 mm guns are also made in South Africa and China.

The first version of the round used a 231 mm case with a longer neck, and shorter, less aerodynamic projectiles. In 1958 the 228 mm case was adopted together with longer projectiles which not only improved the velocity retention, but also increased the HE load from 100 to 120 g.

All current loadings use a low-carbon, drawn-steel cartridge case with the projectiles secured by eight crimping points. These engage in a cannelure around the body of the boat-tailed projectile. Soft-iron drive bands are used. The propellant used with full-calibre shells is from 330 to 333 g of single-based powder. A percussion primer with a Boxer primer cap is threaded into the cartridge case base. Most cartridge cases have an annular groove just below the shoulder to provide a secure location for belt links (some of the guns use belted ammunition, others a linkless feed). Overall round length with full-calibre HE shells is 387 mm, weight 1.58 kg.

Rheinmetall is the principal manufacturer but ammunition is, or has been, also made in Canada, China (whose army has adopted it as an eventual replacement for their 37 mm systems), Finland, France, Germany (Diehl), Greece, Iran, South Korea, Norway, Pakistan, Poland, Romania, South Africa, Spain, Turkey and the UK.

The original ammunition loadings were conventional HEI and SAPHEI with matched ballistics, but the capabilities of the weapon system were greatly widened with the introduction of the AHEAD (Advanced Hit Efficiency and Destruction) system in 1991. This increases the hit probability so much that the system is effective against missiles as well as aircraft. It is in effect a modern version of the old shrapnel round, carrying tungsten-alloy sub-projectiles or pellets which are released by a small ejection charge triggered by a time fuze, forming an expanding pattern of pellets like a huge shotgun blast directed towards the target. AHEAD is also designated KETF for Kinetic Energy Time Fuzed.

As the AHEAD projectile leaves the muzzle it travels through three coils containing electronic sensors. The time taken in passing between the first two coils allows the exact velocity to be measured and fed into the fire-control system computer along with the target location, speed and direction derived from the system sensors. This enables the projectile flying time to be precisely calculated, a value transmitted by electro-induction to the programmable base fuze as the projectile passes through the third coil. At the correct point of its flight the fuze detonates the ejection charge at a given distance in front of the target. At a range of 40 m from detonation, the pellets cover an area 7 m in diameter, at a density of four per square metre, at which time they are still travelling at 75% of the initial ejection velocity. One 25-round burst can produce a 'swarm' containing 3,800 pellets covering the expected target position. If the fuze fails to function for any reason, the shell will self-destruct after 8.19 seconds, an equivalent range of approximately 5,000 m. For use against AFVs, the round may be fired without the fuze being set; it then acts rather like a frangible Armour-Piercing (AP) projectile.

RWM Schweiz has developed three types of this AHEAD ammunition in 35 mm calibre for the engagement of different targets:

AHEAD projectiles were originally loaded with 152 tungsten pellets each 5.85 mm wide and weighing 3.3 g, stacked in eight layers with each layer containing 19 pellets. For anti-personnel use a new loading with a larger number of smaller pellets has been developed. Unlike the AA AHEAD, it is designed to spread the pellets more to the sides to engage targets in defilade (hiding behind walls or in trenches). This PMD330 round contains 407 pellets each weighing 1.24 g – the same pellets as are used in the 30 mm KETF round. A third variant, the PMD375,

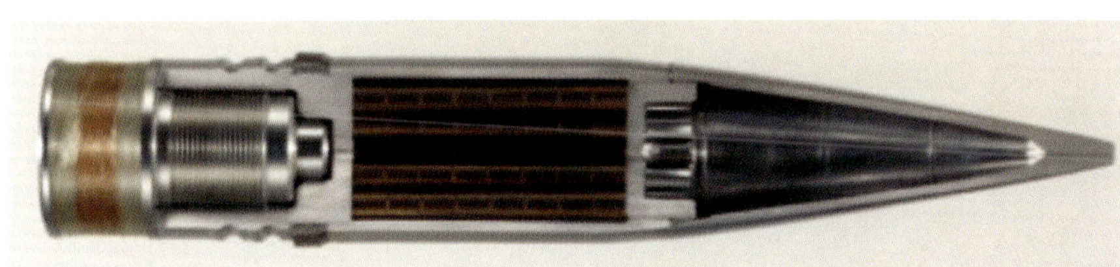

*Sectioned 35 mm AHEAD (Oerlikon Contraves Ag)*

*35 mm APFSDS (RWM)*

was developed for the Counter Rockets and Mortar (C-RAM) mission. This contains 860 pellets each weighing 0.64 g, but it is not clear if it has been adopted.

The use of 35 mm calibre cannon in light AFVs has led to the development of APFSDS ammunition designed to penetrate the armour of most such vehicles.

The following data covers the principal loadings (not all of which are still made by Rheinmetall) with Rheinmetall factory codes (PMD numbers). In addition, there is the usual range of training ammunition with matched ballistics, including a break-up type. For full-calibre, standard weight (550 g) shells made by Rheinmetall and others, maximum average chamber pressure is 380 MPa, muzzle rotation speed is 1,200 rpm, muzzle velocity is 1,175 m/s, and maximum ballistic range is 11,000 m.

- HEI (NF): PMD 042; 535 g; nose-fuzed, filled with 112 g Hexal (or 98 g if tracer fitted), with PD, DA, SD fuze; flight times to 1,000/2,000/3,000 m are 0.96/2.18/3.8 s; tracer burns to 3,100 m; SD range 4,000+ m
- HEI (BF): PMD 074; 550 g; base-fuzed, filled with 65 g Hexal + 20 g incendiary comp; ballistics as HEI (NF); penetration 15 mm RHA/30°/1000 m
- SAPHEI: PMD 044; 550 g; base-fuzed with thick steel body and ballistic cap; filled with 22 g Hexal; ballistics as HEI (NF); penetration 40 mm/90°/1000 m, 20 mm/30°/1000 m
- AHEAD: 750 g; 152 (PMD 062) or 407 (PMD 330) tungsten pellets with base time fuze and burster charge (total pellet weight 500 g); MV 1,050 m/s; flight times to 1,000/2,000/3,000 m are 1.05/2.34/3.98 s; self-destructs in 8.19 s (c.5,000 m)
- APDS-T: PMD 369; 380 g including plastic/light alloy sabot (302 g tungsten-alloy penetrator); MV 1,400 m/s; flight times to 1,000/2,000/3,000 m are 0.73/1.54/2.38 s; tracer burn to 2,000 m; penetration 90 mm/90°/1000 m
- APFSDS-T: PMD 343; 388 g including plastic/light alloy sabot (283 g tungsten-alloy penetrator); MV 1,400 m/s; flight times to 1,000/2,000/3,000 m are 0.70/1.42/2.21 s; penetration 120 mm/90°/1000 m
- FAPDS: PMD 055; 396 g including plastic/light alloy sabot (298 g frangible tungsten-alloy penetrator); MV 1,400 m/s; ballistics as APDS.

## 1 POUNDER Mk III / V (37 × 69R)

This was developed in World War I to produce a lighter version of the Maxim 1-pounder pom-pom (see below), for aircraft use. It did see service, but few were used. The cartridge case was shortened from 94 mm to 69 mm and the propellant load and muzzle velocity decreased accordingly, reducing not only the weight but also the recoil of the cartridge.

The round length is 137 mm and the weight around 550 g. The brass cartridge case is rimmed and has a slight bottleneck. A single segmented crimping groove is used to secure the projectile, which has a copper driving band of complex shape. The standard combat loading was a 454 g HE shell fitted with the very sensitive No.131 impact fuze for use against airship and balloon fabric, the flat nose of which was covered in tin foil. Muzzle velocity was 365 m/s from the 81 cm (L/22) barrel.

## 37 × 87R AAC

The gun was closely based on a Puteaux design in 37 × 94R but was slightly rechambered, presumably to ensure that purchasers of the guns had to buy AAC ammunition. The gun was designated Type M and was offered on a turreted aircraft mounting, and on a wheeled mount as an Automatic Mountain or Infantry Gun. The production status of the gun is not known, but it was advertised for several years and AAC did achieve sales (although exactly how many guns of which types it sold to which countries is unclear). The round fired a 500 g HE projectile at 380 m/s. The total round weight was 635 g, including 24.3 g propellant. The chamber pressure was only 90 MPa.

## 1 POUNDER MAXIM 'pom-pom' (37 × 94R)

Developed in 1885 by Hotchkiss, this round was chosen for the first self-powered automatic shell-firing gun, the Maxim 1-pounder which was introduced by Vickers in the 1890s, and become known (even in official documents) as the 'pom-pom', after the sound of its slow firing; a nick-name later transferred to other guns, particularly the Vickers 2-pounder described later. The ammunition was also widely used in manually-loaded infantry and light tank guns. Many different loadings were developed for this round by many different companies. It remained in front-line service until World War II,

*1–1½-pounder rounds for automatic weapons: 37 × 69R (1-pounder Mk III/V); 37 × 94R (1-pounder Hotchkiss); 37 × 124R (1½-pounder Vickers); 37 × 136R (1-pounder Mk 6, Heavy); 37 × 190 (1½-pounder C.O.W.)*

and in sub-calibre equipment for training purposes for longer than that.

The round has a brass, rimmed, slightly bottlenecked cartridge case, with the projectiles secured in the case by means of a crimping groove. Overall length is normally 165–175 mm. The projectiles typically have two narrow driving bands, but other types were also used.

The 1900 British Handbook on the '1-Pr Maxim Q.F. Gun' described two projectiles: a nose-fuzed Common Shell and a pointed base-fuzed Steel Shell, both weighing 454 g including a 17.5 g black-powder bursting charge. The propellant charge of 40.6 g Cordite generated a chamber pressure of 181 MPa, enough to drive both shells at an MV of 549 m/s from a 107 cm (L/29) barrel. Overall round weight is about 662 g, of which the cartridge case weighs 170 g. Perforation of wrought iron (presumably by the steel shell) was 57 mm at the muzzle, 53 mm at 100 m and 35 mm at 600 m. Time of flight to 450/900/1,800/2,700 m was 0.98/2.33/5.86/10.52 s with remaining velocities being 393/304/225/171 m/s.

Rounds for manually-loaded guns showed great variety, including incendiary and canister types. Projectile weights normally varied between 450–550 g and muzzle velocities were usually around 365–450 m/s; somewhat less powerful than the British loading, possibly because many guns had shorter barrels of around 20 calibres. One notable exception was one of the last combat loadings to be developed; the French Mle 1935 tank gun API round, which fired a 390 g shot consisting of a sub-calibre hardened steel penetrator within a magnesium body which flashed into flame on impact; it was fired at 600 m/s.

## 37 × 101SR 'SOCKELFLAK'

This German development was intended for a cannon for arming airships (proper designation *'3,7 cm Luftschiff Flak'*), which saw limited use in World War I in the ground-based AA role, and again in World War II, firing tracer shells to deter low-level attacks on airfields. The round differs from the normal pom-pom cartridge primarily in having a slightly longer, semi-rimmed case. In other respects it is similar. The ballistic performance was low, with the 465 g shell fired at only 320 m/s, partly because of a very short barrel but also probably to avoid muzzle flash – not a desirable feature in a gun designed to be fired from the top surface of an airship which could be venting or leaking hydrogen.

*37 mm Grabin (courtesy Evgeny Aranov)*

### 37 × 110SR GRABIN V-14–113

This round is for the smallest of a series of Grabin aircraft guns designed in the late 1940s. The OFTZ (HEI-T) projectile weighed 635 g and contained 98 g HEI. The case contained 76 g propellant, enough to generate 602 m/s MV. The round is 255 mm in overall length. Three guns were made before the project was cancelled.

## 37 × 111R Ho-203

This Japanese cartridge was originally designed for the Year 11 light infantry gun adopted in 1922, but was subsequently chosen (in a purpose-designed loading) for the Ho-203 automatic cannon that armed some IJA heavy fighter planes in World War II.

The rimmed, brass cartridge case is unusual in having a very long neck relative to the length of the body. The overall length of the round is 194 mm. The standard Ho 203 projectile was a 475 g nose-fuzed HEI containing 32 g HE plus an incendiary capsule. Muzzle velocity was 570 m/s.

## MAXIM-NORDENFELT Mk III 1¼-pounder 'HEAVY' AND VICKERS 1½-pounder (37 × 124R)

This ammunition was developed late in the 19th Century for the 1¼PR Maxim-Nordenfelt Mk III 'Heavy' of 1898. This saw little use but the cartridge was adopted in modified form (including a different primer) some years later for the Maxim-based Vickers M1915 1½PR, used by the British and Austro-Hungarian navies. This also saw little use with the RN before being replaced by an evolutionary development, the 40 × 158R 2-pounder pom-pom during World War I. The ammunition was also used in the Vickers 1½-pounder Type B semi-automatic aircraft gun.

The ammunition has a rimmed and bottlenecked brass cartridge case wider as well as longer than the 1-pounder. The projectiles initially used in the Vickers copied the Maxim in weighing 1¼ lbs (570 g), but very few of these were made before the shell weight was increased to 1½ lbs (680 g).

Overall round length of the Vickers varies with the projectile but is 210 mm with the APHE shown in photos here; cartridge weight is around 850–900 g. Both base-fuzed steel-pointed and nose-fuzed explosive projectiles (containing 67.5 g explosive) were used, with a shell weight of 680 g and an MV of 640 m/s from an L/42.5 barrel.

There was another cartridge of this size but using a belted case, so has the designation 37 × 124B. It was developed for the Browning-pattern M1924 semi-automatic gun fitted to some US light tanks in the 1920s, but this remained experimental (although it was the ancestor of the M4 aircraft gun described below). Shell weight was 570 g.

## USN 1-pounder Mk VI, HEAVY (37 × 136R)

This cartridge was originally designed for a manually-loaded Hotchkiss gun which entered USN service in about 1890 and was also used in various other guns and sub-calibre training systems. It was adopted for two automatic weapons in the first few years of the 20th century: an enlarged Maxim 'pom-pom' known as the 1-Pounder Maxim-Nordenfelt, Heavy, Mk 1 (later designated 1-Pounder Mk VI), which during World War I was ordered in large numbers for the USN; and another US automatic cannon developed by Samuel N McClean of Lewis

Gun fame. Although the gas-operated McClean gun failed to be adopted by the USA, a sale to Russia (who called it the 'Maklen' gun) during World War I of just over 200 guns subsequently led to them being widely distributed.

Illustrations of ammunition from a Russian catalogue show the following headstamps:

PE&MCo. = Poole Engineering & Machine Co, the cartridge case is dated June, 1916

U.M.C.Co. = Union Metallic Cartridge Company, the cartridge cases appear to be dated June, 1914 and November, 1915

The brass cartridge case is rimmed and slightly bottlenecked. Various projectiles were used, but all seem to have been of the base-fuzed steel pointed type. The overall length of the round is 210 mm and the weight is around 600 g. The standard projectile weight was 476 g and the MV 640 m/s from an L/42.5 barrel.

### 37 × 142R AAC

This round (which is essentially a slightly lengthened version of the 37 × 136R described above) is mainly associated with AAC's manually loaded Rapid Fire Gun, offered on either wheeled or pedestal (naval) mountings. However, it has been included in this section because it was probably the round used in AAC's Automatic Naval Gun. An examination of the illustration of this gun in the AAC catalogue reveals that it was a close copy of the McClean cannon mentioned in the 37 × 136R record. Given AAC's predilection for making their own unique cartridges to ensure that the customers for their guns had to keep buying AAC ammunition, it seems a near certainty that the gun was merely slightly modified to handle the longer case length. Sales of this gun are unknown.

Four different types of ammunition were offered for the Rapid Fire Gun: three of them (AP, HE and canister) shared the same projectile weights and muzzle velocities: 570 g at 610 m/s. The fourth was designated 'Super AP' and had a 660 g projectile fired at 565 m/s and was claimed to penetrate 25 mm of hardened steel 'at all battle ranges and at a considerable angle of impact'. AAC's Automatic Naval Gun is stated in their catalogue to fire a 570 g projectile at 732 m/s; the higher velocity would have been at least partly due to the use of an L/50 barrel rather than the L/40 barrel of the Rapid Fire Gun.

## 37 × 144 IJN Ho-204

This round was developed for an aircraft cannon intended to replace the Ho 203 described above. It saw very little use, emerging right at the end of World War II. The rimless, bottlenecked brass case was developed solely for use in the Ho-204 and has an unusually high, sharp shoulder with a segmented crimping groove. Overall length is 230 mm when loaded with the same HEI shell as used in the Ho 203. Shell weight was therefore 475 g but muzzle velocity was increased to 710 m/s.

## 37 × 145R M4

This round used a slightly lengthened version of the Heavy 1-pounder case described above. It was designed for a Browning aircraft gun that was adopted by the USAAF as the M4 and was subsequently used in a modified version of the gun, the M10. Both guns were used during World War II.

The total length with the standard HE shell is 248 mm (237 mm with the AP) with the overall weight being 900 g for the HE and 1.05 kg with AP. The M38 brass, slightly bottlenecked cartridge case contained 72 g of propellant giving a maximum chamber pressure of 186 MPa and a muzzle velocity of 610 m/s (HE) or 556 m/s (AP).

The most common projectile was the 608 g M54 nose-fuzed HE-T SD shell filled with 45 g Tetryl or Comp A, with a tracer which exploded the HE at the end of its three-second burn. It was also available with the solid steel M80 AP-T shot weighing 753 g, which could penetrate 25 mm/70°/450 m of homogenous plate (20 mm if face-hardened plate) and 15 mm/70°/900 m (10 mm if face-hardened).

## 37 × 155 N-37

The 37 × 155 was developed for the Soviet N-37 aircraft cannon, developed towards the end of World War II, which saw extensive use in the 1950s. Compared with the 37 × 195 NS-37 round described below, the smaller and less powerful cartridge enabled the N-37 to be lighter and faster-firing at the cost of a reduced muzzle velocity. Czechoslovakia, Hungary, Poland and East Germany manufactured this ammunition in addition to the USSR.

The cartridge case (of steel or brass) is rimless and bottlenecked, containing 128 g propellant and with two segmented crimping grooves around the neck. The complete round is 288 mm long and weighs 1.28 kg. Various different loadings were developed, initially using the same projectiles as the NS-37. The muzzle velocity in the N-37 was 690 m/s except where stated, increasing to 725 m/s with the long-barrelled N-37L and NN-37 guns.

- HEI-T SD: designated OZT; 735 g nose-fuzed shell containing 37 g A-IX-2 HE
- HEI SD: designated OFZ; 735 g nose-fuzed shell containing 49 g A-IX-2 HE
- AP-T: designated BZT; 760 g solid steel shot with a ballistic cap; MV 675 m/s

*Comparison between N-37 (37 × 155) below, and HS 606 (37 × 158B, repP) above*

- APHEI SD: designated BR; 734 g base-fuzed shell with a thick steel body and a ballistic cap; contained 36 g A-IX-2 HE.
- Anti-radar: designated PRL; standard HE shell body loaded with chaff, a fixed-time fuze and a small expulsion charge; designed to fool incoming radar-guided missiles.

### 37 × 158B HISPANO HS 606

This round is based on the 30 × 158 HS 603/604 but necked-out and with a belt added for headspace in view of the shallow shoulder. The HS 600-series weapons were the last HS cannon developed in France (the project started in 1951) and they were listed under the 7CG designation, one of a series of French cannon projects in the 1950s.

The overall length of the round is 264 mm, and it fired a 420 g HE shell at 915 m/s. This makes an interesting comparison with the Soviet N-37 round (37 × 155) which has very similar dimensions and ME but fired a much heavier projectile at a much lower MV. It seems that the HS 606 might have been designed for aerial combat in the thin air at high altitude (using the same light projectile/high MV principle as the 30 mm HS 825), whereas the heavy Soviet shell would have retained more of its velocity in the thicker air at low altitudes, especially in ground attack. However, the HS 606 was also apparently intended for ground attack, having a thick-walled shell rather than a thin-walled Mine shell.

## 1½ POUNDER C.O.W. GUN (37 × 190)

Development of a 37 mm automatic cannon at the Coventry Ordnance Works began before World War I. Initially this was designed to use a 1-pounder cartridge very similar in dimensions to the 37 × 94R pom-pom round but with a rimless case. Attention soon switched to a much more powerful 1½-pounder version, with the case doubled in length and widened. The light cannon designed around it was intended for aircraft use but only entered service in very small numbers right at the end of World War I. By the beginning of World War II only a few dozen guns and 10,000 rounds of ammunition 'of doubtful age' were left, but another batch was made as cases exist with 1942 headstamps.

The brass cartridge case is rimless, tapered and with a slight shoulder; it contained 85 gm propellant, generating 178 MPa chamber pressure. Loaded with a 650 g HE shell the overall length of the round is 290 mm and the weight 1.225 kg. Muzzle velocity was 594 m/s.

### 37 × 198 100-K

The 100-K AA gun was a development of Russia's OKB-15 in the late 1930s under Shpitalny; an unsuccessful contender in the competition won by the 37 mm M1939. It used a 37 × 198 cartridge which was a different shape to the later 37 × 198 used in the NS-37, so the designation '100-K' is added to make this clear. Ballistic data for two loadings is known: AP = 665 g at 931 m/s; HE = 630 g at 971 m/s.

## 37 × 198 NS-37

This powerful cartridge was designed for a Soviet aircraft gun that saw service in both fighters and ground attack planes in World War II. The cartridge case is rimless and bottlenecked, and contained 215 g propellant. Overall round length is 332 mm and loaded weight 1.58–1.605 kg. Muzzle velocity was 880 m/s for all of the loadings listed below.

- HEI-T SD: designated OZT; 735 g nose-fuzed shell containing 37 g A-IX-2 HE
- HEI SD: designated OZ; 720 g nose-fuzed shell containing 49 g A-IX-2 HE
- AP-T: designated BZT; 760 g solid steel shot with a ballistic cap; armour penetration 48 mm/90°/500 m.

## 37 × 203R AAC

AAC (the American Armaments Corporation) offered a wide range of different guns and ammunition for various roles in the 1930s, including aircraft and AFV armament, mainly in 37 mm calibre. Two different rounds are known to have been used in their automatic cannon: the 37 × 87R described above, and the 37 × 203R which was used in AAC's Anti-Aircraft Gun and the Type F fixed aircraft gun. A third round, the 37 × 142R, was mainly associated with the manually-loaded Rapid Fire Gun but was almost certainly also used in the Automatic Naval Gun (a copy of the McClean cannon). A fourth round (37 × 178R) was used in the AAC Anti-Tank Gun which was probably manually loaded.

The 37 × 203R was loaded with a 500 g HE projectile fired at 820 m/s from an L/50 barrel; the total round weight was 780 g, including 150 g propellant with a chamber pressure of 160 MPa; all HE rounds were fitted with a tracer and self-destructed on tracer burnout.

The Type F aircraft gun reportedly had an MV of 610 m/s with an L/40 barrel and 820 m/s with an L/50 barrel, but an increase in MV of 34% just by increasing the barrel length by 25% is not credible. Possibly the 610 m/s figure relates to one of the smaller cartridges; either the 37 × 178R or even the 37 × 142R.

### 37 × 208 HOTCHKISS

This almost unknown round, for which only factory drawings and a tiny photograph exist, was developed in the mid-1930s for an automatic AA gun. In the same period, Hotchkiss produced such a gun which competed unsuccessfully in the French naval ACAD competition for a new AA twin mounting. This was probably the same gun which was stated to have been sold to the Romanian Army in single and twin towed mountings, although these never entered service so were presumably acquired for trials purposes. There was also another naval project at around that time: the 37 mm *zénithaux* (zenith), an upward-firing Hotchkiss quadruple mounting intended for use against dive bombers (the guns could not depress below 45 degrees). The mounting would have been set into the deck and loaded from beneath, but this was only at the concept stage by June 1940. It is uncertain what ammunition any of these gun projects might have used, but the 37 × 208 is the only known candidate. One source for the Hotchkiss shell weight gives 550 g, but for the ACAD trials (see below) the Hotchkiss fired a 625 g shell containing 57 g Tetryl; MV in both cases was 875 m/s.

### 37 × 217 M168B3 FOR T250X-2 VIGILANTE

This round was developed towards the end of the 1950s for the T250X-2, the second version of the 37 mm six-barrel rotary cannon intended to replace the 40 mm Bofors L/60 in the US Army (see 37 × 255 for the first version). The M168B3 cartridge selected was based on the Bofors case, shortened, necked-down and with the rim turned down. One example is dated 1960. The mouth, neck and shoulder of the case were similar to the 37 mm M17 cases for the U.S. M1 anti-aircraft weapon, except the neck was shortened. Overall weight of a round was 1.5 kg; the overall length was 397 mm when loaded with a 748 g T324 projectile with a long, tapered ballistic cap or windshield. Muzzle velocity was 915 m/s.

### 37 × 218R Mle 1935

At the start of World War II the standard French naval 37 mm AA guns, the Mle 1925 and Mle 1933 which used a powerful 37 × 277R round, were manually loaded and therefore slow-firing. An attempt was made in the mid-1930s to develop an automatic gun to replace them in the ACAD trials (*Automatic Contre-Avions Double* = automatic twin-AA) of the mid-1930s, which pitched the 40 mm Bofors, 37 mm Schneider, 37 mm Hotchkiss, and a 37 mm design from the *Artillerie Navale* at Ruelle. The A.N. system won the contract as the Mle 1935, but it is unclear which ammunition was used. Official trials data state that it fired an 831 g HE projectile (containing 50 g HMn) at 825 m/s; the 210 g propellant loaded generated a chamber pressure of 294 MPa. The problem is that the 37 × 218R cartridge case, rimmed and slightly bottlenecked, which is generally associated with the Mle 1935 (cases are headstamped 37–1935), seems to have been too small to generate such a high muzzle energy. Just possibly, the A.N. Mle 1935 round might have been the existing 37 × 277R with the 37 × 218R being the less powerful Hotchkiss contender, but that is speculation with no evidence in support. Anyway, the ACAD was still at the service test stage in 1940. The overall length of the round is about 350 mm.

## 37 × 223SR M1 AND M9

This was designed for a Browning AA gun developed in the interwar period and adopted by the US Army as the 37 mm M1, which saw extensive use during World War II. It was also adopted for the M9 aircraft gun, which only saw use in some light naval craft.

The brass M17 cartridge case is basically a semi-rimmed version of the rimmed, bottlenecked M16 case used in ammunition for light tank and anti-tank guns, modified to facilitate feeding in automatic cannon. With the standard HE shell, overall length is 325 mm and weight 1.22 kg. Two different pressure levels were used: for the M1 AA gun, 207 MPa and for the M9, 317 MPa. Penetration figures for the M9 loadings given below assume an additional 156 m/s MV added by the aircraft speed.

Standard warshots were as follows:

- M54 HE: 608 g, filled with 45 g Tetryl or Comp A, fitted with an M56 point-detonating fuze in the nose; MV 792 m/s
- M59 APC shot: 870 g; MV 625 m/s (for self-defence against armoured vehicles); penetration against homogenous plate or face-hardened armour 25 mm/70°/450 m, 15 mm/70°/900 m
- M59 APC shot (for the M9 gun): 870 g; MV 853 m/s; penetration against homogenous plate armour 56 mm/70°/450 m, 33 mm/70°/900 m; against face-hardened armour 48 mm/70°/450 m, 30 mm/70°/900 m
- M80 AP shot (for the M9 gun): 753 g, MV 930 m/s (as used in the 37 × 145R round); penetration against homogenous plate armour 79 mm/70°/450 m, 56 mm/70°/900 m; against face-hardened armour 66 mm/70°/450 m, 51 mm/70°/900 m.

## 37 × 232SR BREDA 37/54

This round was designed for the Breda 37/54, the standard Italian medium automatic AA gun used in World War II which was available in various single and twin mountings, mostly naval. The cartridge is very similar in appearance to the US 37 × 223SR described above. It was loaded with an 828 g HE-T shell containing 30 g MBT explosive. The fuze was an impact-SD type, and the shell was fired at 700 m/s by 200 g of N.A.C. propellant. Tracer duration was 12 s. Overall round length was about 360 mm and weight 1,600 g.

## 37 × 240 TYPE 76

This round was designed for the Chinese Type 76 twin-gun naval AA mounting that emerged in 1976, since when other naval mountings have also been offered. It was clearly intended as an improvement on the old 37 × 250R round, the main difference being a wider rimless case with a much greater propellant capacity in order to increase the muzzle velocity. It bears a close resemblance to the Oerlikon 35 × 228, but fires a heavier projectile at a lower velocity. This may have been derived from the aborted Soviet 'Yenisey' project for advanced 37 mm AA gun systems

*Comparison between 35 × 228 (top) and 37 × 240 (repA)*

that ran during the 1950s and early 1960s, which had an identical ballistic performance.

The projectile is crimped to the steel cartridge case by two annular crimping rings which lock into two cannelures towards the projectile base. It is fitted with a single wide driving band. The complete round is 383 mm long, weighs 1.78 kg and has a chamber pressure of 284 MPa. A 725 g PFHE with a nose proximity fuze fired at a muzzle velocity of 1,010 m/s is the standard combat loading advertised (with self-destruct after 8–11 seconds), although AP rounds have also been observed. Maximum ballistic range is 9,400 m.

## 37 × 250R SK C/36 AND M42

Most German 37 mm Flak automatic cannon used in World War II used the 37 × 264B ammunition described below, but two of the naval guns are exceptions. The 37 × 250R cartridge is almost identical to the German army's 37 × 249R light tank/anti-tank round, only with a slightly thicker rim (3.9 mm rather than 2.5 mm – presumably to strengthen it in view of the more violent extraction forces in an automatic weapon). The thick-rimmed round was developed for the naval Flak 3,7 cm SK C/36 (cartridge cases with C/36 stamps exist); little is known about this gun, but it may have been installed in a number of second-line vessels. The M42 inherited the same ammunition, but itself saw little use, its successor, the M43, using the standard 37 × 264B round.

The thick-rimmed cartridge case is straight tapered. The original C/36 loading fired a 625 g HE-T shell at an MV of 840 m/s. The M42 loadings used 175–185 g of propellant generating 290 MPa. There are two crimping grooves locating the projectiles more firmly in the case. This round used the same projectiles as the 37 × 264B except for the APCR. The complete round is 355 mm long and weighed 1.37 kg when loaded with the standard HEI-T projectile, which weighed 644 g and was fired at 845 m/s. There was also a 680–700 g AP loading fired at 815 m/s and a 550 g *Minengeschoss* fired at 925 m/s, although that last one may not have had time to get into service.

Incidentally, the German navy started the war with a different 37 mm Flak gun, the SK C/30, which used very powerful 37 × 380R ammunition, but this was manually loaded so very slow-firing.

## 37 × 252SR SOVIET M1939

This was a Soviet development inspired by the Bofors long-recoil weapons (as was the 25 mm M40). The gun was first designed around a 45 × 495–510R cartridge in 1939 (exact dimensions uncertain), but this turned out to be too heavy so was dropped in favour of a rapidly-developed 37 mm version; possibly Russia had details of the very similar German naval Flak SK C/36 which fired very similar 37 × 250R ammunition. The rimmed, tapered round has been in continuous production ever since its introduction in 1939 and has been available in a variety of loadings from manufacturers in China, Egypt, Pakistan and Serbia as well as the USSR (where it is no longer made, as it's been out of service for a long time). While it remains in use with many armed forces, it is gradually becoming obsolete as the old guns are retired.

The cartridge cases are lacquered steel or brass and are fitted with percussion primers. The complete round is 383 mm long and weighs around 1.5 kg. The propellent load is usually 200 g of a single-base propellant plus 4 g of decoppering agent and the chamber pressure is 275 MPa. Tracer burn time is 6 seconds and the HEI self-destructs in 8 to 12 s.

Projectiles are fixed to the cartridge cases by two 360° annular crimps which lock into two cannelures towards the projectile base. The forged-steel projectiles are fitted with a single wide driving band. Various projectiles have been developed at different times, the two main ones with typical ballistics being:

- HEI-T SD; Russian designation UOR-167; 735 g shell filled with 37 g A-IX-2; nose fuze is PD, DA, SD; MV 880 m/s
- AP-T; Russian designation UBR-167; 760 g solid-steel shot with ballistic cap; MV 850 m/s; penetration 37 mm/90°/1000 m.

### 37 × 255(SR) T68 FOR T37, T172 AND T250X-1 VIGILANTE

The development of post-World War II US 37 mm ammunition is rather obscure. Information obtained so far suggests that the first model, designated T68, has a semi-rimmed case 255 mm long, with a 57 mm rim and a 54 mm maximum body diameter. This is the most likely candidate for use in the T37 gun (later redesignated T172E2) and also a revolver cannon designated T172E1. A variation of the case, about 263 mm long, is associated with an HEP projectile and was made in 1954.

A further development has the semi-rim turned off to be completely rimless (i.e. with a 54 mm rim diameter) and is designated T68 B1. This was the original cartridge for the T250X-1 Vigilante rotary cannon, with one example dated 1956. It used the same T324 projectile as the 37 × 217 above, giving an overall length of 435 mm, with an MV of 1,100 m/s.

## 37 × 264B FLAK 18

This was the standard German land-based medium automatic AA gun ammunition of World War II, being used in a succession of weapons designated Flak 18, 36, 37 and the M43 which was also used in a naval mounting. It was also used in a modified version of the Flak 18, the BK 3,7 aircraft gun used for ground attack.

The cartridge case is tapered and belted, typical of Rheinmetall-Borsig automatic cannon ammunition of the period. It contained 189 g propellant which generated 255 MPa chamber pressure. The complete round is 367 mm long and weighs about 1.5 kg. Projectiles had two copper driving bands until 1940 when a shortage of copper led to the use of a single soft-iron band instead, which required the gun chamber to be slightly shortened. The usual loadings were:

- HE: 635–644 g nose-fuzed shell, with or without added incendiary element or tracer; HEI-T contained 24 g HEI; SD impact fuze; MV 820 m/s; 6.5 s tracer burn
- AP-T: 680–700 g base-fuzed shell; 790 m/s; 2 s tracer burn; penetration 50 mm/90°/500 m, 40 mm/60°/500 m

*37 × 264B: H-Panzergranate (APCR)*

- APCR: (*H-Panzergranate* for the BK 3.7 only); 380 g tungsten-cored shot, MV 1,170 m/s; penetration 140 mm/ 90°/100 m, 70 mm/60°/ 100 m.
- In addition, a *Minengeschoss* was under development: 550 g, contained 90 g HTA explosive; the tracer burned for 2,000 m (4.1 s); SD took place 2,800–3,500 m.

Russia acquired technology from Rheinmetall in the early 1930s and developed a multi-purpose automatic cannon around the Flak 18 ammunition: the AKT-37. This failed to enter service.

### 37 × 280B GERÄT 341 AND FLEDERMAUS

These two projects were rivals for a contract for a late-World War II German Navy AA gun; the *Gerät 341* was by Rheinmetall-Borsig, the *Fledermaus* by Gustloff and Stübgen. According to one Halstead document both used the same ammunition; however, the *Unterlüss* papers, while confirming the high muzzle velocity of the *Gerät 341*, credits the *Fledermaus* with the same MV as the standard 37 × 264B.

The Rheinmetall ammunition had a cartridge case length of 280 mm, with an overall round length of 425 mm, a case weight (steel) of 975 g, a C/32 mechanical primer, and an 'M.Gr.' (*Minengranate = Minengeschoss*) projectile weighing 725–740 g and containing 125–138 g HTA explosive, fitted with an Lh 12 tracer. Muzzle velocity was 1,000–1,040 m/s (sources vary). No illustrations of the ammunition have so far emerged and the rim/base diameters are not known.

## 37 × 296R SCHNEIDER Mle 1930

This very rare and little-known cartridge was developed for use in the equally little-known Schneider Mle 30 automatic AA gun. Some 700 examples were ordered for the French Army as an interim measure pending the arrival of 40 mm Bofors guns, but only twenty were in service by May 1940, deployed to defend Paris.

The ammunition has a tapered and slightly bottlenecked, rimmed brass case 296 mm long. A drawing of the HE shell indicates an overall round length of c.442 mm. The shell was fired at 800 m/s. Shell weight is usually given as 750 g but for the ACAD (naval automatic twin AA) competition of the mid-1930 s it was 895 g (containing 39 g pentolite) at the same velocity.

# 40–57 mm CARTRIDGES

The UK was the first country to move away from the 37 mm calibre in automatic cannon, by adopting the 40 mm 2-pounder naval pom-pom in World War I. This system saw extensive use in World War II but it was Sweden, with the extremely successful 40 mm Bofors AA gun, that really scored in terms of international sales. The 40 × 311R L/60 Bofors saw ever-increasing use throughout World War II and has since been supplanted by the more powerful 40 × 365R L/70 version.

In contrast to these high-velocity rounds, World War II Ho-301 used a low-velocity caseless 40 mm round. It was installed on some IJA fighters for anti-bomber use, but was unsuccessful. It is described in detail in the Unconventional Systems section.

Few other automatic cannon rounds in this calibre have been developed since, but the Franco-British 40 mm from CTAI (Cased Telescoped Ammunition International) has been selected for service and the ATK Super 40 is at an advanced stage of development.

The other calibre in this group to find significant success in automatic cannon is 57 mm. Achieving a high rate of fire with such large ammunition proved technically difficult and was not perfected until after World War II. Since then only four 57 mm rounds have seen service and only two of them – the 57 × 347SR Russian and 57 × 438R Bofors – can be considered successful. This pair both entered service in the 1950s and both remain in use and under active development to this day.

Automatic cannon in between the 40 mm and 57 mm calibres have seen little use. There have been two relatively unsuccessful 45 mm Soviet rounds, only one of which saw regular service, plus the equally unsuccessful German 50 × 346B for the Flak 41. At the end of World War II Germany was also developing two very different 55 mm systems, but these remained experimental. The future may well see further developments in these larger calibres, as a 50 mm version of the 35 mm Oerlikon round is being developed at the behest of the US Army for arming future AFVs.

## 40–57 mm SERVICE ROUNDS

*20 × 110RB*

*2-pounder No.1 (40 × 158R)*

*40 mm Bofors L/60 (40 × 311SR)*

*40 mm Bofors L/70 (40 × 365R)*

*45 mm SM-7 (45 × 386SR) (repP)*

*50 × 346B Flak 41*

*50 × 420R BK 5*

*20 × 110RB*

*57 × 121R Ho-401 (repP)*

*57 × 230R (Bofors m47) (repP)*

*57 mm S-60 (57 × 348SR)*

*57 × 398 R10*

*Bofors 57 mm L/60 (57 × 438R)*

*57 mm Molins (57 × 441R)*

## 40–57 mm EXPERIMENTAL ROUNDS

*20 × 110RB*

*40 × 180 Super 40 (repA)*

*2-pounder 75 Cal (40 × 277RB) (repA)*

*42 mm Red Queen (42 × 270)*

*42 mm RK420 Red King (42 × 338/348) (repA)*

*NS-45 (45 × 186) (repA)*

*47 mm Class P (47 × 326R)*

*3½-pounder (47 × 428R) (repA)*

*50 × 330 Supershot (repP)*

*501 MK (50 × 355) (repA)*

*20 × 110RB*

*55 × 175RB MK 112 (repA)*

*55 × 205 MK 412 (repA)*

*55 × 450B MK 114 (repA)*

*57 × 165RB N-57 (repA)*

*57 × 350B (RShR-57) (repA)*

*57 x 514R (6$^{PR}$ 6 cwt)*

## 2 POUNDER NO.1 / 40 mm CLASS S (40 × 158R)

Except for the two Bofors, this British round is probably the most famous in this calibre. It was developed during World War I for the naval Vickers 2-pounder pom-pom by necking-out and extending the cartridge case from the short-lived 37 × 124R used in the Vickers 1½-$^{PR}$, and saw extensive service in World War II.

Ammunition included a base-fuzed CP (Common Pointed) shell developed in World War I for firing against smaller surface targets such as MTBs. There were two principal AA loadings: the original one was a 900 g HE shell containing 71 g HE and available with a time fuze in the nose (which could be automatically set by the loading mechanism). Alternatively, percussion fuzes could be fitted. The round had an overall length of 300 mm.

By the end of the 1930s, the muzzle velocity of 610 m/s was felt to be too low, so the shell weight was reduced to 770 g enabling the muzzle velocity to be raised to 730 m/s by means of 123 g Cordite. This HV version was fitted with a flat-nosed point-detonating fuze (initially the No.243 instant action, later the No.246 delayed-action).

*2-pounder No.1: CP; HE (time); HE HV; S Class AP; practice*

The cartridge case was also selected for the Vickers 40 mm Class S aircraft gun, for which special armour-piercing steel projectiles were developed. The Mk.1 AP shot weighs 1,130 g, and 108 g Cordite propellant developed a muzzle velocity of 615 m/s. The total round weight is 1.6 kg. In the Mk.VI shot the weight was increased to 1,360 g to increase the performance by about 9%; MV was reduced to 575 m/s. Penetration figures for the Mk VI (allowing an additional 107 m/s for aircraft speed) were 58 mm at 400 yards (366 m) and 44 mm at 1,000 yards (914 m), both at 20° striking angle. A tracer was added later. HE ammunition was also made for the Class S amounting to about 22% of the 2.4 million Class S rounds made in 1942–3.

There was also an experimental APCNR (or APSV) loading for the aircraft gun, for use with a Littlejohn squeezebore adapter, both of them borrowed from the much more powerful 2-pounder No.2 ammunition (40 × 304R) used in the army AFV/anti-tank guns. The projectile only weighed 567 g so muzzle velocity was increased to 990 m/s and the penetration (allowing an additional 107 m/s for aircraft speed) was 85 mm/400 yards (366 m) and 62 mm/1,000 yards (914 m), both at 20° impact angle. The Littlejohn application caused problems with gun function issues, the light projectile failing to generate enough recoil, so the recoil system had to be modified accordingly. Air firing tests took place in 1944, with

180 rounds fired and a 3.8% stoppage rate. The project was cancelled in favour of using rocket projectiles.

Rolls-Royce developed a rival gun to the Class S using the same ammunition but this was only accepted for service as a modified, semi-automatic naval gun, the Type BH.

During air firing trials, a problem arose with ammunition used in the S gun in that cases often failed to eject properly and suffered torn rims. It was discovered that this problem only affected cases made by Kynoch, and that they were made of much softer brass than Vickers cases. Further trials indicated that oiling the Kynoch case solved the problem, so the Air Staff approved the oiling of S gun cases.

Finally, some 2-PR No.1 cases were shortened to 107 mm and stamped HA on the base. This cartridge was developed for a manually-loaded sub-calibre training gun designed to fit within the barrel of 4' and 4.7' naval guns to provide economical practice. There were two versions, designated Low Angle and High Angle (LA and HA). The LA used the standard 40 × 158R ammunition, the HA the 40 × 107R. They both fired the same projectiles, but the muzzle velocity of the HA was only c.360 m/s instead of c.600 m/s for the LA. The HA was presumably used in the anti-aircraft versions of the 4' and 4.7' guns, whose mountings achieved a higher elevation angle, with the lower velocity being chosen to minimize the range and therefore the range safety requirements. During World War II, these adapters were hastily fitted with their own breech mechanisms and mountings to provide some armament for small naval craft.

*40 mm ALACV Super Forty rounds: 40 x 218 APFSDS (top) and 40 x 165 HE*

#### 40 × 165 ALACV SUPER 40 HE

This was the initial version of the Super 40 (see entry below) with the shorter case length selected for the HE loading.

#### *40 × 180 Super 40*

The term 'Super 40' or S40 refers to a project to extract the maximum performance from guns designed around the 30 × 173 cartridge, by necking-out the case to the largest practicable calibre. The Super 40 cartridge therefore has exactly the same diameter and overall length as the 30 × 173, to fit into the same gun mechanism and ammunition feed. The Super 40 bears the same relationship to the 30 × 173 as the 50 mm Super 50 (described below) does to the 35 × 228. At present the only guns being developed for the Super 40 are the ATK Mk 44 Chain Gun and its XM813 derivative, which require only a new barrel, additional springs in the recoil damper and a modified feed sprocket to handle the more powerful new cartridge. The ammunition was initially developed by both ATK and GD-OTS, but GD-OTS dropped out of the project in 2014.

Development of the Super 40 commenced under the ALACV (Advanced Light Armament for Combat Vehicles) project initiated by the US Army's Armament Research, Development and Engineering Center (ARDEC), between 2000 and 2003. The objectives were to improve the performance of APFSDS projectiles by 30 per cent and to provide a much larger HE shell to increase the lethal area for air-burst by 400 per cent. The result of this programme was unusual in that the case length varied depending on the projectile. For the APFSDS-T round the projectile was mainly buried within the case, which was extended to 218 mm. For the much longer HEAB-T shell, the case was shortened to 165 mm. However, the different case lengths resulted in excessive barrel wear so further development resulted in a common case length of 180 mm. In all versions, the cartridge cases are straight, with no bottleneck, although in the latest 180 mm version the case taper has been increased to aid extraction, resulting in a slight calibre reduction to 39 mm (although it is still designated 40 mm).

The case is made of steel and a standard 30 mm percussion primer and flashtube are used. The propellant is expected to be a hybrid type being developed by GD-OTS' St Marks Powder facility, offering high-temperature stability and reduced barrel wear,. Chamber pressures are stated to be near 400 MPa.

The current HEAB-T design has an incendiary-filled multipurpose nose (to enable the round to be used in airburst and impact modes), an electronic time fuze set while the round is in the feed system and an HE/fragmentation body. Projectile weight is understood to be c.670 g, carrying an HE payload 60 per

*Super 40 display rounds from ATK*

cent greater than the 30 × 173, and the muzzle velocity is c.1,000 m/s.

The current APFSDS flight projectile weighs 230 g (not including the lightweight composite sabot) and is fired at 1,475 m/s. An improvement in penetration of 20–25 per cent over 30 × 173 APFSDS is claimed.

### 40 × 218 ALACV SUPER 40 APFSDS

The top cartridge shown here was the initial version of the Super 40 (see entry above) showing the longer case length selected for the APFSDS loading.

### 40 × 239B HSS 840

This was for a Swiss AA gun project and was designed as a modified version of the HSS 834, which used the existing 34 × 239 ammunition. The steel cartridge case was necked up to 40 mm calibre and also had a belt added. It fired an 820 g shell (containing 100 g HE) at 950 m/s. Overall length is 365 mm and weight 1.72 kg.

### 2-pounder 75 cal (40 × 277RB)

In February 1943 work began on a Bofors replacement for the Royal Navy, intended to deliver a higher performance in terms of both muzzle velocity and rate of fire. B.S.A. was given the job of developing the gun, which was designed around an unusual cartridge from the British Design Department. The case was wide for its length and with a sharply rebated rim, suggesting that the gun may have been an API blowback (in other words a scaled-up 20 mm Oerlikon). The round was officially designated (and headstamped) the '2-pounder 75CAL' and was made at the Royal Laboratory, with 1944-dated cases existing. The project was abandoned in 1946 and while 1,000 cases were ordered it is not clear whether or not a gun was made, or only test barrels.

The round is approximately 425 mm long and weighs around 2.3 kg without propellant; data varied somewhat as development proceeded. The projectile weighed 900 g including 100 g RDX and was designed to be fired at 914 m/s.

## 40 × 311SR BOFORS L/60

In the late 1920s the Swedish Navy, not satisfied with their Vickers 2-pounder pom-poms, requested from Bofors the development of an improved 40 mm AA gun. The result was the famous and widely used 40 mm L/60 Bofors (originally known as the L/56) and its 40 × 311R cartridge. The first gun emerged in 1932, and both naval and mobile field mountings achieved huge export sales from the mid-1930s to the late 1940s. Although long ago replaced by the 40 mm L/70 Bofors for most purposes (see below), the L/60 remains in front-line service with many armed forces in land and naval mountings to this day, and the ammunition accordingly remains in production.

The rounds are fixed, with a rimmed brass cartridge case fitted with a percussion primer. There is a clip groove between the rim and body of the case. The case used in US Army cartridges is designated M25 (brass) or M25B1 (steel). US Navy cases have the Mk 2 brass case or the Mk 3 steel case. US-made ammunition of British design uses the M22 case, threaded for a British primer. The typical charge load is now approximately 320 g of single-base low-erosion propellant, generating a chamber pressure of 300 MPa which produces a muzzle velocity of 880 m/s (early loadings achieved about 850 m/s). Loaded rounds are 447 mm long and weigh around 2.15 kg.

The projectiles typically weigh just under 1 kg and are fixed to the case by one or two annular crimping grooves. Double-crimped ammunition is intended for use in weapons using automatic feed systems or guns with increased rate-of-fire modifications, as were developed after World War II. The projectiles are normally fitted with a copper driving band. The following projectiles are or have been the most significant ones used with this cartridge:

- HE-T: 930 g shell filled with 92 g Hexotonal and fitted with a PD nose fuze
- AP-T: 930 g steel shot, with a soft steel nose cap under a ballistic cap (in British terminology, APCBC); penetrates 40 mm/60°/100 m
- APHC-T: 950 g, with a tungsten alloy penetrator in an aluminium body; at

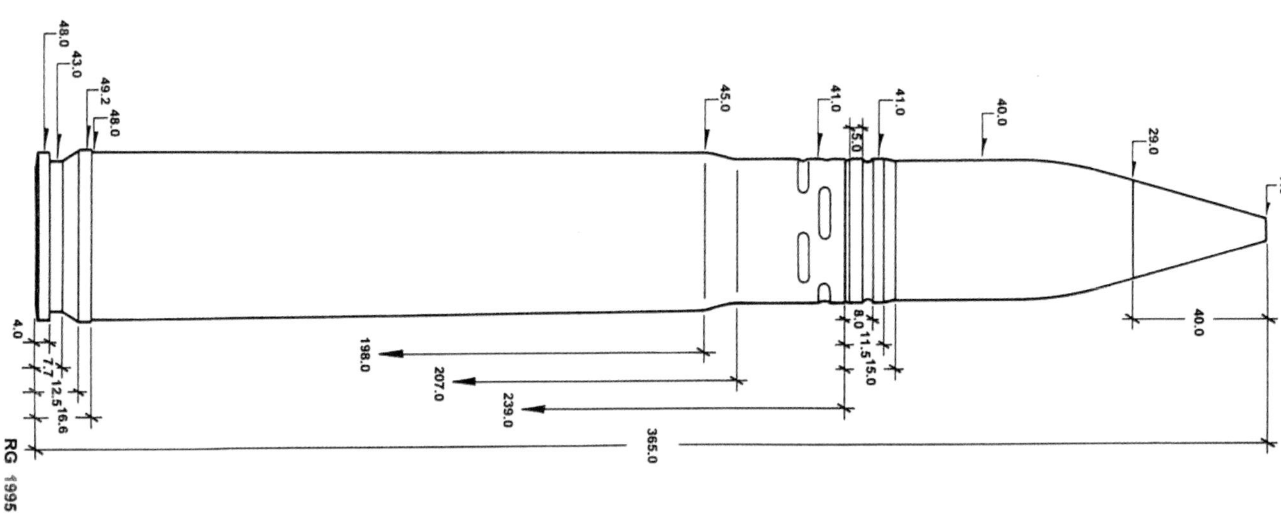


*Drawing of 40 × 239B by Robert Gerber*

least 30% more penetration than AP-T
- PFHE (pre-fragmented HE): modern 980 g shell containing 90 g Octol, lined with 600 spherical tungsten pellets and fitted with a proximity fuze
- MP-T (multi-purpose): modern 980 g shell of Raufoss Multipurpose SAPHEI type

There was one attempt to use an APFSDS shot in 40 × 311R, curiously in aircraft. The 40 mm PGU-31/B was developed for the Bofors gun fitted to the US AC-130U gunship. The business end was a 230 g tungsten alloy penetrator (11 mm diameter with a 15:1 L/D ratio) which had a blunt tip covered by an aluminium windshield containing some incendiary compound to produce a bright flash on impact. Muzzle velocity was 1,335 m/s, giving a muzzle energy of 205,000 joules. However the tests, carried out in the late 1980s/early 1990s, were unsuccessful and the round was not accepted.

## 40 × 365R BOFORS L/70

Despite the great success of the Bofors 40 mm L/60, by the end of World War II it was becoming clear that its performance needed to be improved in order to deal with the increasing speed of aircraft. Bofors therefore designed a new high-velocity cartridge and a new faster-firing gun to fire it – the 40 mm L/70. The first 40 mm L/70 air-defence gun was produced in 1947 and has achieved widespread use in towed AA, naval and AFV mountings. The ammunition is also used in the Bofors-based South Korean S&T Dynamics K40 AFV gun and in the ATK Bushmaster IV Chain Gun, although development of the latter is currently suspended.

The 40 × 365R is a fixed round, with the projectile secured to the brass cartridge case by two annular crimping grooves. The brass case, which is fitted with a percussion primer, is 54 mm longer than the L/60's and is also wider, with a rim diameter of 65 mm compared with 62 mm. It therefore has room for considerably more propellant, weighing about 485 g, which launches the projectiles at just over 1,000 m/s. The complete round is 534 mm long and weighs around 2.5 kg. The projectiles are normally fitted with a copper driving band.

Bofors Explosives has developed Low-Vulnerability Ammunition (LOVA) to increase the safety of rounds through using insensitive components, reducing vulnerability to shock and high temperatures.

The main loadings for the 40 × 365R round are or have been as follows:

- HE-T: 960 g shell containing 100 g Hexotonal and fitted with a PD nose fuze
- AP-T: 930 g steel shot, with a soft steel nose cap under a ballistic cap (in British terminology, APCBC); penetrates 50 mm/60°/100 m
- HCHE: high-capacity 870 g shell filled with 165 g Octonal and fitted with a PD delayed-action nose fuze
- PFHE (pre-fragmented HE): 880 g shell containing 120 g Octol, lined with 650 spherical tungsten pellets measuring 3 mm, and fitted with a proximity fuze
- MP-T (multi-purpose): 940 g shell of Raufoss Multipurpose SAPHEI type, containing 105 g Comp A4
- 3P: 975 g shell similar to PFHE but with the sophisticated multi-function 3 P fuze with impact, time and proximity capabilities; the version for the Trinity air-defence system weighs 1,100 g and is filled with 140 g Octol
- APFSDS: 500 g projectile with 200 g penetrator, fired at around 1,500 m/s; penetrator is 200 mm long and 11 mm in diameter; the Mk 2 version can penetrate more than 150 mm armour at 1,000 m and takes 1.1 s to reach 1,500 m; developed for CV9040 IFV (infantry fighting vehicle). An APFSDS round was earlier developed for the OTO Breda Dardo anti-missile system, which had an MV of 1,350 m/s.

### 42 × 270 RED QUEEN

After World War II Britain started to examine requirements for a high performance automatic AA gun to replace the Bofors. As well as working with Oerlikon on their 421 RK cannon, designated Red King by the British (see below), a parallel development was initiated a couple of years later for a different gun firing different ammunition with a similar performance. This was designated Red Queen. The cartridge case is unusually short, fat and heavy, while the projectile is very long and slender. One technically unusual point was that the 1,090 g HE projectile has a cup-type rubberised base flange which was covered by the case neck until firing. It contained 175 g HE. The project was cancelled after a few years and the L/70 version of the Bofors bought instead. Performance data is assumed to be similar to the Red King. Overall length is 483 mm and round weight about 4.4 kg.

### 42 × 338/348 421 RK / RED KING

In 1950 the UK signed a contract with Oerlikon to develop their 421 RK cannon (42 = calibre in mm; 1 = first model; RK = *Revolver Kanone*), to which the British assigned the project designation Red King, presumably inspired by the Oerlikon initials. It was intended as a high-performance replacement for the wartime 40 mm Bofors L/60. The British lost interest after a few years and Oerlikon stopped work in 1958.

The cartridge used a rimless brass case and was of more conventional shape than the Red Queen's (see above), being longer and more slender. The case was made with two different neck lengths, giving case lengths of 338 mm and 348 mm. The shorter neck was developed first for use with Oerlikon's own projectile design; the longer neck was for use with the British projectiles developed for the Red Queen, in which the driving band was covered by the neck instead of resting on top of it. Overall round length is 542 mm and total weight is 3.06 kg. The HE shells weigh 1,090 g (including 175 g HE) and were fired at a muzzle velocity of 1,070 m/s. Oerlikon produced a limited quantity of cases, primers, HE shells and point detonating fuzes, but surviving cases are British-made; the extent to which Oerlikon actually produced their own 42 mm ammunition is not clear.

### 45 × 186 NS-45

This cartridge was developed from the 37 × 195 round for the Soviet NS-37 aircraft gun by the simple expedient of necking it out to 45 mm calibre and shortening it slightly so that the overall length remained the same. The NS-45 gun which fired it differed from the NS-37 only in the larger calibre barrel. The gun and ammunition saw use only in a few aircraft on an operational trials basis towards the end of World War II.

The rimless cartridge case is of brass. Overall length is 332 mm and round weight 1.9 kg. The only known service round used an OT (HE-T-SD) or OZT (HEI-T-SD) projectile based on a naval shell. This weighs 1,065 g and was fired at 780 m/s. A higher-velocity (820–850 m/s) tungsten-cored AP loading was also tested.

## 45 × 386SR SM-7

This was developed for the Soviet SM-7 naval gun that only saw limited service for a relatively brief period. Both single and quadruple mountings were manufactured in small numbers in the mid-1950s and the last ones left service at the end of the 1970s.

The high-performance ammunition has a heavy, large-capacity brass cartridge case. The complete round is 532 mm long and weighs 4.52 kg. There were two warshot types: the OR-75 HEI-T-SD and the much rarer F-75 APHEI-T. Both fired projectiles weighing 1,410 g at a muzzle velocity of 1,080 m/s.

### 45 × 495–510R M1939

This was for the immediate predecessor of the Russian 37 mm M1939 AA gun and got as far as being formally adopted, with an order placed for 190 guns in 1940, before it was cancelled in favour of the 37 mm system. The appearance and dimensions of the cartridge case are unknown, but the case was presumably rimmed (the system was based on a enlarged Bofors) and the case length is believed to be around 495 to 510 mm, holding 518–545 g of 7/7 propellant. This fired shells weighing 1.44–1.46 kg at an MV of 960 m/s.

### 47 × 326R VICKERS CLASS P

The 40 mm Vickers Class S airborne anti-tank gun was found to be a far more accurate weapon than rocket projectiles in the middle years of World War II, but became increasingly ineffective as tank armour thickened. The Class P was intended as a more powerful replacement, but only reached the test stage after the war was over. The ammunition has a straight, rimmed brass case and was loaded with a steel AP shot. Overall round length is 475 mm and weight is about 3.2 kg. Projectile weight is 2,070 g and muzzle velocity 808 m/s.

### VICKERS 3½-PR (47 × 428R)

All that is known of this late-World War II project for a British naval automatic AA gun is a brief entry in Campbell's *Naval Weapons of World War II*, to the effect that two prototype guns had been made by Vickers Armstrong (one with the barrel recoiling through the water jacket, the other with the water jacket recoiling with the barrel), and the ammunition used 613 g of NF/S 093–031 propellant to drive a 1,590 g projectile at a muzzle velocity of c.1,020 m/s. It appears that the automatic mechanism was never fully developed and the project was cancelled in 1946. No examples of the ammunition are known to exist, but recently a factory drawing of the round came to light and the replica shown in the Experimental group photo was made using that. The overall length of the round is 611 mm.

### 50 × 131 DWM

This is only known from a drawing dated December 1941 which carries the DWM stamp. The projectile is an earlier-pattern M-Geschoss. It was presumably intended for an aircraft cannon, but further information is not known. It looks very similar to the 34 × 149 from the same manufacturer at the same time; presumably DWM were attempting to develop a family of weapons in different calibres.

### *50 mm Supershot / EAPS / Super50*

During the 1980s Rheinmetall developed the Rh 503 AFV gun, initially intended for 35 × 228 Oerlikon ammunition but with the option to upgrade to a new 50 mm Supershot calibre by means of a barrel change. The end of the Cold War meant that the Rh 503 failed to attract any sales and the project lapsed.

More recently the ATK Bushmaster III 35 mm Chain Gun has entered service, and Rheinmetall has indicated its intention to develop a 35 mm version of its new WOTAN30 cannon. Both manufacturers have mentioned the option of introducing a 50 mm version if there is a demand for it.

The Supershot has the same overall dimensions as the 35 mm round (including an overall length of 387 mm) to facilitate conversion to the larger calibre, but uses an elongated 330 mm version of the 35 × 228 cartridge case which is straight rather than bottlenecked, forming a cylinder. The much longer case means that the projectiles are largely contained within it. Two types of 50 × 330 round were under development for the Rh 503:

- HETF-T M-DN 191 (High-Explosive Time-Fuze Tracer), weight 2.1 kg with 192 g HE and an MV of 1,040 m/s.
- APFSDS-T round, weight of penetrator 640 g, MV 1600 m/s, penetrating 180 mm armour at 1000 m.

The cartridge case for the 50 mm Supershot was also adopted for the RDECOM-ARDEC experimental 50 mm EAPS (Extended Area Protection and Survivability) C-RAM (Counter Rocket, Artillery and Mortar) project of the mid-2010s. However, it is loaded with a fin-stabilized, course-corrected projectile that protruded much further from the case than the Supershot, giving an overall round length of 538 mm. This is much longer than could be handled by the Bushmaster III, so a special gun was developed to fire it, combining elements of ATK's Bushmaster III and IV (the latter being designed for the 40 × 365R Bofors round).

At AUSA 2018 the US Army Armament Research Laboratory showed a different, more conventional, configuration of 50 mm round, called the Super50: still based on the 35 × 228 Oerlikon but simply necked out to a straight case, with the case length as well as the overall length of 387 mm remaining the same. General Dynamics is developing the XM1204 High Explosive Airburst with Tracer (HEAB-T) and XM1203 Armor Piercing Fin Stabilized Discarding Sabot with Tracer (APFSDS-T) rounds. The HEAB-T, which is reported to have a projectile weight of 1,540 g, is programmable with settings for Point Detonate (PD),

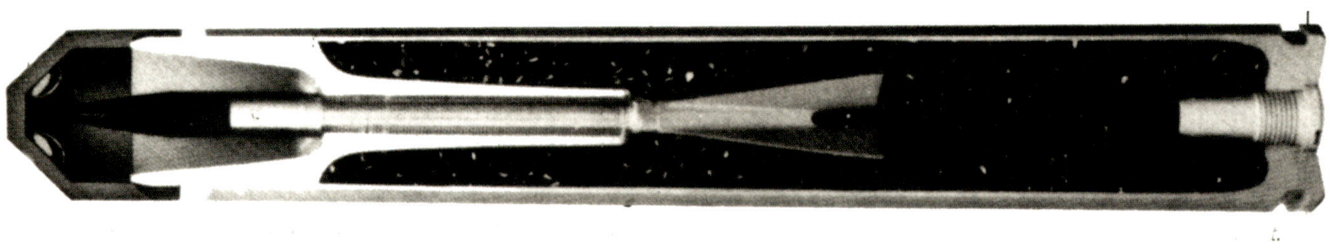

Sectioned 50 mm Supershot APFSDS

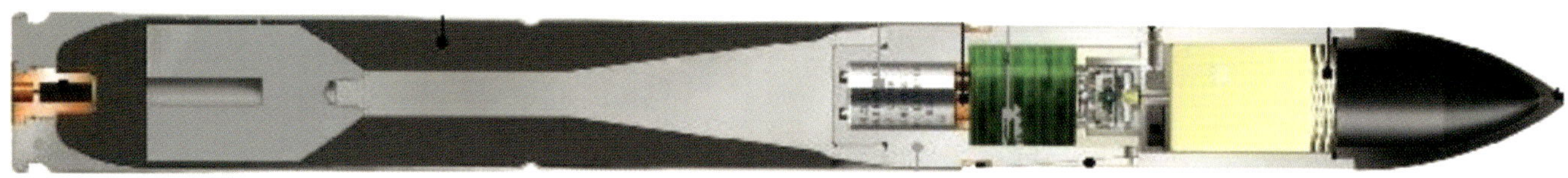

Sectioned 50 mm EAPS

Super50 rounds: APFSDS and TP, next to 35 mm TP for comparison

Point Detonate Delay (PDD), and Airburst (AB). The gun designed to fire this ammunition is based on the Bushmaster III and designated the XM913. Both cartridges were developed by Picatinny Arsenal and produced by General Dynamics. Northrop Grumman are planning to introduce a guided 50 mm projectile which would substantially improve the 4,200 metre range of the XM913.

## 50 × 346B FlaK 41

Even before World War II the Germans realised that there was an anti-aircraft 'engagement gap' between the maximum effective altitude of 37 mm automatic guns and the minimum effective altitude of the smallest large-calibre manually-loaded artillery, in 88 mm calibre. They therefore started development of a 50 mm calibre automatic AA gun, designed around a new cartridge, which entered service early in the war as the 5 cm FlaK 41. The system proved problematic and few were made.

The ammunition was designed around a unique, belted steel case which contained 690 g propellant. The round length is 532 mm and the weight 4.28 kg with the HE shell. Standard HE and HEI shells weigh 2,200 g, were fitted with a tracer and a settable self-destruct fuze, and had a muzzle velocity of 840 m/s. Designations were *5 cm Sprgr Patr 41 L'spur* and *5 cm Br Sprgr Patr 41 L'spur* respectively. For anti-tank use the APHE projectile from the 5 cm PaK anti-tank gun (see below) could be loaded into the case; it was designated 5 cm Pzgr Patr 42.

### 50 × 355 OERLIKON 501 MK

In the early 1950s Oerlikon began to investigate the optimum calibre for future medium AA guns. The USN was interested and supported the development of guns in two calibres: 35 mm and 50 mm. The 35 mm was the eventual winner and has since achieved considerable success (although the USN lost interest). The 50 × 355 ammunition was developed but was only fired in test barrels as the proposed Model 501 MK cannon was not made. Overall round length is 539 mm. The 1,520 g shell was fired at 1,175 m/s.

## 50 × 420R BK 5

This round was developed for the manually-loaded 5 cm L/60 Pak 38 anti-tank gun and KwK 39 AFV gun, in German army service from early in World War II. It features here because it was subsequently adopted for two aircraft guns: the autoloading Rheinmetall-Borsig BK 5 and the more advanced Mauser MK 214A (the latter being just too late to see operational service). The ammunition has a bottlenecked and rimmed steel case weighing 1,150–1,340 g filled with 830–920 g Digl.R.P. propellant. Round length is 607 mm and overall weight 3.5–3.8 kg when loaded with a high-capacity *Minengeschoss* tracer shell developed for air-to-air engagements; this weighs 1,520 g, contains 345 g HE, and was fired at 965 m/s. The weight of each belt link is 1.12 kg.

### 55 × 175RB MK 112

Developed by Rheinmetall-Borsig at the end of World War II, the MK 112 aircraft gun was a response to the USAAF bomber raids over Germany; it was designed to be able to shoot down a heavy bomber with one hit. The gun and its ammunition were virtually double-sized versions of the 30 mm MK 108 and similarly emphasized hitting power over velocity. The round is 355 mm long and weighs 2.4 kg. The short, straight cartridge case,

*57 × 173RB (courtesy of Robert Leiendecker)*

which weighs 650 g, has a highly rebated rim to work with the gun's API blowback mechanism. The case length varied significantly during the development process, with drawings showing cases from 155 mm to 195 mm – these variations being less problematic for straight cases in this type of gun. Priming was electric and the 240 g Nz.R.P. propellant drove the 1,480 g *Minengeschoss* shell at 600 m/s. The shell was reportedly the same as the one used in the MK 412 of which more details are given below. A modified version of the cartridge was made in the USA (see 57 × 173RB below).

### 55 × 180B RHEINMETALL-BORSIG

This is only known from a Rheinmetall-Borsig drawing dated March 1942. The projectile seems to be the usual streamlined 5.5 cm M-Geschoss, common also to the 55 × 175RB MK 112, the 55 × 450B MK 114 and possibly the 55 × 203 MK 412. The round was presumably for an aircraft cannon of some kind.

### 55 × 205RB MK 412

Developed by Krupp at the end of World War II as an unsuccessful rival to the MK 112 above. The case had a rebated rim (although the gun was not an API blowback) and was bottlenecked. Sources differ over the case length, both 205 mm and 210 mm being quoted (possibly, the length varied slightly during development). The overall round length is 385 mm. The steel cartridge case weighs 1,000 g, the total round weight is 2,740 g and the 207 g of Digl.R.P. propellent generated an MV of 700 m/s (one source says 650 m/s). Priming was electric. It used the same 1,480 g *M-Geschoss* tracer shell as the MK 112; the shell was filled with 420 g of HTA 41, an impressive 28% of the total weight. The tracer burned for 2,000 m.

There were a number of other comparable projects for large-calibre aircraft gun ammunition, for example, the 50 × 131 DWM (which seems to have been a scaled-up version of their 34 x 149) and the 55 x 180B Rheinmetall-Borsig, of which only factory drawings are known.

### 55 × 318R MK 318

This semi-combustible round for the recoilless MK 318 cannon is described in the Unconventional Systems section of Part 2.

### 55 × 450B MK 114 / Flak 58

The Rheinmetall-Borsig MK 114 was the high-velocity counterpart of the MK 112, in much the same way that the MK 103 was the HV version of the MK 108 in 30 mm calibre. It was intended to replace the use of the 50 × 420R round in automatic weapons such as the Mauser MK 214A; in fact, a version of that gun in 55 × 450B was being developed under the designation MK 214B. As well as aircraft use, the weapon was intended for AA applications under the Flak 58 designation. Overall length is 682 mm, while the length of the cartridge case is variously quoted as 448 to 452 mm. The steel cartridge case is of the belted, bottlenecked type and electrically primed. Normal chamber pressure was 320 MPa. Given ballistics vary: the most commonly quoted are a 2,030 g HE shell fired at around 1,000 m/s, but a more detailed analysis gives a 1,860 g *M-Geschoss* tracer, containing 480 g HTA 41 explosive, fired at 1,080 m/s.

## 57 × 121R Ho-401

This cartridge was designed for a low-velocity manually-loaded gun for use on Japanese tanks. It was then adopted, in a revised loading, for the Ho-401, an automatic aircraft cannon which saw very little use before the end of the war. The cartridge has a straight-sided and rimmed brass case. A special HE shell with a relatively high capacity was developed for the Ho-401. US sources on this ammunition give the HE shell weight as 1,500 g with a total round weight of 2.1 kg and an overall length of 323 mm. Muzzle velocity was around 500 m/s.

### 57 × 165RB N57

This round was developed for the Soviet N-57 aircraft gun developed at the end of World War II. The initial version used a 57 × 155RB case which was slightly bottlenecked, the base and rim diameters being 58.5 and 51.4 mm respectively. This was later modified to a 165 mm long straight case with base and rim diameters of 56.6 and 50.4 mm. Overall length is approximately 316 mm. The HE shell weighs 2,000 g and was fired at 600 m/s. Only a small number of guns were made, since aircraft designers preferred the lighter N-37.

### 57 × 173RB T78

The 57 mm T78 is the US version of the German 55 mm MK 112 (see above), as modified for testing in the USA. The calibre was increased to 57 mm to use the American T35, 57 mm Practice projectile and the case length was slightly reduced from 175 mm. The brass case, designated T4, was made at the Frankford Arsenal. This projectile was produced in 1948, the same year as the cartridge case. The overall length of the round is 355 mm.

## 57 × 230R BOFORS m/47

This little-known round was developed for a post-World War II aircraft gun, for use in the anti-shipping role. It was much shorter and less powerful than the ammunition for the Bofors 57 mm AA/naval gun. The gun saw relatively little use. Two loadings were developed: an HE shell (2,230 g at 744 m/s) and an AP

(2,450 g at 686 m/s). Overall length is 430 mm.

## 57 × 348SR S-60

As with Bofors in Sweden (see below) the USSR decided after World War II to develop a general-purpose 57 mm automatic cannon for land and naval mountings. The gun was developed in the late 1940s with the army S-60 gun entering service in 1950; the ZSU-57-2 twin-gun SPAAG and the ZIF-31 and ZIF-72 naval mountings followed a few years later. China manufactured all three types and the army guns in particular saw extensive service within the Warsaw Pact. Although some countries carried out updates of the system, Russia preferred to focus on smaller-calibre weapons until recently. However, new Russian naval (A220M) and AFV (AU220M) mountings using modified weapons promise a new lease of life for their 57 mm. Ammunition has been made in Russia, China, Bulgaria, Egypt, Romania and the former Yugoslavia.

The ammunition for the S-60 anti-aircraft gun series uses a semi-rimmed cartridge variously known as the 57 × 347SR, 57 × 348SR or 57 × 350SR. This is a thick and heavy case of lacquered steel or brass, with a percussion primer. The 1.18 kg of nitrocellulose powder propellant drives the projectiles at around 1,000 m/s. The complete round is 536 mm long and weighs about 6.4 kg. The projectiles are firmly attached to the cartridge case by two crimping bands engaging with two prominent cannelures on the projectile. The forged-steel projectiles use a pressed copper or copper-alloy driving band over 26 mm wide with a complex indented outline. The most commonly used ammunition for these guns is of the HEI-T or Fragmentation-Tracer type of traditional design, with armour-piercing ammunition of similar vintage also being made, as follows:

- HEI-T SD: Designated UOR-281 or UOR-281U (slightly modified version) by the army, UFB-71 by the navy; 2.85 kg; filled with 153 g HE and fitted with a delayed-action PD nose fuze with a self-destruct mechanism functioning after 12+ seconds. The tracer burns for 10+ seconds.
- APHE-T: Designated BR-281 or BR-281U, this uses a full-calibre steel shell with very thick walls leaving only a small space for the 13 g bursting charge with its base fuze; a penetrating cap on the nose is covered by a thin ballistic cap, making this an APCBC type in British terminology. Weight is 2.82 kg and armour penetration at 90° is stated to be 106 mm at 500 m and 96 mm at 1,000 m.

In 2013 an advanced HE-ER BB (extended-range, base bleed) shell was displayed by Yugoimport, but no details are so far available. Russia is working on what it calls 'special shrapnel' ammunition in this calibre, probably developed with a view to engaging unmanned aerial vehicles (UAVs). The new ammunition, being developed by NPO Pribor research and industrial association, will probably be similar to the Rheinmetall Air Defence Oerlikon AHEAD (Advanced Hit Efficiency and Destruction) 35 mm ammunition. It may be for use in a new mobile air defence weapon that is currently being developed as the Derivatsiya - PVO anti-aircraft artillery system, or ZAK-57. This is based on a modified BMP-3 infantry fighting vehicle fitted with a brand new turret. The direct-fire range is quoted as 12 km when fired horizontally and up to 8 km when fired vertically. A proximity-fuzed multi-function projectile is being developed.

A guided projectile has also been under development, with the following information from publicity material. It is laser-guided, with a laser sensor in the rear of the missile, implying that it is beam-riding or command-guided rather than using semi-active homing like most such systems. The projectile has a canard configuration with four folding fins. The warhead weighs 2 kg including 400 g HE. It has a minimum range of 200 m and a maximum of 3–5 km against UASs and 6–8 km against aircraft.

### 57 × 350B AND RShR 45/57

This aircraft gun was developed shortly after World War II in two calibres – 45 mm and 57 mm. As well as being belted, the straight/tapered 57 mm cartridge case was much slimmer than the 57 × 348SR S-60 AA gun round. The overall length was 518 mm and the projectile had a double driving band. The round fired a 2,490 g HE shell at 720 m/s. The 45 × 350B version used the same

*57 × 348SR: sectioned BR-281 and UOR-281*

case, necked-down, but nothing more is known about this round.

## 57 × 398 R10 (PLK)

After the end of World War II Czechoslovakia decided to develop its own 57 mm army AA gun and ammunition rather than adopt the Russian S60. This was less than successful, but it did enter service as the R10 gun (although more commonly known in the West as the PLK). The ammunition had a rimless and bottlenecked brass case, was 637 mm long and weighed around 6.6 kg. It was loaded with a 2.56 kg HE shell fired at just over 1,000 m/s.

## 57 × 438R BOFORS

The Bofors 57 mm automatic AA gun was developed during the late 1940s around a new 57 × 438R cartridge. It entered service in 1950 in single and twin naval mountings, and the French Navy developed its own M1951 guns in twin mountings around this ammunition. The same round was used in the Bofors m/54 Army AA gun, which saw service for a while in the Swedish and Belgian armies, but this had a relatively short life. A version of the gun was developed during the late 1970s for installing in an AFV turret, along with an associated Armour-piercing Discarding-Sabot (APDS) round. Both appeared in prototype form only.

The original gun was designated L/60, with development of an updated SAK 57 L/70 version of the naval gun beginning in the 1960s. This used basically the same ammunition but with a higher muzzle velocity aided by the longer barrel. Since then further updates to the gun and mounting have been introduced. It remains in production, and the ammunition continues to be developed.

Early Bofors loadings for the m/54 were HE AA (Spårljusspränggranatm / 501), with tracer and self destruct after 14–20 s; and Armor-piercing (slsgrm / 501): for which the following data applies:

- projectile weight: 2.4–2.7 kg
- HE content: 0.31 kg
- muzzle velocity: 880 m/s.
- maximum ballistic range: 14,500 m
- flight time to a distance of 3000 m: 4.2 sec.
- flight time to a distance of 5000 m: 8.5 sec.

The main loading for the L60 naval gun was a high-capacity HE-T SD shell weighing 2.6 kg (of which 340 g was the bursting charge) driven by 1.0 kg propellant to an MV of 920 m/s, with a 6 secs tracer burn (to c.4,000 m) and self-destruction (via the fuze) after 12.5–15.0 secs (6,700–7,100 m). The maximum horizontal range was 14,500 m. The maximum chamber pressure was 275 MPa. Round weight was 5.9 kg and the weight of a two-round charger was 12.3 kg (including clip). Percussion priming was used.

In modern loadings the rimmed brass cartridge case is fitted with an electrical primer and is loaded with 1.26 kg propellant to deliver muzzle velocities in the 950–1,025 m/s range, depending on the projectile. The overall length of the round is 675 mm and the weight 6.1 to 6.5 kg. The projectiles are fixed by two crimping rings in the case engaging in two cannelures close to the base of the projectiles. Driving bands are of copper. Most rounds have base shapes which are conical (to improve aerodynamics) or hemispherical (to optimize the internal volume).

Recent loadings of this ammunition are as follows:

- HE: 2.4 kg filled with 450 g Hexotonal and a nose PD instant-action fuze (proximity fuzes also available); MV 1,020 m/s, maximum range 13,800 m.
- PFHE: 2.4 kg filled with 380 g Hexotonal and lined with 1,200 tungsten pellets 3 mm in diameter, and fitted with a nose proximity fuze with a triggering distance which varies with the target and with altitude (at 40 m above the surface, it triggers at 7 m against aircraft, 4.5 m against missiles; at 5 m above the surface ir triggers at 3 m); fuze has PD function and can be set to PD only; effective range in proximity mode 7,600 m; self-destructs after 17 s.
- 3P (designated Mk 295 MOD 0 in US service): 2.4 kg filled with 460 g Octonal and lined with 2,400 3 mm tungsten pellets; fitted with 3P nose fuze with six different proximity, time and impact modes as with the 40 mm version.
- HCER(-BB): extended-range shell for use against surface targets; 2.8 kg, loaded with 410 g Octonal and fitted with PD delayed action fuze; MV 950 m/s, maximum range 17,000 m (extendable to 21,000 m in BB – base-bleed – version, which has HE capacity reduced to 310 g).

In 2006 BAE Systems was awarded a contract by the US Naval Surface

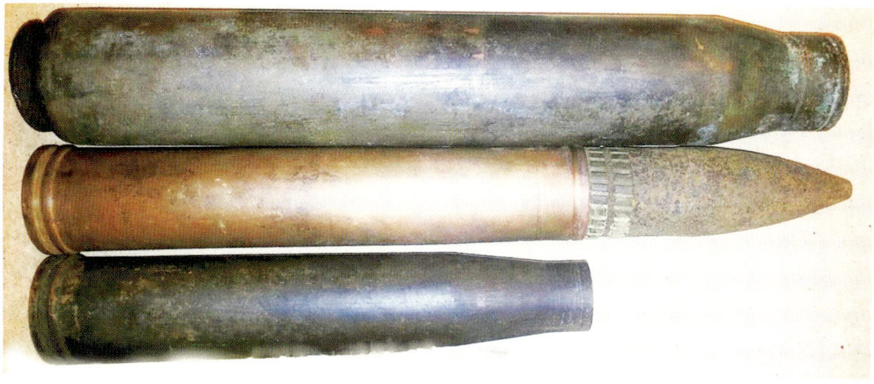

*From top to bottom: 57 × 521RB Ch-21P case (see item below); 57 × 350B round; necked-down 45 × 350B case (courtesy of Evgeny Aranov)*

*57 mm ORKA (BAE Systems)*

Warfare Center for the production and delivery of 57 mm 3P rounds and corresponding Mk 295 MOD 0 cartridges for these gun systems. This incorporates three proximity fuzing modes as well as settings for time, impact, and armor-piercing functions. It is possible to choose ammunition mode at the moment of firing, giving it the ability to switch rapidly between surface targets, air targets, and ground targets.

Bofors has developed Low Vulnerability Ammunition (LOVA, also known as Insensitive Munitions or IM). This includes the propellant and components such as igniters and boosters. This can be applied to all rounds in production.

In April 2015, BAE Systems unveiled a new round for the Mk 110 in the design stage called the Mk 295 Mod 1 Ordnance for Rapid Kill of Attack Craft (ORKA), designed to achieve one-shot kills of surface and air threats. ORKA utilizes technology BAE Systems developed for larger 127 mm and 155 mm guided rounds, using a quad-fin steering system. It is fitted with a multi-mode imaging semi-active seeker that incorporates laser designation or autonomous targeting by downloading an image of the target prior to firing. ORKA retains the 3P multiple fuzing modes. Weight is 3.9 kg, the MV is 740 m/s, the maximum range 10 km and the accuracy claimed to be one metre. The warhead is filled with 212 g PBX and produces 1.4 kg of steel fragments.

In August 2017 L3 Mustang Technology announced the completion of the Critical Design Review (CDR) phase for the U.S. Navy's 57 mm Mk 332 Mod 0 High-Explosive, Guided HE projectile. The round was developed from the Advanced Low Cost Munitions Ordnance program (ALaMO). It is intended for use on the LCS and the new fast Frigate, and the US Coast Guard's Legend and Heritage class cutters. Unlike the much bigger, finned ORKA round, the ALaMO has an external appearance similar to a conventional shell. On March 16 2020 the Mustang Technology Group LP, doing business as L3 Mustang Technology, Plano, Texas, was awarded a contract to manufacture 808 rounds of 57 mm Mk 332.

## 6-pounder 7 Cwt MOLINS (57 × 441R)

This cartridge was developed in the late 1930s for the British Army's 6-pounder 7 cwt tank/anti-tank gun which saw extensive service in World War II. The gun was automated by the Molins company in two versions for different applications: arming naval MGBs (motor gun boats) in which form it used an L/43 barrel; and aircraft (L/50 barrel). It saw limited service during the war.

The ammunition has a brass, rimmed, bottle-necked case. For aircraft use, the intended targets were U-boats so the 2.85 kg solid steel anti-tank AP shot was used, fired at 890 m/s. In this form, the round was 590 mm long and weighed 5.8 kg overall. The naval variant was loaded with an HE fired at the low velocity of 655 m/s, possibly because a flashless loading was wanted for night fighting. The weight of the shell is stated in Campbell to be 2.7 kg, but it seems should be: more likely to have been the already-developed Mk VII (2.85 kg) or Mk X (3 kg) HE-T projectiles which were used. This round was 630 mm long and weighed 5.55 kg overall. There were other loadings for tank/anti-tank guns not given here.

### 6 POUNDER 6 cwt (57 × 514R)

Early in World War II the UK identified a need for an AA gun to cover the performance gap between the 40 mm Bofors and 3.7-inch (94 mm) artillery. The calibre of 57 mm, already used in various 6-pounder guns in British service, was selected and a new cartridge developed, larger and more powerful than any of the others. However the continuing attempts to design a gun of satisfactory performance failed, and the project was officially abandoned after the war.

Curiously, interest in such a weapon seems to have continued for some time since a batch of ammunition was made in 1960, possibly connected with interest in a naval DACR (Direct Action Close Range) system, for which drawings of a twin 57 mm enclosed mounting are known to exist.

The round is 745 mm long and weighs around 6 kg; the 2,720 g HE shell was fired at 945 m/s. There was also an AP shot for self-defence.

### 57 × 521RB Ch-21P

Another large-calibre aircraft cannon designed for ground attack planes, the Ch-21P was designed by Evgeny Charnko the chief of OKBL-46 (Special Design Bureau, Laboratory No. 46 in Leningrad).

The gun was developed in mid-1948 in two calibres – 57 and 76 mm. The cartridge for the 57 mm version (shown above) was based on the 57 × 480R for the ZiS-2 anti-tank gun, slightly extended and with a rebated rim. Overall length is 692 mm and weight 6.5 kg. It fired a 2.485 kg HE shell at 1,087 m/s (figures for the 76 mm version were 9.0 kg, 4.735 kg and 920 m/s respectively).

*Molins 57 mm HE (left) and AP*

# PART IV: THE CANNON

The cannon described in this section are listed under the country of origin or principal use, and within each country are in approximate chronological order. Those made under licence in other countries are not described separately unless they were significantly modified in the process. Entries with the headings in *italics* refer to experimental guns which have not (yet, in some cases) attracted orders. Only a small percentage of experimental guns of particular interest are mentioned below and fewer still are described in any detail. Developments of the former USSR are dealt with under the Russia heading.

Gun data need to be regarded with some caution, as sources can differ. This may be because the specifications change slightly with development, or that different criteria are used: e.g. barrel lengths may be quoted with or without a muzzle brake, and weights with and without various items of related equipment such as belt feeds (these are usually integral but are sometimes separate devices).

The following abbreviations are used as the terms occur so often:

- AA = anti-aircraft
- AFV = armoured fighting vehicle
- L/(number) = the calibre length, which is the barrel length divided by the calibre: so a 20 mm L/70 barrel is 140 cm long
- ME = muzzle energy of the projectile as it leaves the gun, in Joules (J)
- MV = muzzle velocity of the projectile as it leaves the gun, in metres per second (m/s)
- RoF = rate of fire: the cyclic rate at which the gun loading/firing/extracting/ejection sequence operates
- rpm = rounds per minute: the normal way of expressing the cyclic RoF (it does not mean that the gun can fire that number of rounds in one minute – that would be referred to as the practical RoF).

# CANADA

During World War II Canada looked for ways of helping the Allied war effort in various fields, mainly concerning the production of equipment already in use. However, they also tried to develop their own, such as high-velocity versions of the 2 pounder and 6 pounder anti-tank guns. They also developed a lightweight version of the 40 mm Bofors gun dubbed 'Bantam', which was adopted by the War Office for airborne use although too late to see action in World War II.

In 1941 the Canadian Army was experiencing a severe shortage of 20 mm cannon for light AA purposes, so the firm of Inglis developed their own weapon to meet this need.

### 20 mm INGLIS MODEL EX-1–5000: 20 × 110 HS OR 20 × 110RB AMMUNITION.

This project commenced early in 1942 with a proposal from the Canadian Army that the country should develop their own 20 mm AA gun. The initial request was for a gun similar to the Oerlikon Type SS and using the same 20 × 110RB ammunition, but with a quick-change barrel, belt feed (interchangeable between left and right side), and a much higher rate of fire (700–800 rpm). However, the ammunition soon changed to 20 × 110 HS (Hispano) as that was the British Army's preference, and approval was given for Inglis to commence development in the spring of 1942. Progress was rapid, with the first prototype being tested in autumn of that year. After a period of trials and modifications, an initial batch of 58 guns was ordered in spring 1943, of which 41 were delivered. The guns passed their firing tests, albeit with 50 and 100-round magazines; belt feed was expected to be ready in autumn 1944. Single and quad mountings were developed (the latter featuring powered traverse). However, in February 1944 the British Army changed its mind again and decided that all future 20 mm AA guns would use 20 × 110RB Oerlikon ammunition. The Inglis gun was redesigned once more and performed well on test, but by the time it was ready for production, there was no longer a requirement for it. The mountings were adapted to take the Polsten instead.

The Inglis gun had a mixture of the characteristics of other cannon. It took the hybrid gas/blowback operating system from the Hispano, the quick-change barrel from the Oerlikon, and elements of the cartridge feed from Hotchkiss. It was also given an integral cradle to support the gun. Data for a belt-fed gun using HS ammunition include a weight of 68 kg, an L/62 barrel, and a rate of fire of 700+ rpm. In the magazine-fed version, the loaded weight of the scroll-type magazines was 37 kg for the 50-rounder, and 61 kg for the 100-round magazine.

*20 mm Inglis with magazines in quad mounting (Courtesy of Roger Lucy)*

# CHINA

Most of the cannon to have seen service with the People's Republic of China are copies of Russian equipment, but there is an increasing tendency to develop indigenous cannon. With the exception of the aircraft guns, the known designations refer to the mountings rather than the guns. The Type numbers for army and naval guns generally refer to the year of introduction; this does not apply to aircraft guns. More recent developments in army guns have been given different letter designations which still retain the date numbers (e.g. PG87). Modern naval designations such as the Type 730 are western inventions; it simply means a seven-barrel gun of 30 mm calibre. Licence-built cannon include:

- Russian AM-23 aircraft gun (Chinese designation 23 mm Type 2): 23 × 115 ammunition
- Russian 37 mm M39 AA gun (Chinese Type 55, Type 74, PGZ88 SPAAG, Type 61/76 naval): 37 × 252R ammunition
- Russian 2A14 AA gun (Chinese Type 80/85 and naval JW-23 mountings): 23 × 152B ammunition
- Russian 110-PM naval gun (Chinese Type 61 mounting): 25 × 218 ammunition
- Russian 57 mm S60 AA and naval mountings (Chinese PG59, Type 59/66/76 mountings): 57 × 348SR ammunition
- Russian NR-30 aircraft gun (Chinese 30 mm Type 1): 30 × 155B ammunition
- Russian NN-30 naval gun (Chinese Type 69; SRI-206 gun): 30 × 210B ammunition
- Russian GSh-23 aircraft gun (Chinese Type 3–23): 23 × 115 ammunition
- Russian AO18/GSh-6–30K in AK630 naval mounting (Chinese H/PJ-13 – NG-18 for export): 30 × 165 ammunition
- Russian GSh-301 aircraft gun (Chinese designation not known): 30 × 165 ammunition
- Russian 2A72 AFV gun (Chinese designation not known): 30 × 165 ammunition
- Swiss Oerlikon KDF in AF902 Skyguard AA system (Chinese PG99 towed, PGZ07 SPAAG, AN-3 naval): 35 × 228 ammunition.

The guns listed below are Chinese developments.

**37 mm TYPE 76 (H/PJA-76)**: 37 × 240 AMMUNITION.
It is believed that this naval gun was developed from the OTOBreda Fast Forty, using a new round which is unique to Chinese naval guns. As the designation suggests, it was introduced in 1976. The original twin mounting, featuring water-cooled gun barrels and with a fully-automatic radar-directed fire control system, is now designated H/PJ76A (Type 76A). A simpler and less expensive version with a gunner/operator console, possibly for export, is designated H/PJ76F (Type 76F) (4,800 kg). The rate of fire is 380 rpm per gun.

**25 mm PG87**: 25 × 182B AMMUNITION.
This is a gas-operated, belt-fed army gun with a RoF of 600–800 rpm. The ammunition, which resembles a scaled-up 23 × 152B round, is unique to this gun.

First available in the Type 87A twin AA mounting, it has since been used in quad SPAAG mountings: the PGZ95 (Type 95) and the newer PGZ04 (Type 04).

**37 mm AN3**: 37 × 240 AMMUNITION.
This single-gun naval mounting is available with a choice of two guns in different

*H/PJ76F 37 mm naval gun (NORINCO)*

*25 mm Type 87A (Poly Technologies)*

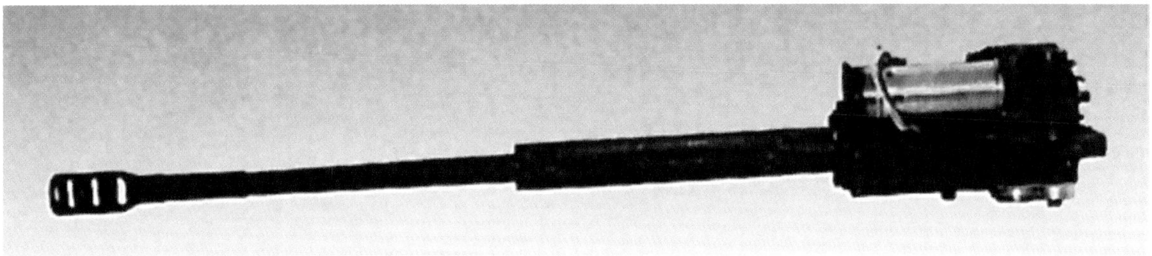

*Chinese 23 mm 'chain gun' (NORINCO)*

calibres: 35 × 228 or the PLAN's unique 37 × 240. The 35 mm gun is the Oerlikon KDF and the 37 mm appears to be derived from that, as it has the same RoF as the Oerlikon: 550 rpm. The mounting weight is c.3,000 kg. The control system is manual or semi-automatic to keep the cost down for use on inexpensive patrol craft.

**23 mm 'CHAIN GUN'**: 23 × 115 AMMUNITION.
This externally-powered gun has been described as a 23 mm copy of the Orbital ATK Chain Gun, although it is not clear whether that is an accurate description or just intended to indicate a general similarity. It presumably uses the same ammunition as the Russian GSh-23L aircraft gun (quoted MV is 770 m/s) and is standard equipment for the Z-10 attack helicopter. It has also been seen fitted to ultra-light combat vehicles and is available in a remotely-controlled mounting, the CS/SP1. It is reported to have a RoF of up to 350 rpm, with the operator able to select single shot, short burst or continous fire. It is belt-fed, from the left side only. No other details of the gun, including the Chinese designation, are currently available.

**30 mm H/PJ-14/15/17 NAVAL MOUNTINGS**: PRESUMABLY 30 × 165 AMMUNITION.
This is a single-barrel gun in three different remotely-controlled naval mountings (H/PJ-17 is a 1,500 kg mounting). The provenance of the gun is not known.

**30 mm TYPE 730 (H/PJ-12) MOUNTING**: 30 × 165 AMMUNITION.
This is fitted with an externally-powered seven-barrel gun which bears a close resemblance to the GAU-8/A and fires at a similar rate (4,200–5,000 rpm – sources vary); the mounting looks similar to that of the experimental French SAMOS system which used that gun. It appears that the gun has been modified to use the Russian electric-primed 30 × 165 rather than the percussion-primed 30 × 173 of the GAU-8/A.

**30 mm TYPE 1130 (H/PJ-11) MOUNTING**: 30 × 165 AMMUNITION.
Developed from the Type 730 with an increase in the barrel count from seven to eleven to boost the RoF to 10,000 rpm (one source says 11,000 rpm), in order to give naval vessels a better chance of stopping supersonic anti-ship missiles. It is also offered in a land-based mounting, the LD3000, for which one source states the RoF is reduced to 8,000 rpm.

**40 mm NORINCO CTAS**: AMMUNITION DIMENSIONS UNKNOWN.
In November 2016 NORINCO showed a 40 mm cased telescoped weapons system and its ammunition. The gun has an adjustable RoF of 100–200 rpm and is intended for AFVs. Ammunition types known so far are HE and APFSDS. The gun is fitted with a muzzle brake and supporting barrel sleeve, and is initially installed in a RWS fitted to an 8 × 8 AFV. There is more information about CTAS in the Unconventional Systems chapter.

*Type 730 CIWS (Rhk111 / Wikimedia Commons / CC-BY-SA-4.0 licence)*

# CZECHOSLOVAKIA & CZECH REPUBLIC

What is now the Czech Republic, or Czechia, has experienced drastic political changes over the last century. Until 1918, it was part of the Austro-Hungarian Empire; after that it became part of the First Czechoslovak Republic, an independent democracy; from 1938 to 1945 it was effectively occupied by Germany; after World War II it became part of the Warsaw Pact and was controlled by the USSR; and in 1989 Czechoslovakia became an independent nation again, before dividing into the Czech Republic and Slovakia in 1993; the Czech Republic joined NATO in 1999 and the EU in 2004.

Throughout all of this, the country had a long tradition of gun design and manufacture; over the last century this has mainly been in small arms (particularly from ZB – Zbrojovka Brno), but Škoda (based in Plzeň = Pilsen) was making heavier weapons from the second half of the 19th Century. Not many designs qualify for inclusion here, either being of less than 20 mm calibre or being manually-loaded, but there have been a few interesting exceptions during and immediately after World War II. However only two of them entered service – the 30 mm M53 and the 57 mm R10 (or PLK), and only the first of these can be counted as moderately successful. Designs which never got past the prototype stage include 30 mm and 55 mm AA systems, and a 50 mm automatic anti-tank gun (Chinn Vol 2 includes full details of these, what follows is a very brief summary).

More recently, the ZPL-20 has been developed and fielded, and what appears to be a copy of the Russian 2A42 is offered as the CZ30.

**30 mm M53**: 30 × 211 AMMUNITION.
This anti-aircraft gun has an unusual history, since it was first developed during World War II at the Brno factory in Czechoslovakia while this was under German control (then known as *Brünner Waffenwerke*). It was apparently initially known as the M 44 but was given the designation MK 303 to acknowledge the involvement of Krieghoff. After the war, the Czechs decided to continue with its development and it emerged as the M53 cannon. The very large cartridge is unique to this weapon.

The gun is gas-operated, fires at 450–500 rpm, has a quick-change L/81 barrel and weighs 200 kg. This was an early application of counter-recoil or floating firing to an automatic cannon. It fires from an open bolt with the barrel held to the rear: after a round is chambered, the piston rod locks the breech block, which simultaneously releases the barrel to start moving forwards; the firing pin is then released and the gun fires as the barrel and piston rod are moving rapidly forwards, spreading the recoil force over a longer period; the gas operation then functions as usual to unlock the breechblock and extract the fired case. Ammunition supply is by a 10-round horizontal strip, limiting the practical RoF to 100 rpm. In the SPAAG installation shown below, the strips are fed into large 50-round boxes on top of the guns, increasing the practical RoF to 150 rpm. Because of their weight (84.5 kg when full) the boxes are reloaded while in place, from the top.

The gun was initially fielded in a twin mounting weighing 1,750 kg in action. The mounting was also later adopted for self-propelled six-wheeled SPAAGs; the M53/59 Praga and M53/70, in which applications it is hydraulically or electrically powered respectively. The fire control arrangements are primitive and the system is effectively obsolete, although YugoImport advertises a laser-based FCS. It has been withdrawn from front-line service, but still may be held in reserve in Cuba, Czech Republic, Libya, Romania, Serbia, Slovakia and Vietnam.

***30 mm SKODA TWIN-BARREL FLAK***: 30 × 248.5RB AMMUNITION.
Work on this unusual weapon commenced in 1937. It was made in two versions, known as A and B. Both have two interconnected barrels and were presumably inspired in part by World War I German Gast MG. It might be a coincidence that the much close spacing of the barrels is more like the later 23 mm GSh-23, but the mechanism, described as retarded blowback plus short-recoil locking, appears to be unique. The bolt of each gun is mounted integrally with the barrel of the opposite gun; when one barrel fires, the bolt recoils by blowback and carries with it the other barrel. The

*Czech M53–59 Praga*

*Skoda twin 30 mm FLAK*

Model A and B differ mainly in the locking of the gun in the recoil position and releasing them at the correct moment: in the Model A this is done hydraulically, in the Model B mechanically.

An ammunition drawing apparently associated with this shows a 30 × 248.5RB case with a 37 mm diameter rim and a 44.9 mm maximum case diameter.

**57 mm SKODA R10 / PLK**:
57 × 398 AMMUNITION.
The origins of this gun extend back to late World War II with a Skoda project for a large-calibre automatic AA gun, initially in 50 mm and then 55 mm calibre (see below). In 1948 this led to an official requirement for a 57 mm AA gun firing at a rate of 180 rpm, with a maximum range of 13,500 m and a ceiling of 5,500 m. Powered traverse and elevation were required, meaning that a petrol engine had to be part of the mounting. Skoda produced three variants, designated R8, R10 and R12, differing in the type of mounting and feed system (3-round clips or 40-round magazines). The R10 was selected, fed by 3-round clips (which fed sideways through oversized trunnions), using an L/62 barrel and with a cross-type mounting. The prototypes proved troublesome but production was authorised; a first batch of eighteen guns was delivered in 1955, but these failed their tests. Despite this, a further 200 were ordered for political and financial reasons, but Russian 57 mm S-60 were bought as well. On paper the R10 (generally known in the West as the PLK) was better than the S-60, with a higher rate of fire, superior ballistics and other advantages, but its problems led to production ceasing in 1956. The R10 was mainly allocated to reserve units, serving until the beginning of the 1990s.

*57 mm PLK (Martin Dubánek)*

***SKODA 50 mm PAK***:
AMMUNITION UNKNOWN.
This was recoil-operated with a vertically-sliding breechblock, fed by either a clip or a metallic belt. The recoil force was used to load the round onto a loading tray and to set a spring-loaded rammer. The weight of the recoiling mass was 250 kg. On test the gun achieved a cyclic RoF of 80–85 rpm.

***SKODA 55 mm FLAK***:
AMMUNITION UNKNOWN.
Development of this was initiated at the request of the German authorities. Like the 50 mm PAK above, this was recoil-operated with a vertical sliding breechblock. Gun weight was 250 kg, the barrel was L/85, and the RoF given as 100–140 rpm using a feed-tray conveyor. It fired a 2.0 kg projectile at 835 m/s, for an ME of 700,000J.

**ZPL-20**: 20 × 102 AMMUNITION.
This is a version of the Russian GSh-23 twin-barrelled aircraft gun, modified in the late 1990s to fire the US electric-primed 20 × 102 ammunition and offered in the *Plamen* (flame) gunpod, which weighs 215 kg when loaded with a 224-round belt. The gun, which is powered by a combination of gas and recoil, weighs 64 kg and fires at a variable rate of 700–2,600 rpm.

# DENMARK

Denmark has had one internationally known manufacturer of automatic weapons: the *Dansk Industri Syndikat, Compagnie on A/S* of Copenhagen, sometimes referred to as DISA but commonly just called Madsen after the inventor. The company's initial focus was on small arms, but in 1926 they announced the prototype of a new 20 mm weapon, primarily for aircraft use. This reportedly fired rimmed ammunition at a cyclic rate of 180 rpm, but over the next few years the design evolved with an increased RoF and the unique rimless 20 × 120 ammunition. Sales had a slow start (prospective customers usually purchasing small numbers for evaluation), but the increasing threat of a major conflict increased interest in the gun, which achieved widespread sales success as an AA and general-purpose cannon with a few fitted to aircraft. It remained in service for a long time after World War II; ammunition for it was still being made in the 1990s.

In the 1930s a 23 mm version of the gun was offered for aircraft use, but only a handful of these saw service with Chile (with their Italian-made Breda aircraft) and possibly China (a couple of Curtiss-Wright Hawk 75s).

**20 mm AND 23 mm MADSEN:**
20 × 120 AND 23 × 106
AMMUNITION.

These two guns are identical except that the 23 mm version received a new barrel of larger calibre with its chamber shaped around a necked-out version of the cartridge case. The short-recoil pivoting or tilting lock action is a scaled-up version of the company's rifle-calibre machine gun. The block is pivoted at its rear end to the barrel extension and during recoil a lug on the right side of the block, acting on cam surfaces in the receiver, raises the forward end of the block to clear the barrel breech and permit case ejection. At the end of recoil the forward end of the block drops below the breech, bringing a curved surface in its left side into alignment with the next round to be fed. The cartridge is forced against this surface by the magazine follower and as counter-recoil begins, the rammer pushes it forward and into the chamber. During the last few millimetres of counter-recoil the block is raised by cam action and brought into battery.

The 20 mm guns were usually magazine-fed; normally by a 40 or 60-round drum, although a 10-round box and a 15-round drum (which weighed 3.6 kg empty and 8.8 kg full) were also available. The 23 mm versions were intended for aircraft use so were fitted with a belt feed; the standard belt length was 200 rounds, a full belt of 100 rounds weighed 38 kg. Belt feed was also available for the 20 mm versions (for both the very few aircraft guns and for AA use), but it is not clear how much use was made of it. Both guns weigh 53 kg; the barrel lengths are the same so the 20 mm version is an L/62, the 23 mm one an L/52. The RoF claimed for both calibres of the gun was 400 rpm (but see below); for the 20 mm on the universal mounting the practical rate (allowing for magazine changes) was quoted at 240 rpm.

Various mountings for the 20 mm gun were available, including a light low-angle field tripod at 65 kg, the same fitted with wheels at 120 kg, and a universal mounting capable of high-angle fire with detachable wheels at 195 kg (changing from legs to wheels or vice versa took about 30 seconds). There was also a lightweight triaxial mount intended for fortifications and naval use, and some were fitted to small numbers of AFVs made in Sweden.

The Danish Navy mounted the guns on a couple of warships, including the training cruiser *Niels Iuel*, which in 1936 was fitted with five twin mountings for the gun, under the designation '20 mm R.K. M/31' (R.K. = Rekyl Kanon = recoil-operated cannon); each gun was said to fire at c.250 rpm, and the mounting weighed 440 kg. In 1941/2 these plus some smaller MGs were replaced by ten single mountings of the '20 mm R.K. L/60 M/41' which was said to fire at c.500 rpm; the mountings weighed 226 kg.

The US Army tested the 20 mm Madsen at the Aberdeen Proving Ground in October 1940, with both an F7 Universal mount (AA and AT, with wheels for travelling) and an F5 light field mount for AT use. Both mountings incorporated a recoil-absorbing mechanism. The gun came with 15-round drum magazines, with which the cyclic rate was measured at 220 rpm at 0° elevation and 174 rpm at 50° – considerably lower than the claimed figure. The gun proved reliable but it was noted that the number of curved surfaces in the mechanism could make it difficult to mass-produce economically.

A later US report by ARF noted that the most distinctive feature of the gun was the breech block which was pivoted around its breech end, and assumed different positions for firing, ejection and loading. It was noted that it was difficult to change the barrel, and to assemble and service the weapon.

*20 mm Madsen on naval mounting (David Castor / Wikimedia Commons / Public Domain)*

# FINLAND

Finland used a variety of cannon in the years before World War II. They obtained a manufacturing licence for the 40 mm Bofors in 1937, purchased a few SEMAG (Oerlikon L) and Madsen 20 mm cannon, and bought 134, Flak 30, 20 mm guns from Germany, fifty of which were delivered just before the Winter War of 1939–40, plus a total of 88 Italian 20 mm Breda M35 cannon, the first of which were being distributed at the end of that war. An order for 195 more Madsen cannon followed before the Continuation War of 1941–44 with more to follow (349 were still in use in 1950, mostly with the navy – which still had 217 of them in 1982). Further help came with deliveries of 20 mm Flak 38 from Germany (a total of around 113 being delivered by the end of the war). However, these 20 mm purchases were mainly due to delays in obtaining enough home-grown cannon, production of which finally delivered some 174 twin-barrelled VKT M40 mountings in service by the end of the Continuation War – by which time there was a total of 720 assorted 20 mm cannon in Finnish service. The Breda cannon remained in service for training purposes until the 1980s, the Flak 30 and 38 until well into the 1990s while the navy retained the Madsen cannon until the end of the century.

The story of indigenous Finnish cannon development is largely concerned with one designer, Aimo Lahti, who was responsible for some successful small arms and went on to produce various cannon designs in the 1930s, based on a variety of different 20 mm ammunition. By far the most successful was the VKT M40 AA gun in the German 20 × 138B calibre; the M39/44 in the same calibre was an unsuccessful automated version of the M39 semi-automatic anti-tank rifle, and the L-34 AA/general-purpose cannon saw minimal use. Various experimental Lahti aircraft cannon saw no use at all: the L-35 (a version of the recoil-operated L-34); the blowback L-37; and the gas-operated L-36, L-37/38 and L-38. Instead, the Mauser MG 151/20 saw considerable use late in the Continuation War, mainly fitted to Messerschmitt Bf 109 aircraft supplied by Germany, but also fitted to some French Morane fighters the Finns possessed (a few of which had the Hispano-Suiza 404 cannon). Even a handful of Soviet 20 mm ShVAK were pressed into service, in the aircraft they were captured with. Thirty of the MG 151/20 guns were subsequently converted as light AA weapons and remained in service until the early 1990s. A few Ikaria MG-FF cannon also served, fitted to bombers supplied by Germany.

**20 mm LAHTI AND TAMPELLA L-34**: 20 × 113 / 20 × 113B AMMUNITION. The first prototype of this cannon emerged in 1934 and after some modifications the Finnish Coast Guard ordered ten L34/II in 1935 to arm new patrol boats. These recoil-operated guns were made by the state-owned VKT and delivered with L/60 barrels and 15-round box magazines, although 32-round flat pan magazine and belt feeds were experimented with. With a weight of 48 kg and a rate of fire of 300–350 rpm the gun was an unremarkable performer. The 20 × 113 ammunition was unique to this gun.

No further sales were achieved by VKT but another Finnish engineering company, Tampella, negotiated

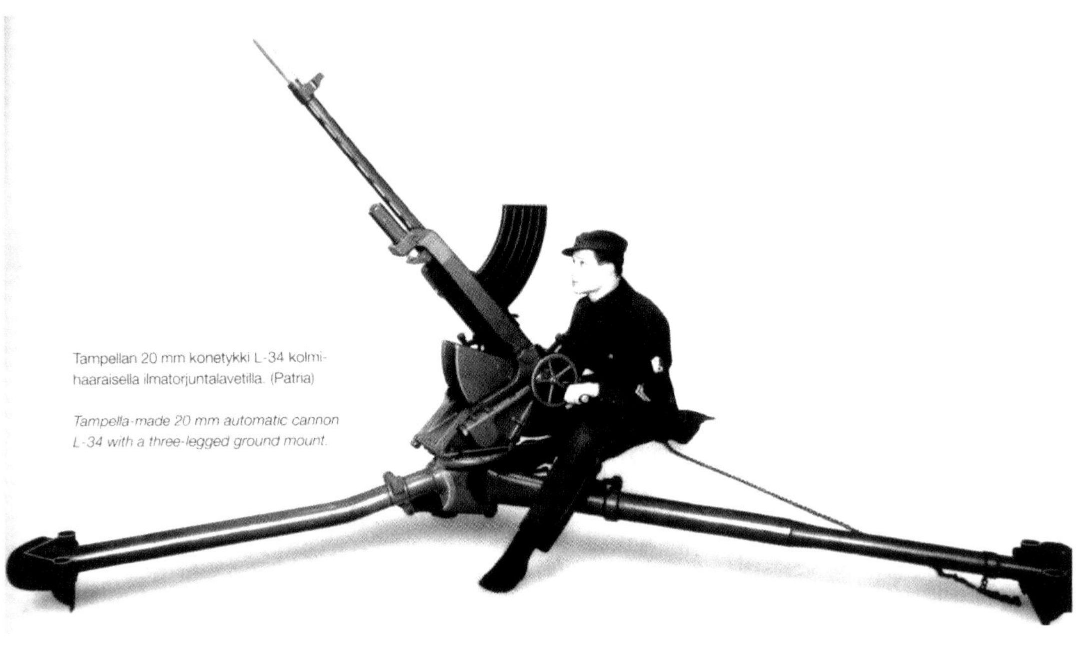


*Lahti L-34 on tripod mounting*

manufacturing and export rights. They slightly modified the gun and its ammunition, designing a belted version of the rimless case, and promoted it abroad. Two examples of the Tampella L-34 were made and saw wartime service with an AA unit of the Civil Guard at Tampere, but no sales resulted. This weighed 40–60 kg, had a 15-round box or 32-round pan magazine and fired at 325–360 rpm.

**20 mm LAHTI VKT M40**:
20 × 138B AMMUNITION.
The prototype of this gun was first designated L-39, being commissioned in 1939. The performance of the gun was satisfactory but the mounting needed work and the German Flakvisier sights were required to deliver the performance – but these were initially not available. Deliveries therefore did not take place until 1943 and continued for a while after the war, with a total of just over 200 built. They remained in service for training until the 1970s. The gun is gas-operated, has an L/65 barrel and is fed by a 20-round box magazine. It uses the same ammunition as the German Flak 30 and 38, fired at 350–375 rpm and weighs 40 kg. A twin mounting was standard.

*20 mm M40 VKT (MKFI / Wikimedia Commons / Public Domain)*

**20 mm LAHTI M39/44**:
20 × 138B AMMUNITION.
Lahti's most successful 20 mm gun in production terms was the semi-automatic gas-operated M39 anti-tank rifle, with over 2,000 being made by 1945. As these became less effective against tanks, a total of 325 rifles were adapted to anti-aircraft use by being given appropriate sights and mountings, plus being modified to fully-automatic fire. However, automatic fire was found to cause an unacceptable level of breakages, so the guns were restricted to semi-auto fire, in which form they remained in service until the 1950s. The gas-operated gun weighs 42 kg, has an L/65 barrel and was fed by a 10-round box magazine.

*20 mm Lahti L-39 anti-tank rifle. This appears to be the 39–44 variant capable of automatic fire, based on the additional recoil springs on both sides of the barrel (MKFI Wikimedia Commons / Public Domain)*

# FRANCE

The French state was traditionally involved in armament development, with the private sector in a supplementary role; relationships between public and private organisations were not always frictionless, with a lot of rivalry involved. The principal private company involved with automatic weapon development in the first decades of the 20th century was Hotchkiss, while the main state organisation working on cannon was the Army's Puteaux Arsenal, generally shortened to A.P.X. (*Atelier de Puteaux* = Puteaux workshop).

Gun designations pre-World War II typically included the calibre and a model number assigned according to the year of introduction (e.g. *Modèle* 1935, often shortened to Mle 35), although this did not generally apply to late-1930s aircraft cannon, which used manufacturers' designations as described below.

The *Service aéronautique* which became the *Armée de l'Air* (it was a branch of the *Armée de Terre* until 1934) had always been keen on large-calibre gun armament, fitting manually-loaded 37 mm and 47 mm guns to twin-engined aircraft from early in World War I. During the war attention was also given to fitting 37 mm cannon in the nose of single-engined fighter aircraft. This was made possible by using Hispano-Suiza V-8 engines with geared propellers: the gun was fixed to the engine and fitted between the cylinder banks, while the raising of the propeller axis from the line of the crankshaft allowed the use of a hollow propeller hub through which the gun could fire. Such installations were known as *moteur canon* and became something of a French speciality. Those used during World War I were manually loaded by the rather busy pilot, but attention understandably switched to developing automatic cannon for this purpose.

The introduction of the Maxim 1-pounder 'pom-pom' at the end of the 19th Century (see UK section) prompted a response from Hotchkiss, in the form of *Le canon automatique de 37 mm Hotchkiss Mle 1900/1901*. A few of these were reportedly sold to Japan and Mexico (a photo exists of one with the Mexican Army). No further progress was made until late in World War I when a *Canon de 37 mm Mle 1918* was being developed by A.P.X.. There was also interest in the British 1½-pounder C.O.W. gun, with reports not only of plans to make it under licence but of A.P.X. producing their own design using its 37 × 190 ammunition. The end of World War I reduced the impetus for these developments and none of them saw service. The same fate befell the little-known Delaunay-Belleville automatic anti-tank gun in its unique 20 × 115SR calibre.

In the interwar period Hotchkiss provided the main focus of automatic cannon work, the first project to emerge being an obscure 25 mm automatic anti-tank gun. Its 25 × 193 rimless ammunition later inspired two different Hotchkiss service rounds: a rimmed version of the case for a manually-loaded anti-tank gun and a version of the rimless case shortened to 163 mm for the 25 mm Mle 38/39, which saw most service with the Imperial Japanese Navy as the Type 96.

In 1936 the Hotchkiss company was nationalised under the name MLS (*Manufacture de Levallois*). Other well-known private companies involved in the interwar period were initially Schneider & Cie, primarily known for 75+ mm calibre artillery, and Hispano-Suiza (Spanish-Swiss, after the two countries initially involved), originally known for luxury cars and then for aero engines. The French branch of Hispano-Suiza became involved in developing automatic cannon in the second half of the 1930s, principally for aircraft. The company was nationalised in 1937 under the name *La Société d'exploitation des matériels Hispano-Suiza*, although it was popularly known as Hispano-Suiza France, or HSF (the Swiss branch of the company – HSS – remaining independent). Guns developed by them were given HS 4xx designations pre-World War II, and HS 6xx post-war.

Various aircraft gun projects of around 30 mm calibre took place interwar, although details of them are incomplete. In 1927 A.P.X. was directed to develop another *moteur canon*, which reached the aircraft test stage in 1932. Known characteristics were: 33 mm calibre, firing a 630 g shell at 650 m/s or 687 m/s. The gun had an L/62 barrel and fired at a RoF of 300 rpm. This was replaced in 1933 by another A.P.X. project for a gun weighing 140 kg with a muzzle velocity of 900 m/s, also firing at 300 rpm and carrying 50 rounds. Next came a *semi-léger* (half-light) cannon from M.A.T. (*Manufacture d'armes de Tulle*, a state factory) using 30 × 167 ammunition which fired 455 g shells at more than 950 m/s. Finally, Hispano-Suiza developed the 30 mm HS 411 which evolved into the post-war Swiss HS 830/Oerlikon KCB weapon series.

Smaller cannon were also developed for aircraft use (still mainly as *moteur canon*), following-on from licence production of the Swiss Oerlikon Type S by HSF

*Hispano Type S9 (copyright Rheinmetall Air Defence AG)*

as the S7 and S9 (the differences between these two being minor, to do with adapting the guns for different installations). In the mid-1930s HSF decided to adopt a new operating system to increase the rate of fire, resulting in the famous HS 404, better known internationally as the 20 mm Hispano. HSF experimental projects begun before World War II included the HS 8 (firing a 23 × 120RB cartridge), HS 401 (23 × 122), HS 405 (20 × 71) and HS 406/7 (23 × 122 – using a larger-diameter case than the HS 401).

The *Armée de Terre* did not initially appreciate the value of automatic cannon in the AA role until experience of the Spanish Civil War of 1936–39 focused attention on the need for AA guns in between machine guns and the larger, manually-loaded artillery. The 37 mm Mle 30 by Schneider & Cie was ignored until then, only obtaining a few export sales. By the late 1930s the army had chosen the 40 mm Bofors, but the Bofors order books were full and only 42 had arrived by May 1940. Licence production was arranged but this would involve a long delay before service use, with 100 expected to be delivered during 1941. An interim order was therefore placed for the 37 mm Schneider Mle 30, but deliveries were very slow, so orders were also placed for the 25 mm Hotchkiss Mle 38/39; these too arrived rather slowly, fewer than 250 making it into service, so a supplementary order was placed for 20 mm Oerlikon Type S, of which some 237 were in service by May 1940.

The Navy (*Marine Nationale*) also became involved in automatic AA cannon development, although again rather late as they were relying on the manually-loaded 37 mm Mle 1925 (single barrel) and Mle 1933 (twin barrel) mountings which were standard on French warships. In the mid-1930s the Navy held a competition to replace these guns with automatic ones in a similar calibre. This was the ACAD (*Automatique Contre-Avions Double* = automatic twin-AA) programme. The competition involved the 40 mm Bofors L/60, 37 mm Schneider Mle 30, 37 mm Hotchkiss, and a 37 mm design from the government *Artillerie Navale* (A.N.) at Ruelle. The A.N. system won the contract and was designated Mle 1935, but by June 1940 only one trial mounting was in use.

The Navy also developed smaller cannon for aircraft use. In 1933, the French Navy HQ ordered D.C.A.N. (*Direction des Constructions et Armes Navales,* which led the naval arsenals) to began the study of new 25 mm weapons for its aircraft. This was carried out by S.T.A.N. (*Section Technique d l'Artillerie Navale*), a branch of D.C.A.N. which planned studies, issued specifications to builders (arsenals or private contractors) and tested prototypes.

An initial prototype used 25 × 161 ammunition (with a much slimmer cartridge case than the 25 × 163 used in Army's Hotchkiss Mle 38/39) and was known as the 'Type 25 Mle 1933'; it was a motor cannon with a 44 round drum. The next model, the Mle 1935, used slightly modified 25 × 159.5 ammunition and was developed for both fixed and flexible mountings with drum magazines of varying capacities from 33 to 56 rounds. Ten of these were made by *Société des Moteurs Salmson* (SMS). Two other 25 mm guns reached the prototype stage: the Mle 1937A and Mle 1937B.

In addition, Chinn describes two other French prewar 25 mm aircraft cannon. One was a Darne *moteur canon*, but apart from the fact that it was belt-fed and could reportedly fire at 750 rpm, nothing more seems to be known. The other was a Hotchkiss aircraft gun, but Chinn's description seems to conflate this with the Mle 38/39 AA gun; Linder, the Mauser engineer who reported on French cannon developments following the German occupation, provided more information (see below).

To sum up: the French armaments industry (both state-owned and private) developed an impressive range of automatic cannon in the 1930s, but with few exceptions were unable to get many into service before such activities ceased following the German invasion of summer 1940. Unfortunately, data on the experimental weapons is patchy.

After World War II the French retained some wartime cannon, notably the Mauser MG 151/20; restarted production by M.A.T. of the HS 404; and also resumed their automatic cannon design and development activities.

At the end of the war, as previously, there were three ministries:

1. The War Ministry was in charge of the French Army, whose Headquarters was (EMAT: *l'état-major de l'Armée de terre*), and of all the State industrial base directed by DEFA (*Direction des Études et Fabrications d'Armement* = Directorate of Arms Study and Manufacturing); before the war it was named DFA (*Direction des Fabrications d'Armement*).
2. The Navy Ministry with the Navy, its HQ and DCAN (*Direction des Constructions et Armes Navales* = Directorate of Naval Shipbuilding and Naval guns manufacture).
3. The Air Ministry with the Air Force and its HQ and, initially, an industrial base similarly controlled by the DCAé (*Direction des Constructions Aéronautiques*).

Control of the State Aircraft industry (resulting from the nationalisation of 1936) soon ended; DCAé became only a staff organisation without any industrial base. When the Air Force needed a new gun, the programme was written by DCAé for the Air Force and DEFA was ordered to develop the gun and to build it. This avoided duplicate mechanical plants. From an industrial viewpoint, to develop and build an aircraft gun was the same as developing a gun for an IFV.

A few years later, a single Ministry was established to control the three departments with a State Secretary for each of them : land, air and sea under the control of the *Minister des Armées*. These State Secretaries were in charge of the industrial branches of their departments: DEFA, DCAé or DCAN.

DEFA (*Direction des Etudes et Fabrication d'Armements*) was the directorate of State Armouries for Army Armaments which became DTAT (*Direction Technique des Armements Terrestres*), then DAT (*Direction de l'Armement Terrestre*); in 1967 this became GIAT (*Groupement des Industries de l'Armée de Terre* = Army Industries Group) and in 2006 Nexter Systems, which is responsible for all automatic cannon, their ammunition and mountings. In 2015, Nexter and Krauss-Maffei of Germany merged to form KNDS (KMW + Nexter Defense Systems), also known as KANT (Krauss-Maffei And Nexter Together).

EFAB (Établissement de Fabrication *d'Armement de Bourges* = Armament Manufacturing Establishment of Bourges) was the State arsenal operated by DEFA/DTAT/DAT/GIAT to manufacture artillery and tank guns. It was in charge of developing and building medium calibre cannon after the closure of AME.

AME, MAC, MAT, MAS were developers and manufacturers of small and medium calibre guns. DEFA's small arms ammunition plants were ALM, ATE, AVE, and ATS (which also manufactured primers).

Today, EFAB and ATS are the only working plants, together with La Chapelle Saint Ursain where all medium and artillery ammunition is manufactured. All State armories dealing with small arms and their ammunition have been closed, and although plans to re-establish a SAA plant were considered, its future is unclear.

The designation system for cannon projects changed twice in the immediate post-war years: initially, the Germanic MG and MK designations were used; then these were replaced by CG and CGF designations (the meaning of CG is no longer known); and finally, the current M designations (usually preceded by the calibre, as in 20M621), in which the first two numbers after them refer to the year in which development started (in this case, 1962), the third to the model number.

For Army use, the main concern post-World War II was to obtain high-performance cannon for AA purposes. Initially, they developed a powered 53T1 mounting for the MG 151/20, as an interim measure pending the development of higher-performance cannon, some 200 mountings being built. For the new AA gun, they started with the same 'million point gun' requirement which led to the German MG 213C revolver cannon: a 20 mm cannon with a muzzle velocity of 1,000 m/s and a cyclic rate of fire of 1,000 rpm. This project was designated 5CG and the cartridge (by A.M.E. – *Atelier Mécanique de Mulhouse*, formerly known as CEAM = Centre d'Études et *d'Armement de Mulhouse*) evolved through various shapes, some with belted cases and some rimless. Both A.M.E. and M.A.C. (*Manufacture d'Armes de Chatellerault*) proposed guns to fire it. However, concerns about lack of effective range led to a switch to the 30 mm calibre at the end of 1954.

Various guns were put forward for testing, including: the 30 mm 5CGF from AME (using ammunition derived from the 3CGF programme); the 30 mm and 37 mm HS 606; a revolver cannon using 30 mm HS 606 ammunition; and the eventual winner, the 30 mm HS 830, which was adopted in 1955 (after some modifications) as the HS 831A. This gun was fed by a box magazine containing eight flat trays each with five rounds, and had a powered mounting featuring a tripod for firing and wheels for towing. Some 500 were bought, from both HSS and MAT. In 1956 a self-propelled system was adopted consisting of an armoured twin-gun turret (the S401A from SAMM) with a tracking radar and fire control system, mounted on an AMX-13 light tank chassis. Only 60 were built, but in 1975 Saudi Arabia bought 57 examples of an updated turret on an AMX-30 tank chassis.

Interest in the 20 mm calibre returned in the FINABEL programme in which France took a leading role (this organisation promotes cooperation in armament development between European Union members: the acronym refers to France, Italy, Netherlands, Germany (*Allemagne*), Belgium and Luxembourg – members in the 1950s). This resulted in the AME 582, using a modified version of the 20 mm 5CGF cartridge, but this was rejected due to the need to develop and field a new round of ammunition. There was now an urgent need to replace the MG 151/20 so the AME 582 was lightened and modified to use the US 20 × 102 cartridge (already in French service with the M39 cannon of the North American F-100 fighter bomber) to create the 20M621. The design was further modified to use 20 × 139 ammunition in the 20M692. Both of these guns proved successful and are still marketed today.

Meanwhile, work was taking place in developing new aircraft cannon. France initially made good use of German experience and technicians from the Mauser factory in order to kick-start French industry (Mauser engineers also went to the UK, Switzerland and the USA to assist with their revolver cannon projects). This led to the famous 30M540/550 series aircraft revolver cannon; total production of the 30M550 series alone was around 14,000 (it was also made in Israel and Brazil).

HSF cannon development briefly restarted in 30 mm and 37 mm calibres using HS 603, 604 and 606 designations, but by the mid-1950s HSF had transferred all automatic cannon work to the Swiss branch (HSS), which used a model numbering system of HS 8xx.

The post-war Navy initially used World War II-era 40 mm Bofors and 20 mm Oerlikon, but replaced these in new-build projects with the same 20 mm and 30 mm cannon as had been adopted by the Army. The one notable addition was the 57 mm Bofors, for which a French mounting was developed.

Other notable post-war guns include the 30M781, 30M791 and 25M811. In addition, there has been an even greater number of experimental guns which failed to make production, many of which are listed below.

Among interesting experimental projects was a joint programme for cannon using caseless ammunition between GIAT (responsible for the 25 mm ammunition) and Rheinmetall (for the gun) which commenced in 1978. EFAB continued with the development in 30 mm calibre, exploring many aspects of the caseless concept, but technical problems led to cancellation in 1984.

Nexter has a Joint Venture with BAE Systems, CTA International, to develop and manufacture case telescoped weapon systems and ammunition of 40 mm calibre.

## PROJECTS UP TO WORLD WAR II

### HOTCHKISS 37 mm MODEL 1901: 37 × 94R AMMUNITION.

As with Maxim, Hotchkiss used a scaled-up version of their machine gun mechanism when developing their first

Hotchkiss 37 mm Model 1901

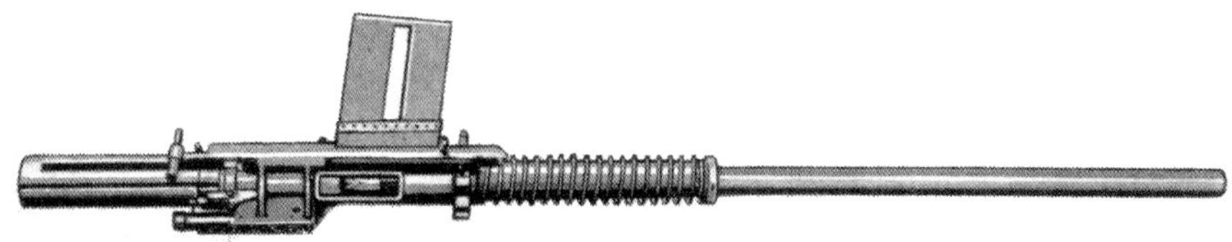

A.P.X. 37 mm Mle 1918 automatic 'moteur canon' (Chinn)

automatic cannon, which in this case was gas-operated. Few details are available, but it fired at 100 rpm and was loaded via horizontal 10-round stripper clips fed into the right side of the gun. It was air-cooled, and had a few of the characteristic Hotchkiss large cooling fins around the rear of the barrel (obscured by the shield in the photo above).

Photos usually show it on a wheeled mounting, with large wooden-spoked wheels. The gun weight was 240 kg and the wheeled mounting weighed 260 kg. The accompanying ammunition trailer contained 288 rounds and weighed 150 kg empty and 450 kg loaded.

The photo shows the cannon on a typical naval pedestal mounting, with the ammunition clips and storage box for 60 rounds.

*DELAUNAY-BELLEVILLE 20 mm AUTOMATIC ANTI-TANK GUN*:
20 × 115SR AMMUNITION.
This was a late-World War I project for an anti-tank machine gun designed by the luxury car-making firm of Delaunay-Belleville to meet a French Army requirement. The performance target was to penetrate a 16 mm armour plate at 300 m. The gun was fed by a 20-round straight strip or an articulated belt (similar to the Hotchkiss M1919 LMG), and used a mixture of gas-operation and long-recoil features; the cyclic rate of fire is not recorded. Weight of the prototype was 30 kg. The ammunition was unique to this gun. The project was abandoned in 1919 after two guns had been made: a film clip survives, showing the gun being manually loaded, so presumably development was incomplete.

*CANON DE 37 mm Mle 1918*:
37 × 94R AMMUNITION.
This was being developed by A.P.X. as a *moteur canon* for fitting to aircraft. It was a long-recoil design (using a mechanism similar to the Chauchat LMG) and fired the usual low-velocity 37 × 94R cartridge at 120–135 rpm from a 5, 7 or 10-round gravity-fed magazine. Weight was 90 kg. A high-velocity version chambered for the British 1½-pounder C.O.W. gun's 37 × 190 ammunition was under development at the end of the war.

*25 mm HOTCHKISS AT MACHINE GUN*: 25 × 193 AMMUNITION.
This little-known automatic anti-tank gun emerged in the early 1930s and predates both the Hotchkiss automatic AA gun and the manually-loaded AT gun. The photo, from a 1934 Spanish catalogue, shows the gun, fitted with a vertical box magazine, on a low-angle mounting.

This gun is principally known from its ammunition: a few cartridge cases survive, plus a factory cartridge drawing dating from 1930. This 25 × 193 should not be confused with the shorter 25 × 163 of the AA gun or the rimmed version of the round (25 × 194R) which saw service in manually-loaded anti-tank guns.

The gun had a slightly longer (L/72) barrel than the AA version and weighed 107.5 kg; the 10-round box magazine weighed 15.5 kg when full. No credible reference has been found to the RoF.

It does not seem to have been adopted by France, but it is possible that some export sales were achieved (Mexico and China being mentioned).

SCHNEIDER 37 mm Mle 1930:
37 × 296R AMMUNITION.
This little-known gun emerged in 1930 but at first the French Army showed no interest, the gun reportedly being made only for an export sale to Romania (but see the note on the 37 mm Hotchkiss below). Official attitudes changed in the late 1930s as a result of the Spanish Civil

25 mm Hotchkiss AT machine gun

Schneider 37 mm Mle 1930

War, during which the importance of automatic cannon against low-flying aircraft became evident. While preferring the 40 mm Bofors, the Army placed an interim order for 700 of the *'Mitrailleur de 37 mm Schneider modèle 1930'*. Even this was too late; production and other difficulties at the factory meant that only twenty mountings were delivered by May 1940 and these were allocated to the defence of Paris so saw no action. None of them appears to have survived the war. The gun had an L/45 barrel and was stated to fire at 175 rpm. The complete land mounting weighed 1,340 kg in action.

*37 mm A.N. Mle 1935*:
37 × 218R AMMUNITION.
This gun was designed by the Artillerie Navale at Ruelle and was selected for the ACAD (*Automatic Contre-Avions Double*) requirement to replace the manually-loaded 37 mm Mle 1925 and Mle 1933 AA guns in service with the French Navy. A twin mounting weighing 1,450 kg (with the guns mounted diagonally) was developed but by May 1940 only one mounting was installed for trial purposes on the Patrol Sloop *Amiens* (the mounting is shown on the right of the photo). This reportedly saw action during the Dunkirk evacuation. In view of the Army's difficulties in obtaining a satisfactory medium-calibre AA gun, the Navy proposed the 37 mm A.N. on a single ground mounting.

Few technical details survive but it is known that the six-round ammunition feed was from the side, through an enlarged trunnion axis. Replenishing the ammunition was possible without interrupting firing. The cyclic RoF was around 165 rpm. It seems clear that a 37 × 218R round was used (these are found with the M.35 headstamp) although the stated performance looks high for such a small round – possibly due to the long (L/70) barrel.

D.C.A.N. 25 mm Mle 1935:
25 × 159.5 AMMUNITION.
According to a report by the Mauser engineer Linder following a visit in August 1940, the French Navy had ordered a former Naval officer to design a 25 mm cannon to be fitted to naval aircraft. The work, including the design of the flexible mount, was done at *Société des Moteurs Salmson*. It was said to be 'designed like a Maxim gun', not belt fed but with a magazine drum with a capacity of 33 rounds. The total length of the cannon was 280 cm and the rate of fire was stated to be 280 rpm (another source says 350 rpm). French sources report that ten guns were made, of which three were installed (one to each plane) on flexible mountings to the three Latécoère 523 heavy flying-boats which entered service in 1938.

Further work on the Mle 1935 took place during World War II, including a 200-round capacity belt-fed version using aluminium belt links, plus an AA version with a large drum magazine, intended for submarines.

*Photo from NavWeaps.com*

*25 mm Mle 1937A & Mle 1937B*: 25 × 159.5 AND 25 × 135.5 AMMUNITION.
The Mle 37A was a lighter gun than the Mle 35 to make it more suitable for flexible mountings. It fired down-loaded ammunition, using the same 159.5 mm case and 300 g projectiles as the Mle 35 but with an MV of 775 m/s rather than 825 m/s. RoF was 400 rpm. The Mle 1937B was a shortened version, modified to fire ammunition with a shorter (135.5 mm) cartridge case: this fired a 200 g shell at 875 m/s. RoF was increased to 500 rpm. The guns were drum-fed, with capacities of 33 and 44 rounds respectively.

*25 mm HOTCHKISS AIRCRAFT CANNON*: 25 × 163 AMMUNITION.
Chinn has a chapter on a 25 mm Hotchkiss aircraft cannon, but this does not draw a clear distinction between the aircraft gun and the Mle 1938/39 AA weapon in 25 × 163 calibre; they appear to be variants of the same gun. Work on an aircraft gun started shortly after World War I and it is clear that that the design evolved over many years, gradually increasing in rate of fire and ammunition capacity and also reducing in weight. At the end of 1937 three guns were being built at the Hotchkiss Levallois factory, able to be fitted in a turret, wing-mounted, or as *moteur canon*. The gun was regarded as secret; a US delegation visiting the factory in February 1937 was told that they could not inspect the gun or engage in negotiations to acquire it.

Linder, mentioned in the description of the 25 mm D.C.A.N. above, provides some details: the gun weighed 65 kg empty (another source says 80 kg), 121 kg with magazines and magazine housing; was 228 cm long overall with a 150 cm (L/60) barrel, and fired at 300 rpm (another source says 300–350 rpm). The ammunition feed was complex, consisting of four horizontal box magazines stacked vertically, with a total capacity of 35 rounds. The magazines collectively weighed 15 kg empty or 40 kg filled with 35 rounds. It appeared to use the same 25 × 163 ammunition as the AA gun, although the loading may have differed: while Linder quotes a 250 g projectile at 912 m/s (the

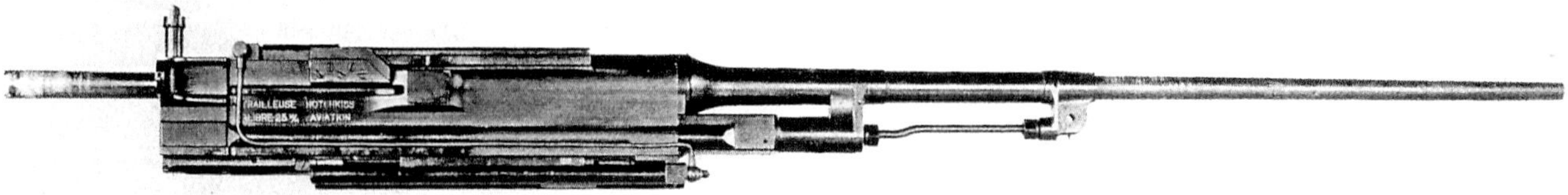

*25 mm Hotchkiss aircraft cannon*

same as the AA gun), Chinn says 300 g at 820 m/s, and another source says 250 g at 850 m/s.

**HOTCHKISS 37 mm**: 37 × 208 AMMUNITION (PROBABLY).
This rather mysterious gun competed with the 37 mm A.N., the 37 mm Schneider and the 40 mm Bofors for the ACAD contract (see introductory notes). This was presumably the same gun as the one being considered during the same period for the Hotchkiss *zénithaux* (zenith), a proposed 37 mm four-barrel naval mounting only capable of high-angle fire, for use against dive bombers. Photos also exist of single and twin land mountings of what is stated to be a 37 mm Hotchkiss sold to Romania, although some sources say that Romania bought the 37 mm Schneider; possibly both were ordered, but numbers delivered of either gun seem to have been small, or possibly fitted to ships, as neither was in Romanian army service in 1941.

*37 mm Hotchkiss twin mounting. Note the second magazine behind the one in use*

The Hotchkiss appears to have been a scaled-up version of the 25 mm Mle 1938/39 AA gun. The usual top-mounted Hotchkiss box magazines were used, with one interesting addition: a second magazine could be installed behind the one feeding the gun, so arranged that when an emptied magazine was removed, the one behind slid down a pair of tracks into the feeding position, thereby minimising the interruption to firing. Presumably this only worked at rather low elevations. In the naval ACAD trials the gun had a cyclic rate of 144 rpm, or 96 rpm for longer bursts allowing for the rapid magazine change. No further details are known, not even the ammunition used, although as far as is known there was only one Hotchkiss round of this calibre from this period; a rimless 37 × 208.

***HOTCHKISS 25 mm Mle 1938/39/40***: 25 × 163 AMMUNITION.
Hotchkiss developed this AA gun in the 1930s for export purposes, but the search for effective close air defence weapons eventually resulted in its adoption by all three French services, in single or twin mountings. The different models were:

- *Modèle 1938*: a tripod-mounted gun intended for Romania, but acquired by France.
- *Modèle 1939*: an improved, heavier tripod.
- *Modèle 1940*: faster-firing and on a fixed mounting.
- *Modèle 1940 jumelée*: twin mobile mounting.

After various modifications, it was adopted by the French Army in June 1938 under the designation *'Matériel de 25 mm CAJ Modèle 1940'* (CAJ = *Contre-Avions, Jumelé* = twin anti-aircraft mount).

Part of an order from Romania (which appears to have had a considerable appetite for French cannon) was seized by the French government and issued to the Army, and an order from the USSR was embargoed. Mountings were also ordered for fitting to some warships, although few if any were installed. Despite all of this effort, it appears that fewer than 250 guns had been delivered by the summer of 1940.

The Imperial Japanese Navy was a far bigger user, having acquired a production licence for the gun and put it into service as the 25 mm Type 96 (indicating adoption in 1936). During World War II it was the IJN's standard automatic AA cannon and was widely fitted to warships in single, twin and triple mountings.

The gun was gas-operated, air-cooled and magazine-fed via 15-round box magazines. It weighed 115 kg with an L/60 barrel. The cyclic RoF was around 200–260 rpm, but the practical rate was about half that.

**HISPANO-SUIZA HS 404**: 20 × 110 HS AMMUNITION.
This is one of the most famous automatic cannon in history, seeing extensive use with the British RAF and also adopted as an aircraft gun by the USN and USAAF. In its modified version, the HS 804, it was widely used post-war and is still in service today.

The HS 404 was designed by Hispano-Suiza's Swiss chief engineer, Marc Birkigt, working at the French branch of the company. Most armament histories suggest that Birkigt started with the Oerlikon Type S and improved it by designing a gas-unlocked delayed-blowback operating system to speed the RoF. However, as noted by Chinn he appears to have been more directly influenced by the Italian designer Scotti's experimental Model 1932; no great surprise, as there were close links between

Scotti and Oerlikon, and between Oerlikon and Hispano-Suiza France (and Scotti's patents had been in the public domain for years). Scotti did start with the Oerlikon and added a gas-unlocked delayed blowback system, albeit using rotary-bolt rather than the flap-locking of the HS 404. Scotti also modified the Oerlikon 20 × 110 cartridge case from rebated-rimmed to rimless, leaving Birkigt only the job of expanding the case at the shoulder. It seems that he did not do this immediately as a drawing from ammunition maker SFM of France shows an 'HS 404' cartridge identical with Scotti's and still with the original sloping shoulder of the Oerlikon; presumably for an early prototype.

The HS 404 was originally introduced into French service as the '20–404' to replace the HS 9 aircraft gun and was in full production by May 1940, equipping all new fighters being built with some being used for airfield AA defence. This was a *moteur canon*, designed to fit between the cylinder banks of a vee-engine and fire through the hollow propeller hub; this accounts for the very long (L/85) barrel, as it needed to extend to the propeller nose, while the ammunition feed had to be behind the engine. The gun was also used in a wing mounting in the Bloch 152 fighters.

The Hispano uses a delayed blowback mechanism; the bolt is locked to the breech at the instant of firing, then unlocked in the same way as a gas-operated gun, after which blowback is used to power the operating cycle. It is therefore similar in principle to the Scotti designs, although the locking mechanism, involving a pair of pivoting flaps, is simpler. The Hispano fired from a rear sear (i.e. an open bolt) and in this form could not be synchronized to fire through the propeller disk. It is not a 'complete gun', in that it does not have a fixed receiver other than the engine it was intended to be mounted on. This caused problems in other installations, particularly in wings, as the supporting structure had to be very rigid. The USA developed special cradles to circumvent this problem; the British adopted a similar expedient for turret mountings.

Like the API blowbacks, the gas-unlocked delayed blowback action did not provide any extraction pull on the fired case, so to prevent the chamber pressure sticking the case to the chamber walls the cartridges were usually lubricated. The rate of fire was initially 700 rpm.

As well as being very long (254 cm) the HS 404 was also heavy, although weight comparisons are difficult due to the modular nature of the gun, so the usually-quoted figure of 50 kg is an approximate minimum.

The HS 404 initially fired from a 60-round drum although smaller magazines were available for flexible mountings. The Swiss developed a 160-round double–drum for fixed installations, although a maximum of only 150 rounds – and much less if the magazine was to be left loaded for any length of time – were normally loaded to avoid problems with deforming the magazine spring.

Development of the HS 404 ground to a halt in its native France with the German invasion. Germany made use of captured weapons, but further development was continued by the UK and USA during World War II (see the British and US sections for more information), and the Swiss branch of Hispano-Suiza adopted the gun thereafter as the HS 804.

*HISPANO-SUIZA HS 405*: 20 × 71 AMMUNITION.

The HS 405 was a significantly shortened and lightened version of the HS 404 intended for mounting in aircraft turrets and flexible mountings. Gun weight was only 25 kg, the length was 176 cm, the calibre length was L/45–50 and the rate of fire was 760 rpm. The cartridge case was shortened also, reducing the MV to 600 m/s. The ammunition feed was presumably by magazine.

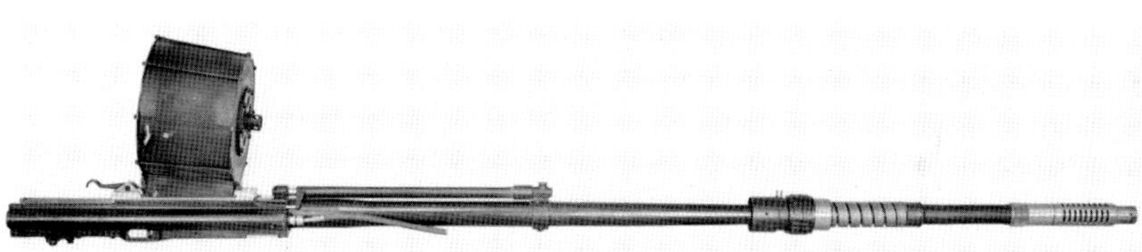

*HS 404 with 60-round drum magazine (Collection Jean-François Legendre)*

*HISPANO-SUIZA HS 406/7*: 23 × 122 AMMUNITION.

These aircraft cannon were scaled-up versions of the HS 404, designed around a new and very powerful 23 × 122 round. The 406 had an L/50–55 barrel, the 407 an L70-/75 one for use as a *moteur canon*. The guns had a protracted development and were still at the experimental stage in May 1940; very little is known about them. It seems that initial French enthusiasm for them faded somewhat, and manufacturing rights were sold to the Russian government at the beginning of 1937. One year later, the Russians started work on developing a 23 mm cartridge of similar power to the 23 × 122; the end result was the 23 × 152B used by the VYa-23 cannon.

## PROJECTS AFTER WORLD WAR II

*HISPANO-SUIZA HS 421/2 AND MAC MLE 1949*: 15 × 113 AND 20 × 96 AMMUNITION.

These guns were developed immediately after World War II to meet the requirements of the *Armée de l'Air* for new fast-firing aircraft guns to replace the HS 404. Both Hispano-Suiza and MAC became involved in designing guns around a common pair of cartridges: 15 × 113 and 20 × 96. The HS gun, about which little is known, was designated HS 421, while the MAC was the Mle 1949. There is also an official drawing of the 20 × 96 round headed as being for the HS 422 (possibly the 15 mm version was the HS 421?), but no further information seems to exist. Some of the characteristics of the MAC gun are known: a weight of 45 kg and a rate of fire of 1,000 rpm. The programme, which was also designated 2CG, was cancelled in favour of the AME 3CGF (see below).

*MG 250*: 20 × 82 AND 15 × 96 AMMUNITION.

This was an AME project for a small revolver cannon (scaled-down from the MG 213C) designed for installation in

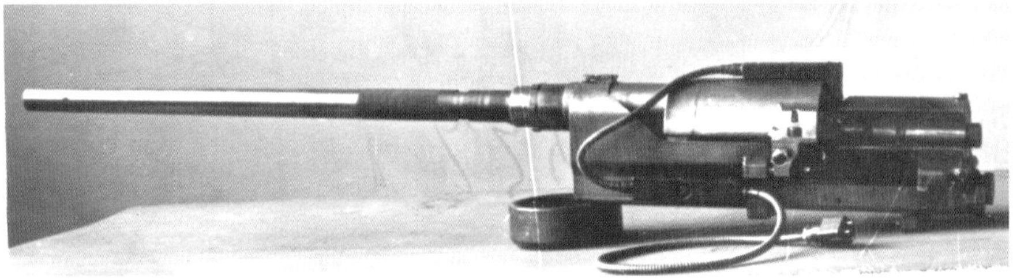

*MG 250 (Collection Jean-François Legendre)*

aircraft turrets, using the electric-primed versions of the Mauser MG 151 ammunition. It was developed in 1946 but was not given a high priority and only a few prototypes were made. The rate of fire was 2,000 rpm.

*MK 253*: 30 × 110B AMMUNITION.
After World War II France was interested in the Rheinmetall-Borsig MK 103 (30 × 184B calibre) and produced a modified version using the designation MK 252. Further work by AME in 1949 produced the MK 253, based on shortened versions of the MK 103 gun and its ammunition. The bottlenecked cartridge case would have held significantly more propellant than the later 30 × 113B DEFA round (see below) so ballistic performance would have been superior. No further details of the gun are known.

*MK 255 / 3CG SERIES*:
30 × 99B, 30 × 116B AND
20 × 122 AMMUNITION.
This was developed from the Mauser MG 213/30 gas-operated five-chamber revolver cannon that was at the prototype stage by 1945. A lot of detailed work by AME was required to produce a service-ready gun, including testing different case lengths (within the same overall round length). The first attempt, which emerged in 1949, was known as the MK 255 and the ammunition had a case 99 mm long, but as the development continued the designations changed to 3CG-F30 and then 3CGF. This work eventually resulted in a service cannon designated the M541 (see below).

In 1952 there was also a 3CG-F20 chambered for a rimless 20 × 122 round which had the same rim diameter as the Oerlikon 20 × 128 case; overall length was 187 mm.

*5CG SERIES*: 20 × 122; 20 × 126(B);
20 × 126; 30 × 97B; 30 × 100B;
AND 30 × 113B AMMUNITION.
These were late 1950s experimental AA guns (originally designated MK 254) using a wide range of ammunition, most of it based on the belted DEFA revolver case.

*7CG SERIES (HISPANO-SUIZA HS 603/4/6)*: 30 × 158,
37 × 158B AMMUNITION.
Developed during the early 1950s, the HS 603 was designed as an aircraft gun using a 30 × 158 cartridge based on a shorter version of the 30 × 170 case. The HS 604 had a lengthened barrel for AA use. A modified version of the gun designated HS 606 was made in two calibres: 30 × 158 and 37 × 158B, the latter cartridge having a narrow belt added to achieve headspace. The HS 606 guns were considered for use in the 7CG programme.

The HS 603 and 606 are believed to have been gas operated and were closely related, both weighing 150 kg and being 225 cm long. They both fired at 750 rpm.

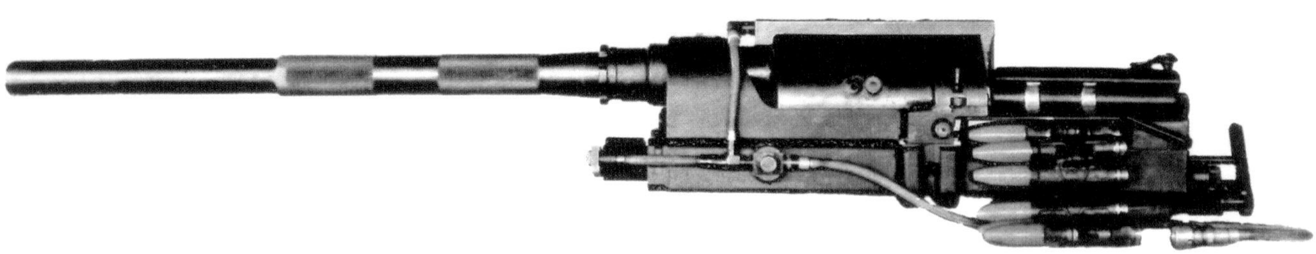

*3CG-F30 (Collection Jean-François Legendre)*

*37 mm HS 606 (Chinn)*

**9CG SERIES**

Very little is known about these cannon, except that they were Gatling type guns initially in 30 mm calibre and later in 25 mm.

***11CG SERIES***: 25 × 133B AMMUNITION.

An aircraft revolver gun based on the usual DEFA cartridge case. An 'oscillating attachment' is referred to, but it is not certain what this means – possibly a steerable mounting?

**DEFA 30M541**: 30 × 97B AMMUNITION.

This was the end result of the 3CG programme described above, which was developed from the Mauser MG 213/30 gas-operated five-chamber revolver cannon that was at the prototype stage by 1945. By the end of this process the case length had changed to 97 mm, and the first version to see service was designated the M541. It was fitted with an L/37 barrel. The rate of fire was 1,200–1,400 rpm and, like all of the other guns based on the MK 213/30, priming was electric and the overall round length was 199 mm. This was soon replaced by the 30M552 cannon.

**DEFA 30M552**: 30 × 113B AMMUNITION.

In 1953 the French agreed to use the same gun chamber dimensions as were established for the Mk 4 version of the British ADEN. This meant that the case lengthened to 113 mm. However, in various other respects such as gas pressure, belt link specification and especially the type of electric priming, the weapon systems were not the same so ammunition was not interchangeable. The gun had an L/47 barrel and fired at 1,100–1,500 rpm.

**GIAT / NEXTER 30M552A**: 30 × 113B AMMUNITION.

This gun was manufactured to replace first generation 30–552 guns when worn out, because 30–553 guns could not be fitted as replacements in the Mirage III series of aircraft due to the different mounting attachments. The 30–551, 552, 552A and 552B were fitted on a removable tray while the 30–553 and 30–554 were attached directly to the airframe. The 552A had a new cradle with all the refinements from the 30–553. They were sold to the French air force and those foreign forces that used Mirage III aircraft.

**GIAT / NEXTER 30M552B SPECIAL 1A/W**: 30 × 113 B AMMUNITION.

This gun is a modified 30–552A using a different electrical circuit to fire 30–550 ammunition fitted with the same insensitive primer (1A/1W) that is used with 30 × 150B ammunition. This gun was used aboard the carrier-based naval *Super-Étendard Modernisé* strike fighter.

**GIAT / NEXTER 30M553**: 30 × 113B AMMUNITION.

The 30M553, which entered service in 1971, is very different gun from the 552 but with the same basic characteristics. It was more strongly built to avoid the jamming problems of the previous gun which enabled it to be fitted directly to the air frame instead of on a removable tray. Its L/47 barrel is fluted and fitted with a decompressor on the muzzle. It is the first gun of the series which could be fitted with an ammunition belt presented either with the hook of the link forward or with the eye forward. This meant that armourers did not need to use different belts for the right-feed and left-feed guns (as in the previous models) which is very important for logistics and servicing in operations. It is fitted with a 6-shot re-cocking system. The 30M553 was also built under licence in Israel and in South Africa, where a 55C530 version was used in its Atlas Cheetah family (upgraded Mirage III).

**GIAT/NEXTER 30M554**: 30 × 113B AMMUNITION.

The final version of the French revolver cannon using this ammunition entered service in 1984. It was modified to increase the rate of fire to 1,800 rpm, by making full use of the five chambers rather than having one always empty. It also has a longer barrel for a higher muzzle velocity, and an automatic 6-shot re-cocking system. Barrel life is said to be 5,000 rounds.

**GIAT/NEXTER 20M621**: 20 × 102 AMMUNITION.

This gun was derived from the experimental FINABEL series (see below), the result of a French Army request in 1957. In the early 1960s the Air Force and Army later combined in the development of the 20M621, but this was protracted and series production did not start until 1973, the French Army finally adopting it in 1986 under the designation CN-MIT M-20 621-F1.

The ammunition chosen was the US electrically-primed 20 × 102 developed for the M39 and M61 aircraft cannon, as this is smaller than the 1950s 20 × 128 Oerlikon and 20 × 139 Hispano-Suiza rounds, enabling the gun to be lighter and more compact, with less recoil. The 20M621 has a gas operated, delayed-blowback action. It fires from a locked breech for single-shot firing and an open breech for bursts.

The L/75 barrel and action recoil into the gun cradle at the same time as two swinging locking pieces are retracted by gas action; the barrel and action are then stopped and the bolt is driven to the rear by residual gas pressure at the same as the muzzle brake exerts a forward pull on the barrel. The gun utilizes a differential or counter-recoil system, firing each round as the barrel and action are moving forwards and thereby further reducing the peak recoil blow. Various recocking devices can be used: electric, hydraulic or mechanical.

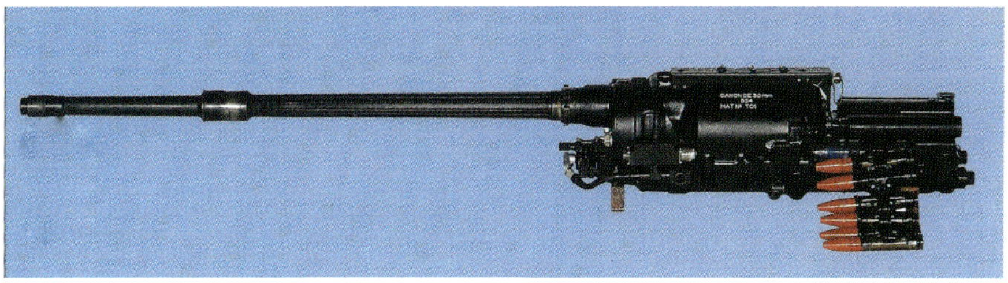

*Nexter 30M544 (30–550 F4) (GIAT)*

*Nexter 20M621 (GIAT)*

The gun has a single belt feed (switchable from one side to the other), weighs 49 kg and fires at 800 rpm. More than 2,000 cannon have been sold and it is available in manual or remotely-controlled vehicle, naval and helicopter mountings, as well as in an aircraft gunpod.

GIAT / NEXTER 20M693:
20 × 139 AMMUNITION.
Development of this gun started in the late 1960s and it is effectively a version of the 20M621 scaled-up to use the longer and more powerful ammunition originally developed for the Hispano-Suiza HS 820. It therefore uses the same gas-operated delayed-blowback operating system with counter-recoil built in. The principal practical differences between the guns are that the 20 × 139 ammunition uses percussion priming (therefore needs no electrical supply to fire) and the 20M693 features a dual belt feed so that two ammunition types can be ready to fire. It is also longer (with an L/98 barrel) and heavier at 71 kg, and it fires at around 750 rpm.

The first prototype appeared in 1969 under the designation Canon Mitrailleur 20 F-1, until the M number designation system came into use. The gun was adopted by France in 1973 under the designation CN-MIT 20 F-1; a couple of years later the F-2 version was introduced, with a modified dual feed system. As with the 20M621, the 20M693 is available in a variety of land, naval and helicopter mountings, both manually and remotely operated. Some 8,000 have been made by EFAB; it is also manufactured in South Africa by Denel Land Systems as the G1–2.

*GIAT 30M741/30M761 AND 30M763*:
30 × 150B AMMUNITION.
These were the initial projects aimed at producing a high-performance revolver gun for the next generation of combat aircraft, which eventually resulted in the 30M791 (see below). The M741 appears to have been a twin gun, featuring two barrels and two five-chamber cylinders. The M761 differed in having a single, ten-chamber cylinder feeding the twin barrels, while the M763 was a single-barrel version of this. The twin-barrel guns fired at up to 3,000 rpm.

GIAT/NEXTER 30M781:
30 × 113B AMMUNITION.
This weapon was originally developed for the Franco-German *Tigre* attack helicopter programme, specifically the French HAP variant (*Hélicoptère d'Appui Protection* = 'Support and Escort Helicopter') which carries the cannon in a THL 30 chin turret. Development started in 1978 but, due to delays with the helicopter project, flight trials did not start until 1996. The 30M781 is marketed in various vehicle and naval mountings, but so far the *Tigre* remains its only application. It is a direct competitor for the US M230 / M230LF.

The 30M781 was designed around the NATO 30 × 113B ammunition. In view of the proximity of the *Tigre* magazine to some powerful electronics, the version of the ammunition used is the same as that developed for the *Super-Étendard Modernisé* naval fighter (SEM), using an electric primer which is less sensitive to radiation. The operating mechanism is

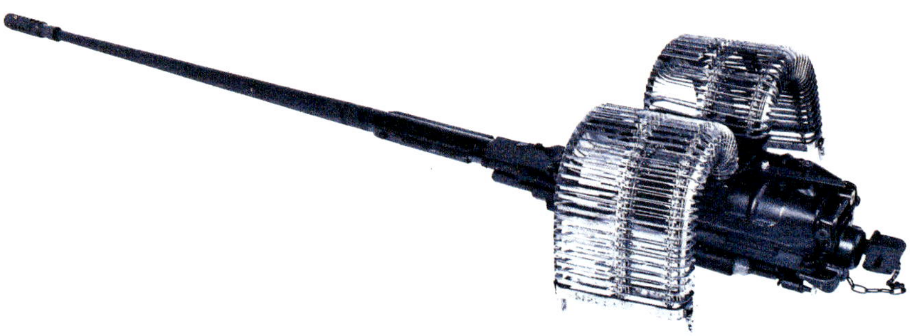

*Nexter 20M693 with dual belt feed (GIAT)*

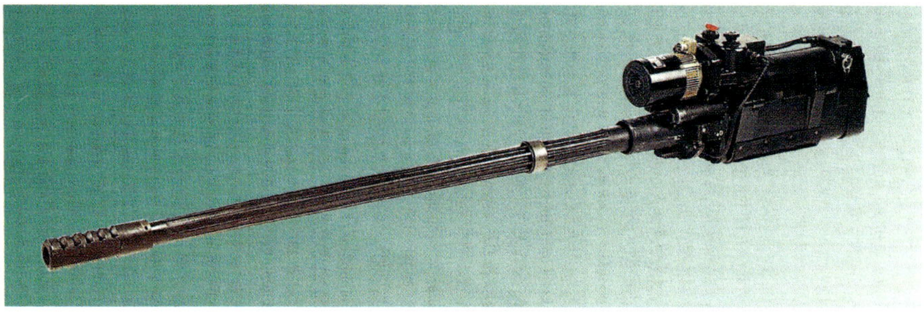

*Nexter 30M781 (GIAT)*

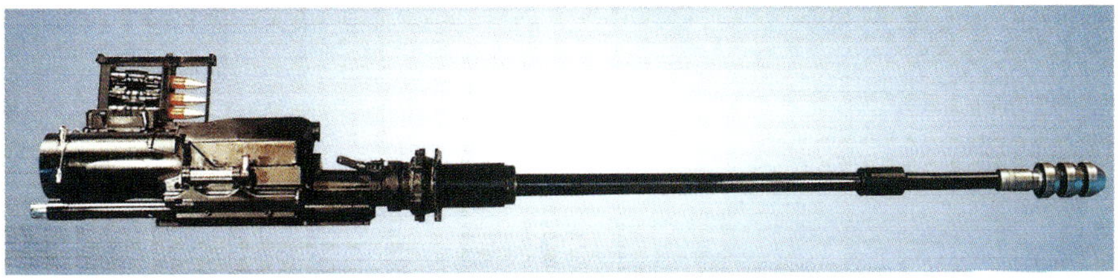

*30M791 (GIAT)*

electrically-powered via a camshaft which has a spiral cam groove that engages with a lug on the bolt. As the shaft revolves, that drives the bolt to and fro. The camshaft also operates the feed mechanism to sychronize the ammunition feed with the bolt movements. A hang-fire safety device is provided. It has a single belt feed and a rate of fire which is variable up to 720 rpm. Single shots or timed bursts can be fired. The gun has an L/47 barrel and weighs 65 kg.

*GIAT/NEXTER 30M791*: 30 × 150B AMMUNITION.

This gun was designed to equip the Dassault *Rafale* jet fighter, the intention being to provide significant improvements in both the muzzle velocity and the rate of fire compared with the 30M554. After initial experiments with the 30M741, 30M761 and 30M763 (see above), the 30M791 design was developed. A higher muzzle velocity was achieved by developing a new, longer, cartridge case to hold more propellant, combined with a longer gun barrel. A higher rate of fire resulted from increasing the number of chambers in the cylinder from five to seven. These changes have put up the size and weight of the gun compared with the 30M554 (120 kg cf 85 kg, and 240 cm cf 201 cm).

Other changes from the 30M554 include automatic recocking and a belt feed (left-hand only) in which the links are stripped before the rounds enter the cannon. Electrical priming is retained. The gun can be set to fire a 0.5 or 1.0 second burst (21 or 42 rounds) or unlimited fire. The rate of fire is up to 2,500 rpm but can alternatively be set to fire at 1,500 rpm.

The first airborne firing trials took place in 1991 in a gunpod, with the 30M791 being cleared for operational use on the Rafale M in 2002.

*GIAT 30M792*: 30 × 150B AMMUNITION.

This was a project for an externally-powered gun firing the same ammunition family as the 30M791; effectively, it was a high-velocity 30M781 intended for helicopter, vehicle and naval applications. This was developed as a result of the Sydia programme, which also investigated 'automechanical' (externally powered) weapons in calibres including 27 × 145B and 35 × 228. The project was cancelled when it was decided that the 30 mm AP ammunition did not have a sufficient performance margin for future AFV use.

*GIAT/NEXTER 25M811*: 25 × 137 AMMUNITION.

Development of the 25M811 began as a private venture by GIAT and the gun was first shown in 1983. It uses the same electrically-driven cam-operated system as the 30M781 (see above), but adapted to the NATO standard 25 × 137 ammunition (which is percussion rather than electric primed) and with a dual belt feed. A range of ammunition in this calibre had already been developed by Manurhin. The rate of fire is 400 or 650 rpm, appropriate for the intended use of the gun in arming AFVs and naval craft, in applications in which accuracy and persistence (in terms of both ammunition use and reduced barrel heating) are more important than a high RoF. The barrel is fluted externally to provide an increased surface area for cooling without excessive weight. The bare gun weighs 93 kg, or 109 kg including the cradle and electric motor, an L/86 barrel is used, and the gun is claimed to be able to achieve a 1.5 m group size at 1,500 m. The gun is in French service on light AFVs and is also made in Turkey for their army.

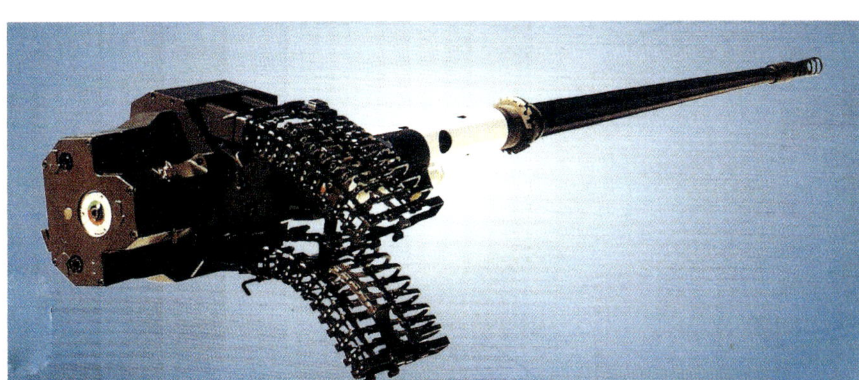

*25M811 (GIAT)*

*GIAT 45M911*: 45 × 305 CASED-TELESCOPED AMMUNITION.

The origin of the Cased Telescoped Armament System (CTAS) project was a proposal to develop the main armament of the future French direct-support vehicle, the *Véhicule d'Appui Direct* (VAD). After experiments in the late 1980s with the XT2000, a design utilising a rotating cylinder with three open chambers, allowing sideways loading of the 35 mm cased-telescoped cartridges, DGA/GIAT contacted Honeywell of the USA who had developed a gun using similar ammunition under the COMVAT programme (Combat Vehicle Armament Technology). This resulted in a decision to develop 45 mm Case-Telescoped Ammunition (CTA), also known as 45 CTA. GIAT and Honeywell (from 1992, Alliant) agreed on a composite cylindrical cartridge measuring 70 mm in diameter and 305 mm long, with a calibre of 45 mm.

A fully functioning automatic prototype, designated the 45M911, was developed by Giat Industries under a

adoption of the MG-FF, a modified version of the Oerlikon FF made by Ikaria (which was initially Oerlikon's German subsidiary). As World War II approached and the Versailles restrictions were increasingly ignored, Rheinmetall-Borsig (as it became in 1936, following the acquisition of August Borsig GmbH in 1933) openly took back the developments previously attributed to Solothurn and continued with them.

A separate effort to develop a 20 mm cannon was made in the late 1920s by Heinrich Lübbe, who had worked at the *Fokker Flugzeugwaffenfabrik* in Berlin during World War I. This was gas-operated, weighed 48.5 kg, used a lightweight skeleton-drum feed and was said by its inventor to fire at 360 rpm and to launch a 124 g projectile at 1,000 m/s. The ammunition it used is unknown, but must have been a considerable size if the MV claim was correct.

During World War II there were two major German manufacturers of automatic cannon: Rheinmetall-Borsig was the established producer, but relative newcomer Mauser provided strong competition, mainly in 20 mm calibre (forays into larger calibres being mostly experimental). Krupp and Krieghoff were also involved, but without much success (Krieghoff put a lot of abortive effort into a series of 20 mm – and eventually 30 mm – cannon). Rheinmetall-Borsig, which was taken over by the state-owned conglomerate *Reichswerke AG Hermann Göring* in 1941, made service automatic cannon in 20 mm, 30 mm, 37 mm, and 50 mm calibres.

At this time automatic weapons of up to and including 20 mm were known as machine guns (*Maschinengewehr*) with larger calibres being known as machine cannon (*Maschinenkanonen*), leading to the MG and MK prefixes. In addition, designations based on function were used for particular applications, whether or not the weapons were automatic. A gun intended for the anti-aircraft role was called *Fliegerabwehrkanone* (sometimes given as *Flugzeugabwehrkanone* or *Flugabwehrkanone*); for the anti-tank role *Panzerabwehrkanone* (renamed *Panzerjägerkanone* during the war); and for mounting in armoured fighting vehicles *Kampfwagenkanone*; Flak, Pak and KwK for short. The BK (*Bordkanone*) designation was used for very large calibre (37+ mm) airborne cannon (although that term is now applied to the 27 mm Mauser BK 27 gun).

After the adoption of the 20 mm MG-FF, a new designation system was introduced for aircraft guns. The first number(s) of the designation indicated the calibre, the next the gun model (e.g. the MG 131 was the first model of 13 mm gun). Other types of automatic cannon used designations that were generally linked to the year they emerged, plus the calibre (e.g. 5 cm Flak 41). In 1942 the system was changed again so that the first of the three numbers indicated the design firm (1=Rheinmetall-Borsig, 2=Mauser, 3=Krieghoff, 4=Krupp) with the next two identifying the model of gun. Gustloff also had some involvement in cannon design, but not enough to merit its own number. There are a couple of minor mysteries over the 30 mm MK 101 designation, which complies with the post-1942 system; when initially used in the early part of the war, it should strictly have been called the MK 301, but this does not appear to have been the case. To add to the confusion, an example of the gun captured in 1940 was reportedly stamped MG 101; however, that was the designation of the 6-round magazine for the MK 101, which possibly caused some confusion. Prototypes were sometimes given a *Gerät* (device or equipment) number, which appears to be random.

There are many unusual World War II experimental guns described in this section and listed in the gun data table, for which most of the information comes from the *Unterlüß* files. These are records resulting from British interviews with German technical staff immediately after the war, held at the Unterlüss site which was (and still is) a Rheinmetall establishment. Other information comes from the HAC (Halstead Exploiting Centre) at Fort Halstead, which had a similar task.

Some unconventional experimental German light Flak guns were developed before (and possibly during) World War II. They consisted of eight barrels, closely packed together either in one horizontal row or in two vertical rows of four. The guns used salvo firing, apparently with a common breech block for all barrels, and therefore resemble the current Spanish Meroka. However, it appears that these guns were recoil-operated. They were known as *Salvenmaschinenkanonen*. An Allied document from 1945 describes five different multi-barrel weapons (*Mehrlaufwaffen*), most of which were in HMG calibres, except for the 2 cm SMK V1 and V2 from Rheinmetall-Borsig.

The most significant development of the late war years was the Mauser MG 213C, the first automatic revolver

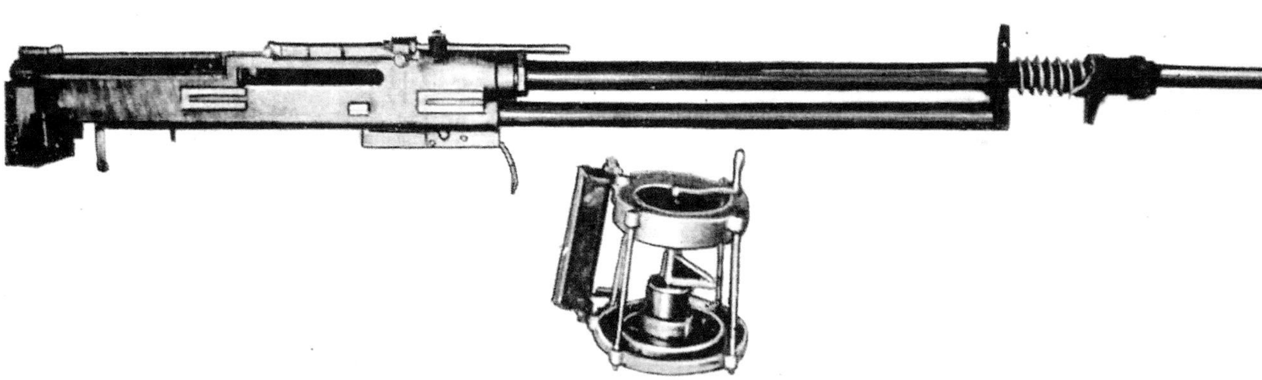

*Lübbe gas-operated 20 mm gun with skeleton-type drum*

cannon which used several chambers to achieve a very high rate of fire. This design principle was developed (with help from former Mauser engineers) in various forms in the 1950s, in France (DEFA), the UK (ADEN), Switzerland (Oerlikon), the USA (M39) and the USSR (NN-30). The most recent revolver cannon to enter service is the Nexter 30M791, installed in the Dassault *Rafale* fighter.

After manufacture of automatic cannon was resumed in the 1950s, Rheinmetall GmbH (as it was renamed in 1956 – further name-changes have taken place since) and Mauser both produced successful designs, Rheinmetall in 20 mm calibre (the Rh 202), Mauser in 30 mm (MK 30F) and later 27 mm (BK 27 revolver cannon).

As usual, for every successful design there were several which failed to be adopted. Mauser's Model B in 20/23 mm and Model E in 25 mm calibre (effectively scaled-down MK 30) were conventional designs offered for some years. Rheinmetall developed the Rh 503 AFV gun in two calibres: 35 mm and 50 mm, but the end of the Cold War effectively brought this to a halt. In contrast, Mauser's RMK series in 30 mm and 35 mm calibres was anything but conventional, combining a recoilless reverse-loading revolver cannon with caseless telescoped ammunition (see the Unconventional Systems section for details). History shows that such radical new designs are rarely rewarded with success.

The rivalry between the two German companies was ended in 1995 when Rheinmetall acquired a majority stake in Mauser. Rheinmetall's dominance of the European cannon market (France's Nexter Systems being the only significant rival) was completed in 1999 when the company acquired a majority stake in the Swiss company Oerlikon Contraves. Since then, Rheinmetall Defence has introduced a new modular multi-calibre externally-powered design, initially in 30 mm then 35 mm calibres, and designated WOTAN.

## WORLD WAR I PROJECTS

**3,7 cm KRUPP 'SOCKELFLAK'**: 37 × 101SR AMMUNITION.

This was a short-barrelled (L/14.5) gun originally intended for installing on World War I airships to defend them against fighters. It was subsequently adapted as a ground-based AA gun, entering service in this role in 1917 and with some 150 surviving to the end of the war, after which some were distributed abroad (a few were retained for airfield defence in World War II, firing tracers to distract attacking planes).

*3,7 cm 'Sockelflak' (Szilas / Wikimedia Commons / Public Domain)*

The term '*Sockelflak*' is a generic one simply describing the type of mounting and the purpose, and is also applied to other calibre AA guns of the period: the gun should properly be called the *3,7 cm Luftschiff Flak* in view of its intended purpose. While this Krupp gun used a Maxim-type short-recoil toggle lock action it was not belt-fed, instead having a 10-round magazine feed mounted on top of the gun; the rate of fire was around 120 rpm. The weight was 260 kg including the tripod mounting.

**2 cm BECKER**: 20 × 70RB AMMUNITION.

This was the original Advanced Primer Ignition (API) blowback design, reportedly developed by the Cönders brothers but patented by Dr. Rheinhold Becker of the *Stahlwerke Becker AG* in the Rhineland in September 1913. At first, the new weapon faced some resistance from the German armament authorities because of the lack of relevant experience of the company, but the German Army Air Service showed interest and firing tests were ordered in mid-1915. These tests were unsatisfactory but the concept showed promise so technical assistance to Becker was provided by the nearby Spandau Arsenal in order to develop a satisfactory weapon.

In June 1916 Becker received a production contract for 120 cannon, while Spandau continued developing the gun. This was followed-up by field trials of five cannon in August of that year, which showed good results. There were still ammunition feed problems which were not resolved until the introduction of the Type 2 Becker in November 1916, which used a slightly different shape of cartridge case. In 1917 it was agreed that Becker would build guns for air service, while Spandau and MAN-Nürnberg would make them for the army, mainly for anti-tank use. According to available information, the guns built by Spandau were heavier – 42 cf 35 kg – and slower-firing – 250 cf 300 rpm – than the Type 2. It is not known whether or not these guns received a different designation ('Type 3' seems to be speculative).

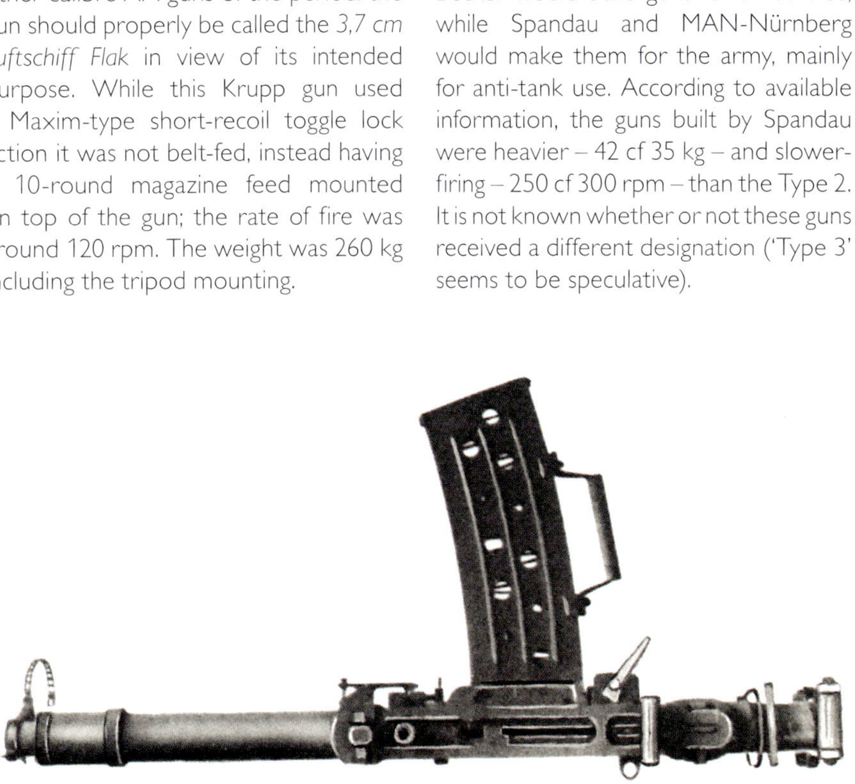

*2 cm Becker (Chinn)*

By May 1918 the development of the gun was considered essentially complete, although work continued on the feed arrangements with various drum and belt feeds being tested. Box magazines mounted on top of the gun were the only ones to see service; the standard 15-round box weighed 5 kg loaded, and a 10-round box was also available.

It is unclear how many Beckers were made. Some 362 were discovered by the Allies after the war, of which approximately one-third were supplied to the Army for anti-aircraft and anti-tank use. However, 428 aircraft Beckers were reportedly made by MAN alone (despite the earlier agreement). Becker's own production was very small until late in 1918; however, 111 cannon were delivered in September of that year. In 1921 the rights to produce and develop the Becker API blowback were transferred to the company's subsidiary, SEMAG of Switzerland.

**2 cm EHRHARDT FLZK**: 20 × 70RB (E) AMMUNITION.

The Ehrhardt FlzK (*Flugzeugkanone*, or aircraft cannon) was a Rheinmetall development which fired a 20 × 70RB cartridge (very similar to the Type 2 Becker's and often mistaken for it). It employed a scaled-up version of the Dreyse short-recoil design, introduced in 1912 in an infantry machine gun of that name. This used a pivoting block lock designed by Louis Schmeisser, a famous gun designer who ironically is best remembered in the popular name of a sub-machine gun with which he was not in fact involved. His mechanism has the bolt supported by a locking piece that pivots downwards under the control of cams in the receiver, releasing the bolt to continue rearwards.

The Ehrhardt weighed 36 kg, had an L/50 barrel and fired at 250–300 rpm from a side-mounted box magazine. The locked-breech mechanism was considered superior to the Becker's by the APK (*Artillerie Prüfungs Kommission* = Artillery Test Commission) but it did produce more recoil and vibration, which led to many stoppages.

The Ehrhardt did make it into production and at least fifty-one were built (one surviving example is stamped '51'), but it appears to have been too late to see service. After successful efforts

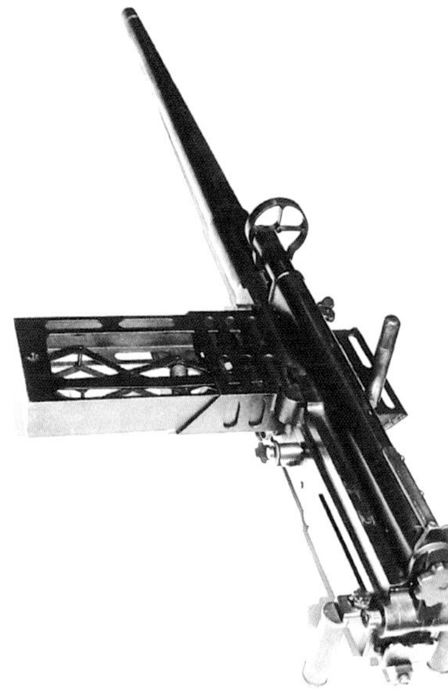

*2 cm Ehrhardt FlzK*

to conceal it from the Inter-Allied Control Commission – the Allies apparently learned of its existence but were unable to locate any guns – the design was further developed in the 1920s into the ST-5 and then ST-52 before further development was (at least nominally) transferred to Solothurn.

## INTERWAR AND WORLD WAR II CANNON

**2 cm S5–100**: 20 × 138B AMMUNITION

The next stage of development following on from the ST-52 was the S5–100 of the early 1930s, in which the short-recoil pivoting or tilting lock design was enlarged to take the powerful 20 × 138B cartridge popularly if erroneously known as the 'Long Solothurn'.

The mountings were designated the S9–100 (pedestal). S9–200 (field – for AA/AT), and S9–300 (twin).

An improved version of the S5–100 was the S5–106 of 1933, two examples of which were bought by the US Army in 1937/8. Rheinmetall further developed this into the Flak 30 service weapon described below. A variant of the S5–100 was the T5–150 intended as an AFV gun: this led to the KwK 30, also described below.

**2 CM FLAK 30, C/30 AND KwK 30**: 20 × 138B AMMUNITION.

These were Rheinmetall designs which formed the bulk of the *Luftwaffe's* light Flak at the start of World War II, plus the standard armament of light tanks and armoured cars respectively. The Flak 30 and its naval version, the C/30, were developed from the S5–100 series, the KwK 30 from the T5–150, a variant of the S5–100. All used the Dreyse pivoting-lock short-recoil design and were fed by side-loading box magazines holding 10 (KwK) or 20 (Flak) rounds. Barrel lengths were L/65 for the Flak and L/55 for the KwK. The guns weighed 63–64 kg and fired at 280 rpm.

The designs were subsequently modified by Mauser to improve the rate of fire, leading to the Flak 38, C/38 and KwK 38 (see below).

*2 cm Flak 30 (Bundesarchiv, via Wikimedia Commons, Bild 101I-301–1953–24 / Kurth / CC-BY-SA 3.0)*

*3,7 cm Flak 36/37 in Italy, 1944 (Bundesarchiv via Wikipedia Commons, Bild 101I-478–2162–26 / Coernig / CC-BY-SA 3.0)*

*BK 3,7 cannon (without fairing) under the wing of a Ju-87G*

**3,7 cm FLAK 18, 36, 37; M36; BK 3,7**: 37 × 264B AMMUNITION.
This series formed the bulk of the *Luftwaffe's* medium-calibre automatic AA guns for the duration of World War II. The prototype of this series was designated the Rheinmetall ST 10, which was nominally handed over to Solothurn for further development as the S10–100, before being formally returned to Rheinmetall-Borsig. The mechanism used a short-recoil rotating-head lock with the cocking handle, attached to the front of the bolt, riding in a long slot in the receiver, which curved sharply as it approached the forward end of its movement, rotating the bolt head locking lugs.

The guns in this series were basically the same, with L/57 barrels, weighing 278–295 kg and firing at 160 rpm. The differences between the Flak 18, 36 and 37 mainly involving the mountings and sights: the Flak 18 had a cumbersome four-wheel mounting and was only made in small numbers, from around 1935, before being replaced by the Flak 36 which had a much improved two-wheel mounting. The Flak 37 differed in being fitted with a new sight, the *Flakvisier 37*, which incorporated a computer with a clockwork drive. Mounting weight in action varied from 1,550 to 1,750 kg. The BK 3,7 cm was adapted to be carried by aircraft, primarily in the anti-tank role.

**2 cm MG C/30L / MG 102**: 20 × 138B AMMUNITION.
This Rheinmetall aircraft gun was intended for engine mounting so the L/90 barrel was extended to reach from behind the engine (where the magazine was located) right to the front of the propeller hub. This allowed it to develop the high muzzle velocity of 950 m/s. The design was derived from the S5–100 series so utilized the same pivoting-lock action as the Flak 30 and KwK 30.

The gun used a huge 100-round drum, unusually mounted underneath the breech, but the RoF was only 300–350 rpm and, following combat testing in the Spanish Civil War on board a prototype He 112, it was not considered to be successful. It was later redesignated MG 102, but not many examples were made, and they ended their lives as Flak guns (with shortened barrels) on single manual mountings.

**2 cm MG 204**: 20 × 108 / 20 × 105B / 20 × 105 AMMUNITION.
Starting in 1934, Rheinmetall-Borsig developed a series of 20 mm cannon for aircraft use, firing smaller cartridges than the 20 × 138B. They competed unsuccessfully with the MG-FF and MG 151 and were made only in small numbers, a few MG 204 seeing service with the *Küstenfliegergruppen* (coastal air service), which had a low priority for equipment so used anything available (including captured 20 mm Hispano HS 404).

The initial prototype of this little-known series was the ST 11, this was followed in the development chain by the S12–100, and then the L or Lb 201a, Lb 202, Lb 204, and finally the MG 204 (the meaning of L or Lb is not clear, but

*The action of the MG C/30L showing the 100-round drum underneath (internet)*

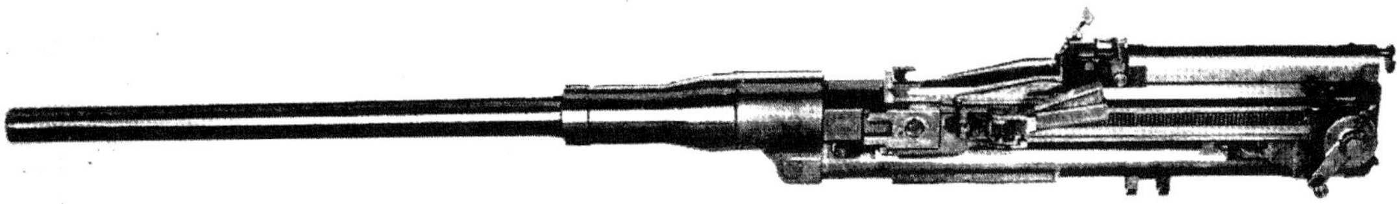

*ST 11 (top) and MG 204 on test stand (Courtesy Evgeny Aranov)*

the most favoured explanation is that they were designed by Heinrich Lübbe, who had previously developed his own gas-operated 20 mm gun as mentioned in the introduction to the German section). All used a short-recoil mechanism with a rotating lock: the Lb 201a used a rotating barrel, while the others used a rotating sleeve to lock the bolt and the barrel together (known as the Stange or Solothurn lock). According to most sources they all weighed around 55 kg (one says 38 kg) and were fed by a 20-round double or saddle-drum for use in flexible mountings (see the ammunition feed section of the chapter on Gun Design). L/45 barrels appeared to be standard. A 100-round drum was also developed for fixed use but not employed, and a subsequent belt-fed MG 204G (*Gurt* = belt) also failed to be adopted. The rate of fire gradually increased from 280 rpm in the ST 11 and Lb 201a, to 350 rpm in the Lb 202, and 400 rpm in the Lb 204. The MG 204 is quoted as firing at around 400–500 rpm.

One curiosity is that three different cartridges were used: the Lb 201a and Lb 202 used 20 × 108 ammunition which was similar in dimensions to the 20 × 110 Hispano HS 404 round; the Lb 204 used a belted 20 × 105B round (sometimes known as the 'Short Solothurn' as it was mainly used in the Solothurn S18–100 series anti-tank rifles); while the MG 204 reverted to a rimless 20 × 105. The development path is obscure but it is believed that the final reversion to a rimless case was prompted by the belted case interfering with smooth feeding from the magazine.

A version of the S12–100 listed in the Unterlüss files is the T12–120, apparently intended for Japan and using MG 204 ammunition. It weighed 43 kg and fired at 300 rpm.

### 3,7 cm FLAK SK C/36: 37 × 250R AMMUNITION.

This almost unknown gun was the *Kriegsmarine's* first attempt to acquire an automatic 37 mm AA cannon. Instead of using the 37 × 264B Flak ammunition (as used in the Flak 18, Flak 36, Flak 37 and Flak M43), Rheinmetall-Borsig designed the gun around a version of the Army 37 mm cartridge used in manually-loaded tank and anti-tank guns. This rimmed cartridge case was modified by giving it a thicker rim, presumably to strengthen it in view of the more violent extraction forces in an automatic mechanism. The HE shells it fired were the same as those used in the Flak 18+ ammunition.

The gun bore a close resemblance to the Bofors guns, including the boxy receiver and a gravity feed above the mechanism taking 5-round clips, and was presumably based on these. It was fitted to a C/36 single mounting. Few technical details are available, beyond a gun weight of 240 kg (930–1,040 kg including the three-axis mounting) and a rate of fire of up to 180 rpm.

It seems that not many C/36 were made, being deployed on only a few vessels, according to a *Handbuch für Admiralstabsoffiziere* which lists the *Aviso Hela*, the two Type 1A submarines U-25 and U-26 (and possibly some Type IX), torpedo boats and fishing protection vessels.

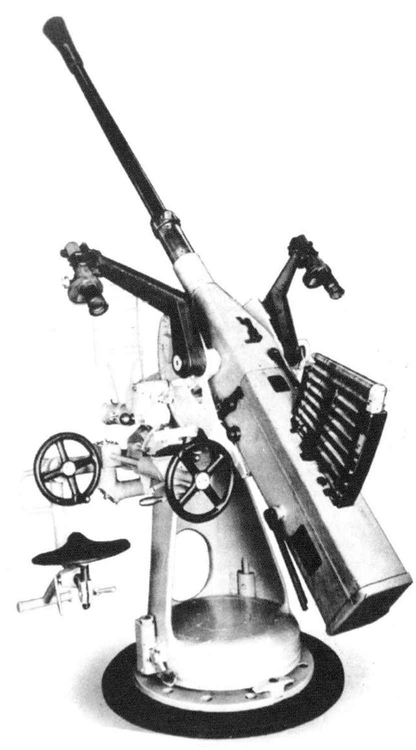

*3,7 cm SK C/36 on C/36 mounting (copyright Rheinmetall Air Defence AG)*

The *Kriegsmarine* had another try some years later, resulting in the 3,7 cm M42 (see below) which used the same ammunition and general design, retaining the boxy receiver and the vertical gravity

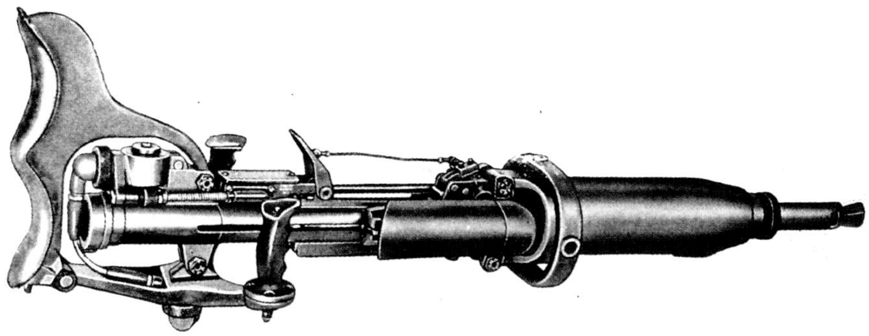

*MG-FF in its flexible mount form (Chinn)*

feed. It is evident that the M42 was an improved SK C/36, and the first mounting used by the M42 was the C/36.

Confusion over the German 37 mm naval AA guns and their designations is widespread, and not only over whether the C/36 term is being used to describe the gun or the mounting (if the latter, the gun could be either the SK C/36 or the M42, as they are hard to tell apart). Also, it is often stated that the SK C/36 is nothing more than a marine version of the *Luftwaffe's* Flak 36, when the two guns are completely different in their design as well as in the ammunition they fired. The situation is even more confused as some Flak 18/36/37 guns were indeed fitted to ships as a quick way of boosting their AA defences, so alertness is necessary in determining which guns are actually being referred to.

**2 cm MG-FF AND MG-FFM:**
20 × 80RB AMMUNITION.
These were developed from the Swiss Oerlikon FF and are therefore in the direct line of descent from the Becker of World War I. The *Luftwaffe* adopted the FF with some modifications (including a longer cartridge case); it was made by Oerlikon subsidiary Ikaria Werke Berlin as the MG-FF (*Flügel Fest* = wing mounted). In mid-1940 the *Luftwaffe* introduced the *Minengeschoss* or *M-Geschoss* (mine shell) HE shell. This thin-walled high-capacity shell was significantly lighter than standard and the resulting reduced recoil required a rebalancing of the gun's bolt weight and return spring strength to operate properly. The revised gun was called the MG-FFM. Ammunition for the MG-FF and MG-FFM was not interchangeable (although there was no obvious difference between them) and most FFs were probably converted to FFM standard; later in the war the terms MG-FF and MG-FFM seem to have been used interchangeably. These guns were used extensively by the *Luftwaffe* but most were eventually replaced by the MG 151/20.

The MG-FF used the Becker's API blowback operating system; the Germans referred to this as *Massenverriegelter Verschluss* (mass-locked). The gun weighed between 24 and 28 kg, had an L/40 barrel, and fired at 520–540 rpm. Almost all of these guns used the standard 60-round drum, although late in the war 30 and 90-round drums were introduced, and a few night fighters carried the GZ-1 electrically-powered belt feed capable of dealing with up to 2,000 rounds. The 90-round drum (used on some Fw 190 fighters) was a clever piece of design, as it was more or less the same size as the standard 60 round one so could be fitted in its place. Weights of the 60 and 90 round drums were respectively 8.2 and 12.0 kg (empty); 20.3 and 30.2 kg full.

One report indicates that at least one attempt was made to substantially upgrade the MG-FF. The 'MG-FF Ausf.L' was modified to use ammunition loaded to a c.20% higher chamber pressure, which in conjunction with a c.50% longer barrel increased the muzzle velocity to 820 m/s. Weight went up to 32 kg. This only reached the prototype stage, and was presumably dropped because the MG 151/20 was a considerably better gun.

**3 cm MK 101**: 30 × 184B
AMMUNITION (PERCUSSION PRIMING).
This Rheinmetall-Borsig aircraft gun used a short-recoil mechanism with Stange's rotary locking sleeve, and fired from a closed bolt. The action was said to be similar to that of the S18–1000. It was magazine fed with a pneumatic charging system, and is stated to have used 6-round box or 30-round drum magazines, although it should be noted that

*The MG-FF was relatively compact and light for a 20 mm cannon (courtesy Tre Colvin)*

*Rheinmetall-Borsig 3 cm MK 101 (Wikimedia Commons / Public Domain)*

a British report concerning one found in a crashed Messerschmitt Bf 110C-6 in August 1940 stated that it had 10-round magazines (changed by the observer); presumably an error. This same report also identified it from stampings as the 'MG 101', but as previously noted that was the designation for the 6-round magazine.

The gun weighed 139 kg empty; 152 kg with a loaded 6-round box magazine; and 185 kg with a loaded 30 round drum (which weighed 18 or 20.5 kg empty; sources differ). It had an L/52 barrel and the RoF was 230–260 rpm.

**2 cm FLAK 38, C/38 AND KwK 38**: 20 × 138B AMMUNITION.

Originally designated the Mauser Flak 35, they were the result of a project by Mauser to increase the rate of fire of Rheinmetall-Borsig's Flak 30, C/30 and KwK 30. The mechanism remained short-recoil but used a rotating bolt-head lock; the same box magazines and barrel lengths were retained. The weights were 56–71 kg, and the cyclic RoF 420–480 rpm (the practical rate being c.220 rpm). The Flak 38 entered production in 1939 followed by the KwK 38. The Mauser guns gradually took over from the Rheinmetall ones during the war.

One particularly effective application of the weapon was in a four-gun mounting: the *Flakvierling*. This was probably the most formidable close-range anti-aircraft system of the war. Naval mountings carrying the Flak 38 were a shipboard version of the four-barrel *Flakvierling* weighing 2,150 kg and designated the *Vierling L/38*, with the *Vierling C38/43* being a stabilized version. Two more specialized mountings were the LM44U powered twin mounting for U-boats, which weighed 3,600 kg, and the L41 single mounting for S-boats at 500 kg.

A lightweight version of the Flak 38 was fielded by mountain troops from around 1942: the *Gebirgsflak* 38. The gun was identical, only the mounting was lighter.

**2 cm MG 151/20**: 20 × 82 AMMUNITION.

Despite all of the activity by Rheinmetall-Borsig, it was their great rival Mauser who produced the MG 151, the most successful German 20 mm aircraft cannon (and arguably the best aircraft gun of World War II to have seen significant service). This was originally developed from 1935 to use high-velocity 15 mm ammunition, but the MG 151/20, a modified version with the barrel shortened to L/55 and increased in bore to fire 20 mm ammunition, entered production in 1940 and saw far more use. About 29,500 of these weapons had been produced by 1943. The cartridge primarily used *M-Geschoss* shells from the MG-FFM for air combat use, although other projectile types were also loaded. The case was only slightly longer than that of the MG-FF but the larger diameter permitted a 40% increase in propellant.

The gun used Mauser's usual short-recoil action with a rotating bolt-head lock and was belt-fed; in some underwing installations, the belt was coiled, with the projectiles pointing inwards, inside a flat pan magazine. The gun weighed 42 kg including not just the integral belt feed but also the electrical charging device and electric trigger, without which the gun weighed 36–37 kg. It had a cyclic rate of 700 rpm.

The MG 151 was available in percussion and electric-primed (MG 151E) forms and both remained in service throughout the war, the latter being used in synchronized installations. The ammunition was not interchangeable so care had to be taken to ensure that no errors occurred. For that reason, if an aircraft had both synchronized and unsynchronized installations of the gun, all of them were of the MG 151E version (the most common example

*2 cm Flak 38 (Bundesarchiv via Wikimedia Commons, Bild 101I-219-0597-15 / CC-BY-SA 3.0)*

*2 cm Flakvierling (Bundesarchiv, via Wikimedia Commons, Bild 183-J08339 / Schwahn / CC-BY-SA 3.0)*

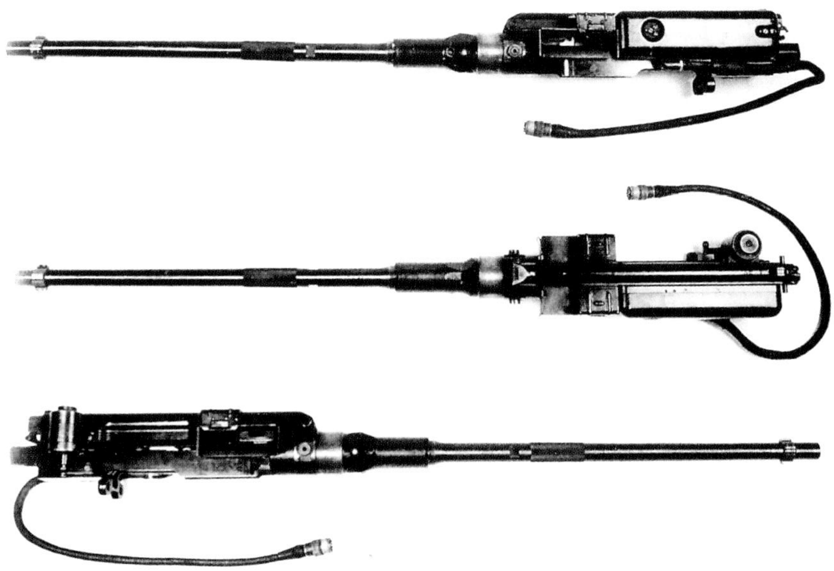

MG 151/20E (internet)

being some Fw 190 variants with wing-root synchronized and outer-wing unsynchronized guns).

When the first examples of this gun were captured the British were reportedly interested in the possibility of copying it, but decided that the precision engineering required to make the rotary lock would be beyond the capabilities of those companies which had surplus production capacity – the Hispano was easier to make.

Later in the war many were converted to the anti-aircraft role on single and triple mountings. After the war, drawings of the gun were acquired from France by South Africa, where it was manufactured from the 1970s onward as the Vektor (now Denel) GA-1; it is still offered today.

**5 cm FLAK 41**: 50 × 346B AMMUNITION.

This gun was the first attempt to produce a heavy automatic AA gun which resulted in a service weapon, although it suffered from technical problems (concerning the unstable mounting, slow traverse and complex sights) which greatly restricted its use, with fewer than 200 guns made in total before the project was cancelled. This was a Rheinmetall-Borsig design made to a *Luftwaffe* contract awarded in 1935 and, unusually for this company, was gas-operated rather than using a short-recoil mechanism. It also differed from the smaller cannon in using an artillery-type vertical sliding breech-block. Ammunition was manually fed in five-round strips onto a loading tray on the left side of the gun; a fresh strip could be loaded without ceasing fire. An L/68 barrel was used, gun weight was 570 kg and the rate of fire 130 rpm.

**3,7 cm FLAK M42**: 37 × 250B AMMUNITION.

This Rheinmetall-Borsig design for the *Kriegsmarine* was a second attempt at an automatic naval 37 mm Flak cannon, following-on from the little-used SK C/36 Flak (see above) which bequeathed its ammunition and Bofors-type feed to the M42, 5-round clips being retained. The M42 was evidently closely based on the C/36, otherwise there would seem to have been no point in retaining its unique ammunition instead of using the standard 37 × 264B of the other Flak guns. The performance was similar to the *Luftwaffe's* 3,7 cm Flak 37 although the mechanism as well as the ammunition differed; the gun was long-recoil operated with an artillery-type wedge lock and continuous feed. An L/69 barrel was used, gun weight was 300 kg and the RoF was 160–180 rpm.

The original mounting for the M42 was the *Flaklafette* C/36 (as used by the SK C/36 cannon) but this was soon replaced by the triaxially stabilized LC/39 (also in a U-boat version) and the LM 42 and LM 43, both developed specifically for the gun and weighing 1,350 kg. There was a twin mounting, the *Doppellafette* LM 42 (DLM 42) which weighed 1,750 kg. Despite the M42 designation, the gun entered service in late 1943, initially for submarines. An installation programme for fitting this gun to smaller surface ships got underway in 1944.

5 cm Flak 41 (US Army)

3,7 cm Flak M42 twin on U-190 (NARA photo, via NavWeaps site)

*Rheinmetall-Borsig 3 cm MK 108 (Courtesy of James Spaw)*

**3 cm MK 108**: 30 × 91RB AMMUNITION.

Once the USAAF's strategic bombing campaign against Germany got underway, the *Luftwaffe* soon discovered that the tough B-17 bombers were difficult to shoot down with 20 mm cannon, German fighters frequently running out of ammunition in the attempt. A decision was made to introduce 30 mm weapons in two classes; low and high-velocity. The MK 108 was Rheinmetall-Borsig's successful contender in the low-velocity class.

This was in effect an enlarged MG-FF using an API blowback system, but organised in a different way from 20 mm weapons of this type. Instead of the recoil spring being wrapped around the barrel and compressed by means of a frame connecting the bolt to the front of the spring, the recoil spring was placed between the back of the bolt and the back of the receiver. Two other changes over the MG-FF were the use of electric priming and an integral belt feed system. The electric priming had the advantage that it only activated when the bolt was travelling forwards at high speed; it would not fire if the bolt was fully forwards and stationary, since the violent recoil which would then occur might wreck the gun.

The result was a very compact and lightweight gun (at 60 kg, it weighed about the same as a 20 mm Hispano), which could be made to fire very quickly by adjusting the bolt weight and spring strength. The service gun fired at 600–650 rpm, but being tested at the end of the war was a modified version, the MK 108 nA (*neuer Art* = new type) reportedly capable of 900 rpm. The big *M-Geschoss* shells were highly destructive but the weakness in the performance was the low muzzle velocity of about 500 m/s (an L/15 barrel was fitted), which required the fighters carrying the gun to get very close to their targets in order to ensure hits.

It is interesting to note that immediately after the war this gun was strongly promoted to the RAF by German engineers as an interim anti-bomber gun while the MG 213/30 (which was eventually developed into the ADEN gun) was perfected, but this was not accepted.

**3 cm MK 103**: 30 × 184B AMMUNITION (ELECTRIC PRIMING).

This was the high-velocity 30 mm gun to complement the low-velocity MK 108 described above. It fired the same *M-Geschoss* HEI projectiles as the MK 108, but at an MV of around 860 m/s, enabling the aircraft which carried it to engage their targets from a longer stand-off distance (or, to put it another way, to score a higher percentage of hits at any given distance). It was also of value in the ground-attack role, since it could penetrate most tanks from the side or rear when firing tungsten-cored *Hartkernmunition* AP projectiles. Finally, it promised to have a useful career as an AA gun (the 3 cm Flak 38) but that was too late to have much impact.

The MK 103 was a direct replacement for the MK 101, both being designed by Rheinmetall-Borsig. The two guns were much the same size and weight and used almost the same ammunition, but there were significant differences: the MK 103 was belt-fed, used electric priming (so ammunition for the two guns was not interchangeable), and was significantly faster firing. A disadvantage was that the different gun action – a combination of the gas operation of horizontally-moving locking pieces, plus a recoiling barrel which drove the belt feed – was not as strong, so the HE ammunition was downloaded somewhat to preserve gun life (the AP ammunition for ground attack was left at full power). This affected the rate of fire, which was 380 rpm firing HE, and 420 rpm firing AP.

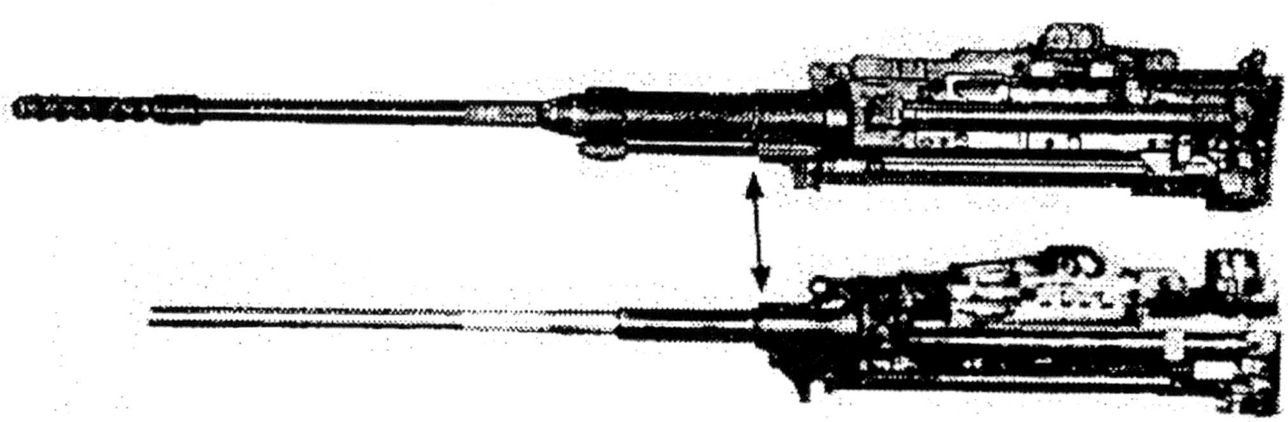

*A comparison between the MK 103 (top) and the MK 103M; note the way the latter had to be slimmed-down to fit into the Bf 109 (Ted Bradstreet)*

The gun weight was 141 kg, each belt link weighed 122 g, and the weight of a belt of 100 rounds was 92.2 kg (*M-Geschoss*) or 94.2 kg (*Hartkern*).

The gun was too big to fit in the space for an engine cannon provided in the Bf 109; the L/45 barrel needed to fit in the 70 mm-diameter tube running between the engine cylinder banks, but the gas tube which ran alongside part of the barrel made this impossible. A modified version was accordingly designed which was slimmed down as much as possible. This was designated MK 103M (also referred to as the MK 103 *mot*) and an example was installed in a Bf 109K for evaluation, but this did not prove to be successful.

*5 cm BK 5, with integral magazine wrapped around the action (National Museum of the United States Air Force)*

**3,7 cm FLAK 43**: 37 × 264B AMMUNITION. ALSO KNOWN AS GERÄT 339.
This was the final successful attempt at an automatic 37 mm Flak cannon, intended for use by both the land forces and the *Kriegsmarine*. It used the same ammunition as the earlier Flak 18, 36 and 37 series and the same L/57 barrel length, but a new gas-operated mechanism derived from that of the 3 cm MK 103 and featuring a diagonal wedge lock boosted the cyclic rate to 250 rpm. Ammunition feed was from the side through enlarged trunnions to minimize the movement of the loading tray and the ammunition came in eight-round strips. Gun weight was 265 kg. There were also twin-gun mountings, with the *Luftwaffe Flakzwilling 43* mounting one gun above the other, while the *Kriegsmarine Flak LM 44* placed them side-by-side.

**5 cm BK 5**: 50 × 420R AMMUNITION.
One of the standard anti-tank guns in the German Army was the 5 cm Pak 38, and this gun was modified for aircraft use; both for engaging bombers and for ground attack. The modifications by Rheinmetall-Borsig included an autoloader using the long-recoil system, and an annular magazine with a capacity of 22 rounds held in an endless belt. The gun had an electrically-controlled firing system with feeding and ramming achieved by pistons driven by compressed air. An L/61 barrel was used, gun weight was 540 kg and the gun fired at 45–50 rpm.

## WORLD WAR II EXPERIMENTAL PROJECTS

There was a large number of experimental cannon projects developed to varying degrees in Germany during World War II. What follows is by no means a comprehensive list, just the ones for which a reasonable amount of information could be found.

***2 cm MG C/35 AND 2 cm Stg 201***: 20 × 138B AMMUNITION.
Rheinmetall-Borsig developed the MG C/35 as a Flak 30 replacement, an unsuccessful rival to Mauser's Flak 38. This had similar performance to the Flak 38 (450 rpm) but used the Solothurn rotary locking sleeve. The weight was 60 kg for the magazine-fed gun or 70 kg for a belt-fed version. Only a small number was made.

There is also a reference to the Stg 201, Rheinmetall-Borsig's proposed replacement for the KwK 30 for arming AFVs. Again, this had similar characteristics to Mauser's successful KwK 38, and weighed 70 kg.

*Zwillings-Flak M43 (Bundesarchiv via Wikimedia Commons, Bild 101I-301-1957-32 / Kurth / CC-BY-SA 3.0)*

***2 cm MG 210:*** 15 × 96 AND 20 × 82 AMMUNITION.
Also known as the GL 15 (for *Gasdrucklader* = gas-operated), this was an attempt by Mauser to improve on the rate of fire of their MG 151 while making it lighter and simpler, using the same ammunition. Development began around 1940 but it was overtaken by the MG 213 programme. It utilized Mauser's usual rotating bolt-head locking system, but was gas-operated via a hollow piston surrounding the barrel. Priming was electric, and the gun weighed 32 kg with a design RoF of 1,000 rpm.

***3 cm MK 212:*** 30 × 91
AMMUNITION.
This was Mauser's unsuccessful rival to Rheinmetall-Borsig's 30 mm MK 108. Unlike the very simple API blowback mechanism of the MK 108, the Mauser was gas-operated with horizontally-moving locking pieces. The weight and performance of this belt-fed gun were very similar to the MK 108, but the muzzle velocity from the L/33 barrel was higher at 580 m/s.

***2 cm MG 301:*** 20 × 135B
AMMUNITION.
Krieghoff reportedly began working on a high-performance 20 mm cannon as early as 1938; their designs evolved through several stages, the final version bearing little resemblance to the original. One version was submitted to meet the Luftwaffe's 1942 requirement for a 'million point gun'; an aircraft gun capable of firing projectiles at 1,000 rpm and 1,000 m/s. The MG 301 failed to achieve the targets set for this, with a rate of fire of only 600–700 rpm (although it is unclear as to which version this refers). At about the same time in a second Luftwaffe project, the ammunition was also developed to fire very long HE projectiles – the *'HT Geschosse'* – but little is known about the performance of these.

The gun was gas-operated with a vertically-sliding wedge lock, controlled by cams cut in the receiver sides. It was belt-fed. According to Chinn, it weighed 52 kg and was 320 cm long with an L/113 barrel (including the muzzle brake); it is unclear which version of the gun this refers to. Other sources indicate a weight of 63–65 kg.

It is unclear at which point the 'MG 301' designation was applied. Further work on the MG 301 was transferred to Mauser, who changed the designation to MG 213A. The project was eclipsed by the success of Mauser's rival MG 213C revolver cannon.

*2.8 cm GERÄT 411*: 28 × 216B
AMMUNITION.
A Krupp project initiated in 1941 for a motor cannon driven by the aircraft engine, and therefore presumably limited to use with vee-engines, with the gun fitting between the cylinder banks and firing through a hollow propeller hub. This used a large and powerful cartridge, but was unsuccessful. It is sometimes confused with the 3 cm MK 411, about which virtually nothing seems to be known except the calibre.

*2.8 cm MK 8202*: 28/20 × 210B
AMMUNITION.
A Mauser anti-tank gun (also referred to as KwK 8202 and the Gerät 231), firing Gerlich squeezebore ammunition. The cartridge is not the same as the 28/20 × 187R used in the PzB 41 manually-loaded AT gun which saw service. The MK 8202 was gas-operated (a recoil-operated version was also tried), via a hollow piston encircling the barrel, had a rotary-locking bolt head and was belt-fed. It appears to have been intended for aircraft as well as AFV use. It weighed 120 kg and fired at 180 rpm (Musgrave says 450 rpm). It appears that this was rejected for ground use in favour of the simpler PzB 41.

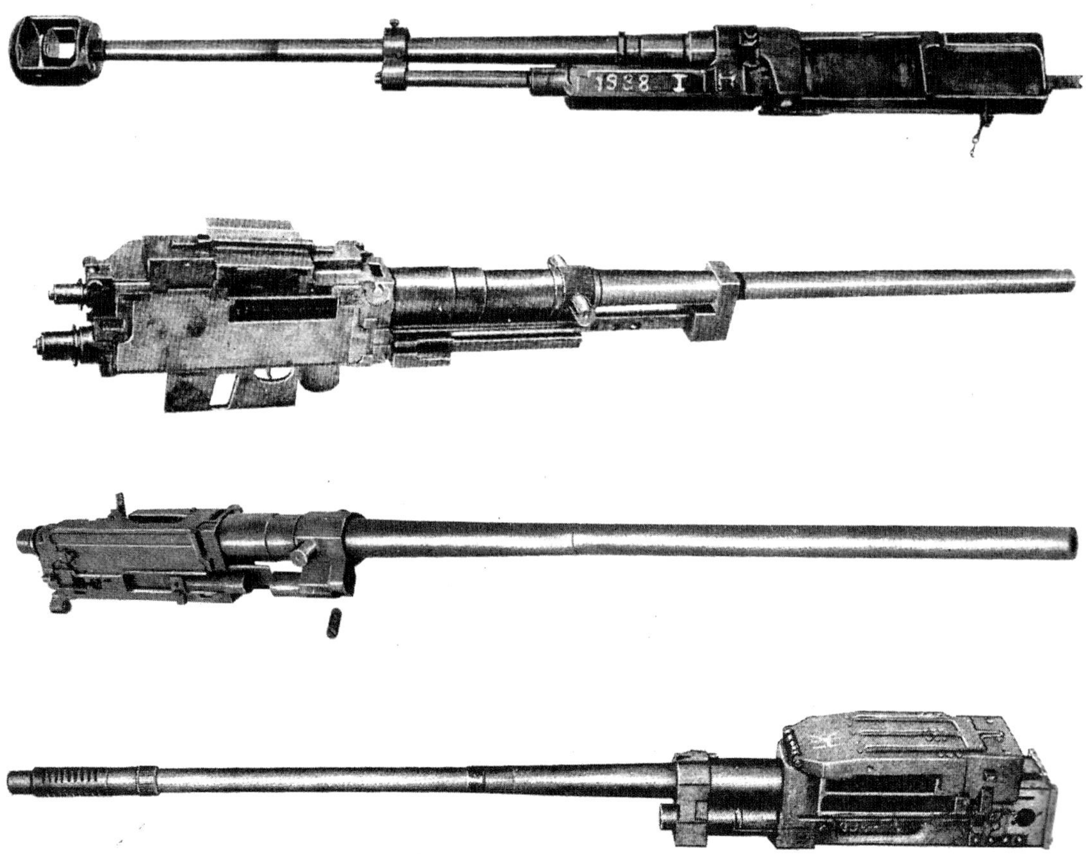

*Krieghoff's 2 cm cannon development, from top to bottom (the second one is referred to as the Model 1941, the fourth one as the MG 301) (Chinn)*

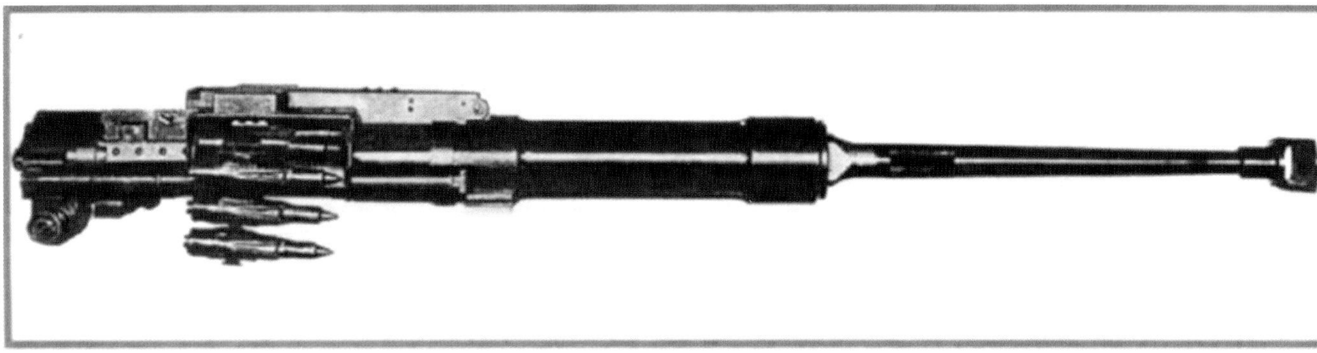

*Mauser MK 8202 (internet)*

*3 cm MK 303*: 30 × 211 AMMUNITION.
This high-velocity gun was developed during World War II at *Brünner Waffenwerke* in Czechoslovakia under German control, with the involvement of Krieghoff. It was a gas-operated design using a horizontal wedge lock and seems to have been a scaled-up version of the 2 cm MG 301 (see above). It used 'frame feeding' (10-round skeletal magazines). It weighed 185 kg, was 295 cm long with an L/73 barrel and fired at 400 rpm. It was intended for naval Flak applications (initially in Type XXI submarines) but was not ready in time. After the war, development was continued by Czechoslovakia and it entered service with the Czech and other armies in a twin AA mounting, designated M53.

*2 cm GERÄT 239 GEBIRGSFLAK*: 20 × 138B AMMUNITION?
A project by Gustloff for a light Flak gun for mountain troops, with a very high rate of fire. The ammunition used is not stated but the muzzle velocity quoted is the same as that for the 20 × 138B. The gun was gas-operated with flap-type locking and a strip feed permitting continuous fire. The weight was 65 kg and the gun achieved a RoF of 700 rpm. A small number was built, but a lack of manufacturing facilities prevented further production.

*2 cm GERÄT 240*: 20 × 138B AMMUNITION?
A Rheinmetall-Borsig project for a very fast-firing Flak gun, possibly a rival for the *GebirgsFlak* described above. It was gas-operated and belt-fed, weighed 50 kg and was credited with a RoF of 900 rpm.

*3,7 cm GERÄT 341 / FLEDERMAUS*: 37 × 280B.
The 3,7 cm *Gerät 341* was a gas-operated, strip-fed naval Flak gun by Rheinmetall-Borsig; the gun weighed 450 kg and fired at 240 rpm. The rival 3,7 cm Flak *Fledermaus* was by 'Gustl./Stübg' (presumably Gustloff and Stübgen) also for the *Kriegsmarine*, weighed 450 kg and fired at 270 rpm. It appears that that the *Gerät 341* and *Fledermaus* were rivals for the same contract. One source (HEC) states that they fired the same high-velocity (1,000+ m/s) ammunition, another (*Unterlüss*) says that only the Rheinmetall gun used the HV round, the other retaining the 37 × 264B.

*5 cm MK 214A*: 50 × 420R AMMUNITION.
This was an attempt by Mauser to improve on the BK 5 described above. The gun used a vertically-sliding breech-block, a long-recoil operating system, a conventional belt feed and electric priming. It weighed 490 kg, had an L/56 barrel, and fired at 160 rpm. The start of serial production had been ordered in 1945.

The gun was also installed on the mounting designed for the 5,5 cm *Gerät 58* (see

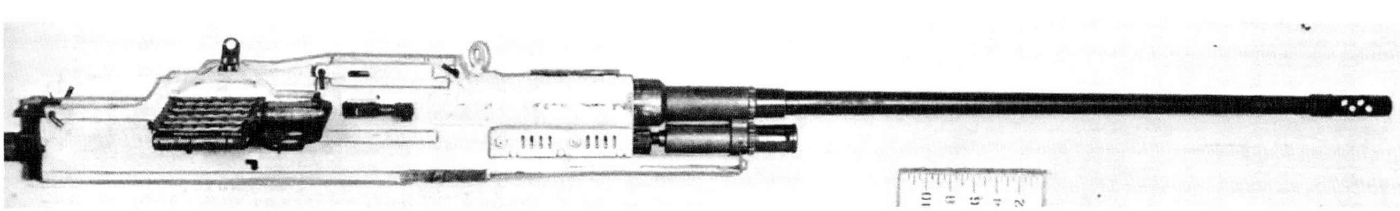

*Krieghoff MK 303 (Chinn)*

*Mauser 5 cm MK 214A (Chinn)*

below) in order to get an effective medium Flak gun into service more quickly, but the resulting Flak 214 had not reached service by the end of the war.

*5,5 cm MK 112*: 55 × 175RB AMMUNITION.
After experience with the 30 mm guns against bombers, the *Luftwaffe* decided that an even bigger calibre would be more effective. Calculations by Rheinmetall-Borsig showed that a single hit with an HEI shell of 55 mm calibre (containing 500 g HE) would normally be enough to bring down a bomber, so both low and high-velocity guns were developed, just as with 30 mm guns. The MK 112 was Rheinmetall-Borsig's low-velocity aircraft gun project; a scaled-up MK 108, using the same API blowback principle and layout, and similarly electrically-primed and belt-fed. It weighed 275 kg, had an L/19 barrel and was reportedly capable of 320–360 rpm, but was still at the prototype stage at the end of the war. It was then tested in the USA under the designation '57 mm T78': the calibre was increased slightly to use the US T35 practice projectile, the case length being 173 mm.

*5,5 cm MK 412*: 55 × 205RB AMMUNITION.
This was a Krupp project for the low-velocity 55 mm aircraft gun. It weighed 230–240 kg (although one Unterlüss report, from Dipl Ing Mix, said it weighed only 180 kg) and fired at 260 rpm. It used a combined gas and recoil operation, with a self-locking rotary bolt head, belt feed from right or left side, an L/40 barrel and electric priming. Daimler Benz planned to install that between the banks of a DB 609/DB 613 V24 (double-V) 2,500 hp engine. Krupp ran into many technical problems so the work was passed to Mauser in mid-1944, but after some development work (resulting in a RoF of 300 rpm) Mauser concluded it was not viable and the project was cancelled later that year.

*5,5 cm GERÄT 58*: 55 × 450B AMMUNITION.
Rheinmetall-Borsig developed this gun as a result of their experience with the 5 cm Flak 41. It was preceded by the 5 cm *Gerät 56* in 50 × 461B, which seems to have been an intermediate development. The gas-operated mechanism had a rotating bolt head and was designed around the new 55 × 450B ammunition. Barrel length was L/77. The recoil was tamed by adopting a 'differential recoil' mounting in which the gun fired as it was moving forwards into battery; the recoil energy was first used to bring the gun to a halt before beginning to drive it rearwards again, thereby smoothing out the recoil impulses. The ammunition was fed in clips of five rounds and the gun fired at 140 rpm. It was still being developed at the end of the war.

*5,5 cm MK 114*: 55 × 450B AMMUNITION.
This was Rheinmetall-Borsig's proposal for the 55 mm high-velocity *Luftwaffe* cannon for aircraft use, developed around the same cartridge as the *Gerät 58*. It was gas-operated, with a gas-driven ammunition belt feed and a very rigid locking system. It weighed 700 kg, had an L/77 barrel and fired at 150–180 rpm.

*5,5 cm MK 214B:* 55 × 450B AMMUNITION.
Mauser's proposal for a rival to the MK 114 (see above), using the same powerful new 55 × 450B ammunition. Despite the similar designation it had no relationship with the MK 214A, employing gas operation and a rocking device to lock the breech. It fired at 300 rpm and weighed 650 kg. It was still under development at the end of the war.

*5,5 cm MK 115*: 55 × 318B AMMUNITION.
This unique gun was a recoilless design by Rheinmetall-Borsig, and had reached the stage of first prototype construction at the end of World War II. The ballistics matched those of the 5,5 cm MK 112 but the cartridge case was much bigger in order to accommodate more propellant. The gas generated by the extra propellant was expelled to the rear of the gun in order to balance the recoil. To add to the complexity, the ammunition used a part-combustible case.

While technically very interesting, combining a combustible case with a recoilless design in a fast-firing automatic weapon was highly ambitious and would probably have resulted in a very protracted and ultimately unsuccessful development process. More details are given in the Unconventional Systems chapter.

*2 cm AND 3 cm MG 213*: 20 × 135 AND 30 × 85B AMMUNITION.
The original 20 mm MG 213 project initiated at the end of 1942 was intended to achieve the same performance targets as the MG 301 and was thereby known as the 'million point gun' (1,000 m/s × 1,000 rpm). Three different designs were produced by Mauser: the 213A (a development of the MG 301 described above), 213B (a further modification of the same design with a

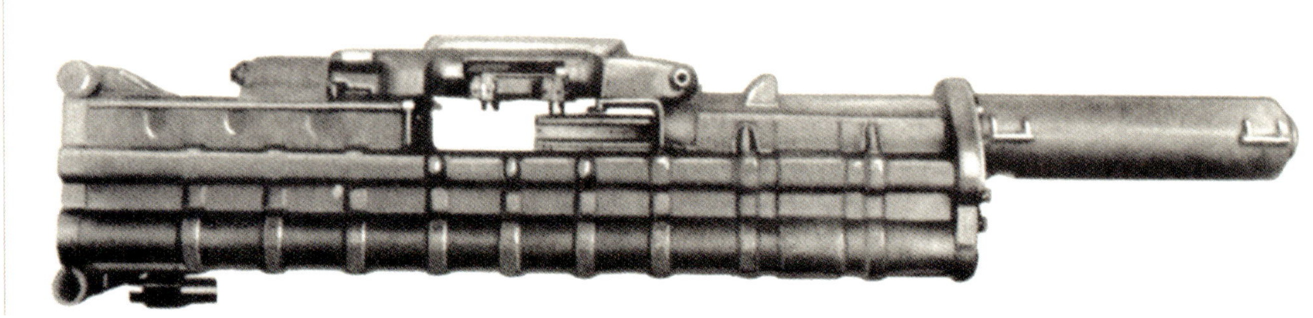

*Rheinmetall-Borsig 5,5 cm MK 112 (Chinn)*

different belt feed) and the 213C (which used a modified cartridge with less case taper and no belt). This last design, by a team led by Anton Politzer, was radically different in that it was the first modern revolver cannon, in which the problems of very fast cartridge handling were solved by loading and extracting the ammunition in stages, using a five-chambered revolving cylinder behind the single barrel. The mechanism is described in the Gun Design section. This gas-operated and electrically-primed gun won the competition; in fact, was the only one to meet the requirements, the others achieving no more than 800 rpm.

Initially developed in 20 mm calibre with an L/80 barrel, attention soon switched to a 30 mm version (L/43 barrel) for the anti-bomber role and the two guns were then designated MG 213/20 and MG 213/30 (some sources state the 30 mm version was called the MK 213, but the official designation never seems have been formally determined). Both guns weighed 75 kg, were belt-fed and gas-operated. The 20 mm version achieved a RoF of 1,200–1,400 rpm, the 30 mm managed 1,100–1,200 rpm. These were only at the test stage by the end of World War II, with about a dozen examples made altogether.

## PROJECTS SINCE 1945

**RH 202 / MK 20**: 20 × 139 AMMUNITION.

This Rheinmetall gun was developed in the 1960s around the ammunition designed for the Swiss Hispano-Suiza HS 820 (which was already in German service). It was intended to deliver reduced costs, improved performance and reliability, and easier maintenance. The mechanism uses a combination of gas and blowback within a differential recoil system. The gun fires from an open bolt; as the bolt moves forward, chambering a fresh round, two locking levers are pushed into recesses in the housing to lock the bolt. The gun then fires while the barrel and bolt assembly is still moving forwards. Gas tapped from the bore is used to power the belt feed drive as well as moving the locking pieces back in again; gas blowback pushes back the fired cartridge case and the bolt. Before recoil is completed the cartridge case is ejected sideways or downwards, depending on the feed type and installation. There are two types of feed: Type 2 double feed and Type 3 single feed; weights are 75 kg with the single feed, 83 kg with double feed. The barrel is L/92 and the rate of fire is 800–1,000 rpm.

In 1967 the Federal Ministry of Defence accepted the Rh 202 for use by all services under the designation MK 20, commencing with adoption by the German Army in a twin-gun *Zwillingsflak* mounting. Deliveries took place to the *Bundeswehr* between 1972 and 1976, also to Argentina, Greece, Indonesia, Portugal, and Turkey.

***20/23 mm MAUSER MODEL B***: 20 × 139 OR 23 × 139 AMMUNITION.

The gas-operated Mauser Model B cannon is interesting in that it incorporated two technical innovations: an hydraulic rate of fire reducer and a piezo-electric bolt. The rate reducer was housed in the recoil buffer and the RoF was continuously variable up to its maximum rate of 900–1,050 rpm: exactly how it worked is not clear. The piezo-electric system in the bolt generated enough electricity during the bolt's forward motion to fire electrically-primed ammunition with no need for a separate power supply. The gun was belt fed and although designed around the standard 20 × 139 HS ammunition it could optionally be modified to use the 23 mm version (which never entered service). The gun weighed 62 kg and was 245 cm long with an L/92 barrel. It appears that only a few prototypes were made and the gun was never completely developed.

**30 mm MK 30**: 30 × 173 AMMUNITION.

The MK 30 cannon was originally developed as a private venture by Mauser-Werke Oberndorf Waffensysteme, now known as Rheinmetall Waffe Munition, Branch Mauser Oberndorf, a subsidiary of Rheinmetall Defence. It was designed around the percussion-primed and aluminium-cased 30 × 173 ammunition developed for use in the GAU-8/A rotary cannon, and was intended for light air defence and naval applications.

The MK30F was subsequently developed to arm light AFVs for ground fighting, using steel-cased ammunition of different natures, particularly APFSDS. This led to the original MK 30F being redesignated the MK 30–1, with the AFV version being the MK 30–2. The latest version is the MK 30–2/ABM, which has an inductive programming fuze-setter fitted to the muzzle, allowing it to utilize air-burst ammunition, initially the Rheinmetall KETF (Kinetic Energy, Time Fuzed) based on the Oerlikon AHEAD 'shrapnel' type, which creates a pattern of tungsten alloy pellets at the predicted location of the target.

The MK 30 is gas-operated, using gas pistons to drive both bolt movement and the ammunition feed. The breech system consists of double locking flaps, and all moving parts are installed along the gun axis. The gun is available with single or double belt ammunition feed, from the sides or top. It is installed in a simple lightweight cradle in which it can recoil, and it can be dismantled and reassembled without tools.

Data for the gun varies according to the model: with a single belt feed, it weighs 154 kg; with a double belt feed the weight goes up to 161 kg (MK30–1) or 173 kg (MK30–2); and reaches 198 kg

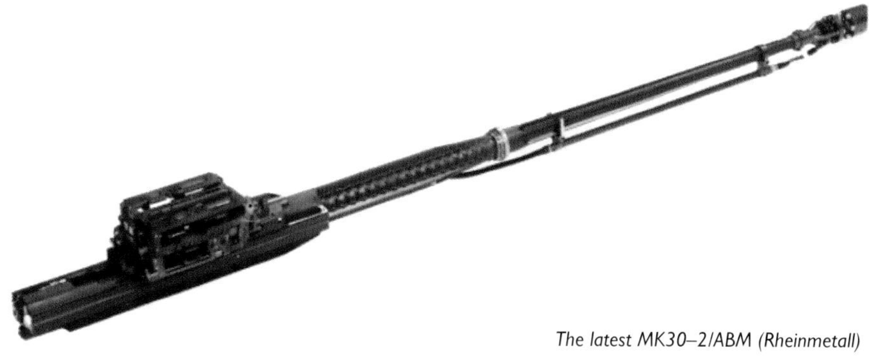

*The latest MK30–2/ABM (Rheinmetall)*

for the MK30–2/ABM. Overall length is 335 cm for the MK30–1, 331 cm for the MK30–2 and 378 cm for the MK30–2/ABM, with the barrel being 258 cm including the muzzle brake (L/82 without). The rate of fire is normally 800 rpm for the MK30–1 and 700 rpm for the MK30–2, but this has been reduced to 200 rpm for the MK30–2/ABM used in the Puma AIFV, in the interests of accuracy.

Some of the early applications of the MK 30 Model F were the Arrow air-defence twin field mounting, the Greek Artemis 30 twin field mount, and the Otobreda twin Sentinel field and single or twin naval mountings, all of which emerged in the 1980s. Various other AA applications were proposed, e.g. the turreted twin AA mounting of the Krauss-Maffei Wildcat 6 × 6 light AFV.

The MK30–2 has been acquired for the Spanish Pizarro Infantry Fighting Vehicle programme; for installation in Austrian Ulan IFVs (the Ulan is the Austrian version of the Spanish Pizarro) and in MK30–2/ABM version for the German Puma AIFVs. It is also offered in the Rheinmetall Lance modular turret system (MTS), which is available in manned and remotely-controlled (RC) versions.

Sales of the MK30–2 have been restricted by the popularity of the highly successful ATK Bushmaster II/MK44/XM813 family, which uses the same range of ammunition and is a direct competitor for AFV installations.

***25 mm MAUSER MODEL E***: 25 × 137 AMMUNITION.

The Mauser MK 25 Model E cannon was developed in the late 1970s as a replacement for the 20 mm MK 20 (Rh 202) fitted to German Marder IFVs. The Model E had the same design and operating principles as the Mauser MK 30 Model F, but was chambered for the NATO standard 25 × 137 ammunition. The MK 25 used a dual belt feed, had an L/84 barrel and weighed 103 kg, plus 52 kg for the cradle within which it recoiled. A counter-recoil or floating firing system evened out the recoil pulses. The unusually high rate of fire of 900 rpm (similar to the MK 20) was required for air defence purposes.

In 1981 the Model E was selected for development by the Bundeswehr in competition with a Rheinmetall design, but it was subsequently decided not to upgun the Marders. With the successful Oerlikon KBA and ATK Bushmaster firing the same ammunition already available, it did not succeed in obtaining any other sales.

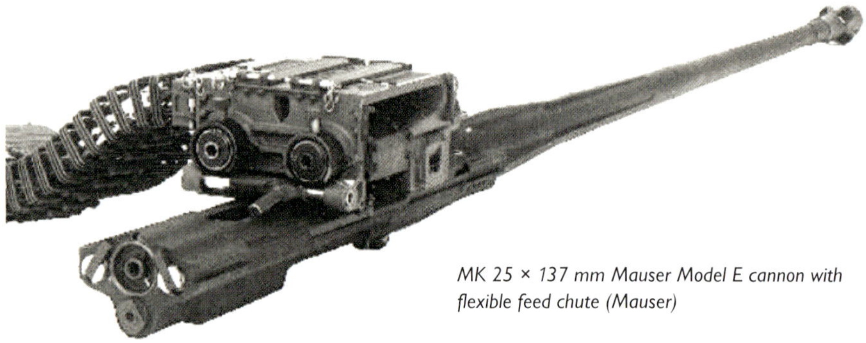

*MK 25 × 137 mm Mauser Model E cannon with flexible feed chute (Mauser)*

**MAUSER BK 27**: 27 × 145B AMMUNITION.

The initial purpose of the Mauser BK 27 was to meet the needs of the European Multi-Role Combat Aircraft (MRCA) programme which led to the Panavia Tornado strike fighter. Development of the BK 27 started in 1971 to meet a requirement for a combination of a high rate of fire, high muzzle velocity and destructive ammunition, which could not be met by existing cannon (the 30 mm DEFA and ADEN guns had too low a muzzle velocity, the US 20 mm M61 fired much smaller shells). Mauser returned to the revolver cannon principle which they had originated late in World War II, and developed a new range of ammunition for the purpose.

The BK 27 is a single (L/63) barrel, gas-operated revolver cannon, which features optional left- or right-hand feed. There are five cartridge chambers and electrical priming. Firing is carried out electrically with automatic re-cocking by a pyrotechnic device. The rate of fire is 1,700–1,800 rpm.

The gun entered service in 1979 in the Tornado; most versions of the plane had two guns, the RAF ADV had one. The BK 27 was subsequently chosen to equip the German version of the Alpha Jet (with 120 rounds) and the SAAB Gripen (120 rounds). The gun was originally belt-fed, but for the Eurofighter Typhoon installation a new linkless-feed system with 150 rounds was developed which also retains the empty cases, avoiding potential problems caused by the ejection of cases and links.

Both of the prime contenders for the US Joint Strike Fighter (JSF), which became the F-35 Lightning II, selected the

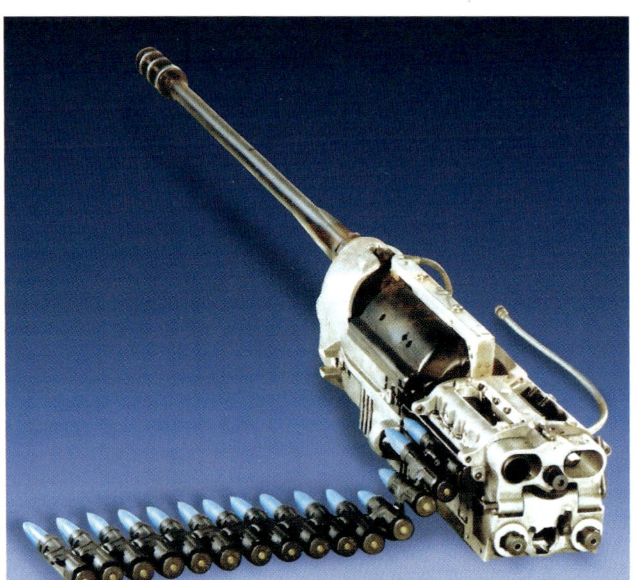

*27 mm BK 27 (Mauser)*

BK 27 as the optimum basis for developing a gun for the JSF, but once the aircraft contract was awarded, Lockheed Martin chose to give the gun installation contract to General Dynamics, who substituted their own GAU-12/U 25 mm five barrel rotary cannon. However, this proved too heavy so, as part of a weight-reduction exercise, GD developed a four-barrel version of the gun, designated GAU-22/A.

The BK 27 has achieved a second life in naval mountings. The initial effort was the Mauser Drakon, a quad mounting developed for the anti-missile role, but this was not adopted by the *Bundesmarine*. Subsequently, a compact, remotely-controlled single mounting, the MLG 27 *(Marine Leichtgeschütz)*, was adopted in 2003 in order to replace 20 mm and 40 mm guns throughout the *Deutsche Marine*, and this has also achieved some success in export sales. Finally, a C-RAM (Counter-Rocket, Artillery, Mortar) land mounting system has also been developed, but has so far not sold.

In August 2016, Rheinmetall stated that more than 3,000 BK 27 were in service worldwide.

***RHEINMETALL Rh 503***: 35 × 228 / 50 × 330 AMMUNITION.

The German Army was one of the first in NATO to introduce cannon-armed infantry fighting vehicles with the Marder 1, but by the 1980s the 20 mm cannon fitted to it were no longer effective against enemy armoured vehicles. A plan to refit the Marders with a new 25 mm gun (see *Mauser E* above) was dropped in favour of adoption of a new IFV, the Marder 2, which was much heavier and better protected. To arm the Marder 2 a major upgrade in gun power was chosen after comparisons with existing armament: a new gun was specified, initially designed around the existing 35 × 228 Oerlikon ammunition (in German service in the Gepard SPAAG) but with the ability to upgrade to a new 50 × 330 'Supershot' round which had the same external dimensions but developed significantly more power. Little more than a barrel change was required to switch calibres.

Prime contractor for what became designated the Rh 503 was Rheinmetall, with Mauser, Diehl, Oerlikon Contraves, Heckler & Koch, and Honeywell all being involved with different aspects of the gun and ammunition development. In 1992 the first protoype of the Marder 2 was ready, fitted with Rheinmetall's TS 503 two-man turret, but at the end of that year the Marder 2 programme was cancelled for financial reasons. Development work on the gun and ammunition continued, with the emphasis on the 35 mm version which was type-classified in 1999 (50 mm ammunition development was mainly focused on the APFSDS, although there was a saboted 35 mm HE air burst shell designated DM31), but hopes that the gun would arm whatever IFV replaced the Marder 1 were dashed by the decision to fit the 30 mm MK30–2 to the new Puma IFV. Given that the ATK Bushmaster III is already in service in 35 mm calibre (with potential for a future upgrade to 50 mm), and that the new Rheinmetall WOTAN modular multicalibre gun (see below) is being offered in 30 mm and 35 mm, the Rh 503 project may be regarded as dead.

The Rh 503 was externally powered with all functional components being positively controlled, making the weapon function reliably regardless of the ammunition gas pressure level. A linkless ammunition feed was provided from both sides allowing an immediate change in the type of ammunition loaded. The empty cartridge case was pushed out of the front in a straight line, minimising the size of the ejection port. In case of ammunition misfire, the cannon automatically stopped in the safe position, with firing recommencing immediately after the misfired round is ejected.

The rate of fire could be selected to within a range of 150 to 400 rpm, and single shots could be fired. If the main power supply failed, then drive could be powered by a battery. In addition, manual operation for single-shot fire was also available. Gun weights were 575 kg for the 35 mm version, 615 kg for the 50 mm, with overall lengths being 473 and 572 cm respectively; barrels were L/90 and L/85.

***MAUSER RMK 30 / 35 SERIES***: 30 × 230 / 30 × 280 / 35 × 300 / 35 × 350 AMMUNITION.

The RMK *(Rückstoßfreie Maschinenkanone* = Recoilless Machine Cannon) family of weapons and ammunition was developed by Mauser-Werke Oberndorf Waffensysteme GmbH in the early 1990s. They incorporated several innovations in the field of automatic cannon: recoilless operation; separate chambers which could be loaded from the front;

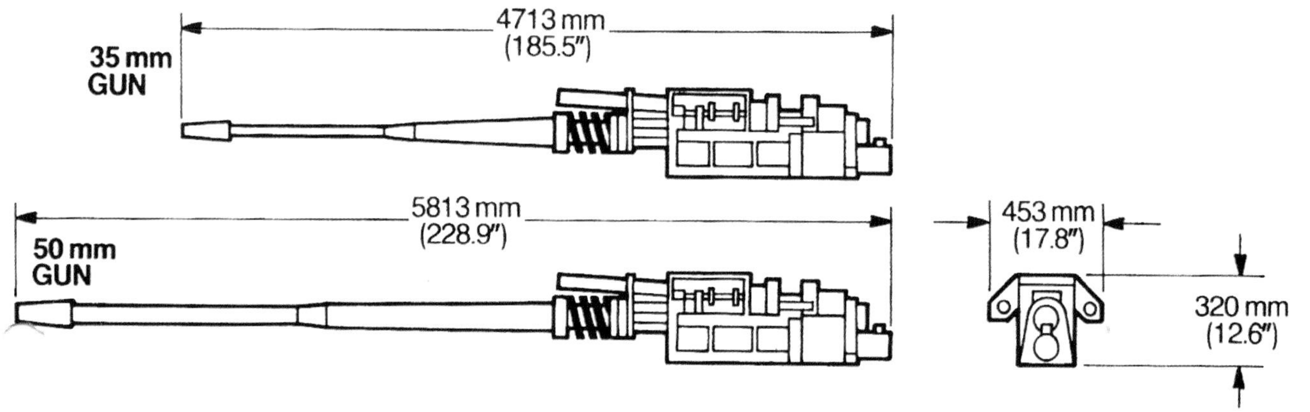


*The two different versions of the Rh 503; the overall lengths varied somewhat during development (Rheinmetall)*

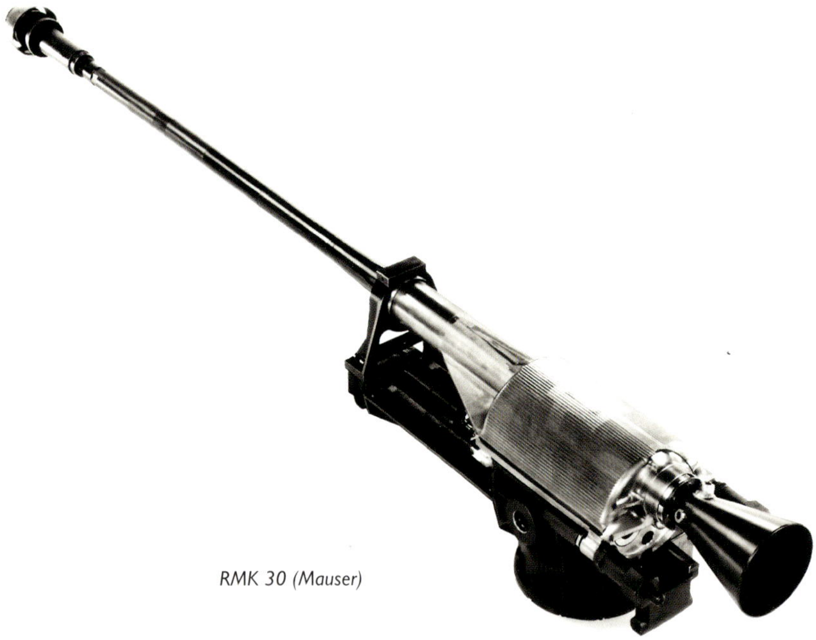

RMK 30 (Mauser)

plus cylindrical, telescoped and caseless (or combustible cased) ammunition using an inductive primer. Four different systems were developed, in two calibres. The RMK 30/1 (30 × 280) was first demonstrated to the German armed forces firing from a Wiesel light airborne tank in November 1996. The RMK 30/2 (30 × 230), with linkless feed, was originally intended to be the main gun armament for the German version of the European Tiger attack helicopter, the UHT (*Unterstützungshubschrauber*), in a chin turret. Another proposed application was the Muraena mast-mounted submarine installation. In the event, no orders were placed so the whole programme was suspended in the mid-2000s.

More information is provided in the Unconventional Systems section.

*RHEINMETALL WOTAN*: 30 × 173 AND 35 × 228 AMMUNITION.

This family of weapons has been designed for AFV and naval applications. While adaptable to various calibres, it has first been developed and tested in 30 × 173. The design is externally powered with a single chamber integral with the barrel, and utilizes a chainless bolt drive system that can accept rounds from two feeds. The gun is link-fed and can use all NATO standard links, including both the US standard MK 44 link and the MK 30–2 link. It features a mechanical hang fire safety, horizontal ammunition feeding and transfer between the bolt and barrel, and the spent cases are ejected forwards. The gun can be fitted with the muzzle-mounted programmable fuze setter required to fire air burst munitions such as the KETF/AHEAD shrapnel type. The rate of fire is adjustable up to 200 rpm. The gun weighs 280 kg and is 367 cm long.

Firing trials commenced in 2010 and the gun has been fitted to Rheinmetall's Lance AFV turret. In mid-2020 an order from Hungary was placed for the Lynx MICV armed with the WOTAN35.

WOTAN 30 (Rheinmetall)

# INTERNATIONAL

Almost all automatic cannon are designed and developed in one country, even if they are subsequently made and further developed elsewhere. However, two AFV cannon were established as international joint projects from the start: the 60 mm HVMS / HVGS from Israel and Italy, and the 40 mm CTAS by CTAI, a joint company established by Nexter and BAE Systems.

*60 mm HVMS / HVGS*:
60 × 410R AMMUNITION.
The inclusion of this cannon is rather marginal, given that its calibre is slightly larger than the general maximum for this book. However, it has been mentioned here as an interesting variation on the typical choices for medium AFV armament, particularly given the recent interest in calibre escalation e.g. the US 50 mm (Super 50) and Russian 57 mm weapons.

The 60 mm system was developed as a result of an agreement between Israeli Military Industries (IMI) and OTO Melara of Italy, now part of Leonardo-Finmeccanica's Defence Systems Division.

This project started in the late 1970s for a new kind of gun for light or medium AFVs, primarily designed around a high performance APFSDS round intended to penetrate earlier generations of main battle tanks and the side armour of contemporary ones. It was proposed for rearming older tanks as well as equipping new vehicles. IMI referred to the gun as the Hypervelocity Medium Support Weapon (HVMS 60) while OTO Melara, who had been thinking along similar lines, called it the High Velocity Gun System (60 HVGS). Initially, IMI were expected to develop the gun and ammunition, and OTO Melara the turret and autoloader. However, in the event both companies went their own way in developing the systems, although they retained the same cartridge case dimensions.

The IMI HVMS 60 has a vertical breech-block, electrical firing and a hydrospring recoil system. The gun weighs 700 kg of which 500 kg is the recoiling mass (recoil distance 27 cm). The barrel is L/70, the overall gun length including autoloader is 510 cm. Two types of loading system were originally being developed for this system, one recoil powered and the other hydraulically power-assisted. The recoil-powered system has two three-round magazines mounted on the gun, permitting firing of a three-round burst in three seconds or semi-automatic fire. IMI also developed a prototype two-person turret for this weapon, weighing about 2,200 kg, which for trials purposes was installed on an M113 APC.

The OTO Melara 60 HVGS also uses a vertical sliding breech block, an electrically actuated firing system and a hydrospring recoil system with a 27 cm recoil movement. The barrel length is the same, but the gun weighs 1,000 kg. It is fitted with a dual automatic loading system capable of loading and firing at any gun elevation with a rate of one round every two seconds. The electrically-operated loading system uses twin magazines holding 32 rounds in total, with reloading possible from inside the vehicle. Oto Melara developed the T 60/70 two-person turret, which can be installed on a wide range of tracked and wheeled vehicles. In mid-2000 the Oto Melara T 60/70 turret was shown fitted to the IVECO/Oto Melara VCI (8 × 8) being developed to meet the operational requirements of the Italian Army.

Incidentally, the 60 × 410R ammunition was developed from the 76.2 × 580R round for the US Army's post-World War II M32 gun, used in the M41 light tank.

After all of this work, the only customer for the system was Chile, which bought a manually-loaded version of the IMI gun for rearming its (ex-Israeli) M50 Sherman tanks (sometimes designated M60), and a version firing a three-round burst to their M24 Chaffee light tanks, which were in service for the last decade of the 20th century. Subsequently, the Chaffee gun was installed in a Chilean-built Piranha 6x6 light AFV, but it is unclear whether or not this saw service.

**CTAS 40**: 40 × 255 AMMUNITION.
The origin of the Cased Telescoped Armament System (CTAS) project is described in the entry for the GIAT 45M911 in the France section. In 1994, Giat Industries (now Nexter Systems) and Royal Ordnance (now BAE Systems) formed a 50:50 joint venture known as CTA International (CTAI). Based at Bourges, CTAI was formed to study, design, develop, industrialise, promote and market all weapons and corresponding ammunition using CTA technology. This resulted in the CT2000 AFV gun in 1997, with a calibre of 40 mm, cartridge dimensions of 255 mm by 65 mm, and a steel case. The complete system was subsequently renamed, with both Cased Telescoped Armament System (CTAS) and Case Telescoped Weapon System (CTWS) being used, as has Cased Telescoped Cannon and Ammunition (CTCA) and, separately, Cased Telescoped Cannon (CTC) and Cased Telescoped Ammunition (CTA). More information about CT weapons and ammunition is included in the Unconventional Systems chapter.

The first British application for the CTAS was planned to be the new Tactical Reconnaissance Armoured Combat Equipment Requirement (TRACER), which was developed alongside the US Army's Future Scout Cavalry System (FSCS) programme. However, both of these projects were cancelled in 2002, being replaced by the UK Future Rapid Effect System (FRES) and US Future Combat System (FCS) programmes (the latter itself being cancelled in 2009).

In 2002 a development contract for the system was awarded to CTAI by the French and British ministries of defence. For the UK, this was intended to replace the 30 mm Rarden gun in future medium

AFV projects, and would initially equip the British upgraded Warrior MICV under the WFLIP (Warrior Fightability and Lethality Improvement Programme – originally WLIP) later known as WCSP (Warrior Capability Sustainment Programme), as well as arming the new Future Rapid Effects System – Special Vehicle (Reconnaissance) (FRES-SV), subsequently SV Scout, and now the Ajax programme. The French were interested in the project to arm their new *Engin Blindé de Reconnaissance et de Combat* (EBRC) 6x6 AFV, since designated Jaguar. The French were initially interested in an unmanned turret, designated TOUTATIS, but later chose their own two-man design. Another application for the CTAS is the Thales RAPIDFire anti-aircraft system, which consists of a 6x6 UNIMOG chassis with a turret mounting the 40 CTAS, and light SAMs being optional extras. A naval mounting is also offered and in autumn 2020 was selected by the French defence procurement agency (DGA) for arming future warships.

In 2008 the British Ministry of Defence announced that the 40 CTAS had won a competition to provide the main gun armament for WCSP and SV Scout, the only rival gun submitted being the ATK 30 mm Mk 44. A separate competition, won by Lockheed Martin, was held to decide which turret the gun would be installed in. In March 2010, it was announced that Lockheed Martin would also be supplying the SV Scout turret, using a version of Rheinmetall's Lance Modular Turret System.

During 2013 the French *Délégation Général pour l'Armement* (DGA) selected the 40 mm CTAS for their new EBRC reconnaissance vehicle, with deliveries from 2020. It is planned to procure 250 EBRC, which could take the total production of the system for French and British armies to more than 800 weapons. CTAS received formal qualification from the French and British ministries of defence in 2014.

In 2015 CTAI was awarded a contract from the UK Ministry of Defence to commence production of 515 CTAS with the first deliveries being due in 2016. The CTAS guns will be made in France, with ammunition being manufactured in both France and the UK. The gun was initially expected to enter British service in the late 2010s, but in 2021 the WCSP was cancelled and technical problems were reported with the Ajax programme; at the time of writing the gun has still not entered UK service. The Jaguar programme does not appear to have experienced such problems and deliveries to the French Army are understood to be underway.

The cannon installation consists of five elements: (1) the externally powered CT Cannon (CTC); (2) the linkless Ammunition Handling System (AHS); (3) the CTAS Controller (CTAS-C); (4) the Gun Control Equipment (GCE); and (5) the gun mount. The chamber is separate from the L/70 barrel, and pivots horizontally through 90° to enable a new round to be pushed in from the side, which simultaneously pushes the fired case out of the other side. The ammunition feed is on the same axis as the trunnions, so the location of the feed is unaffected by gun elevation. This permits the Ammunition Handling System to be fixed in place, both saving on the space normally required to allow belt feeds to move and flex as they connect the gun breech with the magazine, and also removing a major cause of unreliability through failure to feed. Compared with conventional cannon, far less of the turret volume is taken up by the gun mechanism. The loading system incorporates a magnetic induction fuze setter for programming airburst ammunition.

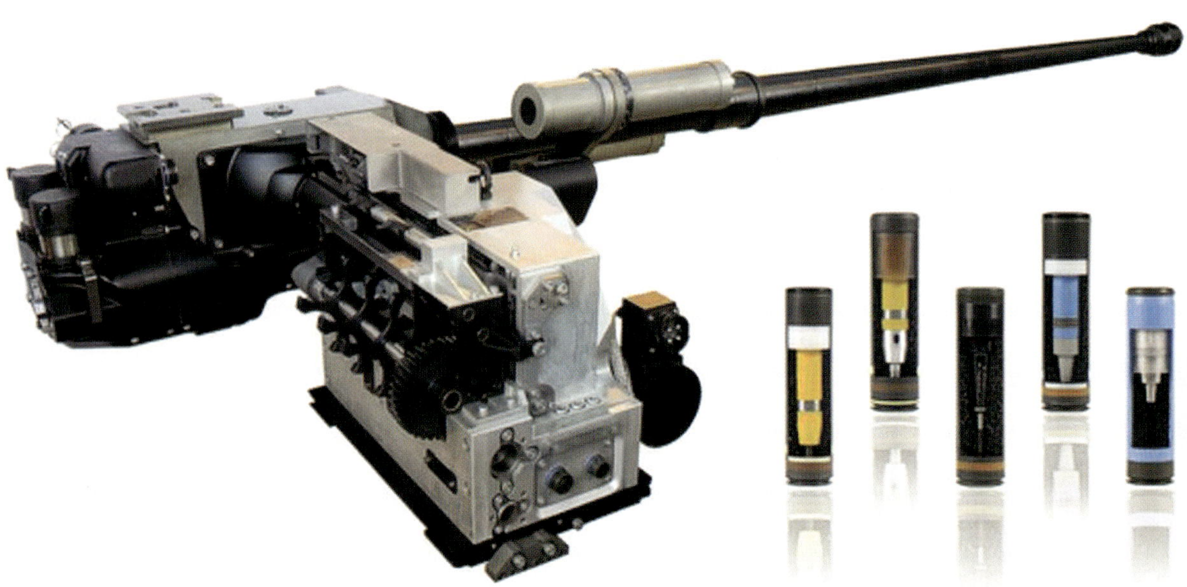

*40 mm CTAS with ammunition and handling system (CTAI)*

# ITALY

During World War I Italy's automatic cannon were largely based on British designs, primarily as a result of the formation in 1905 of Vickers-Terni; a joint project between Vickers and the Terni steelworks in central Italy. In 1911 a joint Army and Navy commission was established to consider the adoption of an AA gun; it remained in existence until 1915. Three British weapons were evaluated, all from Vickers: a 37 mm (presumably the 1½-pounder pom-pom); a 25.4 mm (1-inch); and a 40 mm (2-pounder pom-pom). The 37 mm was not chosen, but the other two did enter production at the Vickers-Terni factory.

The most successful gun was the 40/39 Vickers-Terni (40 mm calibre, L/39 barrel – the standard means of identifying Italian cannon at this time) which was the British 2-pounder Mk II. This entered service in small numbers during World War I, probably from British production initially, and was manufactured in some numbers thereafter. The original version was the Mod. 15–17, followed by the Modif. 30 in 1930. Its full designation was *Cannone-mitragliera da 40/39 Mod. 1915 Modif. 1930* (Automatic gun 40/39 Model 1915 Modified 1930); it wasn't a new build model but a modification of the existing Mod. 15–17. The modifications consisted mainly of a new sight and gunner arrangements, a 50-round rather than 25-round belt feed, and uprated ammunition. It was made until the mid-1930s in single mountings; production numbers are not known, but it remained the Italian navy's standard heavy automatic AA weapon until replaced by the Breda 37/54, and the navy still had 141 in service in 1941. Others were exported as part of the armament of warships sold to Sweden, Greece, Paraguay, Turkey and Argentina.

The other cannon made at Terni was the Vickers 1-inch (described in the UK section), which saw very limited use with Italy, but did inspire designer Revelli (Colonel Abiel Bethel Revelli) to develop an improved weapon using rimless 25.4 × 87 ammunition, which was made by F.I.A.T. (*Fabbrica Italiana Automobili Torino* = Italian car factory of Turin). This Revelli-F.I.A.T. did get into service but production had to be stopped when Vickers threatened to sue for breach of patent.

There was another 25 mm gun which saw Italian service, designated the *cannone 25/41*. This was an old manually-loaded single-shot naval gun which was installed on some aircraft, notably in the nose of a Macchi L flying boat. This probably used 25.4 × 95R Nordenfelt ammunition, although that has not been confirmed.

In the 1930s both Breda and Scotti became involved in designing and making indigenous cannon in both 20 mm and 37 mm calibres. Alfredo Scotti was a designer and had no factory; his guns for Italian use were mostly made by Isotta-Fraschini (owned by the Caproni group) and are sometimes listed under that name, while guns for export were initially made by Oerlikon in Switzerland. Scotti was responsible for a series of machine guns, plus several automatic cannon only one of which saw verified service – a 20 mm AA gun. Breda was the primary manufacturer of service cannon, supplying the standard AA weapons for the army and navy in 20 mm and 37 mm calibre. Small numbers of a 20 mm Grandi cannon were also made; photos exist, but little data except that it used an operating system similar to Scotti's. Ansaldo's 20 mm design seems never to have got off the drawing board, although they did make an experimental 47 mm aircraft gun during World War II.

Other large-calibre automatic cannon projected during World War II were an aircraft version of the 47/48 gun (an experimental anti-tank gun proposed as successor for the 47/40 gun but never adopted); construction is listed as canceled after the armistice. Another mention of a 47/50 mm automatic gun, but for anti-aircraft use; reports that during 1942 the Italian Army and Navy were evaluating a 50/47 mm gun studied by Ansaldo (with the help of Oerlikon) and a project for a 50/62 automatic gun proposed by Caproni.

After the Italian Armistice in September 1943 the Italian armed forces received 40/60 Bofors from the British, and in the post-war period the Army adopted the gun in single mountings, the Navy in single, twin and quadruple mountings, supplied from British and American stocks. Since then, the Italian armaments industry has concentrated on developing mountings – especially naval – for foreign medium-calibre designs, particularly the 30 mm Mauser MK 30F and the 35 mm Oerlikon KDC. They have also produced various mountings for the 40 mm Bofors L/70 (a production licence was bought) plus undertaken some significant modifications to produce the OTO Breda 'Fast Forty'. This resulted from the granting of a licence agreement by Bofors in the 1960s to *Breda Meccanica Bresciana* (later Alenia Difesa OTO Breda Division and, from December 2001, OTO Melara). The focus was mainly on developing large-capacity automatic feed systems in such mountings as the Type 70 twin (also known as the 40 L/70 Compact).

The most recent development is the 'Forty Light'; a lightweight (c.2,000 kg) single naval mounting with dual feed and a carbon composite gun shield, first announced in 2015. The total on-mount magazine capacity is 72 rounds ready for use via a dual-feed system and the gun fires programmable ammunition similar to the Bofors 3P. It normally fires at 300 rpm but the 'Fast Forty' modifications can be applied if required.

Concerning the origin of the names OTO Breda and OTO Melara: OTO (the name dates from 1929) stands for *Odero Terni Orlando*, Odero and Orlando being two of the founders of the company, and Terni being the location of the original factory. Breda and Melara are the names of other factories, located in Brescia and La Spezia respectively. In 1992 the company became part of Finmeccanica S.p.A., and in 2001 the naval gun division of OTO Melara merged with Breda Meccanica Bresciana to form Otobreda. In 2016, the company became a part of Leonardo-Finmeccanica's Defence Systems Division; the company is now known as Leonardo S.p.A..

**25.4 mm F.I.A.T. REVELLI (25,4 F)**: 25.4 × 87 AMMUNITION.

The Italians produced and fielded one of the few cannon to see air service in World War I, the 25.4 mm F.I.A.T. Revelli (*Mitragliera da 25,4 F*). The gun was designed in 1916 by Abiel Bethel Revelli and built by F.I.A.T.. Testing in September/October 1916 was followed by an order for 200 units, of which 133 had been delivered by March 1918. The 25.4 mm cartridge was of similar dimensions to the 1-inch Vickers (described in the British section), but rimless and with a slightly wider case body which allowed the inclusion of a shoulder for headspacing. It fired a similar range of projectiles as the 1-inch Vickers at a slightly reduced MV.

The gun used a short-recoil mechanism which unlocked the bolt after recoiling about 20 mm. It was not belt-fed, instead using an 8-round box magazine mounted vertically above the breech. The gun weighed 45 kg and was 129 cm long with an L/28 barrel.

Despite *Mitragliera* meaning 'automatic gun', this was actually semi-automatic, the gunner having to press the firing button for each shot. It was still possible to empty the 8-round magazine in two seconds, giving a theoretical cyclic rate of 240 rpm, although the need for frequent magazine changes resulted in a very much lower practical RoF.

Despite high expectations, the 25,4 F was disappointing in service due to the low RoF, the poor destructive power of the 25 mm projectiles and an oversensitive percussion fuze which gave the HE shells a tendency to explode prematurely. The ballistic performance of the gun was also regarded as unsatisfactory. As a result of all this, the weapon was never appreciated by the crews, who preferred machine guns. Use of the 25,4 F was therefore relegated to the Savoia-Pomilio S.P.2 second-line air defence aircraft, and it also saw some use in Caproni bombers such as the Ca.3, Ca.4 and Ca. 5 plus two airships (V2 and P7).

In 1917 the 25,4 F was tested as a possible light infantry gun but, in addition to the problems mentioned above, it was rejected for this role because it was too underpowered, the consumption of ammunition was excessively high and the carriage was too large and heavy.

**37 mm BREDA 37/54**: 37 × 232SR AMMUNITION.

In 1928 a joint Army and Navy Commission was set up to evaluate 37–40 mm automatic guns. Guns from several firms were examined over the next few years: Colt-Browning; Rheinmetall/Solothurn; Scotti (Model 31, with an 8-round drum and a RoF of 100 rpm, built by Ansaldo to Scotti's design); 25 mm Hotchkiss and the Breda 37/54. In 1932 the Navy selected the Breda, which used a gas piston-operated gun mechanism with a rising block breech lock and a long bolt-stroke. The Army made no decision for several years, finally adopting the Breda in 1939.

Known formally as the *Cannone-Mitragliera da 37/54*, the Breda was developed in conjunction with a new cartridge that was never used in any other service weapon. The overall gun length was 323 cm and the weight 275 kg. The gun had the usual loading trays fed in this case with six-round clips: these were mounted vertically above the gun in twin mountings, horizontally in the singles. The cyclic RoF was initially variable and could be set to 30, 60, or 90 rpm, or unrestricted; in the latter case, the RoF was around 140–150rpm with the gun horizontal, or 120rpm when elevated. The option of the lower RoFs may have been deleted later. Maximum range was 7,800 m and ceiling 5,000 m.

It was Italy's standard heavy automatic cannon throughout World War II, being used by both the *Regia Marina* and the land forces. There were several mountings, most of them naval.

The first was the M32 twin naval mounting, which had water-cooled barrels and was available either with or without stabilisation; mounting weight was 5,000 kg. The lack of any recoil absorption in the mounting caused severe vibrations.

The M38 was a simplified and lighter version of this, with stabilisation and air-cooling and a mounting weight of 4,300 kg.

Next came the M39 R.M.; a single mounting in three versions. The M39 R.M. *a scomparsa* could collapse into a deck hatch; the M39 R.M. *ad un puntatore* had one aimer, and the M39 R.M. *a due puntatori* had two. All were air-cooled and lacked stabilisation, but did have a recoiling cradle which greatly reduced vibration and enabled lighter mountings to be used – from 1,500 to 2,000 kg. Curiously, the collapsible model was not used in submarines as might be expected, but in the rebuilt *Duilio* class battleships and the newly built *Littorio* class battleships; from photos, it looks as if these cannon may have been mounted

*25.4 mm F.I.A.T.-Revelli (Chinn)*

*Breda 37/54 Mod 38 (Wikimedia Commons / Public Domain)*

in areas affected by the muzzle blast of the big guns.

The first mounting designed for land use was the Mod.39 for the D.I.C.A.T. (the AA territorial defence militia), which differed from the naval M39 only in the sighting arrangements. It was a fixed mounting weighing nearly 3,000 kg. A mobile field mounting version was the Mod.41, but the order was cancelled when the Army opted for the Bofors 40/56 instead.

Another mounting which did not see production was a quadruple turret dating from 1933, effectively two twin mountings side-by-side, which was intended for the initial reconstruction project of the *Cavour* class battleships. It was rejected in favour of the standard twin mountings. Other abortive proposals were a Mod.38 with recoiling cradles, and another proposal for a quad version of this.

The 37/54 also saw some experimentation as an aircraft gun. The Fiat-Cansa FC 20 *bis*, *ter* and *quater* versions (one 37/54 in the nose with 42 rounds, always in clips); the Savoia-Marchetti SM 89 (two 37/54 in the nose with 36 rounds each, always in clips); the AS 14 (one in the nose, a wooden mock- up); one modified BR 20 (it is not known if this plane was completed, the only known photo shows a 37/54 wooden mock-up placed in the nose as for the AS 14). All the guns were fed with a 6-round clip; the twin mountings and the guns installed on planes were fed from above, the single mountings were fed from the left side.

**20 mm BREDA 20/65 AND 20/70:**
20 × 138B AMMUNITION.

This air-cooled gun, properly known as the *Cannone-Mitragliera da 20/65 modello 35 (Breda)* was the culmination of a programme which started with the 20/65 mod.31 prototype; then was put into production for export (notably to China) as the mod.33; and with minor modifications was finally adopted by the Italian Army in 1935 as the 20/65 mod.35 and (in a twin stabilized mounting) by the Navy as the mod.35 R.M.. Both types of mounting were widely exported.

This series of guns was developed around the so-called 'Long Solothurn' ammunition introduced by Rheinmetall's Swiss subsidiary and used extensively by Germany in both AA and light AFV applications. It appears that German assistance with the gun design was also obtained, as there was a contract between Breda and Rheinmetall concerning it. Breda chose a Hotchkiss-type gas piston-operated gun mechanism, an enlarged version of the Breda 13.2 mm M31 machine gun. The mod.35 utilized a rising block breech lock very much like the 37/54 described above, and included a mechanism to replace the fired cases in the 12-round strip feed (which was really more like a tray). The cartridges were clipped into the tray, which was inserted sideways into the gun with the cartridges under the tray rather than on top of it (the 20 mm Scotti used the same tray system). The long bolt-stroke constrained the RoF to about 220–240 rpm (practical rate just 150 rpm). Gun weight was 69 kg, the barrel was L/72, and overall length was 203 cm. As with other Italian automatic cannon, the M35 was often bolted to its mountings without any provision for recoil absorption, which meant that the pedestal mountings needed to be fixed to strong structures, and vibration was considerable.

*Breda 20/65 on mod.35 field mounting in firing position (from Army manual)*

The gun was issued as an AA weapon, although the army hoped it would be effective against light AFVs. The standard M35 army single field mounting was a tripod with widely spaced legs (two wheels were fitted for towing). Single free-swinging mountings were also used by D.I.C.A.T. and the Army (mod.39 and mod.40) and also by the Navy (mod.39 R.M. and mod.40 R.M.). These mountings did allow for a measure of recoil management, in that the guns could recoil 13 mm in their cradle, against the pressure of a spring. Other projects for different mountings (including collapsible ones) were all cancelled due to the armistice.

The mod.35 was slightly modified for AFV installations, and used an 8-round strip to save space.

From 1937 Breda experimented with belt-fed 20 mm aircraft guns, all of them using 20 × 138B ammunition and fitted with L/70 barrels. The first prototype 20/70 looked similar to the 20/65 mod.35 but gradually evolved into the much slimmer-looking mod.41. The only data available for the mod.41 are a weight of 42 kg and a RoF of 600 rpm. All of them were generically known as the Breda AVIO CL.20 (AVIO meaning 'aeronautical' and CL the abbreviation for 'calibre'). It seems that the mod.41 was never formally tested in an aircraft, and it was rejected in favour of the German MG 151/20.

This was not the end of the mod.41, as a magazine-fed version was developed as an AA gun for use by the Army and Navy. This had a shorter but heavier recoiling barrel so was designated the 20/65 mod.43 (or sometimes the mod.41–43). The standard magazine was a top-mounted 50-round open drum with the ammunition exposed (similar to the 20 mm Bofors m/40), but there was also a 25-round box magazine. The mod.43 weighed 63 kg and fired at 350–380 rpm. Installations included free-swinging single and twin mountings, plus a sextuple mount intended for the Navy. Plans were also drawn up for a *Torretta pre-stabilizzata per 4 complessi binati da 20* (Pre-Stabilised turret for 4 twin 20 mm mountings), with the four mountings in a diamond pattern on a stabilised pivoting platform; it had a crew of sixteen.

The Breda 20/65 mod.43 was adopted by both the Army and Navy in 1943, but the armistice soon stopped work.

*Breda 20/65 on mod.35 R.M. Twin mounting for the Navy (from Navy manual)*

***20 mm SCOTTI EXPERIMENTAL CANNON***: 20 × 70RB; 20 × 110 SCOTTI; 20 × 138B.

Alfredo Scotti, who established the *Società Anonima Armi Automatiche Scotti* (Armiscotti) in Brescia near Milan, designed a delayed blowback action which he considered to be suitable for automatic weapons of all calibres. The gun fired from a locked breech which was then unlocked by means of a gas-piston system; locking was by means of a rotating bolt head; extraction was powered by blowback. The lack of primary extraction required the cases to be lubricated.

Several different models of Scotti 20 mm gun were designed and built, but only one series saw service: the 20/70 mod.38, mod.39 and mod 41., which has its own entry below. This entry concerns the other guns, in so far as they are known – there are still various uncertainties.

Scotti started work on a 20 mm gun in the late 1920s and in 1931 formed *Brevetti Scotti AG*, a joint company in partnership with the owner of Oerlikon, Emile George Bührle. In the following year Oerlikon acquired the patent rights to Scotti's products with the exception of Italy, for which Isotta-Fraschini held the rights. Other makers also got involved: there was another production line at the subsidiary company *Fiat – Officine Meccaniche*. This resulted in two sub-designations: Scotti-IF and Scotti-OM. Grandi also made some guns for Scotti, as well as offering their own designs.

The development of Scotti's 20 mm cannon is not entirely understood, so what follows is subject to change if more authoritative sources emerge.

In 1932, the first prototype of Scotti's 20 mm cannon appeared, designated 20/70, as it had a 70-calibre barrel. It was built by Grandi, and is most notable for using a double-drum magazine. Grandi also built a fixed twin mounting in 1934, which presumably used the same ammunition.

The ammunition used in these guns is debatable, with conflicting information from various sources. One of the few certainties is that Scotti developed his own cartridge, a 20 × 110 which was a halfway house between the Oerlikon Type S 20 × 110RB and the Hispano-Suiza 20 × 110 of the HS 404: it combined the sloping shoulder of the Oerlikon with the rimless base of the Hispano. At least one case survives, helpfully headstamped **20 mm Scotti 1933**, thereby providing a date as well as confirming the origin. It seem that Scotti started out using Oerlikon ammunition (not surprising, given his

*Scotti M35 (copyright Rheinmetall Air Defence AG)*

close links with the company) with the balance of probabilities suggesting that his M32 gun may have started development with the Type S ammunition and finished with Scotti's own cartridge.

In 1935 the M35 or Scotti-I.F. 20/40 emerged, a much shorter and lighter gun specifically for wing-mounting in aircraft. Brief details of this weapon are provided in an Oerlikon-Contraves Museum catalogue (the gun still survives, in the Swiss Military Museum at Full). It is stated to weigh 23 kg, with an 80 cm barrel (much smaller and lighter than his other 20 mm guns), fed by a 15-round box or 60-round drum magazine and fired at 350 rpm.

The size and weight of the gun are similar to the Oerlikon FF so the ammunition was also presumably comparable to its 20 × 72RB. This ties in with a drawing from SMI (the ammunition maker *Società Metallurgica Italiana*) of a 20 × 70RB Ehrhardt round, dated 1936 and titled *bossolo modello F per mitragliera Scotti 20 mm* (= model F case for Scotti 20 mm MG). It does seem surprising that the cartridge from the obsolete German Ehrhardt cannon was selected, but that was presumably the most suitable one available with the required performance. The museum catalogue also gives an MV of 850 m/s but that cannot be right and must have related to another cannon: around 670 m/s has been quoted for the Ehrhardt.

At the same time as the M35 was proposed, a third attempt was made at a 20/70 gun, with drum or box magazines, in single or twin AA mountings. This was built initially by Grandi and subsequently by I.F.. In order to compete with the Breda 20/70 for Italian military orders this model adopted both the 20 × 138B ammunition and its 12-round tray feed. In 1938 these efforts succeeded, with both the Navy and the Air Force (for airfield AA protection) adopting the gun as the mod.38, as described in the separate section below.

Scotti continued to develop his ideas: in 1939, following the example of Breda, Scotti proposed another 20 mm gun for aircraft, the Scotti 20/90 (the very long barrel indicating that it was meant for an engine mounting, with the barrel reaching all the way to the propeller hub), but the gun was rejected since at that time there were practically no liquid-cooled vee-engines used in the *Regia Aeronautica* able to take an engine cannon. It is possible that Scotti made other proposals: one source mentions a lightened (56 kg) aircraft version of the M41 AA gun, designated mod.43, but this has not been confirmed.

According to Chinn (but unconfirmed), some of the Scotti aircraft guns were sold to small countries in Europe and South America, and the cannon was ultimately developed to use belt feed and to reliably achieve a RoF of 600 rpm. However, it is not clear to which models he was referring.

It is clear that Scotti's cannon provided inspiration to Marc Birkigt of Hispano Suiza France (which had a close working relationship with Oerlikon), in the development of his HS 404 cannon, which started in 1933. This also uses a delayed-blowback gas-unlocked action (albeit with a different locking system, using pivoting flaps rather than a rotating bolt head) and the cartridge case was clearly developed from the Scotti's, with a sharper shoulder.

**20 mm SCOTTI 20/70 MOD.38, MOD.39 AND MOD.41:**
20 × 138B AMMUNITION.

Scotti's designs for automatic guns were all based around a delayed-blowback operating system as described in the Scotti experimental cannon section (see above). The official name of the guns described here was *cannone mitragliera Scotti da 20/70* for a series of three models, from mod.38 to mod.41. The firing sequence of the *Cannone-Mitragliera da 20/70* started with an open bolt, to aid gun cooling. It used strip feed from 12-round trays, identical to the Breda M35's and similarly replacing the fired cases into the tray (as with the Breda, the rounds were clipped to the underside of the tray – see photo). The RoF was 230 rpm. The gun barrel was an L/70 and the gun was 227.5 cm overall. The weight was 73 kg for the Model 41.

Like other Italian automatic cannon, the Scotti was usually bolted to its mounting without any provision for recoil absorption.

In 1938 Isotta-Fraschini started making the cannon for the AA role, on either a fixed pedestal mounting (Mod.39) or a two-wheeled carriage (Mod.41). About 300 saw service with both the *Regio Esercito* and the *Regia Marina*, the latter also including use in a modified RM35 twin mounting plus an M41 single in a disappearing mounting in some submarines. A quadruple mounting was fitted to both a field mounting and a SPAAG (based on the *Semovente M42*); there are suggestings that the guns were belt-fed, which seems logical given the alternative of a 12-round tray, but that has not been confirmed.

*20 mm Scotti 20/77 AA gun (Bundesarchiv, via Wikimedia Commons, Bild 101I-316–1162–11 / Demmer / CC-BY-SA 3.0)*

***47 mm ANSALDO AUTOMATIC GUN***: (PROBABLY 47 × 195R AMMUNITION).

Chinn describes in detail an experimental Italian 47 mm automatic aircraft cannon in development at the end of World War II, without providing any data or specifying the type of ammunition. However, another source states that this used the same ammunition as the manually-loaded Austrian Böhler light anti-tank/field gun which was in Italian army service as the 47/32, therefore presumably had the same barrel length of 152 cm. The ammunition feed was vertical via strips. The gun used a medium-long-recoil system of operation; the barrel and breechblock recoil together for about 127 mm before unlocking, using an interrupted-thread lock and a rotating breech-block.

There is some confusion over the gun's design history and designation. One unverified source refers to the gun as the C.102, while Chinn states that it was a Breda project. However, it is clear from Italian sources that it was an Ansaldo project; their listof gun projects includes a *Mitragliera da 47 tipo B* (47 mm automatic gun type B).

Various unverified sources state that the gun was installed in a FIAT AS 14 (in the nose) and a Breda 88 (under the fuselage), but the photo of the Ba 88 probably just shows a wooden mock-up of an unidentified gun, while the AS 14 was intended to take the 37/54 (again, the photo probably shows a mockup of the gun). The only aircraft known to have been planned to take 47 mm automatic cannon was the *Cannoniere Santangelo* (literally "Santangelo gunner") a twin-engined plane vaguely resembling the Bf 110, which was to have had two guns in the nose with 35 rounds each. A wooden mockup of this plane was made by Ansaldo in 1943, and two prototypes were under construction at the time of the Italian Armistice.

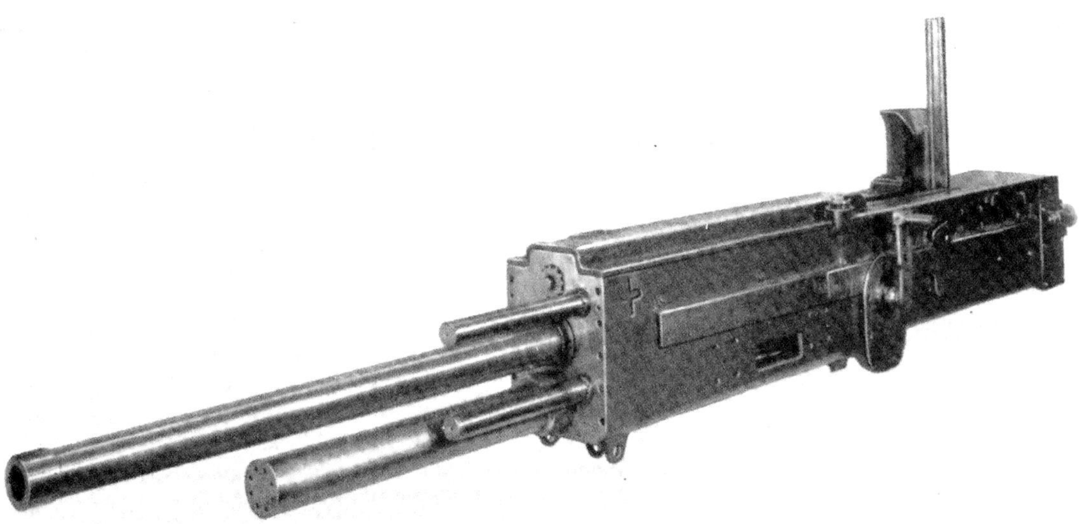

*Ansaldo Mitragliera da 47 tipo B (Chinn)*

**OTOBREDA 'FAST FORTY':**
40 × 365R AMMUNITION.
The particular type of recoil mechanism used in the Bofors has the advantage that the length of the recoil stroke is not determined by the cartridge length. In fact this, like the Oerlikon KDE, is not a 'pure' long-recoil design as the breechblock is unlocked and extraction commences before the barrel has recoiled the full distance required to chamber a fresh cartridge. It is therefore possible to shorten the recoil stroke, and this has been done (from 250 to 110 mm) by OTO Melara in its Fast Forty gun, as a part of a package of measures which have speeded up the rate of fire to 450 rpm. Only the gun barrel and breechblock remain from the original Bofors gun. Ammunition is supplied from a 144-round below-deck magazine via a linkless feed. The Fast Forty first appeared in a naval twin mounting in the mid 1980s, with a single mounting appearing in 1989; this achieved its first export sale in the 1990s.

The Chinese NORINCO 37 mm Type 76A, based indirectly on the Bofors mechanism, has been speeded up in a similar way.

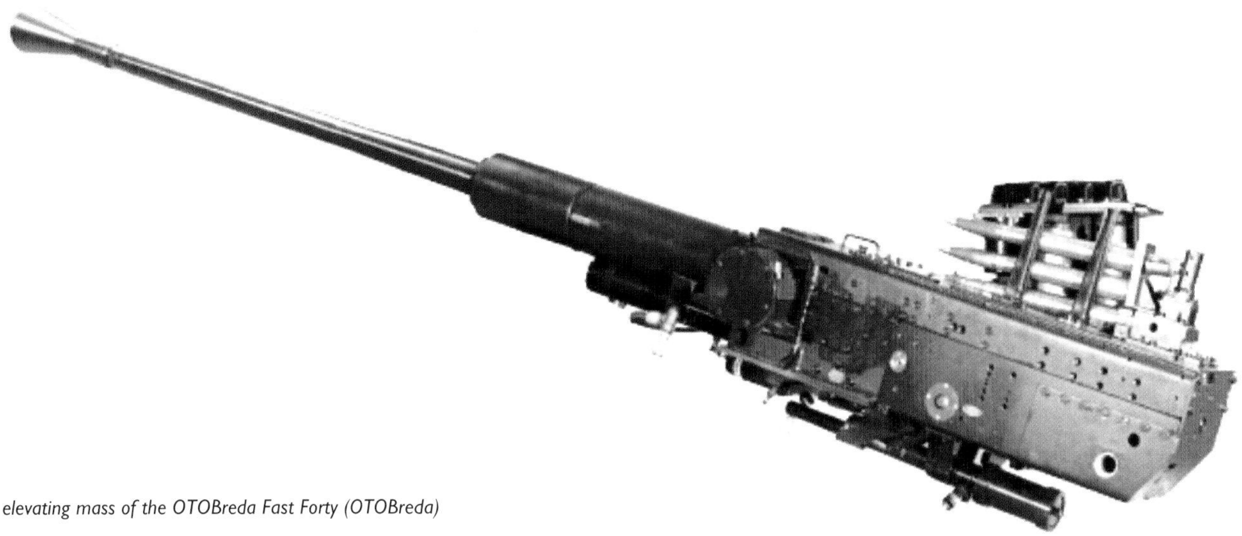

*The elevating mass of the OTOBreda Fast Forty (OTOBreda)*

# JAPAN

The development of Japanese cannon was concentrated into a period between the early 1930s and 1945, but despite this an astonishing variety of guns were produced, particularly for aircraft use. This diversity was multiplied by the segmentation of the market: the Imperial Japanese Army (IJA) air force and Imperial Japanese Navy (IJN) air force were entirely separate organisations which developed their own weapons and ammunition without reference to each other; similarly, the parent IJA and IJN also went their own way in choosing AA cannon.

Two designation systems were used, the standard one being a Type number related to the date of adoption according to the contemporary Japanese Imperial calendar, which counted the years since the first Emperor of Japan in 660 BC. The IJA's 20 mm Type 94 AA gun was completed in 1934, which was 2594 years after 660 BC, and the last two numbers of the date formed the type number (basically, just knock 60 off the Type number and add 19 in front to obtain the western year of adoption). There was potential for confusion with this system; e.g. the IJA had two very different 20 mm Type 94 cannon in service at the same time, one an aircraft gun (designated Flex) and one an AA gun. Starting in about 1940 the IJA started to assign design project Ho-numbers to aircraft guns of more than 11 mm calibre (short for *kikan hou* = machine cannon, or *Tai hō* = big gun). As these were unique designations, they effectively replaced the Type numbers in general use. The IJN stuck with the Type numbers, and called any automatic weapon a machine gun (*Kijyu* = mechanical small gun).

Initially, some foreign cannon designs were acquired or made under licence. One of the first was the Vickers 2-pounder naval AA gun, of which 500 guns and 200 mountings (single or twin) were imported from 1925 to 1935. This gun was obsolete by the start of World War II, having been replaced in the shipborne close-range AA role by the 25 mm Hotchkiss M38/39 (designated 25 mm Type 96 by the IJN, therefore adopted in 1936, earlier than in France), which was made in huge numbers in World War II and installed in single, twin and triple mountings. The first versions, designated *Ho shiki 25 mm Kijyu* and *95 shiki 25 mm Kijyu*, were all imported: the Japanese-made ones were the *96 shiki 25 mm Kijyu*.

Both the IJA and IJN were interested in 20 mm Rheinmetall/Solothurn cannon: the army used the basic design as the basis for their unsuccessful 20 mm Type 94 AA gun (rechambered in 20 × 157), while the the navy designated the 2 cm Flak 30 as the 20 mm Type 2 AA gun (18 guns and 150,000 rounds of 20 × 138B ammunition were supplied to Japan by 1942, with two extra guns and more ammunition the following year). The Japanese were sufficiently impressed with the 40 mm Bofors guns (captured from the British and Dutch) to begin manufacturing the ammunition, and they copied the design as the 40 mm Type 5 but, as the Type number indicates, this only emerged in 1945, too late to have any effect.

Two other firms which had a major influence were Oerlikon and (less directly) Browning. In addition, several cannon were entirely indigenous designs.

Japan began its association with Oerlikon in the early 1930s when the army examined the 20 mm Type F and Type L as candidates for defensive mountings in heavy bombers. Initial experiments in 1932 resulted in a combination of the short barrel from the Type F with the longer action, magazine and ammunition from the Type L. This proved unsatisfactory but 1933 tests of a more mildly modified Type L (from its appearance, the Type AL version) were successful, the outcome being designated the 20 mm Type 94 Flex. These guns were reportedly fitted to the six Mitsubishi Ki-20 – a heavy bomber version of the Junkers G.38. The Type 94 Flex was never fitted to any other aircraft, and seems to have been out of service by 1941.

The IJN became interested in the new Oerlikon FF, FFL and FFS aircraft cannon (lighter and faster-firing than the Type F, L and S) when they emerged in the mid-1930s, acquiring examples of the three models and buying a licence to make them. The IJN adopted the FF and FFL, designating them the 20 mm Type 99–1 and Type 99–2, and these became the IJN's standard aircraft cannon throughout the war (the FFS, which would have been designated the Type 99–3, was envisaged as a motor cannon, but the Japanese made little use of the vee-engines needed for such an installation). The Types 99–1 and –2 did not receive these designations until 1941 and 1942 respectively, being known until then as *E-shiki*, or 'Type Oerlikon' (Japanese does not use the 'L' sound, so Oerlikon sounded rather like 'Erriken'). Different designation systems were used over time for a variety of fixed and flexible sub-types.

Two other Japanese aircraft cannon which also used the API blowback mechanism were the naval 30 mm Type 2, which was just a scaled-up version of the standard Oerlikon design, and the army 40 mm Ho-301, which differed in using caseless ammunition, with the propellant contained within the base of the projectile. Neither saw much use, the promising 30 mm Type 2 apparently being the victim of internal IJN politics. Incidentally, the Type 2 is in principle comparable to the German MK 108 and the ammunition looks very similar, but in fact the MK 108 was radically repackaged and bore no visual resemblance to the Oerlikon. It appears that the German and Japanese weapons were developed in isolation from each other.

The IJA followed a different route with their aircraft cannon, showing a marked preference for the Browning short-recoil designs in three new domestic calibres: 20 × 94, 30 × 115 and 37 × 144. Curiously, only the IJA developed this mechanism to take cannon-calibre ammunition: no other Browning short-recoil guns made elsewhere exceeded 13.2 mm in calibre. It has been suggested that this design of action was not best suited to larger and more powerful ammunition; the ballistics of these Japanese guns were relatively unimpressive, indicating that the ammunition was loaded down to a low pressure.

Japan also had a productive design-and-development capability, the work of engineer Masaya Kawamura being particularly notable. The Ho-301 already mentioned, while adopting the API blowback principle, was radically different from other guns of this type. Other indigenous designs were the Kawamura's 20 mm Type 98 and the similar 20 mm Type 97 for the IJA – originally a semi-automatic anti-tank rifle but adapted as an automatic aircraft gun in two versions, the Ho-1 and Ho-3, for flexible and fixed installations respectively. Then came two related designs, the 37 mm Ho-203 and 57 mm Ho-401, and finally the IJN 30 mm Type 5. These all saw service (albeit only just, in the latter cases); in addition, there were many experimentals about which little is now known.

Post-war there seems to have been little development effort, the Japanese preferring to build foreign cannon under licence (e.g. the 35 mm Oerlikons). There was one experimental effort in the early 2000s: a 40 mm gun which got as far as firing tests while mounted in an AFV turret. Neither the gun nor the ammunition bear any resemblance to the CTAI project, in fact they seem to be closer to the 50 mm Supershot in concept, but no performance data is available.

***20 mm TYPE 94 FLEX***:
20 × 99RB AMMUNITION.

This was a slightly modified version of the Oerlikon Type AL, the changes being made to the trigger and sights. It therefore used the API blowback mechanism. It fed from a compact 15-round drum, had an L/60 barrel (early experiments with an L/40 barrel being unsuccessful) and an overall length of 173 cm, weighed 43 kg (54 kg complete with shoulder stock and magazine) and fired at 350–380 rpm. It was evidently considered unsatisfactory as it saw little use, only being installed as a defensive gun in a few large IJA bombers. It was obsolete by the start of World War II. This gun is often credited with using a unique 20 × 89RB round, but it is now known that this was an error – it used standard Oerlikon Type L ammunition.

***20 mm TYPE 94***: 20 × 157
AMMUNITION.

In their search for a light AA gun the IJA tested a Rheinmetall-Borsig/Solothurn cannon, almost certainly in 20 × 138B calibre. This was chosen as the basis for the IJA's weapon but Lieutenant-Colonel Ishikaza of the Tokyo Arsenal decided to develop a more powerful round of ammunition for the gun. The result was a massive 20 × 157 cartridge. A drum magazine for 15 rounds was developed. No other data on the gun is available. In

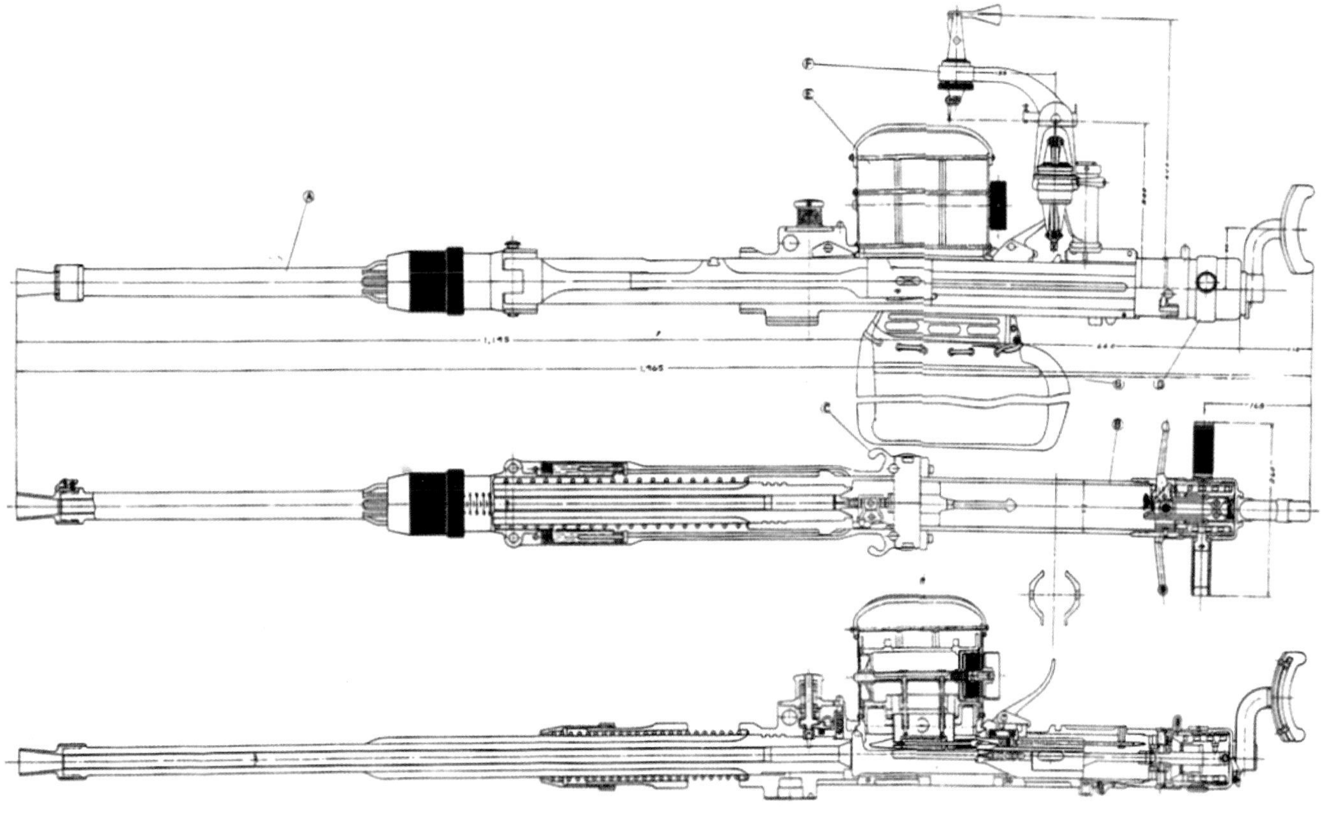

*20 mm Type 94 Flex (Courtesy Yasufumi Kunimoto)*

total, only 28 Type 94 guns were produced, 20 of them being used in military trials in China. These trials revealed shortcomings which led to the development of the Type 98 by Kawamura.

**25 mm TYPE 96**: 25 × 163 AMMUNITION.

This was essentially the same as the gas-operated Hotchkiss M38/39 described and illustrated in the France section. The main changes were the replacement of some parts by castings, the substitution of many stainless steel parts for use in submarines, and the fitting of Rheinmetall type flash eliminators which had a slight muzzle brake action. The RoF could be varied by a gas control valve, up to the cyclic maximum of 220 rpm (the practical RoF was 110 rpm). The top-mounted box magazine held 15 rounds. The gun weighed 115 kg, had an L/60 barrel and a total length of 230 cm. This was the only automatic AA cannon adopted by the IJN during this period and 33,000 were made, for installation in 20,000 single, twin and triple mountings, weighing 785, 1,100 and 1,800 kg respectively. These were mostly unpowered, except for some of the later triple mounts. (the triple mount had a crew of nine).

Despite its ubiquity, the IJN had several complaints about the Type 96: slow training and elevation of the twin and triple mountings (the singles were usually free-swinging); sights inadequate for high-speed targets; excessive vibration; inadequate magazine capacity; and muzzle blast.

**20 mm TYPE 98**: 20 × 143 AMMUNITION.

The Type 98 in 20 × 143 calibre was the standard army light AA gun throughout World War II. It was designed by Kawamura to replace the unsuccessful Type 94 in 20 × 157 calibre (see above), and was essentially a slightly enlarged version of his Type 97 rifle and therefore of the Ho-1 and Ho-3 cannon. The cartridge case was modified to take the same, more conventional, projectiles as the 20 × 124 rounds of these guns.

The operating mechanism was a combination of a gas piston (to unlock the tilting bolt) and blowback (to extract and eject the fired case), and the ammunition was fed via a vertically-mounted 20-round box magazine; the cyclic RoF was 450 rpm.The powerful ammunition made it more useful than most 20 mm guns in the anti-armour role, so it was regarded as a dual-purpose weapon by the IJA. Gun weight was 69 kg, the barrel an L/63 and overall length 201 cm.

In 1944 the IJA introduced the Type 4 Twin AA mounting with two Type 98 guns. Approximately 500 of these were produced.

**20 mm TYPE 99–1**: 20 × 72RB AMMUNITION.

This was the Oerlikon FF described in the Switzerland section, built under licence. Like the German MG-FF, the Type 99–1

*20 mm Type 98 ready to fire, with a magazine in place. (US Army)*

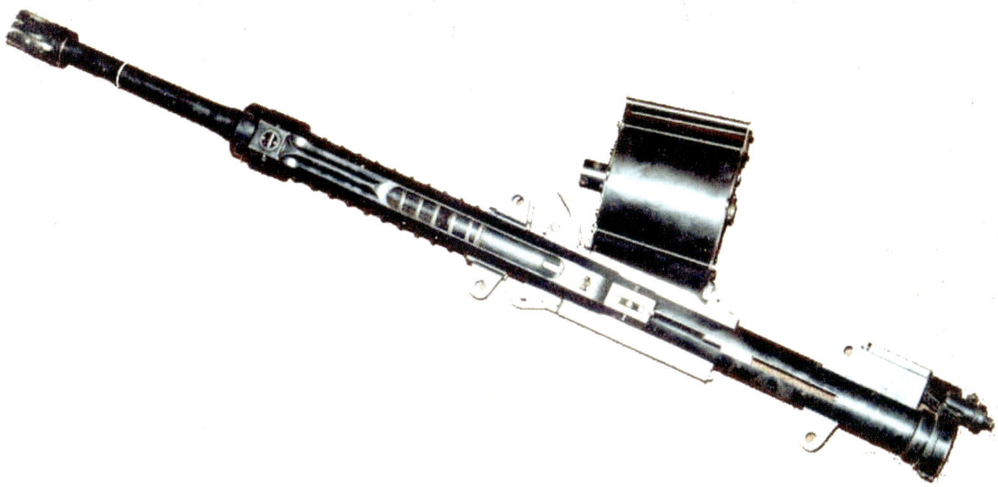

*Type 99–1 Fixed (Courtesy Cliff Carlisle)*

*Type 99–1 Flexible (USA – Wikimedia Commons – Public Domain)*

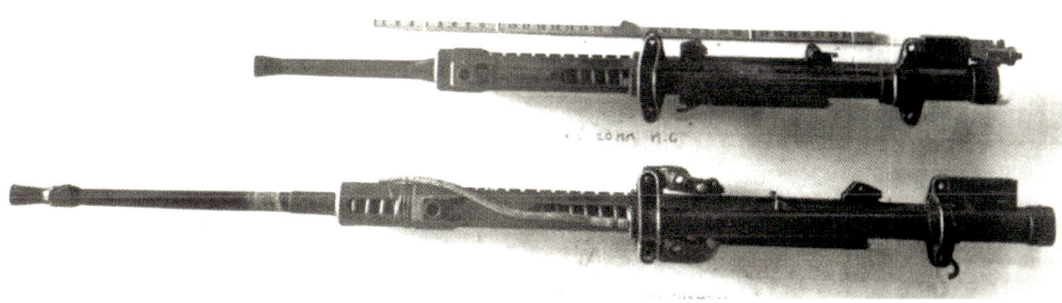

*Navy 20 mm Type 99–1 & 99–2 (USA – Wikimedia Commons – Public Domain)*

was extensively used in both fixed and flexible installations. The flexible gun differed considerably in appearance from the Ikaria weapon, however. The MG-FF had a smooth, bulbous casing covering the reciprocating bolt yoke and return spring, which were wrapped around the barrel. The Type 99–1 had a square and perforated casing, and the gun was mounted upside down so that the magazine was underneath, out of the gunner's line of sight. This required a fired case deflector on the ejection port, directing empties to the right.

Apart from the usual 60-round drums, a 100-round drum was developed in 1942 for the Type 99–1 fixed Model 3 (also usable on earlier models) while the Model 4 had bolt-driven belt feed developed by Dr Kawamura. The standard Oerlikon 45-round and 30-round drums, and possibly 15-round magazines, were available for flexible guns and applications with limited space. The cyclic RoF was 520–550 rpm. The weight varied from 23 to 30 kg depending on the variant, the barrel was an L/40 and the overall length was 133 cm.

**20 mm TYPE 99–2**: 20 × 101RB AMMUNITION.

The Type 99–2 (Oerlikon FFL) was not much used at first, presumably because its extra size and weight were felt to be handicaps, but as combat developed the higher muzzle velocity was more appreciated and it became the preferred fighter armament. Much effort was spent in trying to increase the rate of fire, and this was achieved in two stages: the Type 99–2 Model 4 (which normally fired at 500 rpm) was modified by fitting lighter reciprocating parts and shortening the bolt stroke, resulting in 620 rpm. This was introduced into service from August 1944 onwards. By the end of the war the Model 5, fitted with strong buffer springs at the back of the receiver to kick the bolt forwards more quickly, was running at 720 rpm. Ten prototype and 50 pre-production Model 5 guns had been completed by July 1945 but these were too late to enter service.

As with the Type 99–1, the Model 4 as well as the Model 5 had belt feed, while the Model 3 could take a 100-round drum rather than the original 60-round. The gun weights were 33.5 kg for the Model 3, 37.6 kg for the Model 4 (c. 38 kg for the speeded-up version), and 38.5 kg for the Model 5. The barrel was an L/62 and the overall length was 189 cm. The weight of the full magazines shows the relatively small penalty for moving up to the more powerful ammunition: a full 100-round drum weighed 37 kg for the Type 99–1 and 39.2 kg for the Type 99–2.

**20 mm Ho-1 / Ho-3**:
20 × 124 AMMUNITION.
Kawamura designed the IJA's 20 mm semi-automatic anti-tank gun designated Type 97, using a gas-unlocked blowback operating principle and chambered for its own unique ammunition, contained in a 7-round vertical box magazine. It was subsequently realised that it would be easy to modify the gun to achieve automatic fire and to fit larger-capacity magazines, in which form it could make a powerful aircraft gun.

Two different versions were made, the Ho-1 for flexible mounting and the Ho-3 for fixed installations (although it seems that both were sometimes referred to unofficially as the 'Type 97' leading to some confusion over whether or not the Type 97 rifle was meant to fire automatically – it wasn't). At 30 kg the Ho-1 was significantly lighter than the 45 kg Ho-3 because it omitted the recoil cradle – the mounting incorporated its own recoil management system. While the Ho-1 used a 15 round box magazine, the Ho-3 used a 50-round double-drum. Both guns fired at a rather slow 400 rpm, had L/60 barrels with prominent muzzle brakes and were 174 cm long overall.

**20 mm TYPE 2 (Ho-5)**:
20 × 94 AMMUNITION.
This Browning pattern short-recoil sliding-lock gun became the IJA's standard aircraft cannon in the latter part of the war. It was only marginally increased in size and weight over the .50 Browning it was based on and had a similar RoF, but fired projectiles of almost double the weight, albeit at a lower velocity.

The gun was belt-fed, weighed 35–37 kg and was 145 cm long with an L/45 barrel. Three versions were produced; the Ho-5a for fixed installations (in both wing-mounted and synchronized variants), and the Ho-5b for turret mounting; the latter was only used in later versions of the Ki-67 Hiryu bomber.

The combination of compactness, light weight, a respectable RoF and 20 mm explosive shells should have made the Ho-5 one of the best all-round aircraft guns of the war, were it not for some manufacturing problems. The main weakness was that the 20 × 94 ammunition was less powerful than it looked, firing relatively light projectiles at a modest velocity. This seems to have been due to the Browning action depending on some high-stress components plus the lack of availability of the high-quality steels required; chrome and nickel were in short supply, especially later in the war.

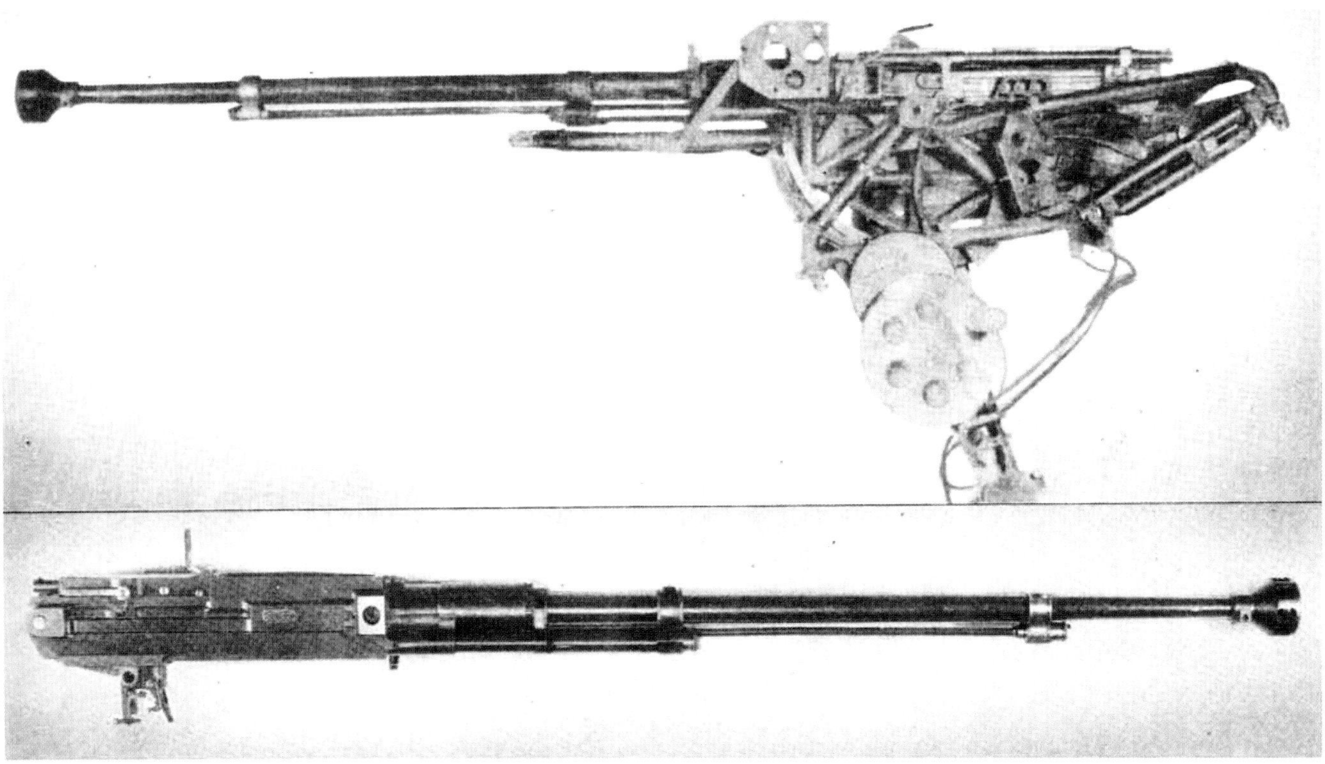

*20 mm Ho-1 flexible (top) and Ho-3 fixed (Office of US Chief of Ordnance)*

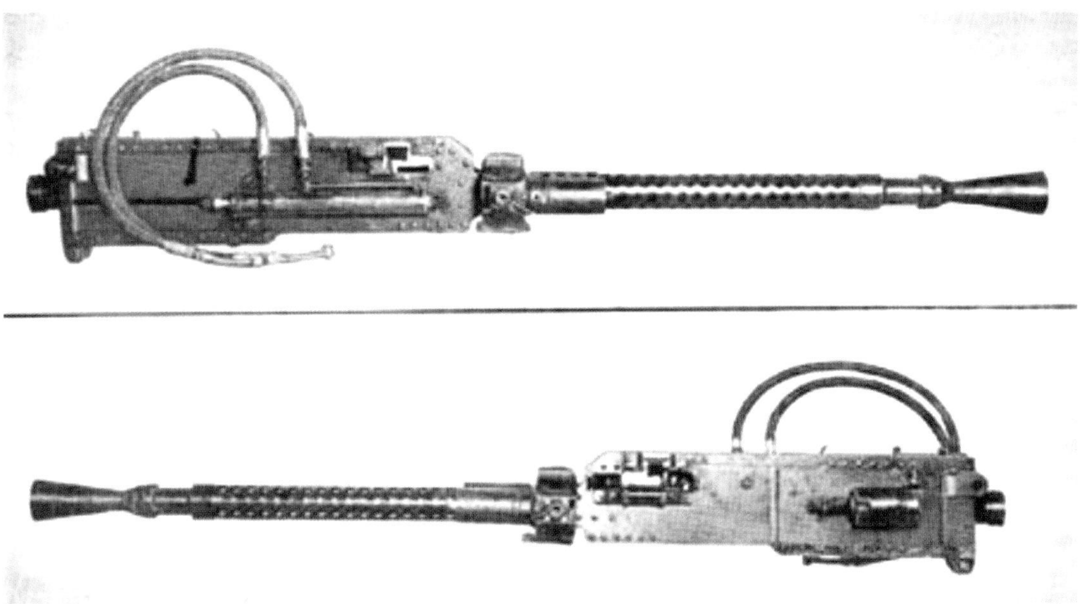

*20 mm Ho-5 (Office of US Chief of Ordnance)*

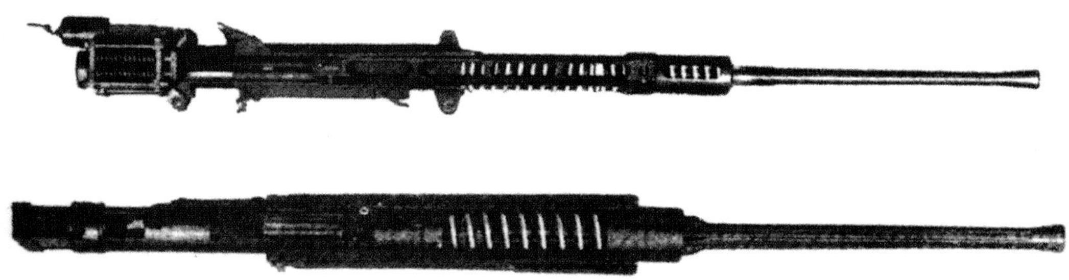

*30 mm Type 2 (lower) compared with 20 mm Type 99–2 (Chinn)*

As a result, the Japanese were compelled to use relatively low-pressure ammunition, and the rate of fire was reduced from 820 rpm to 700–750 rpm.

It is noticeable that the IJN used Oerlikon-pattern 20 mm aircraft cannon with a simpler and more rugged action, the best of which (the belt-fed Type 99–2 Model 5) was similar in weight to the Ho-5 but fired HE shells which were 60% heavier at the same muzzle velocity. However, the Ho-5 had the advantage of being able to be synchronized to fire through the propeller disc (albeit not very well – the RoF was reduced to as low as 500 rpm); synchronisation of the Oerlikon was impossible.

***30 mm TYPE 2**: 30 × 92RB AMMUNITION.*

This was a 'private venture' effort by *Dainihon Heiki*, the Japanese manufacturer of the Oerlikon (Type 99) guns, which had some powerful patrons in the Japanese establishment. Work on a larger version of the Oerlikon seems to have started in 1941, although studies may have begun as early as 1937. They simply scaled-up the 20 mm Type 99 to 30 mm calibre, developing a new cartridge to fire in it.

Initially, the IJN had no interest in such a large-calibre aircraft gun, but this project was given a fresh impetus when a requirement for more powerful armament emerged in August 1942 in the form of the 17-shi specification for a special large-calibre machine gun. *Dainihon Heiki* had advance notice of this requirement in 1941, and promptly offered the 30 mm Type 2. However, this did not comply with the IJN's eventual specifications, mainly because it had a low-capacity (42-round) drum magazine rather than a belt feed. Despite the influence wielded by the company, only a small number of the guns were made early in 1943 and these were sent to Rabaul for operational testing in August 1943, being fitted to at least six A6M3 fighters. Three aircraft were sunk with their transport ship, the others seem to have been shot down before completing the test cycle.

Overall length was 207 cm, the barrel was an L/45, and it was light at only 51 kg. The RoF was 400 rpm. It would have enjoyed the benefit of the relatively smooth recoil characteristics of this kind of gun mechanism. With the benefit of hindsight, the Type 2 could have been a much stronger contender if it had benefitted from the kind of improvements subsequently applied to the 20 mm Type 99–2 Models 4 and 5: belt feed, plus a higher RoF.

**40 mm Ho-301**: 40 mm CASELESS AMMUNITION.
This interesting API blowback gun was unique in that it fired caseless ammunition; the propellant was contained within the rear of the projectile. Since most of the projectile was filled with HE, this greatly limited the quantity of propellant and therefore the muzzle velocity. As a result, the aircraft carrying the gun had to fly extremely close to the target to stand any chance of hitting it, basically limiting the targets to large bombers. Another problem was the box magazine capacity of just ten rounds. The plus points were a light and compact gun for its destructive power, easily able to be mounted in fighter wings, and a high rate of fire of around 450 rpm – although that meant that the magazine was emptied after one burst of 1.3 seconds. As a result of these problems it saw little use.

A report of the British Ordnance Board, dated 5 July 1946, concerned tests of three examples of the Ho-301. This included the following comments: 'guns unreliable throughout…broken side plates, broken strikers and misfeeds…considerable wear noticeable…two premature firings…poor quality of material apparent…shell showed inconsistent detonation…misfires due to defective primers.' The reported concluded: 'The Board do not consider that the Japanese Ho-301 merits any further investigation.' Many Japanese evidently agreed; although the guns were installed in the wings of Nakajima Ki-44-IIc and -III Shoki (Tojo) fighters for anti-bomber purposes from 1942, they were often removed and replaced by 12.7 mm MGs.

There is more detail concerning the ammunition in the Unconventional Systems section.

**37 mm Ho-203**: 37 × 111R AMMUNITION.
Possibly also known as the 37 mm Type 3, this was an aircraft gun designed for shooting down bombers. As such, the destructive effect of the shell was considered more important than a high muzzle velocity or rate of fire. The gun was based on the Year 11 Flat Trajectory Infantry Gun and used the same basic cartridge, but had a long-recoil operating system with a falling-block breech. It was

*40 mm Ho-301 – oblique view, the magazine is horizontal (Chinn)*

*37 mm Ho-203 (US National Archives)*

fed by a closed-loop 'endless belt' magazine holding 15 rounds (shown on top of the gun in the photo). The magazine was so bulky that wing-mounting was not practicable, so the gun was invariably fuselage mounted in twin-engined planes. The gun weighed 89 kg, was 154 cm long with an L/24 barrel, and fired at just 120 rpm. It was installed in the Ki-45KAI (c and d models) and the 20 pre-production Ki-102a.

**37 mm Ho-204**: 37 × 144 AMMUNITION.
Possibly also known as the 37 mm Type 4, this was designed to overcome the disadvantages of the Ho-203 (muzzle velocity and rate of fire both far too low). It achieved both at the cost of extra weight, at 130 kg. It was the largest-calibre example of the Browning-pattern short-recoil sliding lock system to see service, and was 247 cm long with an L/35 barrel. The belt-fed ammunition was unique to this gun, and was fired at about 400 rpm. The gun saw very little use, mainly because it was installed in an obliquely upward-firing mounting in the Ki-46-III-KAI; an aircraft with too low a performance to engage its targets – USAAF heavy bombers.

**30 mm Ho-155**: 30 × 115 AMMUNITION.
The IJA began the search for a more powerful replacement for the 20 mm Ho-5 in 1942, starting with the 25 mm Ho-51 (see below). The Browning pattern sliding lock so favoured by the IJA air force was used. It was felt that 25 mm offered an insufficient advantage over the 20 mm calibre, so attention switched to a 30 mm version. The development path eventually resulted in the Ho-155-I, which seems to have entered production although there are no confirmed examples of its installations.

The Ho-155-I was too big to be a feasible replacement for the 20 mm Ho-5 in fighter wings, so a design review minimized the dimensions and resulted in the Ho-155-II. This was 151 cm long overall instead of 175 cm, had an L/33 barrel instead of L/38, and weighed 44 kg instead of 50 kg. Both right and left-side feed versions were produced. The RoF of both versions was around 500 rpm. The Ho-155-II is known to have been fitted into the wings of thirty Ki-84-Ic fighters which participated in battles above Japan and China. Neither version was formally adopted.

One point worth noting is that an early (and much copied) study on Japanese aircraft guns identified this weapon as the Ho-105; this was an error, there was no such gun.

**57 mm Ho-401**: 57 × 121R AMMUNITION.
This was basically a scaled-up version of the 37 mm Ho-203, and similarly consisted of a tank gun fitted with a long-recoil autoloading mechanism and a

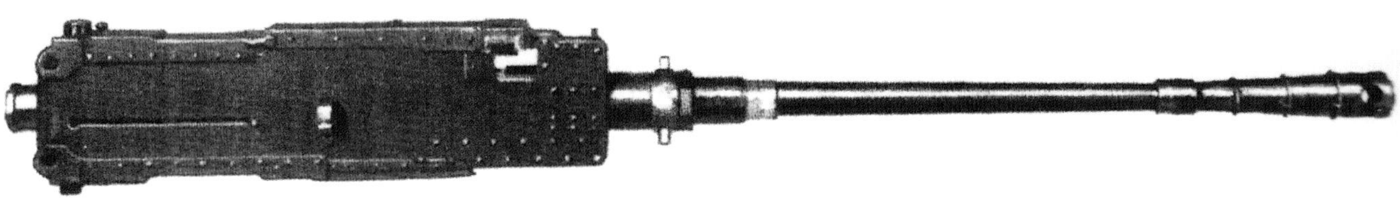

*37 mm Ho-204 (Chinn)*

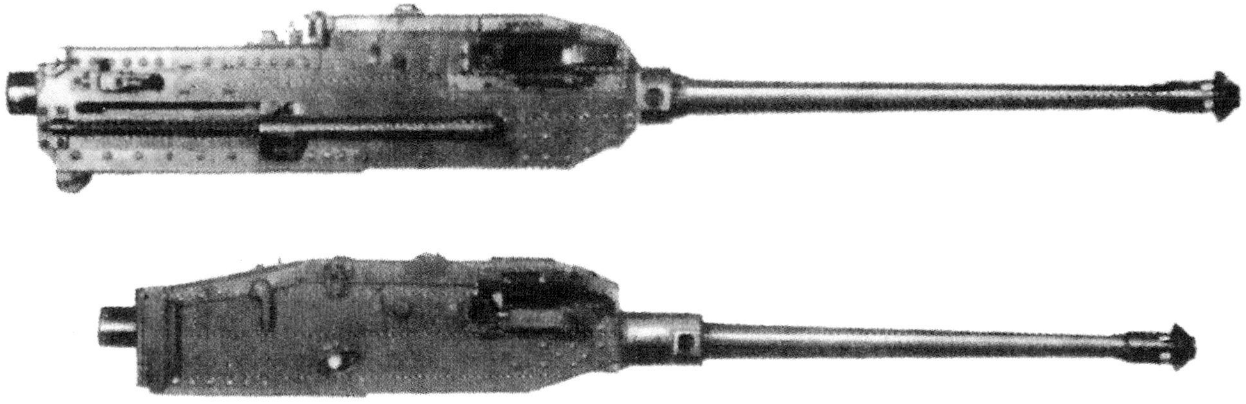

*30 mm Ho-155-I (top) and Ho-155-II (Chinn)*

17-round endless-belt magazine. Length was 160 cm, barrel L/17.5, it weighed 160 kg and fired at 80–100 rpm.

The main purpose of the Ho-401 was for anti-ship operations, and the only model of plane it was fitted to was the Ki-102b 'Randy' attack aircraft. These planes were somtimes used to intercept the B-29, but this was rare. One recorded example was on June 26, 1945, when a Randy from 101st *sentai*, piloted by Lt. Shimamura, attacked a B-29 during a raid on Nagoya. According to Shimamura's report, he hit the right outer engine which was torn off along with a piece of the wing; the inner engine also caught fire. The B-29 lost altitude and the Japanese pilot then lost contact, noting this as a probable kill.

**30 mm TYPE 5**: 30 × 122 AMMUNITION.

This was the IJN air force's equivalent to the IJA's Ho-155, but (unsurprisingly) it used different ammunition and a different gun operating system. It used a gas-unlocked blowback system with locking flaps, and is notable for using 'floating firing', in that the gun was free to move in recoil and fired while moving forwards, so the peak recoil blow was absorbed in bringing the gun to a halt before pushing it back again. The effect of this was to smooth the recoil pulses, putting much less strain on the mounting. This was probably the most successful of the large-calibre Japanese aircraft cannon, but it arrived too late to make any difference. About 2,000 were built. The weight was 66 kg, length 201 cm, barrel L/48, and it was belt-fed at around 500 rpm.

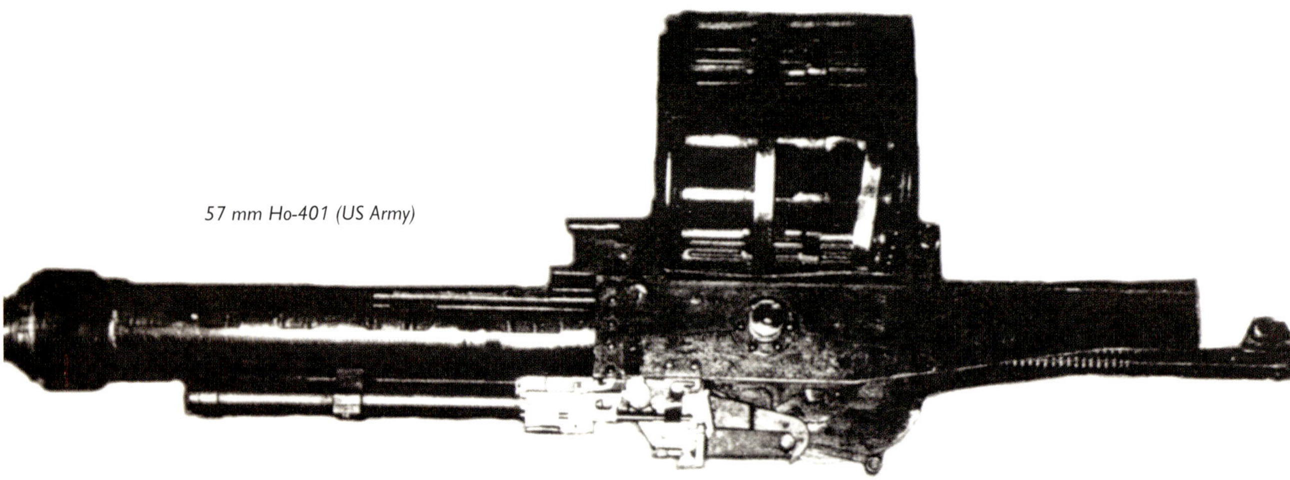

*57 mm Ho-401 (US Army)*

*30 mm Type 5 (Courtesy Cliff Carlisle)*

## EXPERIMENTAL WORLD WAR II GUNS

For completeness, the aircraft guns listed include some in calibres larger than 57 mm.

*Ho-2*: 20 × 124 AMMUNITION.
Nothing known, except that it was designed by Shiro Kayaba and used the same ammunition as the Ho-1/3; it was intended for fixed installations.

*Ho-4*: 20 × 124 AMMUNITION.
A belt-fed version of Ho-1/3 designed by Masaya Kawamura. Its prototype was tested in early 1942, but was rejected by the army. In part, this was due to the fact that the Kawamura couldn't correct the main drawback of these weapons – a low rate of fire.

*Ho-6,7,10,11*: 20 mm AMMUNITION.
20 mm guns mentioned in wartime reports, but not heard of since. Probably designed around 20 × 124 ammunition.

*Ho-12*: 20 × 124 AMMUNITION.
A combination of the Ho-1/3 ammunition with a belt-fed Browning short-recoil action.

*Ho-51*: 25 × 115 AMMUNITION.
The first attempt to scale-up the Browning-pattern 20 mm Ho-5 to a larger calibre. Discontinued in favour of the Ho-155.

*Ho-52*: 25 × 150 AMMUNITION.
Larger version of the Ho-51 to fire more powerful ammunition.

*Ho-201*: 37 × 111 AMMUNITION.
For flexible mountings, used rimless version of Ho-203 ammunition.

*Ho-202*: 37 × 111 AMMUNITION.
Seems to have been a version of the Ho-201 for fixed mountings.

*Ho-205*: 37 × 144 AMMUNITION.
A smaller and lighter version of the Ho-204, using the same ammunition; not built.

*Ho-251*: 47 × 285R AMMUNITION.
This was an automatic fire version of the semi-automatic 47 mm gun Type 1 anti-tank gun, and used the same ammunition. It used a long-recoil operating system and had a box magazine. The development of the gun continued from 1943 to 1945, but never reached the firing stage.

*Ho-402*: 57 × 477R AMMUNITION.
A high-velocity tank/anti-tank gun and ammunition fitted with an autoloader designed by Kawamura. It was intended to achieve 80 rpm, but failed its tests. Two guns were made, intended for installation on the fighter/attack Ki-93 aircraft. The only prototype of the aircraft was tested in July-August 1945, but was destroyed by an Allied air strike. The second, unfinished, was shipped to the United States; its fate is unclear.

*Ho-403*: 57 × 187RB AMMUNITION.
Scaled-down version of the Ho-402. Never got past the design stage.

*Ho-501*: 75 mm AMMUNITION.
Scaled-up version of the Ho-203 and Ho-401, presumably using low-velocity ammunition. This was another gun planned for installation in the Ki-93. Not built.

*Ho-505:* 75 × 497R AMMUNITION.
The basis for development was the anti-aircraft gun Type 88 for 75x497R ammunition. This gun was planned to be used as the main weapon in the Ki-109 heavy interceptors, but the automation system failed and all 22 of the Ki-109 which were built were armed with a manually-loaded standard Type 88 anti-aircraft gun.

*Ho-601*: 120 × 83R AMMUNITION.
An attempt to make an autoloading version of the Type 38 120 mm howitzer. The howitzer fired shells weighing 20 kg with a muzzle velocity of about 300 m/s. The design reached a dead end and work was dropped in the middle of 1944.

*Ho-3057*: 57 × 265 CASELESS AMMUNITION.
Browning action, with 5-round box magazine; one exists in a US museum.

*Ho-3157*: 57 × 187RB AMMUNITION.
Gas-unlocked blowback design, only reached unfinished prototype stage.

*Ho-3075 and Ho-3175:* the equivalents of the Ho-3057 and H0–3157, only in 75 mm calibre. Projects never completed.

# POLAND

Poland was re-established as a separate nation in 1918, so had little opportunity to develop its own weapon systems before the 1939 German and Soviet invasions ended its independence for a further half-century. However, by the mid-1930s they had established a need for 20 mm cannon for both the army and the air force. An early decision, following unsatisfactory comparative testing of foreign cannon, was to pursue a domestic design firing a modified version of the powerful 20 × 138B Rheinmetall ammunition. One subtle point was that the modified cartridge had fractionally larger case dimensions all round, providing only one-way compatibility: Polish guns could fire the German ammunition, but not vice versa. Four different gun designs emerged to use this 20 × 140B ammunition. The ultimate aim was to produce a universal cannon suitable for aircraft, anti-aircraft, tank and anti-tank use, but this proved too ambitious.

The first attempt was the M1938 Model A, also known as the 'nkm wz.38 FK' (nkm = *najciezszy karabin maszynowy* = heavy machine gun; wz.38 = introduced 1938; FK = *Fabryka Karabonow* = the Warsaw company for which designer Bronislaw Jurek worked). This was satisfactory as a ground gun, being rated as superior to the Oerlikon and Madsen guns it was compared with, and it was put into limited production, but it weighed too much for aircraft use. The second attempt, the gas-operated Model B, was a complete failure. The Model C was meant to be the universal gun and much work went into it, in parallel with the development of the Model D which was Jurek's design for an aircraft gun.

The Model D was ready for testing in May 1938 and performed well, but following the failure of the Model C in tests in October 1938, the design of the Model D was modified to make it more suitable as a universal gun. More prototypes were ordered, but it was too late to put the cannon into production.

Despite all of this effort, only about 50–55 of the Model A cannon were completed, of which less than half were fitted to tankettes (light armoured vehicles), the remainder being on ground mounts. They reportedly performed well, with good penetration against the lightly-armoured vehicles in service in 1939.

After World War II Poland primarily used Russian equipment, but in the mid to late 1980s Polish engineers worked on various cannon designs. Work on

*20 mm Nkm wz.38 FK (Wikimedia Commons / Public Domain)*

multi-barrel Gatling type guns was carried out at the Mechanical Equipment Research and Developing Centre (*Ośrodek Badawczo Rozwojowy Sprzętu Mechanicznego* – OBR SM) in Tarnów. In 1983–1990, a 23-mm air cannon named *Sniardwy* was developed, intended to be the armament of the *Iryda I-22* training and combat aircraft, while in 1984 work on another project – the 30-mm Navalnik naval gun – also started in OBR. Neither project led to adoption.

**20 mm M1938 MODEL A**:
20 × 140B AMMUNITION.
The prototype of the Model A gun was completed in November 1937. Initially, 200 were ordered as interim AFV and anti-aircraft weapons with delivery commencing by February 1940, pending the availability of better guns (the Models C and D).

The Model A used the short-recoil method of operation and, for use in a tankette, was fed by relatively small 5-round box magazines, although 15-round and 100-round drums were developed for AA use, and a 10-round box magazine was planned. Gun weight was 58 kg, overall length 201 cm with an L/67.5 barrel, and the RoF was a nominal 320 rpm (it could also fire single shots).

**20 mm M1938 MODEL C**:
20 × 140B AMMUNITION.
The Model C was intended to be the definitive 20 mm gun, designed to be multi-purpose from the start. The gun was to serve as a belt-fed aircraft cannon or, with magazine feed and a lighter barrel, as a ground AT weapon on a light mount or, finally, as a belt or magazine-fed AA gun on a heavy PC 20A mount, derived from the 37 mm Bofors cannon carriage. Feeding of the Mod C was therefore to be by 5 or 10-round box magazines, a 100-round drum magazine or a 200-round disintegrating link belt. Field stripping was very easy and needed no complicated tools. The gun in its ground role could be easily dismantled for carrying by unlatching and turning the barrel by 60°, thus removing it.

The gun was designed by S. Rytwinski and W. Lewandowski; its gas-unlocked blowback action was patterned after the HS 404. In its aircraft version the gun weighed around 50 kg, and had an L/75 barrel. The RoF could be adjusted by means of a gas regulator between 300 and 750 rpm.

Trials of the first prototype were scheduled for October 1938; the trial results are unknown, but the Model C wasn't mentioned again.

**20 mm M1938 MODEL D**:
20 × 140B AMMUNITION.
After completing the Model A, designer Jurek set to work on a second cannon optimized for aircraft use. This was gas-operated and was referred to as the Model D. Despite the different way it operated, many components were interchangeable with the recoil-operated Model A. The prototype was ready in May 1938; the trials gave satisfactory results and orders were placed for a small number of guns for evaluation, but they were scheduled for delivery only a few weeks before the German invasion.

Although originally designed as an aircraft weapon, the failure of the universal Model C led to changes being made to the Model D to enable it to become the universal weapon for air, land and sea service. The RoF of the Model D was regulated by gas between 300 and 600 rpm, and various magazines or a belt were available. The weight was around 40 kg in the aircraft version (very light considering the large and powerful cartridge it fired) with an L/67.5 barrel.

# RUSSIA & USSR

For convenience, 'Russia' will be used here to mean 'Russia and/or the Soviet Union'.

Russia has a long tradition of designing, developing and fielding automatic weapons using a wide range of configurations and operating systems and in a variety of calibres. Initially, some weapons were based on designs obtained from other countries, but most have been indigenous and some quite unique. This applies to the guns which saw service – for every one of those, there were several which never got past the prototype stage. For many of them little information is available, although this is gradually increasing.

In the interwar period an unusual system of design bureaux was established, each focusing on a particular aspect of military technology. For automatic guns, the principal organisation was the Design Bureau for Manual and Automatic Small Arms (PKB), founded in 1927 in the Tula Arms Factory. In 1933 this left the Factory and was renamed the Central Design Bureau of Small Arms (TsKBSV – *Tsyentral'noye Konstruktorskoye Buro Strelkovogo Orujhiya*). In 1936, this became TsKB-14 and was subsequently joined by many design teams that developed all types of artillery and small arms, including the Berezin, Volkov and Yartsev groups and many others.

There were also individual *Opytnoye Konstruktorskoye Buro* (Experimental Design Bureau) or OKB, which focused solely on design and development; they could construct prototypes, but did not have the facilities for mass production. Once a particular design had been approved for production, that task was given to whichever of the large construction plants was most appropriate and had available capacity. This type of organisation facilitated the efficient use of production resources, although in practice relationships tended to build up between OKBs and particular factories.

The TsKB and OKB were given numbers but they were often popularly called by the name of the lead designer, especially if he was well-known. For aircraft cannon development in World War II period, TsKB-14 was involved plus the two main OKBs: OKB-15, led by Shpitalny, and OKB-16, originally under Taubin but later led by Nudelman and including Suranov and Rikhter. Initially, Degtyarev's OKB-2 (whose origin went back to the design office of the Kovrov Weapon Factory in 1921) also featured in cannon projects. Other OKBs focused on related activities, e.g. OKB-43 led by Kondakov mainly specialized in gun mountings although also did some gun design work.

Until the 1930s Russia had few automatic cannon in service, mainly some 40 mm British 2-pounder and 37 mm Maxim guns (the unsuccessful 37 mm McClean or 'Maklen' guns bought from the USA being disposed of as soon as possible).

Initially a high priority was given to naval AA cannon, which in 1932 resulted in the adoption of the 45 mm 21-K (the K designation indicated that this was a product of the Kalinin armament factory), using the same gun and ammunition as the tank/anti-tank gun which had only just entered service. It was fitted to a pedestal mounting, and remained manually-loaded (as were the comparable French 37 mm M1925/1933 and German 3,7 cm SK C/30 naval AA guns). Single and twin turrets using the 21-K gun (designated 40-K and 41-K respectively) were also made for patrol boats.

Initial automatic cannon designs were based on foreign ammunition: from 1934, Rheinmetall's 20 × 138B was favoured and was used in the first version of the cannon designed by Atslega of OKB-2; Oerlikon 20 × 110RB ammunition was also tried but not adopted.

Despite this, the first indigenous automatic cannon to enter service, the ShVAK, used a domestic 20 × 99R cartridge made simply by necking up the ShVAK's original 12.7 × 108R case. This was introduced in 1936 and became the standard aircraft cannon throughout World War II, as well as serving in some light AFVs.

The relatively weak ammunition of the ShVAK led to attempts to produce a

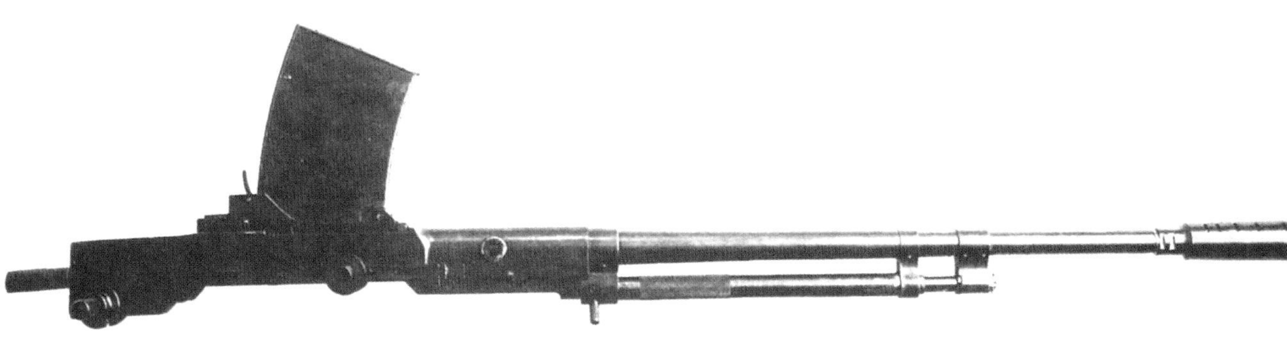

*20 mm AP-20 using Rheinmetall ammunition with DP-based action 1934 (Courtesy Maxim Popenker)*

more powerful 20 mm round: for this, a newly-designed 20 × 110 rimless round was developed, a wider case making it more powerful than the Oerlikon and Hispano 20 × 110 rounds. This was tested in a modified version of the AP-20 Atslega cannon in 1937.

At the same time another new gun, the DShAK (designers Degtyarev and Shpagin, also at OKB-2), was based on the 12.7 mm DShK and used the same 20 × 110 ammunition as the AP-20. In 1941 Berezin modified his 12.7 mm UB to use a unique rimless 20 x 99 round, but at the beginning of the Great Patriotic War was told to redesign it to take the rimmed ShVAK ammunition, resulting in the B-20. This was the only other 20 mm gun to be adopted and produced.

Two other 20 mm guns were developed, each with its own unique round: an enlarged AA version of the ShVAK designated TsKBSV-75 was designed in 1940–41 to use 20 × 110R ammunition, the rim being to facilitate the conversion of the ShVAK design. An AA/anti-tank gun, the TPV Model 1943, used a 20 × 114 round derived from the 14.5 mm anti-tank rifle, and this represented the final effort to make a more powerful 20 mm cannon.

While the manually-loaded 45 mm 21-K (and improved, longer-barrelled 21-KM) remained in production throughout World War II and for a couple of years thereafter, it was clear that automatic cannon were needed for effective air defence of both land and naval forces. While interest was shown in the Rheinmetall/Solothurn 3.7 cm system (and particularly in its 37 × 264B ammunition, which was taken as a basis for development) most attention was paid to indigenous designs. Work on this started very early with the Kovrov Weapon Factory being given an order to develop 25 mm and 40 mm guns in 1926, but progress was very slow due to other priorities. In 1930 the 40 mm gun was dropped and a specification for a 25 mm gun issued. By 1933 a prototype existed based on the Degtyarev MG action, with 7-round box or 12-round drum magazines, but this work was terminated shortly afterwards in favour of lighter 20 mm systems.

In 1932 OKB-43 started development work in 37 mm calibre resulting in the AKT-37, and in 1936 the improved ASKON-37. An interesting feature of both of these guns was an attempt to soften recoil for aircraft use, by increasing the mass of the recoiling parts to include the barrel (and the magazine in the case of the AKT). Tests of both the AKT and ASKON continued until 1938, when both were cancelled, for reasons which might have been political as much as technical (lead designer Kondakov having many enemies).

Recoilless cannon were also under development in the 1930s with Kondakov again involved, resulting in guns such as the ARKON (*Avtomaticheskaya Reaktivnaya Kondakova*), initially in 45 mm and then 37 mm calibre; the RPTR, a 37 mm recoilless gun capable of firing at 30 rpm; and the GK-37 and GK-45 recoilless aircraft guns. More detail is provided in the Unconventional Systems chapter. All work on the recoilless weapon programme was cancelled in 1937.

OKB-43 were also busy in the mid-1930s with another Kondakov design, resulting in the AKON-45 of 1937, a gas-operated aircraft gun in 45 mm calibre. It featured a quick-change L/52.7 barrel and a five-round magazine, and fired at 100–120 rpm. Only one prototype was made in 1937 before cancellation in 1938.

The recoilless gun programme was revived in 1942 resulting in various experimental aircraft gun projects from several development bureaus, including the TsAKB led by Grabin, and OKB-172. While outside the scope of this book, it should also be noted that OKB-43 got involved again, developing in 1953 the DRP-76, a 76 mm RCL gun for small naval craft which was belt fed and remotely controlled. It was not adopted for service.

Meanwhile, from 1935 OKB-15 were also working on a 37 mm gun alongside Kalinin Plant No. 4. This resulted in the gas-operated and magazine-fed 37 mm 100-K, further development being undertaken by OKB-2 in Kovrov. By the autumn of 1937, a prototype gun was achieving a cyclic rate of fire of up to 300 rpm with a new rimless 37 × 198 cartridge (similar but not identical to the later 37 × 198 of the NS-37). Its drum magazine was later reduced in capacity from six rounds to five. Later, and with modified ammunition, the 100-K became the basis for the Sh-37 aircraft gun (ShFK-37).

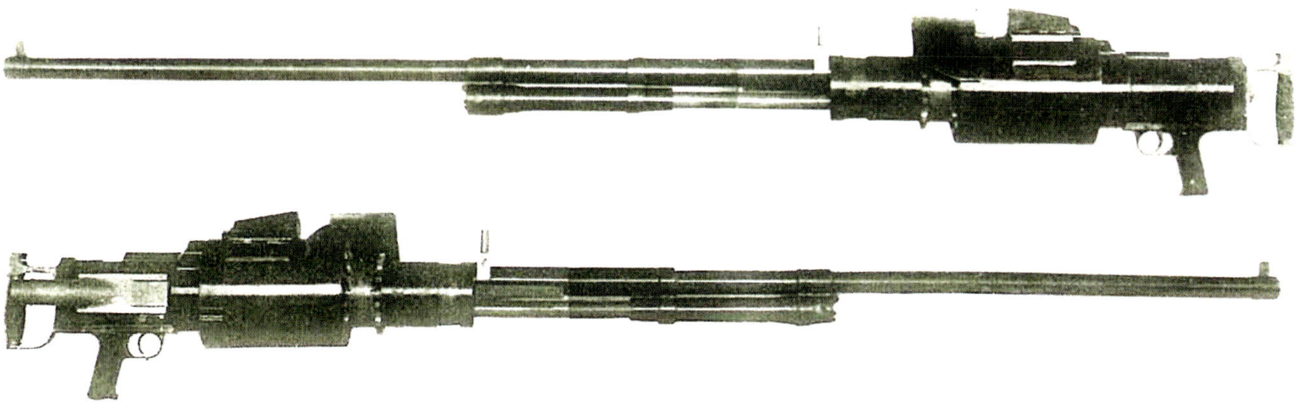

*20 mm TsKBSV-75 (courtesy Evgeny Aranov)*

The Kalinin plant produced a new weapon based on the long-recoil mechanism of the 40 mm M32 Bofors gun, scaled-up to 45 mm calibre. However, it was decided that this was too large for convenient handling so it was rapidly scaled down to 37 mm, leading to the M39 (61-K) in 37 × 252SR calibre, which became the standard Russian medium AA cannon throughout World War II and for some time afterwards. This gun bears a close resemblance to the Rheinmetall-Borsig 3,7 cm SK C/36 and the later, closely related Flak M42, naval guns which were presumably inspired by the Bofors.

The desire for a smaller and handier AA gun led to the adoption of the M1940 in 25 × 218R calibre; it retained the Bofors-type long-recoil action.

Attention then returned to aircraft cannon, prompted by a desire for more destructive ammunition than the ShVAK's, which would also be powerful enough to be effective against armoured vehicles. The first result was the MP-6 from OKB-16 in 23 × 152B calibre. However, not many had been made when it was decided to switch production to the rival VYa-23 using the same ammunition; a successful weapon made in large quantities from 1942 onwards.

Other unsuccessful projects using the same ammunition were the 23 mm Salishchev-Galkin SG-23, the TKB-199 (Silin-Batov-Sivov), and the TKB-200 (Rashkov-Lebedev), all from TsKB-14.

The second result was the Sh-37 in 37 × 198 calibre adopted in 1941–2, but not many had been made before it was replaced by the NS-37 using the same ammunition. Another rival using this ammunition was the Mamontov from OKB-16, while yet another variation of the 37 × 198B – with a belted case – appeared in Taubin's magazine-fed BMA-37, an enlarged copy of the 23-mm MP-6. This cannon was later redesigned by Nudelman to create the NS-37, and the cartridge case lost its belt in the interests of compatibility with the Sh-37.

These 37 mm guns proved to be too powerful for many aircraft applications, so it is not surprising that the NS-45, created by fitting a larger calibre barrel to the NS-37 and necking out the 37 × 198 case to 45 × 186, proved even more problematical and was never adopted for general use.

The VYa-23 was also quite large and weighty and developed heavy recoil, not ideal for the typically small and light Russian fighter aircraft, so a smaller, less powerful 23 × 115 cartridge was developed by necking-out the 14.5 × 114 anti-tank rifle case. This allowed the guns to be considerably smaller and lighter, they also generated much less recoil despite being faster-firing. This enabled more guns and ammunition to be carried for a given weight, further increasing the chance of scoring hits. The 23 × 115 round was introduced in the NS-23 gun of which over 28,000 were made between 1944 and 1953. Eventually a series of single-barrel, twin-barrel and six-barrel guns was developed to fire improved versions of the 23 × 115 ammunition: the NR-23, AM-23, GSh-23 and GSh-6–23 all saw service.

A similar approach was taken with aircraft guns in 37 mm calibre: the projectiles were retained but a smaller cartridge case designed around them, for the same reasons of reducing size, weight and recoil, while boosting the rate of fire. The N-37 in 37 × 155 was the result; it was officially adopted in 1944 but did not enter service until after the war; the modified NN-37 followed on later.

The 23 × 152B aircraft gun ammunition was not ignored post-war, with several new guns designed around it in the 1950s (see also the Tula list below):

- KBP-565: gas operated gun designed by Slostin (OKB-2), tested in 1949; 1,200 rpm; 48 kg; overall length 219 cm
- KBP-585: gas operated gun designed by Vladimirov (OKB-2); tested in 1949; 900 rpm; 48 kg; overall length 208 cm
- KBP-620: gas operated aircraft gun designed by Slostin; tested in 1950
- KBP-780 (APSVV-23): gas operated 5-chamber revolver; designed by Vladimirov; tested in 1953; 1,500 rpm; 58 kg; overall length 191 cm;
- KBP-780A (2-B-O–15): gas operated 5-chamber revolver; designed by Vladimirov; tested in 1954; 2,100 rpm; 72 kg; overall length 205 cm
- KBP-810 (2-B-O–5): gas operated aircraft/AA gun designed by Slostin; tested in 1954; 1,200 rpm; 67 kg; overall length 246 cm
- 2-B-O–30: a Gast-type twin-barrel gun from Krekin in 1957
- TKB-513 single-barrel short-recoil operated gun which fired at 2,100 rpm and weighed 56.5 kg.

None of these were selected for service, although the TKB-513 got as far as a competition with Rikhter's R-23 revolver cannon, which it lost.

Other experimental cannon of the immediate post-war years include:

- 20 mm gas operated aircraft gun designed by Zubov (from OKB-16), tested in 1946–47; 20 × 99R; 850 rpm
- 23 mm Sh-3–23: gas-operated; designed by Shpitalny using 23 × 105 ammunition (23 × 115 with a shorter case) for bomber defensive use; weight 42–48 kg; 1,800–2,000 rpm; suffered cook-offs due to closed-bolt operation
- 30 mm 2-B-0–70: double-barrelled Gast type by Slostin in 1958; 30 × 155B
- 37 mm Grabin V-14–113: short-recoil aircraft gun in 37 × 110SR, from 1946; belt-fed, 65 kg, 420–440 rpm
- 37 mm KBP-775: aircraft gun designed by I. Slostin (from OKB-2); tested in 1953; 37 × 155; gas operated. 1,085 rpm; 78 kg; overall length – 1195 mm
- 45 mm Sh-45: by Shpitalny in 45 × 186: the case shape differed from NS-45
- 45 mm N-45: by Nudelman – the development of a lightweight NS-45 (the same approach as N-23 and N-37) with a shorter case (the exact size is unknown); 1945
- 45 mm Grabin S20: aircraft gun in 45 × 144B, dated to 1945; short recoil; 140 rpm
- 57 mm Sh-57: by Shpitalny using 57 × 155RB, rival to the N-57
- 57 mm N-57: using 57 × 165RB ammunition; described below
- 57 mm Grabin V-14–112: aircraft gun in 57 × 140SR, dated to 1948; short recoil; 555 m/s

- 57 mm Grabin S10: aircraft gun in 57 × 140B, dated to 1946; short recoil; 140 rpm
- 57 mm Grabin V-7031: aircraft gun; dated to 1947; short recoil; 965 m/s. (the ammunition is believed to be similar to the ZiS-2 57x480)
- 57 mm RShR 45/57: 57 × 350B ammunition; described below (there was also a necked-down 45 × 350B version)
- 57 mm Ch-21P: 57 × 521RB ammunition; described below.

KBP at Tula produced a vast number of experimental aircraft guns in the post-war period. The following list is not even complete:

- TKB-403: 23/20 mm air cannon by Chepelev (20 × 99R; 23 × 115)
- TKB-474: 23 mm Volkov and Yartsev (23 × 152B) post-war version of VYa with increased rate of fire
- TKB-494: 23 mm by Volkov (23 × 152B)
- TKB-495: 23 mm gun Volkov (23 × 115)
- TKB-499: 23 mm for experimental cartridge (??)
- TKB-504: 37 mm aircraft gun by Silin for experimental cartridge; 37 × 147SR
- TKB-505: 23 mm aircraft revolver gun by Silin; 23 × 115
- TKB-508: 37 mm aircraft gun Volkov, Yartsev, Afanasyev; 37 × 155
- TKB-513 (AO-7) 23 mm aircraft cannon: Gryazev, Shipunov; 23 × 115
- TKB-515 30 mm aircraft Gatling type by Silin
- TKB-532 23 mm aircraft by Makarov (23 × 115)
- TKB-539 30 mm aircraft revolver type by Makarov (30 × 210B)

Five new cannon cartridges were adopted and saw service in the 1950s: the 30 × 155B for the NR-30 aircraft gun, now obsolete; the 30 × 210B for the NN-30 naval AA gun which saw extensive service and is still in use outside Russia today; the remarkable 23 × 260 used in the Rikhter R-23 in which briefly saw use as a bomber defence gun; the short-lived 45 × 386SR for a naval AA gun; and the 57 × 348SR for land and naval AA guns, which is not only still in service but is having new mountings and ammunition types developed for it.

The 23 × 152B cartridge, in a modified form, also had a second (and much longer) life starting in the 1950s, in the ZU-23 AA system and ZSU-23–4 SPAAG. The SPAAG was adopted in 1960 after a competition with the ZSU-37–2 *'Yenisey'*, which appears to have used different ammunition from the usual 37 × 252SR, in that the quoted muzzle velocity with the same projectiles is considerably higher. It is, in fact, similar to that of the Chinese 37 × 240 round adopted for naval guns in the 1970s, so there could be a connection.

Everything began to change in the 1970s, with most of the older weapons in all three services being replaced in production by a family of cannon all based on a new 30 × 165 round. However, the three services all developed their own versions of the ammunition, with appropriate modifications to the guns. The greatest difference was that the army chose percussion ignition while the navy and air force used electrical priming which was therefore incompatible with the army guns. Furthermore, the navy and air force ammunition also differed in certain respects, e.g. shorter fuzes being used by the air force, so are not necessarily interchangeable.

A family of guns was developed to use the 30 × 165 ammunition, mostly from the design bureau led by Gryazev and Shipunov, who were also responsible for the last of the 23 mm guns, the twin-barrelled GSh-23 and six-barrelled GSh-6–23. In 30 mm calibre they designed the GSh-30 and GSh-6–30, the former in different versions for the army (2A38M) and air force, the latter for the navy (AO-18) and air force. They also produced the single-barrel GSh-301 aircraft cannon. Slower-firing single-barrel guns were produced for army AFVs and helicopters; the 2A42 and 2A72, both by the Instrument Design Bureau KBP at Tula.

A confusing variety of gun designations has been used, with various types often applying to a single gun. For example, the

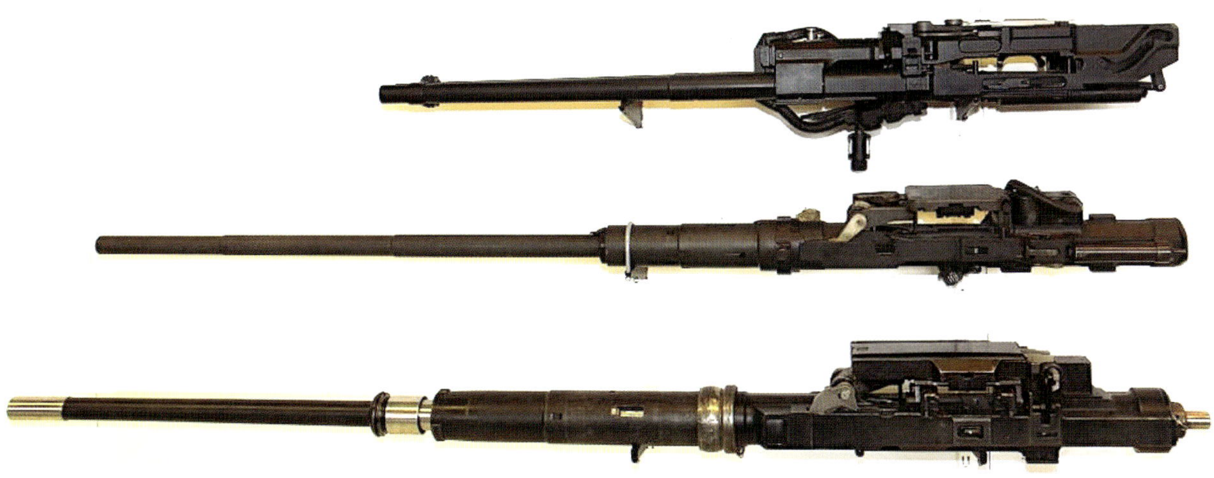

*Three service guns, from the top: GSh-23, NR-23, NR-30 (Courtesy Evgeny Aranov)*

six-barrel 23 mm rotary gun was initially known as the AO-19 or the TKB-613 (both factory internal designations), the 9A-620 (GRAU index – *Glavnoye raketno-artilleriyskoye upravleniye MO RF* = Main Missile and Artillery Directorate of the Ministry of Defense of the Russian Federation, responsible for assigning official index numbers) and finally the GSh-6-23 (public designation, honouring the designers Gryazev and Shipunov).

After World War II the structure of weapon design and production was reorganised several times, resulting in the creation of *Naoochno-proizvodstvyennoye obyedinyeniye* (Scientific Production Organizations), abbreviated to NPO, which included both design and production facilities. There have since been many mergers, amalgamations and splits – a process which continues.

-------------------------------------

***37 mm AKT-37 AND ASKON-37***: 37 × 264B AND 37 × 240 AMMUNITION.

OKB-43 designed a 37 mm cannon, starting in 1932 with the barrel and 37 × 264B ammunition of a Rheinmetall cannon (designated 4K in Russia). They developed their own short-recoil mechanism to create the AKT-37 (Kondakov-Tolochkov automatic, also referred to as the AKT-2 and AAK-37). The recoil length was 153–160 mm. An interesting feature of this gun was an attempt to soften recoil by increasing the mass of the recoiling parts, to include the quick-change barrel and the magazine. A prototype was built in the Artillery Academy in the beginning of 1935. By 1936 OKB-43 had shortened the case to 240 mm and removed the belt. The rifling was progressive, increasing in twist from the chamber to the muzzle. The AKT was 305 cm long with an L/51 barrel and a weight of about 240 kg. It was fed by a 5-round magazine (weighing 12.5 kg full), and demonstrated a cyclic rate of 150–200 rpm.

Firing trials were conducted between May and June 1936 using a temporary ground mounting, followed by successful aerial tests installed in a Tupolev R-6. These resulted in an order for ten guns to try in other aircraft. The AKT-37 was also envisaged as an AA gun to be used in the LAKT and TAKT mounts. The LAKT was a light two-wheeled mount. To deploy for firing, the whole weapon system was lowered and the wheels were pulled away with their frame. The TAKT was a mount for use on the loading platform of a truck (the ZIS-12 being used for trials), or for use on a four wheel mount designated ZU-7 and made by the Bryansk factory. The appearance was similar to the Bofors mount system. Competitive trials held in 1938 resulted in the project being cancelled.

Kondakov did not wait for the results of the AKT trials before starting to modify the design. The result was the 'rapid fire' ASKON-37 gun. The major differences between the AKT and the ASKON were: the addition of a fluted and reinforced barrel, to increase its strength and improve cooling; an improved feeding mechanism, with an added accelerator; a fixed instead of recoiling magazine to speed magazine changes; and a shortened recoil stroke, which raised the rate of fire to 250 rpm. Like the AKT, the ASKON was cancelled in 1938.

**20 mm ShVAK**: 20 × 99R AMMUNITION.

The ShVAK (*Shpitalniy-Vladimirov Aviatsionnaya Krupnokalibernaya*, or heavy-calibre aircraft gun designed by Shpitalniy and Vladimirov), was designed to use the mechanism of the gas-operated and belt-fed 7.62 mm ShKAS but chambered for a purpose-designed 12.7 × 107R cartridge (a different shape from the army's DShK-38 12.7 × 108 heavy machine gun round and not interchangeable with it). The resulting ShVAK was adopted in 1934 with production starting in 1935. Few of the weapons were acquired in this form, however, as in 1936 it was modified to improve the destructive effect by increasing the calibre to 20 mm, simply by straightening and shortening the 12.7 mm case. The guns only required a new barrel to use the 20 mm ammunition.

The rate of fire of 800 rpm (unsynchronized) was among the highest of the wartime 20 mm guns and was adjustable via a gas regulator. The locking system used a tilting breechblock operated by a gas piston. The ShVAK was not without problems: the ammunition feeding (which was from the right side only) and case ejection systems were complex; case extraction sometimes resulted in the case being pulled apart, leaving fragments in the chamber, so the cases had to be lightly lubricated to prevent this; and the HE shells were relatively short and light, limiting their effectiveness.

The barrels of these guns varied from L/62.5 to L/82.5 depending on the installation (there were wing, engine or turret-mounted versions); the overall length was from 168 to 232 cm, and the weight between 38 and 55 kg.

A version for arming light tanks, the TNSH-1 (later, the TNSH-20) standing for Tank gun by Nudelman and Shpitalnyi, was introduced in 1941. It had an L/82.5 barrel which enabled it to achieve an MV of 815 m/s. Rate of fire was 750 rpm. The gun remained belt-fed, with 754 rounds carried in 13 boxes. Some guns were pressed into service in the naval AA role.

The ShVAK remained the standard Soviet aircraft cannon until close to the end of the war, with more than 100,000 made by the time production ended in 1946.

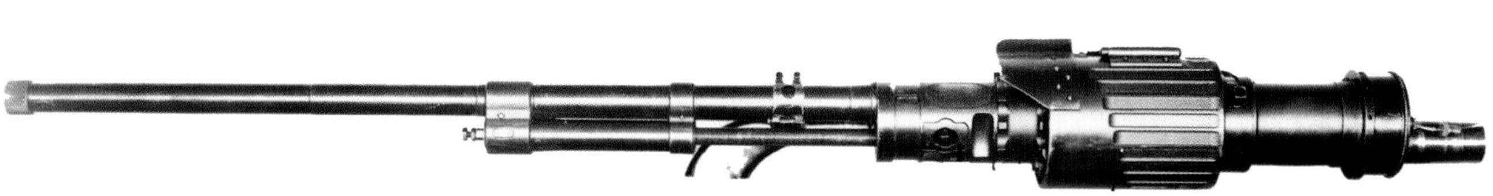

*20 mm ShVAK (Collection Jean-François Legendre)*

***37 mm 100-K***: 37 × 198 100-K AMMUNITION.

An unsuccessful contender for the army's AA gun contract, the 100-K was a product of Shpitalny's OKB-15. Confusingly, the cartridge measured 37 × 198, but it differed in shape from the later NS-37 round and is therefore referred to as the '37 × 198 100-K' cartridge.

Ammunition feeding was from a drum magazine, with a capacity of 6 cartridges (originally), later reduced to 5. Fired cases were tidily returned to the magazine. The weight of the gun was 340 kg, the barrel an L/61. The cyclic RoF was 150–180 rpm and the barrel life about 1,300 rounds. During 1937–38, 20 examples of the 100-K gun were manufactured; these were tested on various carriages and in the turret of a T-26 tank.

***45 mm M1939***: 45 × 495–510R AMMUNITION.

One of the priorities for Russian automatic cannon in the 1930s was a medium-calibre AA gun. After various experiments a 45 mm gun was selected, the operating mechanism being based on that of the Bofors long-recoil guns. The land version, on a four-wheeled mounting, was designated the 49-K (initially known as the ZIK-45), the naval variant the 62-K. The gun was gravity-fed by 3-round clips and fired at 120–140 rpm. An L/70 barrel was used. An order for 190 49-K guns was placed in 1940, but cancelled shortly afterwards in favour of the 37 mm 61-K system.

**37 mm M1939**: 37 × 252SR AMMUNITION (AKA 37 × 250R OR SR).

This gun and its ammunition were designed at short notice in 1938 when it was felt that the 45 mm M1939 was too large; this 37 mm M39 gun, designated ZIK-37 or 61-K, is effectively a scaled-down version of the 45 mm and is therefore a slightly smaller version of the 40 mm Bofors M32. It was fed via 5-round vertical clips and fired at 160–170 rpm. The weight of the army single-barrel mounting was 2,000 kg. A twin-barrelled version of the wheeled land mounting was also made as the 47-V, but this saw more export sales than Russian service; the Su-37 SPAAG also saw only limited service. Nearly 20,000 army guns were produced by the end of World War II, after which Russian production ceased.

The appearance of the M39, with the boxy receiver and vertical feed, is unsurprisingly very similar to that of the Bofors cannon, and also to the German naval 3,7 cm Flak C/36 and Flak M42. Presumably all of these weapons were inspired by the Bofors but the exact development relationship is unclear. The 37 mm M39 was developed with remarkable speed; possibly the Russians already had detailed information about the Flak C/36, as Germany provided various weapons for the Russians to evaluate.

Naval versions of the gun were designated as follows: 70-K (single barrel, air-cooled) over 3,000 made between 1941 and 1956); 66-K (twin barrels, water-cooled) prototype only, cancelled in favour of; V-11 (twin-barrelled, water-cooled, stabilized mounting) and V-11M (improved sights) of which 1,300 were made from 1944 to the early 1980s. A quad mount designated 46-K was developed but not adopted.

These weapons were widely sold abroad, and were made in large numbers by China. They remain in service in various countries, in some cases with upgrades to the mounting and/or fire control system; most recently in Pakistan, with 700 delivered.

*37 mm M39 – 70-K naval mounting (Vikcos75 / Wikimedia Commons / CC-BY-SA-4.0 licence)*

**25 mm M1940**: 25 × 218R AMMUNITION.
The 25 mm M1940 was a follow-on to the 37 mm M1939, intended to be a lighter and handier AA gun, more mobile on the battlefield. It was also developed by the Kalinin plant and was initially designated ZIK-25 before being given the factory designation 72-K. Although ordered in 1940, deliveries did not begin until 1941 with manufacture being spread over several plants as the German invasion affected Kalinin. Some 4,600 of these guns were made with production ceasing at the end of World War II. A twin-gun version using the wheeled mounting of the 37 mm M1939 was developed and given the factory designation 94-KM and service designation 25 mm Model 1944, but only 237 were built between 1944 and 1945.

The army was always somewhat lukewarm about the 25 mm systems, feeling that the lighter weight compared with the 37 mm M1939 wasn't sufficient compensation for the smaller HE shells, and the gun left service shortly after the war. There was a naval version designated 84-KM of which 330 were made; a twin version designed for submarines, designated Z-1, was tested but cancelled at the end of the war.

The 25 mm M1940 retained the long recoil operating mechanism and gravity feeding via vertical clips (of seven rounds in this case) of the Bofors design, the cyclic RoF being 240 rpm. Gun weight is quoted as 445 kg (although this seems far too high) with an L/76 barrel. The gun and ammunition both bore a family resemblance to the Bofors 25 mm systems, but they were not identical.

**23 mm MP-6**: 23 × 152 AMMUNITION.
Development of a 23 mm aircraft gun was initiated in 1938 as a result of dissatisfaction with the destructive effect of the small 20 mm ShVAK HE shells. A projectile weight of 200 g and an MV of 900 m/s were specified, making the ammunition more than twice as powerful as the ShVAK's. The first attempt was a long-recoil gun design by Taubin's OKB-16 bureau designated MP-3, but the RoF of only 300 rpm was insufficient. The design was duly altered into a short-recoil weapon, the MP-6 or PTB-23. One feature which carried over between these guns was the complex magazine feeding system. The magazine contained nine clips of ammunition, each holding nine rounds for a total of 81; the bulky box-shaped magazine was satisfactory in engine mountings but unsuitable for wing-mounting. Despite this, the MP-6 was put into production in 1941 and several hundred were made. The cartridge case was at some point modified from being rimless to belted; a belt-feed system was also developed, although few were made with this.

Meanwhile, TsKB-14 had developed a rival gun, which became the VYa (see below) and which replaced the MP-6 in production. The MP-6 guns were removed from aircraft and used in other roles, mainly AA but also with some experimentally installed in light tanks (PT-23TB). The MP-6 was 231 cm long with an L/59 barrel, weighed 70 kg including its magazine (120 kg with a full magazine) and fired at around 600 rpm.

*25 mm M1940 72-K (One half 3544 / Wikimedia Commons / public domain)*

*25 mm M1940 72-K breech (Saiga20K / Wikimedia Commons / CC-BY-SA-3.0)*

**23 mm VYa**: 23 × 152B AMMUNITION.
In parallel with the MP-6, design bureau TsKB-14 developed a rival gun firing the same ammunition, the TKB-201, later renamed VYa-23 in honour of the designers Volkov and Yartsev. It was essentially a bigger version of the 12.7 mm Berezin MG and retained that gun's gas-operation and disintegrating-link belt feed. It fired from an open bolt with a rising wedge locking mechanism. Cartridges were pulled out of the belt to the rear, rather than pushed-through the belt links.

It was first tested in 1940 but experienced various teething problems and comparative testing with the MP-6 (by then also equipped with belt feed) continued into the spring of 1941. There does not appear to have been much to choose between the two guns, both of which generated considerable recoil due to the powerful ammunition, but the VYa-23 was eventually chosen and over 64,000 were made between 1942 and 1947.

During World War II the VYa was also used in naval AA mountings with one, two or four barrels, but these were removed post-war.

The VYa-23 weighed 66 kg, was 214 cm long with an L/72 barrel. The rate of fire was 550–650 rpm.

**37 mm Sh-37**: 37 × 198 AMMUNITION.
In the late 1930s interest grew in a powerful aircraft cannon which would be able to destroy army tanks, and also bring down an aircraft with one hit, for which tasks the 37 mm calibre was selected. The first gun to meet this requirement was the Sh-37, designed by OKB-15 led by Shpitalny. This was a gas-operated weapon using a 20 or 40-round magazine and firing from a closed bolt. The designer attempted to tame the recoil by allowing the gun to recoil in its mounting: the problem was that the magazine recoiled also, leading to feeding problems; the closed bolt also introduced the risk of cook-offs. In an engine mounting and with a 20-round magazine, the Sh-37 weighed 208.5 kg and fired at 184 rpm; with a 40-round magazine in an under-wing installation, the total load was 302.5 kg and the measured RoF was 169 rpm.

Despite its problems the Sh-37 was put into production but it proved difficult to make and only 236 were made in 1941–42, before it was cancelled in favour of the NS-37.

**37 mm NS-37 AND 45 mm NS-45**: 37 × 198 AND 45 × 186 AMMUNITION.
Meanwhile, OKB-16 produced a rival design designated the PTB-37. This was originally designed for ammunition using a belted case, but this was later changed to the rimless one already in production for the Sh-37. The gun mechanism was based on that of the 23 mm MP-6, featuring a short-recoil rotary lock with a two-piece breechblock (the head rotating to lock/unlock). A bolt-accelerating lever was included. Initially the ammunition supply was also based on the MP-6, with eight clips of five or six rounds each, but this was changed to belt feed. The weight was 171 kg for the engine version and 160 kg for the wing gun; an L/62 barrel was used.

Compared with the Sh-37 this was clearly superior: smaller, lighter, faster-firing and belt-fed. It also proved more reliable. After overcoming some political resistance from the Air Force (which had only just ordered the Sh-37) the 11-P was accepted for service with the designation NS-37, after the lead designers Nudelman and Suranov.

Large-scale production of the NS-37 took place during 1943–45, with just over 8,000 guns made. These were fitted to the Yak-9T single-engined fighter (one gun, engine mounted) and Il-2M (two guns, underwing). About 1,000 Il-2M were made but the recoil caused aiming problems; a muzzle brake was tried but not adopted.

The NS-45 was identical to the NS-37, except that the cartridge case was necked out to the larger calibre to take the HE-T or HEI-T projectiles made for the naval 21-K; the case was also shortened to 186 mm keep the overall round length the same as the 37 × 198, and the gun accordingly had a new, L/56 barrel. The

*VYa-23 (Collection Jean-François Legendre)*

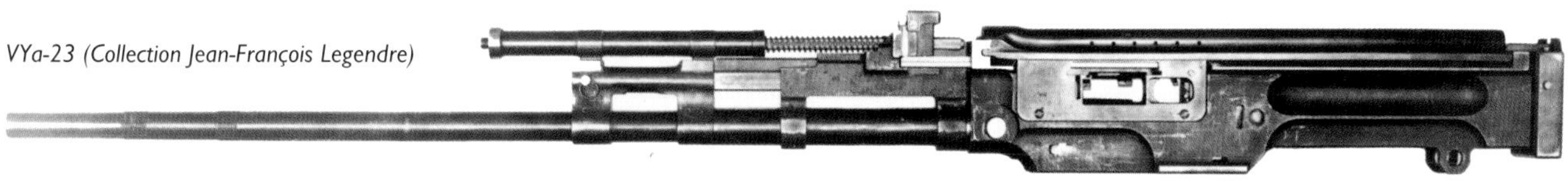

*NS-37 (courtesy Evgeny Aranov)*

gun was installed in an engine mounting in the Yak-9K fighter, but a problem arose: due to restrictions in the diameter of the gun barrel able to fit in the tube through the engine installation, the barrel had to be made very thin, so special steel had to be developed. Although the small numbers of Yak-9K made for trials proved highly effective in skilled hands, the recoil proved too heavy despite a muzzle brake being fitted, so the NS-45 was never accepted for mass production.

**20 mm B-20**: 20 × 99R AMMUNITION.
Since the 20 mm ShVAK had been created basically by replacing the barrel of the 12.7 mm ShVAK with one of 20 mm calibre, it was an obvious move to make a similar modification to the much superior 12.7 mm UB (Berezin) heavy machine gun. This meant that the B-20 shared its basic operating system with the VYa-23 (see above) although there were differences; e.g. the locking wedge slid sideways instead of vertically. Modifying the UB to take the 20 mm ammunition involved slightly more work, because it was originally designed for the army's rimless 12.7 × 108 DShK cartridge rather than the unique, rimmed 12.7 × 107R of the ShVAK which shared the same rim design as the 20 mm round. The UB's breechblock therefore needed to be modified for the B-20. Fortunately, the UB belt feed already used a 'pull out' rather than a 'push through' system which would not have been suitable for a rimmed round.

The B-20, a product of TsKB-14, defeated two rivals in a 1944 competition, all of them using the ShVAK ammunition: the Vladimirov V-20 from OKB-2 and the Sh-20 from OKB-16. The B-20 delivered the same performance as the ShVAK while being substantially lighter, smaller, simpler to make, and more reliable. It weighed a remarkable 25 kg (4 kg less than the aircraft version of the 0.50-inch M2). The rate of fire was 800 rpm unsynchronized or 600 rpm synchronized. Overall length was 148 to 207 cm, depending on the barrel length (L/48.5 to L/78.5). An electrically-seared version, the B-20E, was introduced in 1946 for use in remotely-controlled bomber defense turrets.

The B-20 soon replaced the ShVAK, with production continuing until 1949 by which time over 15,000 guns had been built in three versions: synchronized; unsynchronized (wing-mounted) and turret mounted.

**23 mm NS-23**: 23 × 115 AMMUNITION.
The compact and lightweight 23 × 115 round was developed rapidly at OKB-16, by combining two existing elements: 23 mm projectiles from the VYa, in a smaller cartridge case adapted from that of the 14.5 × 114 anti-tank rifle. The cost was a reduced muzzle velocity compared with the VYa, from around 900 m/s to 700 m/s.

The design bureau obtained approval to proceed with a compact new cannon to fire the new round in mid-1943; this was designated the NS-23 after the OKB-16 designers Nudelman and Suranov. The action was based on that of the NS-37 and therefore had a short-recoil rotary lock with a two-piece breechblock, the head rotating to lock/unlock. Development was rapid, flight-testing being completed only one year after design commenced. There were two rival contenders: the Vladimirov V-23 from OKB-2 (a scaled-up version of the 20 mm V-20) and the Sh-23 from OKB-16 (a 23 mm version of the Sh-20). Approval to begin production of the NS-23 was given in autumn 1944 and a total of over 28,000 guns was made between 1944 and 1953, in two versions: the NS-23KM for wing mounting and the NS-23S for synchronized installations.

The NS-23KM fired from an open bolt (i.e. it had a rear sear) which made it unsuitable for synchronisation, so the NS-23S design was modified by the addition of a front sear, acting on the firing pin rather than the bolt. This added a stage to the firing process, as the pilot pressing the firing button only resulted in the bolt being released, a new round being chambered and the bolt locked – the firing pin was then released by the synchronisation system only when no propeller blade was in front of the muzzle. This added complication reduced gun life from 4,000 to 3,000 rounds.

The NS-23 was belt-fed, from the left side only, and had a cyclic rate of 600 rpm. It weighed 38 kg, and was 198 cm long with an L/63 barrel.

*Berezin B-20 (courtesy Evgeny Aranov)*

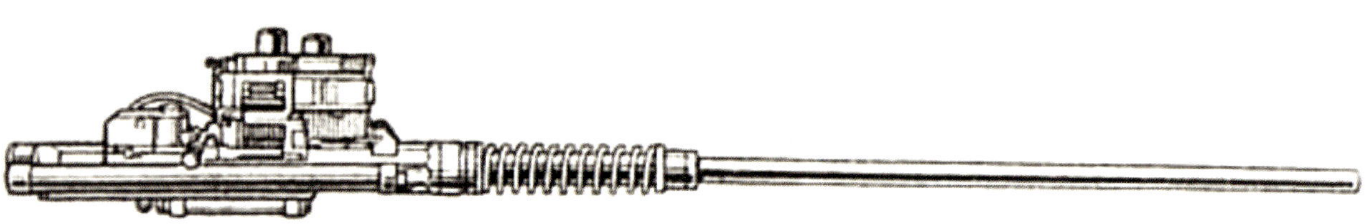

*Nudelman-Suranov NS-23 (courtesy Evgeny Aranov)*

**37 mm N-37 AND NN-37**:
37 × 155 AMMUNITION.
OKB-16 applied the same process to the development of a lightweight 37 mm gun as they had in developing the NS-23: they started in 1943 with developing the ammunition, combining the projectiles from the existing NS-37 with a smaller cartridge case (created by drastically modifying the 37 × 252SR case for the M1939 AA gun). As with the 23 mm pair, the result was a reduction in the muzzle velocity from almost 900 m/s to around 700 m/s. The first N-37 gun (named after Nudelman) was built and officially adopted in 1944 with flight tests in 1945. The initial installation was in the middle of a nose intake of the MiG-9 jet fighter (for which the N-57 was originally intended) and this caused severe problems due to the gun's muzzle blast preventing air from entering the intake, causing the engine to shut down. This problem was not finally resolved until the gun was fitted beneath the nose of the MiG-15, which replaced the MiG-9 between 1948 and 1950. The gun was later fitted with a muzzle brake, then being given the designation N-37D.

The short-recoil gun mechanism of the N-37 was similar to that of the NS-23, and therefore of the larger NS-37. The smaller ammunition allowed the gun weight to be reduced in comparison with the NS-37 from 160 to 103 kg while the rate of fire was increased from c.250 to 400–430 rpm. There were eventually three versions of the N-37:

- N-37D: 245 cm long with an L/32 barrel (plus 16 cm for the muzzle brake);
- N-37L: longer L/53 barrel (without a muzzle brake) giving an overall length of 287 cm, resulting in an increase in MV from 690 to 725 m/s;
- NN-37: N-37L with an added pneumatic counter-recoil accelerator to boost the RoF to 600–700 rpm (at a cost in gun life), used only in a small number of Yak-27 fighters. The extra N referred to designer Nemenov.

The N-37 was in production from 1947 to the early 1960s with over 20,000 guns made. It remained in Russian service until the 1970s (and may still be in use in North Korea).

***57 mm N-57***: 57 × 165RB AMMUNITION (57 × 155RB INITIALLY).
This short-recoil Nudelman design (internal factory index 100-P) emerged from OKB-16 about the end of 1944 and was intended for arming fighter aircraft. The mechanism was a scaled-up version of the N-37 (see above). The weight was 135 kg, length 218 cm with an L/18 barrel and the RoF was 230 rpm. Some 36 examples of the gun were built and it was originally intended for the MiG-9 jet fighter in which it was tested with both

*Two NR-23 and one N-37 from a MiG-15 (Dimitris Kamaras – Wikimedia Commons – CC-BY-2.0)*

ground and aerial firing, but Mikoyan was concerned about the effect of recoil on the small aircraft so chose the N-37 instead.

From the N-57 were developed the 57 mm experimental guns 113-P and 200-P, which were tested in the late 1940s. The I-215 interceptor with two 113-P was developed by Alekseyev in 1947, while another 113-P, with a 51-round belt feed and a longer barrel than the N-57, was used in the experimental ASu-57 self-propelled gun designed by Yakovlev and developed in 1948.

***57 mm RSHR-57***: 57 × 350B AMMUNITION.

This was designed in Nudelman's OKB-16 by Rashkov, Shchentsov and Rozanov. It was essentially an elongated and more powerful version of the short recoil N-57, intended to be available in two calibres. There is no information about the 45-mm version except that it used a 45 × 350B case, necked-down from the 57 mm case. The 57 mm was tested from December 1946 to February 1947 in a TU-2Sh ground attack aircraft. The mass of the high-explosive projectile was 2.49 kg, the muzzle velocity 720 m/s. The gun had a drum magazine for 10 rounds. The rate of fire was 140–160 rpm. The gun weight was 339 kg, the barrel L/51.

***57 mm Ch-21P***: 57 × 521RB AMMUNITION.

Another large-calibre aircraft cannon designed for ground attack planes, the Ch-21P was designed by Evgeny Charnko the chief of OKBL-46 (Special Design Bureau, Laboratory No. 46 in Leningrad). The gun was developed in mid-1948 in two calibres – 57 and 76 mm. The cartridge for the 57 mm version was based on the 57 × 480R for the ZiS-2 anti-tank gun, slightly extended and with a rebated rim. It weighed 6.5 kg and fired a 2.485 kg HE shell at 1,087 m/s (figures for the 76 mm version were 9.0 kg, 4.735 kg and 920 m/s respectively). The gun used a long recoil mechanism with a falling-block action; it had electro-magnetic recharging and used a 15-round vertical magazine. The 57 mm barrel was L/74 and the RoF was 80 rpm (70 for the 76 mm).

Two of the the 57 mm guns were installed on board a TU-2P and tested for several months in 1949, but the gun was not adopted.

**23 mm NR-23**: 23 × 115 AMMUNITION.

It soon became obvious that improvements to the NS-23 would be necessary and design work started at the end of 1946. The key areas requiring change were an increase in the rate of fire (the Air Force asked for 800–900 rpm) and the ability to switch between left and right hand feed to suit twin mountings in defensive turrets.

The result was a new design, in which the RoF was improved partly by using a counter-recoil acceleration lever (boosting the speed of the forward movement of the bolt, which already had a conventional lever to boost the speed of the rearward movement), partly by a general 'tune-up' to increase the speed of movements of the working parts. A universal feed mechanism was adopted permitting easy change of ammunition feed, inspired by the one for the German MK 108. The weight was 39 kg; overall length was 202 cm with an L/63 barrel, and the RoF was between 800 and 950 rpm.

The first air firing tests took place in 1948 but a number of defects emerged, including an excessive rate of wear of some parts, high and inconsistent recoil forces (found to be due to changes in belt pull as the ammunition was used up) and excessive shot dispersion. It took until 1951 to resolve these issues. Even after this, some problems occurred in turreted installations particularly when the guns were firing on the beam, due to the pressure of the slipstream on the side of the gun barrels, which needed to slide to and fro as they recoiled.The NR-23 was in production from 1948 to 1956, with over 80,000 made. In addition, it was made by China as the Type 23–1.

**25 mm 110-PM AND M-110**: 25 × 218 AMMUNITION.

The Navy shared the Army's dissatisfaction with World War II era 25 mm AA guns, designated 84-KM in naval use. Production therefore stopped at the end of the war, but even before then work had begun on defining the specification of an improved 25 mm twin-gun mounting. Nudelman of OKB-16 was given the design role, with the priorities being to reduce size and weight while increasing the rate of fire.

It was clear that the clip feed and long-recoil system of the 84-KM restricted the firing rate, so the new gun, designated 110-P, was designed for a disintegrating belt feed and a short-recoil operating system. Essentially the gun mechanism was a scaled-up NS-23 except for certain technical details, and the cartridge case was redesigned accordingly to be rimless instead of rimmed. The 110-P proved troublesome, being tested and modified repeatedly from 1946 until the final version, designated 110-PM, was approved for service in 1950. This weighed 101 kg and was 284 cm long with an L/80 barrel. The rate of fire was rather disappointing at only 270–300 rpm.

The standard mounting for these guns was designated 2M-3. This carried two guns mounted vertically in separate cradles, and was traversed and elevated by an electro-hydraulic system. The mechanical sights were quite simple, limiting its effectiveness. The on-mount ammunition load was 65 rounds per gun. The 2M-3 was not officially adopted until 1953, with production ceasing in 1957 after just over 1,600 mountings had been made.

*RShR-57 (courtesy Evgeny Aranov)*

*25 mm 2M-3M in an unusual role: on an armoured train (ShinePhantom / Wikimedia Commons CC-BY-SA-3.0 licence)*

The low rate of fire of the 110-PM prompted further work on the design by OKB-43 in the late 1950s, with Sokolov as chief designer. The major change was the introduction of a pneumatic buffer to increase the counter-recoil speed of the breechblock. Propellant gas was tapped from the barrel and fed into the buffer, thrusting the breechblock forwards more violently. This might therefore be considered to be a hybrid gas-assisted short-recoil action. The effect on the rate of fire was dramatic; it increased to 470–480 rpm. The new gun was designated M-110; the length was 281 cm and the weight increased slightly to 110 kg. The mounting was slightly modified and redesignated 2M-3M. It remained in production until 1984, being widely exported and also made by China as the Type 61.

One other application of the 110-PM was a specialized submarine mounting designated 2M-8. This had two guns, vertically stacked as usual, but the gun was marinised (e.g. by the use of stainless steel) and changes to the ammunition feed were made to suit it to its new role. It was officially adopted and fitted to some submarines in the 1950s with 283 being made, but the 2M-8 was withdrawn from service at the end of that decade.

**23 mm 2A14 / 2A7**: 23 × 152B AMMUNITION (MODIFIED).

After World War II the Army adopted a new heavy machine gun for the light AA role, designed around the 14.5 × 114 anti-tank rifle cartridge: the KPV. By the mid-1950s it was becoming apparent that this was no longer effective against armoured ground attack aircraft, and the 37 mm M1939 was also unsuited to dealing with fast jets, due to its inadequate rate of fire and fire control systems. Accordingly, it was decided in 1954 to develop a new anti-aircraft system to replace both existing weapons.

A prototype automatic cannon designed around the 23 × 152B cartridge of the VYa-23 had already been designed at TsKB-14, Tula (now KBP). This was a scaled-up version of Afanasev's A-12.7 aircraft machine gun, and similar to his smaller AM-23 aircraft cannon in 23 × 115. The mechanism is gas-operated, with locking by a vertically-moving wedge breech block which drops to unlock, controlled by cams on the piston extension. The gun is belt-fed from either side, and fires at between 800 and 1,000 rpm. This represents a considerable increase over the VYa-23 and involves more violent ammunition handling, which caused the cartridge to be modified: the case is made from steel rather than brass, and has a thicker rim. It also uses a different primer.

The gun weighs 75 kg and is 255 cm long with an L/75 barrel, plus 15 cm for the cylindrical flash-hider – an instant recognition point compared with the KPV which has a conical flash-hider. The rifling of the quick-change barrels is unusual enough to merit comment: the first 140 mm from the chamber has a fixed rifling twist of 3.6°; in the next 850 mm the twist progressively increases to 6°; for the rest of the barrel, the twist remains constant at 6°.

The gun was designated 2A14 and single, twin and quad towed mountings were developed to use it, although only the twin version was adopted. The towed twin mounting, originally designated ZU-14 (*Zenitnaya Ustanovka* = anti-aircraft system) was formally approved for

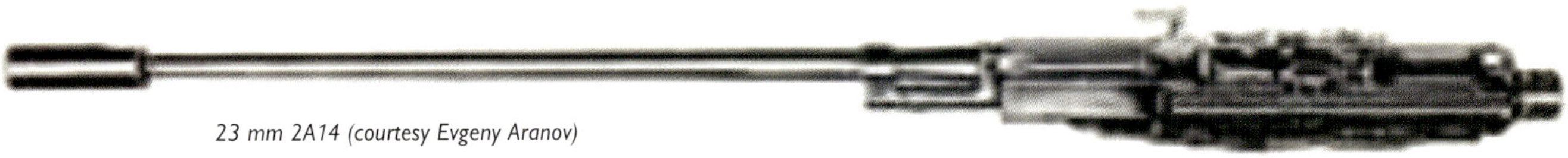

*23 mm 2A14 (courtesy Evgeny Aranov)*

service in 1960 as the ZU-23. It proved highly successful, being widely exported as well as made in several countries (including China, Poland, Bulgaria, Finland and Egypt) and upgrades of various types (mostly concerning the fire control system) are available from several countries.

A modified version of the 2A14 was also selected for use for a self-propelled AA gun, in a quad turreted mounting designated 2A10. The SPAAG is formally known as the ZSU-23–4 (*Zenitnaya Samokhodnaya Ustanovka* = self-propelled anti-aircraft mount), although more popularly as the *Shilka* (Awl). The gun was provided with water cooling to increase the sustained rate of fire (a burst of 150 rounds being permitted), plus an electrical sear, pneumatic charging and a conical flash hider. The gun was designated the 2A7, is 261 cm long and weighs 85 kg. A total of 2,000 ready-use rounds was provided; 480 rounds for each upper gun, 520 for each lower. The SPAAG entered service in 1965 after various problems were overcome, notably a tendency for fired cases to clog the ejection chutes, and it was produced at a rate of 300 per year for the next few years. The system remained in service until replaced by the *Tunguska* in the 1990s.

In the early 1970s both the 2A14 and 2A7 were modified to achieve greater reliability and a longer life (from 4,500 to 8,000 rounds); the 2A14's problematic pneumatic charging was replaced by a pyrotechnic system. The gun designations were accordingly altered to 2A14M and 2A7M.

**23 mm AM-23**: 23 × 115 AMMUNITION.

A need for a faster-firing 23 mm gun for defensive turrets emerged with the arrival of the TU-16 jet bomber in 1953. Shpitalniy's Sh-3–23, briefly described in the introduction to this section, had an impressive performance but the closed-bolt operating system led to cook-offs. Rikhter's 220-P cannon, an uprated NR-23 capable of 1,050 rpm, proved unreliable. So the contract was awarded to TsKB-14 for their TKB-495 gun, designed by Afansev and Makarov and subsequently renamed the AM-23.

*AM-23 (Vitaly V. Kuzmin – Wikimedia Commons – CC-SA-4.0)*

This was based on Afanasev's A-12.7 heavy machine gun, and was developed in parallel with the larger 2A14 in 23 × 152B, so it features belt-feeding (from either side) and gas operation with a vertically-moving wedge breechblock. The AM-23 weighs 43 kg and is 147 cm long with an L/43.5 barrel. There were a couple of variants, both intended for fixed installations in a bomber's nose and featuring longer barrels and muzzle brakes: they were designated AM-23L1 and AM-23L3 depending on the type of muzzle brake. The longer barrel and muzzle brake combined totalled 145 cm. The rate of fire is 1,250 rpm, and the gun has a life of at least 6,000 rounds.

The AM-23 series was in production from 1953 until the late 1970s, and was produced in China as the Type 23–2.

**45 mm SM-21**: 45 × 386SR AMMUNITION.

During the late 1940s the Navy put a lot of effort into developing automatic AA guns which would match the range and power of the manually-loaded 45 mm 21-K, and sufficiently fast-firing to replace the existing 25 mm and 37 mm automatic cannon. Four different guns were developed, for various mountings.

The first effort, from the Central Naval Artillery Design Bureau MATsKB (later TsKB-34) was designated SM-7, a long-recoil gun related to the Bofors design, which fired at 90–100 rpm. It was belt-fed, via 50-round belts stored under the mountings. Twin-gun mountings were developed, both stabilized (SM-16) and unstabilized (SM-17), but were not adopted, although the SM-16 was tested on one destroyer (*Neustrashimiy*) from 1954.

The second attempt was a gun from the design bureau of plant No. 614, designated MIK-1. This used a short-recoil mechanism and could be fed either by charger clips or 130-round belts, and fired at 120 rpm (clips) or 100 rpm (belt). A twin-barrel installation designated

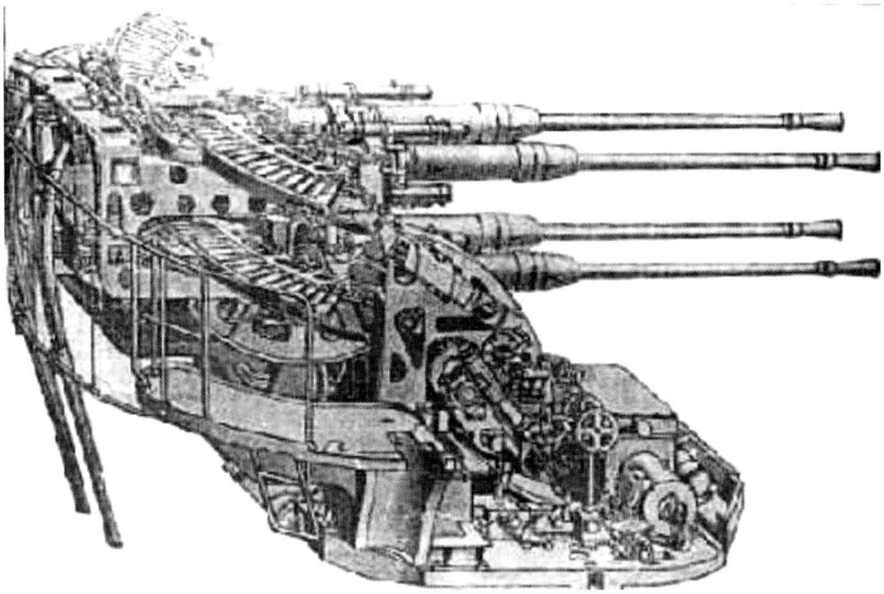

*SM-21 quad 45 mm mounting (from NavWeaps)*

BL-133 was developed, but the project was cancelled in 1950. Another effort, from OKB-16, was a long-recoil gun designated 145-PSh, which fired at 110 rpm.

The most successful effort was a 1950 redesign of the SM-7 gun by TsKB-7 to improve its reliability and increase the rate of fire to 160 rpm by shortening the recoil stroke; this was designated SM-21. Curiously, it reverted to an ammunition feed using four-round clips, which required a lot of manual reloading.

The L/78 barrel had an intermittent cooling feature: every 50 rounds, water was squirted down the barrel. It was produced in two different mountings: a single (SM-21-ZIF), of which few were made, and a quad (SM-20-ZIF), which was formally adopted in 1957 (ZIF referred to the Frunze plant which did the work). 121 quad mountings were produced from 1953–57 and fitted to destroyers, plus another 30, designated ZIF-68, with direct rather than alternating current electricity in order to rearm Type 68bis cruisers which used this type of current.

The 45 mm systems saw relatively short service as the calibre was abandoned in 1959, because the new 30 mm and 57 mm systems were felt to cover the need.

**30 mm NR-30**: 30 × 155B AMMUNITION.

The standard fighter guns in the early 1950s were the NR-23 and N-37, but the air force was dissatisfied with these, wanting something harder-hitting than the 23 mm gun and much lighter and faster-firing than the 37 mm. They decided on 30 mm as the best compromise and a competition was held which involved four different guns:

- Sh-3–30 from Shpitalniy's OKB-15: gas-operated, closed-bolt firing
- TKB-494 (V-30) from KBP Tula: gas-operated, scaled up AM-23
- TKB-500 from KPB Tula: little information available
- 235-P from Nudelman and Rikhter's OKB-16: short-recoil operated, scaled-up NR-23.

It is not clear whether all of the contenders used the same ammunition.

The 235-P won the competition and duly entered service in 1955 under the designation NR-30. It became the standard Russian fighter gun from the mid-1950s and remained in service until at least 1993. It was made in China as the 30 mm Type-1, and may still be in service with North Korea. A photo can be seen in the introduction to the Russia section.

The short-recoil design of the NR-30 utilizes recoil and counter-recoil accelerators to speed the rate of fire and uses a pneumatic barrel brake instead of a recoil spring. Gas is tapped from the barrel and fed into a brake chamber where it is trapped and compressed by the rearward movement of the barrel, bringing it to a halt and pushing it back again; it might therefore be regarded as a hybrid recoil+gas mechanism. The belt feed is driven by barrel recoil, and there is a pneumatic charging mechanism.

The NR-30 weighs just 66.5 kg, is 215 cm long with an L/53 barrel, and is remarkably compact. It fires at 850–1,000 rpm. The main problem with the gun was its short life, not surprising given the performance achieved from such a light weight; this was only 2,000 rounds at the start, but reached 3,000 towards the end of its service life.

What appears to have been an improved version, designated 270-P, appeared in 1957, weighed 78 kg and fired at 1,400 rpm, but was not adopted. Neither was the TKB-515 from Silin's TsKB-14 in Tula, which appeared in the same year and also used 30 × 155B ammunition: an 8-chamber revolver, this weighed 85 kg and fired at 3,000 rpm, but proved unreliable.

**30 mm NN-30 (AK-230)**: 30 × 210B AMMUNITION.

In the second half of the 1950s the navy were looking for a replacement for the 25 mm 2M-3 short-range AA system, which needed to be radar-directed and automated, with a high rate of fire and a good ballistic performance to engage jet fighters. The tasks of developing the gun and its mounting were given to OKB-16 and OKB-43 respectively. The electrically-primed cartridge developed for the gun was similar in general form to OKB-16's 30 × 155B aircraft gun round, being bottlenecked and belted, but was considerably larger to hold the propellant needed to achieve its much higher MV.

A revolver cannon was chosen to obtain the high rate of fire required, but primarily recoil-operated instead of the gas-operation standard for this class of gun. The revolver cylinder also has only four chambers instead of the usual five. The gun, designated 291-P, is belt-fed with separate right and left-side feed versions. With each shot the gun recoils 90 mm and the drive ring is rotated via a cam. Each round is rammed into its chamber by compressed air, and fired cases are blown out by gas tapped from the muzzle. It is water-cooled via two different systems: a conventional external barrel cooling, with a very slim jacket, and an internal evaporation system whereby coolant is sprayed into the bore each time the cylinder is rotated. The gun weighs 155 kg and is 268 cm long with an L/60 barrel. It fires at 1,000 rpm, with a maximum of 100 rounds per burst, after which the barrels have to be cooled.

The KL-302 twin mounting was developed in parallel with the 291-P (requiring some adjustment of both gun and mounting designs to make them fit together). The guns are in one cradle, with the barrel axes 240 mm apart, and are protected by a characteristic dome-shaped steel housing. The mounting extends into the deck below to provide room for the ammunition load of 500 rounds per gun, which is fed with electrical power assistance. The mounting cannot rotate through 360 degrees; it has a 310 degree traverse. The radar control was developed for this system, using the MR-104 Rys radar.

Series production started in 1959, the gun being redesignated NN-30 after the designers Nudelman and Nemenov, the whole mounting being redesignated AK-230; this came in two versions for ships with alternating or direct current power supplies. A modified low-magnetic version for minesweepers was designated AK-230M.

The AK-230 system proved highly successful, with over 1,000 being produced between 1959 and 1970; the final order was placed in 1983. Sixty-seven of the AK-230M were made in 1962–4. The system was made under licence in China as the Type 69.

One other application of the NN-30 was for the Romanian Army, which developed a towed mounting with two guns, designated A436 or Model 80. The guns were modified to provide rates of fire adjustable from 100 to 500 rpm per gun, and 30 rounds of ammunition per gun are held on mount. The original model was manually operated for daylight use but it was followed by a radar-directed version, with the radar on-mount. It appears to be still in service.

*30 mm TKB 539*: AMMUNITION UNCERTAIN.

The Makarov TKB 539 was a 30 mm 4-chamber revolver cannon, with an unusual action in which the cylinder was allowed to recoil via the gas system action in order to unlock from the barrel, rotate 90 degrees and then relock. Obturation was achieved by inserting the case neck into the barrel for about 25 mm. The cylinder was loaded from the front with special cartridges having a reduced-diameter head (a similar arrangement to the R-23). Fired cases were ejected from the front using gas pressure from the next shot. The gun weighed 73 kg (the feed system another 16.8 kg), was 184.5 cm long and fired at 2,000 rpm. The ammunition generated a muzzle velocity of 1,000 m/s.

**57 mm S-60 FAMILY**:
57 × 348SR AMMUNITION.

A project to develop a 57 mm AA system was initiated in 1944, partly to cover the gap in effective range and altitude between the 37 mm cannon and manually-loaded large calibre artillery, partly in recognition of the fact that aircraft were becoming better protected and harder to destroy. Various designs were submitted, the successful one being from the TsAKB Grabin design bureau, which was developed between 1946 and 1949, passing its tests and being officially adopted in 1950 as the S-60.

The S-60 uses a short-recoil action with a two-part bolt locked by a rotating head. Ammunition is held in 4-round clips and fed from the left side, ejecting to the right. Ammunition handling is powered by the barrel recoil. The gun weighs 1,475 kg (3,400 kg including the four-wheel mounting) and uses an L/77 barrel. The cyclic RoF is between 100 and 120 rpm. Fifty to sixty rounds can be fired in one burst before a cooling break is needed; this is aided by a manual system for pumping water through the barrel. Traverse and elevation are electrically powered, with a manual back-up.

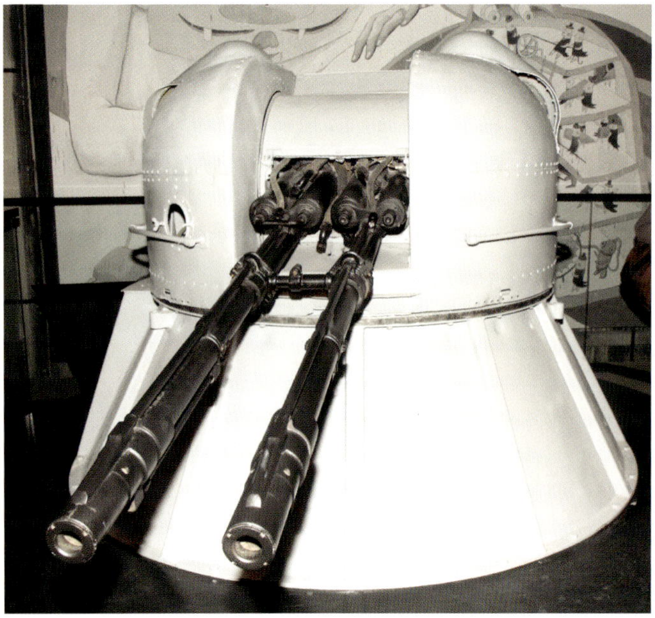

*Two NN-30 in AK-230 mounting (MKFI / Wikimedia Commons / Public Domain)*

*Romanian M80 (courtesy Nic Jenzen-Jones)*

Power is supplied by a generator unit which feeds the whole battery of six to eight guns plus the associated radar and fire control units.

The S-60 went into action during the Korean War, revealing reliability problems due to various shortcomings concerned both gun and ammunition. These issues were resolved and a total of 5,725 guns were made between 1950 and 1957. It was also built in China as the Type 59.

A self-propelled version was developed in parallel with the S-60, consisting of two guns mounted in an open-topped turret on a modified T-54 tank chassis. The gun mounting, with electro-hydraulic traverse and elevation, was known as the S-68, the whole system was designated ZSU-57–2. The right-hand gun was modified to feed from that side; a total of 300 rounds was carried. This was initially successful and was also adopted by China as the Type 80, but the lack of radar and NBC protection led to its early replacement in Russia.

The S-60 has been out of service for decades in Russia, but rather surprisingly it appears to be making a come-back in various forms. One motivation can be to make use of S-60 guns in store, e.g. Valhalla Turrets of Slovenia, which in 2017 revealed that they had commenced a private-venture project to develop the Hildgard 57 Medium Remote Gun System (MRGS) using surplus S-60. It was intended for medium-weight armoured fighting vehicles (AFV) to provide increased firepower for ground fighting compared with the usual 25 to 40 mm cannon. A chain-driven mechanical ammunition handling system was planned, with 33 rounds of ready use ammunition (16/17 each of AP and HE). Both stabilized and non-stabilized models, with various optronics and fire control options, were proposed.

Valhalla Turrets also offer the Heimdal heavy remote controlled turret, the Odin

*ZSU-57–2 (VargaA / Wikimedia Commons / CC-BY-SA-1.0)*

*Hildgard 57, from Valhalla Turrets*

manned turret (both with S-60 cannon), and have developed a novel application announced in 2019. This is the Desert Spider RCWS, which consists of the S-60 mounting with a mechanical ammunition handling system holding 92 rounds of ready-use ammunition, supported by four extendable arms. A 14.5 mm KPV HMG is mounted coaxially. This is intended for the defence of high-value targets.

More significantly, Burevestnik JSC of Russia were offering from about 2015 new-build S-60 guns with various modifications, both for naval general-purpose use and for fitting to AFVs for ground fighting or AA purposes (the modified gun being designated BM-57). The land mounting is designated the AU-220M, an updated version of the AU-220 of c.2000, which was a manned turret intended for rearming the PT-76 light amphibious tank.

The AU-220M turret is remotely-controlled, with mechanical ammunition handling and a rate of fire reduced to 80 rpm; 80 rounds are carried. One application is for the ZAK-57 Derivatsiya-PVO SPAAG, intended for use with a guided artillery projectile developed by the Tochmash Design Bureau.

**57 mm NAVAL GUNS**:
57 × 348SR AMMUNITION.

The Russian navy was also interested in a 57 mm AA gun but instead of adopting the S-60 they developed their own, long-recoil gun designated **SM-15**, with an action based on the 37 mm M1939 but using the same ammunition as the S-60. It fired at a cyclic rate of 100–115 rpm and was fed by 3-round clips; an L/79 barrel gave it a marginally higher muzzle velocity compared with the S-60. It used a barrel cooling system similar to the S-60. A single-barrel mounting with this gun was cancelled in 1950, but the same basic design of gun saw service in various other naval mountings.

A twin mounting of the SM-15, designated **SM-24-ZIF**, was adopted for submarines and entered service in 1955. It remained in service into the 1990s.

This was followed by the **ZIF-31**, an open twin mounting for surface ships which was also adopted in 1955, as usual in different versions depending on the current of the electricity supply. The barrels had modified rifling and were slightly shorter at 427 cm. This was made in China as the Type 66.

Next came a single mounting, adopted as the **ZIF-71** in 1960. The barrel length was 464 cm (L/81) including a muzzle brake, and the cyclic rate was increased to 160–170 rpm, leading to a different gun designation of **ZIF-74**. This also remained in service into the 1990s.

A quad mounting for the ZIF-74 gun was adopted in the early 1960s under the designation **ZIF-75**, but this saw relatively little use with only about 40 mountings made.

The next development involved more radical change, adding a continuous ammunition feed via belts (operated by the recoiling barrel) and continuous water-cooling for the barrel. The rate of fire was increased to 200 rpm per barrel and each gun has a 550-round belt. The new mounting had two modified ZIF-74 guns and was covered by a dome-shaped light-alloy gunshield. This mounting was the **ZIF-72** which was adopted in 1964 under the official designation of **AK-725**. It was very popular, with production continuing until 1988.

Much later came the experimental **A-220**, a lightweight single-barrel mounting for a modified S-60 gun, but with an auger type linkless feed which enabled

*57 mm 2BMP-3 based SPAAG 2S38 Derivaciya PVO (Boevaya mashina / Wikimedia Commons / CC-BY-SA-4.0)*

*57 mm ZIF-31 (ShinePhantom / Wikimedia Commons CC-BY-SA-3.0 licence)*

the gun to fire at 280–310 rpm. It was developed in the late 1970s, but by this time the navy had decided on a 76 mm calibre gun for smaller vessels so was not interested. Despite this, it has recently been revived as the **A-220M** which is a lightweight, remotely-operated mounting with a single S-60 gun, a sea-water barrel cooling system, electrical traverse and elevation, and a shaped (stealth) gun house. The rate of fire is 300 rpm and the mounting has 400 rounds ready to fire; feeding from the magazine to the receiver is by a power-driven 'feed screw', then recoil-operated for chambering the rounds. It will presumably benefit from the guided projectiles developed for the AU-220M described above. It is offered on various light warship designs, so far without any sales being reported.

**23 mm R-23**: 23 × 260 AMMUNITION. The air force had discovered problems with the using the turreted NR-23 as a bomber defence gun because when firing on the beam the pressure of the slipstream against the barrel affected the functioning of the gun. As this was experienced even with the propeller-driven TU-4 it would clearly be a much bigger problem with supersonic jets, so a new gun was needed which would not suffer these problems.

OKB-16's designer Rikhter decided on a radical configuration to meet the requirement, selecting a four-chamber revolver mechanism (with fluted chambers) but with the cartridges loaded backwards into the chambers. This meant that the ammunition belt feed (into the right side of the gun) is sited under the barrel, close to the centre of gravity of the gun, and there is very little of the gun behind the revolving cylinder. As much as possible of the gun barrel is therefore contained within the gun mounting, with only a short length protruding into the slipstream. Altering the elevation of the gun also has a minimal effect on the ammunition feed as it is very close to the trunnions. The gun is gas-operated, with three different gas systems: one to drive the revolver cylinder and feed mechanism, one to blow cartridges into the chambers and a third to blow fired cases out again.

Loading the cartridges backwards means that conventional ammunition cannot be used, as it tapers from the base to the tip; a configuration which is tapered towards the base is needed, and this results in a strange, pipe-shaped case within which the projectile is completely buried. Note that this is not a telescoped cartridge as all of the components are in a line as usual so it is no shorter than a conventional round.

The gun was initially designated the 261-P but on adoption in 1964 it was named the R-23. It initially fired at 2,400–2,600 rpm, but in service this was reduced to 1,800–2,000 rpm to improve reliability. The gun weighs 58.5 kg, is 147 cm long with an L/50 barrel and has a life of 3,000 rounds. It was only ever fitted in a tail mounting to certain variants of the TU-22 supersonic bomber: the A, B, K and R subtypes.

At least 500 guns are believed to have been made, and it remained a secret weapon until the 1990s, possibly because a version (calibre uncertain) was chosen for installation in certain space stations as a self-defence weapon. The first design to incorporate a Rikhter cannon was the Soyuz VI in 1965 but this was cancelled; the only space station which is known to have flown with this armament was the Salyut 3 (also known as Almaz OPS-2) in 1974. At the end of its mission in 1975 the gun was test-fired and reportedly destroyed its targets. It is not known if any later space stations were so armed.

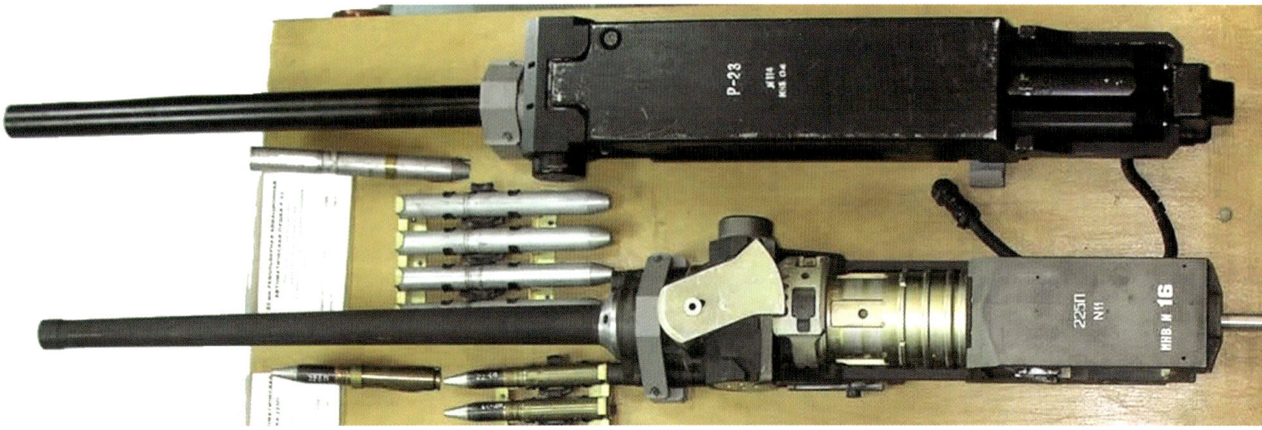


*Rikhter's R-23 (top) and 225-P, each with its own ammunition (courtesy Evgeny Aranov)*

The need to use unique ammunition was a drawback so in the late 1960s Rikhter designed another revolver cannon, using a similar gas-powered system, this time in a conventional layout for standard 23 × 115 ammunition. This was designated 225-P, had a five-chamber cylinder, weighed only 36 kg and fired at an astonishing 4,200 rpm. It was not adopted, possibly because the barrel life would have been very short.

**23 mm GSh-23**: 23 × 115 AMMUNITION.

This was the first of a long series of guns from Gryazev and Shipunov, who took over the mantle of principal cannon designers from Nudelman and Rikhter. Various options for boosting the rate of fire of aircraft guns were tried in the 1950s, with the most successful of them being the twin-barrelled GSh-23.

This system was devised by a German designer, Carl Gast, who developed a machine gun of this type during World War I. It is a twin-barrelled gun with the two bolts linked by a pivoting yoke, so that when one bolt is in battery, the other is held to the rear. The barrels therefore fire alternately, the recoil of one bolt accelerating the forward movement of the other. The GSh-23 differs from the Gast principally in being gas rather than recoil-driven; it also uses rising breech-blocks rather than longitudinally-moving bolts. Gas from each barrel is tapped to drive the operating pistons which are connected to actuating slides; it is divided so that half of the gas goes behind one piston to drive it forwards while the other half goes in front of the other piston to drive it backwards. It is the actuating slides which are coupled and pivoted, effectively acting as accelerators. The gun is belt-fed, from either side. One quirk of the system is that the belts must contain an even number of rounds, to ensure that one isn't left in a chamber after firing the full load. The mechanism is very compact and light, and the twin barrels mean that, for the same rate of barrel wear, the rate of fire can be doubled compared with a single-barrel gun.

Flight tests of what was then called the AO-9 were completed in 1959 and, after modifications by the Kovrov plant to increase the gun life to 4,000 rounds, the cannon was adopted in 1965 as the GSh-23. It has been widely used in fighter aircraft, bombers, helicopters and gunpods, and has also seen some use mounted on AFVs. Most use has been of the GSh-23L, the letter standing for 'localisator', which refers to a long, multi-baffle muzzle brake. The GSh-23 is made under licence in China (as the 23 mm Type 3) and India (where it is known as the *Ghasha*). The Czechs have made a version of the gun, designated ZPL-20, which is converted to fire the US 20 × 102 electric-primed ammunition (the 23 × 115 is percussion-primed).

The GSh-23 weighs 50 kg, is 139 cm long with L/43 barrels, and fires at 3,000–3,400 rpm. The GSh-23L weighs 51 kg and is 154 cm long.

**23 mm GSh-6–23**: 23 × 115 AMMUNITION.

Following experience in the Vietnam War, in which the disappointing performance of guided air-to-air missiles threw fresh emphasis on aircraft guns, Shipunov examined captured US six-barrelled rotary cannon and in 1965 obtained approval to work on an equivalent weapon. This was developed in two versions, both based on the AO-18 naval AA gun which had been developed from

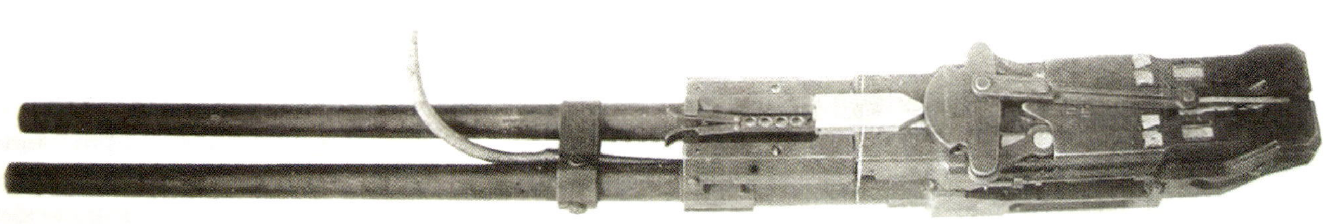

*The first prototype of the AO-9 (courtesy Maxim Popenker)*

1963: one in 30 × 165 calibre became the GSh-6–30 (a relatively simple modification of the AO-18), the other in 23 × 115 was first known as the AO-19 and, apart from being much smaller, also needed to be adapted to fire percussion-primed ammunition. The AO-19 was put into production in 1972 and formally adopted in 1974 under the designation GSh-6–23. Two aircraft types had this gun installed: the MiG-31 interceptor and the Su-24 bomber. It was also used in the 500-round, 500 kg SPPU-6 gunpod carried by the Su-27UB bomber, in which the gun was not only capable of being angled downwards by 45° but also traversed horizontally by the same degree.

The initial version of the gun, designated 9A-620, uses compressed air to spin-up the barrel cluster before the gas-operation takes over; this was replaced by the 9A-768 (GSh-6–23M) with a much faster pyrotechnic spin-up system using PPL blank cartridges (10 were carried). There were also two versions of the 9A-768; the first had a disintegrating-link belt feed (carried over from the 9A-620) but problems with this led to the introduction of a linkless feed.

All of these changes led to some variations in the gun data: the original model weighed 73 kg, was 139 cm long (152 cm with muzzle brake) and had L/43.5 barrels. The belt-fed version of the GSh-6–23M was 140 cm long and weighed 76 kg. Ammunition was held in a 400-round box (designated the NPU-19 system). The linkless feed added substantial weight, raising the total to 148 kg (which presumably included the entire 260-round magazine weight).

The GSh-6–23 was withdrawn from service in the Su-24 because of an unacceptable level of premature detonations in the barrels. After two Su-24 were lost because of shell detonation in 1983, plus other problems (system failures etc., very similar to the situation with GSh-6–30 and MiG-27), use of the GSh-6–23 was stopped by a decision of the Soviet AF Command.

**30 mm AO-10 / GSh-30 / GSh-30K**: 30 × 165 AMMUNITION (ELECTRIC PRIMED).

The first prototypes of the GSh-30, designated AO-10, were developed in parallel with the GSh-23 which it closely resembled, but initially in the percussion-primed 30 × 155B calibre of the NR-30. The AO-10 weighed 87.5 kg, was 210 cm long with L/53 barrels and fired at up to 3,500 rpm. This was ground-tested in 1959 and could have made an outstanding aircraft gun but did not proceed any further due to a premature view that AAMs made guns obsolete. It is worth noting that the comparable Western gun at the time (the ADEN/DEFA 30 mm revolver) weighed the same, yet fired ammunition with only two-thirds of the muzzle energy at less than half of the cyclic rate.

*Two views of the GSh-6–23 (courtesy Maxim Popenker)*

During the 1960s work on the AO-10 continued with the aim of developing an AA gun. This led to the AO-17, which differed mainly in being rechambered for the new 30 × 165 universal round, in its percussion-primed form. Barrel length was increased considerably and the rate of fire was reduced to 2,000–2,500 rpm. This cannon was adopted as the 2A38M and is described below.

In 1971 attention focused to the need to equip the next generation of jet fighters, starting with the Su-27, with a suitable gun. The AO-17 met the performance requirements very well, so KBP shortened the barrels to L/50 (the overall length being 204 cm), switched to electric priming and made other modifications, resulting in a weight of 105 kg and a RoF of 3,000 rpm. This version was initially designated AO-17A but on being formally accepted for service in the first half of the 1970s the name was changed to GSh-30. The advent of the GSh-301, which weighed less than half as much, meant that the GSh-30 was not in fact used as a fighter gun, but it was fitted to the Su-25 ground attack plane in 1981. It is capable of firing 250 rounds (the full capacity of the Su-25's installation) in one burst, and has a life of 4,000 rounds. It is also available in the NNPU-8M gun pod.

A version of this gun designated GSh-30K was developed for the Mi-24P helicopter gunship, in the NPU-30 mount with 250 rounds, the main difference being the L/80 barrels, extended to provide the highest possible impact velocity and therefore armour penetration. This gives an overall length of 294 cm and a weight of 126 kg. The firing mechanism is also different, to permit a variable rate of fire: 300–400 rpm for ground attack, 2,000–2,600 rpm for aerial targets. There is an evaporation cooling system.

**30 mm 2A38M**: 30 × 165 AMMUNITION (PERCUSSION PRIMED).

The development of helicopters and ground attack planes armed with guided anti-tank missiles prompted a search for a replacement for the ZSU-23–4 '*Shilka*' SPAAG as early as 1970, not long after it entered service. A longer range and more effective ammunition were sought, so KBP selected a combination of guided AA missiles and 30 mm cannon, using a common sensor/FCS suite, to provide a wide range envelope. A gun RoF of 5,000 rpm was required to engage high-speed targets, and this was met by fitting two AO-17 guns, adopted in the 1970s as the 2A38 (the 2A38M is a later model, modified to improve the gun life from 6,000 to 8,000 rounds). The resulting SPAAG was designated the 2S6 '*Tunguska*', and was adopted in 1982.

The operating mechanism for 2A38 is basically the same as the GSh-23, and differs from the GSh-30 principally in using percussion-primed ammunition. As installed on *Tunguska*, it looks as if there is only one barrel on each side of the turret, but that is because only one of the two barrels for each gun is fitted with a velocity measurement sensor which adds to the length. The barrels

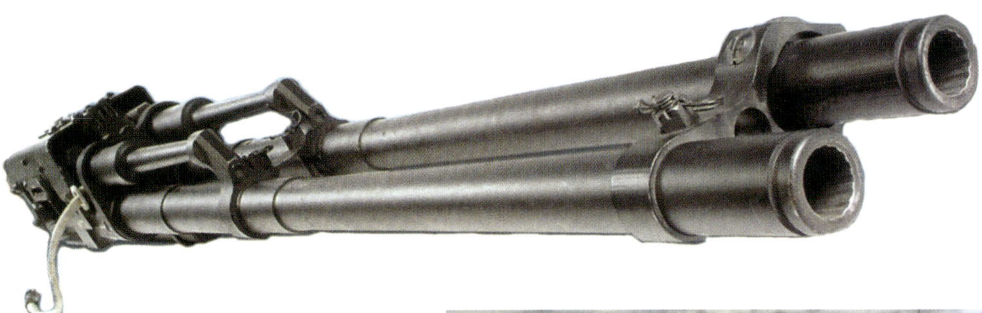

*GSh-30 (courtesy Maxim Popenker)*

*Tunguska at VostoK-2018 military manoeuvres (www.kremlin.ru via Wikimedia Commons – CC-BY-SA-4.0)*

have an evaporative cooling system, with a very slim, high-pressure jacket around virtually the entire barrel and a lower-pressure larger-diameter tank around the back of the barrel. As the water in the high-pressure jackets turns to steam, it is fed into the large tank where it cools down and liquifies, subsequently being fed into the high-pressure jackets again. Ammunition can be fed into the right or left side, fired cases being ejected underneath. The 2A38M weighs 223 kg (complete with water), is 348 cm long with L/80 barrels. It fires at 2,000–2,500 rpm, with the maximum continuous burst length being 100 rounds, after which the gun must be left to cool.

The 2A38M has subsequently been used in various land-based AA plus SAM systems, most notably the Pantsir S-1.

It is worth mentioning an alternative project for an AA gun which also used the Gast principle: the AO-16 from KBP. This was in 57 mm calibre, but no other information about it has yet emerged.

**30 mm AO18 / GSh-6–30K / GSh-6–30P / 6K30GSh**: 30 × 165 AMMUNITION (ELECTRIC PRIMED). The AO18 naval gun was the first to use what is now the standard Russian 30 mm cannon round – indeed the only round for current-production Russian automatic cannon – the 30 × 165. The gun was specifically designed for a naval close-in weapon system able to destroy anti-ship missiles, for which a rate of fire of 5,000 rpm was calculated to be needed. For this, a multi-barrel rotary configuration was chosen; liquid-cooled barrels would also be required to be able to sustain a sufficient engagement time (this was achieved by fitting a coolant-filled cylindrical jacket around the entire barrel cluster).

Another contrast with the contemporary US air-cooled M61 and GAU-8/A rotary cannon is that the Russian gun (initially known as the TKB-025, but on adoption designated the AO-18) is gas-operated instead of externally powered. There is one gas piston which is located in the centre of the six-barrel cluster, on the axis of rotation. Each barrel has its own breechblock. The firing sequence is started by a pneumatic system. Gas tapped from each barrel operates the piston, being directed to the front or the rear of the piston alternately; this movement is converted to a rotary action. The barrels fire at the 6 o'clock position. A cog-wheel around the rear of the cluster drives the ammunition belt feed. The AO-18 weighs 205 kg and is 218 cm long with L/54 barrels. It can fire at 4,000–5,000 rpm and has a life of 8,000 rounds (20,000 in current versions).

In the 1980s the AO-18 was revised in order to improve reliability and gun life (to 12,000 rounds), leading to a change in designation to GSh-6–30K, which replaced the AO-18 in production. Also in the 1980s another variant, the GSh-6–30P, was developed to have a linkless feed instead of a belt feed, which allowed the maximum RoF to be increased to 6,000 rpm. Finally, the 6K30GSh designation was given to a further variant, using evaporative cooling rather than a coolant jacket; this and the GSh-6–30P were intended for use in the big gun/missile CIWS mentioned below.

Just as important as the gun is the mounting, which requires high-speed and very accurate elevation and training mechanisms, plus a sufficient quantity of ammunition. The prototype mounting for a single stabilized AO-18 gun was designated the A-213, but on adoption this became the AK-630. As is usual with Russian naval equipment, different versions of the mounting were made to allow for varying electrical current supplies on warships.

*AK-630 (DoD / Wikimedia Commons / Public Domain)*

The first prototypes were tested in the mid-1960s, but a period of modification and further testing meant that the gun system was not formally adopted until 1976. A modification to the shape of the 2,000-round ammunition magazine was made in order to facilitate its installation in a larger number of vessels, commonly replacing the AK-230 on a one-for-one basis. This was adopted in 1980 as the AK-630M. It is also produced in China as the Type 630.

The AO-18 has also been used in other gun mountings, developed to cope with the increasingly sophisticated anti-ship missiles, especially those with a 'climb and dive' final trajectory. The first such was the AK-630M1–2 'Roy', which was fitted with two AO-18 cannon, one mounted above the other. This was developed during the 1980s but did not attract sales, due to a focus on combined gun/missile CIWS instead. The concept was however revived in the early 1990s in cooperation with China, and the name has changed to 'Duet'; it weighs 3,000 kg without ammunition (4,000 rounds are carried).

In the 1980s KBP developed a system, designated 3M87, which combines two GSh-6–30P or 6K30GSh guns and eight longer-ranged surface-to-air missiles on one mounting using the same sensors and fire control system. The 3M87 mounting was dubbed *Kortik*, and in modified form was offered for international sale as

*'Duet' CIWS (from Tulamashzavod brochure, via Richard Jones)*

*Kortik CIWS (Mike1979 Russia / Wikimedia Commons / CC-BY-SA-3.0 licence)*

the *Kashtan*. This is a formidable system with such features as power reloading of the missiles from under-deck magazines, but is accordingly large and heavy (15,500 kg all-in). Performance improvements (to the mounting, missiles and FCS rather than the guns) have resulted in the *Kortik-M*. This was in part a response to a rival system from Nudelman's KBTM, designated *Palash* (*Palma* being the export version). This is similar but much simpler, using SACLOS missiles and omitting the missile reloading, and weighs only 7,000 kg. The AO-18KG guns use linkless feed. Delays with perfecting the missile system have resulted in a gun-only version being offered.

**30 mm GSh-6–30**: 30 × 165 AMMUNITION (ELECTRIC PRIMED).
The idea of producing an aircraft version of the AO-18 was proposed in the late 1960s with the first prototype made in 1970, approved in the following year and officially adopted in 1974 under the designation GSh-6–30. This speed was due to the relative simplicity of the task, mainly achieved by removing the liquid cooling jacket and shortening the barrels. Even so, there were problems to be overcome concerning barrel overheating and the feeding system.

The GSh-6–30 weighs 149 kg, is 188 cm long and has L/50 barrels. The rate of fire is 5,000 rpm and the maximum continuous burst length is 300 rounds.

Only one type of aircraft had this gun installed in service, the MiG-27 ground-attack fighter. Unfortunately, this small plane had a light structure which proved unable to cope with the fierce recoil and vibration of the big cannon, even after the rate of fire was reduced to 4,000 rpm and the pilots advised to fire only 30 to 40-round bursts. Despite many accidents and other problems the MiG and its cannon were not withdrawn from service until the late 1990s.

**30 mm AO18L**: 30 × 165 AMMUNITION (ELECTRIC PRIMED).
The AO-18L was designed as a lightweight version of the AO-18 for installation in the AK-306 mounting (originally designated the A-219), for fitting to smaller vessels. The most obvious change from the AK-630 is the omission of the barrel coolant jacket; less obvious but more fundamental is the replacement of the gas-driven operating system by an electrically-powered one. The gun weight is 155 kg (the electric drive mechanism weighing a further 35 kg); the dimensions are the same as the AO-18, but the rate of fire is reduced to just 600–1,000 rpm, with a modified version seeing this further reduced to 300–500 rpm (it seems odd that six barrels were retained for such a low rate of fire given that the aim was to lighten the installation). The ammunition capacity is 500 rounds. The fire control system is much simplified, being optical only, and the mounting is not stabilized. The AK-306 weighs just 1,100 kg compared with 1,850 kg (plus radar) for the AK-630. Some 125 AK-306 were built between 1978 and 1986.

**30 mm 2A42**: 30 × 165 AMMUNITION (PERCUSSION PRIMED).
This gun was developed to arm infantry fighting vehicles (IFVs – also known as mechanised infantry combat vehicles or MICVs) with a flat-trajectory weapon primarily intended to engage similar enemy vehicles. It replaced the low-velocity 73 mm smooth-bore gun which armed the BMP-1, as this had a much lower hit probability. A high rate of fire was not normally required, accuracy and high velocity were the priorities. The Instrument Design Bureau KBP at Tula was selected to design the gun, which was designated 2A42. Production started in 1978 with service entry in 1980 as the main armament of the BMP-2 IFV. It has since been used to arm other light AFVs and helicopters.

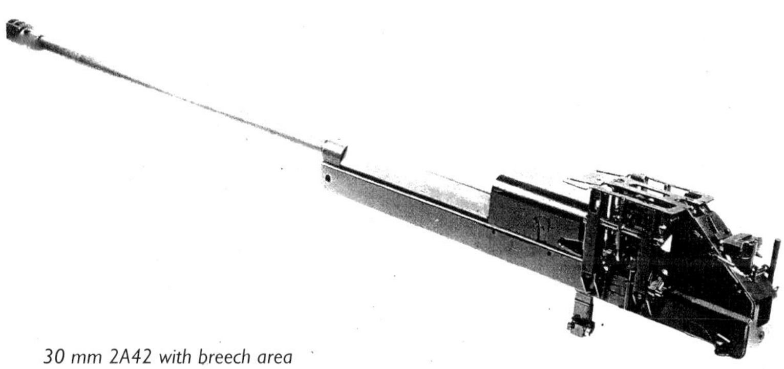

*30 mm 2A42 with breech area of ZTM-2*

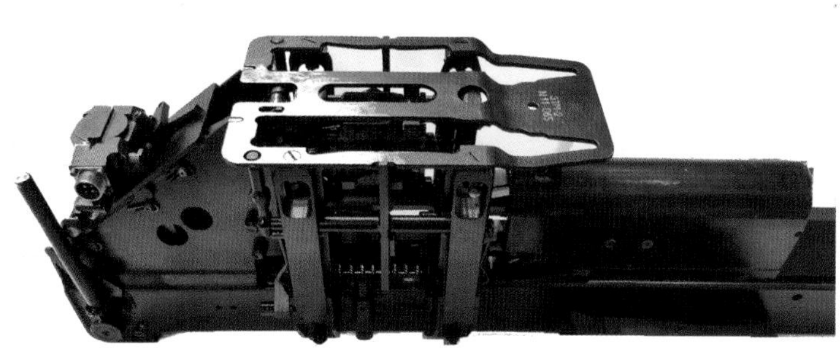

The 30 × 165 round developed for the AO-18 naval gun was chosen first, suitably modified to percussion priming so that the gun would still fire even if there were an electrical failure. The longest of the three barrel lengths developed for 30 × 165 ammunition – 240 cm or L/80 – was selected to maximize the muzzle velocity. The mechanism is gas-operated, with the gas piston driving a breechblock carrier. On chambering a round, the head of the breechblock is pushed forward by the carrier until it is stopped by the rear of the barrel; the carrier keeps going forwards, cams forcing the breechblock head to rotate and lock into the barrel extension. The barrel recoils 30–35 mm within the receiver, which along with the standard muzzle brake helps in controlling the recoil. The gun has a selective rate of fire, 200–300 rpm at the low rate and 550+rpm at the high rate, with 600–800 rpm achieved when the gun is fired mechanically rather than via the solenoid system. There are dual ammunition belt feeds. Overall length is 303 cm and the weight 115 kg. Service life is 6,000 rounds.

In service the 2A42 was found to suffer considerable problems, especially at its higher rate of fire, with fumes and smoke entering the turret. As a result, the 2A72 was developed for use in manned turrets, although the 2A42 is still used for installation in remotely controlled mountings, including helicopters. In 2014 Russia announced the Epoch Almaty stabilized RWS, mounting a 2A42 with 160 rounds of AP ammunition and 340 rounds of HE (plus an MG and four Kornet ATGMs), for installation on three new AFVs: the *Armata* MBT, *Kurganets* MICV and *Boomerang* 8 × 8 APC.

India manufactures this gun under the name Medak for use in the Sarath AFV (based on the BMP-2) and it is also installed in naval mountings which are designated CRN 91. Ukraine also makes a version of this gun under the designation ZTM-2.

**30 mm 2A72**: 30 × 165 AMMUNITION (PERCUSSION PRIMED).
The 2A72 was developed at KBP to overcome the 2A42's problem of propellant gas fumes entering the turret, especially at the high rate of fire, to the considerable discomfort of the crew. The main design feature to achieve this is a change from gas to long-recoil operation; the entire barrel and action recoils some 33 cm before the breech is opened at the end of this recoil stroke, giving far more time for residual gas to escape from the muzzle. It is worth noting that the same method of operation was adopted, presumably for the same reasons, for the British 30 mm Rarden and the Swiss Oerlikon 35 mm KDE, both of which were designed for mounting in manned IFV turrets.

The opportunity was also taken to simplify the design and produce a lighter, more reliable gun. The reduction in weight, plus the omission of the muzzle brake fitted to the 2A42, does not cause recoil problems because long-recoil mechanisms spread out the recoil impulse over a longer period of time, reducing the peak force. The main disadvantage of long-recoil is that it cannot achieve high rates of fire but that is unimportant for engaging ground targets. As with the 2A42, the 2A72 offers variable rates of fire: 200 rpm at the low rate, 330 rpm at the high rate, and 350–390 rpm when fired mechanically rather than electrically. The 2A72 retains the L/80 barrel and is 301 cm long, but weighs only 82 kg. Barrel life was initially 6,000 rounds but this was increased to 9,000 in 1995. Dual ammunition belt-feeds are retained.

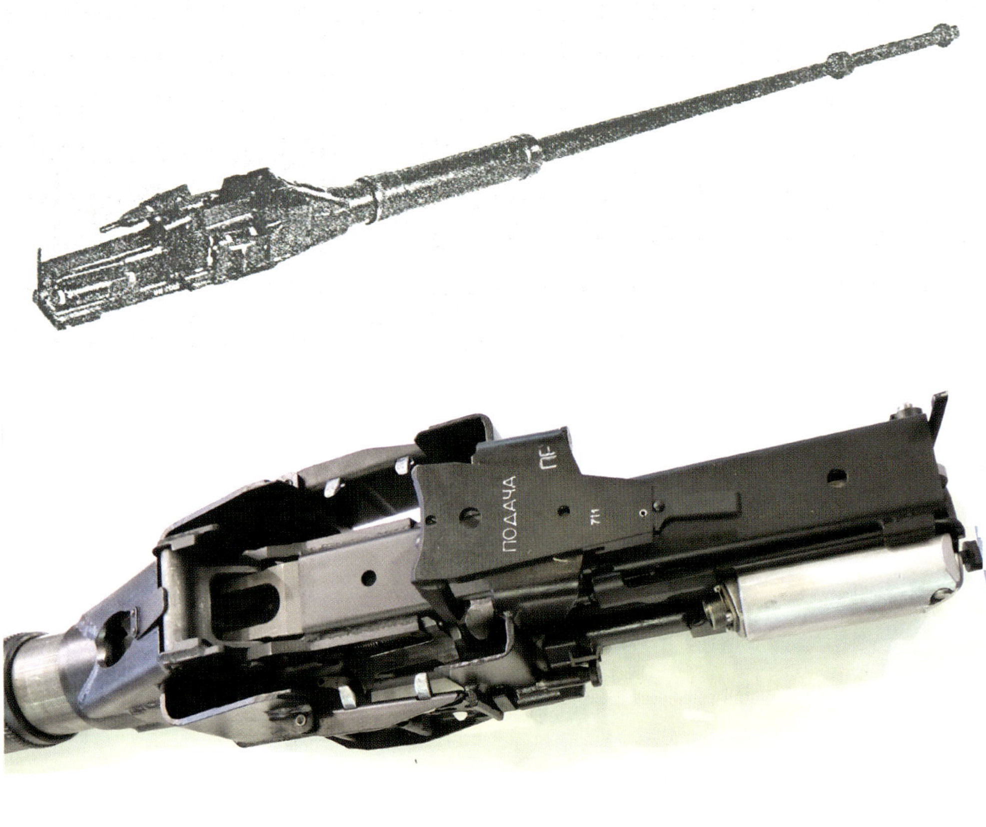


*30 mm 2A72, with breech area of ZTM-1*

It is understood that the 2A72 was accepted for service in 1984, but it first appeared on the new BMP-3 IFV which entered service in 1987 and has since been fitted in other mountings to a range of light AFVs with manned turrets. Uniquely, the BMP-3 solved the dilemma between fitting a high-velocity cannon like the 2A72 or a much larger-calibre low-velocity one, by having both: the 2A72 is mounted alongside the 100 mm 2A70 which can fire guided missiles as well as 'dumb' shells, giving the very busy crew a plethora of choices.

Ukraine also makes a version of this gun under the designation ZTM-1.

In the mid-2010s, western development of modern 30 × 165 ammunition for the 2A42 and 2A72 guns, initially for the 2A42 in Finnish service, was led by Nammo. Natures being developed were MP-T/SD, APDS-T, APFSDS-T and TP-T. This work revealed some interesting contrasts between the two guns. It emerged that the 2A42 is very rough in the way it handles the ammunition, which needs to be particularly rugged to withstand this. Conversely, a lot of recoil is needed to drive the 2A72 mechanism: the Nammo APFSDS projectile is too light for the loading to be able to generate the recoil needed, so they had to revert to APDS instead. This possibly accounts for the continued use of the 2A42 in unmanned installations in preference to the 2A72, including the new-generation T-15 and *KurganetS-25* IFVs.

**30 mm GSh-301**: 30 × 165 AMMUNITION (ELECTRIC PRIMED).
Despite the high performance and moderate weight (by western standards) of the GSh-30, the Russians were looking for an even lighter-weight gun for the next generation of fighter planes being developed in the 1970s. Development by KBP commenced in 1976 and the gun was adopted in 1983, with both right-hand and left-hand belt feed versions to suit the the MiG-29 and Su-27 respectively. After researching alternatives, Gryazev and Shipunov, the chief designers for the Instrument Design Bureau KBP, rather surprisingly selected what is in principle a very old-fashioned design: a single-barrel gun with a short-recoil action. This was developed under the TKB-687 designation and adopted in 1977 as the GSh-301, the GRAU index being 9A4071K. It entered service with the aircraft in the early-to-mid 1980s.

The operating mechanism involves a vertically-moving sliding-wedge breechblock actuated by a pivoting lever. The gun fires so quickly that breech-unlocking begins while the projectile is still in the barrel. While the cartridge case is loosened in the chamber by the ejector, it is blown out of the chamber by residual gas pressure in the barrel. Technically the operating system might therefore be described as a 'recoil-unlocked delayed blowback'. Ground firing tests are spectacular, with large balls of fire coming from the muzzle, clouds of smoke pouring from the breech and fired cases being violently ejected at 100 m/s.

The gun fires from a closed bolt (the first round being manually inserted on the ground) and has an evaporation-cooled barrel to minimize the risk of cook-offs. There is a outer water tank which is heated up by firing, generating a water-steam mixture which is channelled down spiral grooves between the barrel and an inner sleeve; the evaporation of this adds to the cooling effect.

The gun is belt-fed from either side, the normal ammunition load being 150 rounds. The ammunition feed is driven by the recoiling barrel, aided by a pneumatically powered feed assister. The gun has a fluted chamber to ease extraction. If a cartridge fails to fire, there is a rather brute-force wartime-only mechanism to persuade it to do so: a special cartridge next to the chamber is automatically triggered, its penetrator punching a hole in the side of the case via a small port in the chamber wall, the burning gases following through to ignite the propellant.

The GSh-301 is 198 cm long with an L/50 barrel and weighs 45 kg. It is apparently known to the factory as the 'ballerina' due to its weight being ideal for such a dancer. Performance is comparable with Western revolver cannon, while weight is reduced by half and the action is much slimmer. There is a cost to building such a lightweight weapon: gun life is only 2,000 rounds, barrel life normally 1,000 rounds, but if a continuous burst of 100–150 rounds is fired, the barrel must be changed.

One unique aspect of the weapon lies in the planned installations rather than the gun itself: while the gun is installed in a conventional fixed mounting in the MiG-29 family, a flexible mount designated PPU-27 was designed for the Su-27 which permitted the gun to be elevated up to 15° and traversed by +/- 2.5°; studies had shown that a cannon steered automatically by the FCS was three or four times more effective than a fixed one. However, development problems caused this to be abandoned. A flexible mounting is however a feature of the 9A4273 gunpod, in which the GSh-301 can be depressed by 30° and traversed +/- 15°.

***KPVB***: 23 × 115 AMMUNITION.
This is the KPV heavy machine gun adapted from 14.5 × 114 to 23 × 115. The 23 mm case is based on the 14.5 mm so the dimensions are compatible, except obviously the calibre and also the overall cartridge length (the 23 mm being longer). The first mentions of the KPVB date back to around 2006, but progress since then has been slow. Work is apparently continuing, but is far from being complete and no details have been made public.

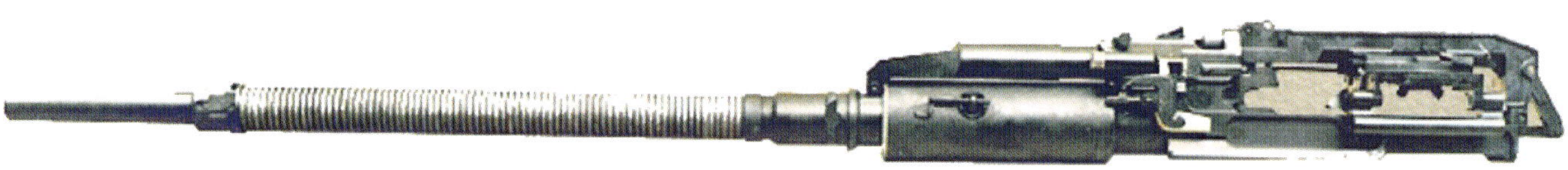

*GSh-301 (courtesy Maxim Popenker)*

# SOUTH AFRICA

South Africa's armaments industry was considerably expanded in response to the United Nations sanctions which (among other things) banned the sale of armaments to the country's apartheid regime. These began in 1967 and were mandatory from 1977 to 1994, during which time the country had to make many of the weapons they needed. The main organisation involved in armament development was originally ARMSCOR (The Armaments Corporation of South Africa SOC Ltd), established in 1968, with products marketed under the Vektor brand name. In 1992 the production facilities were transferred to a new company, Denel, and since 2004 cannon are the responsibility of the Denel Land Systems division. All of these companies were or are state-owned.

Various types of automatic cannon, mostly based on weapons which had already been imported into the country, were manufactured during the sanctions. Work has continued since then on cannon of domestic design, based on standard western 20 mm, 30 mm and 35 mm calibres. Ammunition is manufactured for both domestic and export purposes and, as well as the above-mentioned calibres, this includes the Russian 23 × 152B due to the number of guns captured in the long-running intermittent conflict with other southern African states during the apartheid era.

*Vektor GA-1 (Denel)*

**DENEL LAND SYSTEMS GA-1**: 20 × 82 (83.5) AMMUNITION.

This gun, originally marketed as the Vektor GA-1, is a locally-made version of the Mauser MG 151/20, the principal aircraft cannon of the *Luftwaffe* during World War II. It has retained a somewhat niche place in the market since it is considerably smaller and lighter than other current 20 mm cannon and generates less recoil, so it can be fitted to smaller and lighter vehicles and helicopters, including hand-held mountings. At 39 kg it weighs about the same as a 0.50-inch (12.7 mm) M2HB Browning Machine Gun.

The GA-1 uses a short-recoil mechanism with a rotating bolt head and fires from an open bolt. It features an accelerator within the bolt, an electric charger, and belt-feed with push-out type links. The belt feed can be switched from one side to the other. The rate of fire is in the region of 600–700 rpm. The gun is 177 cm long with an L/55 barrel. The ammunition is often referred to as the 20 × 83.5, but some Denel-made ammunition examined by the author still uses an 82 mm case.

***DENEL GUN AUTOMATIC MULTIPLE AMMUNITION (GAMA)***: VARIOUS AMMUNITION (SEE TEXT).

The GA-1 described above is available with a conversion kit to fire .50 BMG (12.7 × 99) ammunition for low-cost practice; the conversion is relatively simple and quick. In the 1990s Denel extended this idea to offer a multiple-calibre gun based on the GA-1 design but able to use a wider range of cartridges. Qualification of the weapon in 20 × 110 HS, 14.5 × 114, 20 × 82, and 12.7 × 99 mm calibres was completed during the first half of 1999, with 20 × 102 and 12.7 × 108 planned. Weights varied from 47 kg for the 12.7 mm models to 55 kg for a long-barrelled gun in 20 × 102. Rates of fire also varied between around 600 to 800 rpm. There was insufficient interest in the gun to merit production and it is no longer advertised.

**DENEL LAND SYSTEMS GI-2**: 20 × 139 AMMUNITION.

This is a licence-built version of the French *Canon Mitrailleur 20 F-2* (now designated 20M693), the South African

*Vektor GAMA mounted on a M3 heavy machine gun tripod (Denel)*

designation standing for 'Gun, Infantry, 2'. It therefore has a gas-operated delayed-blowback system of operation with a dual belt feed and fires from an open bolt. It is a distant relative of the HS 404 and uses similar side-mounted locking flaps to lock the bolt. There are three main sub-assemblies; the basic gun, the gun cradle, and an electrical fire-control device which enables the user to select single-shot, burst or continuous automatic fire. The gun is 269 cm long with an L/93 barrel, weighs 73 kg and fires at 750 rpm. It is available in a wide variety of mountings for land, naval and helicopter installations.

*VEKTOR 55c5*: 30 × 113B AMMUNITION.
The South African Air Force used the Dassault Mirage III fighter jet between 1963–1990 and a substantial indigenous upgrade of this, the Atlas Cheetah, between 1986–2008. The Mirage came with DEFA 550-series 30 mm aircraft cannon, made under licence, but pilots wanted a higher rate of fire so Vektor began work on an upgrade on the basis of the DEFA 553. The main design change in the XDM (experimental development model) focused on feeding cartridges into the chambers in three phases rather than two (the same principle was adopted for the DEFA 554). Initial firing tests achieved a firing rate of 2,200 rpm, but this caused excessive component wear and reduced reliability, so the action was slowed down. Further development culminated in acceptance flight tests during which the 55c5 achieved a firing rate of 1,900 rpm, and it is claimed that an entire week of evaluation was completed without a single stoppage. Several production cannon were manufactured and integrated into the air force but during the late 1990s major changes took place in the defence force resulting in the replacement of the Cheetah aircraft with the SAAB Gripen. With the phasing out of the Cheetah the 55c5 project was shelved.

The gun is a 5-chamber, gas-operated, belt-fed revolver weighing 82 kg; it is 166 cm long with an L/47 barrel.

There was also a proposal for a version of the gun for helicopters, with the RoF electronically controlled down to 900 rpm. It was intended to be installed in the XC-30 helicopter chin turret developed as an armament proposal for the Rooivalk attack helicopter in the 1990s. A large muzzle brake was fitted, which in conjunction with a constant pressure damper design was claimed to reduce the peak recoil from 29 kN to 3.5 kN.

*AEROTEK 30*: 30 × 113B AMMUNITION.
This was an externally-powered gun intended as a rival to the derated DEFA revolver cannon described above, for installation in the XC-30 chin turret. It weighed 67 kg, was 210 cm long and the rate of fire was selectable between 300 and 600 rpm. The hydraulic power required was up to 4.8kW.

*EMAK 35*: 35 × 228 AMMUNITION.
The EMAK 35 cannon was developed by Denel in order to arm AFVs for ground fighting, the ammunition already being manufactured by Denel for the army's Oerlikon AA guns. A high rate of fire was not required, the main characteristics sought being accuracy, reliability, a controllable rate of fire and minimal levels of gun gas in the turret. The selected design was externally-powered by means of a drum cam, and the ammunition was kept under positive control throughout the firing sequence. The cannon could be fired from an open or closed breech, first-round hit probability being improved from the closed bolt position. The barrel and locking mechanism were taken from the GA 35 cannon.

Different types of feed systems could be installed, depending on the turret requirements. The final part of the feed

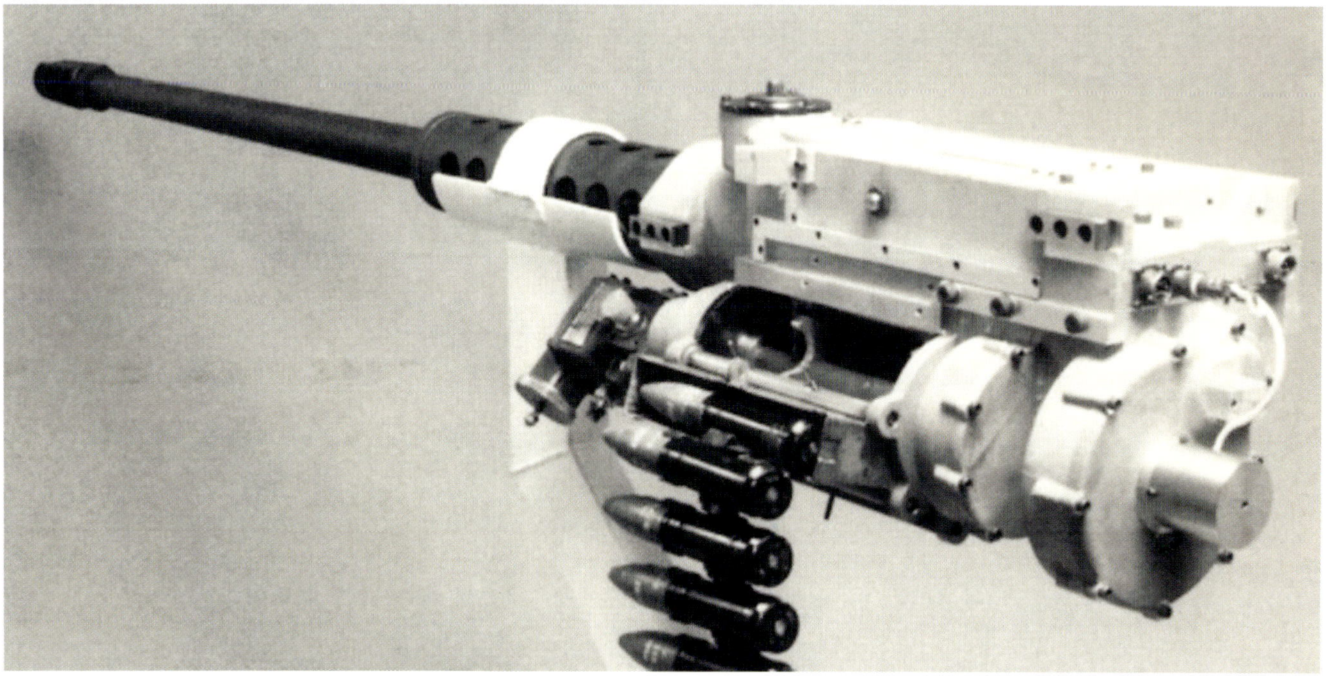

*30 mm Aerotek (Aerotek)*

*EMAK 35*

was linkless, with a first-round selection between right or left-hand feed. The gun was normally expected to use linked ammunition, with a delinker fitted to the main (right hand) feed to deal with a large quantity of ammunition (the prototype installation having 80 rounds); the left-hand magazine had 26 rounds, with a mini-feeder that stripped rounds ouf a clip, and was reloaded from the top. A standard GA 35 type strip-through link was used, with links being discarded sideways and downwards out of the cradle and empty cases ejected clear of the weapon forwards and on top of the cradle. A completely linkless feed was an option. Considerable attention was given to safety: premature activation of the firing pin was prevented by double-locking the firing pin, cocking the pin being performed after bolt locking. Hangfires were prevented by an automatic mechanism, deactivated by at least 20 mm of recoil.

The EMAK 35 weighed 380 kg and was 418 cm long with the usual L/90 barrel length for this calibre. The rate of fire was controllable between 1–120 rpm and the maximum recoil force was 35 kN.

The EMAK 35 was shown for the first time in 1996 and was successfully demonstrated at firing trials in January 1998. It attracted no customers and is no longer advertised, South Africa's own requirements for an AFV gun being met by the GI-30 CamGun.

**GA-35**: 35 × 228 AMMUNITION.
This gun was designed for the proposed ZA-35, a twin 35 mm self-propelled anti-aircraft gun system based on the Rooikat 8x8 vehicle. It was also intended for the Denel Land Systems eGLaS 35 single 35 mm towed anti-aircraft gun system, the first prototype of which was completed in 1992. Neither of these projects were adopted, but the gun is in service in the Vektor 35 DPG (Dual-Purpose Gun) twin-barrel naval gun system – later renamed Naval Air Defence System 35 mm Light Gun Mount (NADS 35 LGM) – which is fitted to the four *Valour* class frigates of the South African Navy.

The GA-35 is a gas-operated gun based on the Oerlikon GDA system, with dimensions, weight and rate of fire therefore a close match for the specification of the Oerlikon GDF cannon which have long been in service with the South African Army. Gas operation is used for both the bolt and the single or dual belt-feed systems, and the gun fires from an open bolt. All 35 × 228 ammunition can be utilized, including AHEAD via an upgrade. The gun weighs 429 kg, is 434 cm long with an L/90 barrel, and fires at 500–600 rpm.

**GI-30 CAMGUN**: 30 × 173 AMMUNITION.
The South African Army wanted a more powerful gun for arming future IFVs than the 20 mm G1–2 but rejected the EMAK 35 for that purpose, preferring instead a gun in what has become the western standard IFV calibre of 30 × 173. It is evident from the description of the gun that it owes a lot to the EMAK: it has an externally driven mechanism that utilizes a drum-cam to cycle the breech, chamber rounds and extract spent cartridge cases. The fired cartridge cases are ejected outside the turret. It fires from a closed breech in the interests of accuracy. Either single shot or a three round burst at a cyclic rate of fire of up to 115 rpm can be selected. There are two linkless feeders with 15 rounds on the left side and 25 rounds on the right. An additional 150 rounds are carried in the turret for manual reloading. The gun weighs 171 kg and is 345 cm long.

Work on the GI-30 started in 2006 with initial firing trials in 2008 and development completed in 2014. The first sales were to Malaysia, who selected the Denel Land Systems LCT 30 two person turret armed with a stabilized GI-30 to arm their new 8x8 IFVs. This is also expected to be used in the new South African Badger IFV.

*GI-30 CamGun (Denel)*

# SPAIN

The armed forces of Spain make use of various cannon of foreign design: for example, the army uses the 35 mm Oerlikon GDF AA system, and the 30 mm Mauser MK30–2 in the Pizarro tracked IFV for ground fighting, while the navy uses the 40 mm Bofors. There is one striking exception, however; the unique 20 mm Meroka naval CIWS – one of the most unusual cannon ever to see service.

**20 mm MEROKA**: 20 × 128 AMMUNITION (ELECTRIC PRIMED).
The Spanish Navy decided in the 1970s to develop a close-in weapon system (CIWS) to defend warships against attack by aircraft and anti-ship missiles. Rather than adopting or developing a fast-firing automatic gun like the 20 mm Phalanx or 30 mm Goalkeeper and AK-630, the Spanish decided to adopt a modern version of the 19th century 'volley gun' (the name Meroka reportedly comes from the German *Mehrrohrkanone* = multi-barrelled gun). Twelve fixed L/120 barrels (from the Oerlikon KAA) are arranged in two rows; all barrels are reloaded simultaneously since there is just one, common breech-block for all twelve barrels. Each barrel weighs 28 kg. The specially-developed ammunition differs from the KAA's in being electrically primed and used only with APDS projectiles.

The power for loading ammunition from the dual belt feeds (one belt for each row; six rounds are loaded simultaneously from each belt), plus unlocking, moving and locking the breechblock comes from compressed air from reservoirs on the mounting, which are recharged by a compressor on the mounting.

The barrels do not quite fire simultaneously; four groups of three shots are fired to reduce recoil forces and to increase accuracy. Electric priming is used; simpler and more precise than a dozen separate firing pin assemblies. At the end of the fourth group the breech-block moves to the rear, extracting spent cartridge cases and any misfires. With the breech-block at the rear, the ammunition belts are moved forward by compressed air to align twelve new rounds with the barrels, which are then loaded.

This volley-firing means that the rate of fire is not a simple number: the slightly staggered firing means that a 12-shot burst is fired at 9,000 rpm (150 rounds per second, thereby taking 0.08 seconds), followed by a 0.42 second pause for reloading before the next volley. Each 12-round volley therefore takes a total of 0.5 seconds, giving a sustained RoF of 24 rps or 1,440 rpm.

The Meroka was developed and produced by the Spanish firm FABA (*Fábrica de Artillería Bazán*), which became known as Bazán and is now Navantia Systems' FABA Factory. The first prototype appeared in 1977 and the system entered service in 1986, but initially almost every mount was unique because of the piecemeal implementation of individual improvements. All the Meroka installations were upgraded and largely standarised in the late 1990s, although variations between the installations on different ship classes persisted. Various improvements have been made to it since, mainly concerning the sensors and fire control system, although in the Mod 2 version the ammunition capacity was increased to 2,160 rounds. Some 25 Meroka systems were acquired and fitted to the *Principe de Asturias* aircraft carrier and the *Baleares* and *Santa María* class frigates. Only the last of these are still in service and no replacement systems have been acquired.

It should be noted that this was not the first 20th century attempt to revive the volley-gun system. In the 1930s several versions of what were variously called *Salvenmaschinenkanonen* or *Mehrlaufwaffe* were made in Germany, but these did not enter service. There is more about them in the Unconventional Systems section.

*Meroka system*

# SWEDEN

Sweden is unusual in this context in that it has only ever had one company developing and making automatic cannon, but it is one which has had continuous world-wide success since the 1930s. AB Bofors started manufacturing arms in the late 19th century, an activity boosted by export demand in World War I, and then cooperated with Krupps in the 1920s, Germany being subject to severe restrictions on armament development at that time. At that time, Bofors production was mainly of large-calibre artillery and naval guns.

In the early 1920s, the Swedish Navy acquired a few Vickers-Terni automatic AA cannon in two calibres: 40 mm (the 2-pounder pom-pom, known as the *40/39 automatkanon (autk) m/22*), and 25.4 mm (the 1-inch Vickers, designated *25 mm kulspruta (ksp) m/22*). The Navy was dissatisfied with the performance of these, in particular the belt feeding (using fabric belts) which couldn't cope with being constantly soaked in water. Initially, Bofors was asked to develop better projectiles for the Vickers-Terni guns and introduced rounds with tracers and more sensitive fuzes; the firm's first involvement with automatic cannon.

In the late 1920s the Navy issued a requirement for a new 40 mm semi-automatic AA gun, but Bofors advised them that a fully-automatic version could be produced instead, and in 1929 the Navy changed its requirement accordingly. The Bofors designation for the new gun was 40 mm L/60. By early 1932 the 40 mm (officially designated M/32) was complete, and two years later the first mobile field mounting for armies emerged as the m/34 (in Swedish terminology a capital 'M' in the designation supposedly indicated a naval weapon, a small 'm' being used for others, although this does not appear to have been consistently applied). These 40 mm weapons achieved huge export sales between the mid-1930s to the late 1940s, with many more being manufactured elsewhere under licence, especially by the USA and UK.

After the successful launch of the 40 mm L/60 an order was placed to develop a 25 mm version, also designated M/32; this was put into production but was never as popular as the 40 mm, only being acquired by the Swedish Navy.

Attention then switched to the 20 mm calibre, initially to meet a 1933 requirement from the Swedish Air Force for an aircraft cannon. Bofors accepted the task and a preliminary designation of *20 mm kulspruta (ksp) L/44 m/34* seems to have been set. The basic operating system was the same as the larger guns, but much modified and lightened. Instead of clips as used on the larger guns the 20 mm L/44 was fed through a 25-round open drum magazine, but the gun was still gravity fed which was not ideal in aircraft. A prototype gun was delivered in 1937, but by then it was considered obsolete due to its low rate of fire (c.360 rpm) and small magazine, and the Swedish air administration adopted the Belgian FN Herstal 13.2 mm Browning and 20 mm HS 404 instead. However this was not the end for the Bofors 20 mm; the company decided to continue work on the gun privately for the AA and AT roles, resulting in the 20 mm L/70 gun described below.

It was also not the end for Bofors 20 mm aircraft guns: by the end of World War II a completely different weapon, the 20 mm *akan* m/45, was developed to replace the HS 404 in the Swedish Air Force; a modified version was known as the m/49. A 30 mm version of the akan m/45 was designed in 1945, but not built; data on the gun and ammunition have so far not emerged.

After World War II, the development of the 40 mm guns switched from those firing the original 40 × 311SR cartridge – the L/56 or L/60 guns – to those firing the significantly larger and more powerful new 40 × 365R round – the L/70 guns – the first model being the m/48. These weapons have also been highly successful in terms of international sales, and are still being developed and marketed today. Apart from this, interest has mainly focused on 57 mm calibre systems in various forms.

Bofors' first attempt at a 57 mm system remained experimental: a recoilless, caseless, aircraft gun intended to arm the T 18B maritime strike plane. It had an MV of 625 m/s and a RoF of 170 rpm. In a 1944 trial, the backblast destroyed the plane, so the idea was scrapped in favour of a more conventional 57 mm gun; the m/47, fitted to the same plane. In addition, a much more powerful 57 mm gun entered service with the Navy (M/50) and Army (m/54).

The Bofors family of cannon, with the exception of the m/45, m/47 and m/49 aircraft guns, used a recoil-powered system of operation which was midway between the short and long-recoil types: the barrel and action remain locked together as they recoil (a distance of around 20 cm for the 40 mm L/60 models, 25 cm for the 40 mm L/70), until the vertically-sliding breechblock opens and the spent case is ejected by a separate mechanism. In most of the early weapons (many of which are still in service) ammunition is fed by gravity, with clips of several cartridges dropped manually into vertical guides above the breech. The guides were generally made tall enough for two full clips to be inserted, thereby permitting continuous fire without the need to pause to change magazines. The Russians studied the Bofors closely (purchasing a few 25 mm guns) and used a very similar design for their 25 mm M1940, 37 mm M1939 and some 57 mm cannon (which were also made by China). So, it seems, did the Germans, as Rheinmetall-Borsig's little-used naval 3,7 cm SK C/36, and the M42 developed from this, bore a very close resemblance to the Bofors guns.

It should be noted that not every cannon which has seen Swedish service came from Bofors: aircraft cannon in particular were imported, not just the HS

404 as mentioned above, but also the 30 mm ADEN (in the SAAB Draken), 30 mm Oerlikon KCA (SAAB Viggen fighters) and now the 27 mm Mauser BK 27 (in the SAAB Gripen).

To attempt to explain the rather complex and changing Swedish designation system: before 1937 the Swedish military used the term *kulspruta* (machine gun) for all automatic weapons, no matter the calibre. With the adoption of the Bofors 25 and 40 mm in the early 1930s the term became outdated and each service chose replacement terms, applying from about 1937. The Navy selected *automatpjäs* (= automatic piece, shortened to *aupjäs, apjäs* or *apj*) for calibres 20+mm, the Army *automatkanon* (shortened to *akan*) for calibers 20+mm and the Air Force *automatkanon* for calibres 12.7+mm. However, only one designation for a gun used by more than one service was allowed; so a naval gun adopted by the army would retain its naval designation.

In addition, some more specific names have been used such as the army *lvakan* (*luftvärnsautomatkanon* = anti-air auto cannon) and *pvkan* (*pansarvärnskanon* = anti tank cannon); and the naval *tdblapj* (*torndubbelautomatpjäs* = twin turret automatic piece). In around 1970 the Swedish navy started to issue new type designations for all their guns using army names. However, the navy subsequently changed the designations to reflect the general-purpose nature of the guns, while still keeping the same abbreviations: so apj now means *allmålspjäs* (all target piece) while the alternative term *akan* means *allmålskanon* (all target cannon). Bofors also use their own SAK designations (*Sjö Automatisk Kanon* = Naval Automatic Cannon).

Later in the 1970s a unique 4-digit code was introduced for everything, e.g. *57 mm apj 7102*, in which the 71 was the randomly-chosen code for the item, while 02 meant it was the second version to enter service. More recently, designations have become somewhat casual and tend to follow the manufacturer, so the example above has now changed to *57 mm apj Mk.2*.

In the early 2000s AB Bofors was acquired by the US firm United Defense, which has since been bought by BAE Systems. Bofors is currently part of BAE Systems Land and Armaments.

**BOFORS 40 mm L/56 OR L/60**:
40 × 311SR AMMUNITION.

As already described, the original L/56 or L/60 (the difference is only in the designation, not the barrel – the actual calibre length for all except the L/43 submarine gun is L/56.25) was introduced as both a mobile army weapon and a naval gun. After the first prototype was tested in 1930, the first naval mounting was delivered in March 1932. This was the compact, folding M/32 for submarines (whose shorter barrel reportedly reduced the muzzle velocity from 900 to about 700 m/s), the first surface-ship version being designated M/36. A land-based version, the m/34, appeared in 1934 with an improved model, the m/36, replacing it in production two years later. The army gun weighed around 2,000 kg, the twin naval mounting about 3,500 kg.

The guns fired at 100–120 rpm initially, later raised to 120–150 (varying with the gun elevation; the higher the angle, the more work the mechanism had to do to push the mass 'uphill' into battery). The ammunition clips held 4 rounds and the guides into which they fitted, 8 rounds.

The army weapons were single-barrelled, air-cooled, and mounted on a four-wheel trailer; elevation and traverse was by hand. The naval mountings for surface ships were mainly twin-barrelled, water-cooled, power-operated in larger vessels and with a stabilized mounting. During World War II some of the weapons fitted to ships (particularly in British service) were just the army guns on fixed mountings and later production formalised this, but the purpose-designed naval guns were fitted in multi-gun mountings. Bofors only produced twin mountings but the USN developed a four-barrel version and the RN one with six barrels.

There were some examples of SPAAGs fitted with this gun, most notably the US M19 and M42 on light tank chassis (M24 and M41 tanks respectively), which had two guns in their turrets. As with all SPAAGs, most of their active service was spent in delivering concentrated fire against ground targets. The gun was used in fixed installations in Sweden, in

*40 mm Bofors L/60 on land mounting (AB Bofors)*

*40 mm Bofors L/60 on twin naval stabilized mounting (AB Bofors)*

both single and twin mounts. The Bofors was even fitted to some Hungarian Messerschmitt Me 210Ca heavy fighters.

The 40 mm L/60 was one of the most successful weapons of World War II, seeing service in huge numbers in both army and naval versions with many nations. Manufacture in other countries was a major contributor to production totals, with licences being granted to Poland and Hungary in 1935, Norway and the UK (with manufacturing also in Canada and Australia) in 1937, and the USA in 1941. Other licences were granted, but did not result in production (e.g. France). The quality of the design and manufacture made it the supreme weapon of its class, but it was also complex and difficult to make; in UK and US production both the gun and mountings were considerably simplified to speed up the process.

The Royal Navy continued to use it into the 21st century (they never adopted the L/70) and it is still fitted to some USAF AC-130 gunships, although now being replaced. Bofors offers an updating service to enable these venerable weapons to benefit from the latest technology in ammunition and fire control systems. Despite being supplanted in production by the L/70, the L/60 has continued in service ever since.

**BOFORS 25 mm M/32**: 25 × 205SR / 25 × 187SR AMMUNITION.

The 25 mm M/32 was designed around a 25 × 205SR cartridge, the feed mechanism being via 6-round clips held in vertical guides above the breech which were capable of taking two clips. The clips were in two halves, which split and fell away as the cartridges entered the guides. The gun, which looked very similar to its 40 mm brother, had an L/64 barrel and fired at 160–180 rpm. It was available in four-wheeled mobile or fixed single mountings, or a fixed twin mounting, although the mobile mounting apparently was only used by the Coastal Artillery. Weights varied from 570 kg for the fixed single mountings (which included one designed for submarines, of which very few were used) to 1,100 kg for both the mobile single and fixed twin mountings. The Swedish Navy fitted these AA guns to a variety of warships from 1934 onwards, but from the late 1930s they were replaced by 40 mm mountings where weight limitations permitted.

One change took place in the late 1930s with the adoption of the M/38 cartridge which had a shorter case, just 187 mm long. The guns were slightly modified to use the new round, but retained the M/32 designation. The reason for this change appears to be that the M/32 cartridge fired relatively blunt-nosed projectiles with poor aerodynamics. The M/38 cartridge utilized longer, more streamlined shells, so the case needed to be shortened to keep the overall round length the same.

**20 mm m/40**: 20 × 145R AMMUNITION.

Originally developed as an L/44 aircraft gun, this was modified for the AA and AT roles, most obviously through fitting a much longer L/70 barrel. The 20 mm m/40 followed the same pattern of long-recoil automatic mechanism as the 25 mm and 40 mm guns. It was chambered for a unique and quite powerful 20 × 145R cartridge, and could fire at 360 rpm. The ammunition feed mechanism consisted of an open 25-round drum magazine above the gun, which in the AT mounting meant that the sights had to be fixed to the side.

The gun was fitted to about six different mountings. On a wheeled AA mounting it weighed 300 kg; on a low tripod for anti-tank use the weight was 65 kg (the same gun was used in both installations, and could be switched between mountings). They were also fitted to about forty PB m/31 armoured

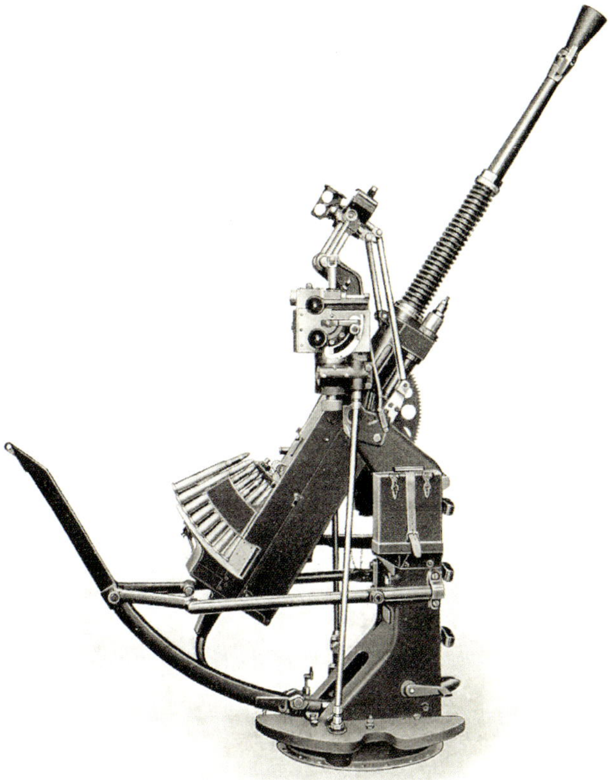

25 mm Bofors gun on single naval mounting (for submarines) (AB Bofors)

Bofors 20 mm m/40: note the open-sided drum magazine

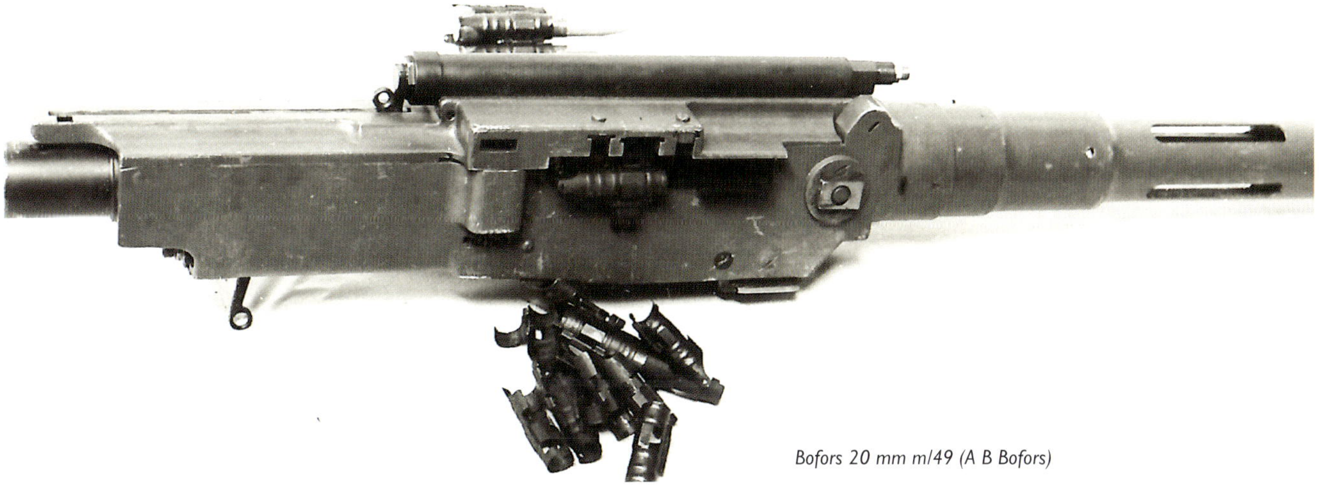

Bofors 20 mm m/49 (A B Bofors)

cars and to fixed AA and 'combination' mountings. In anti-tank form it was reportedly known as the 'grasshopper' as it jumped about so much on firing, even though it had a pneumatic recoil absorber. It was officially the *Pansarvärns/Luftvärnskanon m/40* (PVLV) which translates as 'Anti-tank/anti-aircraft gun'.

The gun was accepted for Swedish Army service as the m/40, with some 2,700 guns being produced, but found no other buyers.

BOFORS 20 mm m/45 AND m/49: 20 × 110 HS AMMUNITION.

With their relatively low rates of fire and small magazine capacities, the typical Bofors recoil-operated guns were unsuited to air combat. As a result, Sweden chose the Hispano Suiza 20 mm HS 404 in the 1930s, but World War II prevented delivery so the Hispano did not enter service until after the war. Concerned about the difficulties in importing weapons in wartime, Sweden decided to make an indigenous design of aircraft cannon, using the 20 × 110 ammunition of the HS 404. The result was the m/45 and the modified m/49 (which had an increased rate of fire, plus the barrel extended from L/65 to L/70).

These differed from most other Bofors cannon in using a short-recoil mechanism and belt feed, which could be switched from one side to the other. The bolt and barrel recoiled together for 20 mm before being unlocked, after which a pivoting lever accelerated the bolt recoil. The fired case was not violently ejected but was pulled to the rear by the bolt, being ejected during counter-recoil as the next round moved into position for chambering. Despite the very different mechanism, the size, weight and rate of fire of these guns were all quite similar to the HS 404.

The Bofors guns saw service in some of the early Swedish jet strike/ground-attack aircraft; the SAAB A21 and A32A respectively. They may also have armed the Pbv 301 and Pbv 302 IFVs from the late 1950s.

BOFORS 57 mm m/47:
57 × 230R AMMUNITION.
Although work on 57 mm guns started during World War II, it did not result in a production weapon until afterwards. The first to emerge was an oddity; the m/47 aircraft gun. The mechanism employed Bofors' usual vertical sliding wedge locking system, with the recoil being tamed by a liquid-filled damping system. The unlocking of the bolt from the barrel takes place after only 7 cm of travel, effectively making this a short-recoil gun.

The gun and ammunition were only fitted to one type of aircraft – the SAAB T 18B. This was a version of a wartime twin-engined attack/torpedo bomber, and it appears that the m/47 was intended for anti-shipping purposes. It was mounted in the lower nose of the plane, with the rotary 41-round magazine (which was fixed to the gun) fitted into the bomb bay. The gun weighed 290 kg, was 3.84 m long with an L/48 barrel and fired at 100–180 rpm (sources differ). The 57 × 230R cartridge was available in both HE and AP loadings.

*Bofors 57 mm m/47 aircraft cannon (Chinn)*

BOFORS 40 mm L/70:
40 × 365R AMMUNITION.
After World War II Bofors sought to build on the success of the 40 mm and developed a new and more powerful version, the L/70, designed around a larger cartridge, the 40 × 365R. The rate of fire was increased, initially to 240, subsequently to 300 and, in the TRINITY version, 330 rpm; a high rate of fire resulted from ramming the new round during recoil. Improved mechanisms for feeding ammunition to the breech also improved the sustained rate of fire, and powered mountings with remote control became standard (although local control and manual operation is still provided for in most cases). As with the L/60, there have been numerous upgrades over the years and with advanced fire-control systems and the latest ammunition technology, the gun system remains highly competitive today.

The prototype first emerged in 1947, with the first model being designated the m/48; this entered service with the Swedish Army in 1951. Like all of the subsequent versions, these mountings (both land and naval, as before) are fitted with a single air-cooled barrel, are power-driven (either by external electricity supply or by an on-mount generator in the case of mobile mountings), and feature extended clip frames holding up to 43 rounds to allow a larger ammunition supply for immediate use (ammunition is still handled in 4-round clips). Weight of a typical four-wheeled land mounting (including on-board generator) is 4,800 kg.

*Bofors 40/70 BOFI (Tereza Sobreira / Brazilian Defense Ministry / Wikimedia Commons / CC-BY-SA-2.0 licence)*

*Bofors 40–70 Mod.58 (Lung / Wikimedia Commons / CC-BY-SA-2.0 licence)*

*Bofors 40–70 Model 1978 (MKFI – Wikimedia Commons – Public Domain)*

Single naval mountings have been produced under licence in Germany and Spain as well as being reported in Brazil, India, and Yugoslavia. OTO Melara of Italy (now Leonardo) have developed a range of mountings and gun improvements, described in the Italy section. Bofors and Bazan (Navantia FABA Systems) have produced more than 650 single mountings and these remain in production while OTO Melara have produced some 130 single and 260 twin mountings,. Bofors has received orders for 15 Mk 3 mountings while Daewoo is believed to have produced 15 Vespa mountings.

In 1976 Bofors introduced a major upgrade for the system; the gun was virtually unchanged, but much more advanced day/night sights were fitted, incorporating a laser rangefinder, a ballistic computer and an optronic fire control, in conjunction with the new proximity-fuzed ammunition. This package of improvements is known as BOFI (Bofors Optronic Fire-control Instrument). By 1979 an on-mount tracking radar was added.

In 1984 TRINITY was introduced; a further refinement of the BOFI, using more advanced ammunition firing a heavier and better-shaped projectile at a slightly higher muzzle velocity and fitted with a 3P fuze (Prefragmented, Programmable, Proximity-fuzed) which provides a wide range of engagement options depending on the target. Intended for naval installations, there are two on-mount power-operated magazines containing 101 rounds of ammunition. TRINITY was not adopted, but elements of it – including the huge magazine – have found their way into other mounts, both new build (as the 40 mm Mk 3) and retrofits. Allied Ordnance of Singapore together with Singapore Technologies have developed a derivative of this mounting as the 40 mm L/70 NADM (Naval Air Defence Mount), plus a mounting for the army: the FADM (Field Air Defence Mounting).

The latest naval mount is the BAE Bofors 40 Mk 4, which weighs less than 2,500 kg complete with 100 3P rounds – 30 in the primary magazine plus another 70 in an intermediate magazine, from which the primary magazine is automatically topped-up. The fully enclosed turret uses electric drive rather than hydraulics and is fully digitized and modular in the modern fashion. Local control is possible using a gun-mounted video sight. The rate of fire is variable between 30 and 300 rpm.

Development was completed with sea trials in 2011–2012, and the launch customer is Brazil, with five mountings ordered for patrol vessels. In 2020 both the 40 mm Mk 4 and the 57 mm Mk 3 were chosen to arm the British Royal Navy's planned Type 31 frigates.

The L/70 guns have found their way into a variety of vehicles, from wheeled and tracked SPAAGs to MICVs. One of the most ambitious efforts was the abortive US M247 Sgt York DIVADS (DIVisional Air Defence) project which featured two guns in a radar-directed turret, all on a tank chassis. Trialled in the early 1980s, this proved a failure for reasons unconnected with the guns, and was cancelled in 1985. One which has entered service with Sweden is the *Luftvärnskanonvagn* (lvkv) 9040: a modified version of the IFV (see below) fitted with PS-95 radar from Thomson CSF Harfang (now Thales Group) and with a high elevation mounting and capable of using programmable ammunition.

Finally, the Swedish CV9040 IFV (infantry fighting vehicle) uses a variant of the Bofors L/70 gun – the 40/70Ba – primarily for engaging land targets, mounted upside down with three 8-round magazines attached to the gun, providing a choice of up to three ammunition types which can be switched between in 2–4 seconds. Fired cases are ejected upwards and forwards through a hatch in the turret roof. South Korea has adopted a similar installation for their K21 IFV, with a K40 gun based on the Bofors.

As with the L/60 model, various upgrade packages for the L/70 are offered by Bofors under the modular REMO programme (Renovation and Modernisation)

**BOFORS 57 mm L/60 AND L/70**:
57 × 438R AMMUNITION.

The true scaled-up 40 mm was not the 57 mm m/47 described above but the naval and army AA guns, all designed around a far more powerful 57 × 438R cartridge. The guns and mountings include: M/50 L/60; French M1951; m/54; SAK 57 L/70 ; BAE Systems 57 mm Mk 3 (BAE Systems Bofors 57 mm Mk 110). These were all naval systems except for the m/54 army AA weapon.

The ***m/54 lvakan*** saw service with the Swedish and Belgian armies (70 and 100 pieces respectively). The barrel was liquid-cooled, and the ammunition was handled in 2-round clips. Cyclic rate was 120 rpm, and the traditional four-wheeled mounting weighed 8,100 kg. Effective range was quoted as 4,000m.

The m/54 had a relatively short life before being replaced by AA missiles.

*Bofors 57 mm m/54 Lvakan*

The 57 mm naval gun has had a much longer run due to its useful multi-purpose capabilities, and it is still in production today. The initial **M/50** versions had water-cooled L/60 barrels, and were available in single or twin mountings. The vertically-sliding breech block opened downwards. The automatic loader had a spring-operated rammer which chambered a fresh round during run-out; the percussion mechanism was released before run-out was completed, so the recoil force first had to bring the gun to a halt before pushing it back again, softening the recoil. Towards the end of the recoil stroke the fired case was first extracted and then deflected towards the ejection chute at the front.

The **M/50 single naval mounting** was not enclosed, and did not penetrate the deck. It had two ready-use ammunition racks, each holding five chargers (10 rounds) to the rear of the gun platform. These were reloaded manually from outside the mounting, and transferred to the loading frame by a crew of two inside. Operation was normally remotely-controlled, but could be locally-controlled by two gun-layers on the mounting. The rate of fire was 120 rpm. It weighed 5,500 kg including a 5 mm steel plate front shield for splinter protection. It appears that this version achieved no sales.

**The M/50 twin mounting** had the two guns mounted close together in the same cradle, and differed from the single in being stabilized and in its very different ammunition handling arrangements. This was a proper turret with a 10 mm steel protection, not only fully enclosed and weatherproof, but also with a magazine buried several decks down, with each gun having individual ammunition supply arrangements. These consisted of a lower hoist to carry two-round chargers from the magazine up into the working chamber immediately below the turret, where the charges were fed by hand into the upper hoist which was fixed to the stabilized part of the mounting (and therefore moved around relative to the working chamber). The upper hoist terminated in the buffer magazine, where the chargers were de-clipped and fed into a fan (a quadrant-shaped hopper) mounted at the trunnions, from which the rounds were fed into the automatic loader. Each mounting had a crew of 10–12, consisting of the commander in the turret (plus two gun-layers if the gun was operated under local control); four men in the working chamber; two to four men in the magazine; and an ordnance maintenance man. Each gun fired at 130 rpm, and the turret weighed 31,000 kg all-in (including the hoists, but not the ammunition).

The French Navy, which for a time regarded the 57 mm Bofors gun as its primary AA weapon for the close-in defence of larger warships, thought that the Bofors twin mounting was somewhat over-specified and developed their own **M1951** twin mounting for the gun which weighed only 15,000 kg. This retained the fan-feed system but this was fed by ready-service boxes on the deck of the mount, containing 80 rounds per gun. These boxes were refilled (at 40 rpm) via a hoist from the handling room below.

In 1966 an improved version of the gun was introduced to meet a Swedish requirement to arm new fast attack craft; it entered service in 1971. Fitted into a single mounting, it was designated the **SAK 57 L/70 Mk 1** (the original model became the Mk 1A when the improved Mk 1B was introduced with new electronics). In

*HSwMS 'Småland', a Swedish destroyer in service between 1956 and 1984, was armed with Bofors automatic guns in 120 mm (forward-most turret), 57 mm (turret ahead of the superstructure) and 40 mm (single mounts) calibres. She is now a museum ship at Gothenburg.*

*57 mm L/70 Mk 1 (57_Mk_1 (Gurun / Wikimedia Commons / public domain)*

contrast to the 40 mm guns, while the barrel was lengthened from the L/60, the cartridge remains the same. There is a plastic gunhouse for weather protection. Water-cooling was retained for the barrel. The magazine is divided into four vertical racks, two on each side of the breech, each with ten rounds in two-round clips. Recoil energy feeds the rounds into the breech alternately from the outer racks; when the outer racks are empty the inner ones are connected automatically. There is a turret crew of three: the operator sits at the front of the mounting on the left and two loaders stand on each side of the gun. The turret has four ammunition racks, each with 32 rounds. Two electrically driven hoists connect the turret to the magazine, where ammunition is stored in racks, each holding seven clips. The two-round clips are inserted manually with one loader for each hoist. The gun mechanism design was modified to increase the rate of fire to 200 rpm.

Further improvements produced the lightweight **SAK 57 L/70 Mk 2**, developed in 1981 and introduced in 1985, with an air-cooled barrel (made from improved, more heat-resistant alloy)

*Bofors 57 mm Mk 2 (gurun / Wikimedia Commons / Public Domain)*

*Bofors 57 mm Mk 3 (MKFI / Wikimedia Commons / Public Domain)*

and the rate of fire further increased to 220–230 rpm. This was much lighter at only 6,500 kg as a result of a strenuous weight-saving effort, which included a much more compact plastic gunhouse which is not manned but controlled by an operator in the Combat Information Centre. An improved stabilisation system is also included. There are 120 rounds in the turret, handled via an intermediate magazine from which two 10-round cassettes are filled, one each side of the gun so that different ammunition types can be selected. In a typical installation, the hoists and below-deck magazine can contain a further 120 rounds.

The current version of the gun is the **SAK 57 L/70 Mk 3**, a modified version of the Mk 2 designed to use 3P (prefragmented, programmable, proximity-fuzed) ammunition. There is also a muzzle-velocity radar to assist the FCS, and an electro-optical manual aiming system. The weight is increased to 7,000 kg (7,500 kg in the US version). The prototype emerged in 1995 and entered service in 2000 in the Swedish Visby class corvettes, this version being in an optional 'stealth' mounting, with a gun barrel which disappears into the mounting when not required. The Mk 3 has been selected (with a conventional mounting) to arm two new US warship projects: Coast Guard cutters, and the Littoral Combat Ship. In US service the gun is designated the **MK 110 Mod 0**. In 2020 the Mk 3 was selected (also with a conventional mounting) to arm the Royal Navy's Type 31 frigates.

*Bofors 57 mm Mk 3 with stealth gun house closed (MKFI / Wikimedia Commons / Public Domain)*

*Bofors 57 mm Mk 3 with stealth gun house open (Poxnar /Wikimedia Commons /CC-BY-SA-1.0 licence)*

# SWITZERLAND

Switzerland made an early start to cannon developments on the back of German work in World War I. Germany was prohibited by the Treaty of Versailles from continuing their development of automatic cannon, so outsourced such work to Swiss-based organisations. In 1921 the patent for the 2 cm Becker was transferred to *Stahlwerk Becker AG's* Swiss development centre SEMAG *(Seebach Maschinenbau Aktien Gesellschaft*, an engineering firm near Zurich), who continued to market the Becker and developed their own, more powerful version, firing a 20 × 99RB cartridge instead of the original 20 × 70RB. By 1923 a 25 mm version was being tested, but was regarded as too heavy.

In 1924 the SEMAG company failed and the rights to the guns were acquired by another Swiss company, Oerlikon (*Werkzeugmaschinenfabrik Oerlikon* = Swiss Machine Tool Factory, Oerlikon), named after the suburb of Zurich in which the factory was based since its founding in 1906. They added a third weapon to the range, chambered for an even more powerful 20 × 110RB cartridge. All of these retained the Becker's Advanced Primer Ignition (API) blowback system of operation. They also had the recoil spring, which was wrapped around the barrel, enclosed in a sleeve or casing. By the late 1920s the three models were known as the Oerlikon Type F (20 × 70RB), Type L (20 × 99RB) and Type S (20 × 110RB) and were offered in various versions as aircraft, anti-aircraft and anti-tank guns but, although these had few rivals at the time, sales were initially slow.

In around 1930 the three Oerlikon models were completely re-engineered for aircraft use, modifications including a lighter bolt and stronger recoil spring to increase the rate of fire, and an overall weight reduction. The appearance also changed, with the sleeve enclosing the recoil spring being skeletonised. The aircraft versions were initially designated AF, AL and AS respectively. By 1935 further refinements to these aircraft

*Early Oerlikon 20 mm Gun on tracked AT mounting*

*Oerlikon Type L*

Early model Oerlikon twin AA

guns resulted in a change in designation to FF, (for *flügelfest* = wing mounted) FFL and FFS. The standard ammunition feed for all weapons was a 60-round drum, although a 15-round box and drums with other capacities were also available, particularly for flexibly-mounted guns. During this period Oerlikon also developed the SSG and SSG-36 semi-automatic anti-tank rifles, in 20 × 70RB and 20 × 110RB respectively.

In 1936 Emil Georg Bührle, who had been managing the factory since 1924, became the sole owner and in 1938 renamed the company *Werkzeugmaschinenfabrik Oerlikon-Bührle & Co.*. Oerlikon sales began to take off with the pre-war rearmament boom. Most sales of aircraft guns were of the FF, but France adopted the big FFS. As the French were interested in engine mounting with the gun firing through the propeller hub, the Oerlikon was modified by Hispano-Suiza and made by them under licence as the HS 7 and 9 (minor differences, for different installations); this was destined to enjoy only a short life, being replaced by the HS 404. The Swiss adopted the FF as the FF-K Oe 37. The major users of the Oerlikon aircraft guns in World War II were however Germany and Japan. In contrast, the UK and USA made extensive use of the big Type SS anti-aircraft gun, primarily for the close-in

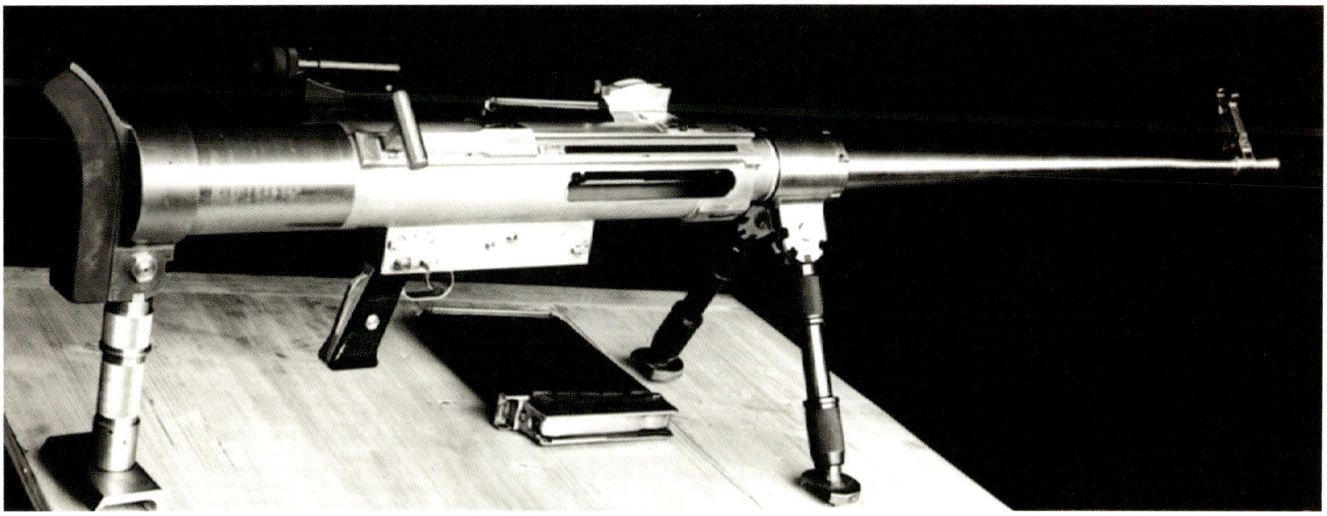

Oerlikon SSG-36 20 mm anti-tank rifle (Oerlikon)

defence of warships. In 1944, Oerlikon Bührle acquired Contraves, a developer specialising in anti-aircraft systems.

The second line of development was based on the 2 cm Ehrhardt cannon made by Rheinmetall, which did not see service although production had started by the end of World War I. Much further development was ostensibly carried out by Rheinmetall's Swiss subsidiary Solothurn, but this was a political fiction concealing the fact that the German company was very much in control of developments.

The third source of cannon developments was *Eidgenössische Waffenfabrik* of Berne, which built guns designed by Colonel Adolf Furrer utilising a version of the Maxim toggle-bolt short-recoil principle. Three basic cannon were developed, varying mainly in calibre: the FM-K 38 aircraft and Flab Kan 38 anti-aircraft guns in 20 × 139 (not quite the same as the 1950s 20 × 139); the Pzw Kan 38 and Tb 41 in 24 × 139 (strictly speaking semi-auto, but they used the same mechanism as the cannon); and the Flab Kan 38 in 34 × 239, all of which only saw service in Switzerland.

Fourth and last, Hispano-Suiza. It should be explained that Hispano-Suiza was a rather complex organisation. It was originally established around 1900 with Spanish investment to support the inventions of its Swiss chief engineer, Marc Birkigt (hence the 'Spanish-Swiss' name). The first factory was in Spain and concentrated on building luxury cars. French production started later, switching to advanced aero engines during World War I. In the 1930s the French factory started work on automatic cannon, first building Oerlikons under licence but in the mid-1930s developing what became the 20 mm HS 404. In the late 1930s, new gun factories were established in Switzerland (Hispano Suiza Switzerland or HSS, in Geneva in 1938), and in the UK (the British Manufacturing and Research Company – BMARCO), while the Spanish factory was taken over by the government during the Civil War. During World War II the French plant was devoted to production for Germany, so gun development switched to the Swiss plant in Geneva where it remained until the company was absorbed by Oerlikon in 1971 (Oerlikon in turn being taken over by Rheinmetall in 1999).

Until the end of World War II, cannon developed by the French Hispano plant (HSF) were identified by a three-number code starting with 4, as in the HS 404. This changed to 6xx for the few developments which briefly took place post-war. HSS used numbers beginning with 8 (so the Swiss development of the HS 404 was designated HS 804). Incidentally, the Swiss also used the MG 151/20 cannon (as well as manufacturing it for Germany), which meant that with the Oerlikon FF and the FM-K 38, they had four different 20 mm cannon from different companies in service at about the same time.

The Type IISS of 1948 took the development of the Oerlikon API blowback cannon as far as it could go. With the development of fast jets making life much harder for AA gunners, significantly higher rates of fire and shorter flight times (i.e. higher muzzle velocities) were priorities in providing an acceptable hit probability. The twin-gun mounting shown here was one of the API's final efforts.

From the late 1940s Oerlikon cannon development continued (assisted by recruiting experienced German gun engineers) with new weapons and ammunition in 20 mm, 25 mm, 30 mm and 35 mm calibres, but the new models abandoned the performance-limiting API blowback operating system. To take advantage of this, Oerlikon replaced their 20 mm ammunition with a new 20 × 128 cartridge which developed 80% more muzzle energy than the 20 × 110RB; the projectile shapes were also more aerodynamic to retain more of their velocity for longer.

Oerlikon developed a number of revolver cannon (given RK designations, as in RK 421 – or 421 RK – meaning first model of 42 mm revolver cannon) with limited success, but also worked on recoil-operated (unsuccessfully) and gas-operated (successfully) models. Since the 1960s, their designation system for service weapons has been K (for Kanone) followed by a letter designating the calibre (A = 20 mm, B = 25 mm, C = 30 mm, D = 35 mm), then another letter to indicate the gun model: so the KCA was the first model of 30 mm gun to be adopted for service.

With the takeover of Hispano-Suiza by Oerlikon Bührle in 1971 (initially known as Hispano-Oerlikon), the HS guns then in production were redesignated using the Oerlikon system, so the 20 mm HS 820 became the Oerlikon KAD and the 30 mm HS 831 became the

*Fliegerabwehr-Zwillingsgeschütz Oerlikon Type 12ZSLa/2SS (copyright Rheinmetall Air Defence AG)*

Oerlikon KCB. In 1989 the Contraves and Oerlikon Bührle military operations were merged to form Oerlikon Contraves. In 1999 the German company Rheinmetall acquired a majority interest in Oerlikon Contraves, ending Swiss ownership of automatic cannon development. In 2009 Oerlikon Contraves was renamed Rheinmetall Air Defence AG, a division of the Rheinmetall Defence Group.

*2 cm Becker in a makeshift AA mounting*

**20 mm BECKER / OERLIKON TYPE F / AF / FF**: 20 × 70/72RB AMMUNITION.
A development of the original German 2 cm Becker cannon, this continued to evolve until the late 1930s, becoming lighter and faster-firing. The AF aircraft gun weighed 25 kg instead of 30 kg for the Becker (mainly due to the removal of the barrel sleeve enclosing the recoil spring), and the rate of fire was increased from 300 to 400 rpm. Ammunition was fed from a 15 or 30-round drum magazine, rather than the Becker's 15-round top-mounted box. The 1935 FF trimmed a further 1 kg from the weight, probably due to further lightening the bolt in order to increase the firing rate to 520 rpm, and larger drum magazines were available: the standard size was 60-rounds, although 45, 75 or 100-round drums were eventually listed. The 45-round drum weighed 7.4 kg, the 60-round 8.3 kg, 75-round 10 kg and 100-round 12 kg (empty weights: a full 60-round magazine weighed 20 kg, a 100-round 37 kg). By this time, the gun was chambered for a slightly lengthened 20 × 72RB cartridge, to make space for a larger extractor groove. The barrel remained at L/40 throughout. The main users of the FF were the Japanese Navy air force (20 mm Type 99–1) , and the German *Luftwaffe*, who designated it the MG-FF (with a modified 20 × 80RB cartridge).

*Oerlikon Type FF with 60-round drum magazine (copyright Rheinmetall Air Defence AG)*

**20 mm SEMAG / OERLIKON TYPE L / AL / FFL**: 20 × 99/101RB AMMUNITION.
The SEMAG (20 × 99RB) was introduced in 1924 but saw little use. When Oerlikon took over after the failure of SEMAG, they renamed the gun the Type L, which also only sold in small numbers. One advertised installation for which data is available was for the Savoia-Marchetti S.62bis flying boat in the early 1930s, which carried one in the bows. A manufacturer's catalogue promoting this installation states that the gun weighed 42 kg and was 182 cm long (the barrel was L/60). The ring-type mounting weighed 55 kg and could traverse through 360 degrees while providing up to 90 degrees of elevation and

*Oerlikon Type L on aircraft flexible mounting, with 15-round drum*

depression (although depression was limited to 60 degrees in the SM.62bis installation). The ammunition feed was via a 15-round drum which weighed 2 kg empty (each round weighed 220 g). Sights weighed another 2 kg and a sack to hold 30 fired cases weighed 1.2 kg (necessary because the SM.62 was a 'pusher', so it was important to prevent fired cases from flying back and striking the propeller). Although the gun had a cyclic rate of 350 rpm, the practical rate, allowing for magazine changes, was 125 rpm. The complete installation weight, including six loaded magazines, was 135 kg.

The AL was drastically lightened to 30 kg (according to one Oerlikon source – another gives 36 kg), but the rate of fire was not increased until the FFL, which saw this raised to 500 rpm. As with the smaller model, an enlarged extractor groove increased the length of the cartridge case, to 101 mm. This saw even less use with the notable exception of the Japanese Navy which adopted it as the 20 mm Type 99–2; it was their principal aircraft gun in the latter part of World War II, and saw more development including belt feed and a higher rate of fire. There was a relatively small penalty in increased magazine weight compared with the FF: an FFL 45-round drum weighed 8.3 kg, 60-round = 9.3 kg, 75-round = 11 kg and 100-round 12.5 kg: all empty weights; a full 100-round FFL drum weighed 39.2 kg compared to 37 kg for the FF (weights from Japanese sources).

**20 mm OERLIKON TYPE S / FFS / SS**: 20 × 110RB AMMUNITION.

The Type S, introduced in 1927, was the third and largest of the Oerlikon 20 mm cannon of the interwar period. In its original form it weighed 62 kg and fired at 280 rpm; the barrel was L/70 in all

*Tankabwehrlafette Oerlikon Type TLA-1S with spare barrel (copyright Rheinmetall Air Defence AG)*

models. The Type SS of 1938 saw the rate of fire increased to 470 rpm, and this is the form in which the gun served with the RN and USN. The gun life was 9,000 rounds. The ammunition was greased before being loaded into the magazines. Development continued, with the Type IISS emerging in 1948 being the ultimate version: this weighed 65 kg and fired at 650 rpm.

The Oerlikon was mostly used as an AA gun, but in its early days it was considered to have considerable potential in the anti-tank role, as were other 20 mm cannon.

The aircraft versions of the Type S developed in a different direction, with the aim being to reduce the weight as much as possible; in the FFS this was just 39 kg. The rate of fire was 470 rpm; no doubt this could have been accelerated to the same rate as the Type IISS had there been a demand for it, but only France made significant use of the FFS; it was made under licence as the HS Type 9 engine cannon.

The combination of sturdiness, reliability, reasonable rate of fire and cartridge ballistics provided by the Type SS made it extremely popular in the light AA role. No surprise, therefore, that it was made in vast numbers during World War II, especially in the UK and USA, principally for fitting to warships in the close-in defence role. The gun entered British service in 1939 and was soon employed in a variety of mounts installed in almost every kind of warship as well as merchantmen. By the end of the war, there were some 55,000 20 mm naval guns in the British inventory. The US Navy approved the weapon in November 1940, with the first American-made weapon being test fired in June 1941. In total, nearly 125,000 Oerlikons were made in the USA during World War II.

*20 mm Oerlikon gun, RN (Beadell, S J (Lt) / IWM_A_7500 / Wikimedia Commons / Public Domain)*

In naval use the Type SS was mainly installed in simple, unpowered single mountings, typically weighing 330 kg without shield. These required no power supply, the standard 60-round drum magazine permitted a much longer engagement time than most of the magazine-fed cannon, and it could if necessary be operated by one man (although there was typically an extra man to change the heavy magazines, plus another to designate which targets to engage). The magazine weighed 14.3 kg empty and 29 kg full. The magazines were asymmetrical, with left- and right-hand versions. An interesting feature of the drum was that it could be kept fully loaded without the operating spring being compressed: in other cannon drums such as the Hispano's, loading the magazine compressed the spring, which shortened its life.

More sophisticated mounting options were developed, including powered twins, but in practice these designs were often modified to take one 2-pounder or 40 mm Bofors gun instead. Late in World War II, the British developed a powered quad mounting in which the guns were belt-fed, but this was cancelled. The USN did make limited use of a powered

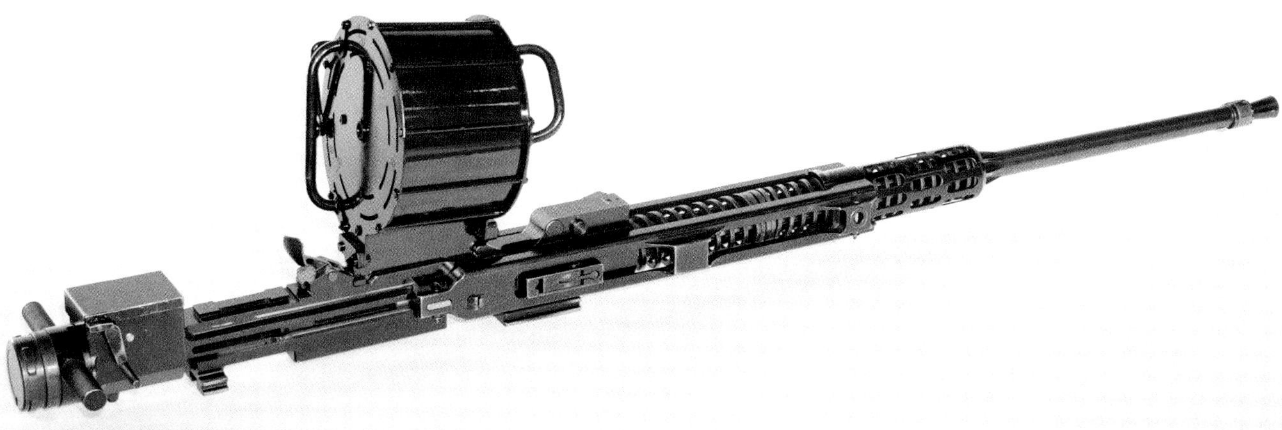

*Oerlikon Type SS (copyright Rheinmetall Air Defence AG)*

quad, but the guns remained magazine-fed. Some Type SS remained in service in secondary roles into the 21st century, and the ammunition is still manufactured.

*25 mm SEMAG*: 25 × *90.5RB* AMMUNITION?

This mysterious gun is referred to in the Oerlikon archives as having been tested in October 1934, using HE shells with explosive charges ranging from 10 g to 22 g. Photographs of the gun in the archive have enabled a surviving example to be identified in the *Schweizerisches Militärmuseum* at Full-Reuenthal, which is now the home of the former Oerlikon-Bührle museum collection.

There are two curious aspects: first, the gun was tested ten years after the collapse of the SEMAG company, and secondly, the external appearance bears no relationship to that of the 20 mm Becker/SEMAG/Oerlikon cannon family. The gun has a long, box-shaped receiver and the ammunition feed is via a horizontal tray extending through both sides of the receiver. This appears to have ten spaces and presumably passes through the gun as it is fired, possibly replacing the fired cases in the tray.

The cartridge used is uncertain, but a previously unknown 25 x 90.5RB round with a straight case and a strongly rebated rim (just 20 mm in diameter) has recently been found, headstamped 1933, which makes it a likely candidate.

This is unverified, however, and generally regarded with some scepticism.

20 mm SOLOTHURN S18–1100:
20 × 138B AMMUNITION.

Rheinmetall's Swiss subsidiary Solothurn gave its name to a pair of semi-automatic anti-tank rifles which used the Stange locking system: the S18–100 series (which went up to S18–500) chambered in the so-called 'short Solothurn' 20 × 105B round, and the S18–1000 series chambered in the 'long Solothurn' 20 × 138B calibre. Weights were 40 and 50 kg respectively, and barrels L/45 and L/65. Magazines could hold up to 10 rounds. This is included here because a selective fire version of the bigger gun was developed as the S18–1100, intended as a 'universal' light cannon available with ground and AA mountings. It saw little use.

*2,54 cm SOLOTHURN S19–100*:
25.4 × 148B AMMUNITION.

This was a 1933 project for a selective-fire anti-tank and AA gun. It had the usual short-recoil action with a pivoting breech-block and was fed by a side-mounted 16-round box magazine. The cyclic rate was 250 rpm, with a practical rate of fire of 150 rpm (allowing for magazine changes). The gun weighed 100 kg and was 240 cm long with an L/60 barrel.

According to Musgrave (German Machineguns) three different mountings were proposed: the S20–100 was a light infantry-accompanying mount, the S20–200 used a portable tripod mount, and the S20–300 was a twin AA mounting (weighing 800 kg). None of these entered production.

20 mm FMK38 AND FLAB K38:
20 × 139 AMMUNITION.

This Furrer toggle-lock short-recoil design was utilized as an aircraft and anti-aircraft gun, seeing Swiss service in both installations. The aircraft gun was designated the FMK38 (*Flugzeugmotor-Kanone*

*2,5 cm SEMAG at the Schweizerisches Militärmuseum Full ( courtesy Robert Gerber)*

*Solothurn S18–1000*

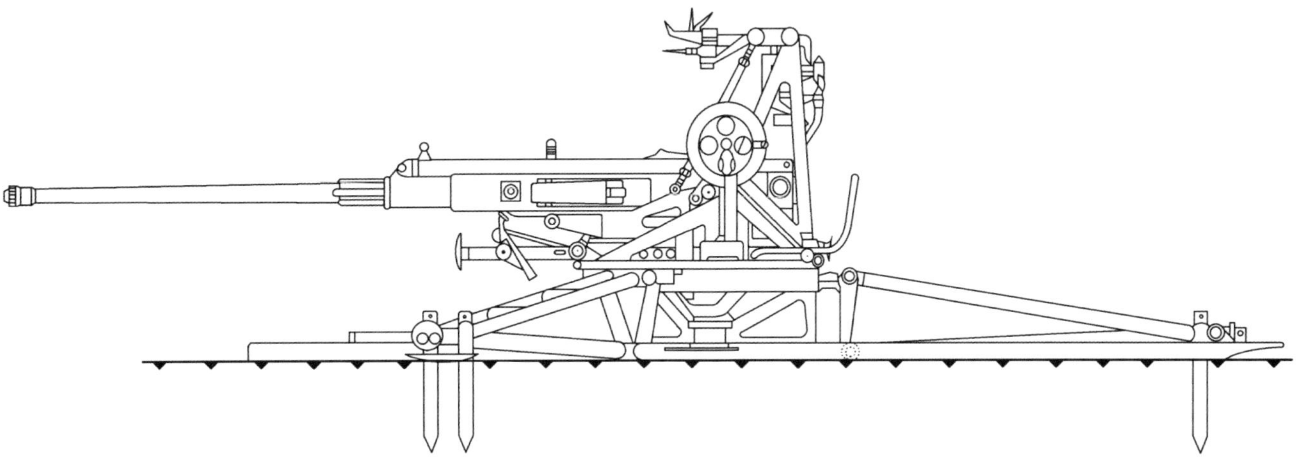

Solothurn S19–100 (courtesy of Robert Berger, redrawn from an original

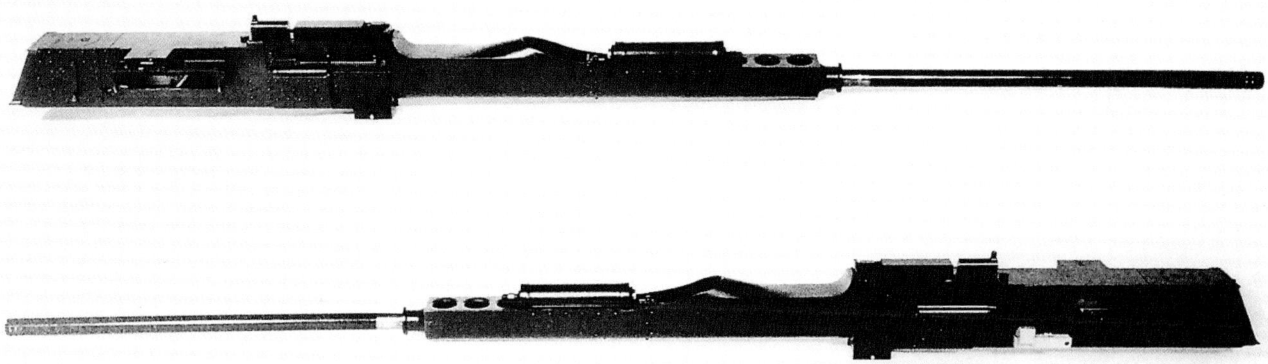

20 mm FMK38 (courtesy Verlag Stocker-Schmidt, Dietikon-Zurich)

1938) and the AA gun the Flab K38 (*Fliegerabwehrkanone* 1938). The gun weighed 57 kg, was 255 cm long with an L/91 barrel, was belt-fed and fired at 400 rpm.

The AA mounting, with a single gun, weighed 390 kg ready to fire, and the ammunition supply was contained in boxes of 50-round belt-lengths, each weighing 30 kg loaded. Some 1,475 AA guns were made between 1939 and 1947; it entered service in 1940 and remained in use in fixed installations until 1977.

The aircraft gun was mounted on the engines of C35, C 3603 and Morane D 3801 fighters to fire through the propeller hub. Two guns were also mounted in the belly of the twin-engined Potez 63. Ammunition capacity was 55 or 110 rounds.

*20 mm Flab Kan 38 (courtesy Verlag Stocker-Schmidt, Dietikon-Zurich)*

**34 mm FLAB K38**: 34 × 239 AMMUNITION.
The biggest of the Furrer short-recoil toggle-lock family, the 34 mm Flab K38 AA gun was unusual in having a water-cooled barrel; a total of 80 litres was carried, 32 in the barrel casing and 48 in a reservoir on the mounting. The gun weighed 280 kg, had an L/80 barrel, was belt-fed and fired at 250–270 rpm (a respectable rate considering the size and power of the cartridge). A total of 205 single-gun mountings was made between 1938 and 1942, and some remained in service until being replaced by the 35 mm Oerlikon in 1968.

**24 mm PZW KAN 38 AND TB 41**: 24 × 139 AMMUNITION.
This Furrer toggle-lock gun only marginally qualifies for inclusion, as it was semi-automatic only, with a rate of fire of 30–40 rpm. However, it used the same short-recoil mechanism as the other Furrer cannon, with the sear control being the main difference, and is of interest due to its unique calibre. It was designed as an anti-tank weapon and saw Swiss service in three versions; vehicle-mounted (Pzw Kan 38 = *Panzerwagen Kanone* 1938); in a fortress mounting (*Panzerabwehr-Befestigungskanone* 38); and as an anti-tank rifle on a low tripod which converted to a wheeled mount (Tb 41 = *Tankbüchse* 1941). Only a small number of vehicle guns were made and installed in LTH light tanks; about 300 guns were installed in fortifications by the end of World War II; and over 3,500 anti-tank rifles were made, some of them subsequently being fitted to lake patrol boats and fortifications. The AT rifle had a short life, being withdrawn from the infantry during 1945, but fortress guns remained in use until the 1960s and one was still mounted on a patrol boat until 1971.

The gun evolved through three versions, the most common being the Type III. Gun weight was 77 kg (the tripod being an extra 53 kg), length 259 cm with an L/78 barrel and a complex (and apparently effective) muzzle brake. Ammunition feed was a box magazine of 6 rounds (AP – the HE was smaller with 5 rounds, to avoid confusion), top mounted in the first two versions but with the whole gun turned on its side in the Tb 41, to reduce its height. The six-round magazine weighed 5.12 kg loaded. The magazine was automatically ejected once the sixth round had been chambered; just inserting a new magazine was all that was required to continue firing.

While the Tb 41 was classified as an anti-tank rifle, it had a crew of no fewer than seven: for transport, the barrel and receiver were separated (weighing 39 and 38 kg respectively) or the gun could be towed on its wheeled mount by two of the crew, others towing an ammunition cart containing 120 AP and 40 HE rounds.

*34 mm Flab K38 (courtesy Verlag Stocker-Schmidt, Dietikon-Zurich)*

*24 mm Tb 41 (courtesy Verlag Stocker-Schmidt, Dietikon-Zurich)*

**20 mm HS 804**: 20 × 110 HS AMMUNITION.
This is basically the Swiss-made version of the HS 404 developed in France (see that section for details). It achieved some sales successes post-war and was made, with some minor modifications, in Yugoslavia, which made towed AA mountings for the gun as the Zastava M1955, available in single (20/1 mm M55) and triple (20/3 mm M55) variants, the latter also being fitted to the BOV-3 SPAAG. Single (M71), triple and quad (M75) naval mountings were also made. Barrels are L/70 – L/75.

TK

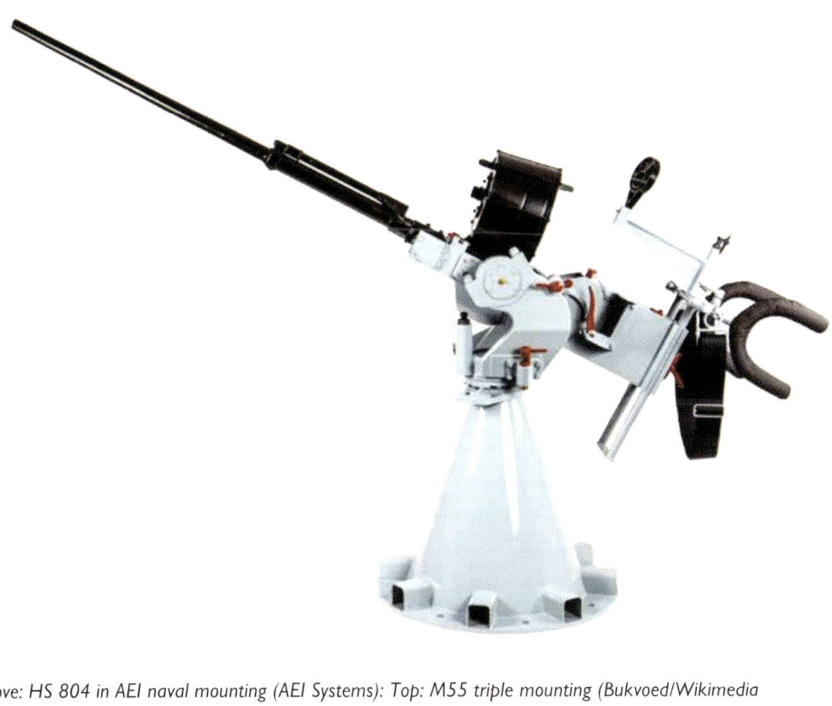

*Above: HS 804 in AEI naval mounting (AEI Systems): Top: M55 triple mounting (Bukvoed/Wikimedia Commons/CC-SA-3.0)*

It is still offered for sale as a new gun, made by AEI of the UK under the AE20 designation. This weighs 43 kg (less magazine) and fires at 620–720 rpm. AEI also offers the AE20-NM naval mount for this gun and, in 2014, announced the lightweight AE20-VM vehicle mount which incorporates a belt-feed mechanism.

**30 mm HSF 411 / HSS 830 & 831 / OERLIKON KCB:**
30 × 170 AMMUNITION.

This gun has a complex history. It commenced as a French project (designated HS 411) in about 1939/40, being designed around a powerful new 30 × 170 cartridge which used a slightly lengthened and necked-up 25 × 163 Hotchkiss case. It appears that little progress was made before the German invasion, but details were transferred to HSS and also to BMARCo, Hispano's British subsidiary at Grantham. By 1942 BMARCo had built an example of the gun (and had some ammunition made for it) and were promoting it to the British Ordnance Board. However, the Board wasn't interested as it considered that such a gun was not needed, since the existing 20 mm and 40 mm guns adequately covered service requirements.

HSS continued the development of this gun from 1942, marketing it as the HS 830 from 1945 to the early 1950s when it was replaced by a slightly modified version, the HS 831. In 1957 the HS 831L was introduced, using modified ammunition. In 1971 the gun was redesignated the Oerlikon KCB and given further modifications, including a fluted chamber.

The KCB is a gas-operated, air-cooled, belt-fed weapon that is available in two versions with left or right feeding; KCB-B02 or KCB-B03 respectively (early versions of the HS 830 family were available with strip or drum feed). The hybrid gas-operated delayed blowback mechanism is similar in principle to that of the HS 404 / 804. The gun fires from an open bolt. On firing, gas is tapped from the bore to unlock the locking plates and release the bolt. Residual gas pressure in the bore then blows the fired case out of the chamber and pushes the bolt to the rear. The gun recoils some 50 mm in its mounting and this is used to drive the belt feed mechanism. Overall length is 352 cm with an L/85 barrel and weight is 159 kg.

The HS 830/831 / KCB family has been primarily utilized in the AA role, in both land-based (towed and self-propelled) and naval mountings, with the latter being much more successful. The French Army was an early user, buying 200 of

*HS 411 in 1942 (Oerlikon, via Robert Gerber)*

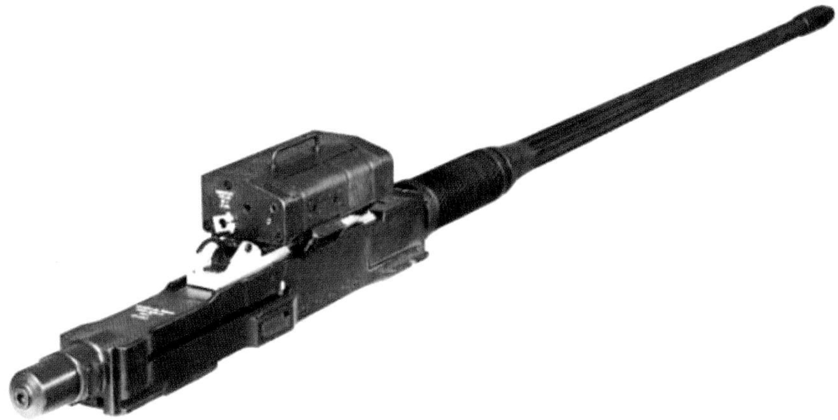

*Oerlikon 30 mm Type KCB cannon (Oerlikon)*

the HSS 831A in the mid-1950s. In the late 1960s, the same gun was fitted in a twin turret mounted on the AMX-13 DCA (*Défense Contre Avions = defence against aircraft*): the French Army bought 60 of these. A few years later a similar turret was mounted on the larger AMX-30 chassis. This was used by Saudi Arabia which bought 53 examples and designated them the AMX-30 SA.

Several naval mountings, both single and twin, were developed. In the early 1970s, the US Navy developed a requirement for a lightweight, low-cost, medium-calibre gun mounting as the primary armament of patrol boats and the secondary armament of larger ships. Emerson produced a twin mounting using the HS 831L in 1972, with the factory name Emerlec-30 and the USN designation Mk 74. It wasn't bought by the USN, but was exported to South Korea and several other minor navies.

In 1971 Oerlikon's British subsidiary BMARCo (sold by Oerlikon in 1987) began marketing the A32, a stabilized twin mounting for the KCB. This was later renamed GAM-A01, with the GCM-A02 being a version featuring an enclosed operator's position. Later still, an improved model was designated GCM-A03. The LS-30B, a single mounting for the KCB, was developed by Laurence Scott whose naval weapons activity continues within a successor company, MSI Defence Systems. The GCM and LS-30 systems were bought by the British Royal Navy as a result of experience in the 1982 Falklands War.

The KCB has not recently been manufactured in Switzerland but is made as required in South Korea, as it is the gun component of the Bi Ho (Flying Tiger) tracked SPAAG system which is in service with the Republic of Korea Army.

*The Oerlikon 421 RK 'Red King' (Oerlikon)*

*42 mm OERLIKON 420 RK SERIES*:
42 × 338/348 AMMUNITION.

In the late 1940s the British Army decided it wanted to replace its 40 mm Bofors L/60 guns with a system with a much better ballistic performance and a higher rate of fire, plus a large magazine capacity. In 1952, the British ARDE approached Oerlikon in Zurich and commissioned them to study a twin AA gun using British 42 mm ammunition. In January 1953 a contract (6/Proj./4624/421RK) was signed between the UK and Oerlikon for the development of this system, which was given the factory designation 421 RK. This project was given the code name 'Red King' by the UK, presumably inspired by the RK designation (at that time British military projects were given colour codes at random to ensure that they didn't indicate the purpose of the system, e.g. the Blue Steel airborne nuclear missile).

The 421 RK (42 mm, 1st model Revolver Kanone) was a revolver, but with a seven-chamber cylinder and two barrels which fired alternately. Ammunition feed was by means of a 75-round pan magazine to the left side of the gun (with the rounds pointing inwards) which was reloaded by means of three-round clips, and a rate of fire of 450 rpm was claimed.

The unusual calibre of 42 mm was selected, with the cartridge case length of 338 mm or 348 mm depending on whether Swiss or British projectiles were fitted. The muzzle energy generated by each round was 1.8x that of the Bofors L/60. The entire weapon on its wheeled mounting weighed over 6 tons (6,500 kg).

The British also began development of a rival weapon of indigenous design, designated Red Queen (see UK section), but in the end abandoned both projects in favour of the new 40 mm Bofors L/70. Oerlikon continued working on the RK 421 until 1958, the final version being the Fla-422.

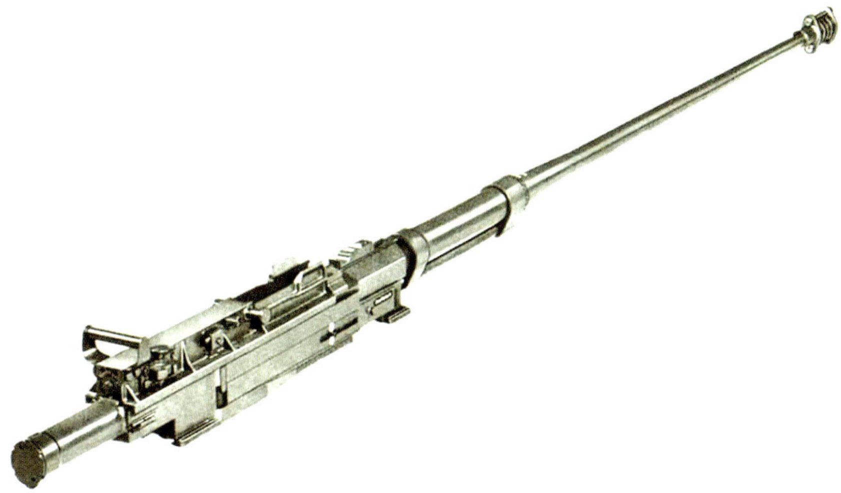

*Oerlikon 5TG / KAB (Oerlikon)*

OERLIKON TYPE 5TG (KAB):
20 × 128 AMMUNITION.

The Type 5TG, later renamed the KAB, was developed in the late 1940s from the experimental 2TMG aircraft gun but seems to have been less successful than the slightly later 204GK described below. It is gas-operated, with an unusual 'power-reserve-loader' system in which energy generated by the bolt's forward movement is stored for a brief period in a spring acting as a power reserve. Once the bolt is in its forward position, the power reserve spring locks the bolt via a pivoting lever below the breech, then releases the firing pin.

Ammunition is fed from a magazine: an 8-round box, or 20 and 50-round drums, the empty/full weights of these being 4.5/8.0, 17/23.5 and 24.5/41.5 kg respectively. It is fired via a mechanical trigger for either single-shot or automatic operation, with a trigger-locking device preventing the bolt from moving forward without feeding a round into the chamber. The barrel has Oerlikon progressive rifling and is fitted with a muzzle brake. It was intended for AA or ground-fighting roles, and given an L/120 barrel to extract the maximum performance from the new, high-velocity ammunition. The gun weighs 109 kg and fires at up to 1,000 rpm.

OERLIKON TYPE 204GK (KAA), TYPE FZK (KAC) AND KAE:
20 × 128 AMMUNITION.

These three are basically the same gun. The Type 204GK, later renamed KAA, was designed by former Mauser engineer Linder who had been responsible for the Mauser MG 151 to which it bears a superficial resemblance, although the locking system is entirely different. The MG 151 uses a short-recoil system, the KAA is gas-piston operated plus blowback case extraction, with two pivoting locking-pieces (similar in principle to contemporary Hispano Suiza cannon). In contrast to the 5TG, the 204GK is fed ammunition via a disintegrating-link belt (which can feed from either side), and it achieves a similar RoF of 1,000 rpm. The barrel is shorter at L/85 (so the muzzle velocity is about 50 m/s lower) and the gun is lighter at 88 kg. It has been widely used in both AFV and AA mountings, including naval installations.

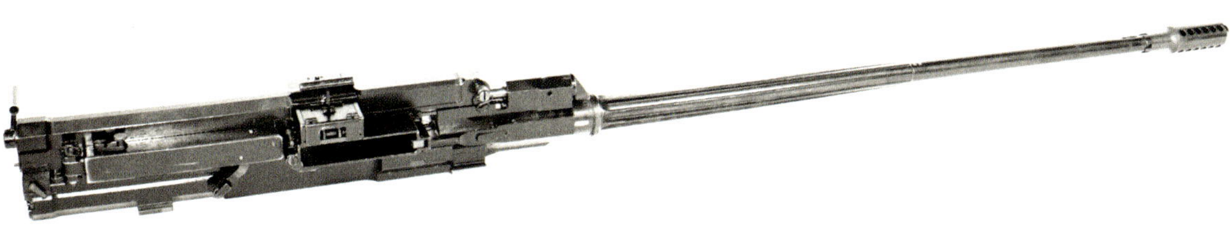

*Oerlikon Type 204 GK (copyright Rheinmetall Air Defence AG)*

*Oerlikon 20 mm Type KAA cannon (Oerlikon)*

An alternative round of ammunition was developed for this gun, a 25 × 116 created by shortening and necking down the 20 × 128 case, but this was not successful.

The Type FZK (*Fahrzeugkanone* = vehicle cannon), later renamed KAC, which was developed from 1965, was a KAA modified to make it better suited for mounting in a turret for ground fighting; modifications included the option of a 12-round box magazine instead of belt feed, and weight increased to 100 kg, presumably due to a heavier barrel. The RoF was reduced to about 800 rpm. It does not appear to have won any orders.

The KAE is a much later (c.2015) update of the KAA, intended for use in the Searanger 20 remote-controlled naval gun system, in combination with new SAPPIE-T (Semi-Armour Piercing Pyrotechnically Induced Effect – Tracer) ammunition to make it more effective in the naval close-defence role. To adapt it to remote control, the KAE has an electrical trigger and control unit. The operator can now fire an exact number of rounds in any one of the three modes: single shot, rapid single shot (100–300 rpm), and burst. A land version of the mounting designated Fieldranger 20 has also been developed.

**20 mm HS 820 / OERLIKON KAD**:
20 × 139 AMMUNITION.

The HS 820 was Hispano-Suiza's post-war replacement for the HS 804, developed in 1947–48 in response to the same need which produced the Oerlikon 5TG and 204GK, and with very similar performance, with a nominal RoF of 1,000 rpm. Like the Oerlikons, the HS 820 was designed around a new cartridge of similarly high power. Hispano-Suiza did start out with an advantage, however; the hybrid gas/blowback mechanism of the HS 804 was capable of being uprated to meet the requirements (and had already been used in larger form for the HS 830 family), so a new mechanism did not need to be developed. The HS 820 (renamed Oerlikon KAD after the 1971 merger) therefore used similar locking flaps to lock the bolt to the breech, unlocked by gas tapped from the barrel, and with blowback pushing the fired case from the chamber and out of the gun.

The bigger cartridge required a larger and sturdier gun with a longer barrel to extract the optimum performance, so it weighed 57 kg (bare gun), with the HS 561/562 BA recoil-driven belt feed mechanism weighing an additional 8.5 kg. A drum magazine feed was an option. The barrel was L/85 for the standard gun, which was 257 cm long overall. The 'HS 820 A' version had a L/105 barrel, was 297 cm overall and weighed 61 kg, or 78 kg with an empty 50-round drum magazine (which weighed 33 kg loaded). The long barrel and magazine feed made this version a direct rival for the Oerlikon 5TG, but the belt-fed standard gun was far more popular and has been used in a wide variety of mountings. Among other successes, it was adopted by the US Army under the designation M139 to arm light armoured vehicles.

*Oerlikon FZK (copyright Rheinmetall Air Defence AG)*

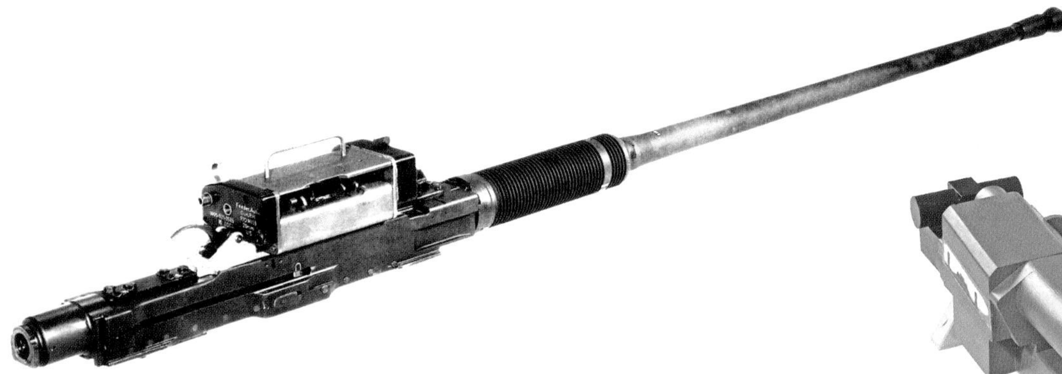

*Oerlikon KAD (Oerlikon)*

Various modifications have been made to the gun, including a dual belt feed system in the KAD-B14. It is no longer offered for sale.

*20 mm HS 821*: VARIOUS AMMUNITION.
This designation was used for two different guns firing different ammunition, one being developed from the other. The original gun, dating from 1960, was designated Heavy 20 mm Infantry MG, and was intended to use a purpose-designed 20 × 70 cartridge. This was soon dropped in favour of an Infantry and Anti-aircraft Machine Cannon which was lengthened to use standard 20 × 139 ammunition.

The guns used essentially the same short-recoil mechanism, fitted within a non-recoiling receiver. The gun was belt-fed and could be rapidly switched between right and left-hand feed. Characteristics of the two versions were: Overall length: 156 and 265 cm; barrel length 70 and 170 cm; gun weight 30 and 59 kg; rate of fire 500 and 600 rpm.

30 mm OERLIKON TYPE 301 FK: 30 × 110B 301 FK AMMUNITION.
This was the first of the Oerlikon 30 mm aircraft cannon, and the project dates from 1948. Only a wooden mockup of the gun was made, plus some ammunition for testing in a pressure barrel, but it is of interest because of the peculiar configuration. Unlike the 302 RK and later models it was not a revolver cannon, but had a short-recoil linear action, with a gas-operated belt feed which could be installed to feed from either side (the fired cases and links were ejected downwards). The gun was electrically primed and designed to fire from an open breech at 1,000 rpm. The gun was estimated to weigh 80 kg and was 158 cm long with an L/30 barrel.

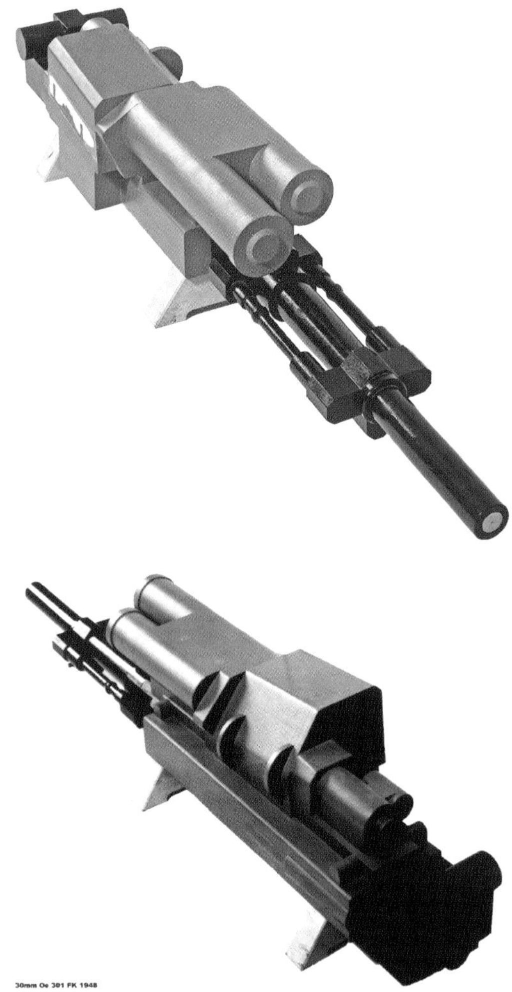


*Oerlikon Type 301 FK (Oerlikon, courtesy Robert Gerber)*

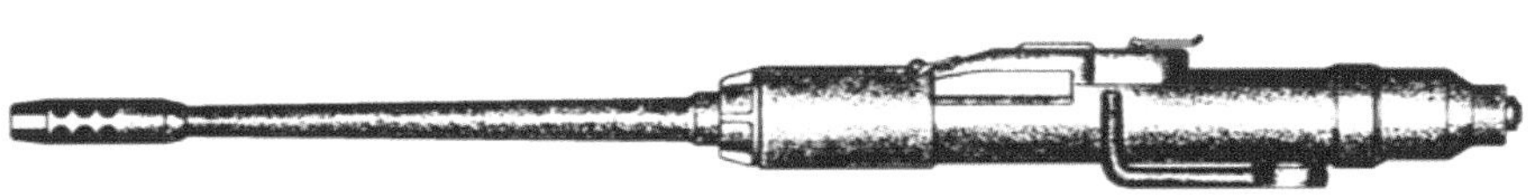

*The two versions of the HS 821 compared (courtesy Robert Gerber)*

***30 mm HS 825***: 30 × 136 AND 30 × 136B AMMUNITION.

This gun and its ammunition were specifically designed at the beginning of the 1950s for aerial combat, the goal being to combine a high hit probability with the substantial destructive effect needed to destroy jet aircraft, in a compact and light weapon. This was achieved by selecting the 30 mm calibre but with lightweight projectiles to enable very high muzzle velocities to be achieved; the light projectile would lose velocity quickly at ground level, but not at the high altitudes at which future aerial combat was expected to take place. This was an efficient solution to the problem: the rival ADEN/DEFA 30 mm guns fired heavier shells but at much lower velocities.

The gun used a version of the usual Hispano-Suiza gas-delayed blowback mechanism. The gun barrel and breech body assemblies recoiled within the receiver, which was rigidly attached to whatever structure it was mounted to. The belt feed was driven by this recoil movement. The mechanism was tuned to achieve the highest possible rate of fire; however, the mechanism was working at the limit of its capabilities which resulted in unacceptable reliability despite much development work. It failed to get into service, although it was considered and frequently tested by the British Fleet Air Arm and was selected for the abortive Swiss P16 ground-attack jet (an ironical choice, given the ammunition's optimisation for high-altitude use). The gun was also initially planned as the armament for the Italian Aerfer Sagittario and Aerfer Ariete, plus the Swedish SAAB 29D fighter aircraft with four guns in the fuselage. However, problems with propellant gas building up in automatic fire and igniting every few rounds were not resolved, and neither the HS 825 nor the 29D model were adopted.

With a short L/53 barrel the gun was 238 cm long overall and weighed 105 kg. An L/70 barrel was also offered, increasing the overall length to 288 cm and the weight to 112 kg. The RoF was 950–1,000 rpm and the belt feed could be to either side of the gun. Two different mounting cradles were available, one being fixed and the other allowing adjustments to be made in azimuth and elevation.

***34 mm HSS 834 AND 40 mm HSS 840***: 34 × 239 AND 40 × 239B AMMUNITION.

In the early 1950s the *Kriegstechnische Abteilung* or KTA (Weapons Technology Section of the Swiss Defence Ministry) requested Hispano-Suiza to develop a new heavy automatic AA gun intended to replace the 34 mm Flab K 38. The resulting gun was was designed in two calibres: the HSS 834 which used an updated loading of the existing 34 mm ammunition; and the HSS 840 which used a 40 mm version of the case, with a belt added to ensure accurate headspacing. Ammunition feeding was via a linked belt or five-round clips and the cyclic rate was an impressive 450–500 rpm.

The project was shelved in the mid-1950s. Oerlikon eventually replaced the Flab K 38 with their KD series guns (see below).

**30 mm OERLIKON TYPE 302 / 303 / 304 RK / KCA / KCE**: 30 × 173 AMMUNITION

This series of revolver cannon was developed in the 1950s and 60s, with the Type 302 emerging in 1952. Although it was clearly inspired by the Mauser MG 213C and developed with the aid of ex-Mauser engineer Linder, Oerlikon selected a much larger and more powerful cartridge in order to combine the characteristics of the 20 mm and 30 mm versions of the Mauser: a destructive 30 mm shell with the high muzzle velocity of the 20 mm. It even managed to retain the same high rate of fire, at 1,100–1,200 rpm. Such a high performance came at a price: the Type 302 was 279–297 cm long with L/66-L/75 barrels and weighed 180 kg, about double that of the Mauser and really too much for an aircraft gun. The gas-powered belt-feed was on the left side of the gun, which was electro-pneumatically charged.

The next step was to rethink the design to reduce the bulk and weight, resulting in the Type 303. The most obvious change was that the barrel was in line with the bottom chamber of the cylinder rather than the top. The cylinder was also smaller, with only four chambers. Weight was reduced to 140 kg with an L/50 barrel, or 150 kg with an L/75 barrel. Overall length was 194 or 270 cm, while the rate of fire was 1,250 rpm.

The Type 304 was a further refinement of this; the barrel was L/65 and the RoF increased to 1,350 rpm, but weight was reduced again to 120–125 kg. The KCA was the production version, firing modified ammunition with a steel rather than brass case. The JA 37 fighter version of the SAAB Viggen was the only

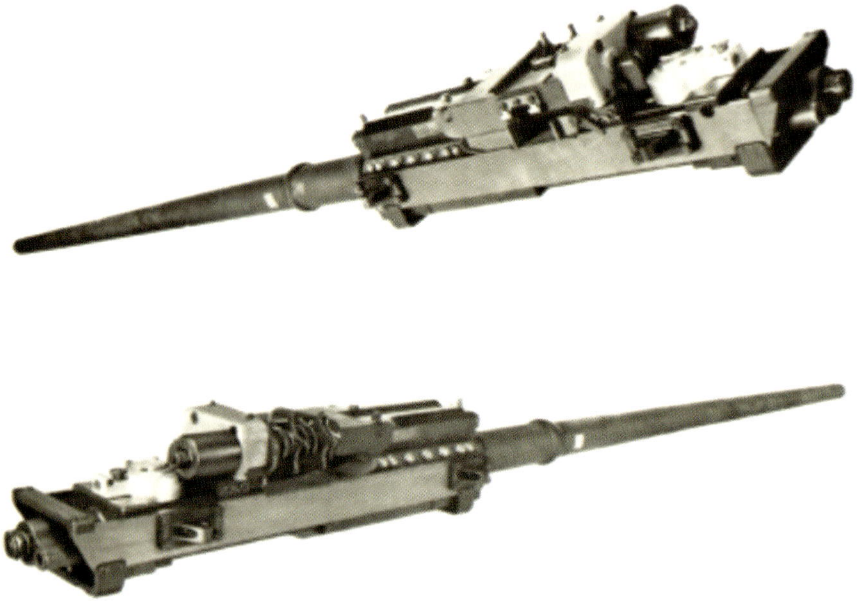

*HS 825 (courtesy of Robert Gerber)*

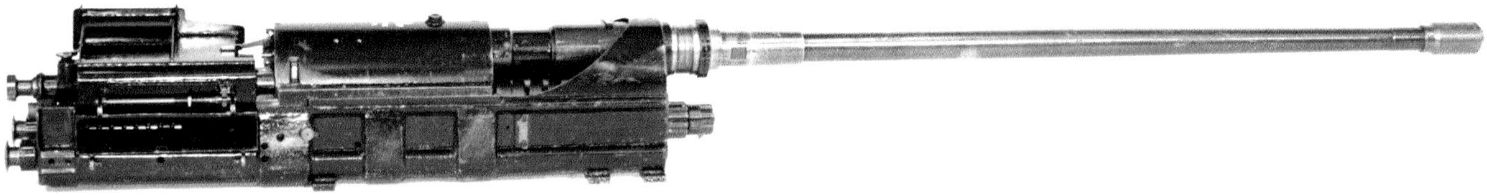

*Oerlikon Type 302 RK (copyright Rheinmetall Air Defence AG)*

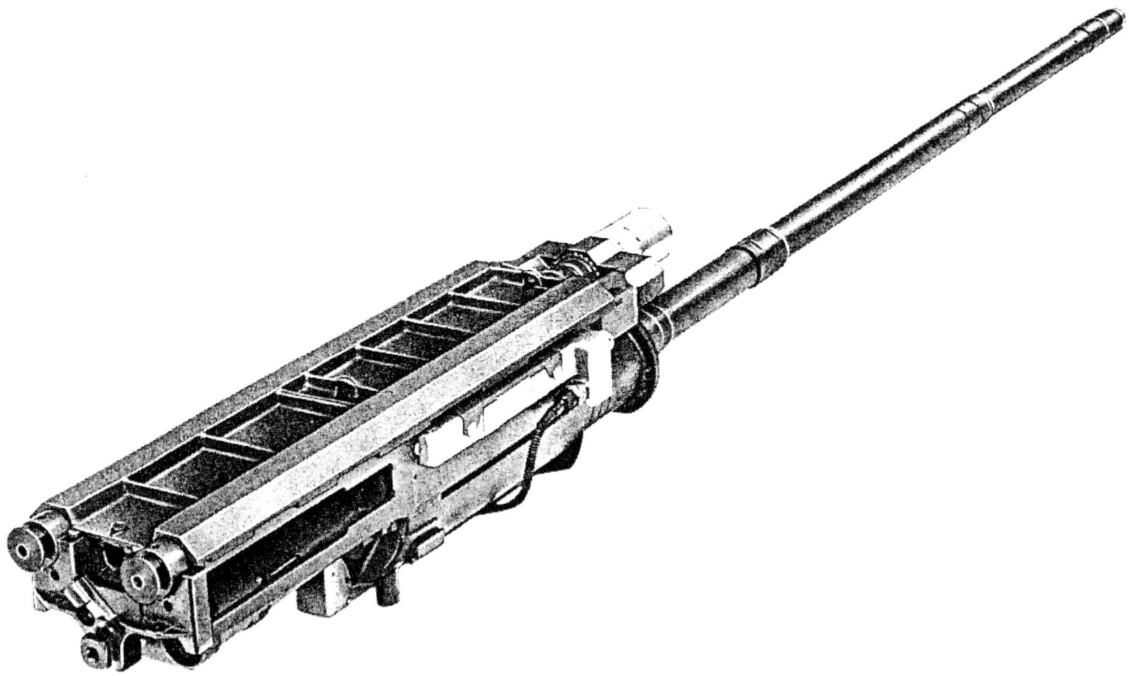

*Oerlikon KCA (Oerlikon)*

application, apart from a Hughes gunpod. The KCA was much admired by the RAF and after performing very well on test it seemed likely to be chosen for the Panavia Tornado European strike fighter, until Mauser developed the BK 27 (Germany was a member of the Panavia consortium, Switzerland was not). One rather curious application the RAF explored was to install it in an aircraft weapon bay, pointing downwards to crater runways, but the length of the gun was too great to be accommodated in the only suitable aircraft – a Buccaneer. It is unclear whether any of the Hughes gunpods were sold and, if so, whether any remain in use, but the Viggen left service in 2005.

The KCA nearly achieved further service in modified forms: as the KCC (see below) and the GAU-9/A (see the USA section), but neither was adopted. The latest version to be offered (in 2020) is the KCE, which is a blend of the KCA and KCC, firing percussion-primed 30 × 173 ammunition. It has a new barrel able to fire APFSDS and KETF ammunition, new sensors and an electric cocking device. It is advertised as suitable for installation in air-defence systems or in remotely controlled weapon systems.

***20 mm OERLIKON TYPE 206 RK***: 20 × 128 AMMUNITION.

This was developed in the early 1950s in parallel with the Type 302 RK, and was essentially a smaller version of it: a gas-operated five-chamber revolver with a belt feed switchable between sides. Belt links were ejected on the opposite side to the belt feed, fired cases to the rear. There was an electro-pneumatic cocking device. The gun weighed 86–90 kg, was 219–239 cm long with L/80-L/90 cm barrels and fired at 1,500–1,800 rpm. It did not achieve any sales.

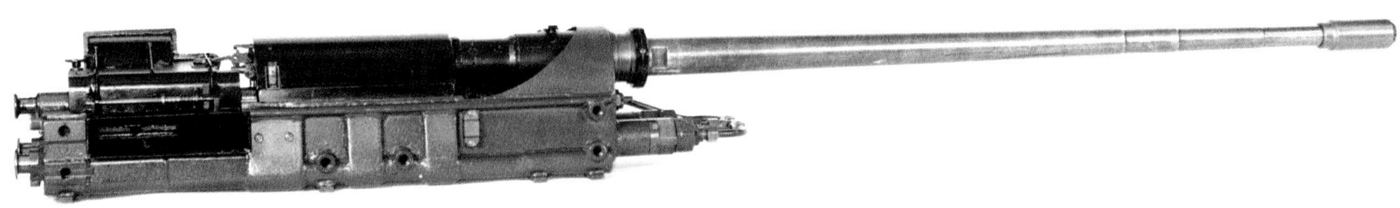

*Oerlikon Type 206 RK (copyright Rheinmetall Air Defence AG)*

***20 mm, 25 mm AND 30 mm OERLIKON 251 RK***: 20 × 128, 25 × 116B, 30 × 113B AMMUNITION.

This was a 'second generation' revolver cannon, reduced in size by the use of a four-chamber cylinder rather than the usual five, and with an increased rate of fire. The 20 × 128 round for the 251/20 gun was the Oerlikon KAA round but with electric priming, the 30 × 113B for the 251/30 from the French DEFA 550 series, but the 25 × 116B for the 251/25 was unique to this gun, being essentially a necked-down version of the 30 × 113B.

The 251/20 weighed about 74 kg, had an L/72 barrel and fired at 1,750 rpm. The corresponding figures for the other guns were: 251/25 – 75 kg, L/70, 1,650 rpm; 251/30 – 75 kg, L/53, 1,400 rpm.

These guns were unfortunate in being launched at the wrong time, at the end of the 1950s when none of the major western powers were looking to replace their aircraft guns (most of them expected guns to be replaced by guided missiles in the near future). The 251/25 in particular appears to have offered a very attractive combination of light weight and high performance, the 165 g projectiles being much harder hitting than the contemporary US 20 mm, while the high muzzle velocity of 1,040 m/s provided a much better hit probability than the 30 mm ADEN/DEFA rounds. Despite these merits the gun failed to attract orders. Oerlikon decided not to pursue a proposal to redesign the 251/25 to use the new 25 × 137 ammunition which they had developed for the US 'Bushmaster' AFV cannon project, although they did produce something similar in the KBC (see below).

**35 mm OERLIKON KDA/B/C/F SERIES**: 35 × 228 AMMUNITION.

At the end of the 1940s Oerlikon started research into the optimum calibre for a future medium AA gun. This led to a contract with the US Navy to develop prototype 35 mm and 50 mm guns (351 MK and 501 MK respectively), with the choice later focusing on 35 mm and the development of a range of 35 × 228 ammunition. The first gun to fire this was the 351 MK of 1955, with the 352 MK and 353 MK progressively incorporating improvements. The first mounting was a twin-gun naval unit featuring water-cooled barrels and with 231-round pan magazines attached to the trunnions, but towards the end of the 1950s this was dropped in favour of focusing on a land-based system. The air-cooled gas-operated guns in this series all have the same rate of fire of 550 rpm, all have L/90 barrels (315 cm without muzzle brake), and are almost invariably used in twin mountings designated GDF in field mountings and GDM for naval systems (GDM-A is a manned mounting with local or remote control and the guns mounted close together; GDM-C is remotely operated and was developed in conjunction with OTO Melara – the guns are on each side of the mounting). The exceptions are a Japanese manually-operated single naval mounting, and a similar Chinese mounting available in 35 mm or 37 mm.

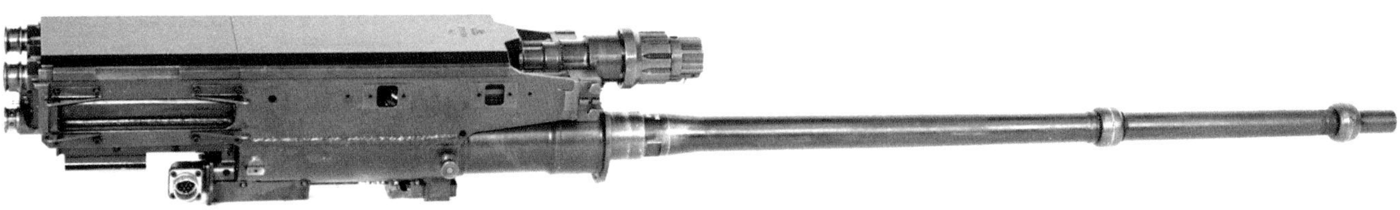

*Oerlikon Type 251 RK (copyright Rheinmetall Air Defence AG)*

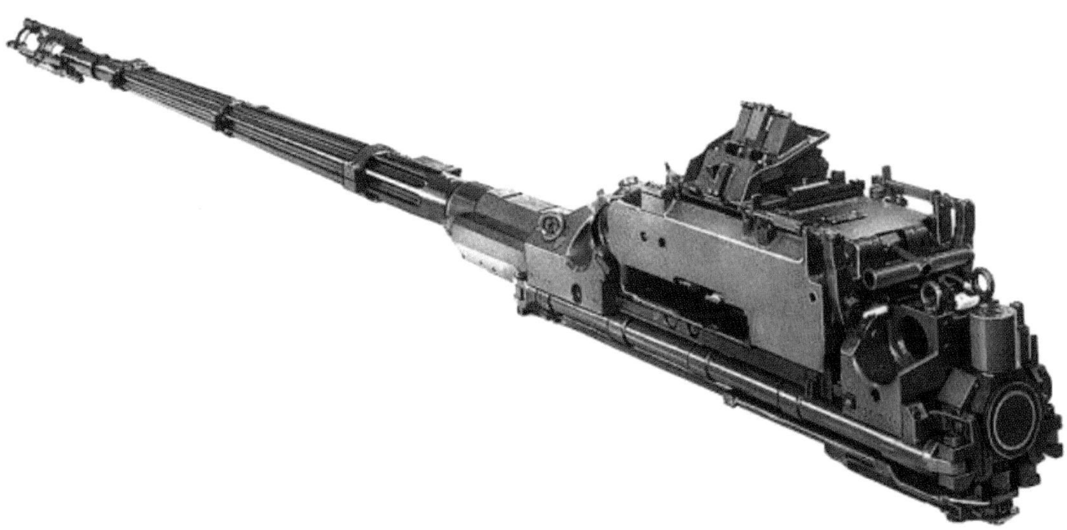

*Oerlikon 35 mm KDA dual belt-fed automatic cannon (Oerlikon)*

The gun housing and barrel slide in the cradle, recoiling 55–60 mm, but the ammunition feed mechanism does not move during recoil. When the trigger is operated the breech lock is released and the round is rammed into the chamber under spring power, the gun firing automatically as soon as the breech is locked. Gas is tapped from the barrel to push back pistons to begin the recoil movement, boosted by blowback. The differences between the gun models are as follows:

KDA: Designed to be operated by means of externally mounted mechanical, hydraulic and electrical devices for handling, monitoring and safety functions. Gun weight 670 kg and length 474 cm. Belt-fed, and typically used in SPAAG (e.g. Gepard) and some naval mountings.

KDB/C: The KDC is an optimized version of the KDB field cannon, developed to accept independently powered linkless ammunition feed systems; it uses seven-round clips from which the rounds are stripped by the feed system conveyer. Weight 430 kg and overall length 521 cm. It is used in both GDF and GDM mountings.

KDF: A 'universal' weapon with a more compact design, adaptable to a variety of ammunition feed mechanisms. Weight 460 kg and overall length 474 cm.

Current Rheinmetall advertising is focused on the KDC model, but the KDA is still in production for the new Polish LOARA-NG SPAAG. China also manufactures field and SPAAG systems for these guns, and Turkey makes GDF mountings. Over 4,000 of these guns have been made.

There has been a wide range of modifications and upgrades to these AA systems, mainly affecting sensors, fire control and ammunition types, the most significant being the use of elements of the 'Skyshield' system (see KDG below) as a successor to the long-established 'Skyguard', including the AHEAD (Advanced Hit Efficiency and Destruction) system.

**35 mm OERLIKON KDE:**
35 × 228 AMMUNITION.

This member of Oerlikon's 35 mm gun family is the only one not to use gas operation. It is recoil-operated with a relatively long recoil stroke of 170 mm, compared with 55–60 mm for the gas-operated single-chamber guns. In combination with the low rate of fire of 200 rpm, this makes the KDE much more suitable for mounting in AFVs to fire against ground targets: the gun is more accurate and the delay before opening the breech after firing minimizes the problem of gun gases entering the turret. Various ammunition feed systems were developed for use with the KDE, in the form of either magazines or automatic ammunition feeders incorporated into the turret, but the standard type has two 17-round magazines fixed to the breech (thereby avoiding the potential feed problems associated with flexible belts). The ammunition feed is selectable, taking less than one second to switch between magazines. The gun weighs 510 kg and has an overall length of 427 cm with an L/90 barrel.

Japan is the only user of the KDE, making it under licence and installing it in the turret of their Type 89 AIFV, which entered service in 1989 and is reportedly still in low-rate production.

***20 mm, 23 mm AND 30 mm HISPANO SUIZA AFV CANNON:***
20 × 139, 23 × 133, 30 × 138 AND 30 × 170 AMMUNITION.

During the late 1960s a family of rotary-locked short-recoil cannon with dual belt feed was developed by Hispano-Suiza for installation in AFV turrets. These consisted of the HS 827-B 'Hurricane' (20 × 139), the HS 827-C MC-23 (23 × 133), the HS 836 'Spitfire' (30 × 138) and the HS 837 'Thunderball' (30 × 170). The 20 × 139 and 30 × 170 were existing Hispano cartridges, but the others were purpose-designed for these guns. Each cannon featured selectable rates of fire with two firing rates as well as single-shot: for the MC-23, which was purpose-designed to meet the US Bushmaster requirement, these were 550 and 1,000 rpm.

The 20 mm Hurricane weighed 56 kg, had an L/85 barrel and was 251 cm long overall. Equivalent data for the others were: MC-23 – 58 kg, L/74, 251 cm; Spitfire – 89 kg, L/60, 274 cm; Thunderball – 105 kg, L/75, 329 cm. None of them was adopted.

**25 mm OERLIKON KBA:**
25 × 137 AMMUNITION.

In the mid-1960s the US Army was looking for a more powerful replacement for the 20 mm M139 (HS 820) light vehicle armament. The requirement was

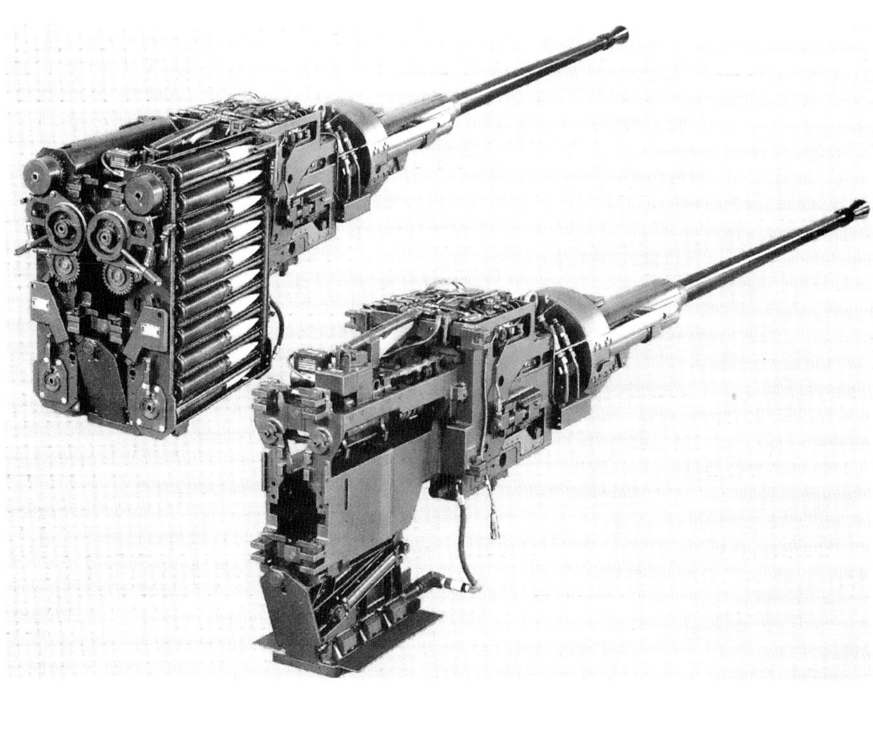

*Oerlikon KDE (Oerlikon)*

officially known as the Vehicle Rapid Fire Weapons System (VRFWS), but is better known as the Bushmaster. Various companies put forward proposals, one being TRW (see USA section) which paired up with Oerlikon; TRW designed the gun, Oerlikon the 25 × 137 ammunition. TRW stopped work in 1969 and Oerlikon decided to design their own gun around their ammunition, resulting in the KBA cannon. This failed to win the Bushmaster contract (although a gun developed around the Oerlikon ammunition did so) but has achieved significant export sales, with more than 5,000 guns delivered. It is still advertised by Rheinmetall Defence as suitable for AFV and naval installations; these are usually in single mountings but there is a quad-mount SPAAG system, the SIDAM 25, in service with the Italian Army.

The KBA is a gas-operated gun with a rotating bolt-head lock and dual belt feed, allowing a rapid switch between two different ammunition types. It fires from an open bolt: single-shot, programmable rapid semi-auto with a rate of fire of up to 200 rds/min, and fully automatic fire with 600 rds/min. It weighs 112 kg, and has an L/87 barrel within the overall length of 289 cm.

*Oerlikon KBA (Oerlikon)*

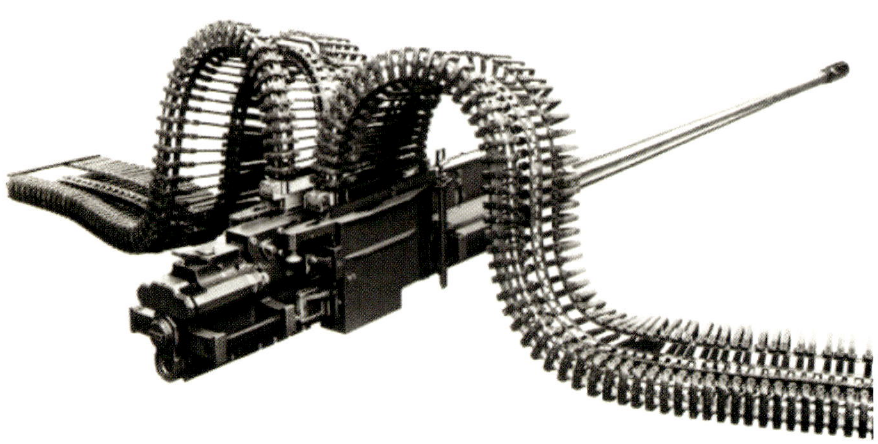

*Oerlikon 25 mm Type KBB cannon with dual-feed system (Oerlikon)*

**25 mm OERLIKON KBB**: 25 × 173/181 AMMUNITION.

This is simply a stretched, higher-performance version of the KBA developed from 1966, the case being lengthened to 173 or 181 mm, depending on the projectile loaded (full-calibre or discarding sabot respectively). The gun action, ammunition feed etc are basically to the same design as the KBA, apart from the dimensional differences to allow for the longer cartridge. The L/92 barrel is also longer, to make the most of the larger propellant charge. The gun weighs 146 kg, is 319 cm long, and fires at 800 rpm.

It was envisaged that the KBB would be installed in a wide variety of mountings, including infantry, AA and AFV purposes, but only one sale was achieved: of the quad Sea Zenith naval CIWS, to Turkey.

***25 mm OERLIKON KBC***: 25 × 137 AMMUNITION (LENGTHENED).

This was a 5-chamber gas-operated revolver aircraft cannon proposal dating from the late 1970s. The ammunition it fired used the standard 25 × 137 NATO cartridge case loaded with a longer and heavier projectile based on the KBB's, increasing the overall round length from 223 to 244 mm. Unlike the other Oerlikon revolver cannon, this used percussion rather than electrical ignition. The gun weight was 90 kg and overall length 236 cm with an L/71 barrel. The rate of fire was 1,700 rpm and a power-boosted linkless ammunition feed was used, which reinserted the fired cases into the continuous conveyor belt. It did not succeed in attracting any orders.

***30 mm OERLIKON KCC***: 30 × 170 AMMUNITION.

This is essentially the KCA revolver cannon adapted to fire 30 × 170 rather than 30 × 173 ammunition, with percussion rather than electric priming. The rate of fire was also increased slightly to 1,500 rpm. The purpose for developing this weapon was to replace the twin naval mountings of the KCB with a much faster-firing twin mounting capable of dealing with anti-ship missiles as well as aircraft. The reason for selecting the 30 × 170 ammunition was that this was already widely in naval service with the KCB in various mountings, although as it turned out the cartridge had to be modified slightly to cope with the violent stresses caused by the higher rate of fire; the resulting KCC round was however backwards-compatible with the KCB. Work began early in the 1980s and reached the firing test stage, when internal company politics led to the project being abandoned for fear that it might harm the hoped-for success of the 25 mm KBB CIWS.

***25 mm OERLIKON KBD***:
25 × 173/181 AMMUNITION.
Uniquely for Oerlikon, this is a rotary gun featuring seven barrels and is driven by external power. It was developed in the latter part of the 1980s with the needs of naval CIWS in mind, the higher rate of fire compared with Sea Zenith (see above) being required to deal with faster and more agile anti-ship missiles. The gun weighs 290 kg and is 276 cm long with L/92 barrels. The first prototypes used hydraulic power, but other systems were envisaged. The rate of fire was 5,000 rpm, with 0.5 secs needed to accelerate the barrel cluster to that rate (firing does not start until that rate has been achieved).

A system using two of these guns, known as Myriad, was developed by an Italian consortium, firing trials being completed in 1995. 1,000 rounds were provided for each gun. There was also a proposal for a single gun to replace the M61 in the Phalanx mounting. No sales were achieved.

*Myriad CIWS, with two Oerlikon KBD cannon*

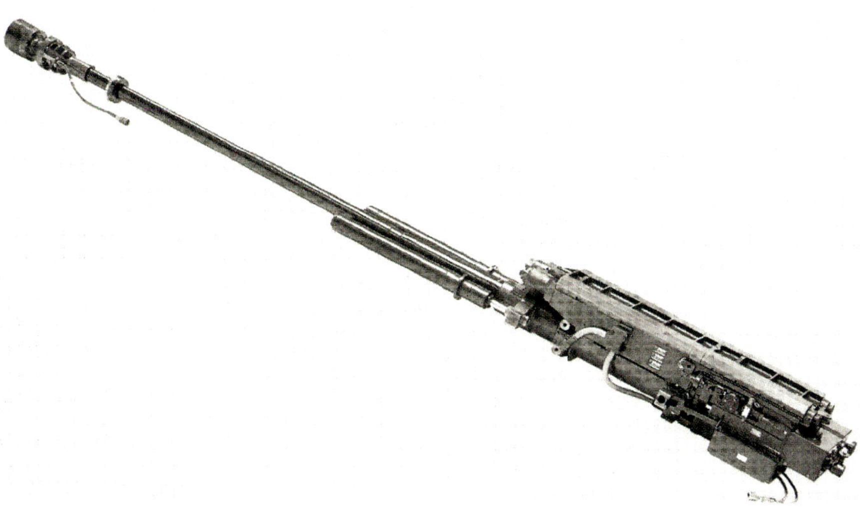
*Oerlikon KDG (Oerlikon)*

**35 mm OERLIKON KDG**:
35 × 228 AMMUNITION.
This was developed by Oerlikon Contraves (now Rheinmetall Air Defence) from 1991 under the designation of the 35/1000. Firing tests started in 1994, and in 1995 the 'Skyshield 35' towed AA system based on this gun was introduced as a successor to the long-established 'Skyguard', together with the AHEAD (Advanced Hit Efficiency and Destruction) system. The key element of AHEAD is the ammunition; the hollow shell is packed with tungsten-alloy pellets released by a small time-fuzed bursting charge just before the projectile reaches the target, resulting in a 'shotgun blast' of pellets covering a significant area and thereby considerably multiplying the hit probability. AHEAD is also designed to be effective in naval mountings against targets ranging from anti-ship missiles to small boats.

A turreted naval mounting, designated Millennium, has also entered service and a similar mounting has been offered on an 8x8 chassis as part of the Skyranger mobile air defence system. Finally, a version of that system has been used since 2012 by the German Air Force in the C-RAM role (Counter-Rocket, Artillery, Mortar) under the 'MANTIS' designation (Modular Automatic and Network-capable Targeting and Interception System).

The KDG is a gas-operated four-chamber revolver with a linkless feed system and is capable of single, selectable burst (to 200 rpm), and automatic fire (1,000 rpm). It has a remote and automatic recocking device, weighs 450 kg and has an L/79 barrel within an overall length of 411 cm. It is mainly used with AHEAD ammunition which is suitable for engaging a wide range of targets.

The KDG is now at the Mk 3 stage: the Mk 1 has a single-side switchable feed, the Mk 2 has a dual feed and other changes. The Mk 3 (introduced in 2016) is designed as part of a remotely controlled, autonomous, network-capable air defence system. Integrated into the 35 mm platform are a tracking/sensor unit with X-band tracking radar and electro-optical sensors as well as electronic warfare components. It has an autonomous surveillance and target engagement capability.

# UNITED KINGDOM

The first automatic cannon originated in the UK in the last decade of the 19th century. This was the 1$^{PR}$ (one pounder) designed by Hiram Maxim, an American who lived in the UK from 1881 onwards and became a British citizen in 1899. It was basically a considerably enlarged version of Maxim's famous (and the world's first) self-powered machine gun which was first demonstrated in 1884 and by 1891 had become the standard machine gun of the British Army, serving in improved form until the 1950s.

The 1-pounder was a scaled-up version of the Maxim, designed around the existing 37 × 94R ammunition which was originally developed for French Hotchkiss manually-operated guns in the 1870s. This calibre was selected by Hotchkiss as a result of the St Petersburg Declaration of 1868 banning the use of exploding projectiles weighing less than 400 g (the lightest 37 mm shell being above this limit). From then on, 37 mm became one of the most popular calibres for guns in this class. The Maxim 1-pounder emerged in the 1890s and saw action in the Boer War, in which it was given its nick-name of 'pom-pom' from the sound it made.

Maxim acquired another armaments company, Nordenfelt, in 1888, the resulting Maxim-Nordenfelt Gun and Ammunition Company merging with Vickers, Sons & Co. in 1897 to form Vickers, Sons & Maxim, by far the most important British company concerned with automatic guns. This company was renamed Vickers Ltd. in 1911, although ammunition made by this company carried the VSM headstamp on cartridge cases until at least 1928.

The basic Maxim recoil-operated mechanism was stretched to accommodate larger and more powerful cartridges; briefly the Vickers 1½-pounder during World War I before the introduction of the largest member of this family, the famous 2-pounder naval AA gun which retained the pom-pom nickname. A few other designs emerged during World War I, notably the much lighter Mk III version of the 1-pounder firing shorter ammunition, and the big 1½-pounder C.O.W. gun (Coventry Ordnance Works) firing its own unique cartridge; both intended for aircraft, but these saw very little use. EOC (Elswick Ordnance Company) were also developing a 1½-pounder automatic aircraft cannon in 1918 but nothing more than a few drawings appears to have survived. The 1-inch Vickers saw no British use, although small numbers were made by Vickers-Terni in Italy.

After World War I, various experimental cannon were made, some as private ventures and others to meet official requirements. There were two separate official British programmes investigating guns and ammunition in 20–25 mm calibres: one by the Ordnance Committee during 1923–25 in order to identify a large-calibre gun (bigger than .5-inch / 12.7 mm) suitable for aircraft use by observers; the other by the Small Arms Committee which was looking for a portable anti-tank weapon from 1925–38. Two British companies were initially involved in these programmes: EOC, a part of Armstrong Whitworth, and Vickers Ltd.. In 1927 Vickers and Armstrong Whitworth merged to form Vickers Armstrong Ltd (VA); ammunition from Elswick was thereafter headstamped VAEL.

In their search for an observer-aimed aircraft gun, the Ordnance Committee tested .8-inch and 1-inch rounds (20.3 mm and 25.4 mm) based on those developed for experimental aiming rifles for field artillery: the 1-inch used a standard aiming rifle case (25.4 × 94R) firing a 227 g projectile at an unstated muzzle velocity. The .8-inch used a necked-down version of the same case (only 10 cases made, in steel) and fired a 128 g projectile at velocities from 850–922 m/s. In both calibres the projectiles were solid AP tracer. In 1925 the .8-inch version was transferred to the Small Arms Committee (see below). The mechanism envisaged for these guns is not known.

The existing .90 inch Elswick (22.9 × 198.5R), a private venture project dating from 1918 about which little is known, was also tested by the Ordnance Committee in 1924. Two steel shot loadings, with projectiles weighing 170 g and 227 g, were supplied. The 170 g loading achieved on test an MV between 850 and 900 m/s. It appears, however, that this gun was intended for manual loading.

In addressing the problem of infantry anti-tank weapons, the Small Arms Committee initially ordered two test cartridges and barrels in .8-inch and 1-inch calibres in order to determine their armour-piercing capabilities. The .8-inch had a chamber length of 138.4 mm (i.e. a case length of c.138 mm) and a cartridge case capacity of 73.74 cc. It fired a 73 mm-long hard-core projectile of 148.6 g with a 90.4 g core, or a 123 g solid steel projectile at an MV of around 850 m/s. The 1-inch had a chamber length of 157.5 mm (c.157 mm case length) and a case capacity of 131 cc. The 91.4 mm long projectile weighed 291 g and was fired at the same 850 m/s MV. At 500 yards and 90 degree impact, the .8-inch penetrated 31 mm armour plate and the 1-inch, 42 mm. Neither of the cartridge cases, nor any illustrations of them, appear to have survived, nor any details of the guns which might have used them.

From 1927, attention focused on the .8-inch calibre. Vickers Ltd had already developed a non-automatic hopper-fed .8-inch gun (further details unknown) and this was acquired for testing by the Committee but proved unsatisfactory. In parallel with this, a new 20.3 × 167SR cartridge was designed to an official requirement. In 1927, 65 and 70 calibre test barrels designed to use this round were ordered from Vickers Ltd., but in 1928 automatic guns made by EOC were received. As a result, this cartridge is generally known as the

.8-inch Elswick. Trials continued until the early 1930s.

The Small Arms Committee minutes of December 1930 includes some information supplied by Vickers-Armstrong concerning their '0.8-inch anti-tank gun'. This states that the particulars are 'almost identical to the 0.8-inch Elswick gun' (which by then was part of the same company), except that the cartridge case was stated to be 'slightly smaller'. It seems likely that the ammunition used was the Vickers 0.8-inch (20.3 × 124.7SR) for an automatic gun.

Ironically, despite all of this research the British Army ended up selecting the .55 inch Boys anti-tank rifle, firing ammunition in the heavy machine gun class – which had (correctly) been rejected as inadequate for the purpose when the Committee started its work in 1925.

The problem of protecting warships against air attack was considered by the Naval Anti-Aircraft Gunnery Committee, which produced a report in April 1932. The Committee went into great technical detail in calculating the effectiveness of different guns. They considered such matters as the nature of the threats, the time during which aircraft could be brought under effective fire by different weapons, the range and rate of fire of the guns, the effectiveness of shells and fuzes, and appropriate methods of fire control.

Three types of attack were considered; precision (i.e. level) bombing, torpedo bombing, and close-range attack with bombs or machine guns. The Committee also identified a further potential risk:

*'The possibility of explosive aircraft being manoeuvred by human pilots to hit ship targets cannot however be ruled out. It is reported that, sooner than accept defeat, ramming other aircraft is a recognised principle among Japanese pilots.*

*Note:- The Air Ministry regard this idea as exceptionally secret and would prefer that it be not generally promulgated.'*

It was clear that the Committee considered such attacks to be potentially extremely difficult to deal with, a concern fully justified by the experience of a dozen years later.

In considering middle-layer defence, the Committee was most concerned about torpedo bombers, calculating that any weapon system able to cope with them would be able to deal with other forms of short-range attack (e.g. torpedo boats) easily enough. Exercises between 1928 and 1931 had shown that the probability of a torpedo bomber hitting a ship was only 10% at 1,250 yards (1,140 m) but rose to 30% at 1,000 yards (910 m), 50% at 750 yards (690 m) then increased very sharply to 85% at 600 yards (550 m).

This led to a demand for a range of 2,500 yards (2,300 m) from automatic AA guns, in order to achieve the aim of certain destruction of an aircraft with 10 seconds of firing at a mean range of 2,000 yards (1,820 m). It was estimated that an aircraft dropping a torpedo from 1,200 yards (1,100 m) would already have been under fire from such weapons for 17 seconds, which was assumed to be more than enough time to shoot it down. That did assume that aircraft would be attacking at about 150 mph (250 kph), appropriate for the period. As aircraft speeds increased, so the available engagement time shortened and the job of the defences became more difficult.

The Committee was not satisfied with the 2-pounder pom-pom as it had too low a muzzle velocity. They were also concerned about its rate of fire, which was felt to be just about adequate in its 8-barrel form. The Committee recommended that no further steps should be taken to develop the 4-barrel version. They expressed a hope that the 2-pounder system might be replaced by an intermediate calibre with a 760 m/s muzzle velocity and a total output of at least 1,250 rpm, and recommended a series of trials to determine the ideal calibre. What emerged from this was the 1½-pounder Mk V of 35 mm calibre, but this was cancelled in 1937, mainly due to its weight, and a High Velocity version of the 2-pounder was developed instead.

Remote power control was introduced for the octuple and quad mountings during World War II. Single mountings for smaller craft were either unpowered (Mk VIII) or powered (Mk XVI) and weighed in the region of 1,100–1,800 kg depending on whether an armoured shield was fitted. About 500 of these guns were sold to Japan, entering service in 1931 as the 40 mm/62 'HI' Type 91, in single or twin mountings.

The eight-barrel (Mk V and Mk VI) mounting was considered the principal type to provide the barrage of AA fire required, and up to eight of them were fitted to some battleships. At 16,000–20,000 kg these were too heavy for smaller ships, so despite the recommendation of the Naval Anti-Aircraft Gunnery Committee, the quad Mk VII mounting, weighing around 9,000–11,000 kg depending on whether unpowered or powered, was introduced in 1936, initially for use in destroyers.

Private venture developments by British companies included 1-inch or 25.4 mm Vickers guns described in more detail below and listed here by cartridge size: 25.4 × 87R; 25.4 × 146R; 25.4 × 189. Armstrong Whitworth also appears to have made a .866 inch (22 mm) gun in around 1928, but this is only known from the ICI drawings of the projectiles.

From all of this, the reader may accurately conclude that the story of British interwar cannon development is rather poorly documented with few physical remains apart from a handful of cartridges.

It is worth noting that various foreign cannon were evaluated in the UK at this time, including: a 37 mm Browning (strip-fed and fired at 100 rpm); the 20 mm Madsen; 20 mm Solothurn (20 × 138B); and 20 mm Oerlikon (Type S and Type L). These 20 mm guns were all effectively dual-purpose weapons, being suitable for AA as well as AT applications, depending on the mountings. Two 20 mm semi-automatic anti-tank rifles were also examined: a Solothurn S18–100 series (20 × 105B) and an Oerlikon SSG (20 × 72RB). The cannon tests resulted in the selection of the Oerlikon Type S, initially for the naval AA role, which was acquired in the improved, Type SS version. Separately from these evaluations, the RAF saw a demonstration in France of the prototype Hispano-Suiza HS 404 aircraft gun, which later become the RAF's standard cannon.

With the later selection of the 40 mm Bofors for the medium AA role, that meant that of the four principal cannon in British service in World War II, only one was of British origin: a modified and updated version of the the naval 2-pounder pom-pom. There were a few other new British-designed automatic cannon which saw service but they were niche products: the Vickers 40 mm Class S, based on the C.O.W. gun (which itself just about remained in service for airfield defence), and the Molins 57 mm Class M guns, fitted (in different versions) to both aircraft and motor gun boats and used to attack surface targets. It would be stretching a point to claim the 20 mm Polsten as a British design, as it was merely an Oerlikon simplified to reduce production costs and the development team was led by Polish engineers who had taken refuge in the UK.

One other cannon is worth mentioning as it was offered to the UK in 1942 but rejected: a new 30 mm Hispano, originally designed in France (designated by them the HS 411) but a least one prototype was assembled by BMARC, Hispano-Suiza's British subsidiary. This cannon is described in more detail in the Swiss section. Ironically, the HS 411 developed into the HS 830 series post-war, which was subsequently renamed the KCB when Oerlikon took over Hispano-Suiza in 1971, and was adopted by the Royal Navy in the 1980s.

The British and Commonwealth forces were not slow in adopting captured foreign weapons when these were useful. A large number of Breda 20 mm Model 35 were captured during Operation Compass in North Africa, enabling the Australian 2/3 Light Anti-Aircraft Regiment, parts of the 4th Anti-Aircraft Brigade (which had a total of 42 Bredas in its Light Anti-Aircraft batteries during the Siege of Tobruk) and one battery of 106th (Lancashire Hussars) Regiment, RHA to be equipped with them. Captured Bredas were also used by the Long Range Desert Group and aboard a number of Royal Australian Navy and Royal Navy vessels including HMAS *Vendetta*, HMAS *Perth* and HMS *Ladybird*, and at least one Marmon-Herrington Mk II armoured car.

A note about nomenclature may be helpful here. British practice for the first half of the 20th century was to describe guns with calibres of 37 mm or more using the 'pounder' system (normally abbreviated to PDR or PR), a 'one pounder' projectile weighing 454 g, although this designation was nominal. Guns in smaller calibres than this were given inch measurements (as were howitzers). However, for guns in the 1-inch calibre, Vickers switched to the metric equivalent of 25.4 mm in around 1932 – or perhaps they used the metric designation for weapons intended for export. The use of inch measurements for British service use began to break down with the adoption of foreign weapons such as the 20 mm Hispano and Oerlikon and the 40 mm Bofors: the RAF was particularly keen to adopt this system, and used it for some British designed guns in World War II, including the 40 mm Class S airborne anti-tank gun and the experimental 47 mm Class P which was intended to replace it. By the end of the war, the use of the metric system for cannon was well established. One further change took place in around 1948: the practice of using Roman numerals was changed to Arabic ones, e.g. Mark V became Mark 5.

Various experimental AA cannon were designed both during and after World War II, including the naval 40 mm HV, 3½-pounder (47 mm), and 34 mm Class T, the army 42 mm 'Red Queen' (alongside the competing Oerlikon 420/1 RK 'Red King'), and the 6$^{PR}$ 6 cwt (57 mm), but none of these got beyond the prototype stage, if that. The final attempt was a little-known naval 57 mm twin mounting at the end of the 1950s which appeared to use ammunition unrelated to the earlier 6$^{PR}$ 6 cwt or to anything else, for that matter. It was the culmination of the naval D.A.C.R. programme (Direct Action, Close Range, meaning the same as the current CIWS), which appears to have commenced with the 34 mm mentioned above. It almost certainly failed to get off the drawing board, and no further details are available.

Despite all of this effort, the 30 mm ADEN aircraft guns and the 30 mm Rarden AFV gun have been the only new British-designed cannon to see service since World War II, and the ADENs were directly developed from the Mauser MG 213C revolver cannon which was at the prototype stage in 1945. The final variant of the design, the ADEN 25 using 25 × 137 ammunition, was developed at the end of the 1990s but the Harrier installation it was principally intended for proved unworkable and it was not adopted.

British companies capable of designing, developing and manufacturing automatic cannon have been few. Vickers Armstrong remained the principal company in this field during World War II and for some decades thereafter. In 1977 the company was reorganised and mostly nationalised, but the armaments department remained independent as Vickers plc. In 1986 Vickers bought the Royal Ordnance Factory, Leeds, and was renamed Vickers Defence Systems. In 1999 the company was bought by Rolls-Royce and the defence department was sold to Alvis, creating Alvis Vickers. This was acquired by BAE Systems in 2004 and became BAE Systems Land Systems. In the following year, BAE Systems bought the US company United Defense Industries (UDI), leading to a further change of name to BAE Systems Land and Armaments. In acquiring UDI, the company also gained Bofors Weapon Systems AB of Sweden (purchased by UDI in 2000), subsequently renamed Bofors Defence, so renewing their involvement with automatic cannon. In 2019 BAE Systems entered a joint venture with Rheinmetall named Rheinmetall BAE Systems Land (RBSL), in which Rheinmetall is the senior partner.

Hispano-Suiza remained the parent company of BMARC, which made about half of the British Hispano cannon produced in World War II as well as making Oerlikons after the war and Rarden cannon in the 1980s. In 1971, the company was taken over by Oerlikon-Bührle, who retained ownership until 1988 when BMARC was sold to Astra Holdings to become Astra Defence Holdings. In 1992 Astra Defence Holdings went into receivership and what was BMARC was bought by British Aerospace.

The Royal Ordnance Factories (ROFs) were the oldest organisations involved in military manufacturing. Government owned, the number of plants saw a dramatic increase in the 1930s and 1940s from three to more than forty to cope with the demand for munitions. In World War II, ROF Poole manufactured 20 mm Hispano cannon, as did BSA. ROF Cardiff was involved in Bofors manufacture. Much of the wartime cannon production was distributed, with factories assembling parts made by many different firms. The end of World War II saw a rapid reduction in the number of ROF sites. In 1985, what was left (sixteen sites) was formed into a government-owned company with the name Royal Ordnance plc. In 1987 the company was sold to BAe (which became BAE Systems in 1999), and its surviving sites became part of BAE Systems Land Systems.

Production of the ADEN and Rarden guns by the original manufacturers has long since ceased, but the UK has joined with France in developing the 40 mm CTAS (Cased Telescoped Armament System – see the International section) which was intended to arm the now-cancelled Warrior IFV upgrade programme, plus the new Ajax reconnaissance vehicle (subject to various problems being overcome), as well as French AFVs.

There is one British firm that continues to manufacture cannon in the UK – AEI Systems Ltd. Starting in the 1960s as a supplier of military aircraft spares and gun parts, they have developed into a design and manufacturing facility producing complete guns including 30 mm ADEN, 20 mm M39 and 20 mm HS 804 (under the designation AE20). They are now introducing their own gun designs, namely Venom and Venom LR (30 × 113B), Inspekta (30 × 173), Alexis (20 × 102) and Aspen (30 × 113B).

Cannon designs purchased from abroad include:

- Swiss Oerlikon Type SS AA gun: 20 × 110RB ammunition
- French Hispano Suiza HS 404 aircraft gun: 20 × 110 ammunition
- Swedish Bofors L/60 AA gun: 40 × 311SR ammunition
- Swedish Bofors L/70 AA gun: 40 × 365R ammunition
- US M61A1 (aircraft gunpod and Phalan x versions): 20 × 102 ammunition
- German Mauser BK 27 (current aircraft use): 27 × 145B ammunition
- Swiss Oerlikon KAA (current naval use): 20 × 128 ammunition
- Swiss Oerlikon KCB (current naval use, being phased out): 30 × 170 ammunition
- US GAU-8/A (in Dutch Goalkeeper CIWS, now phased out): 30 × 173 ammunition
- US Orbital ATK MK44 (current naval use): 30 × 173 ammunition.

It is possible that the podded GAU-22/A using 25 × 137 ammunition may also be adopted, if this option is selected for the F-35B combat aircraft ordered for the RN and RAF. At the time of writing no decision on that had been made.

One other cannon has seen recent British service: the 35 mm Oerlikon GDF twin mounting, of which about a dozen examples were captured in the Falkland Islands following the defeat of the Argentinian invasion in 1982. These remained in service with the RAF Regiment for about a decade, and were reportedly highly regarded.

### MAXIM 1-POUNDER POM-POM: 37 × 94R AMMUNITION.

A 37 mm cannon version of the Maxim machine gun was first tested in 1885. Trial versions were distributed from 1889 resulting in the first sale being achieved in 1895, following which the gun saw extensive international service. It became popularly (and even semi-officially) known as the pom-pom because of its distinctive sound on firing.

The gun was designed around the 37 × 94R cartridge introduced in the Hotchkiss five-barrel manually-cranked *'canon revolver'* and in British service was known as the QF $1^{\text{PDR}}$ or 1-pounder (short for 'quick-firing one pounder'). The Maxim short-recoil mechanism utilized a toggle-lock or 'elbow joint': the hinged bolt was locked to the barrel for an initial recoil distance until the rearward movement of the barrel was stopped and the bolt was forced to bend at the hinge, pulling the front of the bolt to the rear and the fired case out of the chamber. The gun was belt-fed, using a fabric belt from which the rounds were pulled rearwards before being pushed into the breech. It was water-cooled, with a large-diameter water jacket surrounding the barrel, so could sustain a high volume of fire. It was massively constructed, originally using brass for the receiver and water jacket, and weighed 186 kg; it was 187 cm long with an L/30 barrel, and fired at a cyclic rate of 250–300 rpm. It saw extensive international use, both on land (typically on an

*1-pounder Pom Pom on pedestal mounting*

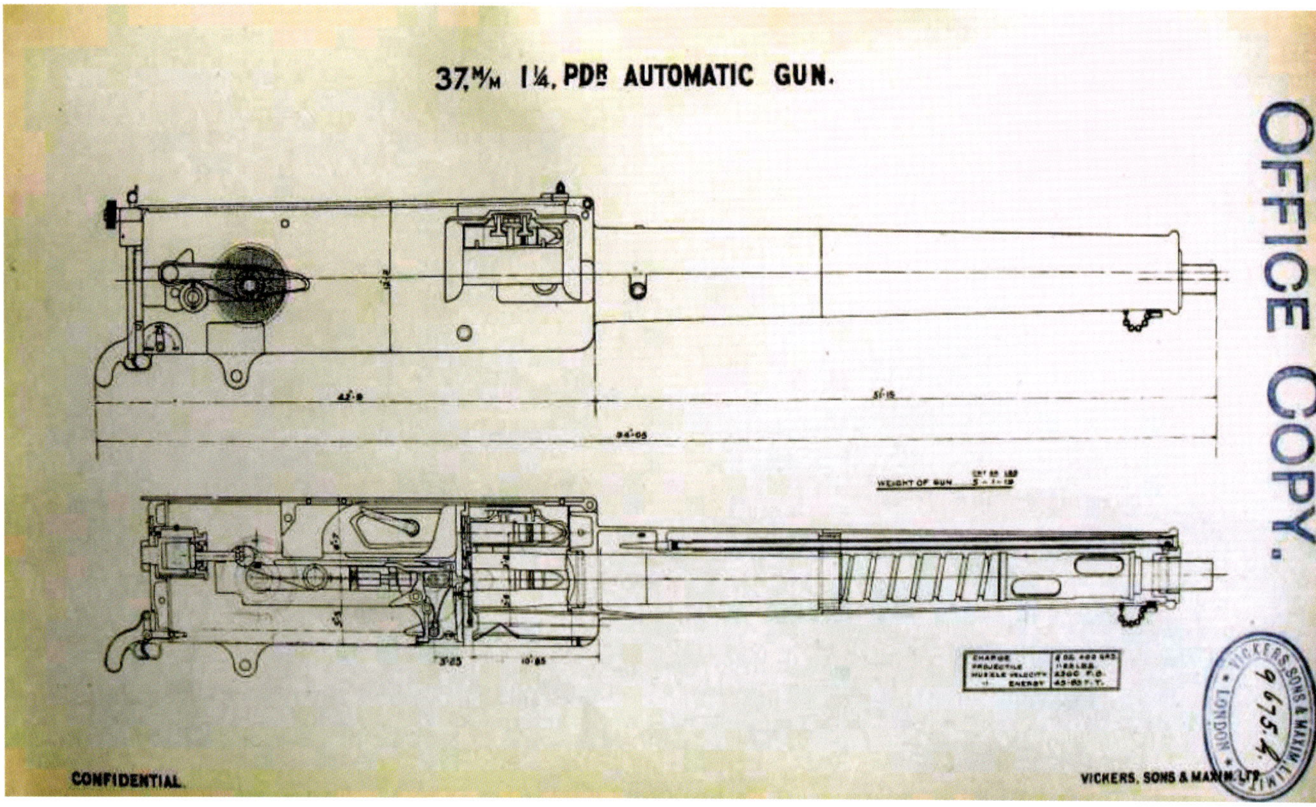


*A Vickers drawing of a rare gun: the 1¼PR (courtesy Norman Bonney)*

artillery-style large-wheeled carriage) and on pedestal-type naval mountings.

### QF 1½-POUNDER VICKERS TYPE A: 37 × 124R AMMUNITION.

The origin of this scaled-up version of the 1-pounder Maxim was the 1¼-pounder Maxim-Nordenfelt Mk III 'Heavy' of 1898. This saw little use but a modified version appeared some years later as the Vickers 1½-pounder Mark 1 (Vickers having by then absorbed Maxim-Nordenfelt), and was used by the Royal Navy in 1914 on board a few cruisers. It was not generally adopted by the RN, being replaced during the following year by an evolutionary development, the QF 2-pounder. Some 1½-pounder Mark 1 were sold to the Austro-Hungarian navy before 1914 and fitted to some classes of cruiser, under the designation '3,7 cm Vickers'. The 1½-pounder Type A weighed 250 kg and was 231 cm long with an L/42 barrel. It fired at 150 rpm.

The Type B was a semi-automatic aircraft gun, used only experimentally.

### 1-INCH VICKERS (25,4 V): 25.4 × 87R AMMUNITION.

The first reference found to the 1-inch Vickers is its consideration by an Italian joint Army and Navy commission of 1911 into acquiring an AA gun. This resulted in 1912 in the purchase of a prototype, but it seems that few, if any, others were made in the UK (in July 1918 just one gun was in service with the RAF; this survives at the NFC).

The Italians designated it the *Mitragliera da 25,4 V* (25.4 Vickers automatic gun) and continued testing it, including on board a Farman Mod. 1914 aircraft in 1914, and in comparison with the 25.4 mm Revelli F.I.A.T. in 1916 (the Revelli was preferred). Production at the Vickers-Terni plant in Italy did not begin until 1918, with 56 guns made in that year and another 8 made in 1919; those received before the end of the war were only used for airfield

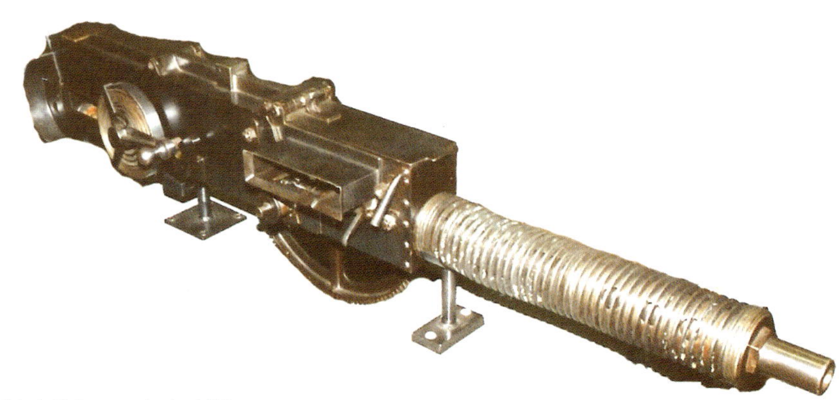

*1-inch Vickers gun in the NFC*

*Prototype 1-inch Vickers gun above, and (with different barrel sleeve) as an AA weapon in the Sino-Japanese War (courtesy Tom Schen)*

AA defence, although some experiments were made with mounting the gun on Caproni bombers. In 1924 some were sold to Sweden (embarked on *Valen* submarine, *Draken* class submarines and *Clas Fleming* mine layers); it was also sold to the Turkish Navy (3 guns) and to the Argentine Navy (24 guns). Curiously, photos exist showing one in Chinese Army AA use in the Sino-Japanese 'Great Wall War' of 1933.

The 25,4 V was belt-fed using fabric belts of 20 or 25 rounds, reportedly with a long-recoil version of the Vickers toggle-lock action (presumably to spread the recoil blow); the barrel group recoiled 19 cm between shots. The weapon was 137 cm long with an L/30 barrel and fired at 150 rpm (one Italian source says 200 rpm). The total weight of gun and mounting was 77 kg, the gun alone weighing 50 kg (another source says 85 kg). The cartridge case was clearly based on that of the 1-inch Nordenfelt or Aiming Rifle. Two different designs of barrel sleeve were used, as shown above and below: on the left is what appears to be a prototype; on the right is the standard version which appears in most photos of the gun (and is also on the NFC survivor).

The 25,4 V inspired Revelli to design an improved weapon using rimless 25 × 87 ammunition, which was made by F.I.A.T. (see Italian section).

QF 1-POUNDER MK III AND MK V: 37 × 69R AMMUNITION.

The Vickers 1-inch design was scaled up to produce the 1-pounder Mk III, firing a shorter and less powerful cartridge (37 × 69R instead of 37 × 94R) than the usual 1-pounder Maxim with which it is often confused. This was developed specifically for use in aircraft during World War I, first appearing in 1915. The belt-fed gun weighed 62 kg and was 143 cm long with an L/22 barrel. Figures for the rate of fire are hard to verify; some sources

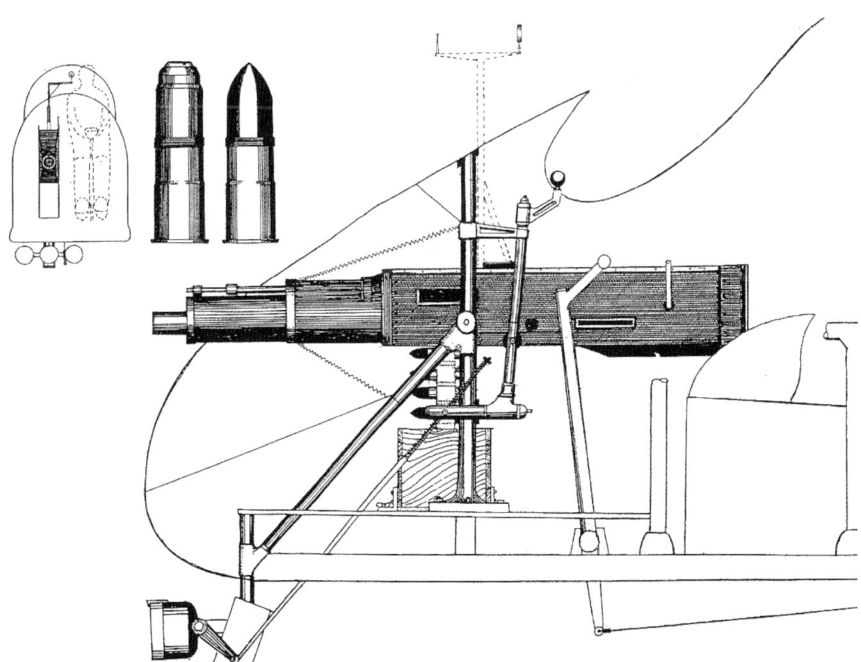

*Vickers 1-pounder Mk III / V, with ammunition (courtesy R. W. Clarke)*

*QF 2-pounder Mk II in 1940 with fabric belt feed (Library & Archives Canada / C-064027 / Wikimedia Commons / Public Domain)*

state that it fired at 300 rpm, others say 150 rpm which seems much more likely. Very few were made but some did see action. Examples of a revised version, the Mk V, were tested on the R31 airship just after the war.

**QF 2-POUNDER VICKERS:**
40 × 158R AMMUNITION.
This was based on the QF 1½-pounder Vickers Type A with the cartridge lengthened and increased in calibre, and the short-recoil toggle-bolt gun action modified accordingly. It was introduced in March 1915 to replace the Vickers 1½-pounder in British naval service and was sold to various other countries in the interwar period (notably Italy, Russia and Japan). Inheriting the pom-pom nickname, it became most famous in World War II being extensively used by the RN in octuple, quadruple and single mountings. Although it was clearly outclassed in performance by the 40 mm Bofors, it was not replaced in service until after the war.

There were two principal service models of the gun, the first being the Mk II of March 1915, the second being the Mk VIII developed for the eight-barrel mounting introduced in 1930 and subsequently becoming standard production for all mountings. Most Mk II were fed ammunition via a 25-round fabric belt, the final Mk II*C and the Mk VIII used a 14-round steel-linked belt, with ten such being able to be hooked together for a 140-round capacity per gun in octuple mountings, enough for about 1¼ to 1½ minutes' continuous firing.

A modified version of the Mk VIII was introduced in 1938 to use a new, higher-velocity loading of the ammunition, which required a different breech mechanism and some changes to the firing gear. Ammunition for the LV and HV versions was not interchangeable; the mounting Mark had a star added to it to indicate that HV ammunition should be used. Both LV and HV guns remained in service throughout the war, which meant that separate supply lines for LV and HV ammunition were required. Other variations in designation were A or C (for automatic or controlled firing – see the explanation in the synchronisation section of the chapter on Design) and a superscript $^{W}$ to indicate a circulating water system (HV guns only), resulting in complete gun designations such as Vickers 2-pounder Mark VIII AHV$^{W}$. In addition to this, the guns varied depending on whether ammunition was fed from the left or the right (both being needed in multiple gun mountings) and their position in the mounting (inner or outer in octuple mounts, upper or lower in the quads), resulting in designations such as RHO and LHI.

The weight quoted for these guns does vary quite considerably, possibly depending on what is included; the weight of the Mk II (gun only) is given as 239 kg (249 kg with cooling water) while that for the Mk VIII is given as 260 kg (LV) or 286 kg (HV). The Mk II was 243 cm long overall, the Mk VIII was 260 cm, and both had

*Quad 2 pdr (IWM D851 / Wikimedia Commons / Public Domain)*

*Eight barrel mounting on HMS Rodney, 1940 (IWM A829 / Wikimedia Commons / Public Domain)*

L/39 barrels. The rate of fire also varies: standard references quote the Mk II as firing at 200 rpm while the Mk VIII fired at 115 rpm in automatic and 90–100 rpm in controlled fire mode. Given that the 1½-pounder was stated to fire at 150 rpm, 200 rpm for the 2-pounder Mk II seems questionable.

A total of 6,691 guns were made in the UK, with another 843 made in Canada.

1½-pounder C.O.W. GUN:
37 × 190 AMMUNITION.

Developed by the Coventry Ordnance Works throughout World War I. The first 1-pounder model, firing a rimless version of the usual 37 × 94 cartridge, was produced in 1913, but effort was soon switched to a 1½-pounder using a much larger (37 × 190) rimless case. The appearance was distinctive in that the recoil spring was wrapped around the barrel in the interest of compactness. The gun utilized a long-recoil mechanism, in which the recoiling parts (weighing 47 kg) recoiled 38 cm; overall it weighed 90 kg and was 233 cm long with an L/41 barrel. Ammunition feed was by gravity, with 5-round clips being loaded, and the RoF was 100–120 rpm.

Development work took a long time and was not until August 1917, following successful trials of the Mk.III gun, that an order was placed for 90 of these weapons; the maximum which could be produced without affecting the output of other guns. The location of a 'shop' for making the C.O.W. gun was identified (somewhat improbably, a lace factory in Basford, Nottingham) and orders increased to 250 for the Air Board, 50 for the Admiralty (cancelled in May 1918) and 50 for France. By August 1918 the orders had increased to 450 and the urgency of the order was stressed but problems continued and only fourteen had been delivered. The end of the war saw the cancellation of the orders with only those under construction being completed, and in all about 76 guns were delivered (planned French production did not have time to commence).

Guns were experimentally mounted in various aircraft in the interwar years, including some on planes which were

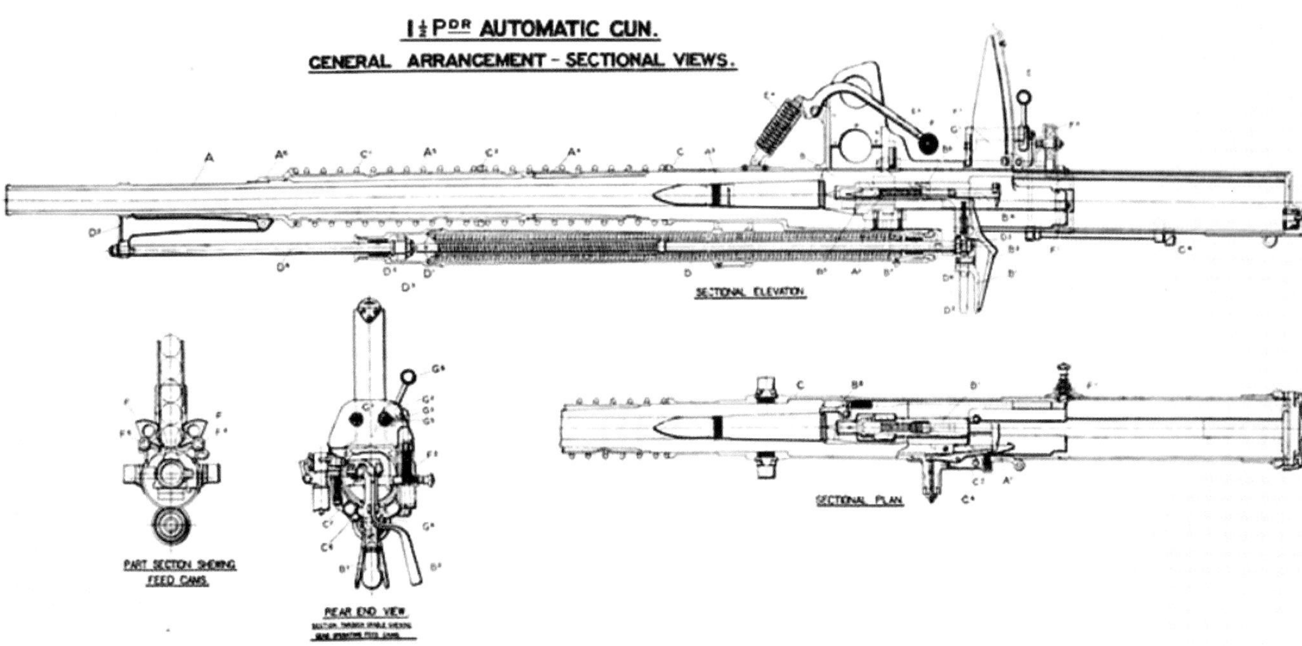


*1½-pounder C.O.W. gun*

*C.O.W. Gun in the nose of a Blackburn Perth flying boat*

specifically designed to mount it, but this never led to anything. In June 1940 with the threat of imminent invasion, methods of arming aircraft with cannon for use against tanks were considered, and the COW gun was suggested. However, that was rejected since there were by then only 45 Mk III guns available plus a limited quantity of ammunition; also, it was noted that the gun would be difficult to mount in an aircraft, presumably because of the lack of a large-capacity magazine or belt feed. The surviving COW guns were employed for airfield defence in World War II, and not finally declared obsolete until 1948.

This weapon was not particularly successful but did lead to the Vickers 40 mm 'Class S' aircraft gun of World War II.

***0.8-INCH ELSWICK***: 20 × 167SR AMMUNITION.

This gun was developed in 1928 to fire a new round of ammunition (actually 20.3 mm calibre) developed to an official specification, primarily for anti-tank purposes. The gun was water-cooled and strip-fed (5 or 10 rounds) and used an automatic mechanism said to be similar to the 37 mm Browning (and therefore presumably a long-recoil type), but no further details or illustrations are available. Trials continued until the early 1930s.

***0.8-INCH VICKERS***: 20 × 125SR AMMUNITION?

Vickers-Armstrong's '0.8-inch anti-tank gun' was mentioned in the Small Arms Committee minutes of December 1930. It was said to use a 'Colt type action with long barrel recoil', fired from a 5 or 10 round strip at 120 rpm, had an L/70 barrel, weighed 76.2 kg if water-cooled or 68.9 kg if air-cooled, and was 192 cm long overall. Some examples of a 20.3 × 125SR Vickers cartridge dating from this time survive, so this was probably the ammunition used.

***1-INCH VICKERS AUTOMATIC GUN***: 25.4 × 146R AMMUNITION?

A '1-inch Vickers Automatic Gun' was mentioned in the Small Arms Committee minutes of November 1930, reported as using 'the usual barrel recoil operated mechanism of the Vickers system with belt or link feed'. The gun was water-cooled, had an L/60 barrel, weighed 95.3 kg and fired at 200 rpm. There is a 1932-stamped VA case in existence (the headstamp also includes '25.4 mm SC', with the meaning of SC not known) which is the most likely candidate for this gun.

**25.4 mm VICKERS**: 25.4 × 189 AMMUNITION.

This gun is simply referred to as the 'Vickers 25.4 mm', presumably to distinguish it from the earlier Vickers 1-inch, although thereby causing confusion with other 25.4 mm Vickers weapons. It appears to have been developed in the early to mid-1930s specifically as a naval AA gun, and was only ever produced in a fixed twin mounting.

The gun looks very different from the traditional Vickers large-calibre automatic guns of the period, which were belt-fed, water-cooled and recoil-operated. This gun is obviously air-cooled, gas-operated, and fed by a box magazine above the breech. Since Vickers was offering the gas-operated,

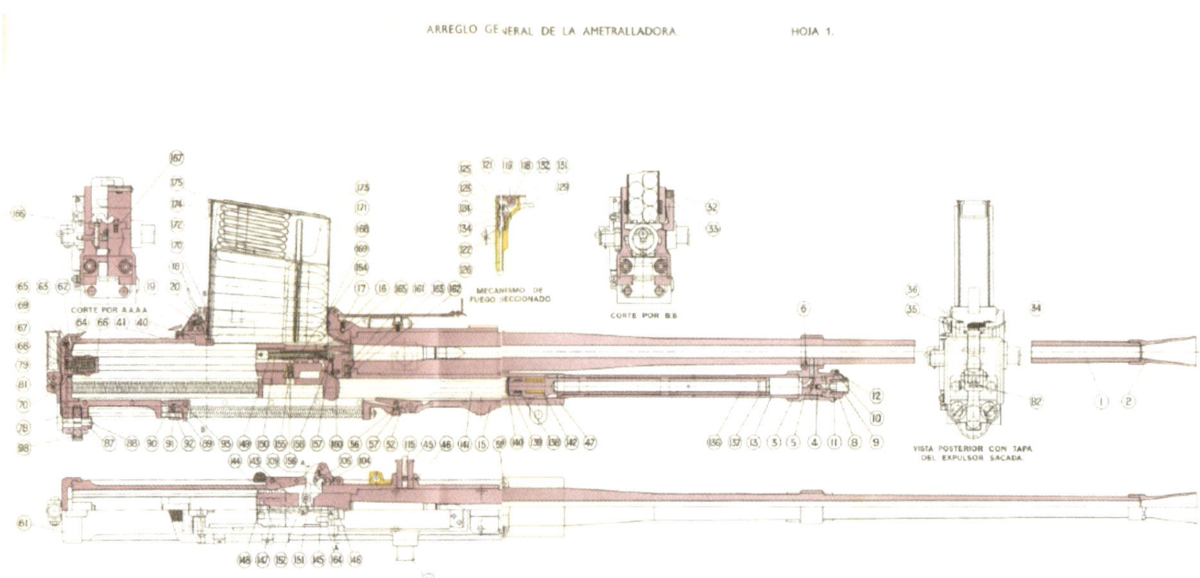


*25.4 mm Vickers, from a Spanish language handbook*

*Vickers 25.4 mm*

magazine-fed Vickers-Berthier .303 LMG at that time (it was used by the Indian Army instead of the Bren), and the Vickers K or VGO aircraft gun derived from it, it seems possible that the 25.4 mm gun used a scaled-up version of this mechanism.

The absence of belt feed and water-cooled barrels would have greatly limited the practical rate of sustained fire.

The gun weighed 127 kg, was 257 cm long with an L/70 barrel, and had a chamber pressure of 18 tons per square inch (278 MPa). The 10-round magazine weighed 5.56 kg empty and 11 kg full, and the rate of fire was 200 rpm.

The main service application for this gun was the Argentinian training cruiser *La Argentina*, built by Vickers at Barrow in 1936–1939, which had six of the twin mountings. However, according to 'Spitfire, the History' by Morgan and Shacklady, a rather unlikely proposal was made in 1938 when considering alternative armament layouts for the Supermarine Spitfire fighter: a pair of 25.4 mm Vickers, equipped with 30-round drum magazines.

The mountings were removed from *La Argentina* in 1949 and replaced by six single 40 mm Bofors mountings. A few of the Vickers guns were then mounted in the *Bahia Thetis* transport/training ship. Further use ended after a fatal accident on board this vessel due to premature detonation of a projectile during testing in the late 1950s. A mounting is on display outside a naval base in Argentina, and one gun (without mounting) still survives in the NFC.

**20 mm OERLIKON**: 20 × 110RB AMMUNITION.

This gun is described in the Switzerland section. The RN decided to acquire the 20 mm Oerlikon Type SS in 1938, and the first prototypes of the 20 mm Mark I were delivered late in 1939. These were successful and Britain placed large orders with Oerlikon as well as obtaining a manufacturing licence. After about 100 guns had been delivered supplies from Oerlikon were cut off by the fall of France and it was not until late 1941 that British production of what was designated the 20 mm Mk II (the same as the US Mk 2) got underway. Similarly, the Mk IV and US Mk 4 were the same, the differences between the various marks concerning details like the arrangement of the buffer springs.

*Twin mounting for Vickers 25.4 mm (courtesy A.A.C.A.M.: Argentinian cartridge collectors' association)*

*Right: Wrens overhauling Oerlikon, 1943 (Tomlin, H W (Lt) / IWM_A_15163 / Wikimedia Commons / Public Domain)*

About 55,000 Oerlikon guns were in British and Dominion naval service by the end of World War II. The great majority of British guns were in single, free-swinging mountings but both manual and power-driven twins saw service. In the planned quad mountings the guns would have been belt-fed, but this was not proceeded with. After World War II, 20 mm naval mountings continued to be produced in the UK by Vickers Armstrong, specifically the single Mk 7A, twin Mk 9 and Mk 12A. A few of the single mountings remained in service on auxiliary vessels into the 21st century.

**40 mm BOFORS L/56**:

40 × 311SR AMMUNITION.

The Bofors Gun is described in the Sweden section. It was the British Army's standard medium automatic AA gun in World War II, and the demand for the gun was such that it took until 1943 for production to meet needs. It also saw increasing use in the Royal Navy although it was never available in large enough numbers to replace the naval 2-pounder. Little was done to alter the gun's performance during the war (most of the developments concerning sights and mountings), but much engineering work was done to make the originally largely hand-finished action more suitable for mass production.

The Army adopted the Bofors gun after trials with some Polish-made guns, received in 1937 and designated the Mk I. After conversion from metric to imperial measurements, the Mk III entered production with 233 completed by the start of World War II. Over 19,000 were produced by British and Dominion factories in World War II. All of these guns were air-cooled and nearly all of them were fitted to single-barrel towed mountings, but some were installed on vehicles for the SPAAG role (typically, on a Morris Commercial 4WD lorry). After the war uprated systems were developed, most notably the 'Bristol Bofors', which incorporated powered high-speed elevation and traverse controlled by one man using a joystick.

The RN, having previously confirmed their choice of the 2-pounder pom-pom for this role, were slower to adopt the Bofors and started in 1940 by fitting Army weapons. In fact, the majority of Bofors guns fitted to RN ships during World War II were army-pattern air-cooled guns on single naval mountings: well over 2,000 such weapons were used.

The first purpose-designed naval mounting (using Mk IV water-cooled guns) adopted by the RN was the Dutch Hazemeyer twin mounting with an improved triaxial stabilisation system. To this the British added an on-mount radar, but the system (introduced in November 1942) proved too complex to be reliable. Even worse in this respect

*Cleaning a twin Bofors 40 mm mounting on an escort carrier (Oulds, D C (Lt) / IWM A19751 / Wikimedia Commons / public domain)*

was the STAAG (Stabilized Tachymetric Anti-Aircraft Gun) which was in effect a Hazemeyer with even more complexity and unreliability. A much simpler system for the Royal Navy was the twin Mk V mounting using Mk XI guns, which saw little use until the very end of the war. Fortunately the USA supplied nearly 800 guns in twin and quad mountings from 1943 onwards; a similar number of British built water-cooled guns was also in service by the end of the war. The Canadians improvised their own mounting in their 'Boffin', based on converting a twin 20 mm Oerlikon mounting to take one 40 mm Bofors and with hydraulic power. Post-war the RN fitted its last battleship (HMS *Vanguard*), plus some aircraft carriers, with six-barrel Mk VI mountings. The single gun mountings weighed 1,200–1,400 kg, the twins 6,500–7,000 kg and the massive six-gun Mk VI 22,000 kg. The heaviest and most sophisticated of the twins, the STAAG (Stabilized Tachymetric Anti-Aircraft Gun) weighed in at 15,000 kg and (like the Mk VI) did not enter service until after the war.

While the Army switched from the L/56 to L/70 Bofors in the early 1950s, the RN continued with the L/56 until the end of its service life, in a modernised version of the air-cooled single-barrel mounting.

*Re-arming two Hispano Mk II on a Hurricane IIC: the two cylindrical objects are the belt feed mechanisms (courtesy Philip Jarrett)*

### 20 mm HISPANO MK II AND MK V: 20 × 110 HS AMMUNITION.

The Hispano-Suiza HS 404 is described in detail in the France section. Here, only the further British developments will be considered.

Initial British production was of the Mark I, built to the original French drawings. This was soon replaced by the Mark II using drawings to British standards so that non-specialized engineering companies could make the guns. Various changes were made during early production: the rate of fire, initially 700 rpm, was slowed to 600 rpm to improve reliability; a triple-wire return spring was fitted to prolong spring life; the firing pin was redesigned; and the Mark II was given a shortened chamber to solve the problem of failures to fire due to light primer strikes.

The delayed blowback mechanism lacked positive extraction of the fired cartridge case, which meant that the cases initially had to be oiled to ensure that they didn't stick in the chamber. This was unacceptable to the RAF so the British apparently solved the problem by applying a thin coating of hard wax in the factory.

The first major change was the development of a belt feed to increase the ammunition capacity. Efforts to develop one in the UK proved unsuccessful, the model eventually adopted being based on the French Chatellerault design. This utilized a spring, kept wound up by gun recoil, to pull the belt into position to load each round in turn. This absorbed about one-third of the gun recoil, so the muzzle brake was removed.

As already observed, the Hispano was long and heavy. The Mark II was 250 cm long with an L/85 barrel. The weight is normally quoted as 50 kg; however, that was just for the gun and front mounting unit. A 60-round magazine weighed 10 kg empty and 25.4 kg full. The separate belt feed unit weighed almost the same as an empty drum (not all belt feeds weighed the same). So installed weights for the guns were in the region of 57–60 kg for the belt-fed version of the Mark II.

The next two gun Marks did not see service: the Mark III was a proposal for a simplified version (comparable with the Polsten) and the Mark IV was the same as the Mark II except for a shorter barrel – it was intended for turret installations. The Mark V was much more significant, however; a major attempt to reduce weight while improving performance. Its weight was reduced by 11.34 kg, partly by shortening the barrel to L/70 but also by eliminating the pneumatic cocking system which was not used in flight anyway. The result was that installed weight was around 46–49 kg, while other changes increased the rate of fire to around 750 rpm.

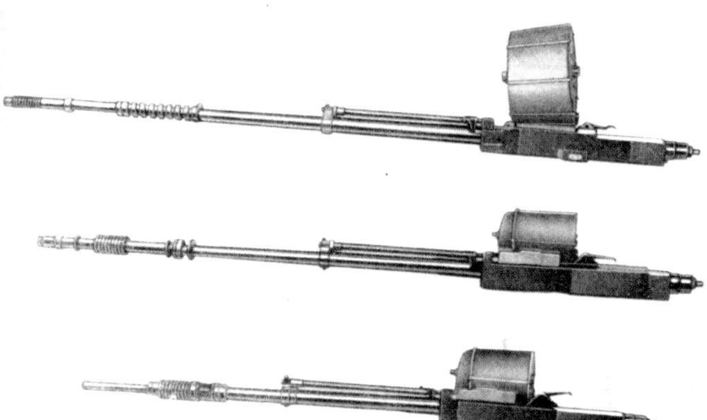

*From the top: Hispano Mk I (magazine and muzzle brake); Mk II (belt feed, no brake); Mk V (shorter barrel) (Chinn)*

In making comparisons with other guns, it is worth remembering that most belt-fed guns had an integral belt feed rather than a separate bolt-on unit. The 42 kg usually quoted for the Mauser MG 151 not only includes the belt feed but also the electrical charging device and electric trigger, without which the gun weighed 36–37 kg. So the difference in normally quoted weights of just 8 kg between the Hispano Mk II and MG 151 was actually about double that, if comparing like with like; even the Mark V was 4–7 kg heavier than the Mauser.

The reliability of the Hispano depended on the conditions. It normally averaged one stoppage every 1,500 rounds, but this rate doubled in the dusty conditions experienced on the temporary airstrips in the Normandy beachheads. American Hispanos were significantly less reliable than British ones due to manufacturing problems, which made them unpopular with US pilots (US-made guns supplied to the UK were never used in aircraft, being relegated to airfield defence). The Hispano was also more affected by extreme cold than the Browning, causing reliability problems at high altitude: one reason given to explain why few Spitfires carried four Hispanos was that the gun heating system, which utilized hot air ducted from the engine, couldn't cope with more than two.

The Hispano Mark V was arguably one of the best aircraft guns of the war for its combination of gun and ammunition performance, rivalled only by the Mauser MG151/20. However, the Mark V came too late to have much of an impact, wartime installations apparently being limited to the RAF Tempest and some late-model FAA Seafires. Molins demonstrated that the Hispano was capable of 1,000 rpm, but the changes required would have interfered with production so were not adopted.

### 40 mm VICKERS CLASS S:
40 × 158R AMMUNITION.

In 1936 the RAF began considering future requirements for guns. Aircraft were becoming faster and stronger, and it was calculated that an HE shell weighing two pounds (900 g) would be required to shoot one down. A draft requirement was written around this, and Vickers Armstrong started on the design early in 1938. A year later, this resulted in a contract to manufacture one 40 mm aircraft gun for trials.

Vickers took the 1½-pounder C.O.W. gun of World War I era as the starting point of the design, the main difference being a change in calibre from 37 × 190 to the 40 × 158R round of the naval 2-pounder Pom-pom. The previously exposed recoil spring was enclosed within a barrel sleeve, and the 5-round clip loading was replaced by a 15-round spring-powered drum (a 16th round could be carried in the chamber, but in practice only 12 rounds were normally loaded, in the interests of reliability). The long-recoil mechanism was retained, with a maximum recoil length of 43 cm; the entire firing and reloading cycle took 0.56 seconds, equivalent to a cyclic rate of 107 rpm. In practice, however, only single shots were fired to allow time to recover the aim after each shot, as the recoil pushed the nose of the aircraft downwards. The gun weighed 175 kg and was 297 cm long with an L/43 barrel.

One technical issue arose concerning sensitivity to ammunition case material; the brass from the two different manufacturers varied in hardness, and while this didn't bother the naval pom-pom, Vickers cases worked much better than Kynoch in the Class S; so the cases were oiled, and this solved the problem. Later, a 30-round drum and also a belt feed (driven by gun recoil) were developed, but did not see service. A Littlejohn squeezebore attachment was also tested, but reliability problems meant that this was not adopted.

*Vickers S Gun with drum magazine, and installed on a Hurricane IID (IWM /cm4957/ Wikimedia Commons / public domain)*

Following successful ground firing tests in March 1940, the initial air-firing tests involved a single gun in a turret mounted on a modified Vickers Wellington bomber in order to assess its value as a bomber defensive weapon. This was not an unqualified success; the muzzle blast stripped the fabric from the fuselage around the turret. In any case, before then the RAF had concluded that 20 mm guns offered more value in air fighting, weight for weight. However, by the beginning of 1940 a new requirement had been identified – attacking tanks from the air.

The 40 mm gun was selected as the best choice for anti-tank use, so ammunition with armour-piercing projectiles was developed. This required some modifications to the oil-filled buffer system as the projectiles were much heavier, with different recoil characteristics. The Hawker Hurricane IID (and later, the Hurricane IV), with a pair of Class S guns underwing, were found to be effective against most tanks of the period, being far more accurate than rocket projectiles. However, they could not reliably penetrate the Tiger and Panther tanks and were withdrawn from European service just before D-day, although they remained useful in the Far East until the end of World War II.

*40 mm ROLLS-ROYCE TYPE BH*:
40 × 158R AMMUNITION.

Rolls-Royce became interested in aircraft gun design and development in 1938, just before the requirement for a 2-pounder aircraft gun was issued. R-R decided to get involved in competition with Vickers, but accepted the Vickers decision to use their 40 × 158R ammunition. R-R had some very talented engineers, particularly Dr Viale who had been recruited from Isotta Fraschini, but none of them had any experience of gun design.

R-R's first attempt at a 40 mm gun was the Type BA, (originally known as the Mk.I), and it included some unorthodox features, such as a cast aluminium body and a hydro-pneumatic recoil system. The breech-block was closed by air from the recuperator cylinder. The locking system was very similar to Degtyarev's, with symmetrical swinging struts pivoting in the body of the breech block, but incorporated some improvements. Like the Class S, the barrel recoiled 43 cm. After two non-automatic guns were complete in late 1939 and early 1940, the first automatic model was made, designated Type BB.

The Type BB (Mk. I*) had a 12-round magazine mounted to the side of the gun (right or left). This consisted of three chargers (trays) of four rounds, so that fresh chargers could be inserted without interrupting firing. This did, however, require a loader as well as a gunner. The gun eventually demonstrated that it could fire a complete load of 13 rounds in one burst, at a cyclic rate of 114 rpm. The complex magazine was eventually swapped for a simpler 8-round hopper above the gun.

The Type BC (Mk. 2) was reduced in size to make it easier to fit into an aircraft turret: it had a recoil length of only 30.5 cm, a smaller cradle and an improved buffer and recoil system. The requirement for a 40 mm turret gun was dropped before this model proceeded beyond the drawing board.

The Type BD (Mk III and Naval Mk. XIV) was the result of a proposal from R-R for a version to arm small naval craft, the Admiralty urgently needing armament substantially more powerful than their machine guns to deal with E-boats and potential invasion craft. As the automatic loading mechanism was not ready for service, the Type BD was devised; a single-shot version of the gun on a simple Oerlikon pedestal mounting. It was modified to fire from an open breech and had a loading tray to enable the loader to maintain a rapid rate of fire – about 30 rpm. It was 320 cm long with an L/50 barrel and weighed 152 kg; the mounting weighed a further 200 kg, or 320 kg with protective shields. The Royal Navy took delivery of 600 of them, pending the availability of an automatic version. The first was delivered in September 1940, and operational experience revealed various problems, especially corrosion and also reliability, which were tackled by a mixture of anti-corrosion measures and better training concerning maintenance. Although it generally performed well and was appreciated by many crews, the gun was still prone to failures of various kinds and the Admiralty lost interest in the automatic version.

The Type BE was a version of the naval BD with the aluminium alloy replaced by bronze, adding 43 kg to the gun weight; only one gun was built. The automatic Type BF aircraft gun had a simple 6-round hopper feed (later increased to 8 rounds); it was intended for the Beaufighter where there was room in the fuselage for a loader to keep it fed. However, firing tests in May 1941 went badly and it was clear that the Vickers Class S was much superior, especially with the belt feed being tested at that time. The nature of the Type BG gun is now lost; it was clearly only an outline proposal.

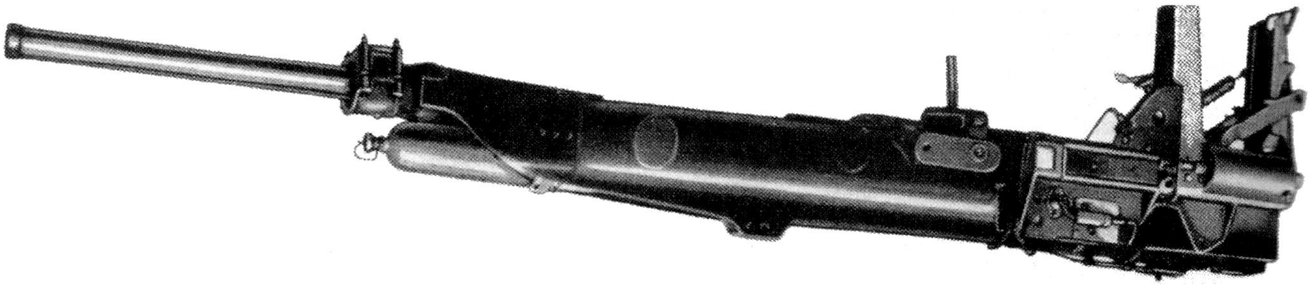

*Rolls-Royce Type BH 40 mm (Chinn)*

This brings us to the definitive automatic version of the R-R 40 mm, the Type BH aircraft gun, which was developed in several Marks: Mk. I intended for the Hurricane, with an 8-round magazine (weight 149 kg); Mk. II with an 11-round magazine and various detailed improvements; and the Mk. IV which was belt-fed, with a 30-round belt. No reference has been found to the MK. III. About 150 of the Mk. II were made before further production was cancelled; the guns all went into store and were never used. By this time, the BH had passed its tests and was a satisfactory weapon, but the Vickers Class S was better, and already in service in significant numbers; there was simply no point in introducing a new weapon to do the same job.

**MOLINS $6^{PR}$ CLASS M:**
57 × 441R AMMUNITION.

In 1942 the British Army fielded the $6^{PR}$ 7 cwt anti-tank gun, an impressive performer which also equipped tanks and was adopted by the US Army as the 57 mm Gun M1. In order to meet a perceived need for a rapid-reaction vehicle-mounted anti-tank gun, the notion was developed of fitting the $6^{PR}$ with an autoloading mechanism and mounting it in a fast, well-armoured, wheeled 4x4 tank destroyer; the Morris Firefly. The Molins Machinery Company (previously mainly noted for making cigarette manufacturing machinery) duly designed and produced a suitable mechanism. By the time it emerged, however, it had been realised that the $6^{PR}$ was not powerful enough to defeat the new Tiger tank so the British Army dropped its requirement.

Despite this setback, the Molins gun was adopted by the RN and was fitted to many MTBs in the power-operated Mark VII mounting. In this form it used the shorter L/43 barrel of the Mk 2 AT gun, had a feed capacity of six rounds in the autoloader – plus one in the breech and another twelve in a ready-use rack on the mounting – and fired semi-automatically only (the gunner had to press the trigger for each shot) at a rate of about 40 rpm. It was probably the most powerful gun fitted to naval craft of this class and was regarded as very successful. Complete in its mounting, it weighed 1,747 kg, and the gun was known to the RN as the $6^{PR}$ QF Mk IIA. Nearly 600 of these were made.

The RAF then became interested in fitting the Molins Gun in the de Havilland Mosquito, to form an airborne anti-tank weapon to replace the Hurricane IID whose Vickers 40 mm Class S guns were ineffective against the latest tanks. The aircraft was duly developed as the Mosquito FB Mk XVIII, popularly known as the 'Tsetse', but by this time the RAF had lost interest in the anti-tank gun role so the aircraft were brought into service by Coastal Command for anti-ship (and specifically anti-U-boat) purposes. The Tsetse, of which about thirty were built, served with No.248 Squadron during 1944 and is credited with sinking a U-boat.

The Molins Aircraft Gun, which was technically known to the RAF as the 'QF $6^{PR}$ Class M Mark I with Auto Loader Mk III' was based on the long-barrelled (L/50) version of the gun. This weighed 487 kg (635 kg with autoloader) and was fully automatic, with a rate of fire of about 55 rounds per minute. The ammunition supply in the autoloader consisted of 21 rounds, held in five racks of unequal length, plus two additional rounds in the feedway. The rounds in each rack were fed by a combination of gravity and a spring-loaded arm and each rack was moved into place in turn by an electric motor. The gun normally used the plain AP shot (that is the only one shown in photographs), so had a high muzzle velocity of 890 m/sec. Against U-boat hulls, it was calculated that it would be able to penetrate the hull when striking at an angle of 45 degrees or more, at a range of about 1,400m, even through 60 cm of water. The gun/aircraft combination was extremely accurate, achieving a hit rate in training of 33% against tank-sized targets – compared with 5% for rocket projectiles. The Tsetse was eventually withdrawn from service when the RAF decided to use rocket projectiles for such roles because, despite their relative lack of accuracy, these were more suited to a variety of purposes and could easily be fitted, or removed, as required.

The Molins Gun in the Mosquito FB Mk XVIII was tested in the USA in 1945, in comparison with the nearest US equivalent, the manually loaded 75 mm

*Two views of the naval $6^{PR}$ Molins gun*

*A Molins $6^{PR}$ aircraft gun and magazine shown next to the Mosquito F.B. Mk XVIII*

AN-Mk 5 in the PBJ-1H. This comparison was more valid than the difference in calibre might suggest, for the 75 × 350R ammunition used in the American gun (the same as was used by the M4 tank gun in the Sherman) was about the same overall size as the 57 × 441R, and the $6^{PR}$ and 75 mm tank guns were effectively interchangeable in the later British tanks. The Molins Gun impressed the Americans with its performance and reliability and was considered superior to the 75 mm as it could achieve a much higher rate of fire. It was noted that fairly violent evasive action and 2.5 positive Gs did not cause stoppages – which could not be said for manual loading! The Americans recommended that the Molins autoloader could be considered as suitable not just for conventional guns but also for recoilless weapons and spin-stabilized rockets.

**20 mm POLSTEN**: 20 × 110RB AMMUNITION.

At the time of the German invasion of Poland in 1939, Polish gun designers were working to simplify the design of the Oerlikon Type SS so that it could be made more quickly and cheaply. These engineers escaped to Britain with details of their work; they were formed into a team including British and Czech engineers and given the task of completing the project.

The resulting gun looked almost identical to the Oerlikon but contained many changes. Most significantly, the number of parts was halved, the total number of machining operations was halved, and the cost to manufacture was reduced to about one-fifth of the Oerlikon. While the gun used the Oerlikon's 60-round drum magazine when installed in an Inglis quad mounting, a top-mounted 30-round quad-stack magazine was normally used in other applications; this had two double columns, each being operated by a separate platform and spring. Like most such complex magazines, this proved troublesome.

The gun weighed 57 kg and a filled 30-round magazine weighed 13.7 kg (compared with 29 kg for the 60-round drum). RoF was 450 rpm.

The gun was named the Polsten: the 'Pol' after the Polish designers, the 'sten' probably a reference to the cheap and simple Sten sub-machine gun. It entered production in 1943 and became the standard British Army light AA cannon from its entry into service in 1944 with British manufacture running at around 1,500 per month. The Polsten was also made in Canada and Australia. The Royal Navy retained the Oerlikon which it already used in huge numbers.

Production data for the Polsten in the UK provides a fascinating insight into wartime practice: scores of different firms were involved, some of them making only one small part, with just one final assembly plant putting all of the parts together.

***VICKERS CLASS P***: 47 × 326R AMMUNITION.

In the summer of 1942 the Panzer VI Tiger tank emerged on the scene and it became evident that the 40 mm Class S gun, fitted to the Hurricane IID then in service in North Africa, was not powerful enough to penetrate the thick armour. The Class S was only capable of penetrating around 50 mm, but it was considered necessary to have a weapon able to penetrate 75 mm of armour plate. After calculating what would be required to achieve this (allowing for an additional 100 m/s velocity for the aircraft speed), the Air Staff issued a request in May 1942

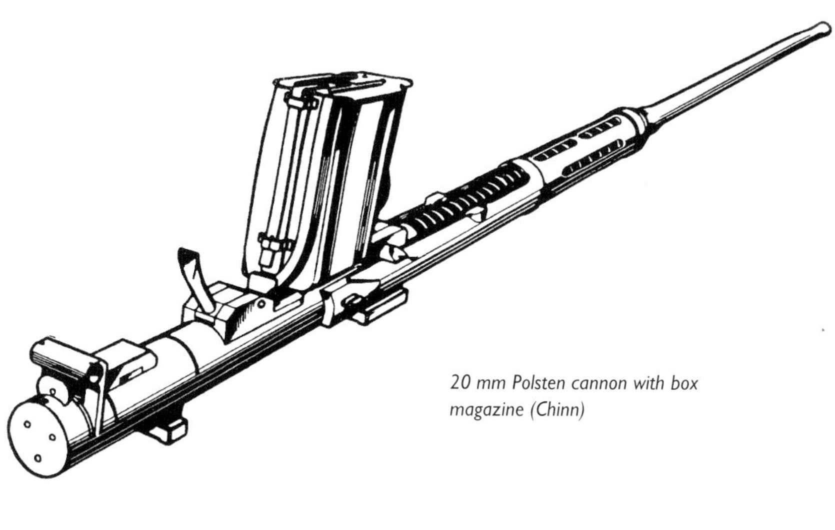

*20 mm Polsten cannon with box magazine (Chinn)*

for a 47 mm airborne anti-tank gun which would fire a 2.07 kg AP shot at 808 m/s. It was required to fire at not less than 60 rounds per minute.

Vickers were contracted to design and manufacture the ammunition. They produced a conventional rimmed straight case with a plain steel shot with no HE content. Three firms were asked to develop guns; Vickers, Rolls-Royce and the War Office Design Department. The project was designated the Class P gun and the competing designs were known as the PV (Vickers), PR (Rolls-Royce) and PC (for the Chief Superintendent of Armament Design in the War Office).

Complete information about the three projects does not appear to have survived. The PR gun was designed with two different barrel lengths, an L/70 (with which the muzzle velocity was calculated to be 815 m/s) which was about 4.3 m long overall, and an L/52 version. The gun was belt-fed and used a long-recoil design; the weapon recoiled 66 cm on firing, generating a peak recoil of just over 2,000 kg. The PC gun weighed 250–270 kg and recoiled only 30 cm, thereby generating a stronger peak recoil of 3,560 kg. This was magazine fed with two alternative designs; a 30-round flat box lying across the gun's body, and a 34–36 round drum wrapped around the rear of the gun. Overall length was 4.5–4.7 m respectively. No information is recorded about the PV gun, although one source mentions a magazine capacity of 36 rpg.

The PV and PC guns were made but were not ready for firing until close to the end of the war. The PR gun was cancelled before completion during the war when Rolls-Royce were directed to cease work on guns. None of the Class P guns proceeded beyond the prototype stage. However a Tempest V was fitted with two of the Vickers guns (which in the one known photo of this installation are usually misidentified as being of 40 mm calibre) in underwing pods just after the war, but no records so far have been found concerning this installation.

***VICKERS 3½-PR NAVAL AA***:
47 × 428R AMMUNITION.
Apart from one factory ammunition drawing, the only information published about this comes from Campbell's Naval Weapons of World War II. The ammunition was developed for a 3½PR naval AA gun, two prototypes of which were made by Vickers Armstrong. Both were water-cooled, the barrel recoiling through the water jacket in one, and with it in the other. The automatic mechanism does not seem to have been fully developed. The barrel was L/75, the chamber volume 820 cc. The project was cancelled in 1946.

***6PR 6 CWT***: 57 × 514R AMMUNITION.
As early as 1940 the UK identified a need for an AA gun to cover the performance gap between the 40 mm Bofors and 3.7-inch artillery. The same process was already taking place in Germany, resulting first in the unsatisfactory 50 mm Flak 41 and much later in the 55 mm *Gerät 58*. After considering a 47 mm gun and a planned Bofors 57 mm, the War Office decided that a British gun would be designed to do the job and a specification was issued in January 1941. The calibre of 57 mm, already used in various 6PR guns in British service, was selected and a new cartridge developed, larger and more powerful than any of the others.

The first prototype twin mounting using semi-automatic guns was manufactured in 1942 but progress was very slow and erratic, affected by requirement changes (including a demand for a single automatic gun mounting) and especially by a call for a much higher performance to cope with the new jet aircaft being planned. To complicate matters further, it was decided early in 1944 that the new gun should also be used in the coast defence role, to replace the manually-loaded 'Twin Six' (6PR 10 cwt); to add further confusion, the Twin Six was also being put into a modified high-angle mounting for the AA role. However the autoloading mechanism of the 6PR 6 cwt was complex, slow and unreliable. Continuing attempts to design a gun of satisfactory performance failed, and the project was officially cancelled some time after the war, the guns being formally declared obsolete in 1948.

The 6PR 6 cwt gun and breech mechanism weighed 318 kg (without the automatic loader), had an L/56 barrel and a vertical sliding breechblock. In this semi-automatic form, it managed 35 rpm per barrel.

***34 mm VICKERS CLASS T***:
34 × 291 AMMUNITION.
The 34 mm Class T was for a late-1940s short-range AA gun system project. The ammunition was designed by C.E.A.D. (Chief Engineer Armament Development) and Vickers-Armstrong were asked to prepare a design study for a gun to fire it. A January 1949 report from the Machine Gun Department, Crayford, summarizes the Vickers-Armstrong Automatic Gun Class T feasibility study.

VA looked at three different gun designs capable of handling this cartridge:

> Design A: a recoil-operated revolver with six chambers.
> Design B: semi-blowback, with the breech unlocked by a gas piston.
> Design C: recoil-operated with a sliding breech block, unlocked by a gas piston.

Design A was very unusual because nearly all service revolver cannon have been gas-operated, not recoil-operated (the Russian NN-30 being an exception). Design B would presumably have used a similar mechanism to the existing 20 mm Hispano aircraft gun. Design C is another strange one – a blend of gas and recoil operation. Design A was selected as the one with most promise and VA were asked to focus on this in completing the design study. However, no other documents have so far been found concerning this project.

One of the intended applications may have been for a 1950s naval close-range AA project which went under the general heading of D.A.C.R. (Direct Acting, Close Range). Like its modern equivalent the CIWS (Close-In Weapon System), this seems to have been a generic term rather than any specific system, but a calibre of 34 mm (among others) is sometimes mentioned in conjunction with D.A.C.R..

***42 mm RED QUEEN***: 42 × 270 AMMUNITION.
In the late 1940s the British Army decided that, in the light of the development of high-performance jet aircraft, it needed

to replace its 40 mm Bofors L/56 guns with a system with a much better ballistic performance and a higher rate of fire, plus a large magazine capacity. In 1950 a contract was signed between the UK and Oerlikon in Zurich to develop this system, which was given the factory designation 421 RK but known to the British as 'Red King'. This system is described in the Switzerland section.

For some reason not yet discovered, in 1952 the UK decided to develop a domestic rival to Red King under the 'Red Queen' designation. The gun which emerged differed in various respects from the Oerlikon, most notably in having a single, water-cooled barrel rather than twin barrels, but it retained the huge, side-mounted pan magazine.

The ammunition was also changed quite significantly; the 42 mm calibre was retained but the cartridge case was made shorter and fatter. It is assumed that the ballistics remained the same as the RK 421, but no firm data survives. To date, no official documentary evidence has been found concerning any aspects of the Red Queen gun or ammunition. It appears that the weapon experienced significant development problems, and Red Queen was cancelled in favour of purchasing the much-improved Bofors 40 mm L/70.

**40 mm BOFORS L/70**:
40 × 365 AMMUNITION.

The British Army was one of the earliest adopters of the L/70 version of the Bofors (see Sweden section for details), the gun entering service in 1953. Well over 1,000 guns were made in Britain and the Bofors provided the Army's close-range air defence until it was replaced by the Rapier missile system in 1979; the Gibraltar Regiment kept theirs until 1983. It was also used by the RAF Regiment. The official Army designation was 'Gun Equipment, 40/70 A.A., L3'.

**30 mm ADEN SERIES**: 30 × 86B AND
30 × 111–113B AMMUNITION.

At the end of World War II Mauser engineers were working on a new concept in aircraft guns, the revolver cannon, which promised much higher rates of fire than conventional mechanisms. The design and development of the MG 213C is covered in the Design and Germany sections, so won't be repeated here. However, examples of the handful of guns actually made fell into the hands of the Allies, and prompted intense interest. As a result, Mauser engineers were recruited by various countries, including the UK, who recognised the value of the new design and proceeded to develop it.

The British weapon was very closely based on the MG 213/30 and the first British gun in service, the ADEN Mark 1 of 1953 (initially designated 3M), even fired cartridges which were almost identical: 30 × 86B instead of 30 × 85B, plus a wider rim, but more obviously with a brass instead of steel case. Like the Mauser, the gun is gas-operated and has a five-chamber cylinder, the barrel being in line with the chamber in the 12 o'clock position; priming is electric. Development of the gun was by the UK Royal Armament Research and Development Establishment at Fort Halstead and the Royal Small Arms Factory at Enfield Lock (the ADEN name being derived from **A**rmament **D**evelopment **En**field).

The ADEN Mk 2 followed a year after the Mk 1, with the Mk 3 in 1958 and Mk 4 in the mid-1960s. The most significant of the modifications during this period was the adoption of ammunition with a longer case: the small propellant load in the original short case resulted in a low muzzle velocity which reduced the hit probability, so the UK sought to improve the MV by shortening the projectile and lengthening the case. In 1955 the UK and France agreed on common case dimensions of 30 × 111–113B for the ADEN and the DEFA/*GIAT M* 550 series (the overall length of these rounds remained at 199 mm to fit within the gun action). This has led to the original short-cased versions being dubbed LV (Low Velocity) with the later ones HV. However, despite the common dimensions with the DEFA there is no standardisation on such matters as the voltage/power of the ignition, the gas pressure characteristics or even the design of the belt links, so the ammunition is not generally interchangeable.

The next stage of development was planned to be a more fundamental change, with a higher rate of fire and a lengthened cylinder to allow for more powerful ammunition with a greater overall length. This was dubbed the 'STRADEN'

*The 42 mm Red Queen (with thanks to Alec Beanse)*

*ADEN 4, 5 and 6 (AEI Systems)*

(stretched ADEN) and would have been adopted as the Mk 5. Both 27 mm and 30 mm calibre rounds were considered (with overall lengths of 222 mm and slightly wider than the ADEN case), but the project was cancelled in 1976 (partly as a result of the frequently-recurring belief that guns would be replaced by missiles). Instead, the Mk 5 designation was given to a more limited upgrade which applied some of the STRADEN's technology to the existing gun.

The design authority and original equipment manufacturer for the ADEN cannon is now AEI Systems Ltd, formerly Aircraft Equipment International Ltd (AEI). They are able to update the HV Mk 3 and Mk 4 guns to Mk 5 standard and also offer guns with further improvements, resulting in the Mk 6. These guns weigh 87 kg, are 159 cm long overall with an L/36 barrel, and fire at 1,200–1,400 rpm.

AEI Systems has also carried out a more thorough redesign of the Mk 5 ADEN 30 gun with numerous improvements and upgrades resulting in Aspen, for aircraft installation. The major improvement is an increase in rate of fire to 1,400 – 1,600 rpm. This weapon is under review for installation on India's indigenous light attack aircraft, the Tejas Mk2. The 'en' within the Aspen name pays tribute to the weapon's heritage (Royal Small Arms Factory, ENfield).

**RARDEN L21A1/2**: 30 × 170 AMMUNITION.

The Rarden gun resulted from an early 1960s study into the requirements for the future armament of light AFVs. The first priority was to defeat armoured personnel carriers (APCs) at a range of 1,000m, with a secondary use against low- and slow-flying aircraft. It was estimated that future enemy APCs would have armour up to 40 mm thick, which set the penetration requirement and resulted in the gun calibre of 30 mm being chosen.

The decision was made to design a new gun, the design work being done at the Royal Small Arms Factory at Enfield Lock. The name Rarden comes from the initials of the Royal Armament Research and Development Establishment (RARDE) and Enfield.

The need for single-shot accuracy, low recoil force, and minimal propellant gas escape into the turret led to the choice of the long-recoil system, with the barrel and mechanism recoiling 33 cm within the receiver: the breech is unlocked after 23 cm of recoil travel. The long period after firing before the breech opens allows most of the propellant gas to escape from the barrel, minimising gas pollution within the enclosed turret. This was further controlled by ejecting the fired cases forwards out of the turret, and enclosing the gun mechanism. The long-recoil system also results in a low rate of fire of about 90 rpm, reduced further by the ammunition feed: manual loading of 3-round clips. A sliding breech-block system was adopted to minimize the length of the gun action within the turret. The gun weighs 113 kg and is 295 cm long with an L/81 barrel.

The 30 × 170 cartridge case was based on the Oerlikon KCB's, but made from brass rather than steel. An APDS loading was developed for it. In the mid-1980s the L21A2 was introduced, with some minor improvements to aid barrel changing and ammunition feeding.

The British Army has used the gun as its standard light and medium AFV cannon, first installing it on the FV107 Scimitar tracked reconnaissance vehicle from 1971, the Fox armoured car from 1975, the Warrior IFV from 1988 and the

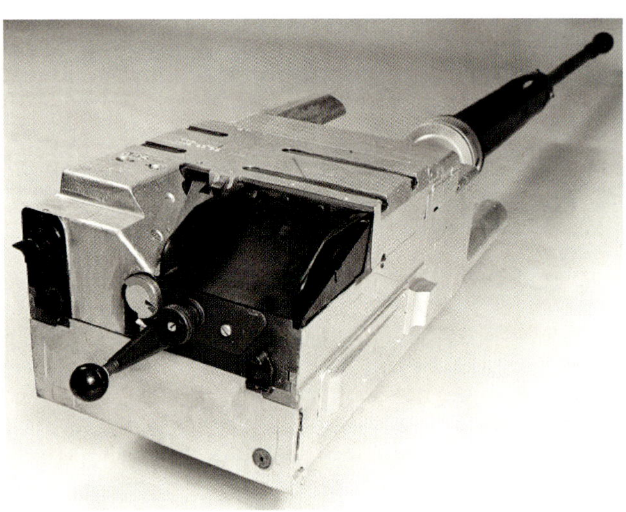

*The L21A1 Rarden cannon, and installed in a CVR(T) Sabre reconnaissance vehicle*

Sabre from 1995. It is due to be replaced by the 40 mm CTAS fitted to the new Ajax reconnaissance vehicle.

*25 mm ADEN 25*: 25 × 137 AMMUNITION.
In the early 1980s the Royal Ordnance Factory at Enfield Lock made another attempt at a fundamental update to the ADEN revolver cannon, this time to adapt the design to fire the high-velocity NATO 25 × 137 ammunition, leading to the designation ADEN 25. A rate of fire of 1,650 to 1,850 rpm was called for, plus dimensions close enough to the 30 mm ADEN to be interchangeable with the existing guns, opening up a potential replacement market.

The 25 mm round has a wider cartridge case as well as being longer overall; furthermore it uses percussion rather than electric priming, so a complete redesign was required, including the use of higher-strength maraging steel to cope with the higher chamber pressures. Both right and left-hand feeds were available, and the gun weighed 113 kg with an overall length, including blast suppressor, of 247 cm and an L/68 barrel. ADEN 25 used the same cradle casting, but the return spring retention tubes protruded a little further to the rear than those for ADEN 30.

In 1984 the new gun was selected to equip the Harrier GR. Mk 5 and Hawk 200, with the first production guns delivered in 1988. There were many teething problems which protracted development, initially connected with the percussion priming system which AEI Systems resolved. However, the major problem was the understrength ammunition links, designed by Royal Small Arms Factory Enfield (latterly Royal Ordnance Enfield). The solution was to redesign and replace the links, but the cost and logistics involved drove the chief designer to look at ammunition feed and link collection. The solutions enlarged the gun pods to unacceptable proportions as they interfered with takeoff from the ski ramp of the Royal Navy's Invincible-class aircraft carriers. This, coupled with the combined weight of 430 kg being considered too great, plus constant ammunition belt breakage stoppages, resulted in the project being cancelled in 1999 after just over 100 guns had been produced and a further 100 guns part manufactured. All except a few guns were scrapped, with some going to museums and a couple retained by AEI Systems for future redevelopment.

*30 mm AEI INSPEKTA*: 30 × 173 OR 30 × 165 AMMUNITION.
This has been developed by AEI Systems Ltd, based on the Oerlikon KCB, for naval and AFV applications. It is gas-operated, with single or dual feed driven by the 5 cm gun recoil, and is available in either the NATO 30 × 173 or Russian 30 × 165 calibres. It weighs 145 kg, is 350 cm long with an L/85 barrel (including the muzzle brake), and fires at 200–300 rpm. This project is about 50% complete, with further development awaiting funding. Inspekta is a derived name using the company initials and initials of the company's managing director and weapon's chief designer.

*20 mm AEI ALEXIS*: 20 × 102 AMMUNITION.
Related to Nexter's 20M621, this is a gas-operated, belt fed gun (with right or left side feeding available). It weighs 49 kg and is 230 cm long with an L/85 barrel (inc. muzzle brake). The rate of fire is selectable at 250–350 or 650–850 rpm. The feed system can be configured for M12 or M14A2 belt links; purpose-designed links have also been developed. Alexis is Greek for 'Defender', the intended use for this weapon, and includes AEIS (AEI Systems) within the name.

*30 mm AEI VENOM AND VENOM LR*: 30 × 113B AMMUNITION
The Venom is based on the ADEN 30 revolver cannon mechanism but with a longer barrel, a muzzle brake, a barrel-aligned recoil buffer for improved accuracy and a fire control offering single-shot or 220–230 rpm bursts. It is 204 cm long with an L/47 barrel and weighs 110 kg. The 'en' within the Venom name pays tribute to the weapon's heritage.

Further modifications have resulted in Venom LR (low recoil) version, in which the mean recoil is reduced from 22 kN to 7 kN; weight is increased to 120 kg which includes a buffered cradle shrouding the weapon, that interfaces directly to a remote weapon station's 'trunnion' arms. Rate of fire is selectable: single shot, 220–230 rpm or 1,200–1,400 rpm. It is marketed as an alternative to the ATK M230LF, with no ITAR restrictions.

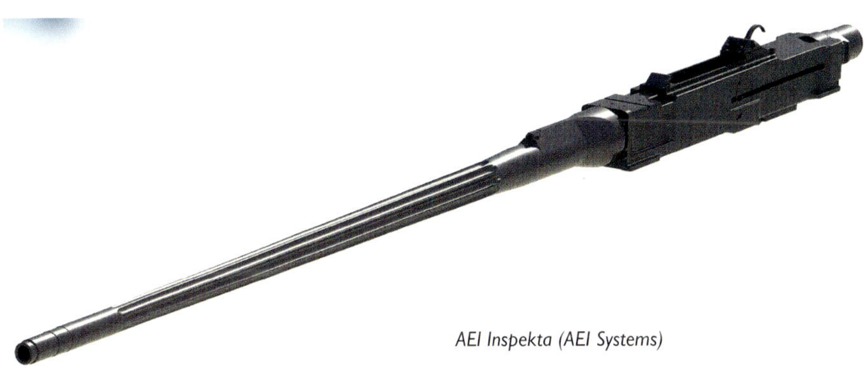
*AEI Inspekta (AEI Systems)*

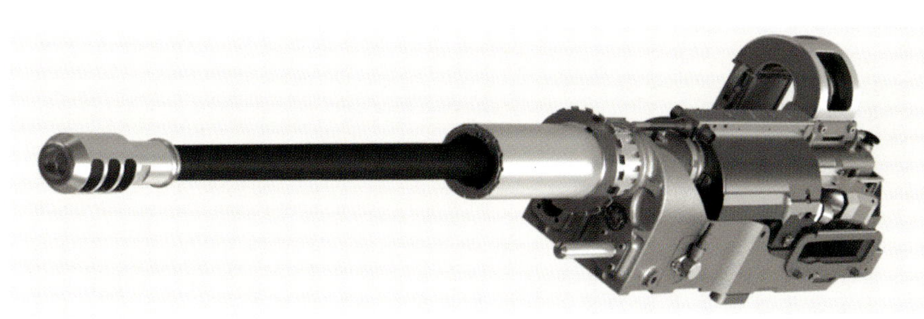
*AEI Venom (AEI Systems)*

# UNITED STATES OF AMERICA

The US Navy adopted the Maxim 1-pounder automatic cannon in 1898, buying 60 examples. In the same year, the USN began negotiations with Vickers for the rights to make the larger Maxim 1¼-PR cannon, but chambered in their 37 × 136R ammunition (already in service in their Hotchkiss quick-fire and revolving cannon) rather than the Vickers 37 × 124R round. The resulting 1 pounder 'Heavy' entered production in 1902, just 60 guns being made. These were soon regarded as obsolete in their surface-fire role but in 1916 were reissued on high-angle mountings for the AA role. Also in 1902, the McClean Ordnance & Arms Co. began the development of a rival, gas-operated gun using the same ammunition; this was rejected for US service but the 200+ guns made were sold to Russia, which in due course dispersed them to other users worldwide (accounting for the surprising number of survivors in museums). A 'USN 1 inch Maxim', known only due to surviving cartridges, was also developed.

Little further progress was made until late in World War I, when the USA began to investigate 37 mm cannon for the aircraft and AA roles. The first attempt was the hapless Baldwin, remembered now only because it prompted John Browning to design something better. A series of 37 mm automatic long-recoil guns chambered for various cartridges followed. John Browning died in 1926, but his work on cannon eventually produced practical results in the form of several weapons which saw service in World War II, as described below: the M4 and M10 aircraft guns, plus the M1 AA gun and its M9 offshoot.

There was initially a weak challenge from the M1924E / M1925E Jervey automatic cannon. Also, the enterprising Miranda brothers who owned AAC (the American Armaments Corporation, established in 1933) put much energy into promoting their somewhat dubious 37 mm cannon, several variants of which were offered in different, unique, chamberings. None of these were adopted by the USA, but some were exported. Much official effort was put into .90 inch (23 mm) cannon series with no practical result, and the 1.1-inch USN AA system proved disappointing in service.

In their search for an aircraft cannon, the USA acquired examples of various foreign types for testing in the mid-1930s, including all three sizes of Oerlikon (Type

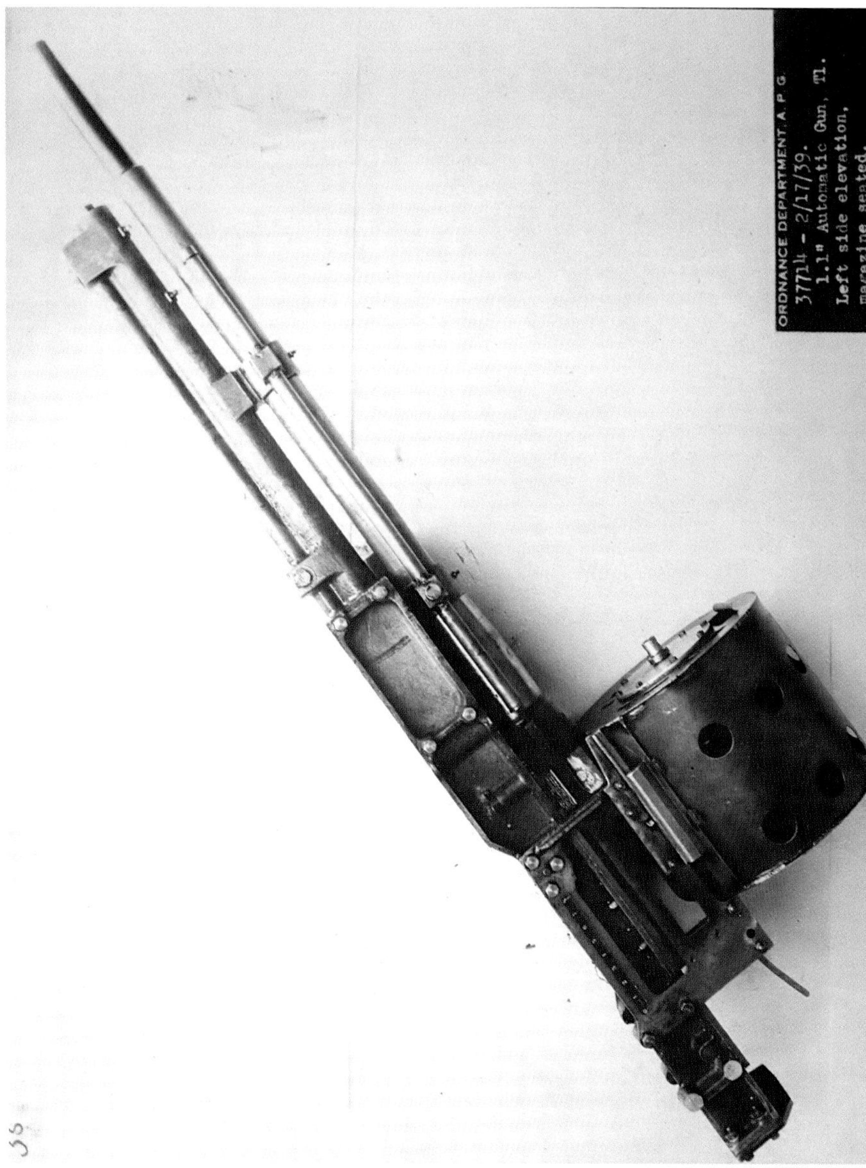


*US Army Ordnance also produced a prototype 1.1-inch gun in 1939, but only a couple of photos have been found in NARA, so it is uncertain what it was for, or even whether or not the USN 1.1-inch cartridge was used. The label on the photo says:*

**Ordnance Department A.P.G.**
**37714 – 2/17/39**
**1.1-inch Automatic Gun, T1**
**Left side elevation, magazine seated**

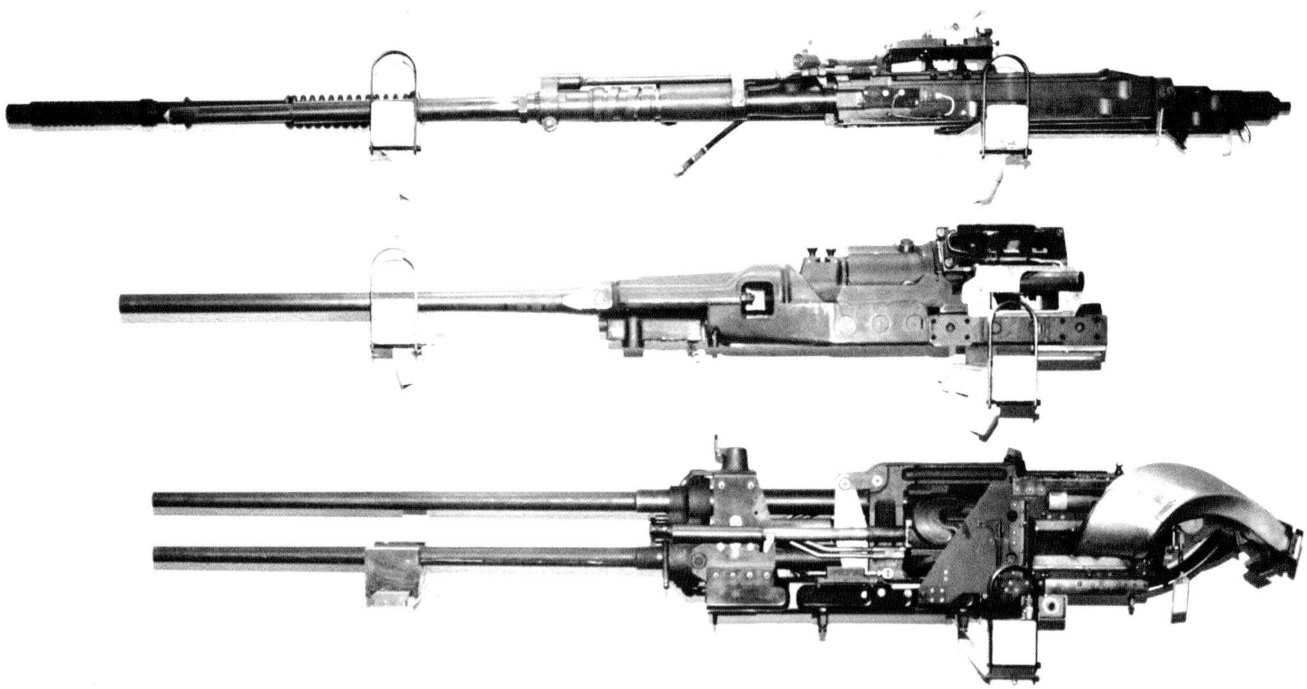

*From top to bottom: 20 mm Mk 12; 20 mm M39; 20 mm Mk 11*

9, Type L and FF); 20 mm Rheinmetall-Borsig (type uncertain) and two 23 mm Madsens. In July 1938 a Joint Bureau of Aeronautics and Bureau of Ordnance Board requested approval to acquire twenty 23 mm Madsens but this was rejected; priority was given to the .90 inch gun developments then underway. However, it was soon recognised that these were a long way from being service ready so some foreign cannon would need to be acquired. By this time, the USN and US Army had both became very interested in the Hispano-Suiza HS 404, in part because of the British were committed to adopting it. The first Hispano acquired for testing arrived in the USA early in 1938 and the first contract to deliver 33 guns was signed in December 1939, with full manufacturing rights being purchased in November 1940.

During World War II the USA did the same as the British in purchasing the best available cannon from abroad: not just the Hispano-Suiza HS 404 for aircraft, but also the 20 mm Oerlikon Type SS and 40 mm Bofors guns for AA use. The USN developed their own mountings for the Oerlikon and Bofors, including quadruple mounts, but the main focus of gun development of both Oerlikon and Bofors was in reducing the complexity and cost of manufacture to meet the need for rapid mass production. The Hispano saw rather more fundamental development during the war, but the US never managed to resolve all of the problems they experienced. Some experimental work during the war, on rechambering the HS 404 and even the MG 151 to use the .60 inch (15.2 mm) ammunition developed for an experimental anti-tank rifle, failed to produce satisfactory results.

The only cannon of foreign design to enter US service since World War II have been the 20 mm HS 820 (arming light AFVs in the 1960s as the M139); the 30 mm ADEN cannon in the USMC's AV-8A aircraft; and currently the 57 mm Bofors naval gun, designated Mk 110.

After World War II the US armament industry demonstrated considerably more enterprise in developing many different cannon of various types. These included further changes to the 20 mm Hispano, ultimately resulting in the USN Mk 12 aircraft gun; developing various revolver cannon based on the German MG 213C technology (the only service results being the 20 mm M39 and, less directly, the naval Mk 11); and reviving the old Gatling multibarrel rotary gun in a new form (resulting in 20 mm, 25 mm and 30 mm service cannon as well as experimentals in other calibres).

Another programme which resulted in a number of competitors varying in calibre and type, was the Army's Bushmaster project, which sought a cannon to replace the 20 mm M139 for arming light armoured fighting vehicles such as the new Bradley MICV. This was won by the 25 mm M242, a success which led to a family of externally-powered single-barrel 'Chain Guns'; cannon of this type have entered service in 25 and 30 mm calibres, plus 35 mm for export. An interim measure to arm the Bradley MICV until the 25 mm gun was available was designated the XM236; it appears to have been an M139 modified to have dual feed.

There were many other post-war experimental projects (especially in the 1950s) of which only a selection of the more significant or interesting are described below.

Responsibility for naval weapons development and procurement was exercized by the Bureau of Ordnance, which existed between 1862 and 1959, when it was briefly replaced by the Bureau of Naval Weapons. In 1966 the various bureaus were replaced by Systems Commands, one of which was Naval Ordnance; this was absorbed by the Naval Sea Systems Command (NAVSEA) in 1974.

The Army equivalent was the United States Army Ordnance Department until 1950, when it became the US Army Ordnance Corps. In 1962 the Ordnance Corps functions came under the Army Materiel Command which assumed responsibility for research, development, procurement, production, storage and technical intelligence. In 1985 the Ordnance Corps was re-established.

Both development and production were usually contracted out to the private sector, although the military departments sometimes became directly involved in development (notably, Springfield Armory and Watervliet Arsenal). Where necessary to meet increased demands in both World Wars, the government established a system of constructing factories but contracting out their operation. For example, towards the end of World War II the Army Ordnance Department was responsible for seven manufacturing arsenals, seven proving grounds, 45 depots, and 77 government-owned, contractor operated (GOCO) plants and works primarily dealing with ammunition.

A few of the private companies were traditional gun-makers such as Colt, which made Browning-designed 37 mm cannon before World War II and continued to be involved after that conflict, e.g. with the USN 20 mm Mk 12 aircraft cannon. Wartime expediency also resulted in the involvement of engineering firms better known for other products, such as the car makers Oldsmobile, Chrysler, Pontiac and Ford (the last of these involved well into the 1970s with the 'Tigerclaws', an improved 20 mm M39). Other specialized companies came and went without any lasting impact, e.g. Shirgun and Johnson, although Armour Research Foundation (ARF) did have success with the M39.

Keeping track of the different private companies involved in US cannon development and production since World War II is not easy, due to the various takeovers, sales and renaming exercises. However, the history of the two major companies still standing is more or less as follows.

Hughes Tool Company was involved in a variety of projects, most notably the Chain Gun family but also including some radical designs such as the Lockless system. Their gun department was acquired by McDonnell Douglas in 1984, which merged with Boeing in 1997; Boeing's ordnance business was sold to Alliant Techsystems (ATK – formerly Honeywell's defence branch) in 2002; ATK became Orbital ATK in 2015, with the range of Bushmaster Chain Guns being the responsibility of Orbital ATK Armament Systems; and Orbital ATK was acquired by Northrop Grumman Corporation in 2018 and renamed Northrop Grumman Innovation Systems. A further reorganisation saw the establishment of Northop Grumman Defense Systems in 2020.

General Electric was responsible for the development of the Gatling-type multibarrel rotary cannon commencing with Project Vulcan. In 1993 it sold GE Armament Systems to Martin Marietta. After Martin's merger with Lockheed in 1995, the cannon became the responsibility of Lockheed Martin Armament Systems, which was acquired by General Dynamics in 1997. The department responsible for the rotary guns was initially General Dynamics Armament and Technical Products; this was absorbed by General Dynamics Ordnance and Tactical Systems (GD-OTS) in 2013.

One previously significant company in this field was Philco-Ford, which developed various experimental weapons in the 1960s and 1970s. In 1975 this was renamed Aeronutronic Ford Corporation, followed a year later by Ford Aerospace & Communications Corporation, which was sold to Loral Corporation in 1990, which sold most of its business to Lockheed Martin in 1996. However, the last significant gun-related activity seems to have been the abortive DIVAD (Division Air Defense), the M247 Sgt. York SPAAG, which mounted a pair of 40 mm Bofors guns.

The Amron and Aerojet companies also became involved in developing ammunition for experimental projects, often working together. Amron focused on manufacturing cartridge cases for Aerojet, and still makes cannon cartridge cases to this day.

Gun designations have (as usual) differed over time and between the services. The US Army Ordnance Corps has always used M (model) numbers for service weapons. Before 1925 each gun was given a designation including the model and year, as in 'Model 1921', but from then on only consecutive model numbers were used for any particular type of equipment. The number alone did not provide much information. There are, for example, dozens of different items from rifles to main battle tanks designated 'M1' so more information is needed; in the case of cannon, the calibre and the purpose, as in '37 mm Automatic Gun, M4'. Modifications are designated by A numbers, as in '37 mm Antiaircraft Gun, M1A1'. While still only at the project stage a 'T' number used to be awarded, which bore no relationship to the ultimate M number: e.g. the 37 mm M4 project was originally designated T9, which sometimes causes confusion with the 37 mm M9, which was a different gun altogether. Experimental modifications or options to T number projects were given E numbers, as in T33E3. The T number system is no longer used. Since the 1950s the Army has used M numbers which are particular to the item of equipment, so further information is unnecessary: e.g. the M230 number is used only for the 30 mm Chain Gun firing 30 × 113B ammunition. Before the gun has been adopted for service it is given an XM rather thanm designation.

FAT numbers have also been associated with experimental ammunition since World War II: these are Frankford Arsenal Test numbers, and may be applied to complete cartridges although more commonly refer to tests of individual components. In 1970 (by which time the numbers had reached FAT 250) these were replaced by FA numbers, with the addition of a letter to indicate the nature of the item tested: C=cartridge, K=case, P=projectile, F=fuze, M=primer. The new series started with FA 251-C, which was a 30 mm dummy round.

The Bureau of Ordnance of the US Navy uses mark rather than model designations, combined with the calibre and often the calibre length (barrel length divided by the calibre), as in 20 mm/70 Mark 2. They may also use a further number to indicate a sub-type, as in Mark 4/1: after World War II this is expressed as 'Mark 4 Mod 1' (Mod 0 being the original). Like the Army's M numbers, the USN now applies unique mark numbers to particular items of equipment, so the Mark 44 is the marinised version of the Bushmaster II Chain Gun. Mark is commonly abbreviated, as in Mk 44; today this often appears as MK44. Experimental USN equipment is given EX numbers, as in EX-29.

Before 1947 what became the USAF was the USAAF (US Army Air Force) so gun projects started before then were given M numbers in accordance with Army practice (and have retained them thereafter). The last cannon to enter USAF service with an Army designation seems to have been the 20 mm M61. Since then, unique GAU designations have been used (Gun, Aircraft, Unit) for new designs, e.g. the GAU-8/A (the '/' meaning first model, the 'A' meaning aircraft, fixed: had there been a second model, it would have been designated GAU-8A/A, the 'A/' meaning second model). The odd one out is the 25 mm GAU-12/U, the 'U' meaning 'Special, Not Otherwise Covered or Combination of Purposes'.

The three services now give their own designations to their versions of some equipment. For example, the 30 mm Bushmaster II (maker's designation) is the Mk 44 in USN service, the XM813 (currently) to the Army and the GAU-23/A when fitted to AC-130 gunships. In contrast, in the past an item of equipment made for both the Army and the Navy was given a common AN designation, e.g. the principal version of the 20 mm Hispano HS 404 was the 'Gun, Automatic, 20 mm AN-M2 (Aircraft)'.

*USN One-pounder Maxim-Nordenfelt 'Heavy' (US Library of Congress / digital ID det.4a14810 / via Wikimedia Commons / public domain)*

## CANNON DEVELOPED UP TO 1945

### ONE POUNDER 'HEAVY':
37 × 136R AMMUNITION.

This was the Maxim-Nordenfelt heavy, automatic 1-pounder short-recoil toggle-lock gun introduced into US Naval service in 1895 and subsequently designated the **1 pdr Mark 6** (the Mark 7 was similar but with a shorter barrel). Considered obsolete by World War I it received a new lease of life in that conflict by being remounted for the anti-aircraft role, but it did not survive to see the next war. The belt-fed gun weighed 259 kg and had an L/42 barrel. The rate of fire is not stated but was probably around 150 rpm.

### MCCLEAN AUTOMATIC CANNON:
37 × 136R AMMUNITION.

This was a private venture project by Samuel Neal McClean. His first patents for automatic weapons were granted in 1902, and he developed both small arms (most famously the prototype of the Lewis LMG) and cannon. The 37 mm was the first gun he made, and this was tested by the US Army Board in 1904. The gun was gas-operated, using a gas-piston type mechanism and a rotating breech-block with an interrupted-thread lock and a compressed-air buffer. It was gravity-fed with 5 or 10-round cartridge clips loaded manually from the top. It had an L/50 barrel, weighed 245 kg, and fired at a maximum of 100 rpm. It was fitted with a muzzle brake that was reportedly effective. The chamber pressure was 170 MPa.

The tests showed up various problems with the gun: there were many springs and other small parts which were inclined to break, and as the gun heated up during intensive firing, parts expanded to differing degrees, causing the mechanism to jam. Although improvements were made and there was further testing at various times by both the US Army and USN, it was never found to be satisfactory so no orders were placed. The only customer was Russia, which had such a weapons shortage during World War I that it could not afford to be too particular, and ordered 16 of these guns in 1916 and a further 200 in 1917 (they referred to it as the 'Maklen' gun). Some were on a light field-gun carriage, the others had pedestal-type static mount for naval and coastal use. Their manufacturer was Poole Engineering Company in Baltimore (Maryland, USA). Russia disposed of the guns as soon as they could and these ended up being distributed around the world, with several still surviving in museums.

Interestingly, an illustration from a 1930s catalogue for the American Armaments Corporation (AAC – see below), showing their 37 mm Automatic Naval Gun, bears an exact resemblance to the McClean cannon.

*McClean 37 mm cannon at the Heugh Battery Museum, UK. This design was scaled up, apparently to 47 mm/3 pounder calibre, but the only evidence for this is a couple of photos. One of these is in Chinn (Vol I page 233), wrongly identified as the 37 mm gun, but it is clearly about twice the size, judging by the men with it.*

*37 mm BALDWIN*: 37 × 94R AMMUNITION.

This was based on the French Puteaux aircraft cannon. Baldwin, an American ordnance engineer, was authorised at the end of World War I to develop an automatic version of the gun. The first gun was completed by Poole Engineering Co. in 1919 and tested several times but proved unreliable, mainly because of the ammunition feed – a simple gravity feed, able to hold eight cartridges. It weighed 63.5 kg, used a long-recoil mechanism with a rotating bolt lock, and fired at a theoretical 120 rpm. The recoil cylinder was mounted above the barrel and was much the same size. The project was cancelled in 1921.

*37 mm EXPERIMENTAL BROWNING GUNS*: VARIOUS AMMUNITION.

Inspired by the failure of the Baldwin, Browning designed a series of 37 mm guns all using a long-recoil mechanism, which were developed in conjunction with Colt.

The first prototype was the M1921 aircraft cannon, which used the standard 37 × 94R ammunition in 5-round stripper clips feeding from the left side. It used a vertical sliding breechblock with the long recoil movement used for feeding the next round and ejecting the fired case, and fired at 150 rpm. This worked well but the Army lacked resources to fund development.

Browning's next effort was the M1924 aircraft gun, made in two versions: one was a low-velocity model which initially fired the usual 37 × 94R, although this was replaced by another round with the same muzzle velocity, the 37 × 120R E1, similar to that used in the heavy revolving cannon. The other was a high-velocity gun chambered for a new 37 × 123B E2 belted case. These two versions are generally referred to by their nominal muzzle velocities, of 1,350 and 2,000 fps respectively (411 and 610 m/s).

The M1925 aircraft gun came next, chambered for a much larger and more powerful cartridge, the 37 × 223R E3 (ancestor of World War II anti-tank gun round), hence the designation '3,000 fps' (914 m/s). This proved to be rather too much for the gun to handle, so in the M1927 this was reduced to 2,800 fps (853 m/s). These versions are also referred to as the Type 1 and Type 2.

The final experimental gun of this series was the M1926 3,000 fps using E3 Type 1 37 × 223R ammunition, which features a longer and heavier watercooled barrel. This was felt to be more appropriate in an anti-aircraft application, so was tested in 1926 as the M1927 AA gun.

Other AA guns developed by the indefatigable John Browning were the M1925/T1 derived from the M1924 2,000 fps, only modified to use a 37 × 145R round and given a much longer barrel and a stronger recoil system. This was quickly followed by a version using the 37 × 223R E3. Further work produced the M1927 which among other things adopted a rimless version of the case, the E3 Type 2. This led directly on to the M1 AA series.

*37 mm JERVEY M1924E / M1925E*: 37 × 94R AMMUNITION (T2).

The M1924E or T2 was designed by Captain Thomas M Jervey, with the main aim of testing various ammunition feed types. The M1925E reportedly was converted to use more powerful ammunition. Few details have survived, except for some specifications for the T2: gun weight 79 kg, with an L/43 barrel, standard 37 × 94R ammunition and an autoloader with a vertical block locking system.

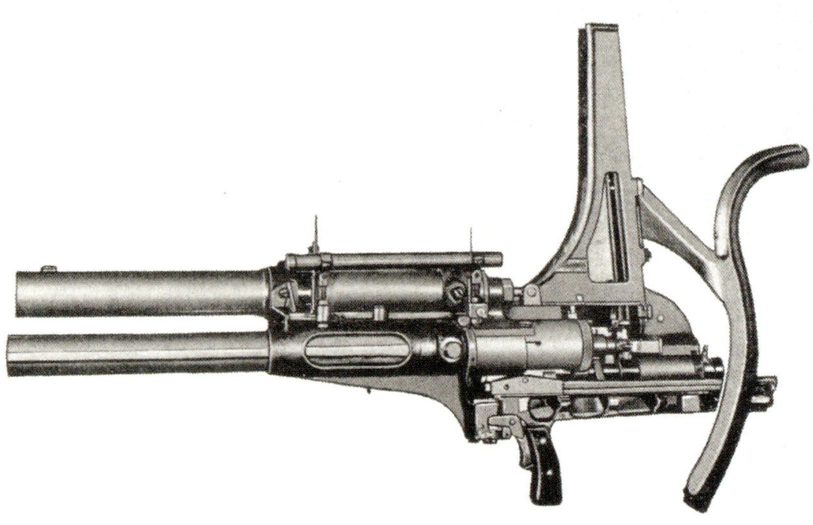

*37 mm Baldwin (Chinn)*

**37 mm AAC SERIES**: 37 × 87R, 37 × 142R, 37 × 178R AND 37 × 203R AMMUNITION.

AAC (the American Armaments Corporation) was a controversial organisation which attracted a lot of publicity (not all of it good) in the mid to late 1930s. They offered a wide range of guns and ammunition for various roles, including aircraft and AFV armament, mostly in 37 mm calibre, some automatic and some manually-loaded. Data on these guns and ammunition is sparse, and since there does not seem to be any AAC document which specifies which cartridge was used with which gun, some of the gun identifications are speculative.

The guns were chambered for four different and unique sizes of 37 mm cartridge. Two of them were certainly for automatic cannon: the 37 × 87R and 37 × 203R. Another (37 × 142R) was certainly used in a manually-loaded Rapid Fire Gun and almost certainly in the AAC Automatic Naval Gun. The fourth round, the 37 × 178R, appears to have been used only in the Anti-Tank Gun and AFV gun, which were therefore likely to have been manually-loaded guns. This may have been used in the Dutch Marmon-Herrington armoured car.

The automatic cannon in the AAC catalogue used a version of the long-recoil system with a rotating bolt head, having the spring wrapped around the barrel and normally covered by a sleeve. The performance of these guns was unremarkable even in their publicity materials (which deserve to be regarded with scepticism). None of the AAC guns was adopted by the USA or any other major army, but some of them were sold to various South American countries and others saw service in obscure corners of the world. There appears to be no reliable information about how many of which guns were sold to which countries.

The guns associated with the ammunition appear to be as follows:

37 × 87R ammunition: L/20 barrel; gun weight 60 kg (100 kg inc mount); recoil length 25 cm; fed by 5-round clips, each clip weighing 3.7 kg; 'hydraulic spring' recoil system; rate of fire variously given as 40 rpm (in 1934 when first tested), 100 rpm and 125 rpm (presumably the design evolved). Applications:

- M1 Automatic Aircraft Gun in dorsal turret; 150 kg including gun
- Automatic Mountain or Infantry Gun on wheeled mounting, 122 kg inc gun

37 × 142R ammunition:

- Rapid Fire Gun (manually loaded, L/40 barrel, on split trail or naval pedestal mounting, 292 and 170 kg respectively)
- Automatic Naval Gun (probably); with an overall length of 244 cm and an L/50 barrel, weighing 340 kg on pedestal mounting, and with a claimed rate of fire of 100–120 rpm. This looks unlike the other AAC guns but appears to be an exact copy of the McClean gun of the pre-World War I period (see above).

37 × 203R ammunition: data for Type F aircraft gun includes: weight 160 kg; L/50 barrel; overall length 244 cm; recoil length 38 cm; fed by 5-round clips; rate of fire 90 rpm (70 rpm for the AA gun). Applications:

- Automatic Anti-Aircraft Gun on a wheeled, towed mounting; 454 or 949 kg (both figures from AAC documents…)
- Type F aircraft; for fixed mountings

As well as the obvious link between the Automatic Naval Gun and the McClean cannon, according to Chinn the smallest of the AAC guns was closely based on the Puteaux cannon being developed in the USA at the end of World War I (and therefore the Baldwin cannon described above).

*American Arms Corp. 37 mm Automatic AA Gun*

**1.1-INCH MARK 1/1**: 28 × 199SR AMMUNITION.

The 1.1 inch USN AA system is unusual for a peculiar reason: a high degree of confusion over who designed it and how it worked.

One account has appeared in various forms in different publications (e.g., Ian Hogg in *Anti-Aircraft Artillery*, Crowood Press, 2002; and *Machine Guns 14th Century to Present*, Krause Publications, 2002; plus works drawing from these: *The Chicago Piano*, by Konrad F Schreier Jr, Naval History, Jul/Aug 1994, United States Naval Institute; and *USN 1.1-in AA gun*, by Richard Worth, Warship International Vol. No. 3, 2003). These state that the gun was designed by Mr. Robert F. Hudson, who was a private inventor, having patented reduced-recoil machine gun designs throughout the 1920s. These accounts and patents describe Hudson's 1.1-inch gun in great detail: It used a unique gas-operated, long-recoil design: the gas cylinder was beneath the barrel, but the gas tapped from the barrel was initially used to blow a weight forward to arrive at the front end of the cylinder and hit a buffer spring just as the projectile was leaving the muzzle, thus providing a counter-recoil effect. As an additional aid to recoil reduction, the weight returning from the front of the gas cylinder compensated for the weight of the bolt moving forwards. This design allegedly achieved a very smooth recoil cycle, at the cost of added complexity and a modest rate of fire.

The problem with the above account is that while it accurately describes the ingenious Hudson gun, it appears that wasn't the one actually chosen for production. The alternative account (e.g. in Norman Friedman's *Naval Anti-Aircraft Guns and Gunnery*) is that the Bureau of Ordnance (BuOrd) designed their own gun, with a conventional recoil-operated system, and this was tested alongside the Hudson gun in 1929. The Hudson gun was found to be unsatisfactory and the BuOrd design was chosen for further development, eventually being manufactured at the Naval Gun Factory in Washington.

This account is supported by the descriptions given in the BuOrd OP 127 Naval Gun Catalogue of 1942. In this document, the 1.1 inch/75 Mark 1 is described as a 'recoil operated' weapon and there is no mention of any gas operation. OP 127 also lists the 'Mark 2 (Experimental) 1.1 inch gun' which is described as 'gas operated' and explains how a hole was drilled 18' from the breech face to allow the gas operation. This is supported by Chinn, who says nothing about the 1.1-inch, except for the following:

> *1.1-inch Machine Guns with Mark and Model numbers*
>
> *1.1-inch Machine Gun Mechanism Mk 1 Mod 0: The recoil operated antiaircraft machine gun designed by the Bureau of Ordnance for shipboard use. Used in World War II and declared obsolete upon demobilisation.*
>
> *1.1-inch Machine Gun Mechanism Mk 1 Mod 1: Variation of Mk 1 Mod 0.*
>
> *1.1-inch Machine Gun Mechanism Mk 1 Mod 2: Another variation of the Mk 1 Mod 0.*
>
> *1.1-inch Machine Gun Mechanism Mk 2 Mod 0: Gas-operated experimental machine gun.*

The Mk 2 was probably the Hudson design, and the fact that it is labeled as 'Experimental' would support the Friedman and Chinn accounts. Unless some contradictory, authoritative source of information is obtained, the conclusion has to be that Mr Hudson was involved only with the gas-operated Mk 2 listed above, and the production recoil-operated gun was designed by BuOrd.

*Quad 1.1-inch on USS Enterprise CV-6 (USN – Wikimedia Commons – Public Domain)*

The gun was first tested in 1933 but development was very slow, eventually resulting in a complex, four-barrel powered mounting which was not put into production until 1939. Twin and single mountings were proposed, but their development was cancelled when the 20 mm Oerlikon was adopted. The complex four-barrel mounting was power-operated, providing maximum elevating and training speeds of 24° and 30°/sec respectively. In early production the guns (which were arranged side-by-side) could also be slewed up to 30° either side without needing to rotate the mounting. The reason for this curious measure was to aid aiming when the guns were pointing vertically for use against dive bombers; traversing the mounting had little effect in these conditions. However, this feature proved too complex to be worthwhile and was dropped during production.

Despite the long gestation period, the 1.1 inch gun was adopted and rushed into service in the late 1930s before the system had been debugged, leading to a reputation for unreliability. It required careful maintenance by a well-trained crew, which became difficult to sustain in wartime conditions. The size of the crew didn't help; a total of fifteen for one quad mounting. Some of these were in a separate 'clipping room', their job being to load the clips with ammunition before these were passed to the gun crews. The ammunition feed for each gun was via two 8-round vertical clips side-by-side, the gun feeding from one clip at a time, so it could keep firing while one clip was changed. The clips resembled box magazines with open sides. A loaded clip weighed about 15.4 kg.

The gun weight was 252–276 kg, the length 304 cm with an L/74 chromium-plated barrel. The mounting weight was between 4,800–6,350 kg. The guns were liquid-cooled, normally by fresh water but this was changed to an ethylene glycol water solution in cold weather.

One problem was that the ammunition used highly sensitive fuzes which sometimes detonated in or close to the muzzle. For this reason, some ships did not exercise with live ammunition, preferring to save this until the ship was under attack. However, the problem was resolved by the introduction of a less-sensitive fuze.

Nearly 1,000 1.1-inch guns were made, with 38 mountings in service by July 1940. The 1.1-inch was gradually replaced by the 40 mm Bofors, although it remained in limited service until 1945.

### 37 mm M4 AND M10:
37 × 145R AMMUNITION.

The first successful US aircraft cannon was a Browning design using a long-recoil mechanism, which was developed and manufactured by the Oldsmobile Division of General Motors as the T9 before being officially adopted as the M4 in December 1939.

The M4 fired from a closed bolt, the barrel and action recoiling together 24 cm before the barrel commenced counter-recoil. The breechblock was unlocked at about half the recoil distance, initial extraction taking place during the recoil stroke, the remainder of the extraction plus ejection, feeding and loading take place during counter-recoil. Both recoil and counter-recoil were controlled hydraulically by a piston and spring combination operating in an oil-filled recuperator cylinder.

Ammunition feed was initially only from the left. There were originally three options: a 5-round clip, a 15-round linked belt held within a box magazine, or a 30-round hoop-shaped endless-belt M6 magazine designed to fit the cowling contours of the aircraft which used it. Eventually, the M6 was the only feed used.

The gun had an L/45 barrel. The weight of the bare gun was 97 kg, the M6 magazine weighed 16 kg empty and 42 kg loaded; the 15-round box weighed 5.7 kg empty and 19 kg full. The rate of fire was 125–150 rpm.

The principal application was in the P-39 fighter aircraft, but the USN acquired some M4s and fitted them to PT boats.

The need to produce a tailor-made endless-belt feed for every installation was inconvenient, so it was decided to develop a conventional disintegrating-link belt feed using M6 links which could feed from either the left or right side. This led to a new designation, the **M10**. It had a slightly higher rate of fire (140–170 rpm) and was slightly heavier at 109 kg, but other data are the same as the M4.

*The layout of the 37 mm M4 cannon installation in the P-39; the engine was located behind the pilot, driving the propeller via a long driveshaft. The gun with its 'endless belt' magazine was in front of the pilot, shooting through the hollow propeller hub*

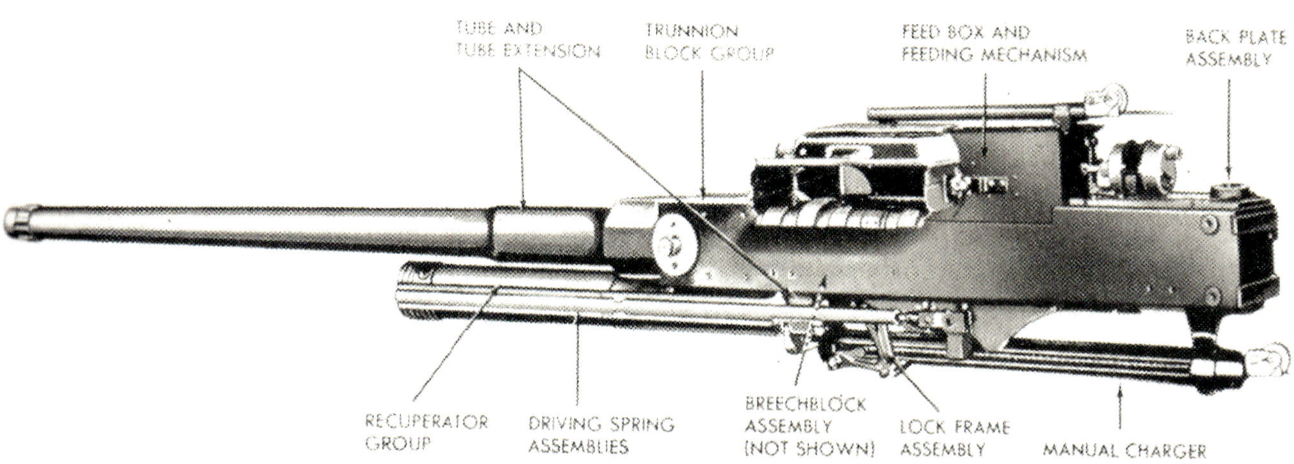


*37 mm M10 (Chinn)*

**37 mm M1 AND M9**:
37 × 223SR AMMUNITION.
The M1 AA gun was essentially a scaled-up version of the M4 described above, using the same design of operating mechanism. Ammunition feeding was again from the left, via 10-round clips in a horizontal feed tray (single rounds could also be loaded), and the cyclic rate was 120 rpm. Gun weight was 166 kg, the barrel length 198 cm.

Design of the M1 gun was essentially completed in 1927, but work on an appropriate AA mounting took longer and it was not standardized until 1938, with manufacture beginning in 1939. The M1A1 added a charging handle, while the M1A2 had modified rifling to reduce the likelihood of the fuze self-destruct mechanism functioning too soon. The mounting weight was nearly 2,800 kg. The 40 mm Bofors was acknowledged to be superior but could not be built quickly enough to meet demand, so the M1 series remained in service throughout World War II with over 7,000 being built. The Mk 3 Mod 0 was a 'wet' version of the gun for use in USN submarines, although it does not appear to have seen service.

The USAAF identified the need for a higher-velocity 37 mm aircraft gun capable of penetrating armoured vehicles, so an aircraft version of the M1A2 was developed by Oldsmobile, initially under the designation T20. The most obvious difference was the use of a disintegrating-link belt feed (which could be from either side). Initially, the gun was intended to be installed in the P-39 aircraft for which a shorter and lighter barrel was developed, but this requirement was dropped so the weapon reverted to the usual M1A2 barrel. The T20 was standardized as the **M9** in January 1943. Despite various experimental installations it appears the M9 was never utilized in aircraft, its only confirmed service being in a few PT boats. Wet versions planned for submarine use were designated 37 mm Automatic Gun Mk 1 and Mk 2 (left or right hand feed).

The M9 weighed 184 kg and had an L/54 barrel. The average rate of fire was 140 rpm. The barrel designed for the P-39 installation was about L/48 and had a slimmer profile, but the difference in weight is not known.

***.90 INCH T1/2/3/4 SERIES***: 23 × 146SR AND 23 × 139/RB/SR AMMUNITION.
By the mid-1930s it was realised that as aircraft became bigger and stronger, automatic cannon of 20+mm calibre would be needed to arm future generations of fighter aircraft. There were difficulties in obtaining samples of such guns then being made in Europe so it was decided that the Ordnance Department should design and develop a gun in collaboration with the Air Corps (USAAC, to which USAAF personnel belonged). In March 1937 the desired characteristics of the weapon were listed by Ordnance, including a calibre of .90 inch (23 mm), a muzzle velocity of at least 2,850 fps (869 m/s), and a rate of fire of 150 rpm (astonishingly low, considering that European cannon in this class were already achieving rates two to three times higher). Gun weight was not to exceed 125 lbs (57 kg) and a 50-round magazine was specified, to weigh 20 lb (9 kg) empty and

*US 37 mm M1A2 AA gun on M3A1 carriage*

82 lb (37 kg) full. There is no comment on the selection of the .90 inch calibre, but it might have been prompted by the existence of the 23 mm Madsen aircraft gun, examples of which were tested in the USA at around this time.

The first effort, the T1, was designed around a 23 × 146SR cartridge, based on a shortened and necked-down 1.1-inch AA gun cartridge case. Chinn states that the gun uses a long-recoil mechanism, has a 'flat drum' (presumably a pan) 50-round magazine, weighs 205 lbs (93 kg) with a loaded magazine (which weighs 37 kg – 9 kg empty – so the gun alone weighs 56 kg) and fired at 150 rpm. An official source refers to the gun action having been based on the Browning 37 mm AA gun, which confirms that it is a long-recoil type (Colt did suggest in October 1938 scaling-up the Browning short-recoil action of the .50 cal MG instead, but this was not accepted). It was made by the Rock Island Gun Arsenal.

The rate of fire of the T1 was considered too low, so a second attempt was made, with initial design studies being completed by January 1938. For this and subsequent designs the required muzzle velocity was reduced to 2,700 fps (823 m/s), enabling a smaller cartridge to be developed, considerably slimmer as well slightly shorter. The barrel length was increased from L/60 to L/75. The basic shape of the T2, T3 and T4 cartridge cases are the same, except that the T2 has a rebated rim while the other two are semi-rimless. There is also a T2E1 case which is identical to the T2 except for having a rimless instead of rebated-rim design. All four T-series rounds were loaded with the same projectile, intended in service use to be an HE-T type weighing 204 g.

The T2 gun design, made by the Watervliet Arsenal, was based on the 20 mm Oerlikon API blowback guns, but there is a direct conflict in the descriptions of the mechanism given by different sources. Chinn states that it is a simple blowback without the Oerlikon's advanced primer ignition feature, and thereby requires recoiling parts weighing 21 kg, leading to a total gun weight of some 109 kg without the 48-round ammunition drum. However, Cabeen (see Unpublished Documents) states that the T2 used a differential recoil mechanism

*.90 inch T4 on a ground mounting, as tested in the anti-tank role*

(a.k.a. API blowback) and he is probably correct, for two reasons: first, the rebated rim used by the T2 cartridge is required for Oerlikon type API blowbacks but not for a straight blowback; second, Cabeen specifies that the cartridge is fired while still 15 mm from the bolt's fully-forwards position, again consistent with API. Furthermore, the rate of fire of 400–450 rpm stated by Chinn would have been impossible for a simple blowback to achieve because of the massive weight of the bolt, so either his description or the rate of fire must be wrong (or possibly both). The T2E1 raises further questions as the rimless case is incompatible with an API blowback gun mechanism.

Of the T3, also made by Watervliet, Chinn only states that it is very similar to the T2 but includes a 'blowback retarding device'. Cabeen states that it fires while the bolt is still 0.125 inches (3.2 mm) from its forward position, and the use of a semi-rimmed case is not inconsistent with this. Cabeen also states that the T3 uses 'counterrecoiling hydraulic buffers' to check the forward motion of the bolt.

The T3 gun was presumably not successful as nothing more is heard of it, while Watervliet continued work on the T2 with a new lightweight version, possibly the one referred to as the T2E2, which Cabeen states was 99% complete by the end of March 1942, at which point work seems to have stopped. The T2E2 was cancelled on 20 October 1942.

Finally, the T4 gun, while using basically the same ammunition as the T3 was, according to Chinn and Cabeen, developed by Colt and based on the design of the Browning 37 mm M1A1 long-recoil cannon (thereby reverting to the T1 operating principle). It was designed to use either flat ammunition clips or a flexible belt in a drum (presumably like the 37 mm M4 aircraft gun). However, by the time it was tested in March 1941 it was being proposed for use as an anti-tank gun. Some further work on T4 seems to have been done after this date, as like the T2E2 it was said to be 99% complete on 30 March 1942, but nothing more was heard thereafter.

The Ordnance Department was responsible for all of the .90 series guns and it appears from official Ordnance Comittee minutes that work on them proceeded in parallel rather than sequentially: design and manufacture of a prototype T1 was approved in April 1937; similar work on both the T2 and the T3 was approved in February 1938; and Colt was discussing with the Ordnance Department work on the T4 in October 1938 (with manufacture of a prototype approved in 1939).

There is clearly a lot of confusion among the various gun descriptions and a complete picture is yet to emerge. However, the entire .90 series seem to have been poor performers compared with European contemporaries, being overweight and slow-firing. They revealed the inexperience of the US Ordnance Department in designing automatic weapons at that time.

**20 mm/70 OERLIKON**:
20 × 110RB AMMUNITION.

For details of the gun, see the Switzerland section. In November 1940 the USN adopted the 20 mm Oerlikon Type SS as a light shipborne AA gun, US production of which had already been requested to meet British needs. The first US-made gun was test-fired in June 1941.

The US-made guns retain the same basic characteristics and performance, but many detailed alterations were made during production. Originally the USN used the Swiss-designed Mk 1/0 gun, followed by the domestically made Mk 2/0, 3/0, 4/0, 4/1, 4/2, and 4/4 guns. The Mk 4 is similar to the Mk 2 but made to different manufacturing tolerances; both have cooling ribs on the barrel. The Mk 2/0 and Mk 4/0 are virtually identical to the British Mk II gun. The Mk 3/0 has no cooling ribs. The Mk 4/4 is of interest in that it has a fluted chamber which eases case extraction and means that brass-cased ammunition does not have to be lubricated, but steel-cased rounds still need lubricating. Most of the differences between versions are relatively minor and did not affect performance, but of much greater significance was a concerted effort to simplify the manufacturing process. This resulted in production time falling from 428.4 man-hours per gun in 1941 to 76.2 man-hours in September 1944.

As with the British versions, nearly all of the US guns were installed in single, free-swinging mountings. There were some twin mountings, the most common being the Mk 24/5 which was also manually-operated, with the first entering service in autumn 1944. Finally, the Mk 15 power-operated quad mounting saw some use in MTBs and on larger ships as a last-ditch defence against suicide bombers.

Around 10,000 single, 2,500 twin (from September 1944) and 700 quad mountings were installed in US warships alone and thousands more were produced for allied navies (including the Royal Navy) and merchantmen, with a total of nearly 125,000 being made in the USA during World War II.

**40 mm/56 BOFORS**: 40 × 311SR AMMUNITION.

A description of this gun is in the Sweden section. The USN acquired an example of the Bofors gun in 1940 – a manual twin mounting with air-cooled guns – and immediately started work on preparing for production, even before a licence was obtained in June 1941. A great deal of work had to be done to suit mass production and to make the changes needed to meet naval requirements, including adding water-cooling to the barrels and power-operation to the twin and quad mountings. The USN also made use of army-pattern guns in Mk 3 single mountings weighing from 1,100 kg for the manual version (powered elevation and training was fitted to some) and large numbers of both twin (4,500 to 6,000 kg) and quad (10,500 to 11,000 kg) mountings, power-operated and with water-cooled guns, which were delivered from early 1942. The first twin was completed in January 1942 and the first quad in the following April. The naval Mark 1 and Mark 2 guns were identical except for being left and right-handed respectively to suit twin mountings (the quad consists of a pair of twins). Curiously, contracts for army and navy weapons were initially awarded to different manufacturers who used different measuring systems, resulting in many parts being non-interchangeable, although this was minimized later. Like the Oerlikon above, the cost of production was reduced significantly during its life – in this case by about 50%.

This was the best medium automatic AA gun available to USA in World War II (as it was in the UK) and saw extensive use in both the Army and Navy with

*Twin Oerlikon mounting on USS Pampanito (SS-383); the asymmetric magazines allowed the guns to be mounted very close together, but meant that two different, non-interchangeable, types of magazines were required (Adam Kliczek / Wikimedia Commons / CC-BY-SA-3.0)*

*Quad 40 mm Bofors (USN – Wikimedia Commons – Public Domain)*

over 42,000 guns being built in World War II by Chrysler Corporation. Nearly 24,000 of these were of the air-cooled M1 for the US Army, with the rest being water-cooled guns divided almost equally between the USN / Coast Guard and exports to allied forces. It had almost entirely replaced the Navy's 1.1-inch gun by the end of the war, although the Army's 37 mm M1A2 remained in use alongside it.

### 20 mm HISPANO SERIES (AN-M2; M3; M24; MK 16): 20 × 110 HS AMMUNITION.

A description of the design and functioning of this gun is included in the France section. The initial US M1 model (designated 20 mm T1 when tested) was made by Bendix Aviation Corp, 1,202 being ordered, including 500 for the USN, with delivery scheduled for May 1941. Before production started, it was decided that all subsequent orders would be for an M2 version, subsequently formally renamed 'Gun, Automatic, 20 mm AN-M2 (Aircraft)' to indicate use by both the Army and Navy. In fact, not all of the initial order was delivered, the change to M2 being accelerated by the USN. The AN-M2 differs from the M1 primarily in its larger and heavier receiver, intended to compensate for the lack of the rigid engine mounting in the originally-intended installation. Weight went up from 47.6 to 50.8 kg.

The detailed weight breakdown for the AN-M2 is interesting, as the gun's construction was unusually modular. It weighs 46.3 kg as a bare gun, but the muzzle brake (not used with belt feed) weighs 2.1 kg, the front mounting adapter between 3.4 and 6.4 kg depending on the model, the electric trigger and sear mechanism 0.8 kg, the hydraulic charger 1.2 kg, and the ammunition feed 10 kg (60-round drum) or 8.4 kg (belt feed drive), giving a total weight of 60–67 kg depending on the feed mechanism and the mounting adapter. The mounting or support cradle would be additional to this.

Despite successful use of the Hispano by the RAF, the US experienced considerable problems with the gun, as related by Chinn. The fundamental problem seems to have been that design authority for the gun lay with the Ordnance Department, which regarded all guns of more than .60 inch (15.2 mm) calibre as 'artillery' with more generous manufacturing tolerances than the Hispano required. The Department also appeared to be reluctant to copy the modifications made by the British (who had been urging the US to make their guns identical to enable them to be interchangeable). In particular, the problem of 'light strikes' – the firing pin not striking the primer hard enough to fire it, which the British solved by shortening the chamber by about 2 mm – was never completely addressed. After initially rejecting any chamber shortening, Ordnance finally agreed in 1943 to shorten it – but only by 1 mm, which did not entirely resolve the problem. The outcome was that the US guns performed badly in most reliability tests compared with the British guns, and although large quantities of US guns were delivered to the UK, they were never installed in RAF aircraft.

Production of the AN-M2 ceased in February 1944 after nearly 135,000 had been made, of which 55,000 were in store.

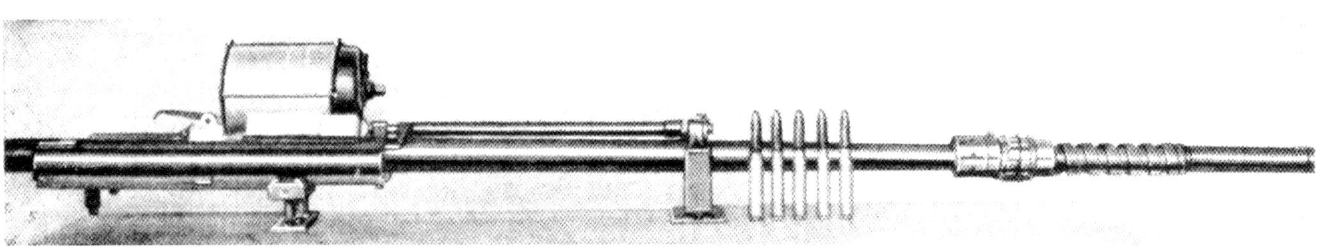

*20 mm AN-M2 with belt feed AN-M1 (Chinn)*

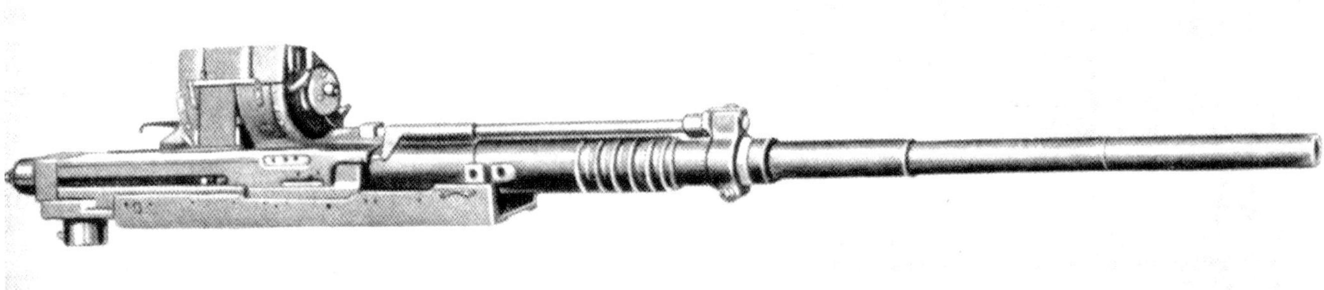

*20 mm T31 (Chinn)*

In 1943, at the request of the USN, the Army began a series of tests to determine the consequences of shortening the barrel by up to 20 inches (51 cm). This determined that it could be shortened by 38 cm without significant extra muzzle flash and with a loss in muzzle velocity of only 24 m/s. This change contributed to a target weight reduction of 11.34 kg; other desired changes were an increase in the rate of fire to 750 rpm, and firing from a closed and locked bolt (as opposed to the standard open-bolt firing); the purpose of the closed-bolt firing was to obtain the precise firing control needed, not just for synchronisation but also to make it possible to ensure that turreted guns could not shoot at their own aircraft. The designation T26 was to be given to the gun with these modifications.

Work by several manufacturers to solve the closed-bolt problem started in October 1943: Oldsmobile went to work on a mechanical closed-bolt system, involving the addition of a front sear rather than the usual rear sear (designated T27); the United Shoe Machinery Co worked on the same problem (T25); Watervliet Arsenal worked on two systems, one retaining percussion priming (T28) the other with a conversion to electric priming (T29). By December 1943 prototypes were being tested and a model incorporating features of both the T28 and T29 was demonstrated (only the bolts had to be changed to switch from percussion to electric priming); this was designated T31, and was promptly selected for further development. It also has the benefit of a recoil system and cradle built in to the gun. The result is a total weight of 52.6 kg including the feeder, cradle and recoil system, whereas the AN-M2 weighs 52.1 kg without any of these. The International Harvester Co. (IHC), followed by Oldsmobile, were given orders to start production of the T31 in January and July 1944 respectively, and achieved this by modifying existing AN-M2 in storage. However, the USN requested that its guns be delivered with a rear sear (i.e. still firing from an open bolt) to expedite deliveries. There were therefore three versions of the T31:

- T31 (standardized as **M3** in June 1945), rear-seared percussion fired
- T31E1 (M23), front seared percussion fired gun (abandoned in 1947)
- T31E2 (standardized as **M24** in November 1947), front seared electrically fired gun (adopted for aircraft turret use)

By May 1945 just over 12,000 guns had been converted by IHC and another 20,000 by Oldsmobile; at that point production stopped.

Work also took place on a more compact ammunition feed mechanism than the bulky M1. The result was designated T14, which was used in conjunction with a new belt link, the T23. This was standardized in May 1944 as the 'Mechanism, Feed, 20 mm, M2'.

Various experiments were made in modifying the Hispano action. Two of them (the T18 and the Colt T23) tried replacing the gas operation with a recoil system. Conversely, another three, the T19 from Draper Corp, the T20 from Colt, and the T32, use gas tapped from the barrel to drive a belt feed mechanism.

After its career as an aircraft weapon came to end, the US Hispano cannon found other employment in the 1960s, on a pedestal mounting as a deck gun for the USN to replace the Oerlikon. The M3 was redesignated **Mk 16 Mod 4** in this role (the similarly modified M24 was designated Mk 16 Mod 5), and was given a recoil-operated chamber lubricator which squirted oil into the chamber before each shot in order to facilitate case extraction. In the 1980s these guns were phased out by the USN but continued in service with the US Coast Guard.

In late 1944 a fast-firing high-velocity 20 mm gun based on the Hispano was being studied by Oldsmobile under the T34 designation. This became overtaken by events, as in April 1945 the USN issued an urgent requirement for for a last-ditch defence against *kamikaze* attacks. It was decided to drop the high-velocity requirement and save time by using standard Hispano ammunition. The rate of fire was increased, mainly by lightening the bolt and installing a Belleville spring buffer. A dozen T34 guns had been completed by the end of the war, but subsequent testing revealed problems, mainly to do with rapid parts wear. Development restarted in 1948 with the electrically-fired T34E1; the T34E2 was modified to fire ammunition which used the standard case loaded with a 104 g projectile instead of the standard 130 g. This led directly on to the Mk 12, which adopted a larger cartridge case for an even higher muzzle velocity.

***20 mm SHIRGUN SERIES***:
20 × 110 HS AMMUNITION.

The Shirgun Corporation was established in New York in 1941 by a European gun engineer named Henri Schirokauer (later changed to Henry Allen Sherwood). In 1942 the Bureau of Ordnance was looking for a 20 mm aircraft cannon compact enough to fit within the thin wings being designed for high-performance

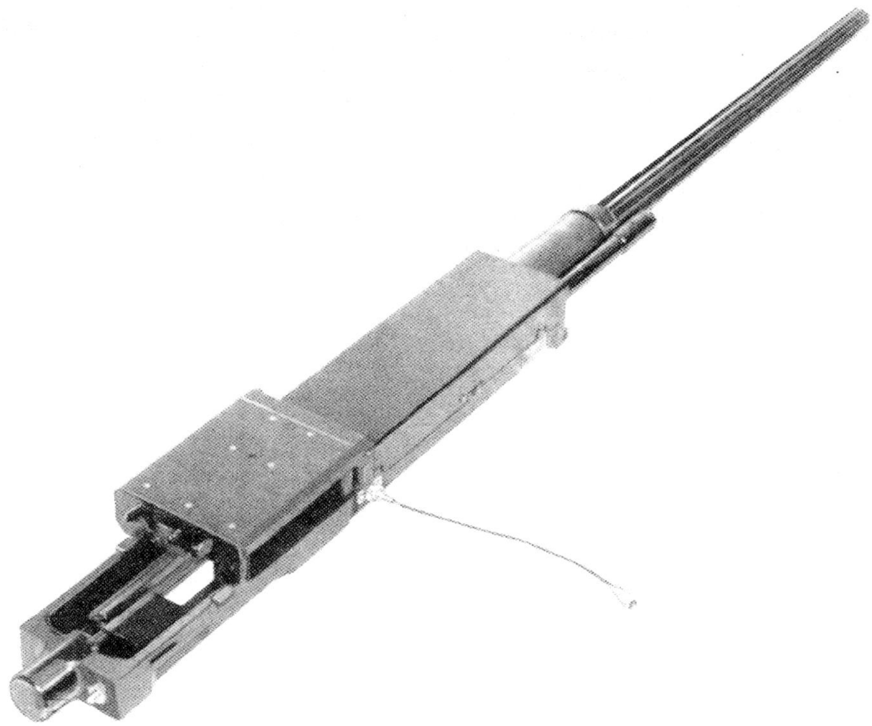

*Shirgun Mk 13 Mod 0 (Chinn)*

aircraft. In 1943 Shirgun succeeded in winning a development project, and in the following year Oldsmobile agreed to manufacture the first prototype, designated EX 1. Shirgun's estimated data for the gun included a cyclic rate of 950 rpm, a weight of 59 kg, and an L/85 barrel. It was a belt-fed gas-unlocked blowback gun with a rising wedge action.

A series of modifications to the EX 1 design (also known by the Army designation of T55) followed, which resulted in the Mk 9. Characteristics of this gun included a weight of 51 kg and a cyclic rate of over 1,200 rpm; the belt feed was driven by bolt travel. The Mk 13 was similar but rechambered for the 20 × 102 round. Little success was achieved and the project ended in 1949. However, the mechanism was revived in the 1950s as the EX 5; in parallel with this a project to convert the Mk 13 to 30 mm calibre began under the designation Mk 3 Mod 0.

A more interesting design was the Mk 10 (Army designation T54) which had twin barrels mounted close together. It was not a Gast-type gun, however, as the barrels were loaded and fired simultaneously from one belt feed. The weight was calculated as 86 kg and the rate of fire as 1,900 rpm. Shirgun's original design was based on the 20 × 110 HS ammunition, but the Naval Gun Factory adapted the project for 20 × 102 rounds. A working gun was never made.

Shirgun (who had changed the company name to Industro-matic Corp. of America in 1947) had earlier proposed scaling-up their EX 1 design to take 37 × 145R, 37 × 223SR, and 40 × 311SR ammunition, but this seems to have got no further.

***20 mm JOHNSON AIRCRAFT CANNON*: 20 × 110 HS AMMUNITION.**

The Bureau of Ordnance started a project early in 1942 for an alternative cannon using 20 × 110 HS ammunition (after an initial preference for the 20 × 110RB Oerlikon round was overturned), intended to supplement or supplant the existing AN-M2. Melvin Johnson, president of Johnson Automatics Inc., was already involved in small-arms design and began working on 20 mm cannon. What followed was a long series of development contracts which, while not resulting in a service weapon, did explore a number of innovations. Four different models of gun were designed, all of them under the collective EX 2 designation.

The First Model (originally made for Oerlikon ammunition, later adapted to Hispano) had a short-stroke (about 40 mm) gas piston action which delivered primary extraction as well as unlocking, before blowback completed the extraction. It was therefore a hybrid gas/blowback-assist action. The belt feed was integral, using closed-loop metal links and an action similar to that of the 0.50-inch Browning. The breechblock contained a breech slide, which moved sideways as the breechblock reciprocated, carrying each round into place for ramming, and there was a sliding wedge lock. The gun weight was 52 kg, the barrel was the usual L/85 and the rate of fire 750–800 rpm.

The Second Model, designed in late 1943, was a paper exercise only, being replaced by the Third Model which appeared early in 1944. This incorporated service requirements including switchable left or right-side feed, bottom eject, and a quick-change barrel.

An increased-diameter piston provided more energy to drive the action and feed, but firing tests ran into problems as the extraction of the rounds from

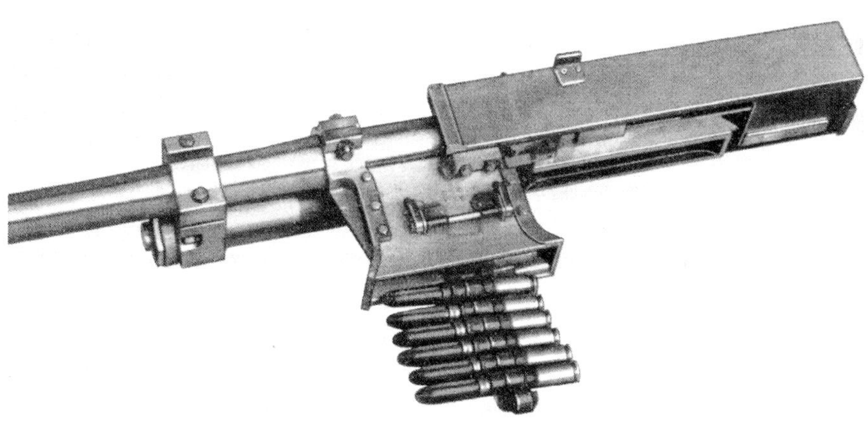

*Johnson EX 2 action (Chinn)*

the belt was so violent that 'de-bulleting' occurred. The gun was also heavy, as attention had been focused on making it work rather than weight-saving. Testing went on into 1945. The Fourth Model was intended to be definitive, featuring considerable weight-saving methods and a cyclic rate estimated to be well over 1,000 rpm, with open-loop belt links as an option.

***20 mm COLT T22***: 20 × 110 HS AMMUNITION.
This was initially based on the Mauser MG 151 design but with a number of modifications, including a longer action to take the significantly longer Hispano ammunition. The action is recoil-operated but uses a pivoting lock instead of a rotating bolt-head, and blowback provides a boost to the extraction/ejection cycle. The gun could be fired from a front or rear sear and was capable of being synchronized. Gun weight was 51 kg, length 198 cm and computed rate of fire was 575–650 rpm. Work commenced in 1943 and was formally cancelled in August 1945, being replaced by the T24 (see below).

***20 mm COLT T24***: 20 × 158RB AMMUNITION.
This used the same mechanism as the T22 above, but substantially redesigned to accommodate the much larger and more powerful ammunition based on the T5 case (as used in the ARF T33). The gun weighed 64 kg and was 254 cm long with an L/80 barrel.

The computed cyclic rate was 700–750 rpm. Development commenced in November 1944 with the firing of the first prototype in February 1945. Test firing went on for several months, but involved single-shot firing only.

## CANNON DEVELOPED FROM 1945 ONWARDS

***20 mm ARMOUR RESEARCH FOUNDATION (ARF) T33***: 20 × 158RB AND 20 × 135 AMMUNITION.
The origin of the T33 dates back to 1944 when Oldsmobile were asked to review the design of a 20 mm T19 gun being worked on by the Draper Corp.. It was soon determined that the design could

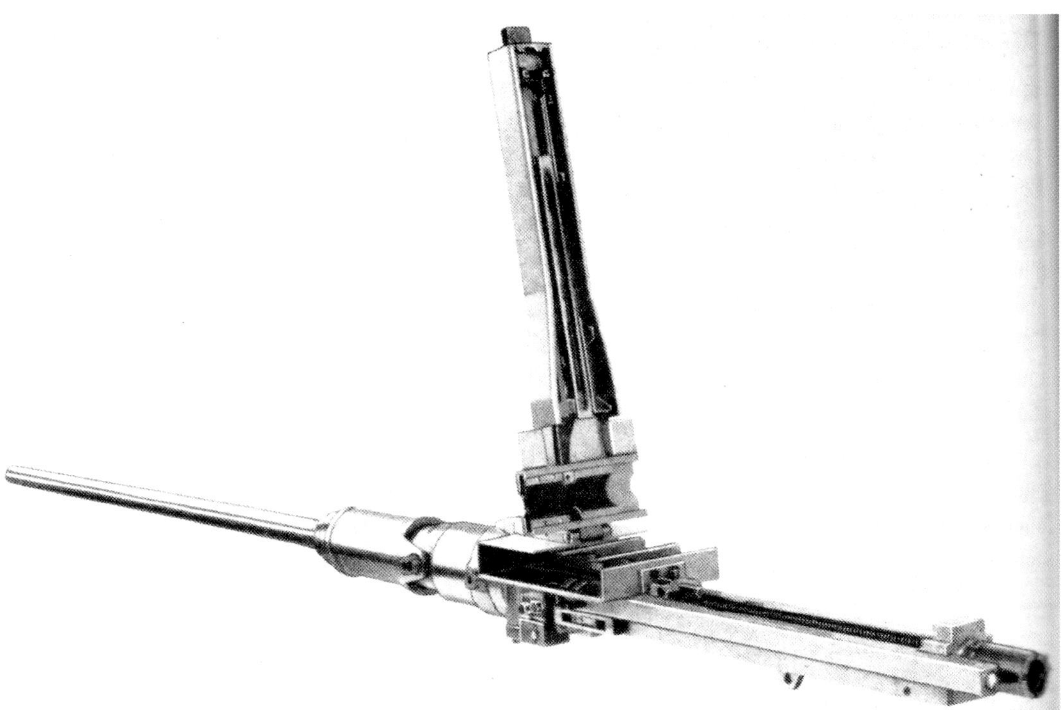

*Colt T24 with cover open (Chinn)*

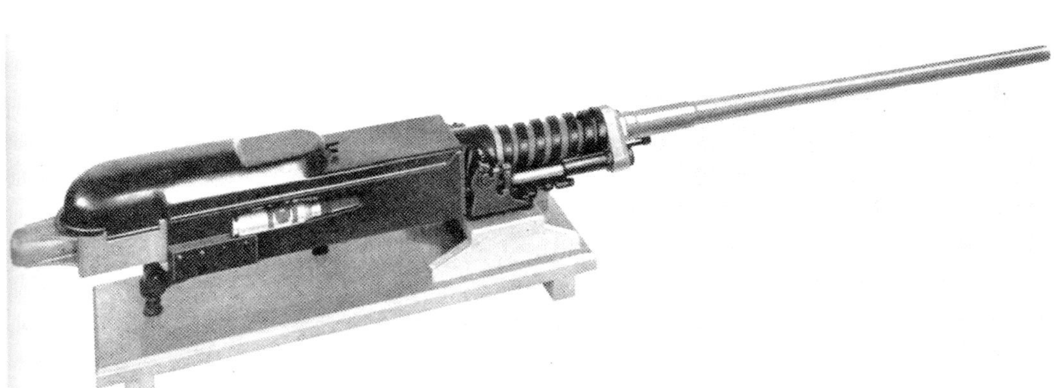

*ARF T33 wooden mockup (Chinn)*

not be made to work at the high rate of fire required because of fundamental feed problems, so in 1945 Oldsmobile started work on an improved design given the T33 designation.

The starting point for this was the 30 mm Rheinmetall-Borsig MK 103, combining a hybrid gas-unlocked blowback action with a belt drive powered by gun recoil via an intermediary spring. The T33, T33E1 and T33E2 (there were minor differences concerned with chrome-plating the barrel) were designed around a very large, high-velocity cartridge, using a 20 × 158RB case designated T5. Data include a gun weight of 80 kg, an L/80 barrel, and a rate of fire computed to be 900–1,200 rpm. Oldsmobile ceased work at the end of the war but in 1946 the Ordnance Corps revived its interest, handing the project over to the Armour Research Foundation (ARF).

ARF worked on the project until 1952, taking the designations to T33E3 Models A and B. The ammunition for these guns used the smaller T7 case as developed for the T74 gun (MG 213C revolver). It appears that the 20 × 135 Mauser case was developed with a new loading for both the T74 and T33E3. The Model A version has the same dimensions and weight as the T33; the rate of fire was 900 rpm. The Model B dropped the weight to 59 kg and raised the cyclic rate to 1,200 rpm. Various bolt locking systems were experimented with.

In 1948 ARF proposed a modified version of the T33; the T109, intended for installation in turrets, so more compact with a much shorter (L/56) barrel. This was never built.

***37 mm T37***: 37 × 255SR AMMUNITION (? ELECTRICALLY PRIMED).
The T37 gun was designed by United Shoe Machinery Corp. It was short-recoil operated with a rotary bolt-head locking and electric priming, and was reportedly based on a Mauser action (presumably the MG 151). The cartridge used is uncertain; the most likely candidate being the 37x255SR T68 (a round with a 263 mm case is also known) The first pilot model was completed in November 1945 and was intended as an aircraft gun. An AA version, apparently redesignated T172E2, was seen as an alternative to the T172E1 revolver cannon. A rate of fire of 400 rpm was reportedly achieved.

The T172E1 was not the only rival to the T37. The burst of creativity in post-war US cannon development saw Colt design the T44; the Bartelt Engineering Co. was assigned the T45 (a radical design using 'a hinge-type sliding shutter to replace the conventional breechblock'); Production Methods Inc. came up with a Solothurn-based design for the T46; and Bartelt Engineering Co tried again with a more conventional T49 based on the Hispano action. All of these were in 37 mm calibre but none was built.

**20 mm Mk 12**: 20 × 110 USN SERIES 100 AMMUNITION.
This gun is essentially the T34E1 (see the Hispano record above) rechambered for a new round of ammunition; the 20 mm Series 100 (initially known as the T130 series). This is the same length as the Hispano round but with a larger-diameter case holding more propellant, and is unique to the USN, only being used in two guns: the Mk 12 and Mk 11.

The design and functioning of the Mk 12 are essentially the same as the M24, that is, a hybrid gas-operated blowback based on the Hispano-Suiza HS 404 but modified to fire from a closed bolt with electrical priming. The externally-powered Mk 7 disintegrating-link belt feed can be set up to feed from either side of the gun. The gun weight is 46 kg with an L/61 barrel. The rate of fire is 1,000–1,100 rpm. This data concerns the Mk 12 Mod 0: there was a short-barrelled version proposed for turret installations, the Mk 12 Mod 1, but this was not used. There was also a very similar gun, the T118, which differed only in being chambered for the USAF's preferred .60/20 mm ammunition (20 × 102), which has the same case diameter as the Series 100 ammunition but is shorter and slightly less powerful. This was tested but not adopted.

The Mk 12, manufactured by Colt, was the USN's standard aircraft cannon from the early 1950s to the early 1970s. It was not an unqualified success, being criticized for inaccuracy and unreliability – particularly in the F-8 Crusader in which there was a long distance between the guns and the ammunition magazines, so violent manoeuvres could cause the ammunition belts to pull apart under the G-forces.

It should be noted that the Mk 12 achieved the 'million point gun' performance (1,000 m/s muzzle velocity and 1,000 rpm rate of fire) which German engineers could only deliver by inventing the MG 213C revolver cannon – a significantly heavier and more complex gun than the Mk 12. After a lacklustre performance just before and during World War

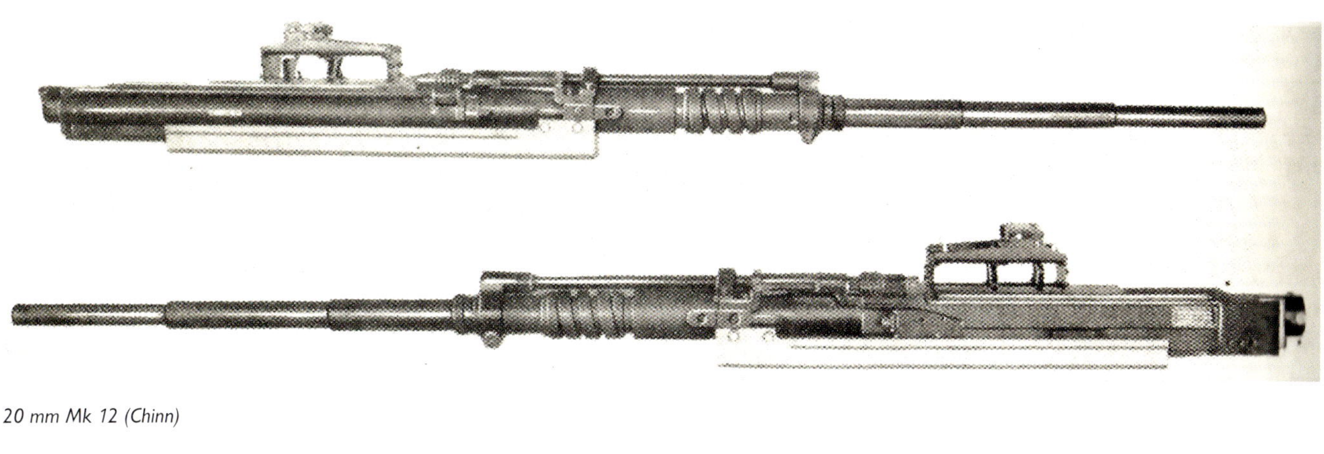

*20 mm Mk 12 (Chinn)*

II, the US gun engineers were beginning to produce some competitive weapons.

*30 mm MK 4*: 30 × 86B AMMUNITION. This was based on the service 20 mm M3 (Hispano), scaled up to take the British ADEN Low Velocity ammunition (T158 in US terminology), very similar to the 30 × 85B Mauser MG 213/30 round. Two rival versions of the 30 mm Mk 4 were developed in the early 1950s, one by Hughes Tool Company, the other by the Naval Gun Factory (NGF). Both long (Mod 0) and short (Mod 1) barrel versions were made, the short ones concerned with the T154 programme for turret-mounted weapons for bomber defence (see the Unconventional Systems section). The guns were designed to be interchangable with the 20 mm Mk 12. Both Hughes and NGF designs used a 30 mm Mk 1 pneumatic feed mechanism, similar to the 20 mm Mk 7. It is not easy to separate the characteristics of the two versions, but it seems the NGF was more of a straight calibre conversion while the Hughes gun included more modifications, particularly concerning how it was mounted. Data available indicates that the two barrel lengths were L/44 and L/27 and the weight (unspecified version) was 52 kg. The guns were belt-fed using 30 mm Mk 1 links and the rates of fire were in the region of 1,050 rpm. This project seems to have been abandoned in around 1957.

*30 mm Mk 5*: 30 × 86B AMMUNITION. This was developed during 1950–51 by the Armour Research Foundation (ARF) and was essentially the 20 mm T33 design (see above) modified to use the T158 (ADEN LV) ammunition. The belt-fed (from either side) gun weighed 61 kg with an L/53 barrel. The gas-unlocked blowback mechanism used swinging locks (after experiments with a rotary locking bolt proved unsatisfactory), and the estimated cyclic rate was 750–800 rpm. The gun fired from a closed bolt, using electric ignition.

*20 mm & 30 mm Mk 22*: 20 × 102 OR 30 × 100B WECOM AMMUNITION. The origins of this project go back to an early 1950s research project carried out by the Naval Ordnance Station in Louisville, Kentucky (NOSL) to determine the optimum characteristics of a 20 mm gun which could also be converted to 30 mm calibre. This resulted in a gun designated EX-6 which went through various tests before being postponed in 1953. The project was revived in 1967, NOSL again receiving the development contract for what became the Mk 22. The USN was primarily interested in this gun for arming patrol boats as a result of their Vietnam War experience.

The gun is a mixture of proven elements: the locking and feeding mechanisms are based on the German MG 42, a short-recoil machine gun which uses a roller-locking system; while the belt link and feeding system were adapted from the .50 Browning. It fired from an open bolt, with ammunition feeding from either side. The gun featured a variable rate of fire, depending on whether or not the muzzle booster was fitted. The ammunition selected for the gun is the electric-primed 20 × 110 USN M100 series instead of the percussion-primed 20 × 110 HS M90 used in the EX-6. The 30 mm round chosen is the percussion-primed 30 × 100B WECOM round developed for the XM-140 helicopter gun. In 1968 a decision was made to change the 20 mm ammunition again, this time to the (electric primed) 20 × 102 M50 series which had been adopted as the tri-service 20 mm cartridge. Several guns were made and test fired, a process continuing until 1972 with various modifications to improve reliability. Data for the Mk 22 include the following (20 mm first, 30 mm in brackets where different): weight 54 kg; barrel L/66 (L/41); rate of fire 400–800 (350–750)rpm.

Interestingly, Chinn mentions a two-stage 'shoot-off' in April 1969 and August 1971: the first between the Mk 22 and the GE 120B (see below); the second involving improved versions of these guns plus the Mauser Model B and the AME 621 (see the German and French sections for details of these guns). It appears that the GE 120B performed the best, although the AME 621 had some attractive features.

*20 mm T74 AND T110*: 20 × 135 AMMUNITION. The USA received one of the prototype Mauser MG 213C 20 mm revolver cannon after World War II, and a research and development contract was awarded to the Armour Research Foundation. ARF analysed the gun in 1946–7 under the designation T74A, making some improvements resulting in the T74B. Further development involved major design changes to reduce the size and

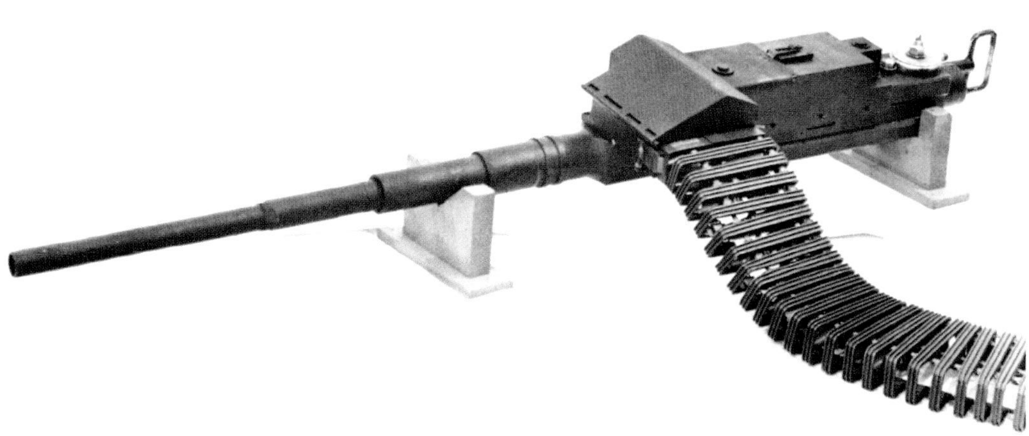

*Mk 22 (Chinn)*

weight, including relocating the barrel to the bottom of the receiver, so the firing chamber was at the 6 o'clock position rather than 12 o'clock; this resulted in the T110 of 1949. Work was also done on the German ammunition; brass rather than steel cases were made, and loaded with US projectiles. This 20 × 135 ammunition also appears to have been used in the T33E3 gun (see above).

The T110 gun was also the basis for the T151, in 27 × 70B and 27 × 96B calibres (see T150 below), and the T160 in 20 × 102, which became the M39.

***57 mm T78***: 57 × 173RB AMMUNITION.
This was the US designation for the Rheinmetall-Borsig 5,5 cm MK 112 which was tested in the USA after modification to take existing 57 mm projectiles (see Germany section).

***30 mm T121***: 30 × 86B AMMUNITION.
This was one of the first results of the work done at Frankford Arsenal on the Mauser MG 213C revolver cannon, and dates from the late 1940s. It led directly on to the T182, the main difference being a switch from the T158 cartridge (ADEN LV) to the longer-cased T204 (30 × 113B).

***27 mm T150 VULCAN***: 27 × 70B AND 27 × 96B AMMUNITION.
An early variation on the Project Vulcan Gatling-style guns which led to the M61, to determine the effect of maximising the destructive power of the HE shells. The 27 mm calibre was chosen as the largest that could be used in a gun sized for the 20 × 102; this resulted in a straight-cased cartridge, so it was given a belted case to ensure accurate headspacing. The first model fired 20 × 70B ammunition with a heavy projectile but the muzzle velocity was felt to be too low, so the projectile was shortened and lightened to permit the use of a longer case holding more propellant. The reduction in shell weight was felt to be worth it in order to achieve the higher hit probability of a higher-velocity round.

***30 mm T168***: 30 × 120RB AMMUNITION.
This was a reverse-loading four-chamber revolver, intended for bomber defence systems. The cartridges were pushed base-first into the front of the chamber. This meant that the ammunition feed was close to the centre of the gun therefore did not move much with elevation. It also shortened the gun overall, providing a more compact installation.

To make such a gun possible, the electric-primed ammunition (designated T268 when loaded with a T328 TP projectile) needed to be repackaged so that it tapered in a reverse direction, with the base being the narrowest part of the case: the rim is tiny.

Test firing started in 1952, and a programme of testing and modification (with various cartridge case lengths and loadings) continued until 1956 when the project was closed down. The ultimate version was the T168E2, a lightweight 'receiverless' gun.

Russian designers faced the same problem and addressed it in the same way with the Rikhter R-23, which did see service.

***37 mm T172***: 37 mm AMMUNITION.
A revolver cannon developed by Armour Research Foundation as an alternative to the reciprocating T37 gun for AA purposes: the designations subsequently changed to T172E1 for the revolver gun and T172E2 for the reciprocating gun. The T172E1 gun was to fire the same (electric primed) ammunition as the T37E2, namely the T181 or T324 37 mm high explosive shell. It is unclear exactly which cartridge was used in these two guns; an electric-primed version of the 37 × 223SR seems the most likely candidate, although it is just possible that the 37 × 256 later used in the T250X-1 was first designed for these guns. The gun weight was quoted as 500 kg, with an L/70 barrel.

Two prototypes were built. Test firing started in 1954 and eventually resulted in a RoF of 1,100 rpm being achieved. Work ceased in late 1956.

***30 mm T182***: 30 × 114B AND 30 × 126B AMMUNITION.
This is the closest US equivalent to the 30 mm ADEN/DEFA, being a five-chamber, gas-operated, electrically fired revolver cannon. The T182 was initially chambered for the T204 round which is dimensionally similar to the ADEN/DEFA 30 × 113B, but the T182E3 onwards were modified for the higher-velocity 30 × 126 B T239 cartridge which uses a longer case but retains the same overall length.

Work on the T182 began in 1951 by Oldsmobile. They developed the gun to the T182E3 stage and in 1953 an order for 3,000 guns was placed (prematurely, as it turned out) with Buffalo Arms Inc.

*T182E3*

However, indecision about the applications for the gun soon led to a reduced order of 450 guns for research and development. Oldsmobile dropped out of the programme in 1954 and were replaced by Ford, under the direction of the Springfield Armory. Further development work improved the reliability and durability of the gun – now the T182E5 – up to production standard, but work ceased in 1957 after approximately 100 guns had been made. The T182E5 weighs 92 kg with an L/36 barrel and fired at 1,200–1,500 rpm.

***30 mm T212 VULCAN II***: 30 × 113B AND 30 × 126B AMMUNITION.
Project Vulcan II was developed in parallel with the T171E1 / M61 to produce a version chambered for the ADEN/DEFA 30 mm ammunition. General Electric started work in 1954 with the first of two prototypes (one electric, one hydraulic) completed at the end of 1956. The US-developed 30 × 126B version of the ammunition was also tried in a version this gun, as well as in the T182 revolver. The contract ended in 1958 when testing was complete. The six-barrelled T212 weighed 155 kg and was 162 cm long. It was belt-fed, the maximum rate of fire for a short burst being measured at 5,500 rpm with hydraulic drive and 6,050 rpm with electric drive: the time to achieve these rates was .50 secs and .78 secs respectively.

***20 mm T220 AND T247***: 20 × 102 (PERCUSSION) AMMUNITION:
In the mid-1950s the US Army was seeking a lightweight All-Arms Anti-Aircraft System (given the code-name Project Octopus in 1956). One of the responses to this requirement was the T220, which was developed at the Springfield Armory in 1954. The gun was designed around the M150 series ammunition, a percussion-primed variant of the M50 series. The T220 had two rates of fire; 1,500 rpm for AA use (operating in short-recoil mode) and 500 rpm for ground fire (operating as a long-recoil gun, with about 25 cm recoil movement). It uses a rotating bolt action, with energy from the recoiling barrel, rather than the bolt, to drive the feed; however, it proved to have insurmountable feeding problems. Attempts to rectify this were the T220E1 from Springfield, the T220E2 from Dixon Research and the T247 from the American Machine and Foundry.

The T220E1 weighs 46 kg and has an L/76 barrel. The rates of fire were 500 and 1,100 rpm. In the low rate of fire, the bolt group was held by a rear sear after each shot, and the return movement was delayed. Ammunition feeding remained a major problem, exacerbated by the two different rates of fire.

The T247 took a different approach to the dual RoF issue, using short recoil for both modes of fire, the rate change being achieved by moving the bolt unlocking cams. However, the mechanism proved very complicated and suffered many problems on test. Weights, dimensions and performance were the same as the T220E1.

The T220E2 differs from the other guns in this series in using the bolt movement to drive the ammunition feed rather than the barrel movement. It also has only one rate of fire. This proved far more satisfactory and largely resolved the problems of the others; it was considered the best of the series. Project Octopus was cancelled in 1958, but in 1961 the T220E2 was recommended for consideration for the Vehicle Rapid Fire Weapon Interim System. However, the HS 820 was selected as it was already in production. The T220E2 weighs 45 kg, has an L/76 barrel and fired at 1,100–1,275 rpm.

***30 mm T221***: 30 × 86B AMMUNITION.
This was known as the Short Drum Revolver Gun, and from 1953 to 1956 explored some substantial changes to the MG 213C pattern used in most revolver cannon, such as the T121 and T182. One principle examined was to drive the operating cycle entirely by gas pressure rather than using springs. Another was the use of a cylinder which is shorter than the cartridges: on firing, the cartridge is supported in a chamber which is divided between the rear part (in the cylinder) and the front (in the barrel). The key advantage is to minimize the slide stroke needed to propel the mechanism, which is also compact and light (helped by the use of aluminium alloy in the receiver). The result is a design calculated to weigh 60 kg and achieve 1,600 rpm (belt fed). At 150 cm long, it was smaller overall than the 20 mm M39. However, it did not progress any further than initial test firing of a few rounds from a firing model.

***37 mm T250***: 37 × 255 AND 37 × 217 AMMUNITION.
The T250 gun, the key element of the T249 Vigilante SPAAG system, was developed in the late 1950s by the Springfield Armory (under the management of the Watervliet Arsenal) to meet a US Army requirement for a 40 mm Bofors replacement. The T250 is the

*20 mm T220E1 (above) and T220E2 (below) (Chinn)*

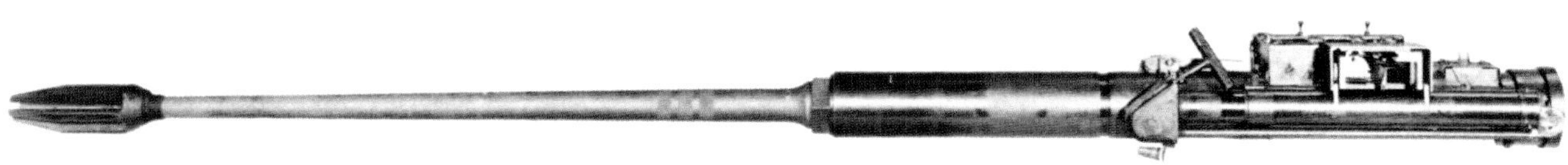

*T250 37 mm rotary cannon*

biggest of the rotary cannon, with six barrels and hydraulic power to achieve a RoF of 3,000 rpm (reduced to 120 rpm for engaging ground targets). The gun has L/69 barrels. Two different prototypes were made: the T250X-1 in 37 × 255 calibre and the T250X-2 in 37 × 217. The T250X-1 was electric-powered and link-fed using 50-round belts: the T250X-2 had a linkless feed system. The 192-round pan magazine was on the left side of the turret and took 2.5 minutes to reload. The system was designed to fire 48 rounds per burst, so could engage four times before reloading.

Two mountings were developed: the T249 tracked carrier, based on lengthened version of the chassis of the M113 APC, and a towed mounting.

Vigilante reached the evaluation stage in 1960 but was cancelled shortly thereafter. Its development was overtaken by an ambitious new American missile system called Mauler, which was intended to operate automatically once switched on. Like many early missile projects, this promised far more than it was able to deliver and was eventually cancelled in 1965, but not before it had terminated the Vigilante.

Sperry modified the gun for the US Army's abortive DIVAD competition of the late 1970s. Their unsuccessful entry used the T250 rechambered for 35 × 228 Oerlikon, mounted on an M-48 chassis.

**20 mm M39**: 20 × 102 AMMUNITION. After working on the T-74 revolver cannon (the German MG 213C), Armour Research Foundation (ARF) were awarded a contract in 1950 to develop a revolver cannon, designated T130, around the .60 inch (15.2 mm) anti-tank rifle cartridge; this was later designated M38 despite not seeing service. This led directly to a 20 mm version, using a necked-out version of the cartridge case and needing little more than a new barrel. This was designated T160 and subsequently adopted as the M39.

The M39 is a single-barrelled, five-chambered revolver type, gas operated, belt fed and electrically fired. It weighs 81 kg and has an L/68 barrel. The rate of fire was initially 1,500 rpm but the final version could achieve 1,700 rpm. In comparison with the parent Mauser MG 213C design, the gun is upside down, with the barrel aligned with the lowest chamber of the cylinder. The complete cannon assembly is contained in an alloy or steel welded cradle unit. It can be mounted in any position on the longitudinal axis and can be fed either right or left-handed. Cocking is pneumatic with provision made for manual operation. The original M39 was superseded in turn by the M39A1, A2 and A3, with only minor differences between them.

Although first tested in action in Korea in 1953, the M39 proved as troublesome as most revolvers to develop and was not completely debugged until 1959. After arming various aircraft in the late 1950s, the M39 saw most use (and is still

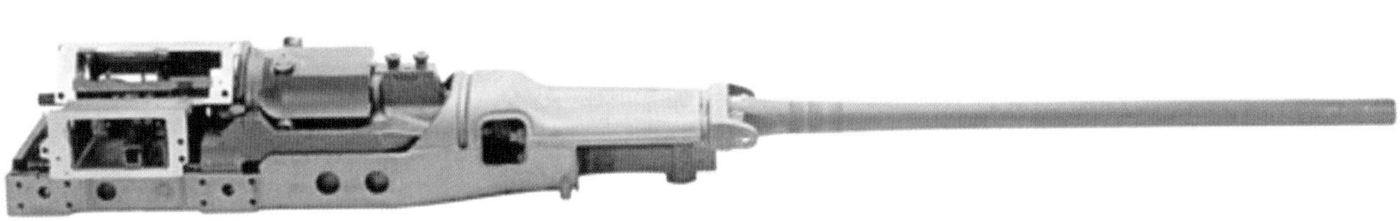

*M39 20 mm revolver gun (AEI Systems)*

in service today) in the F-5 lightweight fighter. More than 35,500 units were produced by the various manufacturers, principally Pontiac and Ford, before production ended in 1988.

As well as its use in aircraft, SMC of the Republic of China (Taiwan) has installed the M39 (designated T-75) in remotely-controlled AA mountings, both single (XTR-101) and twin (XTR-102).

In 1999, the British company Aircraft Equipment International (now AEI Systems Ltd) became an approved manufacturer and supplier of M39 series cannon, spares and support equipment for the F-5 series of aircraft.

**20 mm MK 11**: 20 × 110 SERIES 100 AMMUNITION.

The second American revolver cannon to be adopted was the Navy's remarkable MK 11. The first prototype had some highly experimental features which were soon discarded (see the Unconventional Systems section) but even so the production version was still very unusual.

The production Mk 11 used the 20 × 110 USN ammunition developed for the Mk 12 cannon. It has an eight-chamber cylinder and two barrels (at 12 and 6 o'clock) fed by separate belts and firing simultaneously. The action is part gas, part recoil. The barrels are only rifled for the first few centimetres then smoothbored thereafter. Rate of fire is adjustable, between 700 rpm and 4,200 rpm. Weight is 109 kg including the 20 kg loader, which draws rounds from the belts, rams them into the chamber, fires them and ejects the spent cases.

This peculiar gun was not installed within aircraft, only being available in the Hughes Tool Company Mk 4 gunpod which weighs 357 kg empty and 630 kg with 750 rounds of ammunition. This is referred to as the 'HIPEG' (HIgh Performance External Gun). About 1,200 gunpods were made in the 1960s and saw considerable use against ground targets in Vietnam attached to various types of aircraft, but most commonly the A-4, F-4 and OV-SUU10.

**20 mm M61 SERIES**: 20 × 102 AMMUNITION.

In 1946 the US Army commenced 'Project Vulcan', for which General Electric was contracted to develop a new externally-powered cannon employing the Gatling principle of multiple rotating barrels first used in the 19th century. It is worth noting that an electric motor was coupled up to a Gatling in 1893 and achieved around 3,000 rpm, but no-one could see a use for it.

The first Vulcan prototype, the T45 Model A of 1949, had six barrels and was in .60 inch (15.2 mm) calibre. Ten guns were built and tested (including flight tests) which proved the feasibility of the approach, but many design changes were required (including a switch to the electrically-primed 20 mm calibre). These resulted in the T171E1 or Model C, of which 33 were made during 1953–55. Further exhaustive testing and redesign resulted in the T171E2 (Model D). During this process the weight was reduced from 193 kg (Model A) through 166 kg (Model C) to 119 kg (Model D) and the rate of fire gradually increased to 6,000 rpm. Further testing revealed the need for some changes to avoid case stretch, resulting in the T171E3 which was finalised in December 1956. By the time the E3 was formally adopted as the M61 in December 1957, production was well underway and hundreds had been built.

The gun is externally powered, both electric and hydraulic drives having been used plus, in the SUU-16/A gunpod, a ram-air turbine. When the cannon is fired the barrel group revolves anti-clockwise, looking in the direction of fire. Each barrel has its own bolt, operated by cam followers, which chambers, fires and extracts the rounds as the gun rotates. The multi-barrelled design reduces heating and barrel erosion.

The M61 was belt-fed but in its first installation (F104) problems were encountered with disposing of the links, resulting in a reduced rate of fire of around 4,000 rpm. A Linkless Ammunition Feed System (also powered by an electric motor) was developed for a modified gun, the M61A1, which became the USAF's, and subsequently the USN's, standard aircraft cannon from the 1960s onwards, starting with the F-104C in 1959. This version of the gun not only dispenses with the need for belt links, it also allows for fired cases being fed back to the magazine if required; this is generally preferred, as it not only removes the risk of damage from ejected cases hitting the aircraft, it also helps to maintain the aircraft's centre of gravity.

Each feed system is designed for its particular aircraft. The original systems use helical drum magazines of considerable capacity: the F-16 has a 512-round horizontal drum (the entire weapon

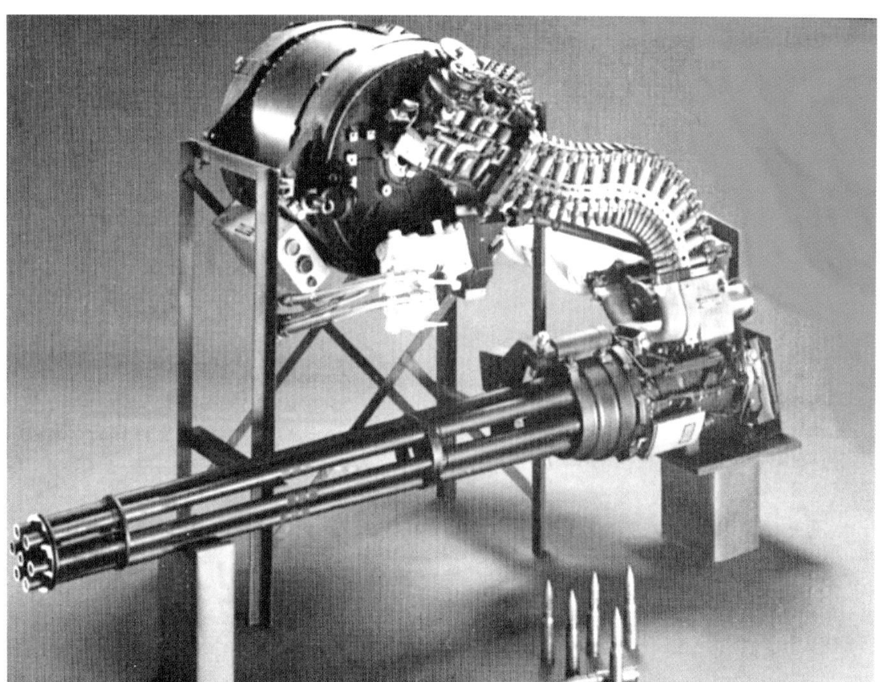

*M61 installation for the AMX (GE)*

installation having a loaded weight of 375 kg); the F/A-18 has a 570-round drum under the gun (381 kg); the F-14 had a 767-round horizontal drum (466 kg); the F-15 has a 940-round drum (526 kg). In more recent installations the big drums have been replaced by more compact linear linkless feed box systems of reduced capacity: the F-15E has a 500-round vertical box in which the feed weight is reduced from 172 to 107 kg; the F-22 has a 480-round box; and while the F/A-18E/F retains a drum (the most efficient shape for its location) it is a lightweight one of 412 rounds with a complete loaded weight of 309 kg.

The installation for the Italian AMX ground-attack aircraft is unusual in using linked ammunition fed from a 403-round drum. An hydraulic drive system provides a rate of fire of 4,000 rpm. Installation weight is 356 kg.

The M61A2 was developed with the aim of achieving a significant weight reduction while still keeping as many parts as possible compatible with the existing model. The main change to achieve this was to fit slimmer gun barrels, reducing total weight from 112 kg to 92 kg. This was made possible by the reduced ammunition capacity of more recent installations, meaning that in a worse-case scenario of a runaway gun, it can still empty the magazine without catastrophic overheating. The reduced inertia of the barrel group also means that it accelerates up to its maximum rate more quickly, achieving 6,000 rpm in 0.25 seconds instead of 0.4. The project was initiated in 1984, completed qualification testing in 1989 and was adopted as the M61A2. In US service, this is fitted to the F-22 and the F/A-18E/F; it is also used in the Taiwanese Ching-kuo fighter.

As well as its aircraft use, the M61 has been adapted to land-based and naval AA purposes. The initial land-based systems were the towed M167 and the self-propelled M163 Vulcan Air Defense System (VADS), the cannon being designated M168. The M167 was made in South Korea (who have since developed an improved Viper version), the M163 (based on the M113 vehicle) was used by Israel and Belgium as well as the US Army. The gun's rate of fire is set to a maximum of 3,000 rpm (with 1,000 rpm available for firing at ground targets). The gunner can select 10-, 30-, 60- or 100-round bursts. A linkless feed system is used, the one in the M163 carrying 1,180 rounds ready to fire with an additional 1,100 rounds inside the vehicle. In 1995 Israel announced the Machbet upgrade to the M163, adding a four-round launcher for the Raytheon Systems Company Stinger fire-and-forget missile launcher, to extend the effective range of the system.

The naval Close-In-Weapon-System (CIWS) designed around the M61 is known as the Mk 15 Phalanx. The development of this system began in 1968, following a realisation that USN warships had no adequate response available to the threat of the new Soviet anti-ship missiles being launched from submarines or surface vessels. General Dynamics were given the contract to develop a quick-reaction gun CIWS using a closed-loop firing system (the on-mount radar not only tracks the targets, it also tracks the outgoing projectiles, calculates whether they will hit the target, and alters the aim for the next burst accordingly). The ammunition fired by the original version is the specially developed Mk 149 APDS (Armour Piercing Discarding Sabot) which initially used DU (Depleted Uranium) penetrators, later changed to tungsten alloy.

The system was originally hydraulically driven with the gun firing at 3,000 rpm, but for the Block 1A this was replaced by a pneumatic drive which achieves 4,500 rpm. The on-mount capacity of the drum magazine was 989 rounds originally and increased to 1,550 rounds from Block 1 onwards; this has the usual linkless feed, returning fired cases to the drum.

The first prototype of the system was available in 1973 and after extensive trials it was approved for service use in July 1977 as the Mk 15 Phalanx. The first production model was completed in August 1979 and it entered service in April 1980. In February 1987 GE Aerospace, Ordnance Systems (now Lockheed Martin, Armament Systems Division) was awarded a second source contract for Phalanx.

Various modifications took place to the Phalanx but in 1994 work started on a package of major improvements which led to the Phalanx Block 1B. This concerned three main development areas: the sensors and fire control systems (now able to engage surface targets as well as incoming air threats); improved Mk 244 APDS ammunition; and longer (L/100 rather than L/76) and heavier gun barrels (OGB = Optimized Gun Barrel) to extract the most from the new ammunition. The first production Block 1B system entered service in October 1999, with all earlier blocks in USN service being upgraded to the new

*Mk 15 Phalanx with M61 Vulcan cannon (USN)*

standard in due course. A total of 900 Phalanx systems have been sold.

In 2004 the conflict in Iraq led to an urgent requirement for a system to defend US military bases from being attacked with rockets and mortar bombs, the task being designated C-RAM (Counter-Rocket, Artillery, Mortar). The most obvious approach was to produce a land-based version of the naval Phalanx Block 1B, and the result is called Centurion, developed by Raytheon and first deployed in 2005. It basically consists of a standard naval Phalanx mounted on a trailer or truck-bed together with a generator. The main difference is that instead of firing APDS projectiles (which can travel a very long way and potentially do a lot of damage when they land) the Centurion fires explosive shells fitted with a self-destruct mechanism so that if they miss the target, they blow up in mid-air. Originally these were the old M246 HEI-T SD from the M168, but these have been replaced by the purpose-designed M940 SAPHEI-T SD, a version of the Raufoss-developed multipurpose ammunition.

In 2015 OTO Melara announced a palletised gunship system for military transport aircraft, consisting of an M61A1 and 750 rounds of ammunition on a single NATO 463L pallet, and designed to fire through a side door at ground targets. It can be installed or removed in about 15 minutes. The total weight is 1,550 kg.

The latest Phalanx development by Raytheon (tested in March 2017) is the 'electric gun'. This replaces the pneumatic motor, compressor and storage tanks with electric power, reducing the system's weight by 82 kg and permitting the use of varying rates of fire to conserve ammunition, surface targets not needing the maximum rate.

**20 mm M195**: 20 × 102 AMMUNITION. A short-barrelled version of the M61 (L/51 barrels) for helicopter use, with the RoF slowed down to about 750 rpm. Six barrels proved an unnecessary weight burden for such a low rate of fire, so it was soon replaced in service by the lighter M197 (see below).

**20 mm M197**: 20 × 102 AMMUNITION. The M197 is a lightweight three-barrel derivative of the M61A1, using many common parts. It was developed in the mid-1960s by General Electric, entering production in 1969 specifically for helicopters, light fixed-wing aircraft, and small naval craft. The cannon is operated by a variable-rate electric motor by which the rate of fire is adaptable to air-to-air, air-to-ground, and ground-to-air applications. The weapon uses the standard L/76 barrels and weighs 66.5 kg. Barrel life is 15,000 rounds. Each barrel has its own bolt, which rams, locks, fires, unlocks and extracts rounds in sequence during one revolution of the barrel cluster. Ammunition can be fed to the cannon as a standard linked belt or by a linkless feed. When using linkless ammunition, as in the GPU-2/A pod, the M197 has a highest rate of fire of 1,500 rds/min; when used with linked ammunition, as with the turret or pintle applications, the rate of fire drops to 750 rds/min. The time to accelerate to the maximum rate of fire is 0.3 seconds.

Over 1,700 M197 have been made, and it is licence-produced in various countries. The principal airborne application has been in the Universal Turret System (UTS), a chin-mounted flexible turret for attack helicopters. It is also employed in the GPU-2A gun pod, and in the Sea Vulcan 20, a manually operated close-defence weapon consisting of an old Oerlikon mounting with an M197, a battery able to drive the gun for 6,000 rounds, and a 300-round magazine. A more advanced model was the remotely-controlled EX-84, actually a universal mounting based on the UTS and adaptable to various light cannon or MGs. It weighed 363 kg empty or 613 kg loaded with 750 rounds.

*20 mm M195*

*Two views of an M197 in a chin turret on an AH-1F Cobra attack helicopter*

20 mm GAU-4/A (ARMY DESIGNATION M130): 20 × 102 AMMUNITION.
This is a version of the M61A1 modified to be gas-operated rather than externally powered, except for an electrically operated inertia starter to initiate gun rotation. Once underway, the action is powered by gases bled from three of the six barrels and transmitted via a gas piston and drive assembly mounted between the barrels. This system also develops enough power to operate the ammunition feed system. This weighs about 4.5 kg more than the externally-powered version.

The GAU-4/A was specifically designed for the SUU-23/A gunpod which weighs 474 kg empty or 780 kg with 1,200 rounds. Like the SUU-16/A the ammunition feed is one way only, the fired cases being ejected from the pod, there being no danger of the cases hitting the aircraft. Gun life is estimated to be 150,000 rounds. ARF carried out initial research work on the system, but GE designed and built it. 1,102 examples were built from when the first contract was signed in 1966 and the end of production in 1972.

*25 mm GAU-7/A*: 25 × 152 COMBUSTIBLE CASE AMMUNITION.
While various unusual gun technologies were worked on in the 25 years following the end of World War II, cannon

ammunition technology changed relatively little. There was one significant exception to this, the GAU-7 programme for a new aircraft gun, intended to arm the USAF's formidable new air superiority fighter, the McDonnell Douglas F-15 Eagle. This was intended to offer a vastly superior performance to the existing 20 mm M61A1 cannon in almost every respect: the projectile weighed almost twice as much and was a far more aerodynamic shape; the projectile velocity at the muzzle was significantly higher at 1,200 m/s, an advantage which increased with range; while the gun was still able to fire at 6,000 rpm and the increase in system weight was to be kept to the minimum. In addition to these performance benefits, the USAF were keen to simplify the operation of the gun by adopting caseless or combustible-cased ammunition technology, which would eliminate the need to extract and eject or store the fired cases; this would also reduce the ammunition weight, and another innovation – telescoped ammunition, with the projectile buried inside the propellant – would reduce its bulk (there is more on the history and characteristics of telescoped ammunition in the Unconventional Systems section).

Work on the new gun was initiated by the USAF at the beginning of 1968, with Philco-Ford and General Electric participating in what was initially a three-phase development programme for a 25 mm system designated GAU-7. For this project, the ammunition development was key to the success; Philco-Ford worked principally with Brunswick, General Electric with Hercules.

Both GE and Philco-Ford chose a multi-barrel Gatling rotary design with hydraulic drive to achieve the performance required; the most obvious difference is that the GE gun has six barrels while the Philco-Ford has five. The GE gun weighs 197 kg and has L/85 barrels. The Emerson Electric linkless feed system (which could also be used with the Philco-Ford gun) is a continuous chain of plastic carrier tubes transporting rounds from the 960-round magazine to the gun. The empty weight of the storage and transfer system is 128 kg empty, or approximately 465 kg when full. The Philco-Ford weighs 193 kg and uses the same L/85 barrel length. Ammunition feed is linkless, using a rotary conveyor.

Both GE and Philco-Ford delivered two guns and 10,000 rounds of ammunition for testing in 1971, after which work on the GE gun ceased and Philco-Ford were awarded the contract for Phase IV. Work continued until 1974 when the GAU-7 programme was terminated due to fundamental problems associated with the combustible-cased ammunition. This proved to deliver performance which varied with climatic conditions, and the need to protect the rounds in the magazine from being ignited by hostile fire complicated ammunition storage and handling.

**30 mm GAU-8/A**: 30 × 173 (30 × 164) AMMUNITION.

Given the success of the M61, General Electric's attention turned to making rotary cannon in other calibres. In 1968 they began studies of a much more powerful gun in 30 mm calibre, and were rewarded in 1971 with a contract to develop the main armament for a new USAF close-support aircraft, which had to be powerful enough to pose a threat to Soviet tanks. The gun was designated GAU-8/A and popularly named 'Avenger'. The first production gun was delivered in 1975 and the A-10A entered service in 1977. Over 700 A-10 aircraft were built.

The ammunition GE chose was based on the 30 × 173 cartridge used in the Oerlikon KCA revolver cannon, with three major changes: the ignition system was changed from electric to percussion; the cartridge case was made from aluminium alloy rather than steel (saving 237 kg or 22% of the weight of a full ammunition load of 1,174 rounds); and the projectiles have polymer rather than metal driving bands, to minimize barrel wear.

The gun follows the same layout and operation system as the M61, except with seven L/77 barrels rather than six. The gun and ammunition feed are hydraulically powered. The GAU-8/A weighs 281 kg, but the complete installation with a full magazine of 1,350 rounds crunches the scales at 1,828 kg (the normal load is 1,174 rounds). Given the huge size of the weapon system (the gun and ammunition tank being 6.4 m long) the aircraft, the Fairchild A-10 or Thunderbolt II (popularly dubbed 'Warthog'), effectively had to be designed around the gun.

The cylindrical ammunition tank contains a linkless ammunition feed system, in which fired cases and any unfired rounds are returned to the storage drum after passing through the cannon. The cannon reaches its maximum firing rate of 4,300 rpm in 0.55 seconds. The system is credited with a minimum rounds before stoppage (MRBS) figure of 10,000, with each barrel having a minimum life of 20,000 rounds.

As well as its use in the A-10 aircraft, the GAU-8/A forms the heart of the Dutch Signaal Goalkeeper close-in weapon system (CIWS) for defending warships from enemy aircraft and anti-ship missiles. The first prototype was completed in 1981, using the General

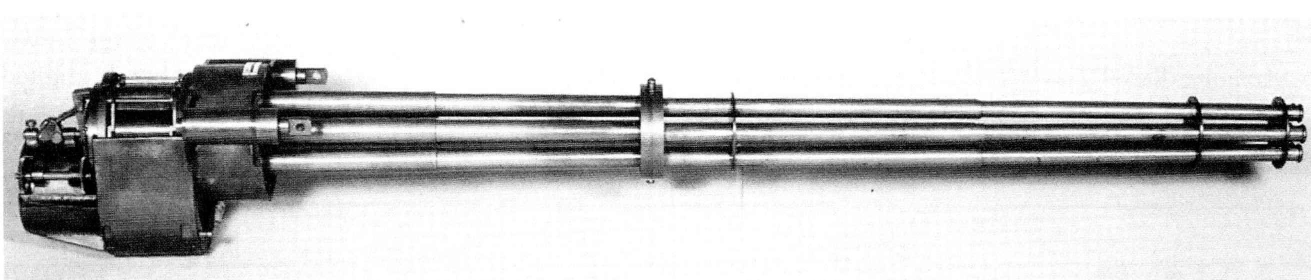

*General Electric GAU-7/A (Chinn)*

*A famous photo illustrating the size of the GAU-8/A and its ammunition tank (USAF)*

Dynamics EX-83 mounting. The first order, for the Dutch navy, was placed in 1983 and since then about 70 systems have been sold.

In order to win the GAU-8 contract GE had to defeat another contender: Philco-Ford. Their version used slightly different purpose-designed 30 × 164 ammunition and had six barrels rather than seven. It weighed 264 kg with feeder and adapters, had L/71 barrels and was capable of firing at up to 4,000 rpm. However, on test it proved less reliable than the GE product.

A 25 mm version of the GAU-8/A was proposed in the mid-1970s (see below).

*30 mm GAU-9/A*: 30 × 173 AMMUNITION.
The GAU-9/A was insurance against the possible failure of the GAU-8/A project intended for installation in the A-10 close-support plane. It was basically the Oerlikon 304 RK revolver cannon (which entered Swedish service, slightly modified, as the KCA) with some changes, most obviously the use of aluminium alloy for the cartridge cases. However, in contrast to the GAU-8/A, the gun and ammunition retained electric rather than percussion priming and a metal rather than plastic driving band. The gun was given pneumatic rather than cartridge charging, and a pneumatic clearing mechanism. The rate of fire was slightly higher than the KCA, at around 1,500 rpm. See the Switzerland section for details of the Oerlikon KCA gun.

The planned installation in the A-10 (developed by Hughes) was self-powered and consisted of two belt-fed guns supplied by two magazines. The system weighed 465 kg empty, and 1,033 kg loaded with 750 rounds. Two system prototypes were made and about 35,000 rounds fired without serious problems, although barrel life was short, averaging about 600 rounds.

*30 mm XM140 WECOM*: 30 × 100B (WECOM) AMMUNITION.
This was the first of three 30 mm US guns to be developed for arming attack helicopters (the other two being the XM188 and the eventually successful M230). The requirement was established by the US Army Weapons Command in 1963 and the proposal from the Springfield Armory was selected. It was known at different times as the Gun Type Aerial System (GTAS), the Area Point Weapon System (APWS) and the Weapons Command, 30 mm (WECOM-30), and

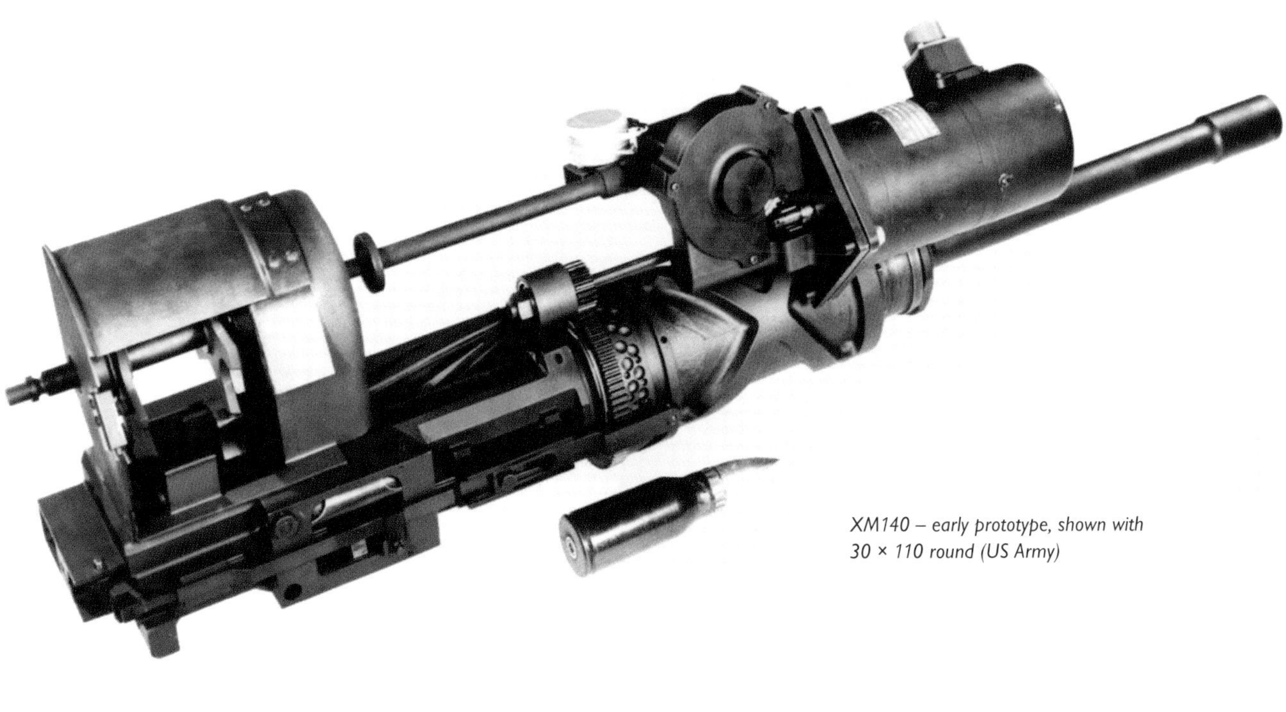

*XM140 – early prototype, shown with 30 × 110 round (US Army)*

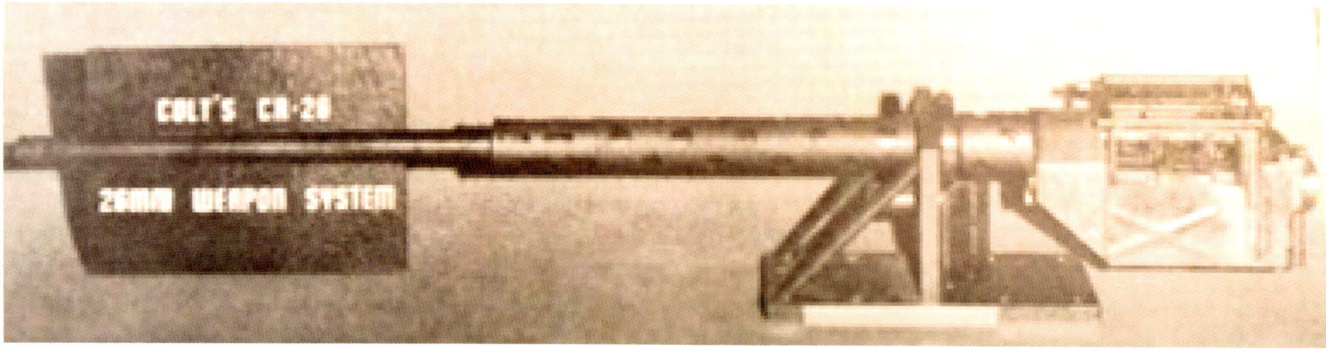


*Colt CR-26 (from Colt brochure, thanks to Dick Fraser)*

impulses and make automatic weapons much more controllable. His work was nearly all focused on small arms of up to .50 calibre but he did design one cannon: the CR-26.

The CR-26 was developed in the late 1960s while Robinson was working at Colt and was intended (to quote from the brochure) to offer 30 mm levels of performance with the dimensions and weight of a 20 mm gun, and the trunnion reaction (maximum recoil thrust) of a 0.50-inch MG. It was designed to be suitable for vehicle mounting, particularly in turrets, and featured recoil operation, delayed extraction and forward ejection in order to minimize the escape of gun gas from the breech. It would have been an interesting competitor in the Bushmaster programme, but by the time that reached the trial stage Robinson had left Colt and there was no-one to continue the development of the gun, of which only two were made.

The bare gun weighed 47 kg, and 54 kg complete with selectable dual feeder, power and manual chargers, and power and manual triggers. It was 201 cm long with an L/64 barrel, and fired at 550 rpm. The barrel reciprocated through 11.4 cm; as the gun fired when the barrel was travelling forwards, this was said to be equivalent to 45 cm long recoil when firing from rest. The peak trunnion reaction was stated to be less than 450 kg.

*27.5 mm GE-127 BUSHMASTER*: 27.5 × 140 INITIALLY, THEN 27.5 × 160 AMMUNITION:
This was GE's competitor for the US Army's Bushmaster requirement to replace the 20 mm M139 in AFV mountings. Work began in 1968 on the GE-127A, but this was soon replaced by the improved GE-127B which was submitted to the Army trials in 1973. The unique family of percussion-primed 27.5 mm ammunition was developed by the Aerojet-General Corporation.

The GE-127 uses a short-recoil operating system based on Browning's action. It has a dual belt feed and achieved a RoF of 550 rpm. For the 127A model, weight with the feed mechanism is 119 kg and overall length is 294 cm. The 127B weighs 106 kg and has a shorter receiver due to repackaging the component parts. The ammunition shape also changed, from a (140 mm) short, fat case to a longer (160 mm) more slender one, but the case capacities and performance remained the same.

***25 mm TRW-6425***: 25 × 137 AMMUNITION:
The US Army's Vehicle Rapid Fire Weapons System (VRFWS – better known as the Bushmaster programme) led to a number of interesting proposals, of which the TRW-6425 was an early contender.

The gun was developed by Thompson, Ramo and Wooldridge Inc, who employed Eugene Stoner (famous for small-arms

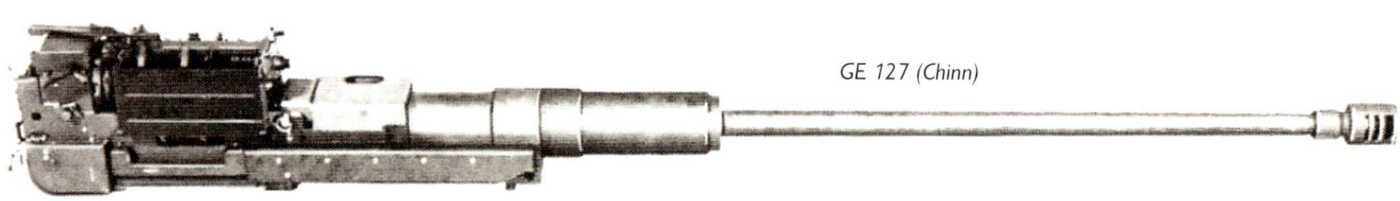

*GE 127 (Chinn)*

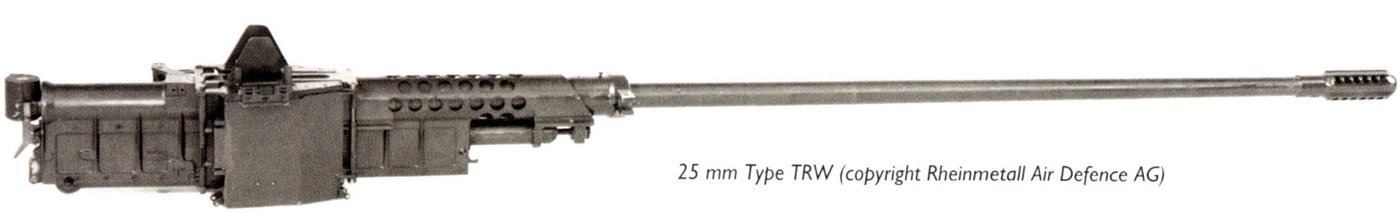

*25 mm Type TRW (copyright Rheinmetall Air Defence AG)*

designs, most notably the 5.56 mm M16 rifle) as a design consultant. Ammunition development was led by Oerlikon. Work commenced in 1964 and the first prototype was test-fired in late 1965. The gun has a short-recoil operating system with a rotating bolt head, and has dual belt feed with a maximum RoF of 550 rpm. Official testing revealed various reliability problems and TRW ceased work on the gun in 1969. Philco-Ford acquired the rights to the gun and developed their PFB-25 from it, with Oerlikon continuing to develop ammunition, both for the PFB-25 and their own gun which became the KBA.

***25 mm PFB-25***: 25 × 137 AMMUNITION.
This was developed by Philco-Ford on the basis of the TRW-6425, the rights to which they had acquired in 1969. It was submitted in the US Army's Bushmaster competition and performed well, but did not enter service. It weighs 88 kg and is 282 cm long, has a dual-belt feed and fires at 550 rpm. It is recoil-operated and has a rotary-locking bolt.

**25 mm M242 BUSHMASTER CHAIN GUN**: 25 × 137 AMMUNITION.
This was the successful competitor in the US Army's VRFWS-Successor (Bushmaster) programme (see the note on the TRW-6425 above), intended to deliver a more powerful cannon to replace the 20 mm M139 in light armoured fighting vehicles such as the Bradley MICV. From 1975 Hughes began to develop a dual-feed, 25 mm version of their new Chain Gun design, and prototypes were delivered to the US Army in February 1978 as the XM-242. It was classified as the M242 in February 1980 and production began in 1981. By December 2014 total production of for all applications had exceeded 11,500 and was still underway for the home and export markets.

The ATK 25 mm M242 cannon uses the same method of operation as other Chain Gun weapons in that all movements and timings are driven and controlled by an external power source. This drives a conventional industrial double-row roller chain which cycles in an oval pattern on four sprockets of which three are idlers and one is driven. Fired cases are ejected forwards out of the turret, while a long dwell-time after firing before the breech opens eliminates gas build-up in turreted installations, as well as minimising the risk of a hang-fire. The action uses a rotating-bolt locking system. The cannon, with a dual (selectable) ammunition feed and internal recoil mechanism, weighs 119 kg and is capable of automatic fire rates of 100, 200, and 500 rpm, or single-shot fire. However, 200 rpm is the normal maximum in US service and various parts have to be changed to attain the highest rate, including a more powerful drive motor. It is fitted with an L/81 barrel.

Improvements have been introduced through the life of the weapon, including an improved triple-spring drive clutch, a high-efficiency muzzle brake, a chrome-lined barrel, an enhanced feeder with removable feed covers, an enhanced receiver, an integral round counter, and a longer recoil stroke. Standard weapons have been upgraded to the enhanced configuration, These improvements resulted in much improved reliability figures, from 6,000 rounds between stoppages originally, up to 22,000 rounds. The barrel life exceeds 13,000 rounds.

The M242 has also found many naval applications. The US Navy took an early interest, issuing in February 1977 an operational requirement for a manually-operated weapon to replace wartime vintage 20 mm Oerlikon weapons. The M242 was designated EX/Mk 38 Automatic Gun System and was installed in a Mk 68 mounting for test firing in August 1980. A different mounting designated Mk 88 was developed for the production system. After production for the USN, the Mk 38 gun system was also chosen by the US Coast Guard in 1991.

In 1994 a remotely-operated version of the Mk 38 was announced by Kollmorgen as the STARC 25 (Stabilized And Remote Controlled 25 mm) naval weapon system, but in 2004 United Defense

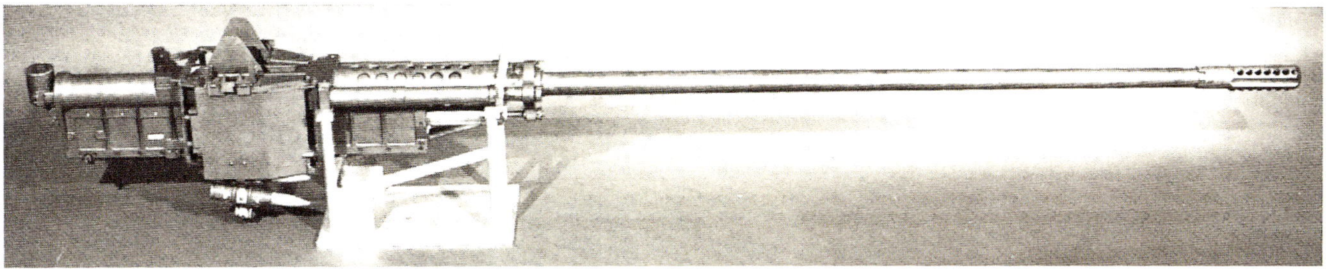

*Philco-Ford PFB-25 (Chinn)*

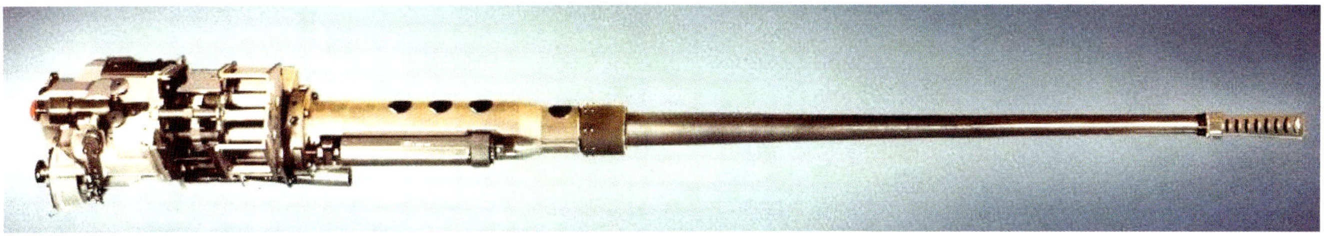

*25 mm M242 'Bushmaster' Chain Gun (ATK)*

(now BAE Systems) and Rafael agreed the former would produce and market a derivative of the latter's Typhoon gun mounting, later designated Mk 38 Mod 2. More than 150 of these systems have been purchased by the US Navy.

**25 mm GAU-12/U**: 25 × 137 AMMUNITION.

The GAU-12/U was a further version of General Electric's rotary cannon family, developed in the late 1970s in conjunction with an external gunpod designed to arm the USMC's AV-8B V/STOL aircraft. Following successful tests, the system entered service with the USMC in 1984. It has since been used in other applications: the AC-130U gunships and the USMC's Light Armoured Vehicle – Air Defense (LAV-AD) for ground-to-air support.

The five-barrelled cannon, also known as the 'Equalizer', fires the standard NATO 25 mm ammunition. Similar in design to the GAU-8/A, it employs the same reversible linkless feed operation to return unfired rounds to the magazine after each burst. The GAU-12/U can be driven by pneumatic, electric or hydraulic means. The gun weighs 123 kg and has L/69 barrels.

The AV-8B Armament System is housed in two detachable pods under the fuselage: the GAU-12/U is housed in the port pod, the 300 round Linear Linkless Feed magazine in the starboard pod; a bridge connects the two. In this installation the gun is driven by a pneumatic motor powered by air bled from the aircraft engine. The cannon can fire at up to 4,200 rpm, and reaches its normal firing rate of 3,600 rpm in 0.4 seconds. The podded system weighs a total of 408 kg empty and 558 kg loaded with 300 rounds.

In 1990 the USAF adapted the GAU-12/U for use in the AC-130U gunship, together with 3,000 rounds of ammunition in a linear linkless feed ammunition system.

**30 mm GAU-13/A**: 30 × 173 AMMUNITION.

This is a version of the GAU-8/A (see above) only with four barrels instead of seven in order to reduce the size and weight so that it could be used in the GPU-5/A gunpod (its only application). The intention was to give a wider range of aircraft (e.g. the F-16) ground attack hitting power comparable with the A-10. The result was a reduction in gun weight from 281 to 154 kg; the GPU-5/A weighs 598 kg or 840 kg with a full load of 353 rounds in its double-ended linkless feed system (which returns fired cases and unused rounds to the magazine). A pneumatic drive system is used with sufficient compressed air stored in the gunpod to fire two ammunition loads (700 rounds) at a maximum rate of 2,400–3,000 rpm. Design commenced in 1976 and deliveries started in 1982, the last of 329 gunpods being delivered in 1986.

In practice, the GPU-5/A proved disappointing; the A-16 (attack variant of the F-16, fitted with the gunpod) was not a success. It appears that due to the heavy recoil the standard attachment points were unable to hold the pod steady enough to achieve the desired accuracy.

***25 mm GE-225***: 25 × 137 AMMUNITION.

The impressive combination of compact size, light weight and high rate of fire demonstrated by the Russian twin-barrel (Gast type) GSh-23 attracted General Electric's attention in the early 1980s. This prompted a design study focused on the standard 25 mm NATO ammunition. The resulting GE-225 was designed to be as modular and flexible as possible to suit it to different purposes, even a choice of externally-driven or gas-operated mechanisms. Overall length is 221 cm. The weight of 82 kg is impressive given the rate of fire of up to 2,500 rpm. Despite this promising performance, this went no further.

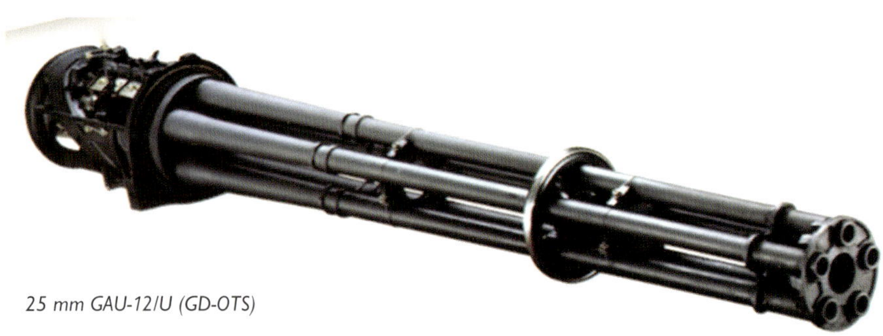

*25 mm GAU-12/U (GD-OTS)*

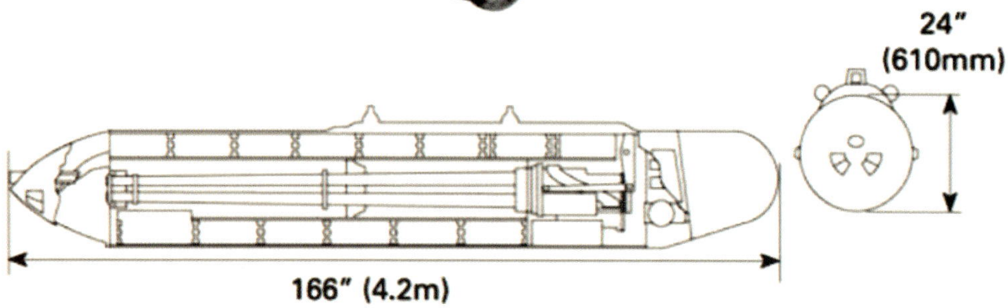


*GPU-5/A gunpod with GAU-13/A cannon (GD-OTS)*

*GE-225 (GE)*

GE carried out an interesting comparison exercise between various different 25 mm gun configurations, both actual and proposed. The table below provides the basic data.

***35 mm ARES TALON***: 35 × 228 AMMUNITION.

The ARES Talon 35 mm lightweight automatic cannon was designed by Gene Stoner initially for the ARES Eagle Air Defence System of 1979 (a twin mounting on an M548 chassis), or for a wide range of other ground combat and AA vehicles. The action was gas-operated, offered either dual feed or single feed powered by propellant gas, had an L/90 barrel and fired at 550 rpm.

***45 mm COMVAT CT***: 45 × 305 CASED-TELESCOPED AMMUNITION.

COMVAT (Combat Vehicle Armament Technology) is one of the few US cannon projects based on unconventional ammunition to have reached an advanced development stage. The cartridge is cased-telescoped: which in this instance means that the case is a cylindrical shape with the projectile buried within it, and most of the propellant surrounding the projectile. This results in a short, if fat, cartridge which provides different handling options to the gun designer.

The programme, which was originally called CVAST (Combat Vehicle Armaments Systems Technology) when it began in the 1980s, was the responsibility of the Close Combat Armaments Center of the US Army's Armament

| GUN | TYPE BARRELS | | RoF | WEIGHT | rpm/kg |
|---|---|---|---|---|---|
| M242 | Chain | 1 | 500 | 119 | 4.2 |
| Oerlikon KBA | gas | 1 | 600 | 112 | 5.4 |
| Mauser Model E | gas | 1 | 900 | 103 | 8.7 |
| **GE 225** | **Gast** | **2** | **2,000** | **82** | **24** |
| CHAG 225 | rotary | 2 | 2,000 | 112 | 18 |
| GE 325 | rotary | 3 | 2,400 | 120 | 20 |
| GAU-12 | rotary | 5 | 4,000 | 160* | 25 |

* = data for prototype at that time: production version is lighter.

*ARES 45 mm XM295*

Research, Development & Engineering Center (ARDEC). The two companies mainly involved were Honeywell's Armaments Systems Division (the prime contractor and responsible for ammunition development) and Ares (responsible for the automatic cannon).

Initially a 35 mm gun firing conventional ammunition was examined, followed by a 30 mm weapon using 30 × 207 Cased Telescoped Ammunition (CTA), before the Army finally decided on a 45 × 305 CTA gun, with a rotating chamber (more details of this ammunition, and about CT systems in general, are in the Unconventional Systems section).

However, in 1996 the Inspector General of the Department of Defense investigated allegations that long search for an effective CT weapon was a waste of money, and concluded that the technology was 'ballistically inefficient with inherent performance, weight, volume and cost problems'. The US accordingly stopped funding CTA programmes. Despite this, the COMVAT work was continued by GIAT (now Nexter) with the addition of BAE Systems, resulting in the 40 CTA which is about to enter service with both the UK and France (see International section).

**30 mm M230 / M230LF / XM914:**
30 × 113B LW AMMUNITION.

In 1972 Hughes Helicopters began researching an electrically powered cannon mechanism, initially to fire the US 20 × 102 M50 series ammunition. The mechanism chosen became known as the 'Chain Gun' on account of its closed-loop chain drive system, and is described in the M242 record above. In the same year the US Army requested proposals for a new 30 mm cannon to arm the Advanced Attack Helicopter (AAH) then under development, so the design was altered to accept the specified 30 × 100B WECOM cartridge (Army WEapons COMmand). The prototype was successfully fired in 1973 but in 1976 the specified cartridge was changed again, this time to the existing NATO 30 × 113B ADEN/DEFA round. The gun was chosen to equip the AAH (which was adopted as the AH-64) and received the M230 designation. The AH-64A equipped with the M230 entered service with the US Army in 1984. The M230 weapon system (including associated sensors, sights and fire control system) is known as the Area Weapon System.

The M230 weighs 56 kg, uses an L/36 barrel, and fires at 625 rpm, a rate it reaches 0.2 secs after start. In the AH-64 a linkless feed ammunition supply is used with a maximum capacity of 1,200 rounds, but in practice this is commonly reduced to 300 rounds as a long-range internal fuel tank is fitted instead. Weight is saved by using aluminium alloy or lightweight steel cartridge cases, with the standard combat round being the M789 High Explosive Dual Purpose (HEDP) cartridge.

The AH-64 family remains by far the principal users of the M230, but there are also special purpose applications such as the SOCOM MH-60L helicopters (which

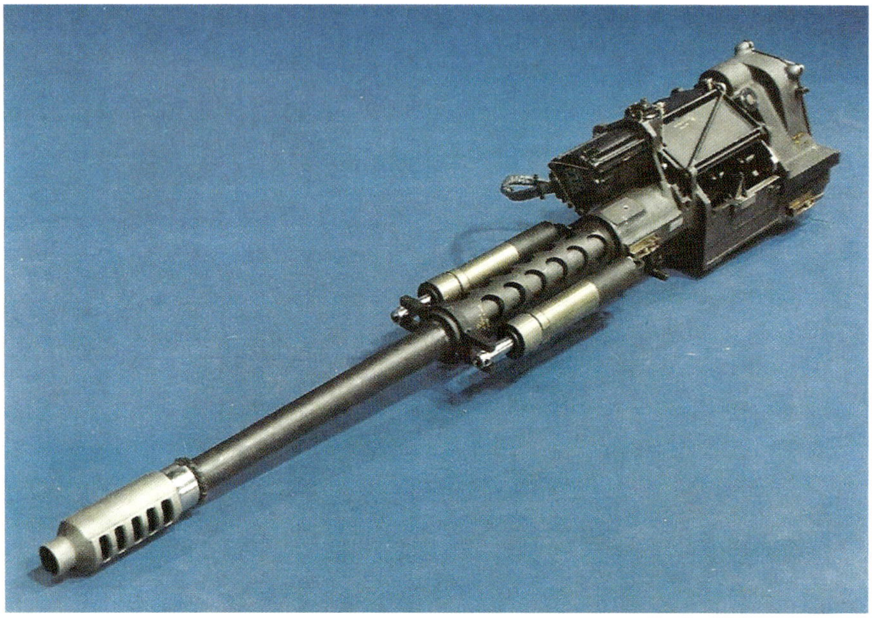

*M230 Chain Gun (ATK)*

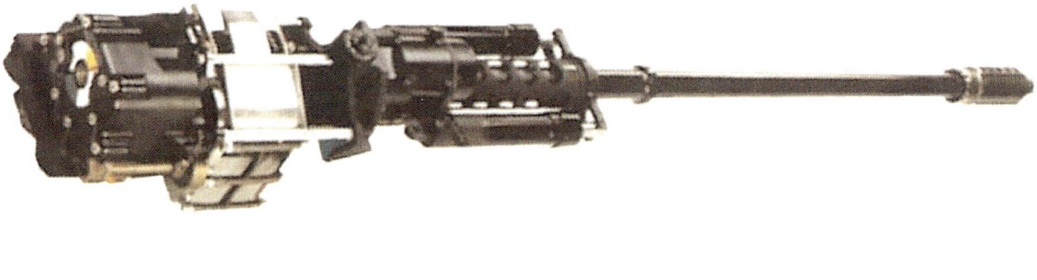

*M230LF (ATK)*

can be fitted with a pylon-mounted fixed forward-firing gun). Most attention as far as new applications are concerned has switched to the M230LF.

The M230LF (Link Fed, indicating that it uses an ammunition belt instead of a linkless feed) differs from the M230 in two other respects: the barrel is lengthened to L/51, increasing muzzle velocity from 800 to 838 m/s, and the rate of fire is reduced from 625 to 200 rpm. Mean time between failures is reported as 22,000 rounds. It is intended primarily for installing on remote weapon stations (RWS) on light vehicles and several different RWS are being marketed for it. The relatively light recoil compared with other, high-velocity, 30 mm or even 25 mm cannon allows it to be fitted to lightweight RWS able to be fitted onto small vehicles. Two versions of the M230LF were developed: LF-1 is intended for installation in heavier vehicles such as an infantry combat vehicle. The LF-2 has hydraulic recoil dampers and a longer recoil stroke, which cut the peak recoil load by almost 75% to enable it to be fitted to light vehicles and small boats.

Other applications for the M230LF have also been developed, e.g. in a side-firing mounting in the AC-235 Light Gunship, a joint project between ATK and KADDB of Jordan. In 2020 the M230LF was redesignated the XM914. In mid-2020 Kongsberg Protech Systems USA were awarded a contract from the USMC to supply XM914E1 Remote Weapon Stations for installation on the Joint Light Tactical Vehicle (JLTV). This will include a number of specialized counter-unmanned aircraft systems (C-UAS) and air defence capabilities.

**30 mm ASP-30 COMBAT SUPPORT WEAPON**: 30 × 113B AMMUNITION.
The ASP-30 (Automatic, Self-Powered, 30 mm cannon) was developed by McDonnell Douglas Helicopter Systems to be readily interchangeable with .50/12.7 mm Browning M2 HB heavy machine guns while providing vastly superior firepower. This is achieved by means of built-in recoil reducers which enable the gun to be used on virtually any existing vehicle mount or, when dismounted, on a standard .50 M3 tripod for use in the direct infantry support role. The gun is gas operated with a rotating bolt. The belt feed is to the left side and the short length of the receiver behind the feed allows simple and easily controlled spade grip operation from any cupola mounting. The ASP has an L/44 barrel and weighs 52 kg. The RoF is 500 rpm.

The ASP was first fired in 1984 and more than 20 examples were built for testing, but the gun failed to achieve any sales.

*30 mm MF*: 30 × 113B AMMUNITION.
The US company Support Weapons Corp (SWC) announced in 1991 that it had developed the MF 30 mm cannon, which fires standard ADEN/DEFA rounds and therefore uses electrical priming. It is recoil-operated, fires from a locked bolt and can operate in semi-automatic (closed bolt) and fully automatic (open bolt) modes, at firing rates up to 450 rpm. Ammunition can be fed from a belt, a 10-round box or a gravity-feed system mounted on top of the gun. Peak recoil was stated to be similar to the .50 M2, achieved by an unspecified recoil-spreading system, and the gun was designed for use on standard .50 M2 and 40 mm MK19 mountings. It was said to weigh less than 50 kg and to be 170 cm long. Successful firing trials were said to have taken place attached to an ultra-light aircraft.

The MF 30 will fire any ADEN/DEFA ammunition including the LW30 range developed for the M230. One interesting variation is that it was also designed to fire a specially-developed 'mortar' round in conjunction with an SWC ground mount which permitted firing at 45°. This round (in development at that time) was intended to contain just enough propellant to cycle the action, allowing a much longer and more destructive HE/Frag or HEDP shell to be fired.

*20 mm XM301 AND XM915*:
20 × 102 AMMUNITION.
This XM301 three-barrel rotary is in effect an updated version of the M197 helicopter gun. It was originally intended to arm the RAH-66 Comanche reconnaissance / attack helicopter developed in the 1990s, which was designed to be a lightweight, stealthy complement to the

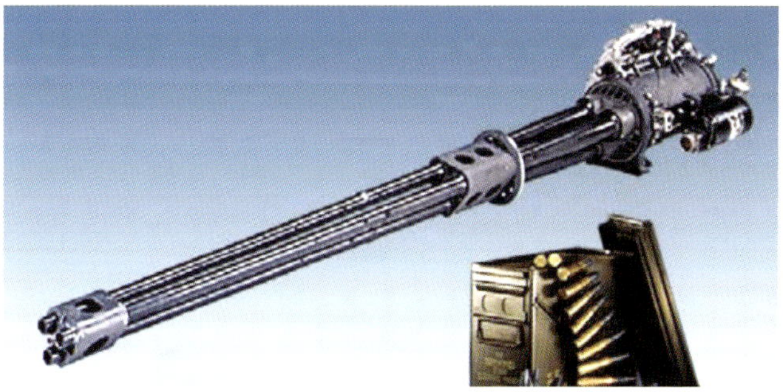
*XM301 (GD-OTS)*

*ASP-30 (MDHC)*

AH-64 Apache. The costs escalated to the point where the RAH-66 was cancelled in 2004 after only two were built, and the XM301 was accordingly shelved for a while, but has since been revived and marketed. Compared with the M197, weight is reduced from 66 to 34.5 kg. The maximum rate of fire is 1,500 rpm (750 rpm also being available) and a 500-round linkless feed ammunition system was installed in the RAH-66.

In June 2020 the US Army announced that the contenders in the Army's Future Attack Reconnaissance Aircraft competition would be working with Picatinny Arsenal to outfit those aircraft with a 20 mm three-barrel Gatling gun, designated XM915. At the time of writing, the only information available was that the gun would weigh less than 52 kg and would fire at up to 1,500 rpm. It is to be mounted in a turreted cradle and have a linkless ammunition feed. At the time of writing, it was unclear exactly what relationship the XM915 has with the XM301.

**30 mm BUSHMASTER II / MK 44 / GAU-23 / XM813**: 30 × 173 AND 40 × 180 AMMUNITION.

The ATK Bushmaster II Chain Gun is closely based on the 25 mm Bushmaster M242, modified as necessary to accommodate the significantly larger and more powerful 30 mm ammunition already in service in the GAU-8/A and the Mauser MK 30. The gun can also fire 30 × 170 ammunition (Rarden or Oerlikon KCB) with a change of the barrel, bolt and rear feed plate. A further development is the potential use in the Mk 44 Stretch and XM813 of 'Super 40' ammunition with little more than a barrel change required.

The Bushmaster II and Mk 44 have a dual ammunition belt feed to allow two types of ammunition to be rapidly selected. The gun has an L/80 barrel and weighs 160 kg. The cyclic rate is a maximum of 200 rpm (variable). The mechanism is essentially the same as described in the M242 entry.

Initial applications of the Bushmaster II were concerned with equipping heavy infantry combat vehicles as a more effective replacement for the 25 mm Bushmaster. It was first chosen by the Norwegian Army for their CV9030 infantry fighting vehicle (IFV), with deliveries beginning in 1995. It has also been selected for similar applications (some remotely controlled) by many other armies including those of Switzerland, Finland, Singapore, Portugal, Poland, the Czech Republic and the Republic of Ireland.

Naval interest in the gun prompted the development of a marinised version featuring greater corrosion resistance and this was qualified as the Mk 44. All production then switched to this version as it is more economical to maintain a single production line. It is now in use on naval mountings on US and British warships, mainly for close defence against small boats. The US naval mounting is designated the Mk 46, the British one is the DS30M Mk2, also known as the Automated Small Calibre Gun (ASCG). By late 2014 the 30 mm Mk 44 had been

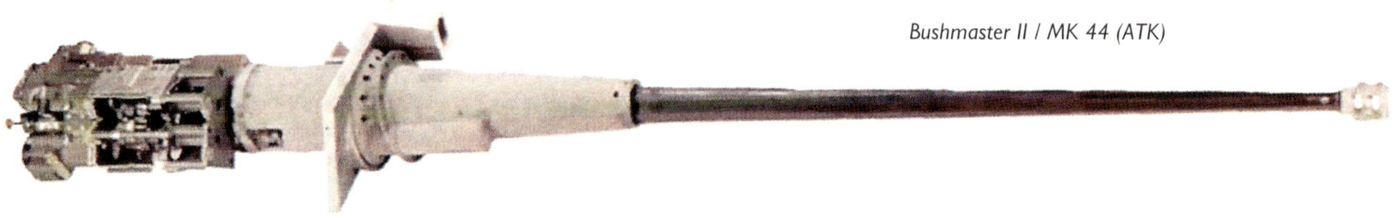

*Bushmaster II / MK 44 (ATK)*

*Two contrasting AFV applications of the Bushmaster II / Mk 44: a Finnish CV9030 with a manned turret (ekeb / Wikimedia Commons / public domain), and a remotely-controlled mounting on a Czech Pandur II (CS92 / Wikimedia Commons / CC-BY-SA-3.0 unported licence)*

*Mk 44 cannon in the US Mk 46 mounting (MC3 Patrick Dionne/U.S. Navy / public domain)*

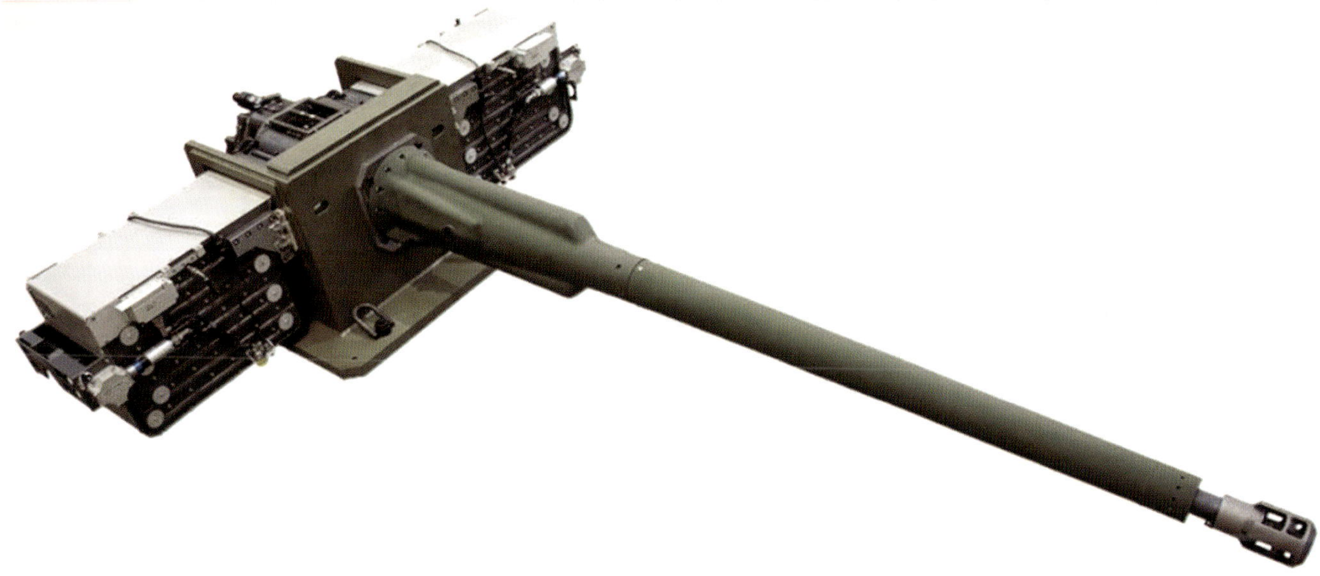

*XM813 cannon with magazines attached (ATK)*

adopted by at least 21 countries for a variety of air, land, and sea applications with well over 2,000 weapons manufactured. The Mk 44 selected for use in US gunships is designated GAU-23.

The latest version of the gun is the XM813, optimized for use in AFVs. The most obvious change is the replacement of the dual belt feed with a linkless feed system with two magazines holding 150 rounds and elevating with the gun, avoiding the reliability problems often caused by the belt feeds flexing through extreme elevation changes. The XM813 also has a heavier barrel and incorporates a fuze setter to fire Air Bursting Munitions (ABM). Finally, it has been designed for rapid conversion to the Super 40 (40 × 180) ammunition if required. The first application of the XM813 is to equip the US Army's M1296 Stryker Dragoon 8x8 AFVs.

**35 mm BUSHMASTER III CHAIN GUN**: 35 × 228 AMMUNITION (POSSIBILITY OF 50 mm VERSION).
A further scaling-up of the ATK Chain Gun design, this time to take standard 35 mm Oerlikon ammunition. Development of this cannon was announced in 1989

with the prototype first firing in 1990. However, there was a long pause before the first cannon entered service in 2007. So far, this cannon has only been used in the CV9035 IFV made for Denmark and the Netherlands, with Estonia receiving some of the Dutch vehicles from 2014.

As with the M242 and Mk 44 this is primarily intended for ground fighting (plus the odd unwary helicopter or UAS that might stray within the parameters of its fire control system) so has the emphasis on accuracy rather than a high rate of fire. One advantage possessed by guns in this calibre is the possibility of using AHEAD ammunition, suitable for engaging a wide range of targets; the gun is available with the muzzle-mounted fuze setter required. The CV9035 has a manned, two-person turret but KMW has developed an RWS for this gun, designated IWS-35.

At least three proposals for necking-out the 35 mm cartridge case to 50 mm calibre have been made in order to generate more power and fire larger projectiles. The original effort dates back to the 1980s for the abortive Rheinmetall Rh 503, another externally powered gun designed from the start to fire either standard 35 mm ammunition or the 50 × 330 'Supershot', with little more than a barrel change. The very long 50 mm cartridge case was optimized for APFSDS with only the tip of the puller sabot protruding from the case – the rest of the projectile was buried within it. Since then, US Army proposals from the late 2010s for acquiring the Bushmaster III at some point in the future have included the possibility of a 50 mm upgrade, which has now emerged as described in the 50 mm XM913 record below. A third variation is the 50 mm EAPS, also described below.

The gun weighs 218 kg, has an L/80 barrel, and fires at up to 200 rpm. Other details are as for the M242.

***40 mm BUSHMASTER IV CHAIN GUN***: 40 × 365R AMMUNITION.

This was the result of a collaboration between ATK and Bofors in the early 2000s, and was designed to use Bofors L/70 APFSDS and 3P ammunition. A prototype was built but it appears that lack of interest has led to it being shelved. Overall weight, including a full dual-feed magazine of 30 rounds, is 400 kg and an L/65 barrel is used. Rate of fire is variable from 1 to 200 rpm.

**20 mm A-50**: 20 × 102 AMMUNITION.

This is a three-barrel version of the M61A2 aircraft cannon, developed specifically for the Korea Aerospace Industries (KAI) A-50 light attack variant of the company's T-50 Golden Eagle supersonic trainer. Airborne trials took place in 2004 and 2005 and the gun is in service. It fires at 3,000 rpm and the A-50 installation has a linear linkless closed-loop feed system holding 204 rounds. The gun is powered by the aircraft's hydraulics system. Weight of the complete system (empty) is 134 kg, of which the gun weighs about 80 kg; it is fitted with L/76 barrels.

**25 mm GAU-22/A**: 25 × 137 AMMUNITION.

The GAU-22/A was developed as a part of the US Joint Strike Fighter (JSF) programme. The two competitors – Boeing offering the X-32 and Lockheed Martin with the X-35 – both provisionally selected the Mauser BK 27 as the most suitable gun. However, after the X-35 had been selected in 2001, Lockheed Martin awarded the gun armament contract to General Dynamics, who instead offered their existing GAU-12/U gun, in service with the USMC's AV-8B aircraft. The F-35 (as it now is) suffered weight problems in development, as a result of which General Dynamics modified the gun design, reducing the number of barrels from five to four, resulting in the GAU-22/A.

The GAU-22/A gun systems completed their critical design reviews in July 2005. The first cannon was completed in January 2006 and in November 2008 the first contract to start production of the guns was awarded. Qualification testing was completed in June 2010. The gun

*35 mm IWS35 (KMW)*

weighs 104 kg and has L/69 barrels. The maximum rate of fire is 3,300 rpm.

Only the USAF's F-35A has the cannon installed internally (together with 181 rounds of ammunition in a linear linkless feed); for the USMC's F-35B and the USN's F-35C the optional cannon is in an external conformal pod with 220 rounds in a helical feed wrapped around the barrel cluster. The system weights are 184 kg for the internal system and 445 kg for the gunpod.

***50 mm EAPS***: 50 × 330 AMMUNITION.
The Extended Area Protection and Survivability programme (EAPS) was proposed in the 2010s to deliver C-RAM protection for military bases at a longer range than the 20 mm Centurion (Phalanx) system could provide. Various missile proposals were made, but the 50 mm gun-based one is interesting in revealing a mix of characteristics. Although the 330 mm long case is similar to the one proposed for the 1980s Supershot (see Bushmaster III record above), it is loaded with a much longer, fin-stabilized course-correcting projectile making the overall cartridge length considerably greater at 538 mm (the same as the 40 × 365R Bofors round). The gun to fire this round has a mix of Bushmaster III and Bushmaster IV elements: it combines the action length of the B.IV with the bolt dimensions of the B.III, and weighs 230 kg. Rate of fire is 200 rpm, and it was proposed to mount two of the guns in a vehicle turret.

***50 mm XM913***: 50 × 228 AMMUNITION.
In October 2018 the US Army Research Laboratory (ARL) revealed that it was working on an AFV gun in 50 mm calibre, designated XM913. The ammunition is based on the 35 × 228 Oerlikon only necked-up to the larger calibre. In contrast with the 50 mm Supershot, the cartridge case is not extended in length, so more of the projectiles are visible when loaded. The XM913 is based on the Bushmaster III from Northrop Grumman Defense Systems. At the same time, General Dynamics announced their interest in developing a gun in this calibre, and orders have been placed by the US Army. The first four of an initial batch of eleven XM913 have been delivered by Northrop Grumman, with an order for another ten expected.

The XM913 is being proposed for arming whatever wins the US Army's Next Generation Combat Vehicle (NGCV) programme, intended to replace the Bradley AFV.

*On show at AUSA 2019: the Super50 XM913 gun in the foreground, compared with the 30/40 mm XM813 Bushmaster in the middle and the 30 mm XM914 (M230LF) in the background (Courtesy G.M. Knowles)*

# OTHER NATIONS

During World War I there was considerable activity in developing cannon in Austria-Hungary. By the end of the war they reportedly had one 20 mm gun (given as the *Szanáts*, but this might have been the *Szakats*) and three different 25 mm cannon (the *Boykow-Czerney*, the *Szebeny* and one designed by the *Technischen Militärkomitee*) under development, but details of how these weapons operated have not emerged.

## UKRAINE

The Ukraine manufactures two cannon which are versions of Russian weapons: the ZTM-1 (2A72) and the ZTM-2 (2A42).

## INDIA

India also manufactures Russian cannon under different names: the Gasha (GSh-23) and the Medak (2A42).

## ISRAEL

***MBT G360***: 20 × 110 HS AMMUNITION.
This multi-purpose light cannon was advertised c.1970s by IAI (Israel Aircraft Industries Ltd) as a low-cost alternative to a heavy machine gun. It appears to have been based on the Russian 14.5 mm KPV, suitably adapted for the 20 mm Hispano ammunition. It was 204 cm long, weighed 50 kg, and had a RoF variable up to 600 rpm.

## ROMANIA

**30 mm A436 MODEL 1980**:
30 × 210B AMMUNITION.
The A436 is based on the Russian NN-30 gun as used in the AK230 naval AA mounting. As such, it is a 4-chamber revolver cannon, powered by recoil rather than the usual gas. The most obvious difference between the NN-30 and A436 is that the Romanian gun is air-cooled, lacking the water+evaporation cooling system installed in the AK230, and it is presumably as a result of this that the RoF is reduced from 1,000 to a maximum of 600 rpm, with the fire control system normally offering a choice of 100, 250 or 500 rpm. The cartridges are linked together but come in 30-round boxes, presumably for ease of handling.

From 1990 the A436 cannon were upgraded to A436M, which had various detailed improvements.

The A436M is installed in a twin-gun, four-wheel mounting designated the Model 80, which is compatible with the Vifor air defense system. As well as seeing Romanian service (in which it was reported in 2016 to be still surviving) the Model 1980 was also exported to Liberia, and from there some ended up being fitted to Libyan coastguard vessels.

## SERBIA AND YUGOSLAVIA

The best-known Yugoslavian-made cannon is the licence-produced Hispano-Suiza HS 804 in 20 × 110 HS calibre. This was installed in two towed AA mountings: the triple-gun M1955 and the single M75. These had very simple sights and were intended for clear-weather use only.

**30 mm ZASTAVA M86/M89 / M97/12**:
30 × 210B / 30 × 173 AMMUNITION.
In the 1980s Zastava developed a purpose-built gun in two versions, intended for arming AFVs for the AA and ground-fighting roles. Unusually, the Russian naval 30 × 210B cartridge was selected as the ammunition. This is a gas-operated gun with an integral gas buffer and spring recuperator to soften the recoil. The

*Two views of the Romanian M80 (Courtesy of N. R. Jenzen-Jones of ARES)*

belt feed is powered by recoil energy which generates enough power to lift a 40-round length of belt; fired cases are ejected forwards. Hydraulic systems perform the bolt cocking and release the sear. Electric priming is used, controlled by a 24 V electric circuit. The gun weighs approximately 200 kg, has an L/70 barrel and fires at 650–750 rpm. The M86 has a single ammunition feed, which can be switched from one side the other. The M89 differs only in providing dual belt feed to permit two different ammunition natures to be available for immediate use. The M97/12 is the latest development, which is available in 30 × 173 as well as 30 × 210B. The two principal applications of these guns are in twin mountings installed on two SPAAGs; the BOV-30 4x4, which is in Serbian service, and the Model 30/2 based on the tracked BVP M80A, which has not been adopted but is being marketed.

## SOUTH KOREA

South Korea has a well-developed armaments industry but until the recent development of the K40 (see below) did not have any indigenous cannon. They do use the 40 mm Bofors as a naval gun, the 35 mm Oerlikon as an AA gun, the 20 mm Vulcan in towed and vehicle-mounted AA systems, and the 30 mm Oerlikon KCB in a twin turret mounting on the K30 Bi Ho (Flying Tiger) SPAAG. The KW2, by Rotem, is another twin-30 mm SPAAG but on an 8x8 rather than tracked chassis; it does not appear to have attracted sales so far.

**S&T DYNAMICS K40**:
40 × 365R AMMUNITION.
This gun was specifically developed to arm the new K21 NIFV (Next Infantry Fighting Vehicle) and is intended primarily for ground fighting, although slow-moving aircraft could also be engaged. The gun is installed in a two-man stabilized turret. It uses the same ammunition as the 40 mm L/70 Bofors and the design is based on the Swedish gun. It is recoil-operated and ejects fired cases from the top of the gun. There is a linkless feed magazine holding 24 rounds, with another 600 being available in the vehicle. Multipurpose fuzes can be programmed by a contactor in the feed mechanism; the country makes its own ammunition, including an APFSDS round. The K40 was designed in the early 2000s and completed its development trials in 2007. The K21 IFV entered service in 2009.

# TABLE I – THE CARTRIDGES

TABLE 1: 20 mm to 57 mm CARTRIDGES FOR AUTOMATIC CANNON
FONT STYLE KEY: **current service;** ***current development;*** obsolete service; *obsolete experimental*
BODY DIAM = the diameter of the main body of the case at its widest point, just above the rim, extractor groove or belt, depending on the case configuration.
NATION = UN 2-letter nationality code (plus INT for international); the current codes are used, e.g. RU is used for both Russia and the USSR.
PROJECTILE = a typical type and weight of projectile is given, with the associated muzzle velocity (MV) and muzzle energy (ME). Other loadings are described in the text.
In ERA / NAME / WEAPONS column: (e) = electric priming; a/c = primarily aircraft use

| METRIC DESIGNATION | RIM DIAM | BODY DIAM | NATION | PROJECTILE | MV m/s | ME Joules | ERA / NAME / WEAPONS |
|---|---|---|---|---|---|---|---|
| 20 x 70RB | 20.2 | 21.8 | DE | HE /T 128–145 | 490 | 15–17,000 | WW1 Becker, 1920s/ early 1930s Oerlikon Typ F |
| *20 x 70RB (E)* | *21.0* | *23.0* | *DE* | *HE / 128* | *670* | *28,700* | *WW1 Ehrhardt a/c, Scotti M35* |
| *20 x 70 HS 821* | *24.8* | *23.8* | *CH* | *HE / 90* | *650* | *19,000* | *1960 HS 821 infantry gun* |
| *20 x 71 HS 405* | *24.8* | *24.8* | *FR* | *HE / 123* | *600* | *22,000* | *WW2 HS 405 flexible a/c* |
| 20 x 72RB | 19.0 | 21.9 | CH/JP | HE / 128 | 605 | 23,400 | 1934+ Oerlikon FF, IJN Type 99-1 a/c |
| 20 x 80RB | 19.0 | 22.0 | DE | HE / 92 | 700 | 22,500 | WW2 Ikaria MG-FFM (M-Geschoss) a/c |
| **20 x 82 (83.5)** | **25.2** | **25.2** | **DE / ZA** | **HE / 92–115** | **800–720** | **29,000** | **WW2 Mauser MG 151/20 (some e) a/c; Denel Land Systems GA-1** |
| 20 x 94 | 24.9 | 25.0 | JP | HEI / 79 | 750 | 22,200 | WW2 Ho-5 / Type 2 a/c |
| *20 x 96* | *26.7* | *26.9* | *FR* | *?* | *863* | *?* | *post-WW2 HS 421/422 and MAC Mle 1949 a/c* |
| 20 x 99R | 25.2 | 21.8 | RU | HE / 96 | 790 | 30,000 | WW2 ShVAK, Berezin a/c |
| 20 x 99RB | 20.0 | 21.9 | CH | HE / 128 | 670–700 | 28,700–31,400 | 1920s/early 1930s SEMAG / Oerlikon Type L, IJA Type 94 Flex |
| 20 x 101RB | 19.0 | 21.9 | CH/JP | HE / 128 | 750 | 36,000 | 1935 Oerlikon FFL, IJN Type 99-2 a/c |
| **20 x 102** | **29.5** | **29.1** | **US** | **HEI / 102** | **1,030** | **53,600** | **1950s M39, M61, M197, A-50, ZPL-20, Nexter 20M621 (e)** |
| 20 x 105B | 25.0 | 25.0 | CH | AP / 146 | 725 | 38,100 | 1930s 'Short Solothurn', Lb 204 |

| METRIC DESIGNATION | RIM DIAM | BODY DIAM | NATION | PROJECTILE | MV m/s | ME Joules | ERA / NAME / WEAPONS |
|---|---|---|---|---|---|---|---|
| 20 x 105 | 24.8 | 26.0 | DE | HE / 134 | 760 | 37,700 | WW2 Rheinmetall-Borsig MG 204 a/c |
| *20 x 108* | *25.2* | *25.7* | *DE* | *HE / 134* | *700* | *32,800* | *WW2 Rheinmetall-Borsig Lb 201a/202 a/c* |
| **20 x 110RB** | **21.9** | **24.9** | **CH** | **HE / 122** | **830** | **42,000** | **1930s Oerlikon F, S, Polsten** |
| *20 x 110 (Scotti)* | *24.8* | *24.9* | *IT* | *HE / 128* | *850?* | *46,200?* | *M1932 (?) Scotti* |
| **20 x 110 (HS)** | **24.8** | **24.8** | **FR/GB/ US** | **HE / 130** | **880** | **50,300** | **WW2 HS 404, Hispano, AN-M2, M3, M24 (e), Mk 16, HS 804** |
| *20 x 110 (AP-20)* | *31.0* | *31.0* | *RU* | *HE / 129* | *940* | *56,500* | *1937 Atslega AP-20, DShAK* |
| *20 x 110R* | *31.0* | *27.0* | *RU* | *HE / 129* | *?* | *?* | *WW2 TsKBSV-75 (AA version of ShVAK)* |
| **20 x 110 (USN)** | **29.5** | **29.3** | **US** | **HE / 110** | **1,010** | **56,100** | **1950s Mk 100 series; Mk 11, Mk 12 (e) a/c** |
| 20 x 113 | 28.0 | 28.0 | FI | AP / 148 | 700 | 36,300 | 1930s Lahti L-34 'boat gun' |
| 20 x 113B | 28.0 | 28.0 | FI | HE / 136 | 800 | 43,500 | 1930s Tampella L-34 |
| *20 x 115SR* | *29.0* | *27.4* | *FR* | *AP / 102* | *c.870* | *c.39,000* | *1918 Delaunay-Belleville AT MG* |
| 20 x 120 | 28.9 | 29.0 | DK | HE / 127 | 840 | 44,800 | 1930s Madsen |
| *20 x 120RB* | *20.0* | *25.2* | *BE?* | *?* | *?* | *?* | *1930s Simoneta* |
| *20 x 122* | *32.3* | *32.3* | *FR* | *HE / 117* | *c.1,000* | *c.58,000* | *1950s 5CG AA* |
| 20 x 124 | 28.5 | 28.7 | JP | HEI / 132 | 820 | 44,400 | WW2 Ho-1, Ho-3 a/c |
| *20 x 125SR* | *30.2* | *27.5* | *GB* | *128.5* | *884* | *50,000* | *1930s Vickers 0.8 inch AT MG* |
| *20 x 126B* | *33.2* | *32.1* | *FR* | *HE / 117* | *c. 1,000* | *c.58,000* | *1950s 5CG AA* |
| **20 x 128** | **32.0** | **32.3** | **CH** | **HE / 125** | **1,100** | **75,600** | **1950s Oerlikon KAA, KAB, KAC, KAE, Meroka CIWS (e)** |
| *20 x 135B* | *30.5* | *30.4* | *DE* | *?* | *1,000* | *?* | *WW2 Krieghoff MG 301* |
| *20 x 135* | *32.0* | *32.0* | *DE* | *HE / 117* | *1,000* | *58,500* | *WW2 Mauser MG 213C / MG 213/20 (e)* |
| *20 x 136B* | *33.2* | *32.0* | *GB* | *HE / 83* | *1,200* | *59,800* | *1950s ADEN (e)* |
| 20 x 138B | 27.0 | 27.0 | CH | HE / 115 | 900 | 46.600 | 1930s FlaK/KwK 30/38 |
| 20 x 139 (FMK) | 28.5 | 28.5 | CH | HEI / 131 | 1,010 | 66,800 | WW2 FMK / Flab Kan 38 (rim 2.6 mm thick) |
| **20 x 139** | **28.5** | **28.5** | **CH** | **HE / 120** | **1,050** | **66,100** | **1950s HS 820, Oe KAD, Rh 202, 20M693, Vektor F2 (rim 3.5 mm thick)** |
| 20 x 140B | 28.0 | 28.0 | PL | 134 | 856 | 49,000 | WW2 wzór 1938 AA/AT/AFV |
| 20 x 143 | 33.4 | 33.5 | JP | HET / 132 | 950 | 59,600 | WW2 Army Type 98 AA/AT |

| METRIC DESIGNATION | RIM DIAM | BODY DIAM | NATION | PROJECTILE | MV m/s | ME Joules | ERA / NAME / WEAPONS |
|---|---|---|---|---|---|---|---|
| 20 x 145R | 29.5 | 25.4 | SE | AP / 145 | 800 | 46,400 | 1930s Bofors m/40 AA/AT |
| 20 x 154B | 33.2 | 32.2 | FR | ? | ? | ? | 1950s AME 20M571 |
| 20 x 157 | 33.4 | 33.5 | JP | ? | ? | ? | 1930s Army Type 94 AA |
| 20 x 158RB | 30.4 | 33.4 | US | TP / 130 | 1,150 | 86,000 | 1945+ T5 case for T33 aircraft gun |
| 20 x 167SR | 36.7 | 34.1 | GB | 146 | 869 | 55,000 | interwar .8 inch Elswick AT |
| 22 x 138B | 27.0 | 27.0 | DE | HE / 150 | 880 | 58,100 | WW2: necked-up 20 x 138 B |
| 23 x 106 | 29.0 | 29.0 | DK | HET / 174 | 720 | 45,100 | 1930s Madsen a/c |
| **23 x 115** | **27.0** | **27.0** | **RU** | **HEI / 184** | **700** | **45,000** | **NS-23, NR-23, GSh-23, GSh-6-23 a/c** |
| 23 x 122 | 34.5 | 34.5 | FR | HE / 200 | 900 | 81,000 | WW2 HS 406, 407 a/c |
| 23 x 133 | 31.5 | 32.5 | CH | HE / 160 | 1,050 | 88,200 | 1960s HS 827 afv |
| 23 x 138B | 27.0 | 27.0 | DE | ? | ? | ? | WW2: necked-up 20 x 138B |
| 23 x 139 HS | 28.5 | 28.5 | CH | HE / 135 | 1,085 | 79,400 | 1950s HS 823 AA |
| 23 x 139 (.90") | 29.1 | 29.1 | US | HE-T / 204 | 823 | 69,100 | late 1930s .90" T2E1 a/c |
| 23 x 139RB (.90") | 24.0 | 29.1 | US | HE-T / 204 | 823 | 69,100 | late 1930s .90" T2 a/c |
| 23 x 139SR (.90") | 30.9 | 29.1 | US | HE-T / 204 | 823 | 69,100 | late 1930s .90" T3, T4 a/c, a/t |
| 23 x 146SR (.90") | 43.5 | 41.0 | US | HE-T / 204 | 869 | 77,000 | late 1930s .90" T1 a/c |
| 23 x 152 | 32.2 | 31.4 | RU | 184–190 | 870–890 | 72,000 | WW2 MP-6 a/c |
| 23 x 152B | 33.4 | 33.2 | RU | API / 199 | 890 | 78,800 | WW2 VYa (brass) a/c |
| **23 x 152B** | **33.4** | **33.2** | **RU** | **HE / 184** | **980** | **88,400** | **1950s ZU (steel) AA** |
| 23 x 260 | 15.0 | 30.0 | RU | HEI / 174 | 850 | 63,000 | 1960s Rikhter R-23 a/c |
| 24 x 139 | 31.4 | 32.0 | CH | AP / 225 | 900 | 91,100 | Pzw.Kan 38, TB.41 Tankbüchse ATR |
| 25 x 90.5RB | 20 | 26.8 | CH | ??? | ??? | ??? | 25 mm SEMAG(?) |
| 25 x 116 | 32.0 | 32.3 | CH | HEI / 165 | 1,050 | 91,000 | early 1960s Oerlikon 204 GK |
| 25 x 116B | 33.3 | 32.3 | CH | HE / 165 | 1,040 | 89,200 | 1958–64 Oerlikon 251/25 RK a/c |
| 25 x 135.5 | 33.8 | 33.8 | FR | HE / 200 | 875 | 76,500 | Mle 1937B a/c |
| 25 x 136 | 44.0 | 43.6 | US | HEI / 194 | 1,220 | 144,000 | 1970s 25 mm GAU-8/A a/c |
| **25 x 137** | **38.0** | **38.2** | **CH/US/FR** | **HEI / 184** | **1,100** | **111,000** | **1970s Oe KBA, M242 Bushmaster, GAU-12/U, GAU-22/A, 25M811** |
| 25 x 145 | 38.2 | 38.2 | US | ? | ? | ? | 1973 AAI polymer/light alloy neckless case |
| 25 x 159.5 | 33.8 | 33.8 | FR | HE / 289 | 825/775 | 102,000/ 90,000 | DCAN Mle 1935, Mle 1937A a/c |

| METRIC DESIGNATION | RIM DIAM | BODY DIAM | NATION | PROJECTILE | MV m/s | ME Joules | ERA / NAME / WEAPONS |
|---|---|---|---|---|---|---|---|
| 25 x 163 | 42.6 | 42.7 | FR/JP | HE / 250 | 900 | 101,000 | WW2 Hotchkiss Mle 1938, Type 96 AA |
| **25 x 173/181** | **38.0** | **38.2** | **CH** | **HE / 230** | **1,160** | **155,000** | **1980s Oerlikon KBB, KBD AA** |
| **25 x 182B** | **40** | **39** | **CN** | **HEI / 250** | **1,050** | **138,000** | **1980s PG87 and PGZ95 army AA** |
| 25 x 187SR | 42.0 | 37.0 | SE | HE / 280 | 850 | 100,000 | Bofors M/38 round for M/32 gun (modified) |
| *25 x 193* | *42.6* | *42.6* | *FR* | *AP / 317* | *823* | *107,500* | *1930s Hotchkiss auto AT gun* |
| 25 x 205SR | 42.0 | 37.0 | SE | HE / 250 | 900 | 100,000 | 1930s Bofors M/32 round for M/32 AA gun |
| 25 x 218SR | 40.0 | 34.7 | RU | HE /288 | 910 | 119,000 | WW2 M1940 AA |
| **25 x 218** | **34.8** | **34.7** | **RU/CN** | **HE / 280** | **900** | **113,000** | **1950s 2-M3, 2-M8 AA** |
| 25.4 x 87R | 31.5 | 28.1 | GB | HE / 200 | 470 | 22,100 | Vickers 1 inch, Vickers-Terni a/c, AA |
| 25.4 x 87 | 30.2 | 30.2 | IT | HE / 200 | 440 | 19,400 | FIAT Revelli M 1917 25.4 mm a/c |
| *25.4 x 111R* | *38.5* | *34.5* | *US* | *???* | *???* | *???* | *1 inch Maxim, USN* |
| *25.4 x 148B* | *c.36* | *c.35* | *CH* | *HE / 280* | *?* | *?* | *1933 Solothurn S19–100* |
| 25.4 x 189 | 34.8 | 34.2 | GB | HE / 260 | 914 | 109,000 | 1930s Vickers naval AA |
| *26 x 130* | *36.0* | *35.9* | *US* | *HE-T / 195* | *945* | *87,100* | *1960s Colt CR-26 AFV* |
| *27 x 70B* | *29.4* | *29.4* | *US* | *HE / 250* | *610* | *46,500* | *1950s T150 Vulcan/ T151 revolver a/c (e)* |
| *27 x 96B* | *29.4* | *29.4* | *US* | *HE / 200* | *840* | *70,600* | *1950s Vulcan/T151 revolver a/c (e)* |
| *27 x 123.5B* | *36.2* | *34.6* | *UK* | *?* | *?* | *?* | *1970s 27 mm STRADEN* |
| **27 x 145B** | **34.0** | **33.1** | **DE** | **HE / 260** | **1,025** | **137,000** | **1970s Mauser BK 27 a/c, MLG 27 naval, (e)** |
| *27 x 173* | *44.0* | *43.6* | *DE* | *TPFRAN-T / 255* | *1,180* | *180,000* | *current Mauser Score sub-calibre system* |
| *27.5 x 140* | *44.8* | *44.3* | *US* | *?* | *?* | *?* | *earlier prototype of the Bushmaster below* |
| *27.5 x 160* | *39.8* | *39.4* | *US* | *APDS / 206* | *1,100* | *125,000* | *1960s Bushmaster programme* |
| 28 x 199SR | 43.5 | 41.4 | US | HE / 416 | 820 | 140,000 | WW2 1.1" naval AA |
| *28/20 x 210B* | *42.0* | *42* | *DE* | *AP / 131* | *1,400* | *128,000* | *WW2 Mauser MK 8202 squeezebore AFV* |
| *28 x 216B* | *41.9* | *38.4* | *DE* | *HE / 300* | *900* | *121,000* | *WW2 Krupp Gerät 411 motor cannon a/c* |
| *30 x 85B* | *31.7* | *32.0* | *DE* | *HE / 330* | *530* | *46,300* | *WW2 Mauser MG 213/30 a/c (e)* |
| 30 x 86B | 33.2 | 32.3 | GB | HE / 273 | 604 | 49,800 | 1950s Aden 3M LV a/c (e) |
| *30 x 91* | *38.0* | *38.1* | *DE* | *HE / 330* | *580* | *55,500* | *WW2 Mauser MK 212 a/c (e)* |

| METRIC DESIGNATION | RIM DIAM | BODY DIAM | NATION | PROJECTILE | MV m/s | ME Joules | ERA / NAME / WEAPONS |
|---|---|---|---|---|---|---|---|
| 30 x 91RB | 26.0 | 32.4 | DE | HE / 330 | 505 | 42,100 | WW2 RhB MK 108 a/c (e) |
| *30 x 92RB* | *29.0* | *32.4* | *JP* | *HE / 273* | *710* | *68,800* | *WW2 IJN Type 2 a/c* |
| 30 x 97B | 33.2 | 32.0 | FR | HE / 297 | 600 | 53,500 | 1950s DEFA 30M540 series a/c (e) |
| *30 x 100B* | *34.5* | *32.5* | *US* | *HE / 195* | *670* | *43,800* | *1960s XM140 WECOM, XM188 a/c* |
| *30 x 107RB* | *29.0* | *33.0* | *DE* | *?* | *?* | *?* | *WW1 Becker type?* |
| *30 x 110B Oe* | *c.35* | *c.37* | *CH* | *HE / 300* | *650* | *63,400* | *1948 Oerlikon 301 FK a/c (e)* |
| *30 x 110B MK 253* | *37.8* | *37.9* | *FR* | *?* | *?* | *?* | *postwar a/c* |
| **30 x 111–113B** | **33.3** | **32.3** | **GB/FR/US** | **HE / 270** | **790** | **84,300** | **1950s ADEN Mk 4, DEFA 30M550, M230, (e)** |
| 30 x 115 | 37.5 | 38.0 | JP | HE / 235 | 700 | 57,600 | WW2 Army Ho-155 a/c |
| *30 x 120RB* | *17.4* | *30.1* | *US* | *TP / 250* | *850* | *90,300* | *1950s T268 for T168 a/c (e)* |
| *30 x 120.5B* | *33.3* | *32.3* | *GB* | *?* | *?* | *?* | *1950s ADEN "C"* |
| 30 x 122 | 40.0 | 41.0 | JP | HE / 350 | 750 | 98,400 | WW2 Navy Type 5 a/c |
| *30 x 126B* | *33.3* | *32.3* | *US* | *TP / 250* | *850* | *90,300* | *1950s T239 round for T182 a/c (e)* |
| *30 x 135.5B* | *36.2* | *34.6* | *UK* | *?* | *?* | *?* | *1970s 30 mm STRADEN* |
| *30 x 136(B)* | *39.4* | *39.4* | *CH* | *HE / 225* | *1,050* | *124,000* | *1950s HS Type 825 a/c* |
| *30 x 138* | *42.7* | *42.7* | *CH* | *HE / 230* | *1,150* | *152,000* | *1960s HS Type 836 Spitfire AFV* |
| **30 x 150B** | **34.6** | **33.8** | **FR** | **SAP / 275** | **1,025** | **144,000** | **1990s GIAT 30M791 a/c (e)** |
| **30 x 155B** | **39.9** | **39.9** | **RU/CN** | **HE / 410** | **780** | **125,000** | **1950s NR30, Type 1 a/c** |
| *30 x 158* | *42.8* | *42.9* | *CH* | *HE / 350*<br>*HE / 400* | *960–1,000*<br>*900* | *161,-175,000*<br>*162,000* | *1950s HS 603 a/c and HS 604 AA* |
| *30 x 164* | *46.3* | *46.3* | *US* | *TP /c. 360* | *c,1020* | *c.187,000* | *1970s Philco-Ford GAU-8/A* |
| **30 x 165** | **40.0** | **40.0** | **RU** | **HE / 390** | **860–960** | **144,000–180,000** | **1970s GSh-30, -301, –6–30 (e): 2A42, 2A72, 2A38M** |
| *30 x 166* | *51.0* | *54.8* | *US* | *?* | *?* | *?* | *1970s AMCAWS CTA* |
| *30 x 167* | *?* | *?* | *FR* | *455* | *950* | *205,000* | *1930s M.A.T. a/c* |
| **30 x 170** | **42.8** | **43.0** | **CH/GB** | **HE / 360** | **1,080** | **210,000** | **1950s HS 831 / Oe KCB AA, Rarden AFV** |
| 30 x 173 | 44.0 | 43.6 | CH | HE / 360 | 1,080 | 210,000 | 1960s Oe KCA, GAU-9/A (e) |

| METRIC DESIGNATION | RIM DIAM | BODY DIAM | NATION | PROJECTILE | MV m/s | ME Joules | ERA / NAME / WEAPONS |
|---|---|---|---|---|---|---|---|
| **30 x 173** | **44.0** | **43.6** | **US** | **HE / 360** | **1,080** | **210,000** | **1970s GAU-8/A, MK 30, Bushmaster II / Mk 44 / XM813** |
| 30 x 184B | 37.8 | 37.9 | DE | HEI-T / 440 | 800 | 141,000 | WW2 RhB MK 101, MK 103 (e) a/c |
| **30 x 210B** | **45.9** | **45.9** | **RU** | **HE / 350** | **1,050** | **193,000** | **1960s NN-30, Zastava M86/89** |
| 30 x 211 | 47.7 | 47.7 | DE/CZ | HE / 435 | 1,000 | 217,000 | WW2 MK 303, 1950s M53 |
| *33 x 190R* | *?* | *?* | *FR* | *HE / 630* | *650* | *133,000* | *1930s APX a/c motor-cannon* |
| *34 x 149* | *50.0* | *50.0* | *DE* | *?* | *?* | *?* | *1941 DWM a/c gun project* |
| *34 x 184* | *38.0* | *38.2* | *DE* | *?* | *?* | *?* | *1980s "mini Supershot" project* |
| 34 x 239 | 47.8 | 47.8 | CH | HE / 720 | 900 | 292,000 | WW2 Flab Kan 38 AA, exp HSS 834 |
| *34 x 291* | *55.9* | *55.9* | *GB* | *?* | *?* | *?* | *Post-WW2 Vickers Class T automatic gun* |
| *34.5 x ?* | *?* | *?* | *GB* | *HE / 680* | *790* | *212,000* | *1930s 1½$^{PR}$ Mk V naval AA* |
| *35 x 173* | *44.0* | *43.6* | *US* | *550* | *?* | *?* | *AMRON experimental* |
| **35 x 228** | **54.5** | **55.1** | **CH** | **HE / 550** | **1,180** | **383,000** | **1960s Oerlikon KD series AA/AFV** |
| 37 x 69R | 44.0 | 39.5 | GB | HE / 454 | 365 | 30,000 | WW1 Vickers 1 pdr Mk III / IV a/c |
| 37 x 87R | 43.9 | 39.6 | US | HE / 500 | 380 | 36,000 | 1930s AAC Type M a/c and mt/infantry |
| *37 x 93* | *41.4* | *?* | *GB* | *?* | *?* | *?* | *1913 1 pr COW gun* |
| 37 x 94R | 44.1 | 39.7 | FR etc | HE / 454 | 549 | 68,400 | 19th century Maxim 'pom-pom' |
| 37 x 101SR | 42.9 | 39.9 | DE | HE / 465 | 320 | 23,800 | WW1 Sockelflak AA |
| *37 x 110SR* | *43.8* | *42.8* | *RU* | *HE / 635* | *602* | *115,000* | *postwar Grabin V14–113 a/c* |
| 37 x 111R | 47.0 | 43.5 | JP | HE / 475 | 570 | 77,200 | WW2 Ho-203 a/c |
| 37 x 124R | 47.6 | 43.7 | GB | HE / 680 | 640 | 139,000 | WW1 Maxim 1¼ and Vickers 1½ pr naval AA |
| 37 x 136R | 46.5 | 40.5 | US | HE / 476 | 640 | 97,500 | pre-WW1 naval 'Heavy 1 pdr', McClean gun |
| 37 x 142R | 44.7 | 40.5 | US | HE / 570 | 610 / 732 | 106,000 / 150,000 | 1930s AAC Rapid Fire Gun; Automatic Naval Gun (probably-732 m/s) |
| 37 x 144 | 44.0 | 44.0 | JP | HE / 475 | 710 | 120,000 | WW2 Ho-204 a/c |
| 37 x 145R | 45.0 | 40.6 | US | HE / 608 | 610 | 113,000 | WW2 M4 / M10 a/c |

| METRIC DESIGNATION | RIM DIAM | BODY DIAM | NATION | PROJECTILE | MV m/s | ME Joules | ERA / NAME / WEAPONS |
|---|---|---|---|---|---|---|---|
| 37 x 155 | 46.0 | 45.8 | RU | HE / 735 | 690 - 725 | 175,000–193,000 | postwar N-37 a/c (725 m/s = N-37L and NN-37) |
| *37 x 158B* | *42.8* | *43* | *CH* | *HE / 420* | *915* | *176,000* | *postwar Hispano-Suiza HS 606 a/c* |
| 37 x 190 | 46.5 | 47.0 | GB | HE / 652 | 594 | 115,000 | WW1 Vickers C.O.W. 1½ pdr |
| *37 x 198 100-K* | | | *RU* | *HE / 630* | *971* | *297,000* | *late-1930s OKB-15 Shpitalny AA* |
| 37 x 198 | 53.3 | 53.3 | RU | HE / 735 | 880 | 285,000 | WW2 NS-37 a/c |
| 37 x 203R | 54.2 | 50 | US | HE / 500 | 820 | 168,000 | 1930s AAC AA and a/c gun |
| 37 x 208 | 53.1 | 53.1 | FR | HE / 625 | 875 | 239,000 | 1930s Hotchkiss AA |
| *37 x 217* | *54.6* | *54.4* | *US* | *HE / 748* | *915* | *312,000* | *1950s T68 for T250X-2 Vigilante AA* |
| *37 x 218R* | *60.0* | *54.0* | *FR* | *HE / 816* | *825* | *278,000* | *WW2 A.N. ACAD Mle 1935 naval AA* |
| 37 x 223SR | 51.1 | 50.3 | US | HE / 608 | 792 | 190,000 | WW2 M1 AA |
| 37 x 232SR | 51.5 | 50.0 | IT | HE / 830 | 700 | 203,000 | WW2 Breda 37/54 AA, Vickers-Terni AFV |
| **37 x 240** | **c.54** | **c.54** | **CN** | **HE / 725** | **1,010** | **370,000** | **c.1970s Type 76 naval AA** |
| 37 x 250R | 51.4 | 45.9 | DE | AP / 644 | 845 | 230,000 | 1930s C/36 and WW2 FlaK M42 (thick rim) |
| **37 x 252SR** | **51.9** | **45.8** | **RU/CN** | **HE / 735** | **880** | **285,000** | **WW2 M1939 AA** |
| *37 x 255 (SR)* | *(57) 54.6* | *54.4* | *US* | *HE / 748* | *1,100* | *450,000* | *1950s T68 for T37/T172 and T250X-1 AA* |
| 37 x 264B | 46.8 | 45.9 | DE | HE / 644 | 820 | 216,000 | WW2 FlaK 18 / 36 / 37 / 43, BK 3,7 |
| *37 x 280B* | *?* | *?* | *DE* | *HE / 725–740* | *1,000–1,040* | *360–400,000* | *WW2 Gerät 341 (+Fledermaus?)* |
| 37 x 296R | 60.0 | 53.8 | FR | HE / 750–895 | 800 | 240–286,000 | WW2 Schneider M1930 AA (895 g for ACAD) |
| 40 x 158R | 48.1 | 43.9 | GB | HE / 900<br>HE / 760<br>AP / 1,130 | 610<br>730<br>615 | 167,000<br>203,000<br>214,000 | WW1&2 2PR naval AA, WW2 2PR Mk XII naval, R-R 40 mm BD, Class S a/c |
| *40 x 165* | *44.0* | *43.6* | *US* | *HE / 680* | *1,000* | *340,000* | *2000s exp Super 40 HE prototype* |
| ***40 x 180*** | ***44.0*** | ***43.6*** | ***US*** | ***HE / c.670*** | ***c.1,050*** | ***370,000*** | ***current exp Super 40 common case*** |
| *40 x 218* | *44.0* | *43.6* | *US* | *APFSDS / 230* | *1,350* | *210,000* | *2000s exp Super 40 APFSDS prototype* |
| *40 x 239B* | *48.0* | *48.0* | *CH* | *HE / 820* | *950* | *370,000* | *Early 1950s HSS 840 AA* |
| *40 x 277RB* | *53* | *60* | *GB* | *HE / 900* | *914* | *376,000* | *WW2 2pr 75cal for BSA naval gun* |

| METRIC DESIGNATION | RIM DIAM | BODY DIAM | NATION | PROJECTILE | MV m/s | ME Joules | ERA / NAME / WEAPONS |
|---|---|---|---|---|---|---|---|
| 40 x 311SR | 61.8 | 55.8 | SE | HE / 930 | 870 | 352,000 | 1930s Bofors L/60 AA |
| 40 x 365R | 65.0 | 57.9 | SE | HE / 960 | 1,030 | 509,000 | 1950+ Bofors L/70 AA/AFV |
| 42 x 270 | 81.2 | 81,7 | GB | HE / 1,090? | 1,070? | 624,000? | 1950s 'Red Queen' AA |
| 42 x 338/348 | 70 | 70 | CH/GB | HE / 1,090 | 1,070 | 624,000 | 1950s Oerlikon RK420 ('Red King') AA |
| 45 x 186 | 53.3 | 53.3 | RU | HE / 1,065 | 780 | 324,000 | WW2 NS-45 a/c |
| 45 x 386SR | 83.0 | 79.0 | RU | HE / 1,410 | 1,080 | 822,000 | 1950s naval AA |
| 45 x 495–510R | ? | ? | RU | HE / 1,450 | 960 | 668,000 | 1939 49-K and 61-K land/naval AA |
| 47 x 326R | 59.7 | 53.3 | GB | AP / 2,070 | 808 | 676,000 | WW2 Class P a/c |
| 47 x 428R | 64.8 | 58.0 | GB | HE / 1,590 | 1,021 | 830,000 | WW2 3½ Pr naval AA |
| 50 x 228 | 54.5 | 55.1 | US | ? | ? | ? | current Super50 AFV |
| 50 x 330 | 54.5 | 55.1 | DE | HETF/ 2,100<br>APFSDS/690 | 1,040<br>1,600 | 1.13K | postwar Rh 503 35/50 Supershot<br>APFSDS flight projectile=440g<br>Case also used for EAPS |
| 50 x 346B | 68 | 68 | DE | HE / 2,200 | 840 | 776,000 | WW2 FlaK 41 AA |
| 50 x 355 | 75.5 | 76 | CH | HE / 1,520 | 1,175 | 1.05K | 1950s Oerlikon 501 MK naval AA |
| 50 x 420R | 78.4 | 70.4 | DE | HE / 1,520 | 965 | 708,000 | WW2 BK 5 a/c (M-Geschoss loading) |
| 55 x 175RB | 45.8 | 59.3 | DE | HE / 1,480 | 600 | 267,000 | WW2 RhB MK 112 a/c (e) |
| 55 x 205RB | 59.9 | 68.2 | DE | HE / 1,480 | 650–700 | 314,000 | WW2 Krupp MK 412 a/c (e) |
| 55 x 450B | 75.0 | 78.2 | DE | HE / 1,860 | 1,080 | 1.085K | WW2 RhB FlaK 58 AA + MK 114 a/c |
| 57 x 121R | 68.4 | 60.8 | JP | HE / 1,500 | 500 | 187,000 | WW2 Ho-401 a/c |
| 57 x 165RB | 50.4 | 56.6 | RU | HE / 2,000 | 600 | 360,000 | late 1940s N-57 a/c |
| 57 x 230R | 80 | 73.0 | SE | HE / 2,230 | 744 | 617,000 | late 1940s Bofors m47 a/c |
| 57 x 348SR | 95.7 | 91.0 | RU | HE / 2,850 | 1,000 | 1.42K | postwar S60 AA |
| 57 x 350B | 67 | 63 | RU | HE / 2,490 | 720 | 645,000 | postwar RShR-57 a/c |
| 57 x 398 | 87.1 | 86.8 | CZ | HE / 2,560 | 1,000 | 1.28K | postwar R10 / PLK AA |
| 57 x 438R | 93.0 | 83.0 | SE | HE / 2,400 | 1,020 | 1.25K | postwar Bofors m54 AA + m50 AT + L/60–70 naval |
| 57 x 441R | 90.0 | 82.0 | GB | AP / 2,850 | 890 | 1.13K | WW2 6 pdr 7 cwt Molins Class M |
| 57 x 514R | 89.5 | 81.5 | GB | HE / 2,720 | 945 | 1.22K | WW2 6 pdr 6 cwt AA |
| 57 x 521RB | 59 | 83 | RU | HE / 2,485 | 1,087 | 1.47K | postwar Ch-21P a/c |

# TABLE 2 – THE CANNON

**TABLE 3: AUTOMATIC CANNON 20 mm - 57 mm**

FONT STYLE KEY: **current service**; ***current development***; obsolete service; *obsolete experimental*

**NAT.** = UN 2-letter nationality code (plus INT for international); the current codes are used, e.g. RU is used for both Russia and the USSR

In the **NOTES** column: (e) = electric priming; (w) = water or evaporation cooled barrel; a/c = primarily aircraft use; AT = anti-tank; AA = anti-aircraft; AFV = armoured fighting vehicle; RhB = Rheinmetall-Borsig.

| NAME | NAT. | METRIC CALIBRE | GUN KG | LENGTH CM | BARREL CM | METHOD OF OPERATION | CARTRIDGE FEED | RATE OF FIRE RPM | NOTES |
|---|---|---|---|---|---|---|---|---|---|
| 2 cm Becker | DE | 20x70RB | 30–42 | 137 | 80 | API blowback | box-15 | 250–300 | WW1 a/c |
| *2 cm Ehrhardt* | *DE* | *20x70RB(E)* | *36* | *150* | *101* | *short recoil* | *box* | *250–300* | *WW1 a/c* |
| *Scotti M35* | *IT* | *20x70RB(E)* | *23* | *?* | *80* | *gas delayed blowback* | *box-15, drum-60* | *350* | *interwar a/c* |
| Oerlikon Type F | CH | 20x70RB | 30 | 135 | 80 | API blowback | box-15, drum-45–100 | 300 | interwar a/c |
| *HS 405* | *FR* | *20x71* | *25* | *176* | *100* | *gas delayed blowback* | *box/drum* | *760* | *WW2 a/c for flexible/ turret mounts* |
| Oerlikon AF / FF/ Type 99–1 | CH / JP | 20x72RB | 23–30 | 133 | 80 | API blowback | drum-30–100 | 400–550 | 1934+ / IJN a/c. Mod 4=belt |
| MG-FF/ MG-FFM | DE | 20x80RB | 24–28 | 134 | 82 | API blowback | drum-60/90 | 520–540 | WW2 Ikaria a/c |
| **MG 151/20** | **DE** | **20x82** | **36–42** | **177** | **110** | **short recoil** | **belt** | **600–700** | **WW2 Mauser a/c: some (e);** |
| *MG 210* | *DE* | *20x82* | *32* | *?* | *?* | *gas* | *belt* | *1,000* | *WW2 Mauser a/c (e)* |
| **Denel GA-1** | **ZA** | **20x82** | **39** | **177** | **110** | **short recoil** | **belt** | **650–750** | **(MG 151); XC-F2 heli version** |
| 20 mm Ho-5 | JP | 20x94 | 35–37 | 145 | 90 | short recoil | belt | 700–800 | WW2 Army a/c (Type 2) |
| *20 mm MAC Mle 1949* | *F* | *20x96* | *45* | | | | *belt* | *1,000* | *ammo also used by HS 421* |
| 20 mm ShVAK | RU | 20x99R | 38–55 | 168–232 | 125–165 | gas | belt | 700–800 | WW2 a/c (barrel length varied) |
| B-20 (Berezin) | RU | 20x99R | 25 | 148–207 | 97–157 | gas | belt | 800 | WW2 a/c (barrel length varied ) |
| SEMAG/ Oerlikon Type L | CH | 20x99RB | 42 | 182 | 120 | API blowback | box or drum-15 | 350–380 | interwar a/c & AT; IJA Type 94 Flex |
| *20 mm Type 94 Flex* | *JP* | *20x99RB* | *43* | *173* | *120* | *API blowback* | *drum-15* | *380* | *interwar Army a/c* |
| Oerlikon FFL / IJN Type 99-2 | JP | 20x101RB | 30–38.5 | 189 | 125 | API blowback | drum-45–100 | 480–620 | Navy a/c. Mod 4=belt |
| **M39A2** | **US** | **20x102** | **81** | **183** | **136** | **gas revolver** | **belt** | **1,700** | **a/c (e)** |
| **M61A1** | **US** | **20x102** | **112** | **187** | **152** | **power rotary** | **linkless** | **3,000–7,200** | **6-bbl a/c (e)** |
| **M168** | **US** | **20x102** | **136** | **?** | **152** | **power rotary** | **belt** | **1,000 or 3,000** | **6-bbl AA (e)** |
| **GAU-4** | **US** | **20x102** | **125** | **187** | **152** | **gas rotary** | **linkless** | **6,000** | **SUU-23/A gunpod (e)** |
| **ZPL-20** | **CZ** | **20x102** | **64** | **200** | **c.150** | **gas+recoil twin bbl** | **belt-224** | **700–2,600** | **a/c (Plamen gunpod) (e) (GSh-23 rechambered)** |

| NAME | NAT. | METRIC CALIBRE | GUN KG | LENGTH CM | BARREL CM | METHOD OF OPERATION | CARTRIDGE FEED | RATE OF FIRE RPM | NOTES |
|---|---|---|---|---|---|---|---|---|---|
| 20M621 GIAT/Nexter | FR | 20x102 | 49 | 223 | 150 | gas delayed blowback | belt | 800 | multiservice (e) |
| M61A2 | US | 20x102 | 92–103 | 187 | 152 | power rotary | linkless | 4,000–6,000 | 6-bbl a/c (e) |
| Phalanx 1B (OGB) | US | 20x102 | c.170 | 235 | 200 | power rotary | linkless | 4,500 | 6-bbl CIWS (e) |
| M195 | U | 20x102 | ? | 137 | 102 | power rotary | belt | 750–850 | postwar 6-bbl heli (e) |
| M197 | US | 20x102 | 66 | 183 | 152 | power rotary | belt or linkless | 750 | 3-bbl: heli (e) |
| A-50 | US | 20x102 | c.80 | 187 | 152 | power rotary | linkless | 3,000 | 3-bbl a/c (e) |
| Ford "Tigerclaws" | US | 20x102 | 58 | 187 | 152 | gas revolver | belt | 2,300 | postwar a/c; developed M39 (e) |
| XM301/ XM915 | US | 20x102 | 34.5 | 183 | 152 | power rotary | linkless | 750–1,500 | postwar 3-bbl: RAH-66 (e) |
| 20 mm Mk 22 NOSL | US | 20x102 | 54.5 | 200 | 132 | short recoil | belt | 400–800 | postwar USN (+30 mm) (e) |
| 20 mm T220E1 | US | 20x102 | 46 | 242 | 152 | short/long recoil | belt | 500–1,100 | 1950s Army AA |
| 20 mm T220E2 | US | 20x102 | 45 | 192 | 152 | short recoil | belt | 1,100–1,275 | 1950s Army AA |
| 20 mm T247 | US | 20x102 | 46 | 242 | 152 | short recoil | belt | 500–1,100 | 1950s Army AA |
| 20 mm Mk 29 Mod 0 | US | 20x102 | 57–59 | 189–193 | 122 | short recoil | belt (dual) | 250–550 | postwar USN |
| 20 mm AEI Alexis | UK | 20 x 102 | 49 | 230 | 170 | gas | belt | 300–750 | current multiservice (e) |
| MG 204 | DE | 20x105 | 38–55 | 157 | 90 | short recoil | drum-20/100 | 400–500 | WW2 a/c. RhB |
| Lb 204 | DE | 20x105B | 55 | ? | ? | short recoil | drum-20/100 | 400 | interwar a/c. RhB |
| Solothurn S18–350 | CH | 20x105B | 40 | 176 | 90 | short recoil | box-5/10 | semi-auto | interwar a/c |
| Lb 201a, Lb 202 | DE | 20x108 | 55 | ? | ? | short recoil | drum-20/100 | 280–350 | interwar a/c RhB |
| Scotti M32 | IT | 20x110 Scotti | ? | ? | 140 | gas delayed blowback | box-15, drum-60, belt feed | 350–600 | interwar M20/70 a/c |
| HS 404 | FR | 20x110 HS | 55–60 | 254 | 160 | gas delayed blowback | drum-60, box | 700 | WW2 a/c fixed mounting |
| Hispano Mk. I/II | GB | 20x110 HS | 50–60 | 250 | 170 | delayed blowback | drum-60/belt | 600 | WW2 a/c (HS 404) |
| 20 mm AN/ M2 | US | 20x110 HS | 51 | 238 | 171 | gas/blowback | drum-60/belt | 600–700 | WW2 a/c (HS 404) |
| Hispano Mk. V | GB | 20x110 HS | 42 | 220 | 140 | " | belt | 750 | WW2+postwar a/c |
| 20 mm M3/ M24/Mk 16 | US | 20x110 HS | 45 | 197 | 133 | gas/blowback | belt | 700–750 | postwar a/c (HS 404). M24 = (e). MK16 = shipboard version |
| 20 mm Shirgun | US | 20x110 HS | 59 | 216 | 171 | gas delayed blowback | belt | 950–1,200 | WW2 EX1/T55/Mk 9 |
| Inglis EX-1–5000 | CA | 20x110 HS | 68 | 190 | 124 | gas delayed blowback | drum 50/100 or belt | 700 | WW2 AA; later in 20x110RB |
| Bofors 20 mm m/45 | SE | 20x110 HS | 49 | 214 | 130 | recoil | belt | 700 | postwar a/c (A21) |
| Bofors 20 mm m/49 | SE | 20x110 HS | 59 | 211 | 140 | recoil | belt | 775 | postwar a/c (A32A) |
| HS 804 | CH | 20x110 HS | 52 | 211 | 140–151 | delayed blowback | various | 750–840 | postwar a/c/AA Zastava M55 |
| 20 mm Johnson | US | 20x110HS | 52 | 246 | 172 | gas+blowback | belt | 750–1,000 | WW2 a/c |

| NAME | NAT. | METRIC CALIBRE | GUN KG | LENGTH CM | BARREL CM | METHOD OF OPERATION | CARTRIDGE FEED | RATE OF FIRE RPM | NOTES |
|---|---|---|---|---|---|---|---|---|---|
| 20 mm Colt T22 | US | 20x110HS | 51 | 198 | ? | short recoil | belt | 575–650 | WW2 a/c |
| 20 mm MBT G360 | IS | 20x110 HS | 50 | 204 | ? | short recoil? | belt | up to 600 | postwar |
| 20 mm TsKBSV-75 | RU | 20x110R | 65 | 220 | ? | gas | belt | 550 | WW2 AA version of the ShVAK |
| Navy Mk 11 | US | 20x110 USN (Mk 100) | 109 | 199 | 143 | gas/recoil twin revolver | twin belt | 700 or 4,200 | postwar a/c: Mk 4 pod (e) |
| Navy Mk 12 | US | 20x110 USN | 46 | 190 | 122 | gas/blowback | belt | 1,000–1,100 | HS404 type (e) |
| Oerlikon Type S/SS/2SS | CH | 20x110RB | 62/65 | 212 | 140 | API blowback | drum-45–100 | 280 / 470 / 650 | interwar/WW2 AA |
| Oerlikon FFS | CH | 20x110RB | 39 | 216 | 140 | API blowback | drum-60 | 470 | 1930s a/c |
| HS Type 9 | FR | 20x110RB | 48 | 207 | 140 | API blowback | drum-60 | 360–420 | interwar a/c (Oerlikon FFS) |
| Polsten | GB | 20x110RB | 57 | 218 | 145 | API blowback | box-30 | 450 | WW2 AA+AFV |
| Lahti L-34 | FI | 20x113 | 48 | 168 | 120 | recoil | box-15 or pan-45 | 300–350 | interwar naval |
| Tampella L-34 | FI | 20x113B | 40–60 | 169 | 120 | recoil | box-15 or pan-32 | 325–360 | interwar general purpose |
| Delaunay-Belville | FR | 20x115SR | 30 | ? | ? | gas/long recoil | strip-20 or articulated belt | ? | WW1 AT |
| 20 mm Madsen | DK | 20x120 | 53 | 200 | 120 | short recoil | box-10, drum-40 or 60 | 400 | WW2 AA/AT and a/c |
| 5CG series | FR | 20x122 | | | | | | | postwar AA |
| 20 mm Ho-1 | JP | 20x124 | 30 | 174 | 120 | gas/blowback | box-15 | 400 | Army a/c (flexible) |
| 20 mm Ho-3 | JP | 20x124 | 45 | 174 | 120 | gas/blowback | double-drum-50 | 400 | Army a/c (fixed) |
| 0.8 inch Vickers | GB | 20x125 | 69–76 | ? | 142 | long recoil | strip 5–10 | 120 | interwar anti-tank |
| 5CG series | FR | 20x126B | | | | | | | postwar AA |
| Oerlikon KAA (204GK) | CH | 20x128 | 88 | 263 | 170 | gas/blowback | belt | 900–1,000 | mainly AA |
| Oerlikon KAB (5TG) | CH | 20x128 | 109 | 339 | 240 | gas | box-8 or drum-20 or 50 | 1,000 | AA & AFV |
| Oerlikon 251/20 RK | CH | 20x128 | 74 | ? | 180 | gas revolver | belt | 1,750 | late 1950s a/c |
| Meroka | ES | 20x128 | ? | ? | 240 | power | belt | salvo-1,440 | 12-bbl naval AA (e) |
| MG 301 | DE | 20x135B | 63 | 320 | 226 | gas | belt | 1,000 | WW2 a/c, Krieghoff |
| MG 213A / 213B | DE | 20x135 | 55–65 | ? | ? | gas | belt | 800 | WW2 Mauser a/c (e) |
| MG 213C (213/20); T74 | DE US | 20x135 | 75 | 193 | 160 | gas revolver | belt | 1,200–1,400 | WW2 Mauser a/c (e) US T74 (T7 case) |
| ARF T33E3 | US | 20x135 | 59 | ? | ? | gas+blowback | belt | 900–1,200 | post-WW2 aircraft (T7 case) |
| S5–100 | DE | 20x138B | 63 | 224 | 130 | short recoil | box-20 | c.300 | interwar AA (Rheinmetall) |
| MG C/30L / MG 102 | DE | 20x138B | 64 | c.280 | 180 | “ | drum-100 | 300–350 | interwar a/c (Rh-B) |
| FlaK 30 | DE | 20x138B | 64 | 230 | 130 | short recoil | box-20 | 280 | WW2 AA (Rh-B) |
| KwK 30 | DE | 20x138B | 63 | 194 | 110 | “ | box-10 | 280 | WW2 AFV (Rh-B) |
| Solothurn S18–1100 | CH | 20x138B | c.50 | 216 | 130 | short recoil | box-10 | ? | 1930s AT/AA |
| FlaK 38 | DE | 20x138B | 57–71 | 225 | 130 | “ | box-20 | 450 | WW2 AA (Mauser) |
| KwK 38 | DE | 20x138B | 56 | 194 | 110 | “ | box-10 | 420–480 | WW2 AFV (Mauser) |

| NAME | NAT. | METRIC CALIBRE | GUN KG | LENGTH CM | BARREL CM | METHOD OF OPERATION | CARTRIDGE FEED | RATE OF FIRE RPM | NOTES |
|---|---|---|---|---|---|---|---|---|---|
| 20 mm Breda 20/65 | IT | 20x138B | 69 | 203 | 144 | gas | strip-12 | 220 | WW2 AA+AT M35 |
| 20 mm Scotti 20/70 | IT | 20x138B | 56–73 | 227 | 154 | gas/blowback | strip-12 | 230 | WW2 AA M39/41 |
| Lahti VKT M40 | FI | 20x138B | 40 | ? | 130 | gas | box-20 | 350–375 | WW2 AA (used in twin mounting) |
| Lahti M39/44 | FI | 20x138B | 42 | 223 | 130 | gas | box-10 | semi-auto | WW2 AA (auto version failed) |
| *MG C/35* | *DE* | *20x138B* | *70* | *?* | *?* | *short recoil* | *box-20* | *450* | *WW2 naval AA (Rh-B)* |
| *St G 201* | *DE* | *20x138B* | *70* | *?* | *?* | *short recoil* | *box-10* | *450* | *WW2 AFV (Rh-B)* |
| *Gerät 239* | *DE* | *20x138B?* | *65* | *?* | *?* | *gas* | *strip feed* | *700* | *WW2 Gebirgsflak (Gustloff)* |
| *Gerät 240* | *DE* | *20x138B?* | *50* | *?* | *?* | *gas* | *belt* | *900* | *WW2 AA (Rh-B)* |
| **HS820/Oe KAD** | **CH** | **20x139** | **66** | **257–297** | **170–210** | **gas/blowback** | **belt or drum-50** | **1,000** | **mainly AA. US=M139** |
| **F2 / 20M693 GIAT/Nexter** | **FR** | **20x139** | **71** | **267** | **197** | **gas delayed blowback** | **dual belt** | **750** | **mainly AA** |
| **Denel G1-2** | **ZA** | **20x139** | **73** | **269** | **186** | **gas/blowback** | **dual belt** | **750** | **(developed from French F2)** |
| **Rh202 / MK 20** | **DE** | **20x139** | **75–83** | **261** | **184** | **gas/blowback** | **single or dual belt** | **800–1,000** | **mainly AA+AFV** |
| *Mauser Model B* | *DE* | *20/23x139* | *62* | *245* | *184* | *gas* | *belt* | *900–1,050* | *postwar* |
| *HS 827B Hurricane* | *CH* | *20x139* | *56* | *251* | *170* | *short recoil* | *dual belt* | *variable* | *postwar AFV* |
| 20 mm FMK 38 | CH | 20x139 FMK | 57 | 255 | 183 | short recoil | belt | 400 | WW2 a/c/ Flab Kan 38 AA |
| 20 mm M1938 Mod A | PL | 20x140B | 58 | 201 | 135 | short recoil | box-5, drum 15–100 | 320 | WW2 AFV/AT/AA |
| *20 mm M1938 Mod C* | *PL* | *20x140B* | *50* | *?* | *150* | *gas* | *box 5–10, drum-100 or belt-200* | *300–750* | *WW2 universal* |
| *20 mm M1938 Mod D* | *PL* | *20x140B* | *40* | *?* | *135* | *gas* | *magazine or belt* | *300–600* | *WW2 a/c* |
| 20 mm Type 98 | JP | 20x143 | 69 | 201 | 126 | gas/blowback | box-20 | 450 | Army AA (some AT use) |
| Bofors 20 mm m/40 | SE | 20x145R | 42 | 205 | 132 | recoil | drum-25 | 360 | WW2 AA/AT |
| *20 mm Type 94* | *JP* | *20x157* | *?* | *?* | *?* | *short recoil* | *drum-15* | *?* | *interwar Army AA (Solothurn)* |
| *20 mm Colt T24* | *US* | *20x158 T5* | *64* | *254* | *160* | *short recoil* | *belt* | *700–750* | *WW2 aircraft, used T5 case* |
| *20 mm Oldsmobile T33* | *US* | *20x158 T5* | *80* | *207* | *160* | *gas+blowback* | *belt* | *900–1,200* | *post-WW2* |
| 23 mm Madsen | DK | 23x106 | 53 | 200 | 120 | short recoil | belt-200 | 400 | interwar a/c |
| NS-23 | RU | 23x115 | 38 | 198 | 145 | short recoil | belt | 600 | postwar a/c |
| NR-23 | RU | 23x115 | 39 | 202 | 145 | short recoil | belt | 800–950 | postwar a/c |
| AM-23 | RU | 23x115 | 43–44 | 147–177 | 100–130 | gas | belt | 1,300 | postwar a/c (also CN Type 2) |
| 23 mm Type 2 | CN | 23x115 | 43 | ? | ? | gas | belt | 1,200 | Soviet AM-23 a/c |
| **GSh-23** | **RU** | **23x115** | **50–51** | **139–154** | **100** | **gas twin bbl** | **belt** | **3,000–3,400** | **Gast type a/c** |
| GSh-6-23 | RU | 23x115 | 73–76 | 140–152 | 100 | gas rotary | belt or linkless | 9,000 | 6-bbl a/c |

| NAME | NAT. | METRIC CALIBRE | GUN KG | LENGTH CM | BARREL CM | METHOD OF OPERATION | CARTRIDGE FEED | RATE OF FIRE RPM | NOTES |
|---|---|---|---|---|---|---|---|---|---|
| 23 mm "chain gun" | CN | 23x115 | ? | ? | ? | power/chain | dual belt | 200–400 | AFV + helo |
| KPVB? | RU | 23x115 | | | | short recoil | belt | 600 | current AFV |
| HS 406 | FR | 23x122 | ? | 207 | 115–127 | gas/blowback | drum-? | 580 | WW2 exp. a/c |
| HS 407 | FR | 23x122 | 56.5 | 252 | 160–172 | gas/blowback | drum-20 | 580 | WW2 exp a/c (motor) & AA |
| HSS 827C MC-23 | CH | 23x133 | 58 | 251 | 170 | short recoil | dual belt | 550–1,000 | postwar AFV |
| HSS 823 | CH | 23x139 | 59.5 | 289 | 223 | | belt or drum-50 | 1,000 | |
| .90" T2 | US | 23x139(RB) | 109 | ? | 172 | API blowback | pan-48 | 400–450? | WW2 a/c |
| .90" T3 | US | 23x139SR | ? | ? | 172 | retarded blowback | strip | ? | WW2 a/c |
| .90" T4 | US | 23x139SR | 77 | 204 | 172 | long recoil | strip or belt | 150 | WW2 a/c / AT |
| .90" T1 | US | 23x146SR | 56 | ? | 138 | long recoil | pan-50 | 150 | WW2 a/c |
| MP-6 (PTB-23) | RU | 23x152 | 70* | 231 | 136 | short recoil | 9x9=81 mag | c.600 | WW2 a/c / AA *inc mag |
| VYa | RU | 23x152B | 66 | 214 | 165 | gas | belt | 550–650 | WW2 a/c |
| 23 mm Vladimirov | RU | 23x152B | 72 | 205 | 139 | gas revolver | belt? | 2,100 | postwar a/c? |
| 23 mm TKB-513 | RU | 23x152B | 56.5 | 216 | ? | gas | belt? | 2,150 | postwar a/c |
| 2A14 | RU | 23x152B* | 75 | 255 | 188 | gas | belt | 800–1,000 | postwar ZU AA *modified |
| 2A7 | RU | 23x152B* | 85 | 261 | 188 | gas | belt | 800–1,000 | postwar Shilka SPAAG (w) |
| R-23 | RU | 23x260 | 58 | 147 | 114 | gas revolver | belt | 2,500>1,900 | postwar a/c |
| 24 mm Pzw.-Kan 38 | CH | 24x139 | 77 | 259 | 188 | short recoil | box-6 | 30–40 | WW2 AFV: Tb41 AT version |
| Oerlikon 251/25 RK | CH | 25x116B | 75 | ? | 175 | gas revolver | belt | 1,650 | late 1950s a/c |
| 25 mm GAU-8/A | US | 25x136 | 240 | 276 | 229 | power rotary | linkless feed | 3,000–6,000 | early 1970s a/c |
| Oerlikon KBA | CH | 25x137 | 112 | 289 | 217 | gas | dual belt | 600 | AA, AFV, naval |
| M242 Bushmaster | US | 25x137 | 119 | 267 | 200 | power | dual belt | 100–500 | Chain Gun |
| GAU-12/U | US | 25x137 | 123 | 211 | 173 | power rotary | linkless | 1,800–4,200 | 5-bbl AA+a/c. |
| Nexter 25M811 | FR | 25x137 | 93–109 | 264 | 215 | electric cam | dual belt | 125–400/650 | AFV |
| GAU-22/A | US | 25x137 | 104 | 211 | 173 | power rotary | linkless | up to 3,300 | 4-bbl a/c |
| TRW-6425 | US | 25x137 | ? | ? | ? | short recoil | belt | 550 | Bushmaster contender |
| Mauser Model E | DE | 25x137 | 103 | 274 | 210 | gas | dual belt | 900 | postwar AA+AFV |
| Aden 25 | GB | 25x137 | 92 | 229 | 170 | gas revolver | belt | 1,650–1,850 | 1990s a/c |
| PFB-25 | US | 25x137 | 88 | 282 | ? | recoil+gas | dual belt | 550 | Bushmaster contender |
| GE-225 | US | 25x137 | 82 | 221 | ? | gas or power | belt | 2,500 or 750 | postwar a/c: Gast type 2-bbl |
| Oerlikon KBC | CH | 25x137 | 90 | 236 | 178 | gas revolver | belt | 1,700 | late 1970s a/c |
| 2,54 cm S19-100 | CH | 25x148B | 100 | 240 | 152 | short recoil | box-16 | 250 | 1933 Solothurn AA/AT |
| GAU-7/A | US | 25x152 | 193 | 229 | 213 | power rotary | linkless feed | 6,000 | postwar 5-bbl a/c (CTA) |

| NAME | NAT. | METRIC CALIBRE | GUN KG | LENGTH CM | BARREL CM | METHOD OF OPERATION | CARTRIDGE FEED | RATE OF FIRE RPM | NOTES |
|---|---|---|---|---|---|---|---|---|---|
| 25 mm Mle 1935 | FR | 25x159.5 | ? | 280 | ? | short recoil | drum-33 or 44 | 280–350 | 1930s naval a/c |
| *25 mm Hotchkiss a/c* | *FR* | *25x163* | *70–80* | *228* | *150* | *gas* | *box-35* | *300–350* | *1930s a/c* |
| 25 mm Mle 1938/9 | FR | 25x163 | 115 | 245 | 150 | gas | box-15 | 220 | WW2 AA (IJN Type 96) |
| 25 mm Type 96 | JP | 25x163 | 115 | 230 | 150 | gas | box-15 | 220 | WW2 Navy AA (Hotchkiss) |
| **Oerlikon KBB** | **CH** | **25x173/181** | **146** | **319** | **230** | **gas** | **dual belt** | **800** | **naval AA** |
| *Oerlikon KBD* | *CH* | *25x173/181* | *290* | *276* | *230* | *power rotary* | *linkless* | *5,000* | *postwar 7-bbl: naval AA* |
| **25 mm PG87** | **CN** | **25x182B** | **?** | **?** | **229** | **gas** | **belt** | **600–800** | **AA** |
| Vickers 25.4 mm | GB | 25x189 | 127 | 259 | 178 | gas | box-10 | 200 | WW2 naval (Argentina) |
| Hotchkiss Auto AT gun | FR | 25x193 | 108 | 244 | 180 | gas | box-10 | ? | 1930s anti-tank |
| Bofors 25 mm M/32 | SE | 25x205SR / 25x187SR | 125 | 180 | 145 | recoil | clip-6 | 160–180 | interwar L/64 AA; converted in 1938 for 25 x 187SR ammunition |
| 25 mm Type 61 | CN | 25x218 | 101 | 281 | 200 | short recoil | belt | 270–300 | Soviet 110-PM |
| 110-PM | RU | 25x218 | 101 | 285 | 200 | short recoil | belt | 270–300 | postwar naval 2M-3/8 AA |
| **M-110 naval** | **RU** | **25x218** | **110** | **281** | **200** | **short recoil** | **belt** | **475** | **postwar naval 2M-3M AA (CN Type 61)** |
| M1940 | RU | 25x218SR | 445(?) | ? | 191 | recoil | clip-7 | 240 | WW2 AA |
| 1" Vickers | GB | 25.4x87R | 50 | 137 | 76 | long recoil | belt-25 | 150 | WW1 (Vickers-Terni) |
| 25.4 mm M1917 | IT | 25.4x87 | 45 | 129 | 70 | recoil | box-8 | 150 | WW1 a/c (FIAT Revelli) |
| 25.4 mm m/22 | SE | 25.4x87–95R | 50 | 137 | 76 | long recoil | belt-25 | 150 | interwar sub. AA (Vickers-Terni) |
| *2,54 cm Solothurn* | *CH* | *25.4x148B* | *100* | *240* | *152* | *short recoil* | *box-16* | *250* | *1933 S19–100 AA/ AT gun* |
| *Colt CR-26* | *US* | *26x130* | *47–54* | *201* | *166* | *long recoil* | *belt* | *550* | *Constant Reaction: postwar AFV* |
| *T150 Vulcan* | *US* | *27x70/96B* | *c.166* | *?* | *?* | *power rotary* | *belt* | *4,000* | *postwar 6-bbl a/c (e)* |
| **Mauser BK 27** | **DE** | **27x145B** | **100** | **231** | **140 (170)** | **gas revolver** | **belt** | **1,700** | **Mauser a/c & naval (MLG 27) (e)** |
| *GE-127 Bushmaster* | *US* | *27.5x160* | *119–106* | *294* | *208* | *short recoil* | *dual belt* | *100–550* | *postwar AFV* |
| USN 1.1" | US | 28x199SR | 252–276 | 304 | 208 | recoil | clip-8 | 150 | WW2 naval AA (w) |
| *Mauser MK 8202* | *DE* | *28/20x210B* | *120* | | | *gas* | *belt* | *180* | *WW2 AFV* |
| *Krupp Gerät 411* | *DE* | *28x216B* | *?* | *?* | *?* | *power (engine)* | *?* | *?* | *WW2 motor a/c* |
| *MG 213/30* | *DE* | *30x85B* | *75* | *163* | *130* | *gas revolver* | *belt* | *1,100–1,200* | *WW2 Mauser a/c (e)* |
| Aden 3M (LV) | GB | 30x86B | 87 | 159 | 108 | gas revolver | belt | 1,200 | postwar a/c (e) |
| *Hughes/NGF 30 mm Mk 4* | *US* | *30x86B* | *52* | *190* | *133* | *gas/blowback* | *belt* | *1,000–1,200* | *postwar USN a/c (e)* |
| *ARF 30 mm Mk 5* | *US* | *30x86B* | *61* | *216* | *160* | *gas+blowback* | *belt* | *750–800* | *postwar a/c (e)* |
| *30 mm T221* | *US* | *30x86B* | *60* | *150* | *?* | *gas revolver* | *belt* | *1,600* | *1950s Short Drum Revolver (e)* |
| MK 108 | DE | 30x91RB | 60 | 105 | 44 | API blowback | belt | 600–650 | WW2 a/c (e) |
| *MK 212* | *DE* | *30x91* | *60–65* | *152* | *100* | *gas* | *belt* | *600* | *WW2 a/c* |

| NAME | NAT. | METRIC CALIBRE | GUN KG | LENGTH CM | BARREL CM | METHOD OF OPERATION | CARTRIDGE FEED | RATE OF FIRE RPM | NOTES |
|---|---|---|---|---|---|---|---|---|---|
| 30 mm Type 2 | JP | 30x92RB | 51 | 207 | 135 | API blowback | drum-42 | 400 | WW2 Navy a/c |
| 30M541 DEFA | FR | 30x97B | 84 | 166 | 110 | gas revolver | belt | 1,200–1,400 | postwar a/c (e) |
| XM140 WECOM | US | 30x100B | 63.5 | 154 | 107 | power | belt | 405 | postwar heli (AH-56) (e) |
| 30 mm Mk 22 NOSL | US | 30x100B | 54 | 191 | 123 | short recoil | belt | 350–750 | postwar naval (20 mm also) |
| XM188 | US | 30x100B | 53 | 156 | 107 | power rotary | linkless | 200–2,000 | postwar 3-bbl heli (e) |
| Oerlikon 301 FK | CH | 30x110B Oe | 80 | 158 | 90 | short recoil | belt | 1,000 | poswar a/c (e) |
| ADEN Mk 4/5/6 | GB | 30x111B | 87 | 159 | 108 | gas revolver | belt | 1,200–1,400 | postwar a/c (e) |
| 30M552/3 DEFA | FR | 30x113B | 80 | 170 | 140 | gas revolver | belt | 1,100–1,500 | postwar a/c (e) |
| T212 Vulcan II | US | 30x113B | 155 | 161 | ? | power rotary | belt | 5,500–6,000 | postwar 6-bbl a/c (e) |
| 30 mm Mk 22 | US | 30x113B | 54.5 | 191 | 123 | short recoil | belt | 350–750 | postwar USN (+20 mm) (e) |
| 30M554 GIAT/Nexter | FR | 30x113B | 85 | 201 | ? | gas revolver | belt | 1,800 | a/c (e) |
| 30M781 GIAT/Nexter | FR | 30x113B | 65 | 192 | 140 | electric cam | belt | up to 720 | heli+AFV (e) |
| Vektor 55C5 | ZA | 30x113B | 82 | 166 | 140 | gas revolver | belt | 1,900 | postwar a/c (DEFA 550) (e); XC-30 is heli version (700–900 rpm) |
| Oerlikon 251/30 RK | CH | 30x113B | 75 | ? | 180 | gas revolver | belt | 1,400 | late 1950s a/c |
| ATK M230 | US | 30x113B | 60 | 168 | 107 | power | linkless | 100–650 | heli chain gun (e) |
| ATK M230LF XM914 | US | 30x113B | 73 | 218 | 152 | power | belt | 200 | AFV (e) |
| ASP-30 | US | 30x113B | 52 | 206 | 132 | gas | belt | 400–450 | postwar multipurpose (e) |
| MF 30 | US | 30x113B | 50 | 170 | ? | recoil | box-10 or belt | 450 | postwar multipurpose (e) |
| AEI Venom (LR) | UK | 30x113B | 110 (122) | 204 | 140 | gas revolver | belt | 230 | multipurpose (e) LR = low recoil version |
| 30 mm Ho-155-I | JP | 30x115 | 50 | 175 | 114 | short recoil | belt | 400 | WW2 a/c (IJA) |
| 30 mm Ho-155-II | JP | 30x115 | 44 | 151 | 98 | short recoil | belt | 500 | WW2 a/c (IJA) |
| 30 mm T168 | US | 30x120RB | 122 | 137 | ? | gas revolver | belt | 1,300 | reverse-loading postwar a/c (e) |
| 30 mm Type 5 | JP | 30x122 | 66 | 201 | 144 | gas+blowback | belt | 500 | WW2 a/c (IJN) |
| 30 mm T182 | US | 30x126B | 93 | 157 | 108 | gas revolver | belt | 1,200–1,500 | postwar a/c (e): initially chambered in 30x113B |
| HS 825 | CH | 30x136 (B) | 103–112 | 238–289 | 174–224 | gas | belt | 950–1,000 | postwar a/c |
| HS 836 Spitfire | CH | 30x138 | 89 | 274 | 180 | short recoil | dual belt | 720 | postwar AFV |
| 30M791 GIAT/Nexter | FR | 30x150B | 120 | 240 | ? | gas revolver | belt | up to 2,500 | a/c (e) |
| NR-30 | RU | 30x155B | 66 | 215 | 160 | short recoil | belt | 850–1,000 | postwar a/c (also CN Type 1) |
| 30 mm Type 1 | CN | 30x155B | 66 | 216 | 160 | short recoil | belt | 850–1,000 | Soviet NR-30 a/c |
| AO-10 | RU | 30x155B | 87.5 | 210 | 160 | gas | belt | 3,500 | postwar 2-bbl a/c |
| 30 mm Hughes Lockless | US | 30x157 CT | 113 | 279 | | recoil | box-50 | 350 | 1972 cased/telescoped |

| NAME | NAT. | METRIC CALIBRE | GUN KG | LENGTH CM | BARREL CM | METHOD OF OPERATION | CARTRIDGE FEED | RATE OF FIRE RPM | NOTES |
|---|---|---|---|---|---|---|---|---|---|
| HS 603/4/6 | CH | 30x158 | 150+ | 225+ | 144+ | gas? | belt | 750 | 1950s a/c |
| Philco-Ford GAU-8/A | US | 30x164 | 264 | 285 | 213 | power rotary | linkless | 2,000–4,000 | 1970s GAU-8/A contender |
| AO18 | RU | 30x165 | 205 | 218 | 162 | gas rotary | belt | 4,000–5,000 | 6-bbl naval AA (w) (e) (AK630) |
| GSh-30 | RU | 30x165 | 105 | 204 | 150 | gas | belt | 3,000 | 2-bbl a/c (e). |
| GSh-30K | RU | 30x165 | 126 | 294 | 240 | gas | belt | 300–2,600 | 2-bbl helo (e) |
| 2A38M | RU | 30x165 | 195–223 | 345 | 240 | gas | belt | 2,000–2,500 | 2-bbl AA (Tunguska) (w) |
| GSh-6–30 | RU | 30x165 | 149 | 188 | 150 | gas rotary | belt | 5,000 | 6-bbl a/c (e) |
| AO18L | RU | 30x165 | 155 | 217 | 162 | power rotary | belt | 600–1,000 | 6-bbl naval AA (e) (AK 306) |
| 2A42 | RU | 30x165 | 115 | 303 | 240 | gas/recoil | dual belt | 200–800 | AFV/heli. ZTM-2 (UA) |
| 2A72 | RU | 30x165 | 84 | 301 | 240 | long recoil | dual belt | 200–370 | AFV. ZTM-1 (UA) |
| GSh-301 | RU | 30x165 | 45 | 198 | 150 | short recoil | belt | 1,500- HS1,800 | a/c and AA (w) (e) |
| 30 mm Type 730 | CN | 30x165 | ? | ? | ? | power rotary | linkless | 4,200 | 7-bbl CIWS (H/PJ12) + LD-2000 (e) |
| 30 mm Type 1130 | CN | 30x165 | ? | ? | ? | power rotary | linkless | 10,000 | 11-bbl CIWS (H/PJ11) (e) |
| AEI Inspekta | GB | 30x165 | 145 | 350 | 255 | gas | belt; single/dual | 200–300 | naval or AFV (developed from KCB). 30x173 opt. |
| HSS 831/Oe KCB | CH | 30x170 | 159 | 352 | 256 | gas/blowback | strip-5 , drum or belt | 600–650 | mainly AA |
| HS 837 Thunderball | CH | 30x170 | 105 | 329 | 225 | short recoil | dual belt | variable | postwar AFV |
| Oerlikon KCC | CH | 30x170 | | | | gas revolver | belt | 1,500 | postwar naval CIWS |
| Rarden L21 | GB | 30x170 | 113 | 295 | 244 | long recoil | clip-3 | 90 | AFV |
| Oerlikon KCA GAU-9/A | CH USA | 30x173 | 136 | 269 | 198 | gas revolver | belt | 1,350 | postwar a/c (e) postwar ground attack |
| GE GAU-8/A | US | 30x173 | 281 | 290 | 230 | power rotary | linkless | 4,200 | a/c, naval AA |
| GAU-I3/A | US | 30xI73 | I54 | 279 | 2I4 | power rotary | linkless | 2,400–3,000 | postwar 4-bbl: a/c GPU-5/A pod |
| CHAG | US | 30x173 | 127 | 309 | 230 | power rotary | linkless | 2,000 | postwar 3-bbl a/c |
| GE 30 mm Rotary Gun | US | 30x173 | 136 | 297 | 226 | various | belt | 300 or 600 | 1970s (30 x 100B also) |
| Mauser MK 30 series | DE | 30x173 | 154–198 | 331–378 | 246 | gas | dual belt | 200–800 | current AA/AFV |
| Bushmaster II / Mk 44 / GAU-23 / XM813 | US | 30x173 or 40x180 | 156 | 340 | 241 | power/chain | dual belt | 200 or 400 | Chain Gun AFV/naval/ gunship |
| GI-30 CamGun | ZA | 30x173 | 171–230 | 345 | ? | electric | dual linkless (40) or belt | 110 (3 rnd burst) | AFV |
| WOTAN 30 | DE | 30x173 | 280 | 367 | ? | power | dual | 200 | AFV and naval |
| AEI Inspekta | GB | 30x173 | 145 | 350 | 255 | gas | belt; single/dual | 200–300 | naval or AFV (developed from KCB). 30x165 opt. |
| Zastava M97/12 | RS | 30x173 or 30x210B | ? | ? | ? | gas | dual belt | ? | AFV (developed from M86) |
| MK 101 | DE | 30x184B | 139 | 259 | 155 | short recoil | box-6 or drum-30 | 230–260 | WW2 a/c |
| MK 103 | DE | 30x184B | 141 | 235 | 134 | recoil+gas | belt | 360–420 | WW2 a/c (e) |
| 3 cm FlaK 38 | DE | 30x184B | 141 | 232 | 134 | recoil+gas | belt | 400 | WW2 AA (MK 103) (e) |

| NAME | NAT. | METRIC CALIBRE | GUN KG | LENGTH CM | BARREL CM | METHOD OF OPERATION | CARTRIDGE FEED | RATE OF FIRE RPM | NOTES |
|---|---|---|---|---|---|---|---|---|---|
| NN-30 | RU | 30x210B | 155 | 268 | 180 | recoil revolver | belt-500 | 1,000 | postwar naval AA (AK230) (e) (w) |
| A436/M | RO | 30x210B | ? | ? | 180 | recoil revolver | belt/box-30 | 100–600 | postwar army AA (Model 80) (e) |
| 30 mm Type 69 | CN | 30x210B | 155 | 268 | 180 | recoil revolver | belt | 1,000 | Soviet AK-230 (NN-30) |
| Zastava M86/89 | RS | 30x210B | 200–215 | 300 | 210 | gas | single or dual belt | 550–750 | AA+AFV (e) M89 has dual belt feed |
| *Skoda 30 mm twin-barrel FLAK* | *CZ* | *30 x 211?* | *625* | *307* | *230* | *blowback+ short recoil* | *magazine and 9-round clip* | *1,000* | *WW2 twin-barrel design* |
| *MK 303* | *DE* | *30x211* | *185* | *295* | *220* | *gas* | *clip-10* | *400* | *WW2 Krieghoff AA* |
| 30 mm M53 | CZ | 30x211 | 200 | 300 | 243 | gas | strip-10 or (SPAAG) box-50 | 450–500 | postwar AA: (development of WW2 MK 303 – see above) |
| *RMK 30/2* | *DE* | *30x230* | *95* | *220* | *170* | *power revolver (4 chamber)* | *linkless* | *300* | *postwar Mauser caseless/ recoilless* |
| *RMK 30/1* | *DE* | *30x280* | *125* | *300* | *240* | *power revolver (3 chamber)* | *dual linkless* | *300* | *postwar Mauser caseless/ recoilless* |
| 34 mm Flab K38 | CH | 34x239 | 280 | ? | 272 | short recoil | belt | 250–270 | WW2+ postwar AA (w) |
| *Vickers Class T* | *GB* | *34x291* | | | | | | | *postwar DACR AA* |
| Oerlikon KDA/B/C/F | CH | 35x228 | 430–670 | 437–474 | 315 | gas | various | 550–600 | army and naval AA |
| Oerlikon KDE | CH | 35x228 | 510 | 409 | 315 | long recoil | various mag | 200 | AFV |
| *35 mm Rh 503* | *DE* | *35x228* | *575* | *473* | *315* | *electric* | *dual linkless* | *150–400* | *1980s AFV* |
| *EMAK 35* | *ZA* | *35x228* | *380* | *418* | *315* | *electric* | *belt* | *1–120* | *postwar AFV* |
| *ARES Talon* | *US* | *35x228* | *270* | *419* | *315* | *gas* | *dual belt* | *550* | *postwar AFV/AA* |
| Vektor GA-35 | ZA | 35x228 | 429 | 434 | 315 | gas | single or dual belt | 500–600 | naval twin mounting |
| Oerlikon KDG | CH | 35x228 | 450 | 411 | 276 | gas revolver | linkless | 1,000 | AA |
| Bushmaster III | US | 35x228 | 218 | 402 | 279 | power | belt | 1 or 200 | Chain Gun AFV |
| *WOTAN 35* | *DE* | *35x228* | *?* | *?* | *?* | *power* | *belt* | *1–200* | *Rheinmetall AFV* |
| *RMK 35/1* | *DE* | *35x300* | *152* | *320* | *280* | *power revolver (3 chamber)* | *linkless* | *300* | *postwar Mauser caseless/ recoilless* |
| *RMK 35/2* | *DE* | *35x350* | *280* | *350* | *280* | *power, rotating chamber* | *linkless* | *200* | *postwar Mauser caseless/ recoilless* |
| Vickers 1$^{PR}$ Mk III/V | GB | 37x69R | 62 | 143 | 81 | recoil | belt | 150 | WW1 a/c |
| 37 mm AAC Type M | US | 37x87R | 61 | ? | 74 | recoil | clip-5 | 100–125 | 1930s a/c and army gun |
| Maxim 1$^{PR}$ | GB | 37x94R | 186 | 187 | 110 | short recoil | belt | 200–300 | 19th Century: (w) |
| 37 mm Hotchkiss M1901 | FR | 37x94R | 240 | ? | ? | gas | clip-8/10 | 100 | pre-WW1 |
| *37 mm M1918* | *FR* | *37x94R* | *90* | *?* | *?* | *long recoil* | *strip-5* | *60* | *interwar Puteaux a/c* |
| 3.7 cm Sockelflak | DE | 37x101SR | ? | ? | 54 | short recoil | mag-10 | 120 | WW1 AA |
| 37 mm Ho-203 | JP | 37x111R | 89 | 154 | 87 | long recoil | belt-15 | 120 | WW2 IJA a/c (37 mm Type 3?) |
| 1½$^{PR}$ Vickers | GB | 37x124R | 250 | 231 | 157 | short recoil | belt | 150 | WW1 naval |
| 1-pounder Heavy | US | 37x136R | 259 | 268 | 157 | short recoil | belt | 150? | pre-WW1 naval |

| NAME | NAT. | METRIC CALIBRE | GUN KG | LENGTH CM | BARREL CM | METHOD OF OPERATION | CARTRIDGE FEED | RATE OF FIRE RPM | NOTES |
|---|---|---|---|---|---|---|---|---|---|
| 37 mm McClean | US | 37x136R | 254 | 259 | 185 | gas | hopper (5) | 100 | pre-WW1 |
| 37 mm AAC Auto Naval Gun | US | 37x142R* | ? | 244 | 185 | gas | hopper (5) | 100–120 | 1930s (copy of McClean cannon) *probably |
| 37 mm Ho-204 | JP | 37x144 | 130 | 247 | 130 | short recoil | belt | 400 | WW2 IJA a/c (37 mm Type 4?) |
| 37 mm M4 | US | 37x145R | 96 | 226 | 165 | long recoil | endless belt - 30 | 140–150 | WW2 a/c (P-39, P-63) + PT boats |
| 37 mm M10 | US | 37x145R | 109 | " | " | " | belt | 150–170 | WW2 a/c (P-63A-9 +) |
| N-37D/N-37L | RU | 37x155 | 103/105 | 245/287 | 136/195 | short recoil | belt | 400–430 | postwar a/c |
| NN-37 | RU | 37x155 | 105 | 287 | 195 | short recoil | belt | 600–700 | postwar a/c |
| *HS 606* | *CH* | *37x158B* | *150* | *225* | *155* | *gas?* | *belt* | *750* | *1950s a/c* |
| 1½^PR^ COW gun | GB | 37x190 | 90 | 234 | 152 | long recoil | clip-5 | 100–120 | WW1+interwar a/c |
| *100-K* | *RU* | *37x198 100-K* | *340* | | *225* | | *drum-5* | *150–180* | *pre-WW2 AA* |
| Sh-37 | RU | 37x198 | 208* | ? | ? | gas | mag-20/40 | 169–184 | WW2 a/c *inc 20-round mag |
| NS-37 | RU | 37x198 | 160–171 | 340 | 230 | short recoil | belt | 240–260 | WW2 a/c |
| 37 mm AAC Type F | US | 37x203R | 163 | 190–238 | 137–185 | recoil | clip | 70–90 | 1930s a/c and AA gun |
| *37 mm Hotchkiss* | *FR* | *37x208 (?)* | *?* | *?* | *178* | *gas* | *box* | *144* | *1930s AA* |
| *37 mm Vigilante* | *US* | *37x217/256* | *?* | *330* | *254* | *power rotary* | *pan-192* | *3,000* | *postwar 6-bbl SPAAG* |
| *37 mm A.N. M1935* | *FR* | *37x218R (?)* | ? | *?* | *259* | *?* | *clip-6* | *165* | *WW2 naval AA* |
| 37 mm M1 | US | 37x223SR | 166 | ? | 198 | long recoil | clip-10 | 120 | WW2 AA |
| 37 mm M9 | US | 37x223SR | 184 | 255–264 | 198 | long recoil | belt | 140 | WW2 exp. a/c + PT boats |
| *ARF T172E1* | *US* | *37x223SR?* | *500* | *350* | *259* | *revolver* | *belt* | *1,100* | *1950s AA gun: 37x256?* |
| 37 mm Breda 37/54 | IT | 37x232SR | 275 | 323 | 200 | gas | clip-6 | 60–140 | WW2 AA M32/38/39/41 |
| **37 mm Type 76A** | **CN** | **37x240** | **?** | **?** | **233** | **recoil** | **belt** | **380** | **naval** |
| **37 mm AN3** | **CN** | **37x240** | **?** | **?** | **233** | **gas** | **belt** | **550** | **naval** |
| 3.7 cm SK C/36 | DE | 37x250R | 240 | ? | ? | long recoil | clip-5 | 160–180 | 1930s naval AA |
| 3.7 cm FlaK M 42 | DE | 37x250R | 300 | ? | 256 | long recoil | clip-5 | 160–180 | WW2 naval AA |
| **37 mm M39** | **RU** | **37x252SR** | **1,000** | **?** | **232** | **long recoil** | **clip-5** | **160–170** | **WW2 AA/naval** |
| 3.7 cm FlaK 18 | DE | 37x264B | 278 | 363 | 211 | short recoil | strip-6 | 160 | WW2 AA |
| BK 3,7 | DE | 37x264B | 295 | 363 | 211 | short recoil | strip-6/8/12 | 160 | WW2 a/c |
| 3.7 cm FlaK 36 | DE | 37x264B | ? | 363 | 211 | short recoil | strip-6 | 160 | WW2 AA. Flak 37 similar |
| 3.7 cm FlaK 43 | DE | 37x264B | 265 | 330 | 211 | gas | strip-8 | 250 | WW2 AA |
| *3.7 cm Gerät 341* | *DE* | *37x280B* | *450* | *?* | *?* | *gas* | *strip* | *240* | *WW2 naval AA* |
| 37 mm Mle 1930 | FR | 37x296R | ? | ? | 166 | ? | strip-8 | 190 | WW2 Schneider AA |
| 40 mm Ho-301 | JP | 40 mm CL | 40 | 139 | 80 | API blowback | box-10 | 450 | WW2 IJA a/c (Ki-44-II Special) |

| NAME | NAT. | METRIC CALIBRE | GUN KG | LENGTH CM | BARREL CM | METHOD OF OPERATION | CARTRIDGE FEED | RATE OF FIRE RPM | NOTES |
|---|---|---|---|---|---|---|---|---|---|
| Vickers 2$^{PR}$ Mk II | GB | 40x158R | 229 | 243 | 157 | short recoil | belt | 200? | WW1 naval AA (w) |
| Vickers 2$^{PR}$ Mk VIII | GB | 40x158R | 259–386 | 260 | 157 | short recoil | belt | 90–115 | WW2 naval AA (w) |
| Vickers 40 mm Class S | GB | 40x158R | 134–175 | 297 | 170 | long recoil | drum-15 | 100 | WW2 a/c |
| Rolls-Royce 40 mm BH | GB | 40x158R | 152–190 | 358 | 203 | long recoil | hopper-8 | 120 | WW2 a/c |
| CTAS 40 (UK/F) | INT | 40x255 | 340 | 343 | 280 | power | linkless | 200 | AFV (cased telescopic) |
| Bofors 40 mm L/60 | SE | 40x311SR | 440 | 340 | 224 | recoil | clip-4 | 120–150 | interwar-postwar AA/naval (w) |
| Bofors 40 mm L/70 | SE | 40x365R | 560 | 405 | 280 | recoil | clip-4 or mags | 240–330 | AA + AFV |
| OtoBreda 40L70N | IT | 40x365R | ? | ? | 280 | recoil | linkless | 450 | AA "Fast Forty" |
| S&T Dynamics K40 | KR | 40x365R | 720 | 389 | 280 | recoil | linkless-24 | 300 | AFV |
| Bushmaster IV | US | 40x365R | 400* | 380 | 260 | power | mag-30 | 200 | AFV *with 30 rounds |
| 42 mm Red Queen | GB | 42x270 | ? | ? | ? | gas revolver | pan(-75?) | (450?) | postwar AA |
| 420/1/2 RK (Red King) | CH | 42x338/348 | ? | ? | ? | gas revolver | pan-75 | 450 | postwar 2-bbl Red King AA |
| NS-45 (OKB-16-45) | RU | 45x186 | 153 | 362? | 252? | short recoil | belt | 260–280 | W2 a/c |
| COMVAT | US | 45x307 | ? | ? | ? | power | linkless | 200–400 | postwar AFV (cased telescopic) |
| 45 mm SM-21 | RU | 45x386SR | ? | ? | 350 | recoil | clip-4 | 160 | postwar naval AA (SM-20/21-ZIF) |
| 45 mm M1939 | RU | 45x495–510R | ? | ? | 315 | long recoil | clip-3 | 120–140 | 49-K and 61-K AA |
| 47 mm Breda C.102 | IT | 47x195R | ? | ? | 152 | long recoil | strip, vertical | ? | WW2 a/c |
| Vickers Class P | GB | 47x326R | ? | ? | ? | recoil | ? | | WW2 a/c-AT |
| Vickers 3½$^{PR}$ | GB | 47x428R | ? | ? | 352 | recoil | ? | ? | late WW2 naval AA |
| 50 mm XM913 | US | 50x228 | ? | ? | ? | power | ? | ? | Super50 AFV |
| 50 mm Rh 503 | DE | 50x330 | 615 | 572 | 425 | electric | dual linkless | 150–400 | postwar AFV (Supershot) |
| 50 mm EAPS | US | 50x330 | ? | ? | ? | power | | 200 rpm | C-RAM |
| 5 cm FlaK 41 | DE | 50x346B | 570 | 469 | 342 | gas | clip-5 | 130 | WW2 RhB AA |
| 5 cm BK 5 | DE | 50x420R | 540 | 434 | 304 | long recoil | belt | 45–50 | WW2 RhB a/c |
| 5 cm MK 214A | DE | 50x420R | 490 | 416 | 282 | long recoil | belt | 150–160 | WW2 Mauser a/c, AA (e) |
| 5.5 cm MK 112 | DE | 55x175RB | 275 | 200 | 105 | API blowback | belt | 300 | WW2 RhB a/c (e) |
| 5.5 cm MK 412 | DE | 55 x 205RB | 229–240 | 309 | 221 | gas + recoil | belt | 260–300 | WW2 Krupp/Mauser a/c (e) |
| 5.5 cm MK 115 | DE | 55x318B | 180–190 | 330 | 120–132 | gas recoilless | belt | 300 | WW2 RhB a/c (e) |
| 5.5 cm MK 114 | DE | 55x450B | 700 | ? | 421 | gas | gas-operated | 150 | WW2 RhB a/c (?) |
| 5.5 cm FlaK 58 | DE | 55x450B | 600 | 615 | 421 | gas | clip-5 | 140 | WW2 RhB AA (?) |
| 5.5 cm MK 214B | DE | 55x450B | 650 | ? | ? | gas | belt | 300 | WW2 Mauser a/c |

| NAME | NAT. | METRIC CALIBRE | GUN KG | LENGTH CM | BARREL CM | METHOD OF OPERATION | CARTRIDGE FEED | RATE OF FIRE RPM | NOTES |
|---|---|---|---|---|---|---|---|---|---|
| 57 mm Ho-401 | JP | 57x121R | 160 | 204 | 100 | long recoil | belt-17 | 80–100 | Army a/c (Ki-102B) |
| *N-57 (OKB-16–57)* | *RU* | *57x165RB* | *135* | *218* | *101* | *short recoil* | *belt* | *230* | *postwar a/c* |
| Bofors 57 mm m/47 | SE | 57x230R | 290 | 384 | 274 | recoil | drum-41 | 100–180 | postwar Saab T18B a/c |
| **S-60** | **RU** | **57x348SR** | **1,475** | **?** | **439** | **short recoil** | **clip-4** | **100–120** | **postwar army AA/ SPAAG** |
| SM-24-ZIF (ZIF-31) | RU | 57x348SR | ? | ? | 449 (427) | long recoil | clip-3 | 100–115 | postwar naval twin AA. CN Type 66 |
| ZIF-71 (ZIF-75) | RU | 57x348SR | ? | ? | 464 | long recoil | clip-3 | 160–170 | postwar naval AA: ZIF-71 single barrel, ZIF-75 quad |
| ZIF-72 | RU | 57x348SR | ? | ? | 427 | long recoil | belt-550 | 200 | postwar naval twin AA (AK-725) (w) |
| ***A-220M*** | ***RU*** | ***57x348SR*** | ***?*** | ***?*** | ***427*** | ***short recoil*** | ***linkless*** | ***300*** | ***naval AA (w)*** |
| ***AU-220M*** | ***RU*** | ***57x348SR*** | ***?*** | ***?*** | ***427*** | ***short recoil*** | ***linkless*** | ***120*** | ***AFV*** |
| *RShR 57* | *RU* | *57x350B* | *339* | *?* | *291* | *short recoil* | *drum-10* | *140–160* | *postwar a/c* |
| 57 mm R10/ PLK | CZ | 57x398 | ? | ? | 353 | ? | 3-round clips | 180 | postwar AA |
| Bofors 57 mm m/54 | SE | 57x438R | ? | ? | 342 | recoil | clip-2 | 160 | postwar AA (w) |
| Bofors M/45 L/60 | SE | 57x438R | ? | ? | 342 | recoil | linkless | 130 | postwar naval (w) |
| **Bofors SAK 57 L/70** | **SE** | **57x438R** | **?** | **?** | **400** | **recoil** | **linkless** | **200–220** | **naval (USN Mk 110)** |
| Molins 6PR | GB | 57x441R | 345 | 256 | 244 | long recoil | tray-6 | 40 (semi-auto) | WW2 naval |
| 57 mm Molins | GB | 57x441R | 486 | 378 | 285 | long recoil | mag-22 | 60 | WW2 a/c |
| *6PR 6cwt* | *GB* | *57x514R* | *318* | *330* | *319* | *recoil* | *hopper-8* | *35* | *WW2 AA* |
| *57 mm Ch-21P* | *RU* | *57x521RB* | *?* | *?* | *423* | *long recoil* | *vertical-15* | *80* | *postwar a/c* |

# GLOSSARY

**AA** anti-aircraft

**ABM** air-bursting munition (see PABM and HEAB)

**AC** armoured car

**accelerator** a pivoting lever inside a short-recoil mechanism which acts to accelerate the rearward movement of the bolt

**advanced primer ignition** see API blowback

**AFV** armoured fighting vehicle

**AGL** automatic grenade launcher

**AHEAD** Advanced Hit Efficiency and Destruction (a proprietary Oerlikon ammunition development)

**AIFV** armoured infantry fighting vehicle; also known as MICV

**air-cooled** a weapon which achieves barrel cooling by radiation direct to the atmosphere

**ALACV** Advanced Light Combat Vehicle Armament (US Army)

**ammunition** collective name for cartridges (or equivalent)

**AP** armour piercing; a projectile designed to penetrate armour, or a cartridge loaded with such a projectile

**APC(BC)** armour piercing, capped, (ballistic capped): an AP shot with a softer cap to aid penetration of face-hardened armour (and a streamlined cap to reduce air resistance)

**APCR** armour piercing, composite, rigid: shot consisting of a penetrating core (usually of tungsten alloy) enclosed within a light-alloy body, which travels as a unit to the target

**APDS** armour piercing, discarding sabot: similar to APCR except that the body (which may be light alloy or plastic) breaks up and falls away from the penetrating core when the shot leaves the muzzle.

**APFSDS** armour piercing, fin stabilized, discarding sabot: similar to APDS except that the penetrating core is long and narrow and is stabilized in flight by fins

**APHC** armour piercing, hard core: an alternative designation for APCR

**APHE** armour piercing, high explosive

**API** armour piercing incendiary: a dual-purpose projectile

**API blowback** advanced primer ignition blowback: a type of blowback mechanism in which the cartridge is fired while still moving forwards within an extended chamber

**APSE** armour piercing, secondary effect

**APSV** armour piercing super velocity; British term for Littlejohn squeezebore ammunition (also initially used to describe APDS)

**arrowhead shot** APCR with reduced-diameter alloy body in centre

**AT** anti-tank

**autocannon** portmanteau word, short for 'automatic cannon'

**automatic** a weapon which continues to fire and reload automatically for as long as the trigger or firing button is pressed

**automatic cannon** fast-firing, high-velocity automatic gun with a calibre between 20 mm and 57 mm

**B** added to a cartridge designation to identify a belted case (e.g. 20x105B)

**ball (round)** an inert projectile, i.e. not AP, I, HE, or -T

**ballistic cap** a streamlined (usually light alloy) nose cone used to improve the ballistic coefficient of a blunt projectile (US: windshield)

**ballistic coefficient** a measurement of the aerodynamic drag of a projectile and therefore the rate at which it loses its velocity

**ballistics** the science concerning the passage of a projectile from the instant of firing to the end of its flight, divided into: internal (within the barrel), external (flying through the air) and terminal (impact on the target)

**barrel** the tube connected to (or integral with) the chamber, down which the projectile is accelerated

**barrel extension** the part of the barrel which extends behind the chamber, usually to accommodate a locking mechanism

**base fuze** a detonating fuze fitted to the base of an HE shell, normally associated with APHE and SAPHE projectiles

**battery** a gun is 'in battery' when it is at rest in its forward position in the mounting

**BC** see ballistic coefficient

**belt** (1) a raised band around a cartridge case, in front of the extractor groove
(2) a strip of fabric (originally) or metal, into which cartridges are fitted to facilitate feeding them into a weapon. Metal belts may be disintegrating, non-disintegrating or a continuous loop

**belt feed** the use of a belt to supply ammunition to a gun mechanism

**belt link** a piece of metal which constitutes a part of a belt

**belted ammunition** cartridges which are held within a cartridge belt

**belted case** a cartridge case with a raised band in front of the extractor groove to aid location in the chamber

**BK** *Bordkanone* (German): a large-calibre airborne cannon for ground attack

**blank** a cartridge which has a primer and propellant but no projectile

**blowback** a type of automatic weapon operating mechanism, in which the bolt is not locked, and is pushed to the rear by expanding gases

**bolt** a part of the operating system, containing the firing mechanism, which slides in line with the barrel, pushing a cartridge into the chamber and holding it there during firing

**bore** the inside of the barrel

**bottlenecked cartridge** a cartridge with a case whose diameter reduces sharply to the neck, creating a shoulder

**box magazine** a type of magazine in which the cartridges are stacked on top of each other (they may be single or double stacked)

***Brandgranate*** (German): incendiary shell

***Brandsprenggranate*** (German): high explosive/incendiary shell

**breech** the opening at the rear of the chamber which allows cartridges to be loaded and fired cases extracted

**breechblock** alternative term for bolt, normally used when its operating movement involves pivoting, or sliding vertically or horizontally

**breech face** the part of the barrel surrounding the breech

**calibre** (1) the diameter of a projectile, or of the inside of a barrel (2) designation of the cartridge a weapon is designed for

**calibre length** the barrel length divided by the calibre, usually prefixed with 'L'; e.g. a 40 mm L/70 indicates a barrel (40x70=) 280 cm long

**cannon** (modern) a large-calibre fast-firing automatic weapon, generally taken to be between 20 mm and around 57 mm calibre

**cartridge** a unit or round of ammunition, normally comprising the cartridge case, projectile, propellant and primer

**cartridge case** the part of the cartridge which contains the propellant and holds the projectile and primer firmly in place

**caseless ammunition** ammunition which does not have a cartridge case but uses solid propellant to hold the projectile and primer

**centrefire** cartridge fired by a primer located in the centre of the base

**chain gun** an externally-powered automatic weapon which uses a chain to drive the operating cycle

**chamber** the space at the rear of the barrel in which the cartridge is positioned and supported during firing

**chamber pressure** maximum pressure generated within the chamber by propellant gases, which forces the projectile down the barrel: see conversion factors at the end of this Glossary

**chemical fuze** a contact fuze on an HE shell which operates by percussion

**chromium plating** used to protect the barrel bore and chamber

**CIWS** close-in weapon system; a naval short-range AA/anti-missile system, usually with automatic aiming and firing

**clip** a piece of metal which holds together several cartridges (normally by the case rims) for feeding into a weapon; or several cartridges held together by a clip

**closed bolt** an automatic weapon designed to commence the firing cycle with the cartridge already loaded into the chamber (see open bolt)

**closed loop** a CIWS control system which uses radar to track both target and projectiles and corrects the aim accordingly

**coaxial mounting** a means of mounting a secondary gun in parallel with a primary gun so that both weapons can be aimed using the same sights

**COIN** counter-insurgency

**combustible case** a cartridge case which is designed to burn with the propellant

**COMVAT** Combat Vehicle Armament Technology (US Army)

**contact fuze** a fuze which is initiated by impact with the target

**cook-off** the unwanted ignition of a cartridge by heat in a gun chamber

**cordite** a type of propellant

**counter-recoil** a type of gun mounting in which the gun is fired as it returns to battery, so the initial recoil force is expended in arresting the forward movement of the gun. See floating mounting, differential recoil and soft recoil

**C-RAM** Counter-Rocket, Artillery, Mortar (for base defence)

**crimp** a depression in the neck of a cartridge case, intended to hold the projectile firmly in place prior to firing

**CTA** cased telescoped ammunition: in which the projectile is contained with the cartridge case (which is often combustible)

**CVAST** Combat Vehicle Armament Systems Technology (US Army)

**D.A.C.R.** Direct Action, Close Range: a post-World War II RN AA project

**dark trace** the initial burn of a tracer, designed to give no light

**delayed blowback** blowback operating system in which the bolt is locked at the moment of firing, unlocked shortly afterwards, then operates as a blowback (see retarded blowback and API blowback)

**detonation** the explosion of an HE projectile

**differential recoil** see counter-recoil

**discarding sabot** a sleeve which encloses a projectile to enable it to be fired in a larger-calibre gun; it disintegrates at the muzzle

**disintegrating link** a type of ammunition belt consisting of metal links, which fall apart as each cartridge is chambered

**double base** a type of propellant

**double drum** a magazine consisting of two drums, with a single, combined ammunition feed into the gun

**DP** dual purpose

**drill round** a cartridge which is totally inert, used to practice loading and unloading drills. It is always made easily recognisable

**driving band** a band of soft metal or plastic around a projectile, which is intended to be gripped by the rifling in order to induce spin

**drum** a type of circular magazine in which the cartridges are held parallel to each other (sometimes used to describe a pan magazine); see helical drum and pan magazine

**DU** depleted uranium; a heavy metal used in some AP shot

**dual belt feed** a gun with two belt feeds to allow a rapid change in the type of ammunition being fired

**dummy round** a cartridge which is completely inert (i.e. no primer or propellant), used in testing weapon cycling

**EAPS** Extended Area Protection and Survivability

**EI** extruded impregnated propellant

**ejection** the act of throwing an extracted cartridge case clear of the gun

**electric ignition** a method of igniting cartridges by passing an electrical current through the primer (see percussion ignition)

**electromagnetic gun** a gun which uses electromagnetic force to accelerate a projectile (also known as a rail gun)

**electrothermal-(chemical)** a system which uses a plasma generator instead of a primer

**elevation** the movement of a gun in its mounting through a vertical arc

**EPVAT** Electronic Pressure Velocity and Action Time: a NATO standard method of testing ammunition, including chamber pressure

**erosion** wear on the inside of a barrel caused by hot propellant gases and friction generated by projectiles

**external ballistics** the science of projectile flight from the muzzle of a gun
**extended chamber** see hooded chamber
**externally powered** a gun mechanism which requires an external source of power to operate; this is usually electric but may be hydraulic or manual, or in aircraft powered by engine gas or the slipstream
**extraction** the act of pulling a fired cartridge case from the chamber
**extractor claw** a hook, attached to the bolt, which fits into the extractor groove in order to pull the cartridge out of the chamber
**extractor groove** a groove around the head of a cartridge case, into which the extractor claw fits

**FAA** (British) Fleet Air Arm (1924 onwards)
**FAPDS** frangible armour piercing discarding sabot
**FCS** fire control system (sights, ballistic computer etc)
**feed** the method of delivering ammunition to the gun
**fin stabilized** a projectile whose flight is stabilized by tail fins rather than by being spun by rifling
**firing cycle** the sequence of loading, firing, extracting, ejecting and reloading
**firing pin** a spring-loaded steel pin which strikes a primer to cause ignition
**Flak** *Fliegerabwehrkanone* (German) AA artillery; also sometimes given as *Flugabwehrkanone* or *Flugzeugabwehrkanone*
**flat trajectory** the flight of a projectile which involves minimal drop due to gravity; associated with high velocity
**flexible mounting** a simple, unpowered gun mounting in which the gun can be moved in traverse and elevation by the gunner's hands
**floating firing** see counter-recoil
**floating mounting** see counter-recoil
**fluted chamber** longitudinal grooves in the chamber to permit gun gas to seep around the cartridge case so it does not stick in the chamber
**FMPDS** frangible missile-piercing discarding sabot
**frangible** a projectile designed to break up on hitting the target
**full-calibre** a projectile which fills the bore of a gun (see sub-calibre)
**fuze** a device for initiating the detonation of an HE shell

**gas operated** a type of gun mechanism using gas tapped from the barrel to drive the firing cycle
**gas-unlocked blowback** a delayed blowback mechanism using gas tapped from the barrel to unlock the breechblock
**Gast** a type of twin-barrel automatic gun mechanism in which the bolts are cross-linked so each barrel fires alternately
**Gatling** an early type of manually powered multi-barrel rotary gun; sometimes used to refer to any rotary weapon
**GAU** (USAF): Gun, Aircraft, Unit: aircraft gun designation
***Glimmspur*** (German): often written *Gl'spur*; a dimly glowing tracer for night use
**GPMG** general purpose machine gun
**grain** measure of weight used in UK and USA for propellant charges and (in smaller calibres) projectiles.
1 gram = 15.432 grains
**gravity feed** the use of gravity to supply ammunition to the gun mechanism
**grooves** the larger interior diameter of a rifled barrel; between the lands
**gun gas** gas generated by ignition of the propellant in a cartridge
**gun pod** a detachable pod, containing a gun (or guns) and ammunition, which is carried by aircraft
**gyro gunsight** a simple type of lead-computing gunsight to aid deflection shooting

**hang fire** an unintended delay in the ignition of a cartridge after the primer has been struck
***Hartkernmunition*** (German, World War II) an APCR shot used in aircraft cannon
**HC** high capacity; a shell with an unusually large (HE) capacity
**HE** high explosive; the normal filling of cannon shells
**HEAB** high explosive, air burst (via time fuze); see PABM
**headspace** the accurate location of a cartridge in the chamber ready for ignition
**headstamp** information about the cartridge, stamped into the base
**HEAT** high explosive, anti-tank; (see hollow charge)
**heavy machine gun** a machine gun with a calibre significantly larger than a military rifle cartridge but smaller than a cannon; in effect, 12.7–15 mm
**HEDP** high-explosive dual-purpose; a modern projectile which combines hollow-charge with general HE effect
**helical drum** a magazine in which all rounds point inwards, following a spiral track
**high-angle mounting** a mounting which permits a weapon to be elevated at or close to the vertical for AA fire
**hollow charge** an explosive projectile with a cone-shaped hollow, lined with metal, in the nose, which generates an intense armour-piercing jet
**hooded chamber** an extended chamber which is longer than necessary to contain a cartridge; associated with API blowback designs
**HMG** heavy machine gun
**HV** high velocity
**HVAP** high velocity, armour piercing. US description for APCR, also used to describe early APDS
**hybrid** a mechanism which uses more than one operating principle
**hydroballistic** a projectile designed to achieve supercavitation in water

**ignition** the igniting of propellant by a primer
**IJA** Imperial Japanese Army
**IJN** Imperial Japanese Navy
**impact fuze** see contact fuze
**inert** a cartridge or explosive which cannot fire or detonate
**internal ballistics** the science of the passage of a projectile down a gun barrel

**joules (J)** unit of measurement of muzzle energy

***kamikaze*** (Japanese) 'Divine Wind': name given to suicide attacks
***Kriegsmarine*** German Navy
**KwK** *Kampfwagenkanone* (German): tank gun

**L/(number)** the calibre length, which is the barrel length divided by the calibre: so a 20 mm L/70 barrel is 140 cm long
***Lafette*** (German): gun mounting
**LAFV** light AFV; typically, an armoured car or light tank
**lands** the smaller interior diameter of a rifled barrel; between the grooves
**leade** unrifled, cone-shaped section of gun bore immediately forward of the chamber
***Leuchtspur*** (German): tracer; often written *L'Spur*
**linear action** a gun mechanism in which the elements reciprocate in line with the gun barrel
**link** an element of an ammunition belt
**linkless feed** a method of supplying ammunition to a gun in which the cartridges are not linked together; uses external power
**lock time** the period of time between pressing the trigger or gun button and the first shot being fired
**long recoil** a type of gun operating mechanism in which the barrel and bolt recoil together for the full length of the cartridge, before being unlocked
***Luftwaffe*** German Air Force
**LV** low velocity

**machine gun** a crew-served automatic weapon of less than 20 mm calibre
**magazine** a container which holds ammunition ready for loading into a gun, in the form of a box, drum or pan
**manually operated** a gun in which the firing cycle is operated by a manual crank or lever
**MG** machine gun (German: *Maschinengewehr*)
**MGB** motor gun boat: a small, fast, naval vessel armed with guns
**MICV** mechanised infantry combat vehicle (also known as AIFV)
***Minengeschoss* or *M-Geschoss*** (German): mine shell
**mine shell** a thin-walled, lightweight type of HE projectile
**MK** (German) *Maschinenkanone*: automatic cannon
***moteur-canon*** (French) motor cannon: an aircraft cannon designed for mounting between the cylinder banks of a vee-engine, firing through a hollow propeller hub
**mounting** the method of supporting a gun ready for firing
**MPa** megapascals: see chamber pressure
**m/s** metres per second, the unit of measurement of projectile velocity
**MTB** motor torpedo boat: a small, fast, naval vessel armed with torpedoes
**multipurpose ammunition** projectiles with a combination of capabilities (AP, HE, I): associated with the Norwegian firm of Raufoss (now nammo)
**muzzle** the end of the barrel from which the projectile emerges
**muzzle blast** the violent escape of gun gas from the muzzle as a projectile leaves the barrel
**muzzle booster** a device fitted to the muzzle to use some of the muzzle blast to increase recoil, to assist the action of recoil-operated guns
**muzzle brake** a device fitted to the muzzle which deflects part of the muzzle blast to the side or rear in order to reduce recoil
**muzzle energy** a calculation of the energy of a projectile as it leaves the muzzle; a function of projectile velocity and weight
**muzzle velocity** the speed of a projectile as it leaves the muzzle

**NATO** North Atlantic Treaty Organization
**neck** the part of a cartridge case which holds the projectile
**necked-down** a cartridge case which has its neck reduced in diameter to accept a smaller calibre projectile than the case was designed for
**necked-up** (or necked-out) a cartridge case which has its neck increased in diameter to accept a larger calibre projectile than the case was designed for
**NFC** National Firearms Centre, Royal Armouries, Leeds, UK

**obturation** the sealing of a gun breech to prevent the escape of gun gas on firing (in automatic weapons, normally achieved by the cartridge case); also forward obturation achieved by the driving bands
**open bolt** an automatic weapon designed to commence the firing cycle without a cartridge loaded into the chamber (see closed bolt)
**optronics** the combination of OPtical and elecTRONIC systems to provide day/night/all-weather sights
**ounce (oz)** UK and USA measure of weight; 1 oz = 28.35 grams.
**over-bore** a cartridge which has a case capacity too large for all of the propellant to be efficiently utilized

**PABM** Programmable Air-Burst Munition. See HEAB
**Pak** *Panzerabwehrkanone* (German): anti-tank gun
**pan magazine** a flat, circular magazine in which the cartridges are held pointing inwards: see drum and helical drum
***Panzergranat*** (German) armour-piercing projectile
***Panzerbrandsprenggranatpatrone L'spur m. Zerl*** (German): armour piercing high-explosive incendiary cartridge with tracer and self-destruct
***Patrone*** (German): cartridge
**PELE** Penetrator with Enhanced Lateral Effect: fragmenting SAP
**penetration** the ability of an AP shot to penetrate armour
**percussion ignition** a method of igniting cartridges by striking a percussion primer with a firing pin (see electric ignition)
**PFHE** pre-fragmented high explosive (a type of shell)
**PGU** Projectile, Gun, Unit: USAF term for ammunition
**pintle mounting** a type of flexible mounting in which the gun is mounted at the top of a vertical rod
**pivoting block** a gun mechanism in which the breechblock movement is pivoted
**pom-pom** unofficial name for a slow-firing automatic weapon, usually of 37 or 40 mm calibre
**primary extraction** when the gun mechanism pulls the fired case out of the chamber
**primer** a percussion cap fitted into the base of a cartridge case, used to ignite the propellant
**primer pocket** the part of a cartridge case into which the primer fits
**progressive or gain rifling** rifling which commences with a gentle or zero twist, which gradually increases down the length of the barrel
**projectile** any bullet shot or shell fired from a gun
**propellant** the chemical which burns rapidly to generate gas which accelerates the projectile up the gun barrel
**propeller disc** the area swept by the blades of an aircraft propeller

**propeller hub or boss** the central part of an aircraft propeller, which may be made hollow to allow an engine-mounted gun to fire through it

**proximity fuze** a shell fuze which is triggered by the close proximity of a target

**PT boat** (American) a small, fast naval vessel

**pyrophoric** gives off sparks on impact and/or ignites spontaneously at high temperature

**Pzgr** *Panzergranate* (German): AP Projectile

**R** rimmed cartridge case (when it occurs at the end of a cartridge designation, as in 20x99R)

**radial engine** an engine in which the cylinders are arranged radially, with the crankshaft in the centre

**rail gun** see electromagnetic gun

**railway gun** a gun mounted on a railway wagon

**rate of fire** the cyclic rate at which the gun loading/firing/extracting/ejection sequence operates, usually measured in rounds per minute

**rebated rim** a cartridge case with a rim of smaller diameter than the case body

**RAF** (British) Royal Air Force (since 1918)

**RAMICS** (USN) Rapid Airborne Mine Clearance System; using helicopter-borne cannon firing hydroballistic ammunition to sink sea mines.

**RAP** rocket assisted projectile

**RB** rebated rim (case description, added to the end of a cartridge designation, also given as RR)

**receiver** the body or chassis of the gun which supports the barrel and operating mechanism

**recoil operated** a gun mechanism operated by the recoiling gun barrel

**reflector sight** a gunsight in which the aiming marks are reflected onto an angled glass plate through which the gunner views the target

**retarded blowback** a blowback operating system in which the bolt is not locked, but its initial rearward movement is slowed by the mechanism

**revolver cannon** a normally single-barrel gun with a rotating cylinder containing several chambers, each of which is loaded and unloaded in stages, and fired as it comes into line with the barrel

**rifling** the spiral grooving within a gun barrel which grips the projectile and spins it in order to ensure its stability (see progressive rifling)

**rimless (case)** a cartridge case in which the rim is the same diameter as the case body, separated from it by an extractor groove

**rimmed (case)** a cartridge case with a rim which has a larger diameter than the case body

**RN** (British) Royal Navy

**RNAS** (British) Royal Naval Air Service (1914–1918)

**rotary cannon** a multi-barrel gun with several parallel barrels rotating around a common axis, each barrel being loaded and fired in turn

**rotary lock** the locking of the bolt to the barrel extension by a rotary movement

**rotating band** see driving band

**round (of ammunition)** a single cartridge (or equivalent)

**RP** rocket projectile

**rpg** rounds per gun

**RR** rebated rim (also given as RB)

**rpm** rounds per minute; the usual measure of rate of fire

**RWS** remotely-controlled weapon system/station

**sabot** a sleeve into which a sub-calibre projectile is fitted, to enable it to be fired from a larger-calibre weapon: the sabot breaks up and falls away after the projectile leaves the muzzle

**saddle drum** a type of drum magazine in which the cartridges are held in two small connected drums on either side of the action, from each of which rounds are fed in turn

**salvo-firing** gun with multiple fixed barrels which are fired in rapid succession

**SAP(HE)** semi-armour piercing (high explosive)

**SD** self destruct: an HE projectile which is designed a detonate a few seconds after firing, to avoid 'friendly fire'

**SDR** sectional density ratio; the ratio between calibre and projectile weight: together with the projectile shape this determines the ballistic coefficient

**selective fire** a gun which can fire semi or fully automatically

**self-powered** gun with an operating cycle powered by the energy of the cartridge (see externally powered)

**semi-automatic** a gun which automatically fires, ejects and reloads each time the trigger is pulled: also used to describe artillery in which the fired case is automatically ejected but a new round is manually loaded

**semi-rimmed (case)** a cartridge case which has a rim only slightly larger in diameter than the case body, separated from it by an extractor groove: also known as semi-rimless

**shaped charge** see hollow charge

**shell** a projectile which is hollow in order to contain HE or other contents

**short recoil** a type of recoil-operated gun mechanism

**shot** any solid armour-piercing projectile (i.e. contains no HE)

**shoulder** the part of a cartridge case where the diameter reduces sharply from the case body to the neck

**shrapnel** a type of anti-personnel artillery ammunition which consists of a shell containing metal balls in the forward part and a small bursting charge at the rear, detonated by a time fuze to explode just before reaching the target: named after its 19th century inventor (nowadays loosely used to describe shell fragments)

**sights** the devices used to aim a gun

**single base** a type of propellant

**slant range** line-of-sight range from an AA position to a target aircraft

**sliding block** a type of gun action locking mechanism which moves across the breech face

**smoothbored** a barrel which is not rifled (used with fin stabilized ammunition)

**soft recoil** see counter-recoil

**SPAAG** self-propelled anti-aircraft gun: usually mounted on an AFV chassis

**spin stabilized** a projectile whose flight is stabilized by being rotated by rifling
***Sprenggranate*** (German): HE shell
**squeeze bore** a gun in which special projectiles are fired down a tapered barrel, or fitment to the end of the barrel, thereby reducing their diameter
**SR** semi-rimmed (or semi-rimless); a type of cartridge case
**stabilized mounting** a mounting in which the gun is kept to a constant point of aim regardless of the movement of the vehicle or ship which carries it
**STANAG** STANdardisation AGreement: applies to NATO equipment
**Stellite** a heat-resistant material used to line gun barrels
**straight-cased cartridge** a cartridge case which little or no taper between the head and the neck; which therefore has no shoulder
**strip** a number of cartridges linked rigidly together for ease of loading
**sub-calibre** a projectile which is smaller than the bore of the gun. For firing, it is supported by a sabot. See APDS, APFSDS and full calibre
**sub-calibre training** a cartridge used for training purposes in a much larger calibre (usually artillery) weapon via an adaptor
**sub-projectiles** projectiles which are carried by a larger projectile
**synchronized** a gun mounting fitted to a piston-engined fighter in which the gun is designed to fire through the propeller disc: each shot has to be precisely timed (synchronized) to avoid hitting a propeller blade

**-T** tracer (when attached to a projectile designation, e.g. HE-T)
**taboo zone** restriction on the freedom of movement of a flexible gun mounting to prevent firing at own ship or aircraft
**taper bore** see squeeze bore
**telescoped ammunition** ammunition in which the projectile is buried within the cartridge case
**terminal ballistics** the science concerning the performance of projectiles on striking the target
**time fuze** a fuze fitted to an HE shell which detonates it a predetermined time after firing
**time of flight** the time taken for a projectile to reach its target
**toggle joint** a type of elbow joint used in some short-recoil gun mechanisms
**TP** target practice: a type of projectile or cartridge used for training
**tracer** a chemical compound in the base of a projectile which burns slowly, giving a visible indication of its trajectory
**trajectory** the curve traced by a projectile in flight
**traverse** the movement of a gun in its mounting through a horizontal arc
**triple base** a type of propellant
**tripod** a type of gun mounting used with light portable weapons
***Trommel*** (German): a drum magazine
**trunnion** Projection from a gun body used to mount the gun; traditionally at the point of balance

***Übungsmunition*** (German): often written *Üb*; practice ammunition
**USAAC** United States Army Air Corps (1926–1947)
**USAAF** United States Army Air Force (1941–1947)
**USAF** United States Air Force (1947 onwards)
**USMC** United States Marine Corps
**USN** United States Navy

**vee-engine** a piston engine in which the cylinders are arranged in two banks, set at an angle to each other, driving a common crankshaft
***Vierling*** quadruple (German): a four-barrel gun mounting
**VT** World War II code letters for a proximity fuze

**warshot** ammunition intended for firing in combat (not training)
**water-cooled** a weapon which achieves barrel cooling by means of a water jacket
**windshield** see ballistic cap
***Wolfram*** (German): tungsten
**World War I** WW1 (1914–18)
**World War II** WW2 (1939–45)

***Zerleger*** (German): self-destroying (fuze), often written *Zerl; m. Zerl* means *mit Zerl* (with self-destruct), *o. Zerl* means *ohne Zerl* (without self-destruct)

## Conversion Factors

The following approximate conversion factors will convert metric to Imperial measures:

| | |
|---|---|
| To convert millimetres to inches | divide by 25.4 |
| To convert centimetres to inches | divide by 2.54 |
| To convert metres to feet | multiply by 3.28 |
| To convert metres to yards | multiply by 1.1 |
| To convert kilometres to miles | divide by 1.61 |
| To convert grams to grains | multiply by 15.432 |
| To convert grams to ounces | divide by 28.35 |
| To convert kilograms to pounds | multiply by 2.2 |

## Chamber Pressure

The following approximate conversion factors apply to the various different units in use:
1 MPa = 10 bar = 10.2 kg/cm$^2$ = 145psi

# SOURCES

A large number of sources have been consulted in preparing this book, including: manufacturers' data (on paper or online); published books and documents; websites and discussion forums focusing on the history of military technology; the back catalogue of The Cartridge Researcher, the Bulletin of the European Cartridge Research Association.

## INTERNET SOURCES

The Acknowledgements section at the start of the book makes reference to some of the internet sources which were particularly useful. The most important internet discussion forums were as follows:

IAA (International Ammunition Association): https://cartridgecollectors.org/?page=forum

BOCN (British Ordnance Collectors Network): http://www.bocn.co.uk/vbforum/forum.php

Military Guns and Ammunition (the author's forum): http://forums.delphiforums.com/autogun

also:

NavWeaps (naval weapons): https://www.tapatalk.com/groups/warships1discussionboards/

Secret Projects (military and aerospace technology): https://www.secretprojects.co.uk

Axis History (includes equipment): https://forum.axishistory.com/index.php

China Defense: http://www.china-defense.com/smf/index.php/

The following websites are also relevant:

NavWeaps (naval guns): http://www.navweaps.com

ECRA (European Cartridge Research Association): http://www.ecra.info/start.php

Military Guns & Ammunition (the author's website): http://quarryhs.co.uk

Wikimedia Commons (for photographs): https://commons.wikimedia.org/wiki/Main_Page

## PUBLISHED BOOKS: KEY TEXTS

For anyone who wants to follow up this subject, the following books are recommended:


Chinn, G.M. *The Machine Gun* (5 volumes) Vols I-IV Bureau of Ordnance, Department of the Navy (USA, 1951–55); Vol V RAMP Inc. 1987 *[Despite the title, this is the standard work on automatic cannon which anyone researching in this field must study. For a large number of experimental projects (mostly American) this is the only available reference. Doesn't say much about ammunition, though]*

Koll, C. *Soviet Cannon: a comprehensive study of Soviet guns and ammunition in calibres 12.7 mm to 57 mm.* Privately published, 2009 *[An absolutely outstanding work which also neatly plugs the biggest gap in Chinn – the lack of anything about Soviet developments post-World War II]*

Jacquinet, C. *The Story of 30 mm DEFA Aircraft Revolver Guns and their ammunition.* Privately published, 2012 *[Immensely detailed, particularly concerning ammunition, and covers a much wider field than the title suggests]*

Elks, K. *Japanese Ammunition 1880–1945 Part 2: 20 mm – 40 mm.* Solo Publications (2009) *[Exceptionally thorough and comprehensive – effectively the sole source on this subject]*

Davis, D.M. *Historical Development Summary of Automatic Cannon Calibre Ammunition: 20–30 Millimeter.* Air Force Armament Laboratory (USA, 1984) *[Written by a man who was there, this provides highly detailed information about the development of USAF cannon ammunition, including many which did not succeed]*


The most comprehensive and authoritative sources of information for all modern weapons and ammunition are the various Janes' annuals, e.g.:

- Artillery and Air Defence
- Armoured Fighting Vehicles: Wheeled
- Armoured Fighting Vehicles: Tracked
- Firepower, Survivability & Mobility
- Janes Weapons: Naval
- Janes Weapons: Air-Launched
- Janes Weapons: Ammunition

## OTHER PUBLISHED BOOK SOURCES

In addition, the following publications, covering current and historical equipment, were consulted in the preparation of this book:


Allsop, D.F. *Cannons.* Brassey's (UK, 1995)

Bradstreet, Theodore E. *Chapter 5. Machine Guns and Machine Cannon,* in Mikesh, R. C. *Japanese Aircraft Equipment 1940–1945.* Schiffer Military History (2004)

Braun, M. *Handbuch der Flugzeug Bordwaffenmunition.* Privately published (1977)

Bürli, A. *Flugzeug-bewaffnung: Die Schusswaffen der Schweizerischen Militärflugzeug.* Verlag Stocker-Schmid (Dietikon-Zürich, 1994)

Campbell, J. *Naval Weapons of World War Two.* Conway Maritime Press (London, 1985)

Friedman, N. *Naval Weapons of World War One.* Seaforth Publishing (2012)

Friedman, N. *The Naval Institute Guide to World Naval Weapon Systems.* US Naval Institute (1989 and 1994 Update)

Friedman, N. *US Naval Weapons.* Naval Institute Press (Annapolis)

Friedman, N. *Naval Anti-Aircraft Guns & Gunnery*. Seaforth Publishing (2013)

Gander, T. *The 40 mm Bofors Gun*. Patrick Stephens (UK, 1990)

Goulding, J. *Interceptor*. Ian Allan (UK, 1986)

Griehl, M. and Dressel, J. *German Anti-Tank Aircraft*. Schiffer (1993).

Gunston, W. *Encyclopedia of Aircraft Armament*. Salamander Books (London, 1987)

Gunston, W. *Encyclopedia of Russian Aircraft 1875–1995*. Osprey (London, 1995)

Hoffschmidt, E.J. *German Aircraft Guns and Cannons*. WE Inc (Greenwich, USA)

Hogg, I.V. (introduction) *The American Arsenal: World War II Official Standard Ordnance Catalogue*. Greenhill Books (London, 1996)

Hogg, I. V. *German Artillery of World War Two*. Arms and Armour Press (London)

Hogg, I. V. *British & American Artillery of World War Two*. Arms and Armour Press (London, 1978)

Hogg, I.V. and Thurston, L.F. *British Artillery Weapons & Ammunition 1914–18*. Ian Allan (London, 1972)

Keith, C.H. *I Hold My Aim*. Tinling & Co. (UK, 1946)

Lenselink, J. and de Hek, W.D. *Military Cartridges Parts 1 to 4*. Privately published (Netherlands, 1995–2003)

Lippert, J. *Deutsche Bordwaffen Munition Bis 1945*. Privately published (München, 1989)

Musgrave, D.D. *German Machineguns*. Greenhill Books (London, 1992)

Petter, F. *Munition für deutsche Flieger Bordwaffen Bis 1945*. Privately published, 2012.

Pitkänenm. and Simpanen T. *20 mm Suomessa Aseet ja ampumatarvikkeet ennen vuotta 1945 (20 mm in Finland: Weapons and Ammunition prior to 1945)*. Apali Oy, Tampere, Finland

Reinhart, C. and Rhyn, m. *Automatwaffen II*. Verlag Stocker-Schmid, Dietikon-Zürich (1983)

Richardson, D. *The History of the Revolver Cannon*. Oerlikon-Contraves (Zurich, 1994)

Schild, H. *Bewaffnung und Ausrüstung der Schweizer Armee seit 1917: Fliegerabwehr*. Verlag Stocker-Schmid (Dietikon-Zürich, 1982)

Schliephake, H. *Die Bordwaffen der Luftwaffe von deb Anfängen bis zur Gegenwart*. Motorbuch Verlag (Stuttgart, 1977)

Tantum, W.H. and Hoffschmidt, E.J., (eds.) *World War II Combat Weapons: Japanese*. WE Inc. (Connecticut, 1968)

Wallace Clarke, R. *British Aircraft Armament*. (two volumes) Patrick Stephens Ltd (UK, 1993 and 1994)

Wallace, G.F. *Guns of the Royal Air Force 1939–1945*. William Kimber and Co. (London, 1972)

Woodman, H. *Early Aircraft Armament*. Arms & Armour Press (London, 1989)


## JOURNAL AND MAGAZINE ARTICLES


Anon. 'Die 3,7-cm-Flakm 42 der Deutschen Kriegsmarine', *Waffen-Revue 106–107*

Anon. 'Paths of Destruction – NWM De Kruithoorn Demonstration of Frangible Medium Calibre Ammunition', *Military Technology* (12/94)

Bustin, I. 'Getting to the Point – Cased Telescopic Ammunition Development in Europe', *Military Technology* (3/96)

Frommer, H. 'Status and Trends in Automatic Cannons', *Military Technology* (3/85)

Gryazev, V. and Zelenko, V. 'Russia's Modern Aircraft Gun Armament' *Military Parade* (July-August 1997)

Kaiser, H and Schmucker, g. 'Military Fuzes', *Military Technology* (4/85)

Labbett, P. 'Cannon Ammunition 15 mm to 35 mm, 1945–1990', *Guns Review* (two parts; November and December 1990)

Labbett, P. 'Russian 20 mm ShVAK Aircraft Ammunition', *Guns Review* (June 1993)

Labbett, P. 'Russian 23x115 mm Aircraft Ammunition', *Guns Review* (July 1993)

Labbett, P. 'Russian 23x152 mm Cannon Ammunition', *Guns Review* (August 1993)

Labbett, P. 'Russian 30 mm and 37 mm Cannon Ammunition', *Guns Review* (September 1993)

Labbett, P. '30 mm Hispano Suiza Type 825 Ammunition', *Guns Review* (June 1989)

Petersen, D. 'Bofors 57 Mk3 Naval All-Target Gun', *Celsius Defence News, Naval Issue 99*

Simpkin, R. 'Light Armoured Vehicle Armaments', *Military Technology* (3/85)

Strandli, K.R. 'Multipurpose Ammunition', *Military Technology* (9/91)

Zhirnikh, g. 'The Development of Soviet Aircraft Armament', *Aviatsiya I Kosmonavtika* (8/1967)

Various contributors: many items in *The Cartridge Researcher*, the Official Bulletin of the European Cartridge Research Association.


### TECHNICAL AMMUNITION GUIDES (PRIVATELY PUBLISHED)


Labbett, P. *German Cannon Ammunition 1938–1945 Service Calibres 13 mm to 3 cm (other than 2 cm)*

Labbett, P. and Brown, F.A. *German 2 cm Cannon Ammunition 1935–1945*

Labbett, P. and Brown, F.A. *British Service 20 mm Oerlikon Ammunition*

Labbett, P. and Mead, P.J.F. *British 20 mm Hispano Ammunition*

Labbett, P. and Brown, F.A. *British 30 mm & 30/20 mm Ammunition*

Labbett, P. and Mead, P.J.F. *British Madsen Ammunition*


## UNPUBLISHED DOCUMENTS

A small sample of those read, many held in the Library of the Royal Armouries Museum, Leeds


Ackers, T. *General Survey of Main Problems Connected with Installation, Functioning and Accuracy of Aircraft Gun Armament During the War*. Technical Note Arm. 383, A&AEE Monograph No. 4.08 (Royal Aircraft Establishment, 1947)

Anon: *World War II 1939–1945 Royal Air Force Armament Volume II: Guns, Gunsights, Turrets, Ammunition and Pyrotechnics*. S.D. 737 Air Ministry (1954)

Anon. *Me 410 A-1/U4 Flugzeug-Handbuch*. (1944)

Anon: *Development of German Fighter Armament*. Air Ministry (UK) 1944.

Anon: *German Aircraft Armament and Ammunition During the War of 1939–45.* A.D.I. (K) Report No. 11/1946

Anon: *New Japanese Aircraft Types and Miscellaneous Armament Development.* Technical Air Intelligence Summary #14, DNI, 1944

Anon: *Japanese Air Weapons and Tactics.* The US Strategic Bombing Survey, 1947

Anon: *Gun Ammunition in 37 mm Calibre.* D. of A. (India) Japanese Ammunition Leaflets Section E, Leaflet E11 (1945).

Anon: *Preliminary Notes on the Ammunition for Vickers Class 'S' Mk.I Gun.* (Official British publication, status unclear)

Anon: *Report of Naval Anti-Aircraft Gunnery Committee 1931* (UK)

Anon: *Renseignments sur les Munitions de Petits Calibres.* P.R.R.E.M. de Bourges (France) 1940.

Bradstreet, T. *The Luftwaffe MG-FF.* 1998

Bradstreet, T. *Notes on Japanese Aircraft Guns and Cartridges Over 13.2 mm.* 1998

Burney, D. *Letter to the Broadway Trust Company to the Ministry of Supply Armaments Design Department* (25th November 1943)

Cabeen, C. *The Calibre Ninety Aircraft Gun.* Lafayette College, Easton, Pennsylvania.

Legendre, J-F. *Atelier de Fabrication de Mulhouse: CEAM-AME 1947–1967: Munitions de Moyen Calibre* (1998)

Various: *Reich Ministry of Armaments and War Production: Interrogation of Speer, Saur, Mommen and Bosch.* Combined Intelligence Objectives Sub-committee, G-8 Division, SHAEF, May/June 1945.

### The Unterlüß Reports

Bühler, Dipl. Ing. and Sörensen, Ing. *Table of German Armament Equipment (Approved and Experimental) Small Arms, Automatic Arms and Guns up to 55 mm Calibre.* Unterlüss Report 295/I, 1948

Bühler, Dipl. Ing. Burgmueller, Dr. and Sörensen, Ing. *Table of German Armament Equipment (Approved and Experimental) Ammunition up to 55 mm Calibre.* Unterlüss Report 295/II, 1948

Mix, Dipl. Ing. (late Director of Aircraft Weapon Development, R.L.M.) *Development of Aircraft Weapons, Munitions and Installations – A comprehensive review of the German position at the end of the War.* Unterlüss Report 376 T, 1948.

N.B. The 'Unterlüss Reports' (named after a Rheinmetall establishment) were written immediately after the end of World War II following interrogation of German technical staff by the British. They provide detailed information about the state of German armament developments at the end of the war. Copies are kept at the Library of the Leeds Royal Armouries Museum.

# INDEX

**CARTRIDGES: METRIC DESIGNATIONS**



**CARTRIDGES: POUNDER DESIGNATIONS**



**CARTRIDGES: INCH DESIGNATIONS**



**CANNON, GROUPED BY NATIONALITY**

**Canada**



**China**